LOGIC, MEANING AND COMPUTATION

SYNTHESE LIBRARY

STUDIES IN EPISTEMOLOGY,

LOGIC, METHODOLOGY, AND PHILOSOPHY OF SCIENCE

VOLUME 305

LOGIC, MEANING AND COMPUTATION

Essays in Memory of Alonzo Church

Edited by

C. ANTHONY ANDERSON
University of California, Santa Barbara, U.S.A.

and

MICHAEL ZELËNY
PTYX, Los Angeles, California, U.S.A.

SPRINGER-SCIENCE+BUSINESS MEDIA, B.V.

A C.I.P. Catalogue record for this book is available from the Library of Congress.

ISBN 978-94-010-3891-1 ISBN 978-94-010-0526-5 (eBook)
DOI 10.1007/978-94-010-0526-5

Printed on acid-free paper

TABLE OF CONTENTS

PREFACE

Alonzo Church was undeniably one of the intellectual giants of the Twentieth Century. These articles are dedicated to his memory and illustrate the tremendous importance his ideas have had in logic, mathematics, computer science and philosophy. Discussions of some of these various contributions have appeared in *The Bulletin of Symbolic Logic*, and the interested reader is invited to seek details there. Here we just try to give some general sense of the scope, depth, and value of his work.

Church is perhaps best known for the theorem, appropriately called "Church's Theorem", that there is no decision procedure for the logical validity of formulas of first-order logic. A decision procedure for that part of logic would have come near to fulfilling Leibniz's dream of a calculus that could be mechanically used to settle logical disputes. It was not to be. It could not be. What Church proved precisely is that there is no lambda-definable function that can in every case provide the right answer, 'yes' or 'no', to the question of whether or not any arbitrarily given formula is valid. To draw the more sweeping conclusion, that there is no "mechanical" procedure, no "algorithm", no "effectively calculable" way of deciding the question requires an identification the class of formally defined methods, with those methods characterized in these more informal ways. The proposal, the hypothesis or conjecture, that this identification is correct is "Church's Thesis".

"Lambda-definability" is a notion defined using another of Church's important contributions—the Lambda Calculus. Church began the construction of this calculus in the hope of showing that Gödel's Incompleteness Theorems are somehow not as conclusive as they seem. The available analyses of the notion of a correct proof all made use of some calculus or other. Yet Gödel had apparently shown that every such formal system, adequate for a certain portion of mathematics, will fail to capture some correct methods of proof and cannot be shown to be consistent except by using assumptions stronger than those it endorses. For Church this must have seemed intolerable. How then can we make the notion of proof precise? Must not we be able to at least give a convincing consistency proof for a formal system adequate for mathematics?

The Lambda Calculus was hoped to be a system that can be proved to be consistent (which it can and has been—by the "Church-Rosser Theorem") and yet somehow escape, or at least mitigate, the limitations on formal systems Gödel's Incompleteness Theorems seemed to impose. (I heard Church

C. Anthony Anderson and M. Zelĕny (eds.), Logic, Meaning and Computation, vii-ix.

say, in later years, that once you understand Gödel's Theorem, it's obvious (!) really that it is correct and there is no escape. But he also said his endeavor was not entirely in vain for it resulted in the Lambda Calculus. This calculus was and continues to be of enormous importance in computer science. So here even Church's failure was a kind of success.) Using the Lambda Calculus and a definition of "lambda-definability" in terms of it, Church proposed to identify this notion or (better) its extension with that of "effectively computable function" and thereby provide an analysis, or definition, of this latter, informal notion. Church's Thesis is certainly not obviously correct. It seems quite amazing that on the basis of a certain amount of experimentation in defining intuitively computable functions, Church's intuition told him that all had been captured. Later work by Turing and Post would provide more intuitively accessible characterizations of effective computability, but it is now virtually universally agreed that Church was correct and that the Thesis is true (whatever exactly that may mean!). This much, his proposed analysis of effective computability and his proof that first-order logic does not yield in these terms a decision procedure, already ensure Church a permanent place as a major figure in the development of symbolic logic. But there was more, much more.

Church contributed early-on to the foundations of theoretical circuit synthesis. He also formulated a theory of weak implication which became part of the basis for work by Alan Ross Anderson, Nuel Belnap, and others, on non-classical conceptions of implication, "entailment" and "relevant implication." He made an important proposal for analyzing the concept of a random sequence. The simple theory of types, a modification of the ramified theory of types proposed by Chwistek and Ramsey, received from Church its first precise syntactical formulation in an elegant version using some of the ideas gleaned from the Lambda Calculus. In later years he would give the first clear formulation of the ramified theory of types itself and show an important relationship between Russell's solution to the semantical paradoxes and Tarski's solution by means of the distinction of object-language and meta-language. And he formulated and gave a relative consistency proof for a set theory that allows for the existence of a universal set.

One would be remiss if *Introduction to Mathematical Logic* were not mentioned for containing important contributions to logic. It contains, for example, what was to become the standard axiomatic formulation of second-order logic (not complete, alas, as follows from Gödel's Incompleteness Theorem.) And Church gives there a correct formulation of the rule of substitution for functional variables, a matter which had eluded Hilbert and Ackermann and others.

In philosophy, besides setting an admirable standard of rigor in philosophical argumentation, Church contributed most of all to defending and developing intensional logic and related matters of general semantics. His own favored approach, the Logic of Sense and Denotation, develops ideas of

Gottlob Frege. But he also gave considerable effort to precisely formulating what he considered to be a viable alternative—Russell's intensional logic as embodied in *Principles of Mathematics* and in *Principia Mathematica.*

He was an able defender of realism in mathematics and logic and a telling critic of various nominalist projects. The sharpening of Frege's argument that sentences denote truth-values, the use of Langford's Translation Test to defeat analyses and arguments about meaning and propositions, an improved formulation of Quine's criterion of ontological commitment, and a refutation of Ayer's formulation of the positivist criterion of empirical significance, are just a few of his distinctively philosophical contributions.

Church wrote several excellent papers on various historical matters concerning logic, for example, on Schröder's partial anticipation of the theory of types and on the history of the notion of a proposition. His explications of the philosophical ideas of Russell and Frege constitute historical scholarship at its very best. Chapter 0 of *Introduction to Mathematical Logic* contains a crystal clear exposition of Frege's ideas about semantics combined with a keen sense of what is worth saving and what ought to be emended.

His monumental *Bibliography of Symbolic Logic* contains every known item on the subject of symbolic logic from the time of Leibniz to 1935. In effect, he continued to work on the *Bibliography* as editor of the reviews section of the *Journal of Symbolic Logic.* The purpose of those reviews was in part to defend symbolic logic against falling into disrepute as a result of misuse and abuse. And he was one of the founders of the Association of Symbolic Logic, playing a large role in the ever increasing respectabilty of the subject.

We rest our case. Church's intellectual legacy plainly establishes for him an honored and permanent place in logic, mathematics, computer science, philosophy, and scholarship about the history of logic.

C. Anthony Anderson

Michael Zeleny

ACKNOWLEDGEMENTS

The editors wish to express their deep gratitude to: Erin Yier Zhu for indispensable assistance with organizing and sustaining this tribute, Neil Nelson for superlative technical typesetting that brought it to completion, and its authors and publishers for patiently bearing with the peripeties of their project along the way.

Alonzo Church

REMEMBERING ALONZO CHURCH

I was a graduate student when Progessor Church visited the University of California at Los Angeles (UCLA) in 1960–61. He lectured on the logic of sense and denotation. I was so captivated by his lectures that I essentially did nothing else that year. I had found an "anomaly" in the axioms; one could prove that there were only two propositions, the Necessary and the Impossible. Clearly this was uninteded. Professor Church was very interested, and I remember distinctly sitting with him as he went over the proof line by line. The opening lines were true, the closing lines clearly false, and he was looking for the line at which things went from true to false. As he checked each line he said "OK", "OK", "OK", until suddenly, "Here it is!". It was an instance of Axiom 16. I also thought that the problem was that axiom, but of course I had spent weeks working on the material. Ultimately, I wrote my dissertation on intensional logic, combining a reformulation of Professor Church's system with the "possible worlds" semantics Carnap had been lecturing on (stemming from *Meaning and Necessity*).[1] Professor Church later published his own revision along somewhat different lines.[2] (He was scheduled to present this work at the American Philosophical Association meeting in Chicago, at which I was scheduled as commentator. When he became ill, I read his paper at the meetings and then gave my commentary entitled "The Church Reformation". I have always thought that my title properly conveyed the importance of a change of view by Professor Church.)

I had already had the opportunity to be overawed by my other teachers Carnap and Montague. But at the time Professor Church visited, Carnap was mainly interested in Inductive Logic and Montague in Set Theory. Logic is a rigorous discipline, but to see the rigor of Professor Church's intellect, the power of insight, and the exactness of his standards applied to *philosophical* problems was a formative experience of my graduate career. When attending his clases or consulting with him privately, I always thought, "This is how it should be done". I believe that I am not the only one of his students who developed a reviews-writing style that was initially in conscious imitation of his.

[1] David Kaplan, *Foundations of Intensional Logic*, University Microfilms (Ann Arbor: 1964).

[2] Alonzo Church, "Outline of a Revised Formalization of the Logic of Sense and Denotation", (Part I), *Nous*, vol. 7 (1973), pp. 24–33; (Part II), *Nous*, vol. 8 (1974), pp. 135–136.

C. Anthony Anderson and M. Zelëny (eds.), Logic, Meaning and Computation, xi-xii.

Despite the awesome intellect, Professor Church was always good-humored and patient with students. At a faculty meeting, after reading the logic prelim of a questionable graduate student, he remarked, "I would be willing to pass him, if he would be willing to promise never to write an article on the philosophical significance of Gödel's Theorem." He loved to laugh, and I have fond recollections of jolly dinner parties at his home and ours. (Although initially the prospect of inviting him scared the daylights out of me.)

One other remark from the early days has especially stayed with me and served me well in a variety of circumstances. I had broken sufficiently free from the Church style to include a few smart-alecky sentences in a review for the Journal of Symbolic Logic. I was delighted with my cleverness. The review came back with the comment that I could include as negative an evaluation of the work under review as I wished, however, sarcasm was not permitted. Of course I obeyed. But I later took the opportunity to ask Professor Church the reason for this policy. His reply was, "Evaluations can be rebutted, and errors can be corrected, but sarcasm can be neither rebutted nor corrected." This has always stayed with me as a lesson about life and human relations.

Professor Church was for me a model of how supreme intellect can combine with supreme objectivity without relinquishing devotion and human kindness. It was my great good fortune to have known him. Throughout the period that we were colleagues, my primary message to our new graduate students was, "Take a class from Alonzo Church. It will change you. And even if you are not interested in pursuing the subjects he teaches, you will be able to tell your grandchildren, 'I was a student of Alonzo Church'."

David Kaplan
Department of Philosophy
University of California
Los Angeles, California 90095-1451 USA
Email: Kaplan@ucla.edu

I found Church to be a very calm, steady, quiet, graciously polite man, without any inclination to carry on normal social conversation, but pleasantly responsive when engaged. In his early years at UCLA, he would often come into the department late at night and depart early in the morning. He frequently talked or sang softly to himself. When I was a young assistant professor, I used to drive him to departmental social events (which he willingly attended but did not contribute much to). Everyone would worry about his standing in a corner smiling at the proceedings but saying nothing. There were elaborate scheduling arrangements to make sure that he was attended to during the parties either by graduate students or by faculty. It fell to me to engage him in converstaion for good portions of several such evenings even though I came to believe he would remain in good spirits whether or not he was talking with anyone. I once asked him about his impressions of other great figures in the history of logic. When we came to Gödel, he said with awe and obvious admiration that Gödel was the only practicing solipsist that he had ever known.

I audited a course by Church on the Gödel theorems. It was Church's practice to begin each class by slowly erasing the blackboard, beginning in the upper lefthand corner, and proceeding in a deliberate, systematic, and orderly fashion to remove every residual chalkmark before the lecture began. Church's class immediately succeeded a class in elementary logic given in the same room by Don Kalish. One day I entered the classroom to find Kalish talking enthusiastically with several students after his class, having forgotten that it was time for Church's class to begin. Kalish's conversation took two or three minutes of Church's time, while Church stood quietly in the corner, with his hands folded over his midriff, swaying slowly back and forth with a benevolent, pastorial smile on his face. Finally, Kalish realized his intrusion, hurriedly made some apologies, quickly erased the board, and bustled out of the room. Church mumbled something gracious and, without moving from his spot, watched Kalish leave. Church stood there, still smiling benevolently, without moving for at least another thirty seconds after Kalish left, before beginning his lecture.

Tyler Burge

C. Anthony Anderson and M. Zelëny (eds.), Logic, Meaning and Computation, xiii.

PART I

LOGIC

PETER APOSTOLI

LOGIC, TRUTH AND NUMBER: THE ELEMENTARY GENESIS OF ARITHMETIC

Abstract. Following Paul Gilmore's LK-style natural deduction based set theories, we present a first order logic of type-free abstraction which employs "truth value gaps" to maintain consistency. By adjoining LK bivalence axioms governing the definitions of numeric concepts, we deduce arithmetic from a combination of first order logic + type-free abstraction.

0. FIRST-ORDER NUMBER-THEORETIC LOGICISM

0.1. Truth and Number

Crispin Wright (*1983*) articulates a position he calls "number-theoretic logicism [1]", the claim that arithmetic is derivable from logic in the strong sense that:

> it is possible to so define arithmetic concepts in terms of logical ones so that every statement of number-theory has a content preserving transcription in terms of a purely logical vocabulary, and every axiom or theorem of number theory can be so transcribed into a theorem of logic. (page 137)

Several authors (Wright *1983,* Boolos *1993,* Demopoulos and Bell *1993*) have reconstructed Frege's *Grundlagen* program so that the required derivations of Peano's postulates are obtained from a consistent fragment of the *Grundlagen* system based upon Frege's principle of equality for the natural numbers, referred to as "Hume's principle" in *Boolos 1993.* However, it is commonly agreed that the reconstruction fails to logicize number-theory in the required sense due to the apparently extra-logical nature of Hume's principle.

This paper demonstrates the truth of number-theoretic logicism by deriving first-order Peano Arithmetic from a purely logical first-order theory. Although the present program is not a reconstruction of Frege's, it deploys some of Frege's central insights concerning the role of definition in the development of arithmetic and the role of the bound variable of "set", or more neutrally, "relational", abstraction in the formalization of mathematical definition. By bringing together techniques from several branches of modern logic, notably analytic proof theory, recursion theory, the lambda calculus, programming language semantics and truth theory, we are able to deploy

C. Anthony Anderson and M. Zelëny (eds.), Logic, Meaning and Computation, 3-59.

relational abstraction in a first-order setting in such a way that the arithmetic of the natural numbers is derived as a consequence of instances of the principle of bivalence applied to the definition of arithmetical concepts.

The idea that the mathematical practice of defining concepts can be analyzed in terms of the operation of *relational abstraction* and the cognate logical device of the *bound variable of abstraction* is a theme which unifies the logicist programs of Frege and Russell. The analysis of mathematical definition as relational abstraction is central to the logicist derivations of mathematics because definition was one of the twin pillars upon which the derivation was to be based; mathematics was to be formally derived from the definition of mathematical concepts and logical laws by logical inference alone. Relational abstraction and conversion were to effect the introduction and elimination, respectively, of definitions in the process of deduction, so that classical truth-preserving inferences could be applied to the conceptual contents of those definitions. Although logicism is now commonly regarded of at most historical interest, the theme of assimilating mathematical definition to the device of the bound variable of abstraction has continued, through the foundational programs of Church (*1941*), Curry (*1930*), Fitch (*1948*) and Gilmore (*1980, 1986*) on into the contemporary analysis of mathematical reasoning in computer languages.

Schütte (*1960*) showed that second-order Peano Arithmetic may be interpreted in second-order logic based upon typed relational abstraction and a single descriptive function symbol. However, it is commonly believed that there is no corresponding derivation of first-order Peano Arithmetic from a purely logical first-order theory based on abstraction. This is a point that could be of moment within the debate over number theoretic logicism, as some philosophers view the existential commitment incurred by second-order quantification on par with the posit of extra-logical set existence principles. The insufficiency of first-order logic to yield a "purely logical" derivation of arithmetic would be taken by most philosophers to be characteristic of the family of first-order languages for which Tarski defined his notion of truth on a relational structure: it is a triviality that these languages admit of finite models in the absence of descriptive vocabulary and extra-logical axioms. However, my notion of a "first-order language" is more liberal than the now proprietary sense in which that term applies to the Tarskian family of languages. By "a first-order theory" I simply mean a logical theory which does not quantify over any objects other than individuals. It is, after all, the resources of second-order *quantification* that are commonly credited with the successful derivation of the induction scheme from the classical logicist definition of the set of natural numbers as the intersection of all zero-successor sets. And, as mentioned, it is the existential commitment of such quantification that purports to undercut that success.

The first-order language L in which this derivation of arithmetic takes place fails to belong to the Tarskian language family by virtue of the type-free conception of predication that underlies it. Because within the long shadow cast by the Tarskian tradition, the concept of a first-order language requires more than the abstinence from quantification over non-individuals. It also requires allegiance to a type-theoretic restriction on the grammar of predication that mirrors a semantic regimentation of the types of objects that can stand in the satisfaction relation. A first-order Tarskian language types syntactic combination so that the semantic function performed by the category of formulas, viz., to be predicated of, or satisfied by, an individual, cannot be performed by the category of singular terms. Singular terms, or rather the individuals they denote, bear identity conditions and existential witness, but they cannot be predicated of, or satisfied by, other individuals.

From the perspective of categorial grammar, a Tarskian first-order language thus assigns two quite distinct semantic tasks or functions to its two syntactic categories *singular term* and *formula*. The semantic value of a *singular term* is a first level individual, where the logical characteristic of a first level individual is its ability to bear identity conditions and serve existential witness, but not to be predicated of other individuals. The distinctive logical role of singular terms is realized primarily through the logic of identity and quantification. In contrast, the semantic value of a *formula* is a set, or characteristic function, of first level individuals (or tuples thereof). The semantic role of a formula is to bear satisfaction conditions, to be satisfied by first level individuals. By stratifying predication, Tarskian languages ensure that the entities that the logic of identity and quantification treat of are distinct from the entities that can bear recursively defined satisfaction conditions.

In the present essay, we interpret first-order Peano Arithmetic in a first-order language, L, designed to express type-free relational abstraction. Although L's surface syntax indeed stratifies syntactic combination, distinguishing between the syntactic categories of *term* and *formula* in the manner of Tarskian languages, this superficial stratification masks the intended interpretation of L, whereupon abstraction terms denote *relations* and "ε" expresses the relation of *satisfaction* or *membership* holding between relations and ordered pairs of first level individuals. Our semantic intentions might be more faithfully represented by using the syntactic combination of singular terms to indicate predication, in the fashion of applicative grammars, viz., $\lambda xy. R(x,y)(a,b)$. Hence, the abstraction terms of L play simultaneously the distinctive semantic roles of Tarskian singular term and formulas: they bear identity conditions and existential witness, but also embed the meta-theoretic satisfaction relation, that is, they are general predicates of individuals.

It is easy to see that the simultaneous deployment of these two logical roles induces numerical existence principles unavailable on the Tarskian conception. For example, by applying the indescernibility of identicals, we see that $0 =_{\text{df}} \{x : x \neq x\}$ is provably distinct from $1 =_{\text{df}} \{x : x = 0\}$, since the latter has a property not shared by the former, namely being satisfied by some object, i.e., having a member. $2 =_{\text{df}} \{x : x = 1\}$ is provably distinct from 0 for the same reason, but also from 1, since it has a property not shared by the latter, namely being satisfied by 1. In this way, the logic of identity combines with type-free abstraction to generate the domain of natural numbers under the encoding of 0 by the empty set and the successor function by the operation of passing from an object to its singleton set. It is crucial to the generation of the sequence

$$0,\ \{0\},\ \{\{0\}\},\ \{\{\{0\}\}\}\ \cdots$$

that the satisfaction relation we embed in L is type-free in the sense that complex singular terms can satisfy other complex singular terms, since every successor term in the sequence is satisfied by its predecessor.

0.2. The grounding of semantics and the semantics of grounding

Church's early hope was to marry type-free *functional* abstraction with full quantificational logic. In the sixties Paul Gilmore showed that type-free *relational* abstraction may be consistently combined with first-order quantification theory if one is willing to give up the principle of excluded middle and work with the notion of a "partial relation". In effect, Gilmore was taking a leaf from the recursion theorists' book; a decade or so later, this leaf would bloom in the philosophical community in the form of theories of a partial truth predicate.

Gilmore (*1980, 1986*) presented his theory as a first-order Gentzen-style sequent calculus, NaDSet I. The deductive rules of NaDSet I are *precisely* those of an *LK* identity calculus with the addition of a pair of *LK*-style introduction and elimination rules which formalize a very general notion of relational abstraction. However, NaDSet I is based upon a proper subset of the *LK* axiom sequents. While *LK* is based upon a complete set of "bivalence axioms" (axiom sequents of the form $A \vdash A$ for any atomic sentence A), NaDSet I's bivalence axioms are restricted to those treating identity statements and those of the form $t\,\varepsilon\, c \vdash t\,\varepsilon\, c$ for non-complex individual constants c only. In effect, consistency is maintained in NaDSet I by the strategy of evaluating only *grounded sentences*, sentences whose truth conditions reduce, under logical decomposition, to the truth conditions of simple sentences which are evaluated on the basis on "non-semantic" or "non-definition-theoretic" facts alone.

Bivalence axioms for identity statements provide a sufficient basis for a NaDSet I derivation of the theory of zero and successor (using the familiar set-theoretic encoding of those operations mentioned in the previous section). However, the absence of bivalence axioms of the form $t \in p \vdash t \in p$ governing free-object variables p undercuts a NaDSet I derivation of the scheme of mathematical induction. In the first part of this paper, we show that, although NaDSet I does not yield an induction scheme, it contains the resources for inductive definition. In section 1, we set out the calculus G, a sub-logic of NaDSet I based on binary abstraction and bivalence axioms for identity statements. The significance of binary abstraction is shown in section 2, where it underwrites the definition of a variable binding "fixed-point" operator fix_y, a relational version of the "paradoxical" Y combinator of the lambda calculus, forming terms which compute the fixed-points of recursive functionals. We are then able to derive LK-style deduction rules which implement a relational version of the second recursion theorem in terms of introduction and elimination conditions for the operator fix_y.

In section 4 we use fix_y to construct a set term $\underline{\omega}$ which represents the set $\mathbb{N}$ of natural numbers in the weak sense that $\underline{\omega}$ enumerates the set of all numerals. In section 6, $\underline{\omega}$ figures in the construction of set terms t_f that represent arbitrary primitive recursive functions $f : \mathbb{N} \to \mathbb{N}$ in the strong sense of enumerating the graph of f. Here, we use the second recursion theorem to simulate primitive recursion. Existential quantification is then used to simulate unbounded search over $\mathbb{N}$, allowing G to enumerate the graphs of all general recursive functions. G is thus "computationally complete", but lacks the resources to *reason* about the relations it computes. Specifically, although $\underline{\omega}$ is infinite, G has no theorem to this effect. Similarly, G does not yield the Peano axioms which recursively specify t_+ and $t_\times$, the set terms that represent the addition and multiplication relations over the natural numbers.

This limitation is entirely due to the failure of bivalence in G for atomic sentences which feature those terms as predicates, and contrasts with the semantic setting of section 3, where term models validate bivalence principles for all primitive recursive terms, hence realizing the full theory of 0, S, $+$ and $\times$. Thus, we pass successively in sections 5 and 7 to extensions G^ω and $G^{+,\times}$ of G by adjoining bivalence axioms governing $\underline{\omega}$ and then t_+ and $t_\times$. For example, G^ω is the extension of G obtained by adjoining as new axioms all sequents of the form $t \varepsilon \underline{\omega} \vdash t \varepsilon \underline{\omega}$ for arbitrary closed terms t. As a result, G^ω yields "the axiom of infinity"—the statement that $\underline{\omega}$ is a 0-successor set—as a theorem, and $G^{+,\times}$ sustains relational versions of the Peano axioms specifying the inductive definition of $+$ and $\times$.

The following foundational paradigm emerges: in the setting of a partial classical logic which maintains unrestricted type-free abstraction by rejecting excluded middle, the posit of bivalence axioms for a primitive recursive set term can *function logically as an existence postulate.* This is seen graphically in the simplicity of the derivation of the axiom of infinity for $\underline{\omega}$ in section 7. Note that adjoining excluded middle axioms to G maintains logical purity: G^ω and $G^{+,\times}$ are obtained from sub-logics of first-order logic by closure under abstraction. And the language of these calculi is just the language L of G. For these reasons, G^ω and $G^{+,\times}$ are referred to as logical extensions of G; the terminology is chosen purposely to contrast with the standard notion of an extra-logical extension of a logic.

Within this paradigm, progressively stronger logics are countenanced by progressively extending the "classical fragment" of G through the addition of bivalence axioms for a select "primitive recursive vocabulary". For example, since G is based upon a complete axiomatization of identity logic, the classical fragment of G contains all formulas and set terms built up from identity statements. Given that the axioms of the theory of 0 and S fall within this fragment, it is not surprising they are derivable in G. By hypothesis, we can reason classically about 0 and S in that fragment and, further, the characteristic properties of 0 and S, viz., that S is an injection from its domain $dom(S)$ into $dom(S) - \{0\}$, are classically entailed by the definitions of 0 and S. This is exactly how logicist reasoning is supposed to work. Passage to the logical extensions G^{ω} and $G^{+,\times}$ in effect extends the classical fragment of G to contain an image of the language of arithmetic.

In section 8, we extend $G^{+,\times}$ to a system G_2 by adding the bivalence axioms governing free-object variables that were lacking from NaDSet I. This addition allows G_2 to simulate enough of the idiom of second-order quantification to underwrite the classical impredicative definition of the set of natural numbers as the intersection of all 0-successor sets. Once this is achieved, the derivation of the first-order induction scheme goes through in G_2 in much the same way as in second-order theories of abstraction, such as NaDSet II.

1. TERM AND FORMULA: THE LANGUAGE OF PAIRWISE RELATIONAL ABSTRACTION

1.1. The language L

We consider a first-order language L containing a three place predicate symbol ε intended to represent set theoretic membership and a binary predicate $=$ intended to represent identity. L has a countably infinite set of individual variables, listed as $v_1, v_2, \ldots$, sentential connectives $\wedge, \vee, \neg$ and universal and existential quantifiers $\forall$ and $\exists$, respectively. In addition, L has a variable-binding term forming operator $[v_i\, v_j : \_\_]$ intended to represent binary, or pairwise, set abstraction. In addition to the individual variables, G contains infinitely many individual *parameters*, listed $p_1, p_2, \ldots$. The parameters are intended to play the role of *eigenvariable* that Gentzen (*1935*) assigned his so-called "free object variables": they provide the arbitrary witness upon which universal generalization is to be performed, but are not bound by any operator.

The set of individual terms and formulas are defined by mutual recursion as follows:

DEFINITION 1.1 **Term and Formula**. All individual variables and parameters are *terms*. If t_1, t_2 and t_3 are terms, then $t_1 = t_2$ and $(t_1, t_2)\, \varepsilon\, t_3$ are (*atomic*) *formulas*. If A is a formula and x, y are variables, then $[x\, y : A]$ is a term. If A and B are formulas and x is a variable, then $\neg A$, $A \wedge B$, $A \vee B$, $\forall x\, A$ and $\exists x\, A$ are *formulas*.

The connectives $\rightarrow$, $\leftrightarrow$ are introduced by their standard classical definitions and used where convenient. As indicated, we let "x", "y", "z" etc. range over variables; "s", "t", "r" range over terms; and "A", "B", "C" etc. range over formulas. In addition, we let "a", "b", "c" etc., and sometimes "p", range over parameters. All these meta-variables may occur with or without numerical subscripts. All logical constants, including "ε", occur autonymously in the meta-language. Atomic formulas of the form $t_1 = t_2$, $(t_1, t_2) \,\varepsilon\, t_3$ are called *identity statements*, respectively, *ε statements*. The latter is sometimes written as $t_3(t_1, t_2)$, to emphasize that t_3 is what we shall call *the predicate term* of the atomic formula $(t_1, t_2) \,\varepsilon\, t_3$. If a term t is of the form $[x\, y : A]$, then t is called an *abstraction term*; x and y are its *variables of abstraction*. A formula or term is *closed* iff it contains no free occurrences of any variable.

Sentences are closed formulas containing no parameters. A closed term is said to be *constant* iff it contains no parameters. Hence, we distinguish closed formulas from sentences. It is the latter that we are interested in evaluating; the former are merely proof-theoretic intermediaries. Let *ctrm* denote the set of constant terms.

For each $k \geq 1$ we use the k-ary substitution operator as follows: for any formula A, distinct variables $\vec{x} = x_1, \ldots, x_k$, and terms $\vec{t} = t_1, \ldots, t_k$, $A_{\vec{x}}[\vec{t}\,]$ denotes the result of simultaneously substituting t_i for x_i uniformly throughout A, with the usual proviso that t_i is free for $x_i (1 \leq i \leq_k)$. The substitution operator will also be used to uniformly replace parameters by arbitrary terms, particularly in the context of performing uniform substitution on ordered sets of formulas. For example, let $\Gamma = \{A^i \mid i \leq n\}$ be a set of formulas and $\vec{p} = p_1, \ldots, p_k$ parameters and $\vec{t} = t_1, \ldots, t_k$ terms. Then, $\Gamma_{\vec{a}}[\vec{t}\,] =_{\mathrm{df}} \{A^i_{\vec{a}}[\vec{t}\,] \mid i \leq n\}$ denotes the ordered set of sentences obtained by uniformly substituting t_i for a_i everywhere in $\Gamma(1 \leq i \leq k)$. An *instance* of a closed L formula is an L-sentence obtained by replacing every parameter occurring in the given formula by a constant term of L. More formally, let $\vec{p} = p_1, \ldots, p_m$ exhaust the parameters occurring in the closed formula A, and let $\vec{t} = t_1, \ldots, t_k$ be constant terms. Then $A_{\vec{p}}[\vec{t}\,]$ is an *instance of A*.

1.2. Syntactic complexity versus semantic complexity

Note that in Definition 1.1, terms and formulas of L are defined by *mutual* recursion. As a result, the language L countenances complex atomic formulas of the form $(t_1, t_2) \,\varepsilon\, [x\, y : A(x, y)]$. In Tarskian first-order languages, the notion of *truth on a structure* is defined by induction on the formation of sentences, a measure which is monotone in the length of the sentence being evaluated. As a result, in a Tarskian language, the measure of the *syntactic complexity* induced by the formation history of sentences coincides with the measure of the *semantic complexity* through which sentences are evaluated. In particular, the sentences which are regarded as *semantically* simple are precisely the atomic sentences. In contrast, the assignment of truth values to sentences of L is defined by an inductive process which is not monotone in the length of the sentence, for in the evaluation of a given sentence, e.g., the atomic sentence above, we are led to evaluate sentences, viz., $A(t_1, t_2)$, which may be longer than the given sentence. This is a familiar phenomenon in theories based on the operation of type-free term reduction, such as the lambda calculus or combinatory logic.

As a result, there are at least two important complexity measures for sentences of L. First, a sentence has a formation history, induced by Definition 1.1, and we can assign the stage at which a given sentence is formed according to 1.1 as a finite complexity measure. This measure ignores the internal structure of set abstract terms, considering all atomic sentences to have 0 complexity, and counts the complexity of Boolean and quantificational complexes as one greater than the sum of that of their principal subformulas. Second, "grounded" sentences can be assigned a "level", the transfinite stage at which they are semantically evaluated in the canonical evaluation process defined in section 3. Roughly speaking, the level of a sentence is the ordinal stage at which it receives a classical truth value, if it receives one at all.

1.3. Proof theory

Next, we introduce a sequent calculus G, in the style of Gentzen's (*1935*) classical calculus *LK*. G is a fragment of NaDSet I. Whereas NaDSet I takes generalized abstraction as a primitive inference rule, we use the rule of simple binary abstraction, which again may be considered a simple instance of the generalized abstraction rule of Gilmore (*1986*).

A *sequent* is a pair of ordered sets of closed formulas Γ, Δ, possibly empty, separated by the turnstile $\vdash$, thus: $\Gamma \vdash \Delta$. The axiom sequents, or axioms, of G are all sequents of the form $t_1 = t_2 \vdash t_1 = t_2$ and $\vdash t_1 = t_1$. The former are referred to as "excluded middle" or "bivalence" axioms (for identity) because sequents are given their classical, disjunctive reading in G. That is, roughly speaking, $\Gamma \vdash \Delta$ may be read as asserting that either some member of Γ is False or some member of Δ is True.

As for NaDSet I, the deduction rules for G are the *LK* structural and operational rules and a congruence rule for identity, =:

Structural Rules

Thinning

$$\Gamma \vdash \Theta \quad / \quad A, \Gamma \vdash \Theta$$
$$\Gamma \vdash \Theta \quad / \quad \Gamma \vdash \Theta, A$$

Contraction

$$A, A, \Gamma \vdash \Theta \quad / \quad A, \Gamma \vdash \Theta$$
$$\Gamma \vdash \Theta, A, A \quad / \quad \Gamma \vdash \Theta, A$$

Interchange

$$\Delta, A, B, \Gamma \vdash \Theta \quad / \quad \Delta, B, A, \Gamma \vdash \Theta$$

$$\Gamma \vdash \Theta, A, B, \Delta \quad / \quad \Gamma \vdash \Theta, B, A, \Delta$$

Cut $\quad \Gamma \vdash \Theta, A \qquad A, \Gamma \vdash \Theta \quad / \quad \Delta, \Gamma \vdash \Theta$

Identity

$=\vdash \quad \Gamma \vdash \Theta, t_2 \,\varepsilon\, s \qquad t_1 \,\varepsilon\, s, \Gamma \vdash \Theta \quad / \quad t_1 = t_2, \Gamma \vdash \Theta$

Operational Rules

$\vdash\wedge$	$\Gamma \vdash \Theta, A \qquad \Gamma \vdash \Theta, B$	/	$\Gamma \vdash \Theta, A \wedge B$
$\wedge\vdash$	$A, \Gamma \vdash \Theta \;/\; A \wedge B, \Gamma \vdash \Theta$		$B, \Gamma \vdash \Theta \;/\; A \wedge B, \Gamma \vdash \Theta$
$\vdash\vee$	$\Gamma \vdash \Theta, A \;/\; \Gamma \vdash \Theta, A \vee B$		$\Gamma \vdash \Theta, B \;/\; \Gamma \vdash \Theta, A \vee B$
$\vee\vdash$	$A, \Gamma \vdash \Theta \qquad B, \Gamma \vdash \Theta$	/	$A \vee B, \Gamma \vdash \Theta$
$\vdash\neg$	$A, \Gamma \vdash \Theta$	/	$\Gamma \vdash \Theta, \neg A$
$\neg\vdash$	$\Gamma \vdash \Theta, A$	/	$\neg A, \Gamma \vdash \Theta$
$\vdash\forall$	$\Gamma \vdash \Theta, A_x[p]$	/	$\Gamma \vdash \Theta, \forall x A$
$\forall\vdash$	$A_x[t], \Gamma \vdash \Theta$	/	$\forall x A, \Gamma \vdash \Theta$
$\exists\vdash$	$A_x[p], \Gamma \vdash \Theta$	/	$\exists x A, \Gamma \vdash \Theta$
$\vdash\exists$	$\Gamma \vdash \Theta, A_x[t]$	/	$\Gamma \vdash \Theta, \exists x A$

where in $\vdash\forall$ and $\exists\vdash$, the parameter p does not occur free in Γ, Θ or A. Note that the use of the substitution operator $\forall\vdash$ requires that t_i be free for x in A. Derived rules of $\vdash\rightarrow$, $\rightarrow\vdash$ will be assumed and used where convenient. Note that we will often use derived variants of the rules, particularly for $\vdash\wedge$ and $\wedge\vdash$. In general, implicit use of Interchange and Contraction will allow us to treat sequents as unordered sets. Also, the form of the identity rule $=\vdash$ that will be used is

$$\Gamma \vdash \Theta, A_x[t_1] \qquad A_x[t_2], \Gamma \vdash \Theta \quad / \quad t_1 = t_2, \Gamma \vdash \Theta,$$

where A is an arbitrary formula with at most x free.

Abstraction Rules. In addition, G has falsity preserving and truth preserving rules for binary relational abstraction, which are very special cases of the generalized abstraction rules of NaDSet I:

$\varepsilon\vdash \quad A_{xy}[t_1, t_2], \Gamma \vdash \Delta \quad / \quad (t_1, t_2) \,\varepsilon\, [x\, y : A], \Gamma \vdash \Delta,$

$\vdash\varepsilon \quad \Gamma \vdash \Delta, A_{xy}[t_1, t_2] \quad / \quad \Gamma \vdash \Delta, (t_1, t_2) \,\varepsilon\, [x\, y : A].$

The use of the substitution operator in these schemas requires that t_i be free for both x and y in A $(i = 1, 2)$.

Derivations take the form of the usual Gentzen trees, though we shall write them in linear fashion to avoid recopying subtrees, where possible. All the cognate proof theoretic notions, e.g., that of *derivable sequent*, are standard.

Note that the axioms and rules of G treating the classical connectives are *classical*. G fails to extend classical logic only because of the absence of the

LK bivalence axioms, for which an adequate basis is $A \vdash A$ for atomic A. G has bivalence axioms for identity statements only. This is a tightening of the policy in *Gilmore 1980, 1986* of restricting NaDSet I's bivalence axioms to those treating identity and $t \,\varepsilon\, c \vdash t \,\varepsilon\, c$ for non-complex individual constants c. In effect, this axiomatic restriction corresponds to the NaDSet I's model theoretic strategy of evaluating only *grounded sentences*, sentences whose truth conditions reduce, under logical decomposition, to the truth conditions of simple sentences which are evaluated on the basis of "non semantic" or "non definition-theoretic" facts alone. Paradoxical atomic sentences such as $\mathsf{R} \,\varepsilon\, \mathsf{R}$, where R is the "Russell" set term $[x : \neg x \,\varepsilon\, x]$, or the set theoretic version of the "truth teller", $\mathsf{T} \,\varepsilon\, \mathsf{T}$, where T is $[x : x \,\varepsilon\, x]$, are not evaluated because they fail to be grounded in this sense.

Thus, the conception of semantic grounding informing NaDSet I is formally implemented by specifying atomic sentences of the form $t_1 = t_2$ and $t \,\varepsilon\, c$ as the only "ground atoms" and allowing only ground atoms to figure in bivalence axioms. G allows only the former to be ground atoms. On this basis, however, excluded middle sequents are derivable for sentences all of whose atomic constituents are identity statements. As we shall see in section 3, this is enough to interpret within G the theory of 0 and the successor function.

1.4. The notion of a logical extension of G

A *logical extension of* G is a calculus for L which is obtained by adjoining to G as new bivalence axioms sequents of the form $s \,\varepsilon\, t \vdash s \,\varepsilon\, t$ for certain closed atomic formulas of the form $s \,\varepsilon\, t$. Here $s \,\varepsilon\, t$ is considered to be a new "ground atom" in the calculus thus obtained. Sometimes we refer to the predicate term t of the new ground atom as a new "ground predicate" of the logical extension.

The ground predicates in the logical extensions studied in this essay are set terms which define number-theoretic concepts that we are interested in reasoning about classically, that is, under bivalence, in the given extensions. The notion of a logical extension contrasts with the standard notions of an extension of a classical theory. We consider a logical extension of G to be a *logic* as opposed to a *theory* (all logics are theories, but not all theories are logics). The simple-minded justification for the "logical purity" of these extensions is the simple observation that their new axioms are instances of the general principle of bivalence.

However, it must be shown why bivalence for numerical concepts counts as a purely logical postulate in a system which rejects the *general* principle of bivalence in order to accommodate unrestricted type-free abstraction.[1] What is needed is a characterization of the class of atomic sentences for which

[1] I owe the present point to Phil Kremer.

the posit of bivalence can be motivated independently of considerations of consistency and the requirements of number-theory. Of course, the class thus characterized will have to be consistent as well as underwrite the derivation of number-theory. The ideal solution is to come up with an analysis of the paradoxes which at once provides for a consistent treatment of type-free abstraction and specifies a class of atomic sentences for which the posit of bivalence is sufficient for number-theory. Such an analysis is provided by a constructivistically motivated analysis of the notion of *semantic grounding.*

On the traditional account, "grounded" sentences are those whose truth-conditions hereditarily reduce under logical analysis and conversion (either set-theoretic or truth-theoretic as *per* convention T) to the truth conditions of "ground atoms"—atomic sentences whose truth-conditions are determined independently of definition-theoretic or semantic facts. This rough intuition is formalized in the inductively constructed term models of NaDSet I and arithmetic models of Kripke (*1975*) in terms of a restriction on the vocabulary that may occur in the atomic formulas that count as ground atoms. Kripke's restriction is the most straightforward: ground atoms must not contain "semantic" vocabulary, such as the truth predicate "T". Gilmore's restriction is slightly more subtle: ground atoms may contain "ε", but not with abstraction terms as predicates. In both settings, however, the motivation for the restrictions is the same: the offending expressions invoke "circular concepts".

In *Gilmore 1986* the motivation for treating identities and ε statements with unstructured predicate terms as ground atoms is that they are *semantically primitive* in the sense that they are not the output of an application of any semantic rule. In particular, they cannot be obtained by abstraction from any sentence; it is in this sense that they are evaluated on the basis of non-definition-theoretic facts alone. So, "semantic primitiveness" converges with a certain notion of "syntactic primitiveness" in NaDSet I.

From the view point provided by logical extensions of G, Gilmore's identification of syntactic simplicity with semantic simplicity is too much to concede to the Tarskian regime. Because, as argued in section 0, the process of semantic evaluation which constitutes the conception of grounding underlying NaDSet I invokes a distinction between syntactic formation and semantic evaluation alien to the Tarskian conception of a first-order language. Once we have broken the identification of semantic complexity with syntactic complexity, there is no reason to think of semantically simple sentences as being syntactically simple as well. Only when this last vestige of the Tarskian stratification of predication is rejected, and syntactically complex ground predicates are countenanced, can first-order theories of type-free relational abstraction bloom.

On our conception of semantic grounding, a given atomic sentence counts as a ground atom iff either it is not the output of a semantic rule, or its truth conditions are such that its semantic evaluation reduces to the evaluation of syntactically *strictly simpler* sentences. This intuition is formalized by identifying ground atoms with *primitive recursive queries.* Primitive recursive notions are precisely those represented by a syntactic structure whose complexity monotonically reduces under evaluation. In section 9 we associate a constant set term t_f with each primitive recursive function f such that t_f enumerates the graph of f in G. It turns out that when term models of G are specified (as in section 3) so that constant terms are individuated under strict syntactic identity, every atomic sentence with t_f as predicate term receives a classical truth value, i.e., t_f represents a "total" set. So all primitive recursive queries are semantically grounded in precisely the original sense of Gilmore (*1986*) and Kripke (*1975*). Again, the motivation for the restrictions is the same as for these earlier semantic proposals: expressions lacking a primitive recursive structure potentially invoke "circular concepts" which lead to non-terminating reduction sequences. More importantly, by allowing the class of logical extension of G to be closed under the adjunction of bivalence axioms for primitive recursive queries, we are adopting *intuitionistically acceptable* instances of bivalence. For a primitive recursive query is a paradigmatic example of the kind of "constructive statement" for which the intuitionist is willing to accept bivalence.

2. EXTENDED SYNTAX FOR L

2.1. K-ary abstraction

In this section we define k-ary abstraction in terms of binary abstraction and derive the corresponding deduction rules of k-ary abstraction for arbitrary logical extensions of G.

DEFINITION 2.1. Our definitions are set up with the intention that 2-ary abstraction is binary abstraction.

Ordered Pairing

$$\langle t_1, t_2\rangle =_{\mathrm{df}} [v_1\, v_2 : t_1 = v_1 \wedge t_2 = v_2],$$
$$\langle t_1, \ldots, t_k\rangle =_{\mathrm{df}} \langle\langle t_1, \ldots, t_{k-1}\rangle, t_k\rangle (k > 2).$$

Vector notation: $\langle \vec{t}\,\rangle$ denotes the term $\langle t_1, \ldots, t_k\rangle$.

Defined Set Terms: Let A be a formula:

Unary abstraction:

$[x : A(x)] =_{\mathrm{df}} [x\, y : A(x)]$, where y is the first variable not occurring in $A(x)$.

Unary membership:

$t \varepsilon s =_{\mathrm{df}} (t, t) \varepsilon s$. So, for example, $t \varepsilon [x : A(x)] =_{\mathrm{df}} (t, t) \varepsilon [x : A(x)]$,

k-ary abstraction: Let $k \geq 3$ and A be a formula. Suppose we have already defined the $(k-1)$-ary abstraction term $[x_1 \dots x_{k-1} : A]$. Then

$$[x_1 \dots x_k : A] =_{\rm df}$$
$$[z\, x_k : \exists y_1 \dots y_{k-1}(z = \langle y_1, \dots, y_{k-1}\rangle \wedge (y_1 \dots y_{k-1})\, \varepsilon\, [x_1 \dots x_{k-1} : A])],$$

where $y_1 \dots y_{k-1}$ are the first variables not occurring in $[x_1 \dots x_{k-1} : A]$ and:

k-ary membership: Let $k \geq 3$,

$$(t_1 \dots t_k)\, \varepsilon\, [x_1 \dots x_k : A] =_{\rm df} (\langle t_1, \dots, t_{k-1}\rangle, t_k)\, \varepsilon\, [x_1 \dots x_k : A],$$

Vector notation: $[\vec{x} : A(\vec{x})] =_{\rm df} [x_1 \dots x_k : A(x_1, \dots, x_k)]$,

Vector notation: $(\vec{t}\,)\varepsilon[\vec{x} : A(\vec{x})] =_{\rm df} (t_1 \dots t_k)\varepsilon[x_1 \dots x_k : A(x_1, \dots, x_k)]$.

Clearly, we may derive rules for unary abstraction. Note that the argument list for $k \geq 3$-ary membership forgoes the comma characteristic of the argument list for binary abstraction. Given $\vec{t} = t_1 \dots t_{k-1}$ and t_k we sometimes preserve the comma, writing "$(\vec{t}, t_k)\varepsilon[\vec{x}\, x_k : A(\vec{x}, x_k)]$" for $(\langle t_1, \dots, t_{k-1}\rangle, x_k)\varepsilon$ $[x_1 \dots x_k : A(x_1, \dots, x_k)]$, when we want to emphasize that $\vec{t}$ is in fact the single argument $\langle t_1, \dots, t_{k-1}\rangle$.

The following corollary is stated for an arbitrary logical extension of G:

COROLLARY 2.2 **The Derived Rules of k-ary**. *Let $k \geq 3$, $\vec{t} = t_1, \dots, t_k$ be any terms, $\vec{x} = x_1, \dots, x_k$ distinct variables and A a formula. Then*

$$\vdash\!\varepsilon_k \qquad \Gamma \vdash \Delta, A_{\vec{x}}[\vec{t}\,] \quad / \quad \Gamma \vdash \Delta, (\vec{t}\,)\, \varepsilon\, [\vec{x} : A],$$
$$\varepsilon_k\!\vdash \qquad A_{\vec{x}}[\vec{t}\,], \Gamma \vdash \Delta \quad / \quad (\vec{t}\,)\, \varepsilon\, [\vec{x} : A], \Gamma \vdash \Delta$$

are derived rules, where, for $k \geq 3$, $\varepsilon_k\!\vdash$ conforms to the restriction that for all closed term $\vec{t} = t_1, \dots, t_k$, $A_{\vec{x}}[\vec{t}\,] \vdash A_{\vec{x}}[\vec{t}\,]$ is a derivable sequent.

2.2. The recursion theorem

We may now present some recursion-theoretic control structures which allow us to calculate fixed-points for formulas in G and logical extensions. The following is a rational reconstruction of a result due to the Scottish logician, Jamie Andrews, who first discovered the distinctive role of *binary* abstraction in the calculation of fixed-point terms in relational settings:

Let $A = A(\vec{x}, y)$ be a formula. Then

$$diag_y A =_{\rm df} [\vec{x}\, y : A(\vec{x}, [\vec{x} : (\vec{x}\, y)\, \varepsilon\, y])],$$
$$fix_y A =_{\rm df} [\vec{x} : (\vec{x}\, diag_y A)\, \varepsilon\, diag_y A].$$

LEMMA 2.3 **Andrews' Recursion Theorem.** *Let* $A = A(\vec{x}, y)$ *be a formula,* $\vec{x} = x_1, \ldots, x_k$ *distinct variables and* $\vec{t} = t_1, \ldots, t_k$ *be terms, where* t_i *is free for* x_j *in* A $(1 \leq i, j \leq_k)$. *Then,*

$$\begin{array}{llcl} \mathit{fix}\vdash & \Gamma, A(\vec{t}, \mathit{fix}_y A) \vdash \Delta & / & \Gamma, (\vec{t}\,)\, \varepsilon\, \mathit{fix}_y A \vdash \Delta, \\ \vdash\mathit{fix} & \Gamma \vdash A(\vec{t}, \mathit{fix}_y A), \Delta & / & \Gamma \vdash (\vec{t}\,)\, \varepsilon\, \mathit{fix}_y A, \Delta. \end{array}$$

are derived rules, where, for $k \geq 3$, $\mathit{fix}\vdash$ *conforms to the restriction that, for all closed term* $\vec{t} = t_1, \ldots, t_k$, $A_{\vec{x}}[\vec{t}\,] \vdash A_{\vec{x}}[\vec{t}\,]$ *is a derivable sequent.*

Proof. For $\vdash\mathit{fix}$:

(1) $\Gamma \vdash A(\vec{t}, \mathit{fix}_y A), \Delta$

(2) $\Gamma \vdash A(\vec{t}, [\vec{x} : (\vec{x}, \mathit{diag}_y A)\, \varepsilon\, \mathit{diag}_y A]), \Delta$ (df. $\mathit{fix}_y A$)

(3) $\Gamma \vdash (\vec{t}, \mathit{diag}_y A)\, \varepsilon\, [\vec{x}\, y : A(\vec{x}, [\vec{x} : (\vec{x}, y)\, \varepsilon\, y]), \Delta$ $(\vdash\varepsilon_{k+1})$

(4) $\Gamma \vdash (\vec{t}, \mathit{diag}_y A)\, \varepsilon\, \mathit{diag}_y A, \Delta$ (df. $\mathit{diag}_y A$)

(5) $\Gamma \vdash (\vec{t}\,)\, \varepsilon\, [\vec{x} : (x, \mathit{diag}_y A)\, \varepsilon\, \mathit{diag}_y A], \Delta$ $(\vdash\varepsilon_k)$

(6) $\Gamma \vdash (\vec{t}\,)\, \varepsilon\, \mathit{fix}_y A, \Delta$ (df. $\mathit{fix}_y A$)

⊣

$\mathit{fix}\vdash$ is treated in a similar manner, using $\varepsilon_{k+1}\vdash$ and $\varepsilon_k\vdash$. In the sequel, we use these two rules to introduce and eliminate recursive definitions in our derivations.

3. TERM MODELS FOR L

3.1. The classical role of 4-valued semantics

In this section we specify a class of structures for L within which we can specify models for G and logical extensions thereof. In this class of "term models", constant terms are posited as their own denotations, individuated under strict syntactic identity. Term models do provide a convenient way of studying G and its extensions, and are here presented to that end. Our term models are based on a simplification of those presented in *Gilmore 1986* for NaDSet I, although we adopt an "algebraic" notational style familiar from the semantical treatment of programming languages (*Scott 1975*), and, more recently, the truth theory literature (*Gupta and Belnap 1993, Fitting 1989, 1993*).

Gilmore (*1986*) utilizes the device of a *signed sentence*, $+A, -A$, to record the assignment of Truth, respectively, Falsity to a sentence A in a model of NaDSet I. Models are then taken to be sets of signed sentences which respect the Kleene strong "three-valued" interpretation of the connectives and quantifiers and the semantic rule corresponding to relational abstraction. There is a subtlety involved here. It is not part of the *definition* of a NaDSet I model that the partial assignment of classical truth values represented by the distribution of + and − among the signed sentences in the model be functional; strictly speaking, "models" may contain both $+A$ and $-A$ for a given sentence. It is rather in the form of a *result* about the structures inductively defined in *Gilmore 1986* that we know that in fact every model of NaDSet I so defined assigns at most one of the classical values to a given sentence. Hence, the device of the signed sentence, and the practice of taking models to be sets of signed sentences, is really a very early representation of a four-valued semantics. Because, here we have the values of classical truth and falsity together with values which represent the situation of a sentence being classically *undefined* and classically *overdefined*.

Of course, the availability of the fourth value, *overdefined*, is utilized by NaDSet I only at the meta-theoretic level, in the construction of three-valued models. For example, in the analytic (or "reductive") style of completeness proof natural to Gentzen calculi, a nested sequence of sets of signed sentences is generated by a branching process that closes branches which come to contain both $+A$ and $-A$ for some atomic sentence A. Here, although we are only interested in producing three-valued sets of signed sentences, we do so by countenancing four-valued sets in the very definition of our canonical reduction process. The present development is contiguous with that of NaDSet I by using four-valued models in the meta-theoretic process of constructing three-valued models. It turns out to be mathematically convenient to first specify the class of four-valued models, and then define the three-valued models to be the structures which only use the three "standard" values.

3.2. Information and truth on a structure

The following presentation borrows from *Fitting 1993*, and *Gupta and Belnap 1993*. Let $\mathbf{4}$ be the set of truth values $\{true, false, \bot, \top\}$. We usually refer to $\top$ as "*both*". Term models are based on the Dunn-Belnap $\mathbf{4}$-valued "bi-lattice", considered under its information ordering, $(\mathbf{4}, \leq_k)$,

both

true *false*

$\bot$

and also under its alethic ordering, $(\mathbf{4}, \leq_t)$,

true

$\bot$ *both*

false

Below, the connectives and quantifiers of L are interpreted as least upper bound operations on $(\mathbf{4}, \leq_t)$.

An *L-structure* is a pair $\mathcal{S} = (ctrm, \varepsilon_{\mathcal{S}})$ where *ctrm*, *the domain of* $\mathcal{S}$, is the collection of all closed set terms and $\varepsilon_{\mathcal{S}} : ctrm^3 \to \mathbf{4}$. $\varepsilon_{\mathcal{S}}$ is intended to be the interpretation that $\mathcal{S}$ assigns the "pairwise membership" predicate ε, and is sometimes called a "membership operator".

As mentioned, the object of this study is the Kleene strong three-valued interpretation of L, based upon the set of values $\mathbf{3} = \{true, false, \bot\}$. We consider $\mathbf{3}$ under the information ordering, $(\mathbf{3}, \leq_k)$,

$$\begin{array}{ccc} false & & true \\ & \bot & \end{array}$$

which gives $\mathbf{3}$ the structure of a complete semi-lattice. The Kleene strong interpretation of the connectives and quantifiers is the least upper bound operation on $\mathbf{3}$ with respect to the corresponding alethic ordering $(\mathbf{3}, \leq_t)$,

$$\begin{array}{c} true \\ \bot \\ false \end{array}$$

which gives $\mathbf{3}$ the structure of a complete lattice. Note that $\mathbf{3}$ is closed under the operation of taking l.u.b's and g.l.b's with respect to $(\mathbf{4}, \leq_t)$. A *3-valued L-structure* is an L-structure $\mathcal{S} = (ctrm, \varepsilon_{\mathcal{S}})$ where $\varepsilon_{\mathcal{S}}(ctrm) \subseteq \mathbf{3}$.

Let A be an L sentence and t_1 and t_2 be constant terms of L. We now proceed to define the notion of the *truth on an L-structure* by induction on the formation of A:

Definition 3.1 Truth on a Structure.

(1) $\mathcal{S}(t_1 = t_2) = true$ if $t_1 = t_2$
$\qquad\qquad\qquad\quad$ *false* else;

(2) $\mathcal{S}((t_1, t_2)\, \varepsilon\, t_3) = \varepsilon_{\mathcal{S}}(t_3, t_1, t_2)$;

(3) $\mathcal{S}(A \wedge B) = \mathcal{S}(A) \wedge \mathcal{S}(B)$, $\mathcal{S}(A \vee B) = \mathcal{S}(A) \vee \mathcal{S}(B)$ and $\mathcal{S}(\neg A) = \neg \mathcal{S}(A)$,

where $\wedge$ is assigned the fixed interpretation of meet on $(\mathbf{4}, \leq_t)$, $\vee$ is treated dually, and $\neg$ is assigned the operation which takes *true* to *false*, *false* to *true*, $\bot$ to $\bot$ and $\top$ to $\top$. Let $A(x)$ be an L-formula containing no parameters. Then,

(4) $\mathcal{S}(\forall x\, A) = \wedge\{\mathcal{S}(A_x[t]) \mid t \in ctrm\}$ where $\wedge$ is arbitrary meet on $(\mathbf{4}, \leq_t)$.

(5) $\mathcal{S}(\exists x\, A) = \vee\{\mathcal{S}(A_x[t]) \mid t \in ctrm\}$ where $\vee$ is arbitrary join on $(\mathbf{4}, \leq_t)$.

We write $A \to_{\mathcal{S}} B$ if $\mathcal{S}(A) \leq_k \mathcal{S}(B)$ and $A \leftrightarrow_{\mathcal{S}} B$ iff $\mathcal{S}(A) = \mathcal{S}(B)$. More usefully, we write

$$\to_{\mathcal{S}} A =_{\mathrm{df}} \mathit{true} \leq_k \mathcal{S}(A),$$

$$\leftarrow_{\mathcal{S}} A =_{\mathrm{df}} \mathcal{S}(A) \geq_k \mathit{false}.$$

We also say that an L sentence A *diverges on* $\mathcal{S}$, and write A, if $\mathcal{S}(A) = \bot$. Intuitively, $\mathcal{S}(A) = \bot$ represents the situation that A is "classically undefined". We say that A *converges on* $\mathcal{S}$, and write $A\downarrow$, otherwise.

Our intention is to take as models of G the L-structures which are 3-valued and validate the rules of binary abstraction. Since 3 is closed under negation and the operation of taking least upper and greatest lower bounds with respect to $(\mathbf{4}, \leq_t)$, if $\mathcal{S}$ is a 3-valued L-structure, then $\mathcal{S}(A) \in 3$ for *all* sentences A, as can be seen by induction on the formation of A. Hence Definition 3.1 specifies the Kleene strong three-valued interpretation of L when $\mathcal{S}$ is restricted to the class of 3-valued L-structures.

We model derivability in G upon a notion of the *validity of a sequent* which a 4-valued generalization of the definition given for NaDSet I in *Gilmore 1986*. This generalization is obtained by replacing the assignment of *true* from 3 in NaDSet I semantics by $\geq_k$ *true* from $\mathbf{4}$.

Let $\Gamma \vdash \Delta$ be a sequent, and let $\vec{a} = a_1, \ldots, a_k$ exhaust all of the parameters occurring in members of Γ or Δ. Let $\vec{t} = t_1, \ldots, t_k$ be constant terms of L. Then $\Gamma_{\vec{a}}[\vec{t}\,] \vdash \Delta_{\vec{a}}[\vec{t}\,]$ is called an instance of $\Gamma \vdash \Delta$.

DEFINITION 3.2. A sequent $\Gamma \vdash \Delta$ is *valid on* $\mathcal{S}$ iff, for every instance Γ' of Γ and Δ' of Δ, either $\mathcal{S}(A) = \mathit{false}$ for some $A \in \Gamma'$, or $\mathcal{S}(A) \geq_t \mathit{true}$ for some $A \in \Delta'$.

An L-structure $\mathcal{S}$ is called *acceptable* iff $\varepsilon\vdash$ and $\vdash\varepsilon$ preserve validity on $\mathcal{S}$. $\mathcal{S}$ is *a model of* G, if in addition to being acceptable, $\mathcal{S}$ is 3-valued. Hence, the notion of the *validity of a sequent* is defined over the entire class of acceptable L-structures, not just the 3-valued structures. Still, the role of the top element of our lattice of truth values is restricted to that of constructing 3-valued L-structures (in particular, models of G).

It is routine to check that the axioms sequents of G are valid on any L-structure $\mathcal{S}$, and that the rules of G, other than $\varepsilon\vdash$ and $\vdash\varepsilon$, preserve validity on $\mathcal{S}$. Hence we have our version of the soundness result for NaDSet I:

THEOREM 3.3 **Soundness of G**. *All sequents which are derivable in* G *are valid on every acceptable L-structure.*

The rest of this section presents some standard definitions and results from the truth theory literature. Here especially, our presentation benefits from *Fitting 1993*:

DEFINITION 3.4. Let $\varepsilon_1, \varepsilon_2 : ctrm^3 \to \mathbf{4}$. We put $\varepsilon_1 \leq_k \varepsilon_2$ iff $\varepsilon_1(t_1,t_2,t_3) \leq_k \varepsilon_2(t_1,t_2,t_3)$. Let $\mathcal{S}_1$ and $\mathcal{S}_2$ be L-structures. Then, we put $\mathcal{S}_1 \leq_k \mathcal{S}_2$ iff $\varepsilon_{\mathcal{S}_1} \leq_k \varepsilon_{\mathcal{S}_2}$.

Note that $\lambda xyz \in ctrm.\ \bot$, which is a representation of the "empty" membership operator, is $\leq_k$ minimal among the membership operators. The space of L-structures is a complete lattice under $\leq_k$; the space of 3-valued L-structures is a complete semi-lattice under $\leq_k$. We let $\mathcal{A}_0$ denote ($ctrm$, $\lambda xyz.\bot$), the $\leq_k$ minimal L-structure.

DEFINITION 3.5. Given L-structures $\mathcal{S}_1$ and $\mathcal{S}_2$, define $\varepsilon_{\mathcal{S}_1} \vee \varepsilon_{\mathcal{S}_2} : ctrm^3 \to \mathbf{4}$ pointwise in $(\mathbf{4}, \leq_k)$ by

$$(\varepsilon_{\mathcal{S}_1} \vee \varepsilon_{\mathcal{S}_2})(t_1,t_2,t_3) = \varepsilon_{\mathcal{S}_1}(t_1,t_2,t_3) \vee \varepsilon_{\mathcal{S}_2}(t_1,t_2,t_3) \qquad (t_1,t_2,t_3 \in ctrm).$$

Then the join $\mathcal{S}_1 \vee \mathcal{S}_2$ of $\mathcal{S}_1$ and $\mathcal{S}_2$ in the $\leq_k$ ordering is the L-structure $(ctrm, \varepsilon_{\mathcal{S}_1} \vee \varepsilon_{\mathcal{S}_2})$.

LEMMA 3.6. *Let $\mathcal{S}_1$ and $\mathcal{S}_2$ be L-structures. If $\mathcal{S}_1 \leq_k \mathcal{S}_2$, then $\mathcal{S}_1(A) \leq_k \mathcal{S}_2(A)$ for all sentences A.*

Proof. By induction on the formation of A, using the monotonicity of the sentential connectives and quantifiers for the induction step. ⊣

DEFINITION 3.7 The Gilmore Revision Operator. For each $\varepsilon_0 : ctrm^3 \to \mathbf{4}$, we may define $\Phi(\varepsilon_0) : ctrm^3 \to \mathbf{4}$ as follows: For all constant terms $[x\,y : A], t_1, t_2$,

$$\Phi(\varepsilon_0)([x\,y : A], t_1, t_2) = (ctrm, \varepsilon_0)A_{x\,y}[t_1, t_2].$$

Φ is called *the Gilmore Revision Operator*. Then, for each L-structure $\mathcal{S}$, let $\Phi(\mathcal{S})$ be the L-structure $(ctrm, \Phi(\varepsilon_{\mathcal{S}}))$. Trivially, if $\mathcal{S}$ is 3-valued, then so is $\Phi(\mathcal{S})$.

THEOREM 3.8 Monotonicity of Φ. *Let $\mathcal{S}_1$ and $\mathcal{S}_2$ be L-structures. Then*

$$\mathcal{S}_1 \leq_k \mathcal{S}_2 \quad \Rightarrow \quad \Phi(\mathcal{S}_1) \leq_k \Phi(\mathcal{S}_2).$$

Proof. Suppose $\mathcal{S}_1 \leq_k \mathcal{S}_2$, i.e., $\varepsilon_{\mathcal{S}_1} \leq_k \varepsilon_{\mathcal{S}_2}$. Then by Lemma 3.6, $(ctrm, \varepsilon_{\mathcal{S}_1})(B) \leq_k (ctrm, \varepsilon_{\mathcal{S}_2})(B)$ for all sentences B, so by definition Φ, 3.7, $\Phi(\varepsilon_{\mathcal{S}_1})(t_1,t_2,t_3) \leq_k \Phi(\varepsilon_{\mathcal{S}_2})([x\,y : A(x,y)], t_1, t_2)$ for all formulas $A(x,y)$ and constant terms t_1, t_2. It follows that

$$\Phi(\varepsilon_{\mathcal{A}})(t_1,t_2,t_3) \leq_k \Phi(\varepsilon_{\mathcal{B}})(t_1,t_2,t_3) \qquad (t_1,t_2,t_3 \in ctrm),$$

i.e., $\Phi(\varepsilon_{\mathcal{S}_1}) \leq_k \Phi(\varepsilon_{\mathcal{S}_2})$. So $\Phi(\mathcal{S}_1) \leq_k \Phi(\mathcal{S}_2)$ as required. ⊣

THEOREM 3.9. *There is an acceptable 3-valued L-structure.*

Proof. By Theorem 3.8, Φ is monotone with respect to the complete semi-lattice that the space of 3-valued L-structures forms under $\leq_k$. Hence, by the Knaster-Tarski lemma, there is a least (and also a greatest) fixed point A of Φ. Its easy to check that for all formulas $A(x, y)$

$$(t_1, t_2)\, \varepsilon\, [x\, y : A(x, y)] \leftrightarrow_{\mathcal{A}} A(t_1, t_2) \qquad (t_1, t_2 \in \mathit{ctrm}, \text{ free for } x \text{ and } y),$$

whence $\varepsilon\vdash$ and $\vdash\varepsilon$ preserve validity on $\mathcal{A}$. ⊣

Hence, there is a model $\mathcal{A}$ of G. It follows that G is syntactically consistent in the sense that the empty sequent, $\varnothing \vdash \varnothing$, is not derivable. Suppose $\varnothing \vdash \varnothing$ is a theorem of G; let A be any sentence of L. Then by Thinning both $\vdash A$ and $\vdash \neg A$ are theorems, whence both are valid on $\mathcal{A}$ by 3.9. Hence, *false* $\leq_t A, \neg A$, and so $\mathcal{A}(A) = \mathcal{A}(\neg A) =$ *both*, contrary to the fact that $\mathcal{A}$ is 3-valued.

In the literature, e.g., *Kripke 1975*, *Gilmore 1984* and *Feferman 1984*, it is usual to explicitly construct the least fixed point of a revision operator "from below", as the $\leq_k$ extension of $(\mathit{ctrm}, \lambda xyz.\bot)$ obtained by closing that initial structure off under Φ by ordinal recursion. We specify a slightly more general construction which will be of use in modeling logical extensions of G.

DEFINITION 3.10. Let $\mathcal{S}$ be an L-structure. We construct an ordinal sequence $\{\mathcal{S}_\alpha \mid \alpha \leq \omega_1\}$ by induction as follows:

0. $\mathcal{S}_0 = \mathcal{S}$,
1. $\mathcal{S}_{\alpha+1} = \mathcal{S}_\alpha \vee \Phi(\mathcal{S}_\alpha)$,
2. $\mathcal{S}_\alpha = \bigvee_{\beta<\alpha} \mathcal{S}_\beta$ for limit ordinals α.

Let $\mathcal{A} =_{\mathrm{df}} \mathcal{A}_{\omega_1}$, where $\mathcal{A}_0 = (\mathit{ctrm}, \lambda xyz.\bot)$. Note it obvious that $\beta \leq \alpha \leq \omega_1 \Rightarrow \mathcal{S}_\beta(A) \leq_k \mathcal{S}_\alpha(A)$ for all sentences A.

OBSERVATION 3.11. $\mathcal{A}$ is the least fixed point of the Gilmore revision operator Φ.

Suppose we are given a sequence of L-structures $\{\mathcal{S}_\alpha \mid \alpha \leq \omega_1\}$ as per Definition 3.10. For each sentence B that converges in $\mathcal{S}_{\omega_1}$ we define *the level of B in $\{\mathcal{S}_\alpha \mid \alpha \leq \omega_1\}$* to be the least α such that B converges on $\mathcal{S}_\alpha$. For each sentence B that converges on $\mathcal{A}$, we define *the level of B* (simpliciter) to be the level of B in the canonical sequence $\{\mathcal{A}_\alpha \mid \alpha \leq \omega_1\}$. In contexts where $\{\mathcal{S}_\alpha \mid \alpha \leq \omega_1\}$ is given, we write simply $\rightarrow_\alpha$ for $\rightarrow_{\mathcal{S}_\alpha}$ and similarly for cognate notation.

3.3. Modeling logical extensions of G

Let $\mathcal{S}$ be an acceptable L-structure. If t_1 is a constant term, we say that $\mathcal{S}$ is t_1-total iff

$$\rightarrow_{\mathcal{S}} \forall x\, y((x,y)\, \varepsilon\, t_1 \vee \neg(x,y)\, \varepsilon\, t_1).$$

Hence, $\mathcal{S}$ is t_1-total just in case for all constant terms t_2, t_3, either $\rightarrow_{\mathcal{S}}(t_2,t_3)\,\varepsilon\, t_1$ or $\leftarrow_{\mathcal{S}} (t_2,t_3)\,\varepsilon\, t_1$. Then it is clear that a t_1-total L-structure $\mathcal{S}$ validates all bivalence sequents of the form

$$(s_2, s_3)\,\varepsilon\, t_1 \vdash (s_2,s_3)\,\varepsilon\, t_1$$

for closed terms s_1 and s_2. Hence, a t_1-total L-structure provides a model of the logical extensions of G obtained by adding the constant term t_1 as a new ground predicate.

In section 7, for each primitive recursive function f, we introduce constant terms t_f computing f and show that $\mathcal{A}$ is t_f-total. Hence, all "primitive recursive terms", i.e., those of the form t_f for some primitive recursive f, may be consistently adjoined as new ground atoms to G. In particular, in sections 8 and 9 we study the logical extension $\mathrm{G}^{+,\times}$ of G obtained by adjoining bivalence axioms treating t_+ and $t_\times$, the constant set terms representing addition and multiplication over the natural numbers, as new ground atoms.

4. SOME SEMANTIC RECURSION THEOREMS

4.1. Recursion introduction and elimination

The purpose of this section is to establish semantic versions of the derived rules $\varepsilon_k\vdash$, $\vdash\varepsilon_k$ and consequently the derived recursion rules $\mathit{fix}_y\vdash$ and $\vdash \mathit{fix}_y$. Let $\mathcal{S} = \mathcal{S}_0$ be an L-structure and $\{\mathcal{S}_\alpha \mid \alpha \leq \omega_1\}$ be the ordinal sequence of extensions of $\mathcal{S}$ obtained by Definition 3.10. Let B be a sentence of the form $(\vec{t}\,)\,\varepsilon\,[\vec{x} : A(\vec{x})]$, for some constant term $\vec{t} = t_1, \ldots, t_k$. Our first task is to see that Φ simulates Φ_k for all $k \geq 2$:

LEMMA 4.1. $\mathcal{S}_\alpha(A(\vec{t}\,)) = \Phi(\mathcal{S}_\alpha)((\vec{t}\,)\,\varepsilon\,[\vec{x} : A(\vec{x})]) \qquad (\alpha \leq \omega_1)$.

LEMMA 4.2 **Abstraction**. *For all* $\alpha \leq \omega_1$,

(a) $\rightarrow_{\mathcal{S}_\alpha} A(\vec{t}\,) \quad \Rightarrow \quad \rightarrow_{\mathcal{S}_{\alpha+1}} (\vec{t}\,)\,\varepsilon\,[\vec{x} : A(\vec{x})]$,

(b) $\leftarrow_{\mathcal{S}_\alpha} A(\vec{t}\,) \quad \Rightarrow \quad \leftarrow_{\mathcal{S}_{\alpha+1}} (\vec{t}\,)\,\varepsilon\,[\vec{x} : A(\vec{x})]$.

Proof. By Lemma 4.1, using Definition 3.10. ⊣

LEMMA 4.3 **Recursion Introduction**. *For all* $\alpha \leq \omega_1$,

(a) $\rightarrow_{\mathcal{S}_\alpha} A(\vec{t}, \mathit{fix}_y A) \quad \Rightarrow \quad \rightarrow_{\mathcal{S}_{\alpha+2}} (\vec{t}\,)\,\varepsilon\, \mathit{fix}_y A$,

(b) $\leftarrow_{\mathcal{S}_\alpha} A(\vec{t}, \mathit{fix}_y A) \quad \Rightarrow \quad \leftarrow_{\mathcal{S}_{\alpha+2}} (\vec{t}\,)\,\varepsilon\, \mathit{fix}_y A$.

Proof. By a "semantic" version of Andrew's recursion theorem. ⊣

Now we represent the converse of abstraction and recursion rules.

LEMMA 4.4 **Conversion**. *Suppose the level of* $B =_{df} (\vec{t}\,) \in [\vec{x} : A(\vec{x})]$ *in* $\{\mathcal{S}_\alpha \mid \alpha \leq \omega_1\}$ *is* α *for some ordinal* $\alpha < \omega_1$ *such that* $\alpha \neq 0$. *Then*

$$\mathcal{S}_\alpha\big((\vec{t}\,) \in [\vec{x} : A(\vec{x})]\big) = \mathcal{S}_{\alpha-1}\big(A(\vec{t}\,)\big).$$

Proof. Since α is a successor ordinal, $\mathcal{S}_\alpha = \mathcal{S}_{\alpha-1} \vee \Phi(\mathcal{S}_{\alpha-1})$ by construction 3.10. So by Definition 3.5,

$$\mathcal{S}_\alpha(B) =$$
$$\varepsilon_{\alpha-1}\big([\vec{x} : A(\vec{x})], \langle t_1, \ldots, t_{k-1}\rangle, t_k\big) \vee \Phi(\varepsilon_{\alpha-1})\big([\vec{x} : A(\vec{x})], \langle t_1, \ldots, t_{k-1}\rangle, t_k\big).$$

By hypothesis $\varepsilon_{\alpha-1}\big([\vec{x} : A(\vec{x})], \langle t_1, \ldots, t_{k-1}\rangle, t_k\big) = \bot$. Hence

$$\mathcal{S}_\alpha(B) = \Phi(\varepsilon_{\alpha-1})\big([\vec{x} : A(\vec{x})], \langle t_1, \ldots, t_{k-1}\rangle, t_k\big) = \text{(by 4.1) } \mathcal{S}_{\alpha-1}\big(A(\vec{t}\,)\big),$$

as claimed. ⊣

LEMMA 4.5 **Recursion Elimination**. *Let* $A = A(\vec{x}, y)$ *be a formula. Suppose the level of* $(\vec{t}\,) \in \mathit{fix}_y A$ *in* $\{\mathcal{S}_\alpha \mid \alpha \leq \omega_1\}$ *is a for some ordinal* α, $2 \leq \alpha < \omega_1$. *Then*

$$\mathcal{S}_\alpha\big((\vec{t}\,) \in \mathit{fix}_y A\big) = \mathcal{S}_{\alpha-2}\big(A(\vec{t}, \mathit{fix}_y A)\big).$$

Proof. Recall $(\vec{t}\,) \in \mathit{fix}_y A =_{\mathrm{df}} (\vec{t}\,) \in [\vec{x} : (\vec{x}\, \mathit{diag}_y A) \in \mathit{diag}_y A]$. Since the level of $(\vec{t}\,) \in \mathit{fix}_y A$ is α, the level of $(\vec{t} \mathit{diag}_y A) \in \mathit{diag}_y A$ is $\alpha - 1$. Then, similarly, the level of $A(\vec{t}, \mathit{fix}_y A)$ is $\alpha - 2$. Then we have

$$\begin{aligned}
\mathcal{S}_\alpha\big((\vec{t}\,) \in \mathit{fix}_y A\big) &= \mathcal{S}_\alpha\big((\vec{t}\,) \in [\vec{x} : (\vec{x}\ \mathit{diag}_y A) \in \mathit{diag}_y A]\big) && (\text{df. } \mathit{fix}_y A)\\
&= \mathcal{S}_{\alpha-1}\big((\vec{t}, \mathit{diag}_y A) \in \big[\vec{x}y : A(\vec{x}, [\vec{x} : (\vec{x}\, y) \in y])\big]\big) && (\text{4.4, df. } \mathit{diag}_y A)\\
&= \mathcal{S}_{\alpha-2}\big(A(\vec{t}, [\vec{x} : (\vec{x}\ \mathit{diag}_y A) \in \mathit{diag}_y A])\big) && (4.4)\\
&= \mathcal{S}_{\alpha-2}\big(A(\vec{t}, \mathit{fix}_y A)\big) && (\text{df. } \mathit{fix}_y A)
\end{aligned}$$

as required. ⊣

4.2. A finite level result for positive sentences

A formula A is *positive* just in case A's only connectives are $\neg$, $\wedge$, $\vee$ and $\exists$, and all occurrences of $\neg$ prefix identity statements in A. Our next result ensures that positive sentences which are verified in the canonical model $\mathcal{A}$ are of finite level. In section 6, we show that positive formulas are sufficient for computational purposes.

LEMMA 4.6. *Let A be a positive sentence. Suppose $\rightarrow_\alpha A$ for some $\alpha \leq \omega_1$. Then A is of finite level.*

Proof. By induction on α. ⊣

5. THE GENESIS OF NUMBER IN G

5.1. *The theory of* $\langle \mathbf{0}, S \rangle$

In this section, we use Andrews' recursion theorem to construct a constant set term $\underline{\omega}$ which enumerates the set of natural numbers in G. First, we define

$$\underline{0} = [v_1 : v_1 \neq v_1],$$
$$St = [v_1 : v_1 = t] \qquad (t \in ctrm).$$

That is, the natural number 0 is represented by the term for the empty set, and the successor function is represented by the operation of taking singleton sets. Let *Num*, the set of *Gilmore numerals*, be the smallest subset of *ctrm* that contains $\underline{0}$ and contains St whenever it contains the term t. Write $\underline{n}$ for $S^n\underline{0}$ ($n \in \mathbb{N}$); $\underline{n}$ is the constant term corresponding to the natural number n. The following derivation of the theory of 0 and the successor function follows that of Gilmore (private correspondence) for NaDSet I:

A1: $\forall x\,(Sx \neq \underline{0})$ ("0 is not the successor of any number"),
A2: $\forall x_1\, x_2\,(Sx_1 = Sx_2 \rightarrow x_1 = x_2)$ ("Successor is 1 - 1").

LEMMA 5.1. $\vdash \forall x\,(Sx \neq \underline{0})$.

Proof.

(1) $\vdash a = a$
(2) $\vdash a \,\varepsilon\, [v_1 : v_1 = a]$
(3) $\vdash a \,\varepsilon\, Sa$
(4) $a \neq a \vdash$
(5) $a \,\varepsilon\, [v_1 : v_1 \neq v_1] \vdash$
(6) $a \,\varepsilon\, \underline{0} \vdash$
(7) $Sa = \underline{0} \vdash$ (3, 5, by $=\vdash$)
(8) $\vdash Sa \neq \underline{0}$
(9) $\vdash \forall x\,(Sx \neq \underline{0})$ ⊣

LEMMA 5.2. $\vdash \forall x_1\, x_2\,(Sx_1 = Sx_2 \rightarrow x_1 = x_2)$.

Proof.

(1) $a_2 = a_1 \vdash a_1 = a_2$

(2) $a_2 \,\varepsilon\, [v_1 : v_1 = a_1] \vdash a_1 = a_2$
(3) $a_2 \,\varepsilon\, Sa_1 \vdash a_1 = a_2$
(4) $\vdash a_2 \,\varepsilon\, Sa_2$
(5) $Sa_1 = Sa_2 \vdash a_1 = a_2$ (3, 4, by $=\vdash$)
(6) $\vdash Sa_1 = Sa_2 \rightarrow a_1 = a_2$
(7) $\vdash \forall x_1\, x_2\, (Sx_1 = Sx_2 \rightarrow x_1 = x_2)$ $\dashv$

COROLLARY 5.3. *If* $m \neq n$, *then* $\underline{m} = \underline{n} \vdash$ $(n \in \mathbb{N})$.

5.2. The natural numbers object

DEFINITION 5.4. Let $\Gamma \subseteq ctrm^k$ and t be a constant k-ary abstraction term. We say that t *enumerates* Γ (*in* G) iff

$$\Gamma = \{(s_1, \ldots, s_k) \mid \vdash (s_1 \ldots s_k) \,\varepsilon\, t \text{ is derivable}\}.$$

We say that Γ is *G-enumerable* iff there is a constant k-ary abstraction term t such that t enumerates Γ. We say that $S \subseteq \mathbb{N}^k$ is *G-enumerable* iff $\{(\underline{n_1} \ldots \underline{n_k}) \mid (n1, \ldots, nk) \in S\}$ is G-enumerable.

DEFINITION 5.5 **The Natural Numbers Object**. Let $A(x, y)$ be the formula

$$x = \underline{0} \vee \exists z\, (x = Sz \wedge z \,\varepsilon\, y),$$

and let $\underline{\omega}$ be the set term $fix_y A$.

The main goal of this section is to show that $\underline{\omega}$ enumerates $\mathbb{N}$ in G. However, since $\underline{\omega}$ is our first example of a syntactically complex ground predicate, and our notion of *grounding* is semantically motivated, we first establish that $\underline{\omega}$ represents a "total" set in $\mathcal{A}$.

LEMMA 5.6. $\rightarrow_{2(n+1)}\underline{n} \,\varepsilon\, \underline{\omega}$ $(n \in \mathbb{N})$.

Proof. By induction on n. For the basis, let $n = 0$:

(1) $\rightarrow_0 \underline{0} = \underline{0}$ (df. 3.1)
(2) $\rightarrow_0 \underline{0} = \underline{0} \vee \exists z\, (x = Sz \wedge z \,\varepsilon\, \underline{\omega})$ (df. 3.1)
(3) $\rightarrow_2 \underline{0} \,\varepsilon\, \underline{\omega}$ (4.3)

For the induction step, assume $n = k + 1$ and that $\rightarrow_{2(k+1)}\underline{k} \,\varepsilon\, \underline{\omega}$ for some $k \in \mathbb{N}$. Then we have

(1) $\rightarrow_{2(k+1)}\underline{k} \,\varepsilon\, \underline{\omega}$ (assumption)
(2) $\rightarrow_0 S\underline{k} = S\underline{k}$ (df. 3.1)
(3) $\rightarrow_{2(k+1)} S\underline{k} = S\underline{k} \wedge \underline{k} \,\varepsilon\, \underline{\omega}$ (1, 2, df. 3.1)
(4) $\rightarrow_{2(k+1)} \exists z\, (S\underline{k} = Sz \wedge z \,\varepsilon\, \underline{\omega})$ (3, df. 3.1)

(5) $\rightarrow_{2(k+1)} S\underline{k} = \underline{0} \vee \exists z\,(S\underline{k} = Sz \wedge z \,\varepsilon\, \underline{\omega})$ (4, df. 3.1)
(6) $\rightarrow_{2(k+1)+2} S\underline{k} \,\varepsilon\, \underline{\omega}$ (4, 4.3, df. $\underline{\omega}$)
(7) $\rightarrow_{2(k+2)} \underline{n} \,\varepsilon\, \underline{\omega}$

Hence, $\rightarrow_{2(n+1)} \underline{n} \,\varepsilon\, \underline{\omega}$, as required to complete the proof of Lemma 5.6. ⊣

LEMMA 5.7. *Let t_1 be a constant term which is not a Gilmore numeral. Then* $\leftarrow_{\mathcal{A}} t_1 \,\varepsilon\, \underline{\omega}$.

Proof. Let the *rank*, $R(t)$, of a term t be the maximum $n \geq 0$ such that $t = S^n s$ for some term s. Let $R(t_1) = n$; we show by induction on n that $\leftarrow_{2n} t_1 \,\varepsilon\, \underline{\omega}$. ⊣

COROLLARY 5.8. $\rightarrow_\omega \forall x\,(x \,\varepsilon\, \underline{\omega} \vee \neg\, x \,\varepsilon\, \underline{\omega})$.

THEOREM 5.9 **The Omega Theorem.** $\underline{\omega}$ *enumerates* $\mathbb{N}$.

Proof. It suffices to show that:
(1) $\vdash \underline{n} \,\varepsilon\, \underline{\omega}$ $(n \in \mathbb{N})$, and
(2) $\vdash t \,\varepsilon\, \underline{\omega} \Rightarrow t \in Num$ (for all terms t).

(2) follows from Lemma 5.7 and the soundness of G with respect to $\mathcal{A}$: Let $t \notin Num$ be a term. Then t has a constant instance t' which is likewise not a numeral. By 5.7, $\leftarrow_{\mathcal{A}} t' \,\varepsilon\, \underline{\omega}$; since $\mathcal{A}$ is 3-valued, it follows that it is not the case that $\rightarrow_{\mathcal{A}} t' \,\varepsilon\, \underline{\omega}$. Hence, by soundness, $\vdash t' \,\varepsilon\, \underline{\omega}$ is not derivable in G. Since theorems of G are closed under the uniform substitution of arbitrary closed terms for parameters, it follows that $\vdash t \,\varepsilon\, \underline{\omega}$ is not a theorem of G, as required.

The proof of (1) is by induction on n. For $n = 0$, we have

(1) $\vdash \underline{0} = \underline{0}$ (= ax.)
(2) $\vdash \underline{0} = \underline{0} \vee \exists z\,(\underline{0} = Sz \wedge z \,\varepsilon\, \underline{\omega})$ (1, $\vdash\vee$)
(3) $\vdash \underline{0} \,\varepsilon\, \underline{\omega}$ (rec. thm.)

For the induction step, assume $n = k + 1$ and that $\vdash \underline{k} \,\varepsilon\, \underline{\omega}$ has been derived. Then we have

(1) $\vdash \underline{k} \,\varepsilon\, \underline{\omega}$ (assumption)
(2) $\vdash S\underline{k} = S\underline{k}$ (= axiom)
(3) $\vdash S\underline{k} = S\underline{k} \wedge \underline{k} \,\varepsilon\, \underline{\omega}$ (2, 3, $\vdash\wedge$)
(4) $\vdash \exists z\,(S\underline{k} = Sz \wedge z \,\varepsilon\, \underline{\omega})$ ($\vdash\exists$)
(5) $\vdash S\underline{k} = \underline{0} \vee \exists z\,(S\underline{k} = Sz \wedge z \,\varepsilon\, \underline{\omega})$ ($\vee\vdash$)
(6) $\vdash S\underline{k} \,\varepsilon\, \underline{\omega}$ (rec. thm.)

⊣

It is time to consider our first logical extension of G.

6. G^{ω} AND A DERIVATION OF THE AXIOM OF INFINITY

6.1. *The axiom of infinity*

The purpose of this section is to draw attention to a connection in a logical extension of G between the principle of bivalence for atomic sentences featuring $\underline{\omega}$ as a predicate term and the Axiom of Infinity, asserting that $\underline{\omega}$ is a 0-successor set. Roughly put, the former entails the latter in short order. This connection is surprising, since bivalence is often taken as the paradigm of a *logical* principle, whereas critics of logicism have taken the Axiom of Infinity to be unarguably *extra logical.* The main point of this section is that the coherence of the distinction between the "logical" and the "extralogical" may be harder to defend than previously thought.

Let G^{ω} be the logical extension of G obtained by adjoining as new bivalence axioms all sequents of the form

$$t \,\varepsilon\, \underline{\omega} \vdash t \,\varepsilon\, \underline{\omega}$$

for closed terms t. The idea here is that, while G is sufficient to *compute* all r.e. sets, it is too weak to *reason* about the sets it computes. By adding bivalence axioms for atomic formulas whose predicate term is $\underline{\omega}$, we extend G's ability to reason about the first infinite collection. In particular, G^{ω} can prove that $\underline{\omega}$ is, in fact, a 0-successor set.

THEOREM 6.1 *The Axiom of Infinity.* $\vdash \exists x\,(\underline{0} \,\varepsilon\, x \wedge \forall y\,(y \,\varepsilon\, x \to Sy \,\varepsilon\, x))$ *is derivable in* G^{ω}.

Proof. It suffices to derive $\vdash \underline{0} \,\varepsilon\, \underline{\omega} \wedge \forall y\,(y \,\varepsilon\, \underline{\omega} \to Sy \,\varepsilon\, \underline{\omega})$.

(1) $p \,\varepsilon\, \underline{\omega} \vdash p \,\varepsilon\, \underline{\omega}$	(G^{ω} axiom)
(2) $p \,\varepsilon\, \underline{\omega} \vdash Sp = Sp \wedge p \,\varepsilon\, \underline{\omega}$	(= axiom, $\vdash\wedge$)
(3) $p \,\varepsilon\, \underline{\omega} \vdash \exists z\,(Sp = Sz \wedge z \,\varepsilon\, \underline{\omega})$	($\vdash\exists$)
(4) $p \,\varepsilon\, \underline{\omega} \vdash Sp = \underline{0} \vee \exists z\,(Sp = Sz \wedge z \,\varepsilon\, \underline{\omega})$	($\vdash\vee$)
(5) $p \,\varepsilon\, \underline{\omega} \vdash Sp \,\varepsilon\, \underline{\omega}$	(rec. thm.)
(6) $\vdash p \,\varepsilon\, \underline{\omega} \to Sp \,\varepsilon\, \underline{\omega}$	($\vdash\to$)
(7) $\vdash \forall y\,(y \,\varepsilon\, \underline{\omega} \to Sy \,\varepsilon\, \underline{\omega})$	($\vdash\forall$)
(8) $\vdash \underline{0} \,\varepsilon\, \underline{\omega}$	(omega thm.)
(9) $\vdash \underline{0} \,\varepsilon\, \underline{\omega} \wedge \forall y\,(y \,\varepsilon\, \underline{\omega} \to Sy \,\varepsilon\, \underline{\omega})$	(7, 8, $\vdash\wedge$)

as required. ⊣

6.2. *Semantics for* G^{ω}

We need only observe that $\mathcal{A}$ is a model of G^{ω} to see that G^{ω} is consistent.

THEOREM 6.2. G^{ω} *is consistent.*

Proof. By 5.9, $\mathcal{A}$ is an ω-total L-structure. Hence, $\mathcal{A}$ is a model of G^{ω}. ⊣

7. HOW TO COMPUTE IN G

Before we can explore the space of logical extensions of G, we need to return to G and finish an important, though pedestrian, logicist task. We need to show the reader how to compute in G. We associate a positive set abstraction term with each primitive recursive function of the natural numbers. Intuitively, existential quantification is used to effect the composition of primitive recursive functions, and fix_y is used to simulate primitive recursion. Then, the existential quantifier will simulate unbounded search over $\mathbb{N}$ to yield representations of all partial recursive functions.

Let $[\vec{x}\,\varepsilon\,\underline{\omega} : A]$ abbreviate the set term $[\vec{x} : x_1\,\varepsilon\,\underline{\omega} \wedge \ldots \wedge x_1\,\varepsilon\,\underline{\omega} \wedge A]$.

Definition 7.1. We associate with each primitive recursive function $f : \mathbb{N}^k \to \mathbb{N}$ a constant k+1-ary set abstraction term t_f, intended to enumerate the graph of f, by induction on the formation history of f:

Initial Functions:

(1) Suppose f is the 0-function $\lambda m.0$. Then t_f is the binary abstraction term $[x\,y : x\,\varepsilon\,\underline{\omega} \wedge y = \underline{0}]$.
(2) Suppose f is the successor function $\lambda m.\, m+1$. Then t_f is the binary term $[x\,y : x\,\varepsilon\,\underline{\omega} \wedge y = Sx]$.
(3) Suppose f is the projection function $\lambda m_1, \ldots, m_k . m_i$ for some $k \geq 0$ and i, $1 \leq i \leq_k$. Then t_f is the k+1-ary term $[x_1 \ldots x_k\,\varepsilon\,\underline{\omega} : y = x_i]$.

Composition:

(4) Let $k \geq 0, n > 0$. Suppose $h : \mathbb{N}^n \to \mathbb{N}$ and $g_i : \mathbb{N}^k \to \mathbb{N}$ are primitive recursive functions $(1 \leq i \leq n)$ and f is $\lambda\overline{m}h(g_1(\overline{m}), \ldots, g_k(\overline{m}))$, where $\overline{m} = m_1, \ldots, m_k$. Then t_f is the k+1-ary term

$$[\vec{x}y : (\exists z_1 \ldots z_n)\{(\vec{x}, z_1)\,\varepsilon\,t_{g_1} \wedge \ldots \wedge (\vec{x}, z_n)\,\varepsilon\,t_{g_n} \wedge (z_1 \ldots z_n, y)\,\varepsilon\,t_h\}].$$

Primitive Recursion:

(5) Suppose $g : \mathbb{N}^{k-1} \to \mathbb{N}$ and $h : \mathbb{N}^{k+1} \to \mathbb{N}$ are primitive recursive and $f : \mathbb{N}^k \to \mathbb{N}$ is given by

$$f(0, m_2, \ldots, m_k) = g(m_2, \ldots, m_k),$$
$$f(j+1, m_2, \ldots, m_k) = h(j, f(j, m_2, \ldots, m_k), m_2, \ldots, m_k,).$$

Inductively, we have k-ary, k+2-ary, terms t_g, t_h representing g, h, respectively. Let $\vec{x} = x_1, \ldots, x_k$. Let $A(\vec{x}, y, z)$ be the formula

$$(x_1 = \underline{0} \wedge (x_2 \ldots x_k, y)\,\varepsilon\,t_g) \;.\!\vee. \quad \exists u\,v\,(x_1 = Su \wedge (u\,x_2 \ldots x_k, v)\,\varepsilon\,z \wedge (u\,v\,x_2 \ldots x_k, y)\,\varepsilon\,t_h).$$

Then t_f is the k+1-ary set term $fix_z A(\vec{x}, y, z)$.

THEOREM 7.2. *Let* $f : \mathbb{N}^k \to \mathbb{N}$ *be primitive recursive and* $\vec{m} = m_1, \ldots, m_k \in \mathbb{N}$. *Then*

$$\vdash (\underline{m_1} \ldots \underline{m_k}, f(\underline{\vec{m}}))\, \varepsilon\, t_f$$

is derivable.

Proof. By induction on the formation history of f, using an inner induction on m_1 to establish the primitive recursion case of the induction step. ⊣

THEOREM 7.3. *Let* $f : \mathbb{N}^k \to \mathbb{N}$ *be primitive recursive. Let* $t_1, \ldots, t_k, s$ *be constant terms such that it is not the case that there are* $m_1, \ldots, m_k, n_s \in \mathbb{N}$ *such that* $t_i = \underline{m_i}$ $(0 \leq i \leq_k), s = \underline{n_s}$ *and* $f(m_1, \ldots, m_k) = n_s$. *Then* $\leftarrow_{\mathcal{A}} (t_1 \ldots t_k, s)\, \varepsilon\, t_f$.

Proof. By induction on the formation history of f. The primitive recursion case of the induction step is established on the basis of the following three claims, the first of which is established by induction on n_1:

Claim 1: Let $n_1, \ldots, n_k, n_{k+1} \in \mathbb{N}$ with $f(n_1, \ldots, n_k) \neq n_{k+1}$.
Then $\leftarrow (\underline{n_1} \ldots \underline{n_k}, \underline{n_{k+1}})\, \varepsilon\, t_f$.

Claim 2: $\leftarrow t_1 = \underline{0} \wedge (t_2 \ldots t_k, s)\, \varepsilon\, t_g$.

Claim 3: $\leftarrow \exists u\, v\, (t_1 = Su \wedge (u\, t_2 \ldots t_k, v)\, \varepsilon\, t_f \wedge (u\, v\, t_2 \ldots t_k, s)\, \varepsilon\, t_h)$. ⊣

COROLLARY 7.4. *Let* $f : \mathbb{N}^k \to \mathbb{N}$ *be primitive recursive. Let* $t_1, \ldots, t_k, s$ *be terms such that*

$$\vdash (t_1 \ldots t_k, s)\, \varepsilon\, t_f.$$

Then there are $m_1, \ldots, m_k, n_s \in \mathbb{N}$ *such that* $t_i = \underline{m_i}$ $(0 \leq i \leq_k), s = \underline{n}_s$ *and* $f(m_1, \ldots, m_k) = n_s$.

Proof. Suppose there are no $m_1, \ldots, m_k, n_s \in \mathbb{N}$ such that $t_i = \underline{m_i}$ $(0 \leq i \leq_k), s = \underline{n_s}$ and $f(m_1, \ldots, m_k) = n_s$. Then there are constant instances $t'_1, \ldots, t'_k, s'$ of $t_1, \ldots, t_k, s$ which likewise fail the conclusion of the Corollary. Then, by 7.3, $\leftarrow_{\mathcal{A}} (t'_1 \ldots t'_k, s')\, \varepsilon\, t_f$, whence $\rightarrow_{\mathcal{A}} (t'_1 \ldots t'_k, s')\, \varepsilon\, t_f$ fails since $\mathcal{A}$ is 3-valued. Then by soundness, $\vdash (t'_1 \ldots t'_k, s')\, \varepsilon\, t_f$ is not a theorem of G. Since theorems of G are closed under the uniform substitution of closed terms for parameters, it follows that $\vdash (t_1 \ldots t_k, s)\, \varepsilon\, t_f$ is not a theorem of G either. ⊣

COROLLARY 7.5. *Let* $f : \mathbb{N}^k \to \mathbb{N}$ *be primitive recursive. Then* t_f *enumerates* f *in* G.

Proof. By Theorem 7.2 and Corollary 7.4. ⊣

THEOREM 7.6. *All r.e. sets of numbers are* G *enumerable.*

Proof. Let $P : \mathbb{N}^k \to \{0,1\}$ be a primitive recursive predicate and t_P is a $k{+}1$-ary abstract term which enumerates the graph of P, as given by 7.5. Then the r.e. predicate

$$\{m_1, \ldots, m_{k-1} : \exists n\, P(m_1, \ldots, m_{k-1}, n) = 0\}$$

is enumerated by the $(k-1)$-ary abstract term

$$[x_1 \ldots x_{k-1} : \exists y\, ((x_1 \ldots x_{k-1}\, y\, \underline{0})\, \varepsilon\, t_P))].$$

Since every r.e. set of numbers can be obtained as the existential projection of the primitive recursive Kleene T-predicate, this is sufficient. ⊣

COROLLARY 7.7. *Let* $f : \mathbb{N}^k \to \mathbb{N}$ *be primitive recursive. Then*

$$\to_A \forall x\, (x\, \varepsilon\, t_f \vee \neg\, x\, \varepsilon\, t_f).$$

Proof. By 7.3 and 7.6. ⊣

8. TOWARDS A LOGICAL DERIVATION OF THE THEORY OF $\langle 0, S, +, \times\rangle$

8.1. The logical extension $\mathrm{G}^{+,\times}$

We pass to the logical extension $\mathrm{G}^{+,\times}$ by adjoining to G^{ω} excluded middle axiom sequents for the set terms representing the addition and multiplication functions over the natural numbers. Although Definition 7.1 provides set terms which enumerate the graphs of those functions, in this section we must reason formally over the content of these terms, so it is convenient to define them more simply, afresh. Let $t_+, t_\times$ be

$$\mathit{fix}_u B^+(x,y,z,u) \text{ and } \mathit{fix}_u B^\times(x,y,z,u),$$

respectively, where

$$B^+(x,y,z,u) =_{\mathrm{df}}$$
$$x = \underline{0} \wedge y = z \wedge y\, \varepsilon\, \underline{\omega} \;.\!\vee\!.\; \exists x'\, z'\, (x = Sx' \wedge z = Sz' \wedge (x'\, y\, z')\, \varepsilon\, u),$$
$$B^\times(x,y,z,u) =_{\mathrm{df}}$$
$$x = \underline{0} \wedge z = \underline{0} \wedge y\, \varepsilon\, \underline{\omega} \;.\!\vee\!.\; \exists x'\, v\, (x = Sx' \wedge (v\, y\, z)\, \varepsilon\, t_+ \wedge (x'\, y\, v)\, \varepsilon\, u).$$

Let $\mathrm{G}^{+,\times}$ be the logical extension of G^{ω} obtained by adjoining as new bivalence axioms all sequents of the form

$$(t_1\, t_2\, t_3)\, \varepsilon\, t_+ \vdash (t_1\, t_2\, t_3)\, \varepsilon\, t_+, \qquad (t_1\, t_2\, t_3)\, \varepsilon\, t_\times \vdash (t_1\, t_2\, t_3)\, \varepsilon\, t_\times$$

for closed terms t_1, t_2, t_3. So in $\mathrm{G}^{+,\times}$, t_+ and $t_\times$ count as new ground predicates. The reason why these arithmetic set terms are well behaved in their new role is that they have the syntactic structure of primitive recursive definitions. Infinite descending reduction sequences are ruled out because each application of a semantic rule in the semantic evaluation process monotonically reduces the syntactic complexity of the sentences requiring evaluation. However, absent from the above definitions of B^+ and $B^\times$ are conjuncts specified in 7.1 for primitive recursive terms which restrict the variables "x", "y" and "z" and the quantifier "$\exists v$" to $\underline{\omega}$. These conjuncts are required if $t_+, t_\times$ are to bear the *falsification* conditions asserted in 7.3 for primitive recursive terms, which underlie 7.7. The consistency of $\mathrm{G}^{+,\times}$ will therefore be established by modifying the construction of $\mathcal{A}$ in 3.1 to obtain a model in which $t_+, t_\times$ are total after all. In effect, we trade simplicity in the statement and derivation of the Peano Axioms for a slightly more complex consistency proof.

Let us pause to reflect upon developments so far. We've seen that G's bivalence axioms for identity statements underwrite a "classical fragment" of L containing all closed formulas whose atomic constituents are restricted to identity statements. On this basis, G yields the theory of zero and successor and also the Omega theorem stating that $\underline{\omega}$ enumerates the set of natural numbers. Thus, it is a metatheoretic fact about G that $\underline{\omega}$ is infinite. By adding bivalence axioms for all atomic formulas featuring $\underline{\omega}$ as predicate, and thus extending the classical fragment of L to include all closed formulas whose atomic constituents feature $\underline{\omega}$ as predicate, G^ω embeds this fact in the form a theorem which asserts that $\underline{\omega}$ is a 0-successor set. In this section, we further extend the classical fragment of L to include all sentences with "arithmetic syntactic structure" by adjoining bivalence axioms for atomic formulas having t_+ or $t_\times$ as predicate terms. We thus obtain a logical extension $\mathrm{G}^{+,\times}$ that underwrites the bivalence of a sublanguage of L containing a relational image of the arithmetic language based on 0, S, $+$ and $\times$. This sublanguage—the so called "L fragment"—will be explicitly defined in the next section. For now however, it is enough to note:

LEMMA 8.1. *Let A be a closed L formula all of whose atomic subformulas are either identity statements, or ε statements with one of $\underline{\omega}$, t_+ or $t_\times$ in predicate position. Then, $A \vdash A$ is derivable in $\mathrm{G}^{+,\times}$.*

Proof. By induction on the formation of A. The proof is formally identical to the proof that the classical Gentzen calculus LK can be based on sequents of the form $A \vdash A$ for *atomic* A. ⊣

Thus, Lemma 8.1 justifies the applications of $\mathit{fix}_y\vdash$ to formulas constructed from $\mathrm{G}^{+,\times}$ ground atoms.

8.2. Axioms for + and ×

We present the following "relational" versions of the Peano axioms which specify the recursive definitions of addition and multiplication over the natural numbers:

A3. $\forall x\,(\underline{0}\,x, x)\,\varepsilon\, t_+$.

A4. $\forall x\,y\,z\,((Sx\,y, Sz)\,\varepsilon\, t_+ \leftrightarrow (x\,y, z)\,\varepsilon\, t_+))$.

A5. $\forall x\,(\underline{0}\,x, \underline{0})\,\varepsilon\, t_\times$.

A6. $\forall x\,y\,z\,((Sx\,y, z)\,\varepsilon\, t_\times \leftrightarrow (\exists v)((v\,y, z)\,\varepsilon\, t_+ \wedge (x\,y, v)\,\varepsilon\, t_\times))$.

THEOREM 8.2. A3–A6 *are derivable in* $G^{+,\times}$.

Proof. The sequents $\vdash \forall x\,(\underline{0}\,x, x)\,\varepsilon\, t_+$ and $\vdash \forall x\,(\underline{0}\,x, \underline{0})\,\varepsilon\, t_\times$ are clearly derivable in G. So it suffices to show that we can derive A4 and A6. Note that the justification for applications of $\vdash \mathit{fix}_y$ and $\mathit{fix}_y \vdash$ below simply refer to "the recursion theorem". Also, we also use prefix notation, "$t_1(t_2\,t_3, t_4)$", suppressing "ε", in contexts where space is at a premium:

A4.1: $\forall x\,y\,z\,\big((Sx\,y, Sz)\,\varepsilon\, t_+ \rightarrow (x\,y, z)\,\varepsilon\, t_+\big)$.

(1) $a = a',\ c = c',\ (a'\,b, c')\,\varepsilon\, t_+ \vdash (a\,b, c)\,\varepsilon\, t_+$ ($G^{+,\times}$ ax, $=\vdash$)
(2) $a = a' \wedge c = c',\ (a'\,b, c')\,\varepsilon\, t_+ \vdash (a\,b, c)\,\varepsilon\, t_+$ (1, $\wedge\vdash$)
(3) $Sa = Sa' \wedge Sc = Sc' \vdash a = a' \wedge c = c'$ (A2, $\vdash\wedge$)
(4) $Sa = Sa' \wedge Sc = Sc' \wedge (a'\,b, c')\,\varepsilon\, t_+ \vdash (a\,b, c)\,\varepsilon\, t_+$ (3, 2, Cut)
(5) $\exists x'\,z'\,(Sa = Sx' \wedge Sc = Sz' \wedge (x'\,b, z')\,\varepsilon\, t_+) \vdash (a\,b, c)\,\varepsilon\, t_+$ (4, $\exists\vdash$)
(6) $Sa = \underline{0} \vdash$ (A1)
(7) $Sa = \underline{0} \wedge b = c\,\varepsilon\,\underline{\omega} \vdash$ (6, $\wedge\vdash$)
(8) $Sa = \underline{0} \wedge b = c\,\varepsilon\,\underline{\omega}\ .\!\vee\!.\ \exists x'\,z'\,(Sa = Sx' \wedge Sc = Sz' \wedge (x'\,b, z')\,\varepsilon\, t_+)$
$\vdash (a\,b, c)\,\varepsilon\, t_+$ (5, 7, $\vee\vdash$)
(9) $(Sa\,b, Sc)\,\varepsilon\, t_+ \vdash (a\,b, c)\,\varepsilon\, t_+$ (8, rec. thm.)
(10) $\vdash (Sa\,b, Sc)\,\varepsilon\, t_+ \rightarrow (a\,b, c)\,\varepsilon\, t_+$ (9, $\rightarrow\vdash$)
(11) $\vdash \forall x\,y\,z\,((Sx\,y, Sz)\,\varepsilon\, t_+ \rightarrow (x\,y, z)\,\varepsilon\, t_+)$ (10, $\vdash\forall$)

Line (1) is obtained by repeated applications of $=\vdash$ taking t_+ bivalence axioms as premises. Note the use of A2 at line 3.

A4.2: $\forall x\,y\,z\,\big((x\,y, z)\,\varepsilon\, t_+ \rightarrow (Sx\,y, Sz)\,\varepsilon\, t_+\big)$.

(1) $(a\,b, c)\,\varepsilon\, t_+ \vdash Sa = Sa \wedge Sc = Sc \wedge (a\,b, c)\,\varepsilon\, t_+$ ($G^{+,\times}$ ax, $=$ ax, $\vdash\wedge$)
(2) $(a\,b, c)\,\varepsilon\, t_+ \vdash \exists x'\,z'\,(Sa = Sx' \wedge Sc = Sz' \wedge (x'\,b, z')\,\varepsilon\, t_+)$ (1, $\vdash\exists$)
(3) $(a\,b, c)\,\varepsilon\, t_+ \vdash Sa = \underline{0} \wedge b = Sc\,\varepsilon\,\underline{\omega}\ .\!\vee\!.$
$\exists x'\,z'\,(Sa = Sx' \wedge Sc = Sz' \wedge (x'\,b, z')\,\varepsilon\, t_+)$ (2, $\vdash\vee$)
(4) $(a\,b, c)\,\varepsilon\, t_+ \vdash (Sa\,b, Sc)\,\varepsilon\, t_+$ (3, rec. thm.)
(5) $\vdash (a\,b, c)\,\varepsilon\, t_+ \rightarrow (Sa\,b, Sc)\,\varepsilon\, t_+$ (4, $\rightarrow\vdash$)
(6) $\vdash \forall x\,y\,z\,((x\,y, z)\,\varepsilon\, t_+ \rightarrow (Sx\,y, Sz)\,\varepsilon\, t_+)$ (5, $\vdash\forall$)

A6.1: $\forall x\,y\,z\,\big((Sx\,y, z)\,\varepsilon\, t_\times \rightarrow (\exists v)((v\,y, z)\,\varepsilon\, t_+ \wedge (x\,y, v)\,\varepsilon\, t_\times)\big)$.

(1) $a = a',\ (d\,b, c)\,\varepsilon\, t_+ \wedge (a'\,b, d)\,\varepsilon\, t_\times \vdash (d\,b, c)\,\varepsilon\, t_+ \wedge (a\,b, d)\,\varepsilon\, t_\times$ (8.1, $=\vdash$)

(2) $Sa = Sa' \vdash a = a'$ (A2)

(3) $Sa = Sa' \wedge (db, c) \,\varepsilon\, t_+ \wedge (a'b, d) \,\varepsilon\, t_\times \vdash (db, c) \,\varepsilon\, t_+ \wedge (ab, d) \,\varepsilon\, t_\times$ (1, 2, Cut, $\vdash\wedge$)

(4) $Sa = Sa' \wedge (db, c) \,\varepsilon\, t_+ \wedge (a'b, d) \,\varepsilon\, t_\times \vdash \exists v\, ((vb, c) \,\varepsilon\, t_+ \wedge (ab, v) \,\varepsilon\, t_\times)$ (3, $\vdash\exists$)

(5) $\exists x'\, v\, (Sa = Sx' \wedge (vb, c) \varepsilon t_+ \wedge (x'b, v) \varepsilon t_\times) \vdash \exists v\, ((vb, c) \varepsilon t_+ \wedge (ab, v) \varepsilon t_\times)$ (4, $\exists\vdash$)

(6) $Sa = \underline{0} \vdash$ (A1)

(7) $Sa = \underline{0} \wedge c = \underline{0} \wedge b \,\varepsilon\, \underline{\omega} \vdash$ (6, $\wedge\vdash$)

(8) $Sa = \underline{0} \wedge c = \underline{0} \wedge b \,\varepsilon\, \underline{\omega}$ .V. $\exists x'\, v\, (Sa = Sx' \wedge t_+(vb, c) \wedge t_\times(x'b, v))$
$\vdash \exists v\, (t_+(vb, c) \wedge t_\times(ab, v))$ (4, 6, $\vee\vdash$)

(9) $(Sab, c) \,\varepsilon\, t_\times \vdash \exists v\, (vb, c) \,\varepsilon\, t_+ \wedge (ab, v) \,\varepsilon\, t_\times)$ (8, rec. thm.)

(10) $\vdash (Sab, c) \,\varepsilon\, t_\times \rightarrow \exists v\, ((vb, c) \,\varepsilon\, t_+ \wedge (ab, v) \,\varepsilon\, t_\times)$ (9, $\vdash\rightarrow$)

(11) $\vdash \forall x y z\, ((Sxy, z) \,\varepsilon\, t_\times \rightarrow (\exists v)((vy, z) \,\varepsilon\, t_+ \wedge (xy, v) \,\varepsilon\, t_\times))$ (10, $\vdash\forall$)

A6.2: $\forall x y z\, \big((\exists v)((vy, z) \,\varepsilon\, t_+ \wedge (xy, v) \,\varepsilon\, t_\times) \rightarrow (Sxy, z) \,\varepsilon\, t_\times\big)$.

(1) $(db, c) \,\varepsilon\, t_+ \wedge (ab, d) \,\varepsilon\, t_\times \vdash Sa = Sa \wedge (db, c) \,\varepsilon\, t_+ \wedge (ab, d) \,\varepsilon\, t_\times$ (8.1, = ax, $\vdash\wedge$)

(2) $(db, c) \,\varepsilon\, t_+ \wedge (ab, d) \,\varepsilon\, t_\times \vdash \exists x'\, v\, (Sa = Sx' \wedge (vb, c) \,\varepsilon\, t_+ \wedge (x'b, v) \,\varepsilon\, t_\times)$ (1, $\vdash\exists$)

(3) $\exists v\, ((vb, c) \varepsilon t_+ \wedge (ab, v) \varepsilon t_\times) \vdash \exists x'\, v\, (Sa = Sx' \wedge (vb, c) \varepsilon t_+ \wedge (x'b, v) \varepsilon t_\times)$ (2, $\exists\vdash$)

(4) $\exists v\, (t_+(vb, c) \wedge t_\times(ab, v)) \vdash Sa = \underline{0} \wedge c = \underline{0} \wedge b \,\varepsilon\, \underline{\omega}$ .V.
$\exists x'\, v\, (Sa = Sx' \wedge t_+(vb, c) \wedge t_\times(x'b, v))$ (3, $\vdash\vee$)

(5) $\exists v\, ((vb, c) \,\varepsilon\, t_+ \wedge (ab, v) \,\varepsilon\, t_\times) \vdash (Sab, c) \,\varepsilon\, t_\times$ (4, rec. thm.)

(6) $\vdash \exists v\, ((vb, c) \,\varepsilon\, t_+ \wedge (ab, v) \,\varepsilon\, t_\times) \rightarrow (Sab, c) \,\varepsilon\, t_\times$ (5, $\vdash\rightarrow$)

(7) $\vdash \forall x y z\, ((\exists v)((vy, z) \,\varepsilon\, t_+ \wedge (xy, v) \,\varepsilon\, t_\times) \rightarrow (Sxy, z) \,\varepsilon\, t_\times)$ (6, $\vdash\forall$)

This concluses the proof of Theorem 8.2. ⊣

We must now show that $G^{+,\times}$ is not too powerful!

9. SEMANTICS FOR $G^{+,\times}$

9.1. The semantics of grounding in models of $G^{+,\times}$

By 7.7 we know that $\mathcal{A}$ is "almost" a model of $G^{+,\times}$. Thus, we will take the opportunity to extend $\mathcal{A}$ to a model $\mathcal{B}$ of $G^{+,\times}$.

We say that an $\mathcal{L}$-structure $\mathcal{S}$ is $+, \times$*-total* iff $\mathcal{S}$ is $\underline{\omega}$-total, t_+-total and $t_\times$-total. Evidently, $+, \times$-total $\mathcal{L}$-structures are models of $G^{+,\times}$.

Since our present definition of t_+ and $t_\times$ omit some of the clauses restricting the variables "x", "y", "z" and "v" to $\underline{\omega}$ as per 7.1, these terms are not guaranteed to be *false* of triples of non-numerals in $\mathcal{A}$. They are, however, *false* of all triples of numerals representing triples of numbers not belonging to the graphs of their respective number-theoretic functions. Hence, to construct an $+, \times$-total L-structure, it suffices to construct an extension of $\mathcal{A}$ on which t_+ and $t_\times$ are "forced" to be *false* of all triples containing non-numerals.

9.2. Consistency

We extend the construction of $\mathcal{A}$ in Definition 3.10 to produce an acceptable L-structure $\mathcal{B} \geq_k \mathcal{A}$ which holds t_+ and $t_\times$ *false* of all constant terms which are not Gilmore numerals. Then we give a reductive argument showing that the resulting acceptable L-structure $\mathcal{B}$ is 3-valued. This argument can be viewed as analogous to a "semantic" version of a cut elimination argument.

DEFINITION 9.1. Define $\varepsilon_0 : ctrm^3 \to 4$ by

$\varepsilon_0(t_1, t_2, t_3) = false$ if at least one of the following conditions hold:

(1) $t_1 = t_+$ and it is not the case that there are $m, n, j \in \mathbb{N}$ such that $t_2 = \langle \underline{m}, \underline{n} \rangle$ and $t_3 = \underline{j}$,

(2) $t_1 = t_\times$ and it is not the case that there are $m, n, j \in \mathbb{N}$ such that $t_2 = \langle \underline{m}, \underline{n} \rangle$ and $t_3 = \underline{j}$;

and $\varepsilon_0(t_1, t_2, t_3) = \perp$ otherwise.

Let $\mathcal{B}_0 =_{\text{df}} (ctrm, \varepsilon_0)$. Define an ordinal sequence $\{\mathcal{B}_\alpha \mid \alpha \leq \omega_1\}$ of L-structures as per 3.10.

THEOREM 9.2. *$\mathcal{B}_\alpha(t_1 \,\varepsilon\, \underline{\omega}) \neq true$ for all constant terms t_1 which are not Gilmore numerals ($\alpha \leq \omega_1$).*

Proof. By transfinite induction on α. Suppose, towards a contradiction, that $\mathcal{B}_\alpha(t_1 \,\varepsilon\, \underline{\omega}) = true$ for some constant term t_1, and let α be the least such ordinal. Evidently, α is the level of $t_1 \,\varepsilon\, \underline{\omega}$. Since α is a successor ordinal, $\mathcal{B}_\alpha = \mathcal{B}_{\alpha-1} \vee \Phi(\mathcal{B}_{\alpha-1})$ by construction 3.10. So by Definition 3.5,

$$\mathcal{B}_\alpha(t_1 \,\varepsilon\, \underline{\omega}) = \varepsilon_{\alpha-1}(\underline{\omega}, t_1, t_1) \vee \Phi(\varepsilon_{\alpha-1})(\underline{\omega}, t_1, t_1).$$

By hypothesis, $\varepsilon_{\alpha-1}(\underline{\omega}, t_1, t_1) \neq true$. Hence,

$true =$

$$\mathcal{B}_\alpha(t_1 \,\varepsilon\, \underline{\omega}) = \Phi(\varepsilon_{\alpha-1})(\underline{\omega}, t_1, t_1) = (\text{by df. } \Phi)\ \mathcal{B}_{\alpha-1}((t_1, diag_y A) \,\varepsilon\, diag_y A),$$

where A is the formula "$x = \underline{0} \vee \exists z(x = Sz \wedge z \,\varepsilon\, y)$", since $\underline{\omega} =_{\text{df}} [xv : (x, diag_y A) \,\varepsilon\, diag_y A]$. It follows that $\alpha - 1 > 0$, since $\mathcal{B}_0$ makes no ε-sentence *true*. Also, since α is the level of the sentence $t_1 \varepsilon \underline{\omega}$, $\alpha - 1$ is the level of $(t_2, diag_y A) \varepsilon diag_y A$. So by 4.4 we have:

$$\mathcal{B}_{\alpha-2}\big(t_2 = \underline{0} \vee \exists z\,(t_2 = Sz \wedge z\,\varepsilon\,\underline{\omega})\big) = \mathit{true},$$

but, since $t_2 \notin \mathit{Num}$, $t_2 \neq \underline{0}$, whence $\mathcal{B}(t_2 \neq \underline{0}) = \mathit{false}$, so

$$\mathcal{B}_{\alpha-2}\big(\exists z\,(t_2 = Sz \wedge z\,\varepsilon\,\underline{\omega})\big) = \mathit{true}.$$

Then there is an $s \in \mathit{ctrm}$ such that $\mathcal{B}_{\alpha-2}(t_2 = Ss \wedge s\,\varepsilon\,\underline{\omega}) = \mathit{true}$. Hence, $t_2 = Ss$ and $\mathcal{B}_{\alpha-2}(s\,\varepsilon\,\underline{\omega}) = \mathit{true}$. By the least ordinal assumption, $s \in \mathit{Num}$, so t_2 is also a numeral, contrary to the original assumption. ⊣

THEOREM 9.3. *α is coherent* $(\alpha \leq \omega_1)$.

Proof. By transfinite induction on $\alpha \leq \omega_1$. Suppose, towards a contradiction, that α is not coherent; we may assume that α is the least ordinal which is not coherent, and hence, by construction, that $\alpha = \beta+1$ for some coherent ordinal $\beta < \omega_1$. Then there are constant terms t_1, t_2, t_3 such that $\varepsilon_\alpha(t_1, t_2, t_3) = \mathit{both}$. But by construction,

$$\varepsilon_\alpha = \varepsilon_\beta \vee \Phi(\varepsilon_\beta).$$

Since β is coherent, the range of $\Phi(\varepsilon_\beta)$ is contained in 3. Since neither ε_β nor $\Phi(\varepsilon_\beta)$ assign the value *both*, it follows that

(a) $\varepsilon_\beta b(t_1, t_2, t_3) = \mathit{false}\ (\mathit{true})$ and (b) $\Phi(\varepsilon_\beta)(t_1, t_2, t_3) = \mathit{true}\ (\mathit{false})$.

Let t_1 be written $[x\,y : C]$ for some formula C. So (b) yields

(c) $\Phi(\varepsilon_\beta)(t_1, t_2, t_3) = \mathcal{B}_\beta(C_{xy}[t_2, t_3]) = \mathit{true}\ (\mathit{false})$

by definition of Φ. With (a) in mind, let $\gamma \leq \beta$ be the least ordinal such that $\varepsilon_\gamma(t_1, t_2, t_3) = \mathit{false}\ (\mathit{true})$. We separate two cases:

Case 1: $\gamma > 0$. Then $\varepsilon_\gamma(t_1, t_2, t_3) = \mathcal{B}_{\gamma-1}(C_{xy}[t_2, t_3]) = \mathit{false}\ (\mathit{true})$, contrary to (c), since $\mathcal{B}_{\gamma-1} \leq_k \mathcal{B}_\beta$ by construction.

Case 2: $\gamma = 0$. Then by definition of ε_0, 9.1 above, $\varepsilon_\gamma(t_1, t_2, t_3) = \mathit{false}$, whence $\Phi(\varepsilon_\beta)(t_1, t_2, t_3) = \mathit{true}$, and there are two cases, corresponding to clauses 1 and 2.

Case 2.1: Clause 1 holds. Then $t_1 = t_+$ and it is not the case that there are $m, n, j \in \mathbb{N}$ such that $t_2 = \langle \underline{m}, \underline{n} \rangle$ and $t_3 = \underline{j}$. Recall that

$$t_+ = \mathit{fix}_u B^+(x, y, z, u) = [x\,y\,z : (x\,y\,z\ \mathit{diag}_u B^+)\ \varepsilon\ \mathit{diag}_u B^+],$$

where $B^+(x, y, z, u) =_{\mathrm{df}} x = \underline{0} \wedge y = z \wedge y\,\varepsilon\,\underline{\omega}$ .V. $\exists x'\,z'\,\big(x = Sx' \wedge z = Sz' \wedge (x'\,y\,z')\,\varepsilon\,u\big)$. Also,

$$[x\,y\,z : (x\,y\,z\; \mathit{diag}_u B^+)\, \varepsilon\, \mathit{diag}_u B^+] =_{\mathrm{df}} [v\,z : \exists x\,y\,(v = \langle x, y\rangle \wedge (x\,y\,z\; \mathit{diag}_u B^+)\, \varepsilon\, \mathit{diag}_u B^+)]$$

by Definition 2.1 of k-ary abstraction. Then,

$$\Phi(\varepsilon_\beta)(t_+, t_2, t_3) = \mathcal{B}_\beta(\exists x\,y\,(t_2 = \langle x, y\rangle \wedge (x\,y\,t_3\; \mathit{diag}_u B^+)\, \varepsilon\, \mathit{diag}_u B^+)) = \mathit{true}$$

by (c) above. So we have constant terms s_1, s_2 such that $\rightarrow_\beta t_2 = \langle s_1, s_2\rangle \wedge (s_1\, s_2\, t_3\; \mathit{diag}_u B^+)\, \varepsilon\, \mathit{diag}_u B^+$. We may assume that β is the least ordinal verifying $(s_1\, s_2\, t_3, \mathit{diag}_u B^+)\, \varepsilon\, \mathit{diag}_u B^+$; also, since $\mathcal{B}_0$ makes no ε-atoms *true*, we know $\beta > 0$. Then by df. $\mathit{diag}_y B^+$ and then 4.4,

$$\rightarrow_\beta (s_1\, s_2\, t_3\; \mathit{diag}_u B^+)\, \varepsilon\, [x\,y\,z\,u : x = \underline{0} \wedge y = z\, \varepsilon\, \underline{\omega}\ .\!\vee\!.\ \exists x'\,z'\,(x = Sx' \wedge z = Sz' \wedge (x'\,y\,z')\, \varepsilon\, [xy\,z : (xy\,z\,u)\, \varepsilon\, u])],$$

$$\rightarrow_{\beta-1} s_1 = \underline{0} \wedge s_2 = t_3\, \varepsilon\, \underline{\omega}\ .\!\vee\!.\ \exists x'\,z'\,(s_1 = Sx' \wedge t_3 = Sz' \wedge (x'\,s_2\,z')\, \varepsilon\, [xy\,z : (xy\,z\; \mathit{diag}_y B^+)\, \varepsilon\, \mathit{diag}_y B^+]).$$

Now, let the left disjunct of the last line, $s_1 = \underline{0} \wedge s_2 = t_3 \wedge s_2 \varepsilon \underline{\omega}$, be denoted L, and let the right disjunct be R. We show that $\mathcal{B}_{\beta-1}(L) \neq \mathit{true}$, from which it follows that by Definition 3.1 that $\mathcal{B}_{\beta-1}(R) = \mathit{true}$, since $\beta - 1$ is coherent. For this, we may assume that $s_2 = t_3$, since otherwise $\mathcal{B}_{\beta-1}(L) = \mathit{false}$ trivially. Since $\rightarrow_\beta t_2 = \langle s_1, s_2\rangle$, $t_2 = \langle s_1, s_2\rangle$, so by assumption, it is not the case that both s_1 and s_2 are numerals. If $s_1 \notin \mathit{Num}$, then $s_1 \neq \underline{0}$ with the desired result. So assume that $s_2 \notin \mathit{Num}$. Then by 9.2, $\mathcal{B}_{\beta-1}(s_2\, \varepsilon\, \underline{\omega}) \neq \mathit{true}$, whence $\mathcal{B}_{\beta-1}(L) \neq \mathit{true}$, as required. Hence:

$$\rightarrow_{\beta-1} \exists x'\,z'\,(s_1 = Sx' \wedge t_3 = Sz' \wedge (x'\,s_2\,z')\, \varepsilon\, [xy\,z : (xy\,z\; \mathit{diag}_y B^+)\, \varepsilon\, \mathit{diag}_y B^+])$$

$$\rightarrow_{\beta-1} \exists x'\,z'\,(s_1 = Sx' \wedge t_3 = Sz' \wedge (x'\,s_2\,z')\, \varepsilon\, t_+) \qquad (\text{df. } t_+)$$

$$\rightarrow_{\beta-1} s_1 = Sr_1 \wedge t_3 = Sr_2 \wedge (r_1\, s_2\, r_2)\, \varepsilon\, t_+, \text{ for some } r_1, r_2 \in \mathit{ctrm};$$

$$\rightarrow_{\beta-1} s_1 = Sr_1 \wedge t_3 = Sr_2$$

$$\rightarrow_{\beta-1} (r_1\, s_2\, r_2)\, \varepsilon\, t_+.$$

So, $\mathcal{B}_{\beta-1}((r_1\, s_2\, r_2) \varepsilon t_+) = \mathit{true}$. Then, since $\varepsilon_0 \leq_k \varepsilon_{\beta-1}$, $\varepsilon_0(t_+, \langle r_1, s_2\rangle, r_2) \neq \mathit{false}$. Hence, by definition of ε_0, all of r_1, s_2, r_2 belong to *Num*. But, since $s_1 = Sr_1$ and $t_3 = Sr_2$, all of $s_1, s_2, t_3 \in \mathit{Num}$, contrary to the assumption of this case.

Case 2.2: Clause (2) holds. Then $t_1 = t_\times$. Recall that, by definition of $t_\times$,

$$t_\times = \mathit{fix}_u B^\times(x, y, z, u) = [x\,y\,z : (x\,y\,z\; \mathit{diag}_y B^\times)\, \varepsilon\, \mathit{diag}_y B^\times],$$

where $B^{\times}(x,y,z,u) =_{\mathrm{df}} x = \underline{0} \wedge z = \underline{0} \wedge y \,\varepsilon\, \underline{\omega}$ .V. $\exists x' v \big(x = Sx' \wedge (v\,y\,z) \,\varepsilon\, t_+ \wedge (x'\,y\,v) \,\varepsilon\, u\big)$. Then

$$\rightarrow_\beta \exists x\,y \big(t_2 = \langle x, y\rangle \wedge (x\,y\,t_3\, \mathit{diag}_y B^{\times}) \,\varepsilon\, \mathit{diag}_y B^{\times}\big)$$

by (c) above, since $\Phi(\varepsilon_\beta)(t_\times, t_2, t_3) = \mathit{true}$. So we have constant terms s_1, s_2 such that:

$$\rightarrow_\beta t_2 = \langle s_1, s_2\rangle \wedge (s_1\,s_2\,t_3, \mathit{diag}_u B^{\times}) \,\varepsilon\, \mathit{diag}_u B^{\times}.$$

We may assume that β is the least ordinal verifying $(s_1\,s_2\,t_3, \mathit{diag}_u B^{\times}) \,\varepsilon\, \mathit{diag}_u B^{\times}$. Then $\beta > 0$. Then, by definition of $\mathit{diag}_y B^{\times}$ and 4.4

$$\begin{aligned}\rightarrow_\beta (s_1\,s_2\,t_3, \mathit{diag}_u B^{\times}) \,\varepsilon\, [x\,y\,z\,u : x = \underline{0} \wedge z = \underline{0} \wedge y \,\varepsilon\, \underline{\omega} \text{ .V. } \\ \exists x' v \big(x = Sx' \wedge (v\,y\,z) \,\varepsilon\, t_+ \wedge (x'\,y\,v) \,\varepsilon\, u\big)],\end{aligned}$$

$$\begin{aligned}\rightarrow_{\beta-1} s_1 = \underline{0} \wedge t_3 = \underline{0} \wedge s_2 \,\varepsilon\, \underline{\omega} \text{ .V. } \exists x' v \big(s_1 = Sx' \wedge (v\,s_2\,t_3) \,\varepsilon\, t_+ \wedge \\ (x'\,s_2\,v) \,\varepsilon\, [xy\,z : (xy\,z\, \mathit{diag}_y B^{\times}) \,\varepsilon\, \mathit{diag}_y B^{\times}]\big).\end{aligned}$$

Now, let the left disjunct of the last line be denoted L. We show that $\mathcal{B}_{\beta-1}(L) \neq \mathit{true}$, from which it follows that by Definition 3.1 that $\mathcal{B}_{\beta-1}$ verifies the right disjunct. For this, we may assume that $s_1 = t_3 = \underline{0}$, since otherwise $\mathcal{B}_{\beta-1}$ *falsifies* L trivially. Since $\rightarrow_\beta t_2 = \langle s_1, s_2\rangle$, we have $t_2 = \langle s_1, s_2\rangle$, whence, by assumption, s_2 is not a numeral. But $s_2 \notin \mathit{Num}$ yields $\mathcal{B}_{\beta-1}(s_2 \varepsilon \underline{\omega}) = \mathit{false}$ by 9.2, whence $\mathcal{B}_{\beta-1}(L) \neq \mathit{true}$, as required. Then:

(1) $\rightarrow_{\beta-1} \exists x' v \big(s_1 = Sx' \wedge (v\,s_2\,t_3) \,\varepsilon\, t_+ \wedge (x'\,s_2\,v) \,\varepsilon\, [xy\,z : (xy\,z\, \mathit{diag}_y\, B^{\times}) \,\varepsilon\, \mathit{diag}_y B^{\times}]\big)$

(2) $\rightarrow_{\beta-1} \exists x' v \big(s_1 = Sx' \wedge (v\,s_2\,t_3) \,\varepsilon\, t_+ \wedge (x'\,s_2\,v) \,\varepsilon\, t_\times\big)$ (df. $t_\times$)

(3) $\rightarrow_{\beta-1} s_1 = Sr_1 \wedge (r_2\,s_2\,t_3) \,\varepsilon\, t_+ \wedge (r_1\,s_2\,r_2) \,\varepsilon\, t_\times$, for some $r_1, r_2 \in \mathit{ctrm}$

(4) $\rightarrow_{\beta-1} s_1 = Sr_1$

(5) $\rightarrow_{\beta-1} (r_2\,s_2\,t_3) \,\varepsilon\, t_+$

(6) $\rightarrow_{\beta-1} (r_1\,s_2\,r_2) \,\varepsilon\, t_\times$

Since $t_2 = \langle s_1, s_2\rangle$, not all of s_1, s_2, t_3 are numerals. From line 4 above, $s_1 = Sr_1$. Since $\varepsilon_0 \leq_k \varepsilon_{\beta-1}$ and $\beta-1$ is coherent, we have $\varepsilon_0(t_+, \langle r_2, s_2\rangle, t_3) \neq$ *false* and $\varepsilon_0(t_\times, \langle r_1, s_2\rangle, r_2) \neq$ *false* from 5 and 6. Then by clauses 1 and 2 of df. ε_0, all of $r_1, s_2, t_3 \in \mathit{Num}$. Then, since $s_1 = Sr_1$, s_1 is a numeral too. This contradiction concludes Case 2, and thus the theorem is established. ⊣

LEMMA 9.4. *Let* $m_1, m_2, m_3 \in \mathbb{N}$. *Then,*

(a) *If* $m_1 + m_2 \neq m_3$, *then* $\leftarrow_\omega (\underline{m_1}\,\underline{m_2}, \underline{m_3}) \,\varepsilon\, t_+$, *and*

(b) *If* $m_1 \times m_2 \neq m_3$, *then* $\leftarrow_\omega (\underline{m_1}\,\underline{m_2}, \underline{m_3}) \,\varepsilon\, t_\times$.

Proof. By induction on m_1, using (a) to establish (b). This proof is a special case of that of Claim 1 in the primitive recursion case of the proof of 7.3. ⊣

COROLLARY 9.5. *$\mathcal{B}$ is $+, \times$-total L-structure.*

Proof. $\mathcal{B}$ is acceptable by construction. By Theorem 9.3, $\mathcal{B}$ is a 3-valued L-structure. $\mathcal{B}$ is ω-total since $\mathcal{A}$ is and $\mathcal{A} \leq_k \mathcal{B}$. So let $* \in \{+, \times\}$ and t_1, t_2, t_3 any constant terms; it suffices to show that one of the following holds:

(1) $\rightarrow_{\mathcal{B}} (t_1\, t_2\, t_3)\, \varepsilon\, t_*$,

(2) $\leftarrow_{\mathcal{B}} (t_1\, t_2\, t_3)\, \varepsilon\, t_*$.

We may suppose that all of t_1, t_2 and t_3 are Gilmore numerals, since otherwise (2) holds by construction 9.1. So there are $m_1, m_2, m_3 \in \mathbb{N}$ such $t_i = \underline{m_i}$ $(i = 1, 2, 3)$. If $m_1 * m_2 = m_3$, then (1) holds by Theorem 7.2 and the soundness of G with respect to acceptable L-structures. If, on the other hand, $m_1 * m_2 \neq m_3$, then 2 follows by Lemma 9.4, (a) or (b), respectively as $*$ is $+$ or $\times$. ⊣

COROLLARY 9.6. $\mathrm{G}^{+,\times}$ *is consistent.*

This completes our analysis of how much arithmetic may be derived within $\mathrm{G}^{+,\times}$. To go further, we need a first-order scheme of mathematical induction.

10. THE PRINCIPLE OF MATHEMATICAL INDUCTION FOR THE NATURAL NUMBERS

10.1. Varieties of ungroundedness

As mentioned in section 0, Tarskian first-order languages obey two kinds of restrictions: a restriction on the range of the first-order quantifiers, and a stratification of the logical grammar of predication. These two restrictions can be seen as talismans intended to ward off, respectively, the two varieties of definitional impredicativity that Russell was reacting to when he formulated his ramified theory of types.

The first kind of impredicativity, quantificational in nature, is featured by the classical logicist definition of the set $\mathbb{N}$ natural numbers as the intersection of *all* zero-successor sets. Here, the source of the anxiety is the fact that the target of the definition, viz., $\mathbb{N}$, is itself a member of the range of the universal quantifier occurring in the definition. This kind of impredicative definition is ruled out by Russell's stratification of the range of the quantifiers. The second kind of impredicativity is due to self-application and self-membership, or more generally, the possibility of non-well-founded membership relations. Russell ruled out this kind of impredicativity by restricting the permissible level of an argument to a propositional function, of which the Tarskian stratification of predication may be seen as a descendant. Since fix_y introduces the self-application control structure $diag_y$ into the logical form of $\underline{\omega}$, this definition of the natural numbers can be charged with the second, but not the first, kind of impredicativity.

In this section, we introduce a constant term $\mathbb{N}$ intended to formalize Frege's classical definition of the natural numbers as the intersection of all zero-successor sets. The derivations of 10.11 and 10.12 in this section demonstrate the startling result that the posit of bivalence for $\mathbb{N}$ is sufficient to yield a schema of first-order induction for arithmetic formulas expressed in terms of $\mathbb{N}$. Although the variety of ungroundedness exhibited by $\mathbb{N}$ can be described in the same general terms as that exhibited by $\underline{\omega}$ (both definitions are liable to generate infinitely descending evaluation sequences), it emerges in the consistency proof of the following section that the quantificational variety of ungroundedness exhibited by $\mathbb{N}$ is of a less tractable nature than that associated with $\underline{\omega}$. Whereas $\underline{\omega}$ is total in $\mathcal{A}$, it appears that there are fixed points of Φ in which the extension of $\mathbb{N}$ is empty (although $\mathbb{N}$'s anti-extension in $\mathcal{A}$ contains the non-numerals; as we shall see, this is a property that $\mathbb{N}$ inherits from that distinguished zero-successor set, $\underline{\omega}$).

10.2. Articulating an image of arithmetic in L

It is important for the consistency of NaDSet I that axioms sequents of the form $t\varepsilon p \vdash t\varepsilon p$ are disallowed. Gilmore (*1986*) showed that Russell's paradox could be reintroduced in the extension of NaDSet I that contains such axioms by considering the following version of the Russell set term R in which an existential quantifier binds a variable in predicate position in an ε statement:

$$[x : \exists y\, (y = x \wedge \neg\, x \,\varepsilon\, y)].$$

This is to be expected, since parameters of NaDSet I are, *qua eigenvariables*, understood as standing for an arbitrary constant set term; if we countenance bivalence for atoms with parameters as predicate terms, then intuitively, we are *ipso facto* countenancing bivalence for atoms featuring R as predicate term.

However, it is precisely sequents of the form $t\,\varepsilon\, p \vdash t\,\varepsilon\, p$ that secure the bivalence of $\mathbb{N}$ atoms and thus underwrite the classical derivations of the axiom of infinity and the principle of induction for this impredicative definition of the natural numbers. Gilmore (*1986*) therefore opted to stratify predication, moving on to a second-order version of NaDSet I, NaDSet II, where axioms of the form $t\,\varepsilon\, P \vdash t\,\varepsilon\, P$ for second-order parameters P are introduced. In the characteristic idiom of second-order quantification, a second-order quantifier can happily bind a second-order variable occurring as the predicate term of an ε statement.

Our solution avoids stratifying predication by restricting the range of the quantifiers instead. We pass to the logical extension G_2 by adjoining to $G^{+,\times}$, as new bivalence axioms, all sequents of the form $t \mathrel{\varepsilon} p \vdash t \mathrel{\varepsilon} p$ for constant terms t and parameters p. Consistency is maintained by retracting the range of the quantifiers in G_2 to a syntactically defined subset of *ctrm*, which, it will turn out, contains terms representing "total" objects only. The range of the G_2 quantifiers is the set of all constant terms of the form $[x : \phi]$, where ϕ is a formula all of whose atomic subformulas are $G^{+,\times}$ ground atoms or $\mathbb{N}$ atoms (ϕ is, as we shall say, "arithmetic"). Thus, as in G, we may quantify over first level representations of predicates in G_2, but not arbitrarily.

The bivalence of arithmetic formulas provides both the philosophical and the technical motivation for thus restricting the range of the G_2 quantifiers. Intuitively, we think of the logical extensions of G as concerning a domain of partial objects (partial "sets" or "concepts") ordered under an information ordering $\leq_k$. In G_2 we intend the quantifiers to range over total objects only, those which are maximal with respect to $\leq_k$. From the perspective of G_2, only total objects—those having "sharp edges" as Frege would say—represent legitimate concepts capable of sustaining the ontic commitment incurred by quantification. Since the collection of set terms which represent total concepts is not recursively enumerable, the intended range of the G_2 quantifiers is not syntactically specifiable. Hence, the notion of an arithmetic term functions as a *syntactically specifiable approximation* of the notion of a total concept. As a result, G_2 has the desirable technical property, not shared by NaDSet II, that its derivations are closed under the substitution of arbitrary closed terms for parameters. This property allows for a straightforward "analytic" proof of the consistency of G_2 in section 12. Since the dependence on the notion of an "arithmetic term" could be replaced by the notion of a "primitive recursive term", the specification of the range of the G_2 quantifiers does not import prior number-theoretic notions in a way that undercuts the purely logical nature of our specification of the calculus.

First define $ZS(p)$ to be the closed formula

$$\underline{0} \mathrel{\varepsilon} p \wedge \forall y \, (y \mathrel{\varepsilon} p \rightarrow Sy \mathrel{\varepsilon} p),$$

which "says" that p is a 0-successor set. We now define a set term intended to represent the set of natural numbers as "the smallest 0-successor set":

$$\mathbb{N} =_{\mathrm{df}} [x : \forall z \left(ZS(z) \rightarrow x \mathrel{\varepsilon} z \right)].$$

The first observation we should make is that $\mathbb{N}$ inherits from $\underline{\omega}$ the desirable property of being false of all non-numerals:

OBSERVATION 10.1. Let t_1 be a constant term which is not a Gilmore numeral. Then $\leftarrow_{\mathcal{A}} t_1 \mathrel{\varepsilon} \mathbb{N}$.

Proof. By 5.7, $\leftarrow_{\mathcal{A}} t_1 \varepsilon \underline{\omega}$. By 6.1 (the axiom of infinity for $\underline{\omega}$) and the fact that $\mathcal{A}$ is a model of G^{ω}, we have $\rightarrow_{\mathcal{A}} ZS(\underline{\omega})$. So, by 3.1, $\leftarrow_{\mathcal{A}} (ZS(\underline{\omega}) \rightarrow t_1 \varepsilon \underline{\omega})$, whence $\leftarrow_{\mathcal{A}} \forall z\,(ZS(z) \rightarrow t_1 \varepsilon z)$. Then by abstraction (4.2), $\leftarrow_{\mathcal{A}} t_1 \varepsilon [x : \forall z\,(ZS(z) \rightarrow x \varepsilon z)]$, i.e., $\leftarrow_{\mathcal{A}} t_1 \varepsilon \mathbb{N}$. ⊣

Now, we formally define the "classical" sublanguage of L which, roughly speaking, contains a relational image of the language of Peano Arithmetic.

DEFINITION 10.2. We define a sublanguage Λ of L recursively as follows:

0. All parameters are *terms* of Λ. $\mathbb{N}$ is a term of Λ. If ϕ is a formula of Λ and x a variable, then $[x : \phi]$ is a *term* of Λ.

1. If t_1, t_2 and t_3 are terms of L, p is a parameter and ϕ a formula of Λ, then

   i. $t_1 = t_2$, ii. $t_1 \varepsilon [x : \phi]$, iii. $t_1 \varepsilon \mathbb{N}$, iv. $t_1 \varepsilon \underline{\omega}$, v. $(t_1\, t_2, t_3) \varepsilon t_+$, and vi. $(t_1\, t_2, t_3) \varepsilon t_\times$ are (atomic) formulas of Λ.

2. If ϕ and ψ are formulas of Λ and x a variable, then $\neg\phi$, $\phi \vee \psi$, $\phi \wedge \psi$, $\exists x\, \phi$ and $\forall x\, \phi$ are formulas of Λ.

The connectives $\rightarrow$ and $\leftrightarrow$ are introduced into Λ by their usual definitions, as in L.

A formula or term of Λ is called an *arithmetic formula* or *term*. Note that we use ϕ, ψ, κ, χ to range over formulas of Λ. We let "s", "t", "r", ..., with or without subscripts, range over terms of Λ.

So the arithmetic formulas are precisely those L formulas whose atomic subformulas are ground atoms of $G^{+,\times}$ or atomic formulas of L featuring arithmetic predicate terms, that is $\mathbb{N}$ or set terms of the form $t_1 \varepsilon [x : \phi]$ where ϕ is arithmetic. We define a notion of the *complexity*, *comp*(ϕ), of a formula ϕ of Λ inductively on the formation of f as follows:

1. If ϕ is atomic and one of the forms i, iii, iv, v or vi, then $comp(\phi) = 0$.

2. If ϕ is atomic and of the form $t_1 \varepsilon [x : \psi]$ for some formula ψ of Λ, then $comp(\phi) = 1 + comp(\psi)$.

3. If ϕ is non-atomic, the $comp(\phi) = 1+\max\{$the proper subformulas of $\psi\}$.

10.3. The logical extension G_2

Let G_2 be the logical extension of $G^{+,\times}$ obtained by adjoining as new bivalence axioms all sequents of the form $t \varepsilon p \vdash t \varepsilon p$, where t is any closed term and p is any parameter, and replacing the $\forall\vdash$ and $\vdash\exists$ rules by

$\forall_2\vdash$	$A_x[t], \Gamma \vdash \Theta$	/	$\forall x A, \Gamma \vdash \Theta,$
$\vdash\exists_2$	$\Gamma \vdash \Theta, A_x[t]$	/	$\Gamma \vdash \Theta, \exists x\, A$

respectively, where t is an arithmetic term.

G_2's parametric bivalence axioms yield a derivation of bivalence for our impredicative representative of the natural numbers, $\mathbb{N}$. First,

LEMMA 10.3. *Let p be a parameter. Then $ZS(p) \vdash ZS(p)$ is derivable in* G_2.

Proof.

(1) $b \,\varepsilon\, p \vdash b \,\varepsilon\, p$ (G_2 axiom)
(2) $Sb \,\varepsilon\, p \vdash Sb \,\varepsilon\, p$ (G_2 axiom)
(3) $b \,\varepsilon\, p \rightarrow Sb \,\varepsilon\, p,\ b \,\varepsilon\, p \vdash Sb \,\varepsilon\, p$ (1, 2, $\rightarrow\vdash$)
(4) $b \,\varepsilon\, p \rightarrow Sb \,\varepsilon\, p \vdash b \,\varepsilon\, p \rightarrow Sb \,\varepsilon\, p$ (3, $\vdash\rightarrow$)
(5) $\forall y\,(y \,\varepsilon\, p \rightarrow Sy \,\varepsilon\, p) \vdash b \,\varepsilon\, p \rightarrow Sb \,\varepsilon\, p$ (4, $\forall\vdash$)
(6) $\forall y\,(y \,\varepsilon\, p \rightarrow Sy \,\varepsilon\, p) \vdash \forall y\,(y \,\varepsilon\, p \rightarrow Sy \,\varepsilon\, p)$ (5, $\vdash\forall$)
(7) $\underline{0} \,\varepsilon\, p \vdash \underline{0} \,\varepsilon\, p$ (G_2 axiom)
(8) $\underline{0} \,\varepsilon\, p,\ \forall y\,(y \,\varepsilon\, p \rightarrow Sy \,\varepsilon\, p) \vdash \underline{0} \,\varepsilon\, p \wedge \forall y\,(y \,\varepsilon\, p \rightarrow Sy \,\varepsilon\, p)$ (6, 7, $\vdash\wedge$)
(9) $\underline{0} \,\varepsilon\, p \wedge \forall y\,(y \,\varepsilon\, p \rightarrow Sy \,\varepsilon\, p) \vdash \underline{0} \,\varepsilon\, p \wedge \forall y\,(y \,\varepsilon\, p \rightarrow Sy \,\varepsilon\, p)$ (7, $\wedge\vdash$)

⊣

Finally,

LEMMA 10.4. *Let t be a closed term. Then $t \,\varepsilon\, \mathbb{N} \vdash t \,\varepsilon\, \mathbb{N}$ is derivable in* G_2.

Proof.

(1) $ZS(p) \vdash ZS(p)$ (10.3)
(2) $t \,\varepsilon\, p \vdash t \,\varepsilon\, p$ (G_2 axiom)
(3) $ZS(p) \rightarrow t \,\varepsilon\, p,\ ZS(p) \vdash t \,\varepsilon\, p$ (1, 2, $\rightarrow\vdash$)
(4) $ZS(p) \rightarrow t \,\varepsilon\, p \vdash ZS(p) \rightarrow t \,\varepsilon\, p$ (3, $\vdash\rightarrow$)
(5) $\forall z\,(ZS(z) \rightarrow t \,\varepsilon\, z) \vdash ZS(p) \rightarrow t \,\varepsilon\, p$ (4, $\forall_2\vdash$)
(6) $\forall z\,(ZS(z) \rightarrow t \,\varepsilon\, z) \vdash \forall z(ZS(z) \rightarrow t \,\varepsilon\, z)$ (5, $\vdash\forall$)
(7) $t \,\varepsilon\, [x : \forall z\,(ZS(z) \rightarrow x \,\varepsilon\, z)] \vdash t \,\varepsilon\, [x : \forall z\,(ZS(z) \rightarrow x \,\varepsilon\, z)]$ (6, $\varepsilon\vdash$, $\vdash\varepsilon$)
(8) $t \,\varepsilon\, \mathbb{N} \vdash t \,\varepsilon\, \mathbb{N}$ (7, df. $\mathbb{N}$)

⊣

LEMMA 10.5. *Let ϕ be an arithmetic formula, $\vec{s} = s_1, \ldots, s_k$ terms and $\vec{x} = x_1, \ldots, x_k$ variables. Then $\phi_{\vec{x}}[\vec{s}]$ is also an arithmetic formula, and of the same complexity as ϕ.*

Proof. By induction on *comp*(ϕ). ⊣

LEMMA 10.6. *Let ϕ be a Λ formula with free variables among $\vec{x} = x_1, \ldots, x_k$, and let $\vec{x} = s_1, \ldots, s_k$ be closed terms. Then $\phi_{\vec{x}}[\vec{s}] \vdash \phi_{\vec{x}}[\vec{s}]$ is derivable in* G_2.

Proof. By induction on *comp*(ϕ). ⊣

LEMMA 10.7. *Let t be a closed term and* $[x : \phi]$ *be a closed arithmetic term. Then*

$$t \,\varepsilon\, [x : \phi] \vdash t \,\varepsilon\, [x : \phi]$$

is derivable in G_2.

Proof. By Lemma 10.6, $\phi_x[t] \vdash \phi_x[t]$ is derivable in G_2. Since t is free for x in ϕ, two applications of unary abstraction yield $t \,\varepsilon\, [x : \phi] \vdash t \,\varepsilon\, [x : \phi]$, as required. ⊣

LEMMA 10.8. *Let t be any closed arithmetic term and p a parameter. Suppose* $\Gamma \vdash \Delta$ *is derivable in* G_2. *Then so is* $\Gamma_p[t] \vdash \Delta_p[t]$.

Proof. By induction on the depth of G_2 derivations. ⊣

10.4. Deriving induction from the impredicative definition of the natural numbers

Thus far we have delineated a fragment Λ of L on the basis of two desiderata. First, the language Λ contains a relational image of the language of Peano Arithmetic. Second, we can reason classically in G_2 about the relations represented in Λ by virtue of the fact that Λ is bivalent in G_2. Λ is based upon the $G^{+,\times}$ ground atoms and the set term $\mathbb{N} =_{df} [x : \forall z\,(ZS(z) \rightarrow x \,\varepsilon\, z)]$ which has been given the impredicative syntactic structure of the classical definition of the natural numbers as "the least 0-successor set". Because parameters are not ground predicate terms of $G^{+,\times}$, $\mathbb{N}$ is deductively inert in that calculus. However with the introduction of parameters as ground predicates, G_2 can simulate enough of the idiom of second-order quantification to reason classically about $\mathbb{N}$. In particular, an axiom of infinity and a first-order scheme of induction for $\mathbb{N}$ are forthcoming.

Note that by Lemma 10.4 $\mathbb{N}$ is "total" in G_2 in the sense that all closed atomic formulas with $\mathbb{N}$ as predicate terms are bivalent.

Given the term model semantics of section 3, the reader may object to reading $\mathbb{N}$ as "the least 0-successor set" since the quantifier $\forall z$ does not range over sets at all, let alone 0-successor sets. Indeed, in G_2 the quantifiers range over closed arithmetic *terms*, of the form $[x : \phi(x)]$ where $\phi(x)$ is a Λ formula. The justification is given by extending the nominalistic viewpoint presupposed by our term model semantics to the reading of $\mathbb{N}$. Here, we consider $\phi(x)$ to be the Λ-representation of an arbitrary formula $F(x)$ of the language of arithmetic. We may think of quantification over the countably many closed arithmetic terms $[x : \phi(x)]$ as simulating quantification over the countably many formulas $F(x)$ of arithmetic. So, in general, in G_2 a sentence involving quantification can in fact specify an infinite first-order schema in F

for the language of arithmetic. A more accurate reading of ℕ would be "the set of objects satisfying every 0-successor formula", where a "0-successor formula" is a formula of arithmetic with at most one free variable which holds of 0 and the successor of any object it holds of. This reading suffices to underwrite a first-order principle of induction for ℕ.

First, we establish that ℕ itself represents a 0-successor set:

LEMMA 10.9. $\vdash \underline{0}\,\varepsilon\,\mathbb{N} \wedge \forall y\,(y\,\varepsilon\,\mathbb{N} \rightarrow Sy\,\varepsilon\,\mathbb{N})$ *is derivable in* G_2.

Proof. First, we show $\vdash \underline{0}\,\varepsilon\,\mathbb{N}$:

(1) $\underline{0}\,\varepsilon\,p \vdash \underline{0}\,\varepsilon\,p$
(2) $\underline{0}\,\varepsilon\,p \wedge \forall y\,(y\,\varepsilon\,p \rightarrow Sy\,\varepsilon\,p) \vdash \underline{0}\,\varepsilon\,p$
(3) $ZS(p) \vdash \underline{0}\,\varepsilon\,p$
(4) $\vdash ZS(p) \rightarrow \underline{0}\,\varepsilon\,p$
(5) $\vdash \forall z\,(ZS(z) \rightarrow \underline{0}\,\varepsilon\,z)$
(6) $\vdash \underline{0}\,\varepsilon\,[x : \forall z\,(ZS(z) \rightarrow x\,\varepsilon\,z)]$
(7) $\vdash \underline{0}\,\varepsilon\,\mathbb{N}$

Now we show $\forall y\,(y\,\varepsilon\,\mathbb{N} \rightarrow Sy \in \mathbb{N})$:

(1)	$Sb\,\varepsilon\,p \vdash Sb\,\varepsilon\,p$	(G_2 axiom)
(2)	$b\,\varepsilon\,p \vdash b\,\varepsilon\,p$	(G_2 axiom)
(3)	$b\,\varepsilon\,p,\ b\,\varepsilon\,p \rightarrow Sb\,\varepsilon\,p \vdash Sb\,\varepsilon\,p$	(1, 2, $\rightarrow\vdash$)
(4)	$b\,\varepsilon\,p,\ \forall y\,(y\,\varepsilon\,X \rightarrow Sy\,\varepsilon\,p) \vdash Sb\,\varepsilon\,p$	(3, $\forall_2\vdash$)
(5)	$b\,\varepsilon\,p,\ \underline{0}\,\varepsilon\,p \wedge \forall y\,(y\,\varepsilon\,p \rightarrow Sy\,\varepsilon\,p) \vdash Sb\,\varepsilon\,p$	(4, $\vdash\wedge$)
(6)	$b\,\varepsilon\,p,\ ZS(p) \vdash Sb\,\varepsilon\,p$	(df. ZS)
(7)	$ZS(p) \vdash ZS(p)$	(Lemma 10.3)
(8)	$ZS(p) \rightarrow b\,\varepsilon\,p,\ ZS(p) \vdash Sb\,\varepsilon\,p$	(6, 7, $\rightarrow\vdash$)
(9)	$ZS(p) \rightarrow b\,\varepsilon\,p \vdash ZS(p) \rightarrow Sb\,\varepsilon\,p$	(8, $\vdash\rightarrow$)
(10)	$\forall z\,(ZS(z) \rightarrow b\,\varepsilon\,z) \vdash ZS(p) \rightarrow Sb\,\varepsilon\,p$	(9, $\forall_2\vdash$)
(11)	$\forall z\,(ZS(z) \rightarrow b\,\varepsilon\,z) \vdash \forall z\,(ZS(z) \rightarrow Sb\,\varepsilon\,z)$	(10, $\vdash\forall$)
(12)	$\forall z\,(ZS(z) \rightarrow b\,\varepsilon\,z) \vdash Sb\,\varepsilon\,[x : \forall z\,(ZS(z) \rightarrow x\,\varepsilon\,z)]$	(11, $\vdash\varepsilon$)
(13)	$b\,\varepsilon\,[x : \forall z\,(ZS(z) \rightarrow x\,\varepsilon\,z)] \vdash Sb\,\varepsilon\,[x : \forall z\,(ZS(z) \rightarrow x\,\varepsilon\,z)]$	(12, $\varepsilon\vdash$)
(14)	$b\,\varepsilon\,\mathbb{N} \vdash Sb\,\varepsilon\,\mathbb{N}$	(13, df. ℕ)
(15)	$\vdash b\,\varepsilon\,\mathbb{N} \rightarrow Sb\,\varepsilon\,\mathbb{N}$	(14, $\vdash\rightarrow$)
(16)	$\vdash \forall y\,(y\,\varepsilon\,\mathbb{N} \rightarrow Sy\,\varepsilon\,\mathbb{N})$	(15, $\vdash\forall$)

Hence, ℕ is itself an arithmetic term which represents a 0-successor set, and so falls within the range of the quantifier $\forall y$ that occurs in its own definition binding a variable in predicate position. ⊣

LEMMA 10.10. *The axiom of induction for* $\mathbb{N}$,

$$\vdash \forall x\,\big(ZS(x) \rightarrow (\forall y\,\varepsilon\,\mathbb{N})y\,\varepsilon\,x\big),$$

is derivable in G_2.

Proof.

(1)	$b\,\varepsilon\,p \vdash b\,\varepsilon\,p$	(G_2 axiom)
(2)	$ZS(p) \vdash ZS(p)$	(Lemma 10.3)
(3)	$ZS(p),\ ZS(p) \rightarrow b\,\varepsilon\,p \vdash b\,\varepsilon\,p$	(1, 2, $\rightarrow\vdash$)
(4)	$ZS(p),\ \forall z\,(ZS(z) \rightarrow b\,\varepsilon\,z) \vdash b\,\varepsilon\,p$	(3, $\forall_2\vdash$)
(5)	$ZS(p),\ b\,\varepsilon\,[x : \forall z\,(ZS(z) \rightarrow x\,\varepsilon\,z)] \vdash b\,\varepsilon\,p$	(4, $\varepsilon\vdash$)
(6)	$ZS(p),\ b\,\varepsilon\,\mathbb{N} \vdash b\,\varepsilon\,p$	(5, df. $\mathbb{N}$)
(7)	$ZS(p) \vdash b\,\varepsilon\,\mathbb{N} \rightarrow b\,\varepsilon\,p$	(6, $\vdash\rightarrow$)
(8)	$ZS(p) \vdash \forall y\,(y\,\varepsilon\,\mathbb{N} \rightarrow y\,\varepsilon\,p)$	(7, $\vdash\forall$)
(9)	$\vdash ZS(p) \rightarrow \forall y\,(y\,\varepsilon\,\mathbb{N} \rightarrow y\,\varepsilon\,p)$	(8, $\vdash\rightarrow$)
(10)	$\vdash \forall x\,(ZS(x) \rightarrow \forall y\,(y\,\varepsilon\,\mathbb{N} \rightarrow y\,\varepsilon\,x))$	(9, $\vdash\forall$)
(11)	$\vdash \forall x\,(ZS(x) \rightarrow (\forall y\,\varepsilon\,\mathbb{N})y\,\varepsilon\,x)$	(10, df. $\forall y\,\varepsilon\,\mathbb{N}$)

$\dashv$

THEOREM 10.11 **The Schema of Pre-Induction for N**. *Let* $\phi(x)$ *be an arithmetic formula,* Γ, Δ *sets of formulas, and* p *a parameter such that the sequent* $\phi(p), \Gamma \vdash \Delta$ *is derivable in* G_2. *Then,*

$$\phi(\underline{0}) \wedge \forall y\,\big(\phi(y) \rightarrow \phi(Sy)\big),\ p\,\varepsilon\,\mathbb{N},\ \Gamma \vdash \Delta$$

is derivable in G_2.

Proof. Since $\phi(x)$ is in Λ, by Lemma 10.5 so are the closed L_2 formulas $\phi(\underline{0})$, $\phi(p)$ and $\phi(Sp)$. Then

(1)	$\phi(\underline{0}) \vdash \phi(\underline{0})$	(10.4)
(2)	$\phi(\underline{0}) \vdash \underline{0}\,\varepsilon\,[x : \phi(x)]$	(1, $\vdash\varepsilon$)
(3)	$\phi(p) \vdash \phi(p)$	(10.5)
(4)	$\phi(Sp) \vdash \phi(Sp)$	(10.5)
(5)	$\phi(p) \rightarrow \phi(Sp),\ \phi(p) \vdash \phi(Sp)$	(3, 4, $\rightarrow\vdash$)
(6)	$\phi(p) \rightarrow \phi(Sp),\ p\,\varepsilon\,[x : \phi(x)] \vdash Sp\,\varepsilon\,[x : \phi(x)]$	(5, $\varepsilon\vdash$, $\vdash\varepsilon$)
(7)	$\phi(p) \rightarrow \phi(Sp) \vdash p\,\varepsilon\,[x : \phi(x)] \rightarrow Sp\,\varepsilon\,[x : \phi(x)]$	(6, $\rightarrow\vdash$)
(8)	$\forall y\,(\phi(y) \rightarrow \phi(Sy)) \vdash p\,\varepsilon\,[x : \phi(x)] \rightarrow Sp\,\varepsilon\,[x : \phi(x)]$	(7, $\forall_2\vdash$)
(9)	$\forall y\,(\phi(y) \rightarrow \phi(Sy)) \vdash \forall y\,(y\,\varepsilon\,[x : \phi(x)] \rightarrow Sy\,\varepsilon\,[x : \phi(x)])$	(8, $\vdash\forall$)
(10)	$\phi(\underline{0}) \wedge \forall y\,(\phi(y) \rightarrow \phi(Sy)) \vdash \forall y\,(y\,\varepsilon\,[x : \phi(x)] \rightarrow Sy\,\varepsilon\,[x : \phi(x)])$	(9, $\wedge\vdash$)
(11)	$\phi(\underline{0}) \wedge \forall y\,(\phi(y) \rightarrow \phi(Sy)) \vdash \underline{0}\,\varepsilon\,[x : \phi(x)]$	(2, $\wedge\vdash$)

(12) $\phi(\underline{0}) \wedge \forall y\,(\phi(y) \to \phi(Sy))$
$\vdash \underline{0}\,\varepsilon\,[x : \phi(x)] \;.\wedge.\; \forall y\,(y\,\varepsilon\,[x : \phi(x)] \to Sy\,\varepsilon\,[x : \phi(x)])$ (10, 11, $\vdash\wedge$)
(13) $\phi(\underline{0}) \wedge \forall y\,(\phi(y) \to \phi(Sy)) \vdash ZS([x : \phi(x)])$ (12, df. ZS)
(14) $\phi(p), \Gamma \vdash \Delta$ (assumption)
(15) $p\,\varepsilon\,[x : \phi(x)], \Gamma \vdash \Delta$ (14, $\varepsilon\vdash$)
(16) $\phi(\underline{0}) \wedge \forall y\,(\phi(y) \to \phi(Sy)),\, ZS([x : \phi(x)]) \to p\,\varepsilon\,[x : \phi(x)], \Gamma \vdash \Delta$
(13, 15, $\to\vdash$)
(17) $\phi(\underline{0}) \wedge \forall y\,(\phi(y) \to \phi(Sy)),\, \forall z\,(ZS(z) \to p\,\varepsilon\,z), \Gamma \vdash \Delta$ (16, $\forall_2\vdash$)
(18) $\phi(\underline{0}) \wedge \forall y\,(\phi(y) \to \phi(Sy)),\, p\,\varepsilon\,[x : \forall z\,(ZS(z) \to x\,\varepsilon\,z)], \Gamma \vdash \Delta$ (17, $\varepsilon\vdash$)
(19) $\phi(\underline{0}) \wedge \forall y\,(\phi(y) \to \phi(Sy)),\, p\,\varepsilon\,\mathbb{N}, \Gamma \vdash \Delta$ (18, df. $\mathbb{N}$)

⊣

THEOREM 10.12 **The Schema of First-Order Induction for N.** *Let $\phi(x)$ be an arithmetic formula. Then*

$$\vdash \phi(\underline{0}) \wedge \forall y\,(\phi(y) \to \phi(Sy)) \;.\!\to.\; (\forall y\,\varepsilon\,\mathbb{N})\phi(y)$$

is derivable in G_2.

Proof. Since $\phi(x)$ is in Λ, by Lemma 10.5 so are the closed formulas $\phi(\underline{0})$ and $\phi(p)$. By Lemma 10.6,

$$\phi(p) \vdash \phi(p)$$

is derivable in G_2. Then, taking Γ to be the empty set of formulas and Δ to be $\{\phi(p)\}$, by the schema of pre-induction (10.11), the sequent

(1) $\phi(\underline{0}) \wedge \forall y\,(\phi(y) \to \phi(Sy)),\, p\,\varepsilon\,\mathbb{N} \vdash \phi(p)$

is derivable in G_2. So, we have

(2) $\phi(\underline{0}) \wedge \forall y\,(\phi(y) \to \phi(Sy)) \vdash p\,\varepsilon\,\mathbb{N} \to \phi(p)$ (1, $\vdash\to$)
(3) $\phi(\underline{0}) \wedge \forall y\,(\phi(y) \to \phi(Sy)) \vdash \forall y\,(y\,\varepsilon\,\mathbb{N} \to \phi(y))$ (2, $\vdash\forall$)
(4) $\vdash \phi(\underline{0}) \wedge \forall y\,(\phi(y) \to \phi(Sy)) \;.\!\to.\; \forall y\,(y\,\varepsilon\,\mathbb{N} \to \phi(y))$ (3, $\vdash\to$)
(5) $\vdash \phi(\underline{0}) \wedge \forall y\,(\phi(y) \to \phi(Sy)) \;.\!\to.\; (\forall y\,\varepsilon\,\mathbb{N})\phi(y)$ (4, df. $\forall y\,\varepsilon\,\mathbb{N}$)

as required. ⊣

11. PEANO ARITHMETIC

11.1. Interpreting Peano's axioms in Λ

It is apparent that A7 and A8 are stated in a form more general than that required to sustain an interpretation of Peano Arithmetic in $G^{+,\times}$. Because, the range of the quantifiers occurring in those axioms includes constant set terms which are not Gilmore numerals and are therefore not intended to denote natural numbers. When giving an interpretation of a first-order theory

T_1 in another such theory T_2, in addition to providing an L_2 formula to interpret each predicate of L_1 and an L_2 term or formula to interpret each function symbol of L_1, we provide an L_2 formula $F(x)$ to carve out the intended domain of the interpretation and then we restrict the quantifiers occurring in the L_2 image of L_1 to $F(x)$. In the following interpretation of Peano Arithmetic in G_2, $\mathbb{N}$ will play the role of the quantifier restriction formula. Then, the resulting versions of A7 and A8 may be derived using the schema of induction for $\mathbb{N}$ to formalize proofs by induction over the natural numbers.

Our new versions of the Peano Axioms are obtained by simply restricting the universal quantifiers in A1–A8 to $\mathbb{N}$, as follows:

PA1. $(\forall x \,\varepsilon\, \mathbb{N})(Sx \neq \underline{0})$.

PA2. $(\forall x_1\, x_2 \,\varepsilon\, \mathbb{N})(Sx_1 = Sx_2 \rightarrow x_1 = x_2)$.

PA3. $(\forall x \,\varepsilon\, \mathbb{N})(\underline{0}x, x) \,\varepsilon\, t_+$.

PA4. $(\forall x\, y\, z \,\varepsilon\, \mathbb{N})\big((Sx\, y, Sz) \,\varepsilon\, t_+ \leftrightarrow (x\, y, z) \,\varepsilon\, t_+)\big)$.

PA5. $(\forall x \,\varepsilon\, \mathbb{N})(\underline{0}x, \underline{0}) \,\varepsilon\, t_\times$.

PA6. $(\forall x\, y\, z \,\varepsilon\, \mathbb{N})\big((Sx\, y, z) \,\varepsilon\, t_\times \leftrightarrow (\exists v \,\varepsilon\, \mathbb{N})((v\, y, z) \,\varepsilon\, t_+ \wedge (x\, y, v) \,\varepsilon\, t_\times)\big)$.

PA7. $(\forall x\, y \,\varepsilon\, \mathbb{N})(\exists! z \,\varepsilon\, \mathbb{N})(x\, y\, z) \,\varepsilon\, t_+$.

PA8. $(\forall x\, y \,\varepsilon\, \mathbb{N})(\exists! z \,\varepsilon\, \mathbb{N})(x\, y\, z) \,\varepsilon\, t_\times$.

The previous $G^{+,\times}$ derivations of A1–A5 may be trivially transformed into G_2 derivations of PA1–PA6, respectively, using just Thinning and $\rightarrow\vdash$. However PA6 is not derivable in G_2 given the current definition of the set term t_+. This is because the quantifier $(\exists v \,\varepsilon\, \mathbb{N})$ that effects the composition of $\times$ with $+$ in A6 is not matched by an $\mathbb{N}$-restriction on the corresponding existentially quantified variable that effects composition in the matrix $B^\times(x, y, z, u)$ of $t_\times$. We redefine $B^\times$ thus:

$$B^\times(x, y, z, u) =_{\mathrm{df}}$$
$$x = \underline{0} \wedge z = \underline{0} \wedge y \,\varepsilon\, \underline{\omega} \;.\!\vee\!.\; \exists x' \,(\exists v \,\varepsilon\, \mathbb{N})\big(x = Sx' \wedge (v\, y\, z) \,\varepsilon\, t_+ \wedge (x'\, y\, v) \,\varepsilon\, u\big).$$

Now the derivation of PA6 goes through nicely. Here, we make essential use of the fact, guaranteed by 10.4, that atoms of the form $p \,\varepsilon\, \mathbb{N}$ are bivalent in G_2. We start at line 3 of the derivation of A6.1:

(1) $Sa = Sa' \wedge (d b, c) \,\varepsilon\, t_+ \wedge (a' b, d) \,\varepsilon\, t_\times \vdash (d b, c) \,\varepsilon\, t_+ \wedge (a b, d) \,\varepsilon\, t_\times$ (line 3, A6.1)

(2) $d \,\varepsilon\, \mathbb{N} \vdash d \,\varepsilon\, \mathbb{N}$ (10.3)

(3) $d\varepsilon\mathbb{N},\; Sa = Sa' \wedge (d b, c)\varepsilon t_+ \wedge (a' b, d)\varepsilon t_\times \vdash d\varepsilon\mathbb{N} \wedge (d b, c)\varepsilon t_+ \wedge (a b, d)\varepsilon t_\times$ (1, 2, $\vdash\wedge$)

(4) $d\varepsilon\mathbb{N}\wedge Sa = Sa' \wedge (db,c)\varepsilon t_+ \wedge (a'b,d)\varepsilon t_\times \vdash d\varepsilon\mathbb{N}\wedge(db,c)\varepsilon t_+ \wedge (ab,d)\varepsilon t_\times$ (3, $\wedge\vdash$)

(5) $d\,\varepsilon\,\mathbb{N} \wedge Sa = Sa' \wedge (db,c)\,\varepsilon\, t_+ \wedge (a'b,d)\,\varepsilon\, t_\times$
$\vdash (\exists v\,\varepsilon\,\mathbb{N})((vb,c)\,\varepsilon\, t_+ \wedge (ab,v)\,\varepsilon\, t_\times)$ (4, $\vdash\exists_2$)

(6) $(\exists v\,\varepsilon\,\mathbb{N})(Sa = Sa' \wedge (vb,c)\,\varepsilon\, t_+ \wedge (a'b,v)\,\varepsilon\, t_\times)$
$\vdash (\exists v\,\varepsilon\,\mathbb{N})((vb,c)\,\varepsilon\, t_+ \wedge (ab,v)\,\varepsilon\, t_\times)$ (5, $\exists\vdash$)

(7) $\exists x'\,(\exists v\,\varepsilon\,\mathbb{N})(Sa = Sx' \wedge (vb,c)\,\varepsilon\, t_+ \wedge (x'b,v)\,\varepsilon\, t_\times)$
$\vdash (\exists v\,\varepsilon\,\mathbb{N})((vb,c)\,\varepsilon\, t_+ \wedge (ab,v)\,\varepsilon\, t_\times)$ (6, $\exists\vdash$)

(8) $Sa = \underline{0} \vdash$ (A1)

(9) $Sa = \underline{0} \wedge c = \underline{0} \wedge b\,\varepsilon\,\underline{\omega} \vdash$ (8, $\wedge\vdash$)

(10) $Sa = \underline{0} \wedge c = \underline{0} \wedge b\,\varepsilon\,\underline{\omega}$.∨. $\exists x'\,(\exists v\,\varepsilon\,\mathbb{N})(Sa = Sx' \wedge t_+(vb,c) \wedge t_\times(x'bv))$
$\vdash (\exists v\,\varepsilon\,\mathbb{N})(t_+(vb,c) \wedge t_\times(abv))$ (4, 6, $\vee\vdash$)

(11) $(Sab,c)\,\varepsilon\, t_\times \vdash (\exists v\,\varepsilon\,\mathbb{N})((vb,c)\,\varepsilon\, t_+ \wedge (ab,v)\,\varepsilon\, t_\times)$ (10, rec. thm.)

(12) $\vdash (Sab,c)\,\varepsilon\, t_\times \rightarrow (\exists v\,\varepsilon\,\mathbb{N})((vb,c)\,\varepsilon\, t_+ \wedge (ab,v)\varepsilon t_\times)$ (11, $\rightarrow\vdash$)

(13) $a\,\varepsilon\,\mathbb{N},\ b\,\varepsilon\,\mathbb{N} c\,\varepsilon\,\mathbb{N} \vdash (Sab,c)\,\varepsilon\, t_\times \rightarrow (\exists v\,\varepsilon\,\mathbb{N})((vb,c)\,\varepsilon\, t_+ \wedge (ab,v)\,\varepsilon\, t_\times)$ (12, Thin)

(14) $\vdash (\forall xyz\,\varepsilon\,\mathbb{N})((Sxy,z)\,\varepsilon\, t_\times \rightarrow (\exists v\,\varepsilon\,\mathbb{N})((vy,z)\,\varepsilon\, t_+ \wedge (xy,v)\,\varepsilon\, t_\times))$ (13, $\rightarrow\vdash$, $\vdash\forall$)

Here, line 11 is an application of the recursion theorem using the alternative definition of $t_\times$ as $\mathit{fix}_u B^\times(x,y,z,u)$. We similarly modify the derivation of A6.2 to obtain:

$$\vdash (\forall xyz\,\varepsilon\,\mathbb{N})\big((\exists v\,\varepsilon\,\mathbb{N})\big((vy,z)\,\varepsilon\, t_+ \wedge (xy,v)\,\varepsilon\, t_\times\big) \rightarrow (Sxy,z)\,\varepsilon\, t_\times\big),$$

as required to complete the derivation of PA6. ⊣

11.2. Deriving the functionality axioms in G_2

We now proceed with G_2's main task, that of yielding derivations of PA7 and PA8.

LEMMA 11.1. $\vdash (\forall y\,\varepsilon\,\mathbb{N})(\exists! z\,\varepsilon\,\mathbb{N})(\underline{0}y,z)\,\varepsilon\, t_+$.

Proof. First, we derive $b\,\varepsilon\,\mathbb{N} \vdash (\exists z\,\varepsilon\,\mathbb{N})(\underline{0}b,z)\,\varepsilon\, t_+$:

(1) $b\,\varepsilon\,\mathbb{N} \vdash b\,\varepsilon\,\underline{\omega}$ (9.11)

(2) $b\,\varepsilon\,\mathbb{N} \vdash \underline{0} = \underline{0} \wedge b = b \wedge b\,\varepsilon\,\underline{\omega}$ (1, = ax, $\vdash\wedge$)

(3) $b\,\varepsilon\,\mathbb{N} \vdash \underline{0} = \underline{0} \wedge b = b \wedge b\,\varepsilon\,\underline{\omega}$.∨. $\exists x'\,z'\,(\underline{0} = Sx' \wedge b = Sz' \wedge (x'b,z')\,\varepsilon\, t_+)$ (2, $\vee\vdash$)

(4) $b\,\varepsilon\,\mathbb{N} \vdash (\underline{0}b,b)\,\varepsilon\, t_+$ (3, rec. thm.)

(5) $b\,\varepsilon\,\mathbb{N},\ b\,\varepsilon\,\mathbb{N} \vdash (\underline{0}b,b)\,\varepsilon\, t_+$ (4, Thin)

(6) $b\,\varepsilon\,\mathbb{N} \vdash b\,\varepsilon\,\mathbb{N} \rightarrow (\underline{0}b,b)\,\varepsilon\, t_+$ (5, $\vdash\rightarrow$)

(7) $b \,\varepsilon\, \mathbb{N} \vdash \exists z\,(z \,\varepsilon\, \mathbb{N} \to (\underline{0}\,b, b) \,\varepsilon\, t_+)$ (6, $\exists\vdash$)

(8) $b \,\varepsilon\, \mathbb{N} \vdash (\exists z \,\varepsilon\, \mathbb{N})(\underline{0}\,b, z) \,\varepsilon\, t_+$ (df. $\exists z \,\varepsilon\, \mathbb{N}$)

Next, $\vdash \forall z_1\, z_2 \big((\underline{0}\,b, z_1) \,\varepsilon\, t_+ \wedge (\underline{0}\,b, z_2) \,\varepsilon\, t_+ \,.\!\!\to\!\!.\; z_1 = z_2\big)$:

(1) $b = a_1 \vdash b = a_1$ (= ax)

(2) $\underline{0} = \underline{0} \wedge b = a_1 \wedge b \,\varepsilon\, \underline{\omega} \vdash b = a_1$ (1, $\wedge\vdash$)

(3) $\underline{0} = Sq_1 \vdash$ (4.1)

(4) $\underline{0} = Sq_1 \wedge a_1 = Sq_2 \wedge (q_1 b, q_2) \,\varepsilon\, t_+ \vdash$ (3, $\wedge\vdash$)

(5) $\exists x'\, z'\,(\underline{0} = Sx' \wedge a_1 = Sz' \wedge (x'\,b, z') \,\varepsilon\, t_+) \vdash$ (4, $\exists\vdash$)

(6) $\underline{0} = \underline{0} \wedge b = a_1 \wedge b \varepsilon \underline{\omega} \;.\vee.\; \exists x'\, z'\,(\underline{0} = Sx' \wedge a_1 = Sz' \wedge (x'\,b, z') \varepsilon t_+) \vdash b = a_1$
(2, 5, $\vee\vdash$)

(7) $(\underline{0}\,b, a_1) \,\varepsilon\, t_+ \vdash b = a_1$ (6, rec. thm.)

By a similar derivation, we obtain:

(8) $(\underline{0}\,b, a_2) \,\varepsilon\, t_+ \vdash b = a_2$ (7[a_1/a_2])

(9) $(\underline{0}\,b, a_1) \,\varepsilon\, t_+ \wedge (\underline{0}\,b, a_2) \,\varepsilon\, t_+ \vdash b = a_1 \wedge b = a_2$ (7, 8, $\vdash\wedge$, $\wedge\vdash$)

(10) $b = a_1 \wedge b = a_2 \vdash a_1 = a_2$ (= logic)

(11) $(\underline{0}\,b, a_1) \,\varepsilon\, t_+ \wedge (\underline{0}\,b, a_2) \,\varepsilon\, t_+ \vdash a_1 = a_2$ (9, 10, Cut)

(12) $\vdash (\underline{0}\,b, a_1) \,\varepsilon\, t_+ \wedge (\underline{0}\,b, a_2) \,\varepsilon\, t_+) \,.\!\!\to\!\!.\; a_1 = a_2$ (11, $\vdash\to$)

(13) $\vdash \forall z_1\, z_2\,((\underline{0}\,b, z_1) \,\varepsilon\, t_+ \wedge (\underline{0}\,b, z_2) \,\varepsilon\, t_+ \,.\!\!\to\!\!.\; z_1 = z_2)$ (12, $\vdash\forall$)

Putting these two results together

$b \,\varepsilon\, \mathbb{N} \vdash (\exists z \,\varepsilon\, \mathbb{N})(\underline{0}\,b, z) \,\varepsilon\, t_+ \wedge \forall z_1\, z_2\,((\underline{0}\,b, z_1) \,\varepsilon\, t_+ \wedge (\underline{0}\,b, z_2) \,\varepsilon\, t_+ \,.\!\!\to\!\!.\; z_1 = z_2)$

$\vdash b \varepsilon \mathbb{N} \to ((\exists z \varepsilon \mathbb{N})(\underline{0}b, z) \varepsilon t_+ \wedge \forall z_1\, z_2\,((\underline{0}\,b, z_1) \varepsilon t_+ \wedge (\underline{0}\,b, z_2) \varepsilon t_+ \,.\!\!\to\!\!.\; z_1 = z_2))$

$\vdash \forall y\,(y \,\varepsilon\, \mathbb{N} \to (\exists z \,\varepsilon\, \mathbb{N})(\underline{0}\,y, z) \,\varepsilon\, t_+ \wedge$
$\forall z_1\, z_2\,((\underline{0}\,y, z_1) \,\varepsilon\, t_+ \wedge (\underline{0}\,y, z_2) \,\varepsilon\, t_+ \,.\!\!\to\!\!.\; z_1 = z_2))$

$\vdash (\forall y \varepsilon \mathbb{N})((\exists z \varepsilon \mathbb{N})(\underline{0}\,y, z) \varepsilon t_+ \wedge \forall z_1\, z_2\,((\underline{0}\,y, z_1) \varepsilon t_+ \wedge (\underline{0}\,y, z_2) \varepsilon t_+ \,.\!\!\to\!\!.\; z_1 = z_2))$

$\vdash (\forall y \,\varepsilon\, \mathbb{N})(\exists! z \,\varepsilon\, \mathbb{N})(\underline{0}\,y, z) \,\varepsilon\, t_+$ (df. $\exists!$)

as required. ⊣

LEMMA 11.2. *Let a and b be any parameters of L. Then,*

$$\exists z\,\big(z \,\varepsilon\, \mathbb{N} \wedge (a\,b, z) \,\varepsilon\, t_+\big) \vdash \exists z\,\big(z \,\varepsilon\, \mathbb{N} \wedge (Sa\,b, z) \,\varepsilon\, t_+\big),$$

is a theorem of G_2.

Proof. Let c be a parameter distinct from a and b.

(1) $(a\,b, c) \,\varepsilon\, t_+ \vdash (Sa\,b, Sc) \,\varepsilon\, t_+$ (A4.2)

(2) $c \,\varepsilon\, \mathbb{N} \vdash Sc \,\varepsilon\, \mathbb{N}$ (10.8)

(3) $c \,\varepsilon\, \mathbb{N},\ (a\,b, c) \,\varepsilon\, t_+ \vdash Sc \,\varepsilon\, \mathbb{N} \wedge (Sa\,b, Sc) \,\varepsilon\, t_+$ (1, 2 $\vdash\wedge$)

(4) $c \varepsilon \mathbb{N} \wedge (ab, c) \varepsilon t_+ \vdash Sc \varepsilon \mathbb{N} \wedge (Sab, Sc) \varepsilon t_+$ (3, $\vdash\wedge$)
(5) $c \varepsilon \mathbb{N} \wedge (ab, c) \varepsilon t_+ \vdash \exists z (z \varepsilon \mathbb{N} \wedge (Sab, z) \varepsilon t_+)$ (4, $\vdash\exists_2$)
(6) $\exists z (z \varepsilon \mathbb{N} \wedge (ab, z) \varepsilon t_+) \vdash \exists z (z \varepsilon \mathbb{N} \wedge (Sab, z) \varepsilon t_+)$ (5, $\exists\vdash$)
(7) $(\exists z \varepsilon \mathbb{N})(ab, z) \varepsilon t_+ \vdash (\exists z \varepsilon \mathbb{N})(Sab, z) \varepsilon t_+$ (6, df. $\exists z \varepsilon \mathbb{N}$)

⊣

LEMMA 11.3.

$$\vdash \forall x \left((\forall y \varepsilon \mathbb{N})(\exists! z \varepsilon \mathbb{N})(xy, z) \varepsilon t_+ \rightarrow (\forall y \varepsilon \mathbb{N})(\exists! z \varepsilon \mathbb{N})(Sxy, z) \varepsilon t_+\right).$$

Proof. Let a, b and c be any parameters of L. Let

$$\phi(x, y) =_{\text{df}} (\forall z_1\, z_2)\left((xy, z_1) \varepsilon t_+ \wedge (xy, z_2) \varepsilon t_+ \,.\!\rightarrow\!.\, z_1 = z_2\right).$$

By a lengthy derivation, we can show that $\phi(a, b) \vdash \phi(Sa, b)$. Then from 11.2 by $\vdash\wedge$ and $\wedge\vdash$,

$$(\exists z \varepsilon \mathbb{N})(ab, z) \varepsilon t_+ \wedge \phi(a, b) \vdash (\exists z \varepsilon \mathbb{N})(Sab, z) \varepsilon t_+ \wedge \phi(Sa, b)$$

from which the desired sequent is obtained by $b \varepsilon \mathbb{N} \vdash b \varepsilon \mathbb{N}$ (10.4) using $\rightarrow\vdash$, $\vdash\rightarrow$, $\vdash\forall$ and $\vdash\forall$. ⊣

COROLLARY 11.4. *A7 is derivable in* G_2.

Proof. Let $\phi(x)$ be the arithmetic formula $(\forall y \varepsilon \mathbb{N})(\exists! z \varepsilon \mathbb{N})(xy, z) \varepsilon t_+$. By 11.1 and 11.3,

$$\vdash \phi(\underline{0}) \wedge \forall x \left(\phi(x) \rightarrow \phi(Sx)\right)$$

is a theorem of G_2. By 10.12, the induction schema for $\mathbb{N}$, $\vdash (\forall x \varepsilon \mathbb{N})\phi(x)$ is derivable in G_2. That is,

$$\vdash (\forall x \varepsilon \mathbb{N})(\forall y \varepsilon \mathbb{N})(\exists! z \varepsilon \mathbb{N})(xy, z) \varepsilon t_+$$

is a theorem of G_2, as required. ⊣

LEMMA 11.5. $\vdash (\forall y \varepsilon \mathbb{N})(\exists! z \varepsilon \mathbb{N})(\underline{0}y, z) \varepsilon t_\times$.

Proof. Similar to the treatment of the 0-clause for A7 in 11.1. ⊣

LEMMA 11.6. *Let* d, b, c_1, c_2 *be distinct parameters. Then*

$$d \varepsilon \mathbb{N},\ (db, c_1) \varepsilon t_+,\ (db, c_2) \varepsilon t_+ \vdash c_1 = c_2$$

is a theorem of G_2.

Proof. Let $\phi(x)$ be the Λ formula $\forall z_1 z_2 \big((x\,b\,z_1)\,\varepsilon\,t_+ \wedge (x\,b\,z_2)\,\varepsilon\,t_+ \ .{\rightarrow}.\ z_1 = z_2)\big)$. We can derive

$$\phi(d),\ (db, c_1)\,\varepsilon\,t_+,\ (db, c_2)\,\varepsilon\,t_+ \vdash c_1 = c_2$$

using G_2 bivalence axioms. By 10.11, the schema of pre-induction,

$$\phi(\underline{0}) \wedge \forall x\,(\phi(x) \rightarrow \phi(Sx)),\ d\,\varepsilon\,\mathbb{N},\ (db, c_1)\,\varepsilon\,t_+,\ (db, c_2)\,\varepsilon\,t_+ \vdash c_1 = c_2$$

is a theorem of G_2. Now, by line (13) of the derivation of Lemma 11.1 and the lengthy derivation mentioned in the proof of Lemma 11.3,

$$\vdash \phi(\underline{0}) \wedge \forall x\,(\phi(x) \rightarrow \phi(Sx))$$

is a theorem of G_2. Hence, we obtain a G_2 derivation of the desired sequent by Cut. ⊣

LEMMA 11.7. *Let a and b be arbitrary parameters of L.*

$$(\forall z_1 z_2)\big((a\,b\,z_1)\,\varepsilon\,t_\times \wedge (a\,b\,z_2)\,\varepsilon\,t_\times \ .{\rightarrow}.\ z_1 = z_2\big)$$
$$\vdash (\forall z_1 z_2)\big((Sa\,b\,z_1)\,\varepsilon\,t_\times \wedge (Sa\,b\,z_2)\,\varepsilon\,t_\times \ .{\rightarrow}.\ z_1 = z_2\big).$$

Proof. Let

$$\phi(x, y, z_1, z_2) =_{\text{df}} (x\,y\,z_1)\,\varepsilon\,t_\times \wedge (x\,y\,z_2)\,\varepsilon\,t_\times \ .{\rightarrow}.\ z_1 = z_2,$$
$$\forall\phi =_{\text{df}} (\forall z_1 z_2)\phi(a, b, z_1, z_2).$$

(1) $d_1\,\varepsilon\,\mathbb{N},\ (d_1\,b, c_1)\,\varepsilon\,t_+,\ (d_1\,b, c_2)\,\varepsilon\,t_+ \vdash c_1 = c_2$ (11.6)

(2) $d_1 = d_2,\ (d_2\,b, c_2)\,\varepsilon\,t_+ \vdash (d_1\,b, c_2)\,\varepsilon\,t_+$ ($G^{+,\times}$ ax, $=\vdash$)

(3) $d_1\,\varepsilon\,\mathbb{N},\ d_1 = d_2,\ (d_1\,bc_1)\,\varepsilon\,t_+,\ (d_2\,bc_2)\,\varepsilon\,t_+ \vdash c_1 = c_2$ (2, 3 Cut)

(4) $Sa = Sp,\ Sa = Sp' \vdash p = a \wedge p' = a$ (A2, $\vdash\wedge$)

(5) $p = a \wedge p' = a,\ (pbd_1)\,\varepsilon\,t_\times,\ (p'\,bd_2)\,\varepsilon\,t_\times \vdash (abd_1)\,\varepsilon\,t_\times \wedge (abd_2)\,\varepsilon\,t_\times$ ($G^{+,\times}$ ax, $=\vdash$, $\vdash\wedge$)

(6) $Sa = Sp,\ (pbd_1)\varepsilon t_\times,\ Sa = Sp',\ (p'\,bd_2)\varepsilon t_\times \vdash (abd_1)\varepsilon t_\times \wedge (abd_2)\varepsilon t_\times$ (4, 5, Cut, Inter)

(7) $\phi(a, b, d_1, d_2),\ d_1\,\varepsilon\,\mathbb{N},\ Sa = Sp,\ t_+(d_1\,bc_1),\ t_\times(pbd_1),\ Sa = Sp',\ t_+(d_2\,bc_2),\ t_\times(p'\,bd_2) \vdash c_1 = c_2$ (3, 6, $=\vdash$)

(8) $\forall\phi,\ d_1\,\varepsilon\,\mathbb{N},\ Sa = Sp,\ t_+(d_1\,bc_1),\ t_\times(pbd_1),\ Sa = Sp',\ t_+(d_2\,bc_2),\ t_\times(p'\,bd_2) \vdash c_1 = c_2$ (7, $\forall_2\vdash$)

(9) $\forall\phi, d_1\varepsilon\mathbb{N}, Sa = Sp, t_+(d_1\,bc_1), t_\times(pbd_1), d_2\varepsilon\mathbb{N}\wedge Sa = Sp'\wedge t_+(d_2\,bc_2)\wedge t_\times(p'\,bd_2) \vdash c_1 = c_2$ (8, $\wedge\vdash$)

(10) $\forall\phi,\ d_1\,\varepsilon\,\mathbb{N},\ Sa = Sp,\ t_+(d_1\,bc_1),\ t_\times(pbd_1),\ \exists x'\,(\exists v\,\varepsilon\,\mathbb{N})(Sa = Sx' \wedge t_+(v\,bc_2) \wedge t_\times(x'\,bv)) \vdash c_1 = c_2$ (9, $\exists\vdash$)

(11) $Sa = \underline{0} \vdash$ (A1)

(12) $Sa = \underline{0} \wedge c_2 = \underline{0} \wedge b\,\varepsilon\,\underline{\omega} \vdash$ (11, $\wedge\vdash$)

(13) $\forall\phi,\ d_1 \,\varepsilon\, \mathbb{N},\ Sa = Sp,\ t_+(d_1bc_1),\ t_\times(pbd_1),\ B^\times(Sa,b,c_2,t_\times) \vdash c_1 = c_2$ (10, 12, $\vee\vdash$)

(14) $\forall f,\ d_1 \,\varepsilon\, \mathbb{N},\ Sa = Sp,\ t_+(d_1\,b\,c_1),\ t_\times(p\,b\,d_1),\ t_\times(Sa\,b\,c_2) \vdash c_1 = c_2$ (13, rec. thm.)

(15) $\forall\phi,\ d_1 \,\varepsilon\, \mathbb{N} \wedge Sa = Sp \wedge t_+(d_1\,b\,c_1) \wedge t_\times(p\,b\,d_1),\ t_\times(Sa\,b\,c_2) \vdash c_1 = c_2$ (14, $\wedge\vdash$)

(16) $\forall\phi,\ \exists x'(\exists v \,\varepsilon\, \mathbb{N})(Sa = Sx' \wedge t_+(v\,b\,c_1) \wedge t_\times(x'\,b\,v)),\ t_\times(Sa\,b\,c_2) \vdash c_1 = c_2$ (15, $\exists\vdash$)

(17) $Sa = \underline{0} \wedge c_1 = \underline{0} \wedge b \,\varepsilon\, \underline{\omega} \vdash$ (A1, $\wedge\vdash$)

(18) $\forall\phi,\ B^\times(Sa,b,c_1,t_\times),\ t_\times(Sa\,b\,c_2) \vdash c_1 = c_2$ (16, 17, $\vee\vdash$)

(19) $\forall\phi,\ (Sa\,b\,c_1) \,\varepsilon\, t_\times,\ (Sa\,b\,c_2) \,\varepsilon\, t_\times \vdash c_1 = c_2$ (18, rec. thm.)

(20) $\forall\phi,\ (Sa\,b\,c_1) \,\varepsilon\, t_\times \wedge (Sa\,b\,c_2) \,\varepsilon\, t_\times \vdash c_1 = c_2$ (19, $\wedge\vdash$)

(21) $\forall\phi \vdash (Sa\,b\,c_1) \,\varepsilon\, t_\times \wedge (Sa\,b\,c_2) \,\varepsilon\, t_\times \;.\!\!\rightarrow\!\!.\; c_1 = c_2$ (20, $\vdash\rightarrow$)

(22) $(\forall z_1\,z_2)\phi(a,b,z_1,z_2) \vdash (Sa\,b\,c_1) \,\varepsilon\, t_\times \wedge (Sa\,b\,c_2) \,\varepsilon\, t_\times \;.\!\!\rightarrow\!\!.\; c_1 = c_2$ (21, df. $\forall\phi$)

(23) $(\forall z_1\,z_2)\phi(a,b,z_1,z_2) \vdash (\forall z_1\,z_2)((Sa\,b\,z_1)\varepsilon t_\times \wedge (Sa\,b\,z_2)\varepsilon t_\times \;.\!\!\rightarrow\!\!.\; z_1 = z_2)$ (22, $\vdash\forall$)

(24) $(\forall z_1\,z_2)((a\,b\,z_1) \,\varepsilon\, t_\times \wedge (a\,b\,z_2) \,\varepsilon\, t_\times \;.\!\!\rightarrow\!\!.\; z_1 = z_2)$
$\vdash (\forall z_1\,z_2)((Sa\,b\,z_1) \,\varepsilon\, t_+ \wedge (Sa\,b\,z_2) \,\varepsilon\, t_\times \;.\!\!\rightarrow\!\!.\; z_1 = z_2)$

Note that $B^\times(Sa,b,c_2,t_\times)$ at line 13 abbreviates

$$Sa = \underline{0} \wedge c_2 = \underline{0} \wedge b \,\varepsilon\, \underline{\omega} \;.\!\vee\!.\; \exists x'\,(\exists v \,\varepsilon\, \mathbb{N})\big(Sa = Sx' \wedge t_+(v\,b\,c_2) \wedge t_\times(x'\,b\,v)\big),$$

with similar remarks for $B^\times(Sa,b,c_1,t_\times)$ at 18. ⊣

LEMMA 11.8. $c \,\varepsilon\, \mathbb{N},\ b \,\varepsilon\, \mathbb{N} \vdash (\exists z \,\varepsilon\, \mathbb{N})(cb,z) \,\varepsilon\, t_+$ *is a theorem of* G_2.

Proof. Let $\phi(x) =_{\text{df}} (\forall y \,\varepsilon\, \mathbb{N})(\exists z \,\varepsilon\, \mathbb{N})(xy,z) \,\varepsilon\, t_+$. The G_2 bivalence axioms, 10.4 and df. $\exists z \,\varepsilon\, \mathbb{N}$ yield:

$$(\forall y \,\varepsilon\, \mathbb{N})(\exists z \,\varepsilon\, \mathbb{N})(cy,z) \,\varepsilon\, t_+, b \,\varepsilon\, \mathbb{N} \;\vdash (\exists z \,\varepsilon\, \mathbb{N})(cb,z) \,\varepsilon\, t_+,$$

that is, $\phi(c), b \varepsilon \mathbb{N} \;\vdash (\exists z \varepsilon \mathbb{N})(cb,z) \varepsilon t_+$, so by 10.11, the schema of pre-induction,

$$\phi(\underline{0}) \wedge \forall y\,(\phi(y) \rightarrow \phi(Sy)), c \,\varepsilon\, \mathbb{N}, b \,\varepsilon\, \mathbb{N} \vdash (\exists z \,\varepsilon\, \mathbb{N})(cb,z) \,\varepsilon\, t_+.$$

From 11.2, we have

(1) $(\exists z \,\varepsilon\, \mathbb{N})(a\,b,z) \,\varepsilon\, t_+ \vdash (\exists z \,\varepsilon\, \mathbb{N})(Sa\,b,z) \,\varepsilon\, t_+$ (11.2)

(2) $\vdash (\exists z \,\varepsilon\, \mathbb{N})(a\,b,z) \,\varepsilon\, t_+ \rightarrow (\exists z \,\varepsilon\, \mathbb{N})(Sa\,b,z) \,\varepsilon\, t_+$ (1, $\rightarrow\vdash$)

(3) $\vdash \forall x((\exists z \,\varepsilon\, \mathbb{N})(x\,b,z) \,\varepsilon\, t_+ \rightarrow (\exists z \,\varepsilon\, \mathbb{N})(Sx\,b,z) \,\varepsilon\, t_+)$ (2, $\vdash\forall$)

And the first derivation in Lemma 11.1 yields, $b \,\varepsilon\, \mathbb{N} \vdash (\exists z \,\varepsilon\, \mathbb{N})(\underline{0}\, b, z) \,\varepsilon\, t_+$, so $\vdash\wedge$ yields a G_2 derivation of

$$b \,\varepsilon\, \mathbb{N} \vdash \phi(\underline{0}) \wedge \forall y\, (\phi(y) \to \phi(Sy)).$$

Hence, we obtain the desired G_2 derivation of

$$c \,\varepsilon\, \mathbb{N},\, b \,\varepsilon\, \mathbb{N} \vdash (\exists z \,\varepsilon\, \mathbb{N})(c b, z) \,\varepsilon\, t_+$$

by Cut. ⊣

LEMMA 11.9.

$$\vdash \forall x\, ((\forall y \,\varepsilon\, \mathbb{N})(\exists! z \,\varepsilon\, \mathbb{N})(x\, y, z) \,\varepsilon\, t_\times \to (\forall y \,\varepsilon\, \mathbb{N})(\exists! z \,\varepsilon\, \mathbb{N})(Sx\, y, z) \,\varepsilon\, t_\times).$$

Proof. Let a, b and c be any parameters of L. Let

$$\phi(x, y) =_{\text{df}} (\forall z_1\, z_2)((x\, y\, z_1) \,\varepsilon\, t_\times \wedge (x\, y\, z_2) \,\varepsilon\, t_\times \mathrel{.\!\!\to\!\!.} z_1 = z_2).$$

Note Lemma 11.7 is a G_2 derivation of $\phi(a, b) \vdash \phi(Sa, b)$.

(1) $(c b, d) \,\varepsilon\, t_+, (a b, c) \,\varepsilon\, t_\times \vdash (c b, d) \,\varepsilon\, t_+ \wedge (a b, c) \,\varepsilon\, t_\times$ ($G^{+,\times}$ ax, $\vdash\wedge$)
(2) $(c b, d) \,\varepsilon\, t_+, (a b, c) \,\varepsilon\, t_\times \vdash Sa = Sa \wedge (c b, d) \,\varepsilon\, t_+ \wedge (a\beta, c) \,\varepsilon\, t_\times$
(1, = ax, $\vdash\wedge$)
(3) $(c b, d) \,\varepsilon\, t_+, (a b, c) \,\varepsilon\, t_\times \vdash \exists x'\, v\, (Sa = Sx' \wedge (v\, b, d) \,\varepsilon\, t_+ \wedge (x'\, b, v) \,\varepsilon\, t_\times$
(2, $\vdash\exists_2$)
(4) $(c b, d) \,\varepsilon\, t_+, (a b, c) \,\varepsilon\, t_\times \vdash Sa = \underline{0} \wedge d = \underline{0} \wedge b \,\varepsilon\, \underline{\omega} \mathrel{.\vee.}$
$\exists x'\, v\, (Sa = Sx' \wedge (v\, b, d) \,\varepsilon\, t_+ \wedge (x'\, b, v) \,\varepsilon\, t_\times)$ (3, $\vdash\vee$)
(5) $(c b, d) \,\varepsilon\, t_+, (a b, c) \,\varepsilon\, t_\times \vdash (Sa b, d) \,\varepsilon\, t_\times$ (4, rec. thm.)
(6) $d \,\varepsilon\, \mathbb{N} \vdash d \,\varepsilon\, \mathbb{N}$ (10.3)
(7) $d \,\varepsilon\, \mathbb{N} \wedge (c b, d) \,\varepsilon\, t_+, (a b, c) \,\varepsilon\, t_\times \vdash d \,\varepsilon\, \mathbb{N} \wedge (Sa b, d) \,\varepsilon\, t_\times$ (5, 6, $\vdash\wedge$)
(8) $d \,\varepsilon\, \mathbb{N} \wedge (c b, d) \,\varepsilon\, t_+, (a b, c) \,\varepsilon\, t_\times \vdash (\exists z \,\varepsilon\, \mathbb{N})(Sa b, z) \,\varepsilon\, t_\times$ (7, $\vdash\exists_2$, df. $\exists z \,\varepsilon\, \mathbb{N}$)
(9) $(\exists z \,\varepsilon\, \mathbb{N})(c b, z) \,\varepsilon\, t_+, (a b, c) \,\varepsilon\, t_\times \vdash (\exists z \,\varepsilon\, \mathbb{N})(Sa b, z) \,\varepsilon\, t_\times$ (8, $\exists\vdash$, df. $\exists z \,\varepsilon\, \mathbb{N}$)
(10) $c \,\varepsilon\, \mathbb{N}, b \,\varepsilon\, \mathbb{N} \vdash (\exists z \,\varepsilon\, \mathbb{N})(c b, z) \,\varepsilon\, t_+$ (11.7)
(11) $b \,\varepsilon\, \mathbb{N}, c \,\varepsilon\, \mathbb{N} \wedge (a b, c) \,\varepsilon\, t_\times \vdash (\exists z \,\varepsilon\, \mathbb{N})(Sa b, z) \,\varepsilon\, t_\times$ (9, 10, Cut, $\wedge\vdash$)
(12) $b \,\varepsilon\, \mathbb{N}, (\exists z \,\varepsilon\, \mathbb{N})(a b, z) \,\varepsilon\, t_\times \vdash (\exists z \,\varepsilon\, \mathbb{N})(Sa b, z) \,\varepsilon\, t_\times$ (11, $\exists\vdash$, df. $\exists z \,\varepsilon\, \mathbb{N}$)
(13) $\phi(a, b) \vdash \phi(Sa, b)$ (11.6)
(14) $b \,\varepsilon\, \mathbb{N}A, (\exists z \,\varepsilon\, \mathbb{N})(a b, z) \,\varepsilon\, t_\times \wedge \phi(a, b) \vdash (\exists z \,\varepsilon\, \mathbb{N})(Sa b, z) \,\varepsilon\, t_\times \wedge \phi(Sa, b)$
(12, 13, $\vdash\wedge$)
(15) $b \,\varepsilon\, \mathbb{N}, (\exists! z \,\varepsilon\, \mathbb{N})(a b, z) \,\varepsilon\, t_\times \vdash (\exists! z \,\varepsilon\, \mathbb{N})(Sa b, z) \,\varepsilon\, t_\times$ (14, df. $\exists! z \,\varepsilon\, \mathbb{N}$)
(16) $b \,\varepsilon\, \mathbb{N} \vdash b \,\varepsilon\, \mathbb{N}$ (10.3)
(17) $b \,\varepsilon\, \mathbb{N} \to (\exists! z \,\varepsilon\, \mathbb{N})(a b, z) \,\varepsilon\, t_\times, b \,\varepsilon\, \mathbb{N} \vdash (\exists! z \,\varepsilon\, \mathbb{N})(Sa b, z) \,\varepsilon\, t_\times$ (15, 16, $\to\vdash$)
(18) $b \,\varepsilon\, \mathbb{N} \to (\exists! z \,\varepsilon\, \mathbb{N})(ab, z) \,\varepsilon\, t_\times \vdash b \,\varepsilon\, \mathbb{N} \to (\exists! z \,\varepsilon\, \mathbb{N})(Sab, z) \,\varepsilon\, t_\times$ (17, $\vdash\to$)

(19) $\forall y\,(y \,\varepsilon\, \mathbb{N} \;\rightarrow (\exists! z \,\varepsilon\, \mathbb{N})(ay, z) \,\varepsilon\, t_\times) \vdash b \,\varepsilon\, \mathbb{N} \;\rightarrow (\exists! z \,\varepsilon\, \mathbb{N})(Sab, z) \,\varepsilon\, t_\times$
(18, $\forall_2\vdash$)

(20) $\forall y\,(y \,\varepsilon\, \mathbb{N} \;\rightarrow (\exists! z \,\varepsilon\, \mathbb{N})(ay, z) \,\varepsilon\, t_\times) \vdash \forall y\,(y \,\varepsilon\, \mathbb{N} \rightarrow (\exists! z \,\varepsilon\, \mathbb{N})(Say, z) \,\varepsilon\, t_\times)$
(19, $\vdash\forall$)

(21) $(\forall y \,\varepsilon\, \mathbb{N})(\exists! z \,\varepsilon\, \mathbb{N})(ay, z) \,\varepsilon\, t_\times \vdash (\forall y \,\varepsilon\, \mathbb{N})(\exists! z \,\varepsilon\, \mathbb{N})(Say, z) \,\varepsilon\, t_\times$
(10, df. $\forall y \,\varepsilon\, \mathbb{N}$)

(22) $\vdash (\forall y \varepsilon \mathbb{N})(\exists! z \varepsilon \mathbb{N})(ay, z) \varepsilon t_\times \rightarrow (\forall y \varepsilon \mathbb{N})(\exists! z \varepsilon \mathbb{N})(Say, z) \varepsilon t_\times$ (21, $\vdash\rightarrow$)

(23) $\vdash \forall x\,((\forall y \,\varepsilon\, \mathbb{N})(\exists! z \,\varepsilon\, \mathbb{N})(xy, z) \,\varepsilon\, t_\times \rightarrow (\forall y \,\varepsilon\, \mathbb{N})(\exists! z \,\varepsilon\, \mathbb{N})(Sxy, z) \,\varepsilon\, t_\times)$
(22, $\vdash\forall$)

⊣

COROLLARY 11.10. A6 *is derivable in* G_2.

Proof. Let $\phi(x)$ be the arithmetic formula $(\forall y \,\varepsilon\, \mathbb{N})(\exists! z \,\varepsilon\, \mathbb{N})(x\,y\,z) \,\varepsilon\, t_\times$. By 11.5 and 11.9,

$$\vdash \phi(\underline{0}) \wedge \forall x\,(\phi(x) \rightarrow \phi(Sx))$$

is a theorem of G_2. By the 10.12, induction schema for $\mathbb{N}, \vdash (\forall x \,\varepsilon\, \mathbb{N})\phi(x)$ is derivable in G_2. That is,

$$\vdash (\forall x \,\varepsilon\, \mathbb{N})(\forall y \,\varepsilon\, \mathbb{N})(\exists! z \,\varepsilon\, \mathbb{N})(x\,y\,z) \,\varepsilon\, t_\times$$

is a theorem of G_2, as required. ⊣

This completes our interpretation of Peano Arithmetic in G_2. Now we must show that G_2 is not *too* powerful.

12. SEMANTICS FOR G_2

12.1. Term models for G_2

In this section we modify the Definition 3.1 of *truth in an L-structure* to accommodate the new syntactic features of G_2. Evidently, the challenge here is the validation of G_2's bivalence axioms $t \,\varepsilon\, p \vdash t \,\varepsilon\, p$, which requires that the given structure be t-total for all terms t comprising the range of the G_2 variables, viz., the constant arithmetic terms. Now, the $+, \times$-total L-structure $\mathcal{B}$ defined by 9.1 comes very close to providing such a structure. Our strategy is to modify the Definition 3.1 of truth on a structure to accommodate the intuitive interpretation of the G_2 quantifiers and variables and then show that $\mathcal{B}$ can be extended to produce an $+, \times$-total L-structure on which every instance of an arithmetic formula converges.

DEFINITION 12.1. The G_2 inductive definition of *truth on an L-structure* is given by clauses 1–4 of Definition 3.1 and, for the G_2 quantifiers, by replacing clauses 4 and 5 by

$$\mathcal{S}(\forall x\, A) \;= \bigwedge\{\mathcal{S}(A_x[\mathtt{t}]) \mid \mathtt{t} \text{ is a constant arithmetic term}\},$$
$$\mathcal{S}(\exists x\, A) \;= \bigvee\{\mathcal{S}(A_x[\mathtt{t}]) \mid \mathtt{t} \text{ is a constant arithmetic term}\}$$

respectively, where, as in those clauses, $\bigwedge$ is arbitrary meet and $\bigvee$ is arbitrary join with respect to $(4, \leq_t)$. Hence, the quantifiers conform in G_2 semantics to the same $\leq_k$-monotone interpretive scheme as the quantifiers in G.

All cognate semantic definitions are the same as those of section 3, except for the definition of an *instance* of a sequent, which needs to be modified to accord with our intention that the range of permissible substitutends for the G_2 parameters are constant *arithmetic* terms only. An *instance* of a closed $\mathcal{L}$-formula is an $\mathcal{L}$-sentence obtained by uniformly replacing every parameter occurring in the given formula by a constant term of Λ. If $\Gamma \vdash \Delta$ is a sequent, $\vec{a} = a_1, \ldots, a_k$ exhaust all of the parameters occurring in members of Γ or Δ, and $\vec{t} = t_1, \ldots, t_k$ are constant terms of Λ, then $\Gamma_{\vec{a}}[\vec{t}\,] \vdash \Delta_{\vec{a}}[\vec{t}\,]$ is *an instance of* $\Gamma \vdash \Delta$. Since G_2 semantics is $\leq_k$-monotone, it is clear that the definition of the Gilmore revision operator extends to yield $\leq_k$-monotone revisions of $\mathcal{L}$-structures.

12.2. Consistency

Our strategy is to modify the construction of $\mathcal{B}$ in Definition 9.1 to produce an $\mathcal{L}$-structure $\mathcal{C} \geq_k \mathcal{B}$ which totalizes our impredicative representation of the natural numbers, N. Then as in the proof of Theorem 9.3, we will show that $\mathcal{C}$ is in fact 3-valued.

DEFINITION 12.2. Define $\varepsilon_0 : ctrm^3 \to 4$ by:

1. $\varepsilon_0(t_1, t_2, t_3) = \mathit{false}$ if at least one of the following two conditions hold:

   (i) $t_1 = t_+$ and it is not the case that there are $m, n, j \in \mathbb{N}$ such that $t_2 = \langle \underline{m}, \underline{n} \rangle$ and $t_3 = \underline{j}$,

   (ii) $t_1 = t_\times$ and it is not the case that there are $m, n, j \in \mathbb{N}$ such that $t_2 = \langle \underline{m}, \underline{n} \rangle$ and $t_3 = \underline{j}$,

   otherwise,

2. $\varepsilon_0(t_1, t_2, t_3) = \mathit{true}$ if $t_1 = \mathrm{N}$ and $t_2 \in \mathit{Num}$, otherwise,

3. $\varepsilon_0(t_1, t_2, t_3) = \bot$.

Let $\mathcal{C}_0 =_{\mathrm{df}} (ctrm, \varepsilon_0)$. Using our new Φ, define an ordinal sequence $\{\mathcal{C}_\alpha \mid \alpha \leq \omega_1\}$ of $\mathcal{L}$-structures as per 3.10. We write ε_α for $\varepsilon_{\mathcal{C}_\alpha}$. For the rest of this section, we write simply $\to_\alpha$ for $\to_{\mathcal{C}_\alpha}$. Also, by "the level of a sentence B" we mean the level of B in $\{\mathcal{C}_\alpha \mid \alpha \leq \omega_1\}$.

THEOREM 12.3. *Let A be any arithmetic sentence. Then A converges on $\mathcal{C}$.*

Proof. We show by induction on $comp(A)$ that $A\downarrow$. For the basis, assume A is of one of the forms

i. $t_1 = t_2$, iii. $t_1 \,\varepsilon\, \mathbb{N}$, iv. $t_1 \,\varepsilon\, \underline{\omega}$, v. $(t_1 t_2, t_3) \,\varepsilon\, t_+$, or vi. $(t_1 t_2, t_3) \,\varepsilon\, t_\times$

as per Definition 10.2. All identities converge by clause 2 of Definition 12.2 and Observation 10.1 together with the fact that $\mathcal{A} \leq_k \mathcal{C}$, $t_1 \,\varepsilon\, \mathbb{N}\downarrow$. Of course, $t_1 \,\varepsilon\, \underline{\omega}$ similarly converges on $\mathcal{C}$ since $\mathcal{A}$ is ω-total and $\mathcal{A} \leq_k \mathcal{C}$. By the same argument as in 9.3 and 9.4, $\mathcal{C}$ is $+$, $\times$-total, so $(t_1\, t_2, t_3)\varepsilon t_+$ and $(t_1\, t_2, t_3)\varepsilon t_\times$ converge on $\mathcal{C}$. For the induction, suppose A is of form 10.2.ii: $t_1 \,\varepsilon\, [x : \phi]$ for some arithmetic formula ϕ with at most x free. By Lemma 10.5, the sentence $\phi_x[t_1]$ is arithmetic and of the same complexity as ϕ; $comp(\phi_x[t_1]) = comp(A) - 1$. Then by the hypothesis of induction, $\phi_x[t_1]$ converges on $\mathcal{C}$, whence A converges on $\mathcal{C}$ by definition of Φ (or abstraction, 4.2). If, on the other hand, A is a Boolean compound, then by the hypothesis of induction, A's principle subformulas converge on $\mathcal{C}$, so A converges on $\mathcal{C}$ since the interpretation of the connectives preserves convergence on an L-structure. Finally, suppose A is of the form $\forall x\, \phi$ for some arithmetic formula ϕ with at most x free. Let t be any arithmetic constant term. By 10.5, $\phi_x[\mathrm{t}]$ is an arithmetic sentence of the same complexity as ϕ, so by the hypothesis of induction, $\phi_x[\mathrm{t}]\downarrow$. Then by Definition 12.1, $A\downarrow$, as required. This completes the proof of Theorem 12.3. ⊣

THEOREM 12.4. *α is coherent ($\alpha \leq \omega_1$).*

Proof. Suppose, towards a contradiction, that α is not coherent. Then as in the proof of Theorem 9.3, there is a coherent $\beta < \alpha$ and constant terms t_1, t_2, t_3 such that both of the following hold:

(a) $\varepsilon_\beta(t_1, t_2, t_3) = \mathit{false}\ (\mathit{true})$, (b) $\Phi(\varepsilon_\beta)(t_1, t_2, t_3) = \mathit{true}\ (\mathit{false})$.

Let $\gamma \leq \beta$ be the least ordinal such that $\varepsilon_\gamma(t_1, t_2, t_3) = \mathit{false}\ (\mathit{true})$. As in 9.3, the case $\gamma > 0$ is excluded. Then, if $\varepsilon_0(t_1, t_2, t_3) = \mathit{false}$, one of clauses 1 or 2 of the definition of ε_0 (12.2) holds and $t_1 \in \{t_+, t_\times\}$. This case is handled exactly in the manner of the proof of Theorem 9.3 (in particular, subcase 2.2 is not affected by the fact that we are now operating with a modified definition of the matrix $B^\times(x, y, z, u)$ of that term). So suppose $\varepsilon_0(t_1, t_2, t_3) = \mathit{true}$. Then by clause 2 of Definition 12.2, t_1 is $\mathbb{N}$ and there is an $m \in \mathbb{N}$ such that $t_2 = \underline{m}$. By (b) and definition of Φ,

$$\Phi(\varepsilon_\beta)(\mathbb{N}, t_2, t_3) = \mathcal{C}_\beta(\forall z\, (ZS(z) \rightarrow t_2 \,\varepsilon\, z)) = \mathit{false},$$

so there is a constant arithmetic term s such that $\mathcal{C}_\beta(ZS(\mathrm{s}) \rightarrow t_2 \varepsilon \mathrm{s}) = \mathit{false}$. Since β is coherent, it follows that $ZS(\mathrm{s})$ is *true*, but $t_2 \,\varepsilon\, \mathrm{s}$ is *false*, on $\mathcal{C}_\beta$. But given $\mathcal{C}_\beta(ZS(\mathrm{s})) = \mathit{true}$, it is easy to show by induction on n that $\mathcal{C}_\beta(\underline{n} \,\varepsilon\, \mathrm{s}) = \mathit{true}$ ($n \in \mathbb{N}$), which contradicts $\mathcal{C}_\beta(t_2 \,\varepsilon\, \mathrm{s}) = \mathit{false}$. This contradiction establishes Theorem 12.4. ⊣

LEMMA 12.5. *$\mathcal{C}$ is a model of* G_2.

Proof. $\mathcal{C}$ is acceptable by construction and 3-valued by 12.4. So it suffices to show that $\mathcal{C}$ validates the G_2 bivalence axioms for ε atoms having parameters as predicate terms. So suppose we a given an arbitrary such axiom $t \varepsilon p \vdash t \varepsilon p$ for some closed term t and parameter p. Let $t' \varepsilon \mathbf{s} \vdash t' \varepsilon \mathbf{s}$ be an arbitrary instance of the given axiom sequent, where $\mathbf{s}$ a constant arithmetic term and t' is an instance of t obtained by uniformly replacing all parameters in t by constant arithmetic terms. Then $t' \varepsilon \mathbf{s}$ is an arithmetic sentence by Definition 10.2.2, and so converges on $\mathcal{C}$ by Theorem 12.3. Since $\mathcal{C}$ is 3-valued, it follows that

$$\mathcal{C}(t' \varepsilon \mathbf{s}) \in \textit{true}, \textit{false}.$$

This is sufficient. ⊣

COROLLARY 12.6. G_2 *is consistent.*

REFERENCES

Apostoli, P. J.

1993 The alethic genesis of number, Department of Philosophy, University of Toronto, unpublished monograph.

Belnap, N. D.

1977 A useful four-valued logic, *Modern uses of multiple-valued logic* (J. M. Dunn and G. Epstein, editors), Reidel, Dordrecht.

Boolos, G.

1990 (editor), The standard of equality of numbers, *Meaning and method: Essays in honor of Hilary Putnam,* Cambridge University Press, Cambridge, England.

Church, A.

1941 *The calculi of lambda conversion,* Princeton University Press, Princeton, New Jersey; reprinted by University Microfilms Inc., Ann Arbor, 1963.

Curry, H. B.

1930 Grundlagen der kombinatorischen Logik, *American Journal of Mathematics,* vol. 52, pp. 509–536, 789–834.

1958 *Combinatory logic,* vol. 1, North-Holland, Amsterdam.

Demopoulos, W. G., and J. L. Bell

1993 Frege's theory of concepts and objects and the interpretation of second-order logic, submitted to *The Journal of Philosophy.*

Feferman, S.

1984 Towards useful type-free theories I, *The Journal of Symbolic Logic,* vol. 49, pp. 75–111.

Fitch, F. B.

1948 An extension of basic logic, *The Journal of Symbolic Logic*, vol. 13, pp. 95–106.

1955 *Symbolic logic: An introduction*, Ronald Press, New York.

1967 A complete and consistent modal set theory, *The Journal of Symbolic Logic*, vol. 49, pp. 75–111.

1974 *Elements of combinatory logic*, Yale University Press, New Haven.

Fitting, M.

1986 Notes on the mathematical aspects of Kripke's theory of truth, *Notre Dame Journal of Formal Logic*, vol. 27, pp. 75–88.

1989 Bilattices and the theory of truth, *Journal of Philosophical Logic*, vol. 18, pp. 225–256.

1993 A theory of truth that prefers falsehood, *Journal of Philosophical Logic*, vol. 26 (1997), pp. 477–500.

Frege, G.

1959 *Die grundlagen der arithmetic*; translated by J. L. Austinas as "The foundations of arithmetic", Blackwell, Oxford.

1964 *Grundgesetze der arithmetik*; translated by M. Furth as "The basic laws of arithmetic", University of California Press, Berkeley and Los Angeles, California.

Gabbay, D., and F. Guenthner

1983 (editors), *Handbook of philosophical logic*, four volumes, Reidel, Dordrecht, 1983-1989.

Gentzen, G.

1935 Untersuchungen über das logiische schliessen, *Mathematissche Annalen*, vol. 39, pp. 176–210, 405–431; translated as "Investigations into logical deduction," in *Szabo 1969*.

Gilmore, P. C.

1971 A consistent naive set theory: Foundations for a formal theory of computation, IBM Research, RC 3413 (June 22).

1974 The consistency of partial set theory without extensionality, in *Jech 1974*.

1980 Combining unrestricted abstraction with universal quantification, in *Hindley and Seldin 1986*.

1986 Natural deduction based set theories: A new resolution of the old paradoxes, *The Journal of Symbolic Logic*, vol. 51, pp. 393–411.

Gupta, A. K.

1982 Truth and paradox, *Journal of Philosophical Logic*, vol. 11, pp. 1–60.

1988 Remarks on definition and the concept of truth, *Proceedings of the Aristotelian Society*, vol. 89 (1988-89), pp. 227–246.

Gupta, A. K., and N. D. Belnap

1993 *The revision theory of truth,* MIT Press, Cambridge, Massachusetts.

Herzberger, H. G.

1982 Naive semantics and the Liar paradox, *The Journal of Philosophy*, vol. 79, pp. 690–716.

Hindley, J. R., and J. P. Seldin

1986 *Introduction to combinators and lambda-calculus*, London Mathematical Society Texts, 1.

Jech, T.

1974 (editor), Axiomatic set theory, *Proceedings of symposia in pure mathematics,* vol. 13, Part II, American Mathematics Society.

Kripke, S.

1975 Outline of a theory of truth, *The Journal of Philosophy,* vol. 72, pp. 690–716.

Schütte, K.

1977 *Proof theory* (Beweisetheorie), Springer-Verlag, Berlin.

Scott, D. S.

1971 Models for various type-free calculi, *Logic and Methodology and Philosophy of Science IV* (P. Suppes and others, editors), North-Holland, Amsterdam.

1975 Combinators and classes, *Lambda-calculus and computer science,* Lecture Notes in Computer Science, vol. 37, Springer-Verlag, Berlin, pp. 1–26.

Szabo, M. E.

1969 (editor), *The collected papers of Gerhard Gentzen,* North-Holland, Amsterdam.

Seldin, J. P., and J. R. Hindley

1980 (editors), *To: H. B. Curry: Essays in combinatory logic, lambda calculus and formalism,* Academic Press, New York.

Smullyan, R.

1986 *First-order logic,* Springer-Verlag, Berlin.

Wright, C.

1983 Frege's conception of numbers as objects, *Scots Philosophical Monographs,* vol. 2, Aberdeen University Press, Aberdeen.

Visser, A.

1984 Four valued semantics and the Liar paradox, in *Gabbay and Guenthner 1983,* vol. 4.

JOHN CORCORAN

SECOND-ORDER LOGIC

Abstract. This expository article focuses on the fundamental differences between first-order logic and second-order logic. It employs second-order propositions and second-order reasoning in a natural way to illustrate the fact that second-order logic is actually a familiar part of our traditional intuitive logical framework and that it is not an artificial formalism created by specialists for technical purposes. To illustrate some of the main relationships between first-order logic and second-order logic, this paper introduces basic logic, a kind of zero-order logic, which is more rudimentary than first-order and which is transcended by first-order in the same way that first-order is transcended by second-order. The heuristic effectiveness and the historical importance of second-order logic are reviewed in the context of the contemporary debate over the legitimacy of second-order logic. Rejection of second-order logic is viewed as involving radical repudiation of part of our scientific tradition. But even if genuine logic comes to be regarded as excluding second-order reasoning, which is a real possibility, its effectiveness as a heuristic instrument will remain and its importance for understanding the history of logic and mathematics will not be diminished. Second-order logic may some day be gone, but it will never be forgotten. Technical formalisms have been avoided entirely in an effort to reach an interdisciplinary audience, but every effort has been made to limit the inevitable sacrifice of rigor.

No matter what human action you consider, if everyone does it to everyone doing it to them, then everyone has it done to them by everyone to whom they do it. For example, if everyone teaches everyone who teaches them, then everyone is taught by everyone they teach. Likewise, if everyone helps everyone who helps them, then everyone is helped by everyone they help. The same holds for "encourages", "hinders", "supports", "opposes", "ignores", and the rest.

Each of the above propositions is actually a *tautology*, a proposition implied by its own negation. In fact, each of them can be proved to be true by logical reasoning alone; e.g., by deducing them from their own negations. Since every proposition in the same form as a tautology is again a tautology, a discourse formally similar to that expressed above obtains in every universe of discourse, not just in the universe of humans.

In metalogic, for example, we often discuss the universe of propositions in so far as various logical relations are concerned. By a *logical relation* I mean relations such as implication, consequence, contradiction, compatibility, independence, etc. More specifically, I mean what are called binary relations on the universe of propositions. If R indicates such a relation and if a and b are each individual propositions, then aRb can be used to express the propo-

C. Anthony Anderson and M. Zelëny (eds.), Logic, Meaning and Computation, 61-75.

sition that the first proposition a is related by R to the second proposition b. It is not excluded, of course, that a and b are the same proposition. For example, every proposition implies itself and some but not every proposition contradicts itself.

Now logical relations are certainly not actions. Saccheri's Postulate contradicts the Parallel Postulate but there is no action that Saccheri's Postulate could perform. Nevertheless, as we have just seen, *relation verbs* function grammatically in certain contexts in a manner similar to the function of *action verbs.* The relation verbs significant in the universe of humans include the following: outweighs, outlives, succeeds (in several senses), precedes (in several senses), equals (in many senses), and many others. The action verbs significant in the universe of humans include the following: calls (in at least one sense), serves (in at least one sense), teaches, commands, obeys, and many others.

In normal English some of the logical relations are expressed by relation verbs, as we have seen. For example, implication is expressed by 'implies' and contradiction is expressed by 'contradicts'. However, some of them are expressed by *relation nouns.* For example, consequence is expressed by the relation noun 'consequence'. The law of transitivity of consequence is that every consequence of a consequence of a proposition is again a consequence of that proposition. Moreover, there are logical relations expressed by *relation adjectives.* Compatibility and independence are expressed by 'compatible' and 'independent'. Aristotle's fundamental law of compatibility of truth is that every two true propositions are compatible with each other. Using 'independent' in the most widely accepted sense we can say that every proposition which is independent of a given proposition is neither implied by nor contradicted by the given proposition, and conversely, every proposition which is neither implied by nor contradicted by a given proposition is independent of the given proposition.

One reason for reviewing the various ways that logical relations are expressed in English is to point out what all creative writers already know, viz. that knowledge of the conventional rules of English should enhance but not inhibit English writing. For example, my very first sentence uses the plural pronoun 'them' as coreferential with 'everyone', which is singular. Even worse, from the point of view of conventional rules, is my use of the filler 'you consider'. The proposition being expressed is not a prediction of what will happen if *you* consider something. The proposition is not about you *per se* at all. The sentence expresses a *general proposition* predicating a certain complex property of every action on the universe of humans. The phrase 'no matter what human action you consider' is just a heuristically effective way of expressing a universal quantifier. From a logical point of view the following would do just as well: 'every human action is one such that', 'every human action is one where', 'with every human action', etc. At any rate, a sentence that violates the conventional rules of English applicable to the expression of a given proposition is sometimes nevertheless a perfectly acceptable and effective way of expressing that very proposition.

Whether a sentence is an acceptable and effective expression of a given proposition is a matter of how readers take it, and not a matter of conventions established in the past. Now we are ready to present a discourse obtaining in the universe of propositions and formally similar to the one which began this essay.

No matter what logical relation you consider, if every given proposition bears it to every proposition bearing it to the given proposition, then every given proposition is borne it by every proposition the given proposition bears it to. For example, if every proposition contradicts every proposition contradicting it, then every proposition is contradicted by every proposition it contradicts. Likewise, if every proposition implies every proposition implying it, then every proposition is implied by every proposition it implies. The same holds for "is a consequence of", "is compatible with", "is logically equivalent to", "is independent of", "is a contradictory opposite of", and the rest.

The propositions expressed in the above paragraph are all tautologies *and* they are all laws of logic. The propositions in the first paragraph of this essay are all tautologies but none of them are laws of logic because they are not about a logical subject-matter. The proposition "Every proposition implies every proposition implying it" is about a logical subject-matter but it is not a law of logic because it is false. For example, "Every proposition is true" implies "Every false proposition is true", but not conversely. The proposition "Every proposition contradicts every proposition contradicting it" is a law of logic, of course, but it is *not* a tautology because it is in the same form as a proposition considered just above and found to be false. By a law of logic I mean a true proposition about a logical subject-matter, e.g., about propositions, about arguments, about argumentations, etc.

The two properties, being tautologous and being a law of logic, are *orthogonal* in the sense that each of the four combinations of the two is exemplified. We have seen above that some but not every tautology is a law of logic and that some but not every non-tautology is a law of logic. There is much confusion concerning this elementary point. Some but not all of the confusion is more or less deliberately nurtured in the service of various dogmas which, happily, are waning in popularity.

The proposition, "Every proposition contradicts every proposition contradicting it", is the law of symmetry (or reciprocity) of contradiction and "Some proposition implies some proposition not implying it in return" is the law of non-symmetry (or non-reciprocity) of implication. "No contradictory opposite of a contradictory opposite of a proposition is a contradictory opposite of that proposition" is the law of antitransitivity of contradictory opposition. A *contradictory opposite* of a proposition is, of course, a proposition logically equivalent to the negation of that proposition. For example,

"Some true proposition is not tautologous" and "Not every true proposition is tautologous" are both contradictory opposites of "Every true proposition is tautologous". In order to avoid confusion it should be noted that, although every two propositions that are contradictory opposites of each other contradict each other, not every two propositions that contradict each other are contradictory opposites. To take an extreme example, "No proposition implies itself" contradicts "Some proposition implies every proposition". The same example illustrates another point that clarifies things and helps to avoid confusion, viz. that although no true proposition contradicts a true proposition, some false propositions contradict false propositions. In fact, some false propositions contradict themselves. Thus although no two contradicting propositions are both true, some two contradicting propositions are both false. In such cases, i.e., when two contradicting propositions are both false, they are not contradictory opposites because every two contradictory opposites have different truth-values.

A *contradiction* (or a self-contradiction) is a proposition that contradicts itself, i.e., that implies its own negation. Every contradiction is a contradictory opposite of a tautology and every tautology is a contradictory opposite of a contradiction. A proposition is said to be *contradictory* (or self-contradictory) if it is a contradiction. Every two contradictory propositions contradict each other but no two contradictory propositions are contradictory opposites of each other. The expression 'two contradictory propositions' means "two propositions each of which is self-contradictory" whereas 'two contradicting propositions' means "two propositions contradicting each other" which, in view of the symmetrical nature of "contradicts", amounts to "two propositions one of which contradicts the other".

In ordinary technical English, 'is contrary to' and 'is a contrary to' are ambiguous. Sometimes, "contradicts" is meant and sometimes "is a contradictory opposite of" is meant. Surprisingly, the ambiguity does not seem to be troublesome. However, in former times logicians had attached a third technical meaning that did lead to confusion. Two propositions were said to be *contraries* (sc of each other) if they contradict each other but their negations do not contradict each other. For example, "Every number is prime" and "Every number is non-prime" are contraries. It is easy to prove that every two contradicting propositions that are not contradictory opposites are contraries, and vice versa. In modern logic, 'contrary' is rarely used in the obsolete technical sense.

We have had occasion just now to state several laws of logic and to mention (or talk about) several laws of logic. As indicated above, by a *law of logic* I mean a true proposition about a logical subject-matter (propositions, arguments, argumentations, etc.). The most basic laws of logic are the laws of excluded middle, non-contradiction, and truth and consequence: "Every proposition is either true or false", "No proposition is both true and false" and "Every proposition implied by a true proposition is true".

The laws of logic, in fact all propositions about logical subject-matter are in some sense *second-level* (or *meta-level*) propositions in the sense that they are about things that are themselves about things (usually, of course, non-logical things). Some people, either ignorant of or in opposition to logical tradition, call such propositions 'second-order'. This is *not* how 'second-order' is used in this essay although some second-level propositions are also second-order. A proposition is classified as *basic, first-order, second-order*, etc., not on the basis of what it is about but rather on the basis of its logical structure. The very first proposition of this essay is second-order. The second proposition is first-order. Every proposition to the effect that one named proposition is in a mentioned logical relation to another named proposition is basic, e.g., "Saccheri's Postulate contradicts the Parallel Postulate". It will become obvious that the two properties, being second-level and being second-order, are orthogonal.

The *basic propositions*, very roughly speaking, are those without common nouns. It is perhaps easiest to describe the *basic propositions of arithmetic* (*BPA*). Actually, instead of describing the *BPA* outright, it is convenient to describe the *basic sentences of arithmetic* (*BSA*) and then to say that the basic propositions of arithmetic are the propositions expressed by the basic sentences when the sentences are understood in their intended interpretations. Now, the substantives of the basic sentences of arithmetic are exclusively *numerals* (number-names) in the wide sense: 'zero', 'one', 'two', 'three', ..., 'zero plus one', 'zero plus two', ..., 'two plus (zero times one)', ... Among the numerals I also intend: 'two-squared', 'two-cubed', etc. The *atomic* sentences of arithmetic include, in the first place, all so-called equations: 'one plus one is two', 'one plus two is one', etc., in other words any sentence in the pattern *numeral is numeral.* The 'is' here, of course, is intended to express numerical identity which is often improperly called *equality* and expressed by 'equals'. Next we have the sentences that normally attribute a quality to a number, e.g., 'one is even', 'one is odd', 'two is prime', 'five is perfect'; and so on. Next we have the sentences that normally relate one number to another, e.g., 'two exceeds three', 'three divides two', etc. These include the identities (or equalities, equations) already mentioned. Next we have the sentences that normally indicate that three numbers are in a ternary relation, e.g., 'two is between one and three', etc.

There are also quaternary relational sentences, e.g., 'one is to three as three is to nine'. And so on. Anything of this sort is countenanced as long as there are no common nouns. Even common nouns are allowed as long as they are understood as nominalized adjectives (e.g., 'two is a prime' means "two is prime") or as nominalized relatives (e.g., 'two is a divisor of four' means "two divides four"), etc. Once the atomic basic sentences of arithmetic have been determined, the basic sentences can be defined as the so-called truth-functional combinations of atomic sentences, the atomic sentences plus what can be obtained from atomic sentences by negations, conjunctions, disjunctions, conditionals, bi-conditionals, etc. It should be explicitly mentioned in this connection that passives (or converses) of binary relation verbs are again binary relational verbs and thus sentences such as 'two is divided by four', 'two is exceeded by four', etc. are included. Likewise included are sentences involving the so-called modified relation verbs: 'properly divides', 'immediately precedes', 'immediately exceeds', etc.

Basic logic is the logic of basic propositions. Basic logic is concerned fundamentally with the question of how we determine the validity or invalidity of an argument whose premises and conclusion are exclusively basic propositions. As you know, an argument is determined to be valid by giving a derivation (or a deduction) of its conclusion from its premises. This means giving an extended discourse, normally much longer than the premises-plus-conclusion which shows step-by-step how the conclusion can be seen to be true were the premises true. The rules for making up these derivations are obtained by looking at what people do with basic propositions when they are reasoning correctly. In order to deduce from any set of basic premises any basic conclusion that actually follows, it is sufficient to use rules from a very small set. These include the usual rules of propositional logic, the rule of substitution of identities, the rule of conversion (the active and passive are interdeducible) and the logical axioms of identity ("one is one", etc.). To show that a given basic conclusion does not follow from a given basic premise-set, it is sufficient to produce a *counterargument*, i.e., a conclusion and a premise-set together in the same form and having false conclusion and true premises. For example, to show that the argument on the left below is invalid it is sufficient to notice that the argument on the right is in the same form and has true premises and false conclusion.

Two is not three.	One is not two.
Three is not two plus two.	Two is not one times one.
? Two is not two plus two.	? One is not one times one.

The argument on the right is obtained in three steps from the argument on the left. First 'one' is substituted everywhere for 'two' on the left. Then in the "new" left argument (in which 'two' no longer occurs), 'two' is substituted everywhere for 'three'. Then in the "second new" left argument, 'times' is substituted for 'plus'.

Strictly speaking an *argument* (more properly, *premise-conclusion argument*) is a two part system composed of a set of propositions called the *premise-set* and a single proposition called the *conclusion.* To represent or express an argument we use an argument-text which is a list of *sentences* (not propositions) followed by a single sentence somehow marked as the conclusion-sentence. Some logic books use a line above the conclusion-sentence, but it is easier and less messy to use a question-mark as above. The method outlined above of transforming one argument-text into another argument-text in such a way that the argument represented by the second is in the same form as the argument represented by the first works only when the argument-texts are written in a so-called *logically perfect language* in which the outer grammatical form of the sentences mirrors exactly the inner logical form of the propositions. When logical issues are important, the language in question is *regimented* (*normalized*) so that it becomes logically perfect (or approximately so). This is why I write 'two plus (zero times one)' instead of 'two plus zero times one'. Logicians typically go immediately to a symbolic language carefully constructed to be logically perfect but for many purposes, especially that of exposition, this method, though virtually essential for some purposes, can be counter-productive.

Basic logic can be called *finite logic* because every finite invalid argument of basic logic is refutable by a counterargument whose propositions have reference only to a finite number of individuals. By a finite argument I mean an argument having only a finite number of premises and by reference only to a finite number of individuals I mean not only that the propositions refer only to finitely many individuals (which is obvious) but also that the functions referred to are all defined on one and the same finite universe of discourse.

By the way, this includes the so-called zero-premise arguments (arguments having the null premise-set) which are valid when and only when the conclusion is a tautology. Some examples follow.

? One is one.

? If one is two then two is one.

? If (if one is not two then one is two) then one is two.

? If one exceeds two then two is exceeded by one.

It follows from what was said above that every basic proposition that is not a contradiction is in the same logical form as a true basic proposition having reference only to a finite number of objects. This means that among the basic propositions there are no so-called *infinity propositions.* In order for a proposition to be an *infinity proposition* it is necessary and sufficient that it be non-contradictory and for every proposition in the same form having reference only to a finite number of individuals to be false. In other words an infinity proposition is a proposition expressed by a sentence which is "satisfiable" only in infinite universes of discourse.

Basic logic covers most of the arithmetic reasoning done by school children, all of the logic "done" by computers (though in a sense computers can *simulate* finite stretches of higher logics), and much of the logic on the normal aptitude test.

In a sense, *first-order logic* (*FOL*) begins when we generalize basic propositions. In fact, it is not stretching things to say that basic tautologies are tautologies *because* they are instances of first-order tautologies. When you prove a basic tautology you feel that you have not exhausted your reasoning in that direction. To illustrate this I will give a basic tautology and then give four first-order generalizations.

? If three exceeds two then two is exceeded by three.

? Every number exceeding two is a number that two is exceeded by.

? Every number that three exceeds is exceeded by three.

? Every number exceeding a given number is a number the given number is exceeded by.

? Every number that a given number exceeds is exceeded by the given number.

We are inclined to think that the basic tautology is logically derived from its generalization, e.g., that "if three exceeds two then two is exceeded by three" is true because "every number exceeding two is one that two is exceeded by" is true ... thus emphasizing the fact that the former is no peculiarity of three. Likewise we are inclined to think that the latter generalization is true because of the truth of *its* generalization, viz. "Every number exceeding an arbitrary number is one that the arbitrary number is exceeded by" ... thus emphasizing that no peculiarity of two is involved.

The *first order sentences of arithmetic* (*FOSA*) are the sentences obtainable from the basic sentences by quantification and taking truth-functional combinations. It is important that these operations are taken recursively, e.g., a basic sentence can be generalized and then combined with other generalizations by truth-functional combinations and then generalized again before taking further truth-functional combinations. The *first-order propositions of arithmetic* (*FOPA*) are the propositions expressed by the first-order sentences interpreted in the usual way. Below is an example of one of the simplest valid arguments in first-order logic.

Every number is either even or odd.

No number is both even and odd.

Every number which is odd is one whose square is odd.

? Every number whose square is even is itself even.

There is a radical increase in expressive power of first-order languages as compared to basic languages. For example, even the first premise in the above argument implies infinitely many basic consequences but it is not implied by any number of its basic consequences, not even by all of them together. The idea that a generalization is logically equivalent to the set of its singular instances is but one of the fallacies that is to be confronted by those seeking to reduce first-order logic to basic logic. Below are a few of the singular instances of the proposition under discussion.

One is either odd or even.

Two is either odd or even.

Three is either odd or even.

As mentioned above basic logic is sometimes called finite logic because each of its consistent (or non-contradictory) propositions is finitely satisfiable. This is no longer true of first-order logic. Indeed, the conjunction of the following two propositions is not satisfiable in any finite universe of discourse.

Zero is not the successor of any number.

Every two numbers which are successors respectively of distinct numbers are themselves distinct.

It is known, however, that every first-order proposition which is consistent is satisfiable in a countable universe of discourse. In fact, every consistent first-order proposition that is not satisfiable in a finite universe of discourse is, like the above conjunction, satisfiable in the universe of natural numbers. For this reason, first-order logic can be called *countable logic.*

Just as every valid basic argument is deducible using a small set of axioms and rules of inference, the same is true of valid first-order arguments. This means that as far as knowledge of validity of first-order arguments is concerned, human knowing faculties are equal to the task. The so-called principle of sufficiency of reason, viz. that every true proposition can be known to be true, can be shown to be false. Human faculties of knowing truth are not equal to the task of knowing truth—truth outruns knowledge. With *validity of first-order arguments*, reason is sufficient—every valid first-order argument can be known to be valid. Whether *every* valid argument (whatever the order) can be known to be valid is a question of considerable complexity and well beyond the scope of this elementary exposition.

There is another much less important fact about first-order and basic logic that is worth mentioning. For this we have to divide the logical concepts into *positive* and *negative.* Without going into the details, let me say that there are no surprises here. "Every", "Some", "Is", "And", "Or", "If", etc. are positive. "Not", "No", "Distinct", "Nor", etc. are negative. The result is that every contradictory first-order proposition involves at least one negative logical concept.

Just as we motivated the transition from basic logic to first-order logic by reflecting on the fact that the reasoning used to establish a basic tautology seems stronger than needed for that purpose and indeed is sufficient (or virtually so) to establish all generalizations of the basic tautology, we use the same sort of insight to transcend first-order logic. Consider the following first-order propositions.

> No number divides exactly the numbers that do not divide themselves.
>
> No number precedes exactly the numbers that do not precede themselves.
>
> No number exceeds exactly the numbers that do not exceed themselves.
>
> No number perfects exactly the numbers that do not perfect themselves.

The relation of perfecting arises in connection with the so-called perfect numbers. Every number having proper divisors is perfected only by the successor of the sum of its proper divisors. The other numbers, viz. zero, one and the prime numbers, are not perfected by any numbers at all. Thus four is perfected by three since two is the only proper divisor of four. But six is perfected by itself. In fact, as you may have seen already, every perfect number perfects itself and, conversely, every number perfecting itself is perfect. Now, the reason for introducing the perfecting relation is to give an example of a tautology in the same form as the first three of the above set but not as mathematically trivial. The first of the above propositions is mathematically trivial because zero, which is the only number that does not divide itself, is divided by every other number. The second is trivial because every number precedes other numbers but not itself. The third is trivial for similar reasons.

Now, as you know, each of the above can be deduced from their own respective negations by familiar (but intricate) reasoning. The fact is that the following premise-conclusion argument is valid.

> Some number perfects exactly the numbers that do not perfect themselves.
>
> ? No number perfects exactly the numbers that do not perfect themselves.

A deduction of this argument, i.e., a deduction of its conclusion from its premise, can easily be transformed into an indirect *proof* of its conclusion. The reason that a deduction of a conclusion from the null-premise set is a *proof* (i.e., a deduction whose premises are known to be true) is because universal propositions with null "subjects" are vacuously true. Every member of the null set of premises is known to be true ... there being no counterexamples.

Once one of these tautologies has been proved to be true by a deduction from the null set of premises the others are also virtually proved to be true also. The reason for this is the principle of form for deductions: every argumentation in the same form as a deduction is again a deduction. Thus a proof of, say, the fourth can be obtained from a proof of, say, the first by substituting in the latter the concept "perfects" for the concept "divides". So it is clear that the reasoning establishing one of the four *virtually* establishes much more.

Now we move to *the* second-order generalization of the above. Actually, the following second-order proposition is at once a generalization of each of the above four first order tautologies and, in a certain reasonable sense, the only generalization.

> No matter which numerical relation you consider, no number bears it to exactly the numbers that do not bear it to themselves.

Once you have seen that this is true you will feel that it is the *ground* of the truth of the previous four propositions, e.g., that the truth of the fourth of them depends on no peculiarity of the perfecting relation.

My main point in this essay is that the reasoning in a given logic achieves more than can be expressed in that logic and that the transcending of a given logic by going to a higher order is one way of reaping the full fruit of one's reasoning in a given logic. This vague principle applies not just to first-order in relation to basic logic and to second-order logic in relation to first-order but in general to any logic in relation to the next lower order.

In basic sentences, there are no common nouns. In first-order sentences, there are common nouns, but no "second-order nouns" such as 'property', 'relation', 'function', etc. The presence of nouns inevitably and automatically entails the presence of quantifiers because nouns require articles and articles express quantifiers. For example, the following sentences express the same proposition.

> Every false proposition implies a true proposition.
>
> Every proposition which is false implies some proposition which is true.
>
> For every proposition which is false there exists a proposition which is true and which is implied by the false proposition.
>
> The same phenomenon can be exemplified in the universe of natural numbers (beginning with zero).
>
> Every odd number exceeds an even number.
>
> Every number which is odd exceeds some number which is even.
>
> For every number which is odd there exists a number which is even and which is exceeded by the odd number.

When we move to second-order by adding second-order nouns we also add second-order adjectives whose ranges of significance are the second-order objects denoted by the second-order nouns. Examples of second-order adjectives are the familiar terms indicating properties of relations: reflexive, symmetrical, transitive, dense, etc. The following are typical second-order sentences involving such expressions.

> Every reflexive relation relates every object to itself.
>
> Every relation that relates every object to itself is reflexive.
>
> Every symmetric relation relates to each other every two objects one of which it relates to the other.
>
> Every relation that relates to each other every two objects one of which it relates to the other is symmetric.

Orthogonality is a second-order relation between properties. In order for one property to be orthogonal to another it is necessary and sufficient that there be four objects, one having both properties, one having the first but lacking the second, one lacking the first but having the second, and one lacking both. These examples show that much of this essay has been written using a second-order language.

Since basic logic is finite and since first-order logic is countable, neither is adequate to axiomatize theories whose universes of discourse are uncountable. The most familiar examples of such theories are calculus and geometry. Now just as first-order logic is not finite, second-order logic is not countable. There are consistent second-order propositions which are not satisfiable in any countable universe. One example is from Hilbert's axiom set for the theory of real numbers (which is foundational for calculus). Another is from Veblen's axiom set for Euclidean geometry. Naturally, second-order logic can be called *uncountable logic*.

First-order logic is not even adequate to axiomatize theories whose universes of discourse are countably infinite. The paradigm case of such a theory is number theory, or the arithmetic of natural numbers, which requires *the principle of mathematical induction* (*PMI*).

> Every property belonging to zero and to the successor of every number to which it belongs also belongs to every number without exception.
>
> In order for a property to belong to every number it is sufficient for that property to belong to the successor of every number having it and also that zero have it.

Mathematical induction is the second-order generalization of each of the following propositions which are among its first-order instances.

> *If* zero is even and the successor of every even number is even, *then* every number is even.
>
> *If* zero is perfect and the successor of every perfect number is perfect, *then* every number is perfect.

In first-order axiomatizations of arithmetic *PMI*, induction, is replaced by the infinite set of its first-order instances, a set which is insufficient to imply mathematical induction. In fact, no set of true first-order propositions is sufficient to imply *PMI* and therefore no first-order axiomatization of arithmetic adequately codifies our knowledge of arithmetic. Moreover, the ground of our knowledge of the instances is our knowledge of *PMI* itself. Thus infinitely many of the so-called axioms of first-order arithmetic are not axiomatic in the traditional sense. Nevertheless, there are able logicians and mathematicians who reject the traditional second-order axiomatizations due to Dedekind and Peano in favor of first-order axiomatizations which date from the 1930's.

Just as second-order logic is necessary to fully exploit first-order reasoning as well as to understand the ground of first-order tautologies, likewise second-order logic is necessary to fully exploit first-order knowledge in arithmetic as well as to understand the ground of acceptance of first-order axiomatizations of arithmetic. Even logicians who reject second-order axiomatizations of arithmetic admit their historic importance and make heuristic and pedagogical use of such axiomatizations. By the way, the same thing may be said of axiomatizations of set theory, but the technical details involved in set theory require distinctions and principles which go beyond the scope of this essay.

In the case of basic logic, as well as that of first-order logic, a small set of simple rules of inferences suffices to enable every valid argument to be deduced. This is no longer the case with second-order logic. In fact, it is a corollary to the famous Gödel Incompleteness Theorem that no simple set of rules is sufficient for this purpose. This means that the principle of sufficiency of reason when applied to second-order validity is false. To be explicit, there are finite valid arguments in second-order logic whose conclusions can not be deduced (in a finite number of steps using simple rules) from their premise-sets. This result is known as the *incompleteness of second-order logic.*

There are logicians who feel that human reasoning must be equal to the task of determining the validity of valid arguments. In most cases such logicians are empiricistically oriented and are fully willing to accept the fact that there are true propositions about the material universe that can not be known to be true. But they feel that validity is intrinsically amenable to analytic *a priori* methods and, in particular, that every valid argument must be deducible. One way out of this quandary is to deny that second-order logic is really logic.

Incidentally, second-order axiomatizations do not evade the incompletability of arithmetic. First-order axiomatizations are deficient because first-order languages are too weak to express our knowledge of arithmetic even though first-order reasoning is adequate to first-order validity. Within second-order the situation is reversed. Second-order axiomatizations are deficient because second-order reasoning is too weak to deduce all of the consequences of second-order axioms even though second-order language is strong enough to express our knowledge of arithmetic. In fact, our second-order arithmetic knowledge implies absolutely every true second-order arithmetic proposition even those that we are powerless to deduce (using any given simple set of rules fixed in advance).

Another phenomenon that gives some logicians doubts about second-order logic is existence of contradictory propositions devoid of negative logical concepts. Recall that in first-order logic every contradictory proposition involves at least one negative logical concept. Below are two second-order propositions the first of which is tautological and the second of which is contradictory, neither of which involve negative logical concepts.

Every object has at least one property.
Every property belongs to at least one object.

The reason that the second proposition is self-contradictory is that it contradicts the following tautology.

No object has the property of being distinct from itself.

We have seen that second-order logic differs radically from first-order. First-order is a logic of countability; second-order is a logic of uncountability. First-order is deductively complete; second-order is deductively incomplete. In first-order every contradiction is negative; in second-order there are self-contradictory propositions which are exclusively positive. The above-mentioned historic examples of axiomatized sciences remind us that higher-order reasoning is not a recent innovation but rather a feature of human thought having a long history. Moreover, it is not the case that logicians started out studying basic logic and then moved on to first-order and then to second-order, etc. In the first place, Aristotle's logic is a fragment of first-order and fundamental aspects of basic logic were not to be discovered for some centuries later. In the second place, in modern times higher-order logics were studied *before* first-order logic was isolated as a system worthy of study in its own right.

After Aristotle's logic had been assimilated by later thinkers, people emerged who could not accept the idea that Aristotle's logic was not comprehensive. These conservative logicians attempted to "reduce" all logically cogent reasoning to Aristotle's syllogistic logic. Likewise, after first-order logic had been isolated and had been assimilated by the logic community, people emerged who could not accept the idea that first-order logic was not comprehensive. These logicians can be viewed not as conservatives who want to reinstate an outmoded tradition but rather as radicals who want to overthrow an established tradition. It remains to be seen whether higher-order logic will ever regain the degree of acceptance that it enjoyed between 1910 and 1930. But there has never been a serious doubt concerning its heuristic and historic importance. In fact, people who do not know second-order logic can not understand the modern debate over its legitimacy and they are cut-off from the heuristic advantages of second-order logic. And, what may be worse, they are cut-off from an understanding of the history of logic and the history of mathematics, and thus are constrained to have distorted views of the nature of the two subjects. As Aristotle first said, we do not understand a discipline until we have seen its development. It is a truism that a person's conceptions of what a discipline is and of what it can become are predicated on a conception of what it has been.

ACKNOWLEDGEMENTS

This essay is based on a tutorial that I led at the Ohio University Inference Conference, October 9–11, 1986. I am indebted to Professor Richard Butrick not only for organizing the conference but also for organizing me. In the year prior to the conference he supplied me, by phone and in writing, with dozens of questions, hypotheses, suggestions and requests from which the basic content and goals of my tutorial emerged. James Gasser, Woosuk Park and Ronald Rudnicki helped with the editing and proof-reading. I am also indebted to the students and colleagues with whom I studied second-order logic over the years, especially George Weaver, Stewart Shapiro, Michael Scanlan and Edward Keenan. Very little in this essay is original. Most of what is here is already in the writings of Alfred Tarski, Leon Henkin, Alonzo Church, Georg Kreisel and George Boolos. Almost every humanistically oriented essay on modern logic is indebted to Tarski, Church, and Quine. Their technical competence, their objectivity, their creativity and, above all, their constant attention to the human importance of logic are largely responsible for preserving, transforming and revitalizing a rich tradition that traces back to Aristotle. One sign of the vitality of the tradition is the fact that not one of the above-mentioned logicians agrees with everything written in this essay, even when the heuristic over-simplifications are emended.

J. MICHAEL DUNN

A REPRESENTATION OF RELATION ALGEBRAS USING ROUTLEY-MEYER FRAMES*

Abstract. A representation of relation algebras is given by modifying the Routley-Meyer semantics for relevance logic. This semantics is similar to the Kripke semantics for intutionistic logic, but its frames use a ternary accessibility relation instead of a binary one. The representation is foreshadowed in the work of Lyndon as well as by the representation of Boolean algebras with operators by Jónsson and Tarski, but the aim here is to make it explicit and to connect it to the related representations of algebras arising in the study of relevance logic as well as other "substructural logics," e.g., linear logic. A philosophical interpretation is given of the representation, showing that an element of a relation algebra can be understood as a set of relations, rather than as the intended interpretation as a single relation. It is shown how a Routley-Meyer frame can be represented so its states are relations, and an interpretation is provided in terms of a "relational database". Connections are mentioned to recent work by Jon Barwise on "information channels" between "sites" and by J. M. Dunn and R. K. Meyer on a ternary frame semantics for combinatory logic.

DEDICATION

This paper was originally dedicated to Alonzo Church on the occasion of his ninetieth birthday and to Robert K. Meyer on the occasion of his sixtieth birthday. The fundamental inspiration for this paper comes from these two. The most direct debt is to Meyer or, more obviously, to Meyer and Richard Routley (who less obviously changed his name to Richard Sylvan). I shall here be extending the "Routley–Meyer" semantics for relevance logic (*Routley and Meyer 1973*) to provide a representation for relation algebras.

Less directly but no less importantly an influence is Church. His "weak theory of implication" (*Church 1951* was not only a seminal paper in the development of relevance logic, but more precisely, his notion that there is an appropriate deduction theorem which holds for "relevant deducibility" influenced my treatment (*Dunn 1966*) of the algebra of relevance logic as based on residuation, as did also his definition of a non-extensional "conjunction" (variously labeled "intensional conjunction", "consistency", "cotenability", and "fusion" in the subsequent relevance logic literature).

* A version of this paper was presented at the Australasian Association for Logic's annual meeting, Canberra, July, 1992, and essentially the present version was distributed in the Indiana University Logic Group Preprint Series in August, 1993 (IULG-93-28). I wish to thank Gerry Allwein, Jon Barwise, Dirk van Gucht, Anil Gupta, Roger Maddux, and Vaughan Pratt for helpful comments and/or information.

C. Anthony Anderson and M. Zelëny (eds.), *Logic, Meaning and Computation*, 77-108.

At the end of this paper, Church, Meyer, and Routley come together in a kind of grand finale, where I describe how the ideas of the representation of relation algebras have been modified (in a recent paper I have written with Meyer) so as to provide Routley-Meyer frame models for Church's λ-calculus.

Since the original dedication, regrettably both Church and Sylvan have died. I now dedicate this paper to both their memories, and of course to Meyer in happier circumstances.

1. INTRODUCTION

The calculus of relatives (relations) originated in the last half of the 19th century and was developed by De Morgan, Peirce, and Schröder. Good sources for both mathematical and historical information are *Tarski and Givant 1985,* and *Maddux 1991.* Cf. also *Pratt 1990* and *van Benthem 1991.* Tarski (*1941*) provided a set of postulates for a "relation algebra"[1] and asked whether every relation algebra could be represented as a set of relations on some domain. Lyndon (*1950*) provided a counter-example. Relation algebras have received renewed attention in recent years because of an interest in a more dynamic conception of logic which incorporates "actions", i.e., relations between states (cf., e.g., *Pratt 1991* and *van Benthem 1991*). Sometimes this is understood as "information based" with relations serving as "channels" for information (cf. *Barwise 1993*). Relation algebras have also been of interest to database theorists (cf. *K.-D. Kraegeloh and P. C. Lockemann 1976*, and *M. Gyssens, L. V. Saxton, and D. van Gucht 1993*).

Routley and Meyer (*1973*) provided a "Kripke-style" semantics for relevance logic using a ternary accessibility relation. De Morgan monoids had been introduced in *Dunn 1966* to provide an algebraic counterpart of the paradigm relevance logic **R**, and Routley and Meyer (*1973*) give a representation of De Morgan monoids using their ternary relation. Routley and Meyer (*1973,* cf. also *Routley and Meyer 1972* and *Meyer and Routley 1972*) indicate that one can vary features of their frames and corresponding features of either the logic **R** or its De Morgan monoid models. A relation algebra is very similar to a De Morgan monoid, and we show here that relation algebras can be represented with only slight modifications of the Routley-Meyer frames used for **R.**

We do, however, have to modify the construction of Routley-Meyer (*1973*), since that construction leads only to immediate representations of what they call "prime" De Morgan monoids,[2] with "representations" of general De Morgan monoids obtained as subdirect products of these representations. The same would hold true of relation algebras. We, however, obtain a direct rep-

[1] And other sets of axioms produced by Tarski and his students which are discussed by Maddux (*1991*).

[2] A De Morgan monoid is *prime*, when for the identity element e, $e \leq a \vee b$ implies $e \leq a$ or $e \leq b$.

resentation of all relation algebras (and the same can be done for De Morgan monoids) by dropping the requirement on a Routley-Meyer frame that there be a distinguished "zero state" 0, and replacing it with a set Z that may contain many "zero states". In *Dunn 1933a* there are parallel discussions that help provide a context for understanding this and other aspects of the current paper.

I discovered only after working out the representation results of this paper that these results were foreshadowed by Lyndon, and by Jónsson and Tarski, though the treatment here is more explicit and also has the point of locating relation algebras and their representation within the geography of substructural logics.

Lyndon (*1950*) showed how every finite relation algebra can be analyzed so that each element is decomposed into a join of atoms with composition and conversion defined on the atoms. In this representation, a natural ternary relation arises on the atoms. It is well-known that the representation of a finite Boolean algebra as a field of sets corresponds to its decomposition into joins of atoms (let each such join be replaced with the set of atoms occurring in it), and that atoms correspond to the maximal filters that appear in the representation of arbitrary Boolean algebras due to Stone (*1936*). Relations on such filters then correspond to relations on atoms. Cf. also *Maddux 1991*.[3]

Another foreshadow is cast by Jónsson and Tarski (*1951–52*). The first part of their paper shows, in general, how to represent a Boolean algebra with n-ary operators using $n+1$ placed relations. These operators must meet the conditions that they distribute over join in each of their positions and that if any operand is the Boolean **0** then the operator must output **0**. In the second part of their paper they point out that relation algebras meet these conditions (with relative product and converse as the operators) and go on to establish a number of results concerning relation algebras. Curiously, they never detail how their general representation relates to these results on relation algebras, and in fact, never state first-order conditions on the relations used in representing the abstract operations corresponding to relative product and converse to assure that they satisfy the postulates for relation algebras.

These foreshadows fall "after the fact" in terms of the development of the present paper, which has as its immediate cause my own work on "gaggle theory" (*Dunn 1991, 1993a*), which generalizes the framework of Jónsson and Tarski (*1951–52*). Relation algebras are gaggles because of residuation and the fact that each of their operations distributes over meet or join in some way or another (and not just preserving join, as was required by Jónsson and Tarski). While the work here has been heavily influenced by the more general gaggle-theoretic approach, it is presented in an independent way.

---

[3] Maddux replaces the ternary relation $a \circ b \geq \check{c}$ of Lyndon with $a \circ b \geq c$, which is more in synchrony with the Routley-Meyer relation on filters.

Also it should be pointed out that the official gaggle-theory representation (as well as that of Jónsson and Tarski) would employ a binary relation in interpreting the algebraic operation corresponding to converse, whereas the representation here uses instead a unary operation. The relationship between these two styles of representations is explored in *Dunn 1993b.*

Relation algebras can be thought of as a kind of substructural logic related to the calculus of Lambek (*1958*) by the addition of the Boolean operators and converse.[4] Cf. *P. Schröder-Heister and K. Dŏsen 1992* for the general framework of substructural logics, which includes relevance logic, linear logic, and Lambek calculus among others. The present representation of relation algebras places them firmly in the general framework of representations for substructural logics. The Routley-Meyer ternary semantics for the first of these has been modified and extended to linear logic by Allwein and Dunn (*1993*) and to Lambek calculus by Dŏsen (*1992*). Cf. also *Dunn 1993a.*

Relation algebras have been proposed as models for non-commutative linear logic by several people, with linear negation understood as complement of the converse.[5] To the best of my knowledge the first to do this was Vaughan Pratt at the Edinburgh Jumelage in 1989.[6]

2. THE ALGEBRAS

We will introduce a standard definition of *relation algebras* after first introducing a few more general definitions with the aim of setting relation algebras in an abstract context that makes clear their relationship to the algebras that arise in studying substructural logics.

DEFINITION 2.1. A *lattice-ordered groupoid* is a structure $(L, \wedge, \vee, \circ)$, where $(L, \wedge, \vee)$ is a lattice, and $\circ$ is a binary operation on L satisfying:

$$x \circ (y \vee z) = (x \circ y) \vee (x \circ z), \tag{1}$$

$$(y \vee z) \circ x = (y \circ x) \vee (z \circ x). \tag{2}$$

When $\circ$ is associative, i.e.,

$$x \circ (y \circ z) = (x \circ y) \circ z, \tag{3}$$

we speak of a *lattice-ordered semi-group.*

[4]This Gentzen system was introduced by Lambek as a way of formalizing certain linguistic constructions.

[5]While writing my dissertation in 1966 I considered relation algebras as models for the system **R** of relevance logic, but they were too weak. Not only is relative product not commutative, it also does not satisfy square increasingness ($R \circ R \subseteq R$), the algebraic counterpart of the logical principle of *contraction* $(\phi \to (\phi \to \psi)) \to (\phi \to \psi)$, rejected by linear logic, and some relevance logics, but a principle of the system **R**. I was able to obtain a representation of De Morgan *lattices* (the *extensional* or, in "linearese," *additive* fragment of both **R** and linear logic) in terms of sets of relations closed under intersection, union, and complement of converse. See *Dunn 1982.*

[6]Cf. also *Brown and Gurr 1995.*

It follows from (1) and (2) that $\circ$ is isotone in each of its arguments.

DEFINITION 2.2. $(L, \wedge, \vee, \circ, e)$ is a lattice-ordered monoid when $(L, \wedge, \vee, \circ)$ is a lattice-ordered semigroup and e is an "*identity element*" satisfying

$$e \circ x = x \circ e = x. \tag{4}$$

DEFINITION 2.3. Right and left residuals are characterized as follows:[7]

$$x \circ y \leq z \textit{ iff } y \leq x \rightarrow z \textit{ iff } x \leq z \leftarrow y. \tag{5}$$

DEFINITION 2.4 **Peirce Groupoid**. A *Peirce groupoid (semi-group, monoid)* is a distributive-lattice ordered groupoid (semi-group, monoid) that is *doubly residuated,* i.e., has both right and left residuals.

Peirce groupoids arise naturally in the structure of substructural logics, since residuation corresponds naturally to the Deduction Theorem and its converse, with the two residuals $x \rightarrow y$ and $y \leftarrow x$ read as implications. There are two implications because the "premiss bunching" operator $\circ$ is not necessarily commutative.[8] When Gentzen sequents are understood in the usual way as formed by anadic constructions rather than pairing, associativity is a natural property. The identity element e is a way of indicating "theorems", with $e \leq a$ read as "a is a theorem". Cf. *Dunn 1993a* and other papers in *Schröder-Heister and Dŏsen 1992.* Peirce monoids correspond to the (associative) Lambek calculus which permits empty left-hand sides, but Peirce monoids have the additional connectives of conjunction and disjunction with the requirement that they distribute over each other. This requirement makes the name "Peirce groupoid" even more appropriate because of the emphasis that Peirce placed upon the distributive laws.

De Morgan and Peirce observed in the last century that the structures we call "Peirce monoids" arise naturally in the context of the algebra of relations (cf., e.g., *Maddux 1991*). We think of relations in the usual way as sets of ordered pairs. Given a set X and two binary relations R and S on X, the *relative product* $R \circ S = \{(x, y) : \exists z((x, z) \in R \,\&\, (z, y) \in R)\}$ is an associative operation, and e is the identity relation (restricted to X). Then $R \rightarrow S = \{(x, y) : \forall z((z, x) \in R \textit{ implies } (z, y) \in R)\}$, and $S \leftarrow R = \{(x, y) : \forall z((x, z) \in R \textit{ implies } (y, z) \in R)\}$.

DEFINITION 2.5 **Proto De Morgan Monoid**. $(L, \wedge, \vee, \circ, \rightarrow, \leftarrow, \sim, e)$ is a *proto De Morgan monoid* just when $(L, \wedge, \vee, \circ, \rightarrow, \leftarrow, e)$ is a Peirce monoid, and $\sim$ is a unary operation satisfying:

$$a \circ b \leq c \textit{ iff } b \circ \sim c \leq \sim a, \tag{6}$$

$$\sim\sim a = a \quad (\textit{period two}) \tag{7}$$

[7] Cf. *Ward and Dilworth 1939.*

[8] The idea of "premiss bunching" can be found in *Meyer and McRobbie 1982.*

The reader can easily see that in the presence of (7) we may restate (6) in symmetric forms as follows.

PROPOSITION 2.6. (6) *may be replaced in the definition of a proto De Morgan monoid by* (8) *below, and also by* (9):

$$a \circ b \leq c \textit{ implies } \sim c \circ a \leq \sim b, \tag{8}$$

$$a \circ b \leq c \textit{ iff } b \circ \sim c \leq \sim a \textit{ iff } \sim c \circ a \leq \sim b. \tag{9}$$

Note that $\sim$ is a *dual automorphism,* i.e., 1-1, which follows from (7), and order inverting, since by substituting e for b we obtain

$$a \leq c \textit{ iff } \sim c \leq \sim a. \tag{10}$$

Dual automorphisms of period two are in the relevance logic literature called De Morgan complements since they satisfy the De Morgan laws:

$$\sim(a \vee b) = \sim a \wedge \sim b, \tag{11}$$

$$\sim(a \wedge b) = \sim a \vee \sim b. \tag{12}$$

A De Morgan monoid is just a proto De Morgan monoid with the following additional properties:

$$x \circ y = y \circ x \textit{ (commutativity)}, \tag{13}$$

$$x \leq x \circ x \textit{ (square} - \textit{increasingness)}. \tag{14}$$

In the presence of commutativity the two arrows collapse to just one, say $\rightarrow$, and this is the algebraic counterpart to relevant implication in the system **R**.[9] A proto De Morgan monoid can also be described as a distributive, non-associative, non-commutative *Girard monoid* (cf. *Allwein and Dunn 1993*) since the algebraic models for the multiplicative-additive fragment of linear logic differ from those for the relevance logic **R** just by lacking distribution and square-increasingness.

Given a lattice, a unary operation x^{-1} (which we shall think of as "converse") is an *automorphism of period two* when it is a 1-1, order preserving, and

$$(x^{-1})^{-1} = x \quad \textit{(period two)}. \tag{15}$$

An alternative definition is that $^{-1}$ is of period two and satisfies:

$$(x \vee y)^{-1} = x^{-1} \vee y^{-1}, \tag{16}$$

$$(x \wedge y)^{-1} = x^{-1} \wedge y^{-1}. \tag{17}$$

Note that if L is a Boolean algebra it can also be shown that

$$(-x)^{-1} = -(x^{-1}). \tag{18}$$

(This follows easily from the well-known fact that in a Boolean algebra complements are unique.)

---

[9] See *Dunn 1966*.

Meyer and Routley (*1973, 1974*) study the result of adding to relevance logic a "Boolean negation" in addition to the usual "De Morgan negation". Meyer (*1979*) extends this even further by studying "Boolean-valued models" for relevance logic, in which the underlying lattices are Boolean algebras (in place of the usual De Morgan lattices). In addition to the usual operations of relevance logic he also introduces an explicit unary connective corresponding to the "star operator", and from it he defines De Morgan negation: $\sim x = -(x^*)$. (We will use the notation x^{-1} in place of x^*.) Meyer and Routley show that Boolean negation can be added conservatively in the sense that these Boolean algebraic models validate the same theorems in the "relevant" vocabulary (obviously excluding Boolean negation except in the context $-(x^*)$) as do the De Morgan monoids.[10] These results of Meyer and Routley can be easily generalized to proto De Morgan monoids and the following algebras, which we name in honor of Meyer:[11]

DEFINITION 2.7 **Meyer Monoid.** A structure $(L, \wedge, \vee, \circ, \rightarrow, \leftarrow, -, ^{-1}, e)$ is a *Meyer monoid* just when

$$(L, \wedge, \vee, -) \text{ is a Boolean algebra,} \tag{19}$$

$$x^{-1} \text{ is an automorphism of period two,} \tag{20}$$

$$(L, \wedge, \vee, \circ, \rightarrow, \leftarrow, e) \text{ is a Peirce monoid,} \tag{21}$$

$$a \rightarrow b = \sim(\sim b \circ a), \quad b \leftarrow a = \sim(a \circ \sim b) \tag{22}$$
$$\text{when } \sim x \text{ is defined as } -(x^{-1}).$$

A Meyer monoid is essentially a proto De Morgan monoid where the underlying distributive lattice is a Boolean algebra and where $\sim$ has been replaced with Boolean complement and converse, from which it can be recovered by way of the definition:

$$\sim x = -(x^{-1}). \tag{23}$$

An alternative way to view a Meyer monoid is to take both Boolean and De Morgan complements as primitive and then define:

$$x^{-1} = -(\sim x). \tag{24}$$

DEFINITION 2.8 **Boolean Proto De Morgan Monoid.** A structure $(L, \wedge, \vee, \circ, \rightarrow, \leftarrow, -, ^{-1}, e)$ is a *Boolean proto De Morgan monoid* just when

$$(L, \wedge, \vee, -) \text{ is a Boolean algebra,} \tag{25}$$

$$(L, \wedge, \vee, \circ, \rightarrow, \leftarrow, \sim, e) \text{ is a De Morgan monoid,} \tag{26}$$

$$- \sim x = \sim - x. \tag{27}$$

[10] This seems not quite an embedding result, except at the level of the free algebras, the issue being preservation of the identity elements.

[11] It is fitting that Meyer have a class of algebras named after him, since he has populated the literature with Church monoids, Dunn monoids, etc. Cf. *Anderson and Belnap et al. 1975.*

THEOREM 2.9. *A Meyer monoid becomes a Boolean proto De Morgan monoid given the definition* (23).

Proof.

$$-\sim x = -[-(x^{-1})] = -[(-x)^{-1}] = \sim -x.$$
$$\sim\sim x = -([-(x^{-1})]^{-1}) = - - x^{-1-1} = x.$$
$$a \circ b \leq c \text{ iff } b \leq a \to c \text{ iff } b \leq -[(-c^{-1} \circ a)^{-1}] \text{ iff}$$
$$(-c^{-1} \circ a)^{-1} \leq -b \text{ iff } -c^{-1} \circ a \leq -b^{-1} \text{ iff } \sim c \circ a \leq \sim b.$$

⊣

THEOREM 2.10. *A Boolean proto De Morgan monoid becomes a Meyer monoid given the definition* (24).

Proof. That $^{-1}$ is an automorphism follows from the fact that each of $-$ and $\sim$ are dual automorphisms. That $^{-1}$ is of period two follows from (27) and the fact that each of $-$ and $\sim$ are of period two. We leave to the reader the proof of $a \to b = \sim(\sim b \circ a)$ and its dual, the hint being to use (6), (7), and (5). ⊣

DEFINITION 2.11 **Pre Relation Algebra**. A *pre relation algebra* is a structure $(L, \wedge, \vee, \circ, ^{-1}, e)$, where the structure $(L, \wedge, \vee, \circ, e)$ is a distributive-lattice ordered monoid and x^{-1} is an automorphism of period two satisfying

$$e^{-1} = e \tag{28}$$
$$(x \circ y)^{-1} = y^{-1} \circ x^{-1}. \tag{29}$$

The following definition comes ultimately from Tarski, though it is essentially in a form due to Birkhoff (*1967*), which emphasizes residuation.[12]

DEFINITION 2.12 **Relation algebra**. A structure $(L, \wedge, \vee, \circ, \to, \leftarrow, -, ^{-1}, e)$ is a *relation algebra* just when

$$(L, \wedge, \vee, -) \textit{ is a Boolean algebra,} \tag{30}$$
$$(L, \wedge, \vee, \circ, ^{-1}, e) \textit{ is a pre relation algebra,} \tag{31}$$
$$(L, \wedge, \vee, \circ, \to, \leftarrow, e) \textit{ is a Peirce monoid,} \tag{32}$$
$$a \to b = -(a^{-1} \circ -b), \quad b \leftarrow a = -(-b \circ a^{-1}). \tag{33}$$

REMARK. Note that a pre relation algebra does not become a relation algebra simply by adding the requirement that L be a Boolean algebra. We must also add that the addition of Boolean complement to relative product and converse allows us to define residuals as in (33) above. Note the intriguing similarity of the earlier (22) with (33) above. To eyes trained in relevance logic, the latter seems a "typo" version of the former. We show next that these are in fact the same. This is the key link between relation algebras and relevance logic.

[12] Cf. the definition of Jónsson and Tarski and other definitions discussed by Maddux (*1991*). The definition below does not make it obvious that relation algebras have an equational definition, but all of the algebras studied here do.

THEOREM 2.13. *A structure* $(L, \wedge, \vee, \circ, \rightarrow, \leftarrow, -, ^{-1}, e)$ *is a relation algebra if and only if* (i) *it is a Meyer monoid, and* (ii) $(L, \wedge, \vee, \circ, ^{-1}, e)$ *is a pre relation algebra.*

Proof. Upon comparing the definitions, it is clear that it suffices to show that the two "definitions" of the arrows in (22) and (33) are equivalent in a pre relation algebra, i.e.,

$$-[(-(b^{-1}) \circ a)^{-1}] = -(a^{-1} \circ -b), \text{ i.e.,} \tag{34}$$

$$(-(b^{-1}) \circ a)^{-1} = a^{-1} \circ -b. \tag{35}$$

But (35) follows from (15), (18), and (29). ⊣

3. THE FRAMES

We next define the structures out of which the representations will be constructed.

DEFINITION 3.1. By a *frame* we shall mean a pair (U, R), where U is a non-empty set and R is a ternary relation on U, i.e., $R \subseteq U^3$.

We augment these frames with an explicit partial order.[13]

DEFINITION 3.2. By an *articulated frame* (for short, *a-frame*), we shall mean a triple $(U, \sqsubseteq, R)$, where $(U, \sqsubseteq)$ is a non-empty poset and R is a ternary relation on U, subject to the following conditions:

$$R\alpha\beta\gamma \ \& \ \gamma \sqsubseteq \gamma' \textit{ implies } R\alpha\beta\gamma', \tag{36}$$

$$R\alpha\beta\gamma \ \& \ \alpha' \sqsubseteq \alpha \textit{ implies } R\alpha'\beta\gamma, \tag{37}$$

$$R\alpha\beta\gamma \ \& \ \beta' \sqsubseteq \beta \textit{ implies } R\alpha\beta'\gamma. \tag{38}$$

We shall refer to the elements of U as "information states" (or sometimes "points") and we shall think of $\sqsubseteq$ as the "information order," interpreting $\alpha \sqsubseteq \beta$ as saying that "β contains all of the information in α".

Note that any frame can be viewed as a special case of an a-frame where $\sqsubseteq$ is just the identity relation restricted to U. Adapting the notion of Meyer and Routley *1973, 1974*, we shall call such a-frames *classical.* In this paper we shall often tacitly assume when we talk of a frame (*simpliciter*) that it has been made augmented in this trivial way as context requires.

Every a-frame gives rise to a Peirce groupoid which we shall call a *framed Peirce groupoid.* Thus let $\mathcal{P}(U)^{\uparrow}$ be the set of *cones* of U, where a *cone* A is a member of the powerset $\mathcal{P}(U)$ such that $\forall a, b(a \in A$ and $a \sqsubseteq b$ implies $b \in A)$. For $A, B \in \mathcal{P}(U)^{\uparrow}$, we define operations as follows:

$$A \odot B = \{\chi : \exists\alpha \in A, \beta \in B : R\alpha\beta\chi\} \tag{39}$$

$$A \Rightarrow B = \{\chi : R\alpha\chi\beta \ \& \ \alpha \in A \textit{ implies } \beta \in B\} \tag{40}$$

$$B \Leftarrow A = \{\chi : R\chi\alpha\beta \ \& \ \alpha \in A \textit{ implies } \beta \in B\}. \tag{41}$$

[13]For Routley and Meyer this was implicit, defined as $R0\alpha\beta$.

DEFINITION 3.3. We shall call $(\mathcal{P}(U)^{\uparrow}, \cap, \cup, \odot, \Rightarrow, \Leftarrow)$, with $\odot, \Rightarrow, \Leftarrow$ defined as in (39)–(41), the *full framed Peirce groupoid* on the a-frame $(U, \sqsubseteq, R)$. Any collection of members of $\mathcal{P}(U)^{\uparrow}$ that is closed under these operations (including the lattice ones) will be called a *framed Peirce groupoid.*

The name "framed Peirce groupoid" was chosen in anticipation of the fact that is easy to verify that this is in fact a distributive lattice-ordered residuated groupoid (with $\subseteq$ as the lattice order, and $\cap$ and $\cup$ as the meet and join), and so we have the following.

PROPOSITION 3.4. *A "framed Peirce groupoid" is a Peirce groupoid.*

In order to get a *framed Peirce monoid* we need both associativity and a two-sided identity, and we must impose more conditions on an a-frame. We develop these by stages so that if anyone desires they can abstract out a number of intermediate structures.

We leave to the reader the verification that one gets associativity (and hence a semi-group) by making the following requirement (as in *Routley and Meyer 1973*) on the accessibility relation:

$$\exists\chi(R\alpha\beta\chi \,\&\, R\chi\gamma\delta) \textit{ iff } \exists\chi(R\alpha\chi\delta \,\&\, R\beta\gamma\chi), \tag{42}$$

or in the notation introduced by Routley and Meyer,

$$R^2(\alpha\beta)\gamma\delta \textit{ iff } R^2\alpha(\beta\gamma)\delta \quad (R^2\textit{-associativity}). \tag{43}$$

We shall call any frame satisfying (42) *associative.*

PROPOSITION 3.5. *A framed Peirce groupoid defined on an associative a-frame is a Peirce semi-group.*

In order to get an identity element (and so a monoid), we need a set $Z \subseteq U$, so that

$$Z \odot A = A, \tag{44}$$

$$A \odot Z = A. \tag{45}$$

Given a non-empty cone Z of U, an a-frame will be called *ordered in its 1st position by* Z when

$$\exists\zeta \in Z(R\zeta\alpha\beta) \textit{ iff } \alpha \sqsubseteq \beta, \tag{46}$$

and *ordered in its 2nd position by* Z when

$$\exists\zeta \in Z(R\alpha\zeta\beta) \textit{ iff } \alpha \sqsubseteq \beta. \tag{47}$$

DEFINITION 3.6. An *assertional a-frame* is a structure $(U, \sqsubseteq, R, Z)$ where $Z \subseteq U$ satisfies both (46) and (47), i.e.,

$$\exists\zeta \in Z(R\zeta\alpha\beta) \textit{ iff } \alpha \sqsubseteq \beta \textit{ iff } \exists\zeta \in Z(R\alpha\zeta\beta) \quad (\textit{ordered}). \tag{48}$$

We shall also say that an a-frame $(U, \sqsubseteq, R)$ is *assertional* if $\exists Z \subseteq U$ such that $(U, \sqsubseteq, R, Z)$ is an assertional a-frame.

With respect to a subset Z, an a-frame will be called *1st position weakly reflexive*[14] if

$$\forall \alpha \in U, \exists \zeta \in Z, R\zeta\alpha\alpha, \tag{49}$$

and *2nd position weakly reflexive* if

$$\forall \alpha \in U, \exists \zeta \in Z, R\alpha\zeta\alpha. \tag{50}$$

LEMMA 3.7. *An a-frame ordered in its 1st position by Z is 1st position weakly reflexive, and similarly, for the second position.*[15]

Proof. Substitute α for β in (46) and (47). ⊣

REMARK. Note that given our conventions, a frame *simpliciter* (U, R) is assertional just when $\exists Z \subseteq U$ such that:

$$\exists \zeta \in Z(R\zeta\alpha\beta) \text{ iff } \alpha = \beta \text{ iff } \exists \zeta \in Z(R\alpha\zeta\beta) \quad (\textit{equality} - \textit{ordered}). \tag{51}$$

PROPOSITION 3.8. *A framed Peirce groupoid defined on an assertional a-frame has both left and right identity elements.*

Proof. We first show (44). Suppose that $\chi \in Z \odot A$, i.e., $\exists \zeta \in Z$, $\alpha \in A$, $R\zeta\alpha\chi$. We know from ordering in the first position (48) that $\alpha \sqsubseteq \chi$. But since A is hereditary, then $\chi \in A$. For the other direction, let us suppose that $\chi \in A$. For $x \in Z \odot A$, we need that $\exists \zeta \in Z$, $\exists \alpha \in A$, $R\zeta\alpha\chi$. But by 1st-position reflexivity (49), we have $\exists \zeta \in Z$, $R\zeta\chi\chi$.

The argument for (45) is similar, using the corresponding properties for the second position. ⊣

The following follows from the previous propositions.

COROLLARY 3.9. *The framed Peirce groupoid defined on an assertional associative a-frame is a Peirce monoid.*

4. INVOLUTED FRAMES

We next put conditions on a frame so as to accommodate converse. Let us suppose there is a map $^\cup$ on U into itself such that for $\chi \in U$,

$$\chi^{\cup\cup} = \chi \quad (\textit{period two}). \tag{52}$$

[14] Note the order of the quantifiers. If they were reversed we would talk of "*strong reflexivity*" in each position. They can be so reversed in a De Morgan monoid, indeed in any Peirce monoid satisfying square increasingness (14). Moreover in De Morgan monoids, with their commutativity (13), there is no need to distinguish reflexivity in the two positions.

[15] Indeed, 1st position weak reflexivity (49) can replace the conditional from right to left in the definition of 1st position ordering (46), and similarly for the 2nd position. Cf. *Dunn 1993.*

Such a map is commonly called an *involution.*[16] We also add (writing $\breve{\chi}$ in place of $\chi^{\cup}$):

$$R\alpha\beta\gamma \text{ only if } R\breve{\beta}\breve{\alpha}\breve{\gamma} \quad (\textit{tagging}). \tag{53}$$

Let us define for $A \in \mathcal{P}(U)^{\dagger}$,

$$A^{-1} = \{\breve{\alpha} : \alpha \in A\}. \tag{54}$$

By virtue of (52), this is equivalent to:

$$A^{-1} = \{\alpha : \breve{\alpha} \in A\}. \tag{55}$$

DEFINITION 4.1. A structure $(U, \sqsubseteq, R, {}^{\cup}, Z)$ is a *pre-relation-algebra a-frame* just when $(U, \sqsubseteq, R, Z)$ is an associative assertional a-frame, ${}^{\cup}$ is an involution satisfying tagging, and Z is *converse closed*, i.e., *if* $\zeta \in Z$ then $\breve{\zeta} \in Z$. We shall call $(\mathcal{P}(U)^{\dagger}, \cap, \cup, \odot, {}^{-1}, Z)$, with $\odot$, ${}^{-1}$ defined as in (39), (54), the *full framed pre relation algebra* on the pre-relation-algebra a-frame $(U, \sqsubseteq, R, {}^{\cup}, Z)$. Any collection of members of $\mathcal{P}(U)^{\dagger}$ that is closed under these operations (including the lattice ones) will be called a *framed pre relation algebra.*

PROPOSITION 4.2. *A "framed pre relation algebra" is a pre relation algebra.*

Proof. It is practically carried by the notation that $(A \cup B)^{-1} = A^{-1} \cup B^{-1}$ (and similarly for $\cap$ and complement relative to U). And clearly $Z^{-1} = Z$ (because of the requirements that Z be closed under ${}^{\cup}$, and ${}^{\cup}$ be of period two). So we only need verify:

$$(A \odot B)^{-1} = (B^{-1} \odot A^{-1}), \tag{56}$$

using tagging and period two (which last allows us to treat any arbitrary point as having a breve over it). Thus:

$$\breve{\chi} \in (A \odot B)^{-1} \textit{ iff } \chi \in (A \odot B) \textit{ iff } \exists \alpha \in A, \beta \in B, R\alpha\beta\chi \textit{ iff}$$
$$\exists \beta \in B^{-1}, \breve{\alpha} \in A^{-1}, R\breve{\beta}\breve{\alpha}\breve{\chi} \textit{ iff } \breve{\chi} \in B^{-1} \odot A^{-1}.$$

⊣

To obtain a-frames suitable for proto De Morgan monoids we add to an involution (52), instead of tagging:

$$R\alpha\beta\gamma \textit{ iff } R\breve{\gamma}\alpha\breve{\beta} \quad (\textit{antilogism}). \tag{57}$$

REMARK. Routley and Meyer (*1973*) can state (57) simply as (57′) $R\alpha\beta\gamma$ *only if* $R\alpha\breve{\gamma}\breve{\beta}$, because in a De Morgan monoid $\circ$ is commutative, and this is reflected by postulating that the first two positions of R permute. But without this postulate we must be more subtle. Using the fact that ${}^{\cup}$ is of period two, (57) can be analyzed as two conditionals (the two "correct" forms

[16]Following Routley and Meyer, this involution is customarily denoted by * in the relevance logic literature. Here we mostly use ${}^{\cup}$ because in the literature on relations * is customarily used for the "ancestral".

of (57′) that arise when one is careful not to build in permutation):

$$R\alpha\beta\gamma \textit{ only if } R\breve{\gamma}\alpha\breve{\beta}, \tag{57a}$$

$$R\alpha\beta\gamma \textit{ only if } R\beta\breve{\gamma}\breve{\alpha}. \tag{57b}$$

DEFINITION 4.3. A structure $(U, \sqsubseteq, R, ^{\cup}, Z)$ is a *proto De Morgan monoid a-frame* just when $(U, \sqsubseteq, R, Z)$ is an associative assertional a-frame and $^{\cup}$ is an involution satisfying antilogism. $(\mathcal{P}(U)^{\uparrow}, \cap, \cup, \odot, \Rightarrow, \Leftarrow, \sim, Z)$, with $\odot, \Rightarrow, \Leftarrow$ defined as in (39), (40), (41), and $\sim A$ defined as $\{\alpha : \breve{\alpha} \notin A\}$,[17] is the *full framed proto De Morgan monoid* on the proto-De-Morgan-monoid a-frame $(U, \sqsubseteq, R, ^{\cup}, Z)$. Any collection of members of $\mathcal{P}(U)^{\uparrow}$ that is closed under these operations (including the lattice ones) will be called a *framed proto De Morgan monoid.*

PROPOSITION 4.4. *A "framed proto De Morgan monoid" is a proto De Morgan monoid.*

Proof. We show that $A \circ B \subseteq C$ *iff* $\sim C \circ A \subseteq \sim B$.

$A \circ B \subseteq C$ means that for arbitrary α, β, γ:

$$\alpha \in A \,\&\, \beta \in B \,\&\, R\alpha\beta\gamma \textit{ implies } \gamma \in C. \tag{58}$$

$\sim C \circ A \subseteq \sim B$ means that for arbitrary γ', α, β' :

$$\gamma' \in \sim C \,\&\, \alpha \in A \,\&\, R\gamma'\alpha\beta' \textit{ implies } \beta' \in \sim B, \text{ i.e.,} \tag{59}$$

$$\breve{\gamma}' \notin C \,\&\, \alpha \in A \,\&\, R\gamma'\alpha\beta' \textit{ implies } \breve{\beta}' \notin B. \tag{60}$$

Substituting $\breve{\gamma}$ for γ', $\breve{\beta}$ for β' and using period two, (60) is equivalent to

$$\gamma \notin C \,\&\, \alpha \in A \,\&\, R\breve{\gamma}\alpha\breve{\beta} \textit{ implies } \beta \notin B. \tag{61}$$

It is easy to see that (61) and (58) are equivalent by using antilogism and the logical principle of contraposition. ⊣

The reader can easily verify that one gets a De Morgan monoid by adding the following requirements:[18]

$$R\alpha\beta\gamma \textit{ implies } R\beta\alpha\gamma \; (\textit{R-commutativity}), \tag{62}$$

$$R\alpha\alpha\alpha \; (\textit{R-reflexivity}). \tag{63}$$

To obtain a-frames suitable for Meyer monoids we in effect add the requirement that they be classical. The reason is that, in general, the complement of a cone is not a cone. But by requiring that $\sqsubseteq$ be simply $=_U$ (the identity rela-

[17] This definition of $\sim$ was first due to A. Białynicki-Birula and H. Rasiowa (*1957*) and was studied in my dissertation (cf. *Dunn 1966*) as part of the algbraic semantics of relevance logic. It was explicitly introduced into the philosophical semantics of relevance logic by R. Routley and V. Routley, and made a part of the semantics for **R** by R. Routley and R. K. Meyer. For history, see *Dunn 1986*, or *Anderson, Belnap, and Dunn et al. 1992.*

[18] See *Routley and Meyer 1973.*

tion restricted to U), every subset (and its complement) becomes a cone since every subset is closed under the identity relation. Then $\mathcal{P}(U)^{\uparrow} = \mathcal{P}(U)$. We get the same effect of working with classical a-frames by simply not requiring that the frames be articulated. Finally, Z satisfies (51).

DEFINITION 4.5. A structure $(U, R, {}^{\cup}, Z)$ is a *Meyer-monoid frame* exactly when $(U, =_U, R, Z)$ is an associative assertional classical a-frame and ${}^{\cup}$ is an involution satisfying antilogism. A *framed full Meyer monoid* is then a structure $(\mathcal{P}(U), \cap, \cup, \odot, \Rightarrow, \Leftarrow, -, {}^{-1}, Z)$, with $\odot, \Rightarrow, \Leftarrow$ defined as in (39), (40), (41) and ${}^{-1}$ defined as in (54). A *framed Meyer monoid* is any collection of subsets of U closed under the above operations.

PROPOSITION 4.6. *A "framed Meyer monoid" is a Meyer monoid.*

Finally we get the required frames for representing relation algebras by simply adding tagging to the requirements for Meyer-monoid a-frame.

DEFINITION 4.7. A structure $(U, R, {}^{\cup}, Z)$ is a *relation-algebra frame* just when $(U, R, {}^{\cup}, Z)$ is a Meyer-monoid frame with ${}^{\cup}$ satisfying tagging, i.e., the structure $(U, =_U, R, {}^{\cup}, Z)$ is an associative assertional classical a-frame and ${}^{\cup}$ is an involution satisfying both antilogism and tagging. A *framed full relation-algebra* on a relation-algebra frame $(U, =_U, R, Z)$ is a structure $(\mathcal{P}(U), \cap, \cup, \odot, \Rightarrow, \Leftarrow, -, {}^{-1}, Z)$, with $\odot, \Rightarrow, \Leftarrow$ defined as in (39), (40), (41) and ${}^{-1}$ defined as in (54). A *framed relation-algebra* is any collection of subsets of U closed under these operations.

PROPOSITION 4.8. *A "framed relation algebra" is a relation-algebra.*

Proof. We must verify

$$A \Rightarrow B = -(A^{-1} \odot -B), \; B \Leftarrow A = -(-B \odot A^{-1}). \tag{64}$$

We show the first identity, the second following in a similar fashion. It clearly suffices to show that

$$-(A \Rightarrow B) = (A^{-1} \odot -B). \tag{65}$$

We note that

$$\chi \in -(A \Rightarrow B) \text{ iff } \exists \alpha, \beta \; R\alpha\chi\beta \;\&\; \alpha \in A \;\&\; \beta \notin B, \text{ and}$$
$$\chi \in (A^{-1} \odot -B) \text{ iff } \exists \breve{\alpha}, \beta \; R\breve{\alpha}\beta\chi \;\&\; \breve{\alpha} \in A^{-1} \;\&\; \beta \notin B.$$

Since $\alpha \in A$ *iff* $\breve{\alpha} \in A^{-1}$, the proof boils down to showing that

$$R\alpha\chi\beta \text{ iff } R\breve{\alpha}\beta\chi. \tag{66}$$

This follows by first applying antilogism and then tagging (with some period two thrown in):

$$R\alpha\chi\beta \text{ iff } R\breve{\beta}\alpha\breve{\chi} \text{ iff } R\breve{\alpha}\beta\chi. \tag{67}$$

⊣

5. REPRESENTATION RESULTS USING ROUTLEY-MEYER FRAMES

It turns out that not only can one produce framed relation algebras given an associative assertional classical a-frame, but up to isomorphism, every relation algebra can be identified with such a framed relation algebra.

We define the *canonical a-frame* as follows. The points of the a-frame are (proper) prime filters. The following definitions of the *canonical accessibility relation* R can all be shown equivalent using residuation:[19]

$$R_\circ\alpha\beta\gamma \text{ iff } \forall a,b, \text{ if } a\in\alpha \,\&\, b\in\beta \text{ then } a\circ b\in\gamma, \tag{68}$$

$$R_\rightarrow\alpha\beta\gamma \text{ iff } \forall a,b, \text{ if } a\in\alpha \,\&\, a\rightarrow b\in\beta \text{ then } b\in\gamma, \tag{69}$$

$$R_\leftarrow\alpha\beta\gamma \text{ iff } \forall a,b, \text{ if } b\leftarrow a\in\alpha \,\&\, a\in\beta \text{ then } b\in\gamma. \tag{70}$$

We define the *canonical information order* $\alpha \sqsubseteq \beta$ *iff* $\alpha \subseteq \beta$. It is easy to verify the "tonicity conditions" (36)–(38)

Before proceeding, it is convenient to pause and state a couple of lemmas.[20]

THEOREM 5.1 **Extension Lemma**. *Let F be a filter, I be an ideal, with $F\cap G=\emptyset$. Then there exists a prime filter $F'\supseteq F$ and a prime ideal $I'\supseteq I$, so that $F'\cap I'=\emptyset$.*

THEOREM 5.2 **Squeeze Lemma**. *Let γ be a prime filter, and let α_0 and β_0 be filters. Suppose $R_\circ\alpha_0\beta_0\gamma$. Then there exist prime filters α and β such that $\alpha_0\subseteq\alpha$, $\beta_0\subseteq\beta$ and $R_\circ\alpha\beta_0\gamma$, $R_\circ\alpha_0\beta\gamma$.*

THEOREM 5.3 **Exclusion Lemma**. *Let χ be a prime filter, and let α_0 and β_0 be filters. Suppose $R_\circ\chi\alpha_0\beta_0$, and suppose that $b\notin\beta_0$. Then there exists a prime filter β such that $\beta_0\subseteq\beta$ (and so $R_\circ\chi\alpha_0\beta$) with $b\notin\beta$, and similarly, when $R_\circ\alpha_0\chi\beta_0$.*

Proofs of these may in effect be found in *Dunn 1986*. The first of these derives from Belnap. The second derives from *Routley and Meyer 1972*, Lemma 4, where it is called the "Primeness Lemma." The name "Squeeze Lemma" is due to Routley, I believe. The third is an easy consequence of the Extension Lemma, letting $F=\beta$ and $I=(b]=\{x : x\le b\}$. The proofs presented in the referenced literature deal with theories (sets of sentences) rather than with filters, but it is easy to translate between the two.

We are now in a position to prove the following.

THEOREM 5.4. *Every Peirce groupoid is isomorphic to a framed Peirce groupoid on an a-frame.*

Proof. For the representation, we take the canonical a-frame $(U,R,\sqsubseteq)$ (whose points are prime filters) and define the *canonical isomorphism*,

$$h(x)=\{\alpha\in U : x\in\alpha\}. \tag{71}$$

[19] Cf. *Dunn 1993a*.

[20] Basic terminology from lattice theory may be found in *Birkhoff 1967*.

It is well-known from Stone (*1937*) that this is a 1-1 map preserving $\wedge$ and $\vee$, carrying them into intersection and union. It is also known, as a special case, from Jónsson and Tarski (*1951-52*) that h preserves $\circ$, and Dunn (*1991*) has provided a generalization of the their work which contains as a special case that h preserves $\rightarrow$ and $\leftarrow$ as well.

But it is difficult to "see the trees for the forest" in this earlier work, and so we prove

$$h(a \circ b) = h(a) \odot h(b), \tag{72}$$

$$h(a \rightarrow b) = h(a) \Rightarrow h(b), \tag{73}$$

$$h(a \leftarrow b) = h(a) \Leftarrow h(b). \tag{74}$$

Thus we show that $a \circ b \in \chi$ *iff* $\exists \alpha, \beta\ R_\circ \alpha\beta\chi \& a \in \alpha \& b \in \beta$. Right-to-left is immediate. For left-to-right, let us suppose that $a \circ b \in \chi$. The principal filter determined by x, $[x) = \{y : x \leq y\}$. Let $\alpha_0 = [a)$, $\beta_0 = [b)$. Because $\circ$ is isotone, $R_\circ \alpha_0 \beta_0 \chi$. Applying the Squeeze Lemma twice gives us prime filters α, β such that $R\alpha\beta\chi$.

We next show $a \rightarrow b \in \chi$ *iff* $\forall \alpha, \beta, R_\rightarrow \alpha\chi\beta \,\&\, a \in \alpha$ *only if* $b \in \beta$. Again, half is immediate, this time from left-to-right. We prove the other half by proving the contrapositive: $a \rightarrow b \notin \chi$ *only if* $\exists \alpha, \beta, R_\circ \alpha\chi\beta \,\&\, a \in \alpha \,\&\, b \notin \beta$. Let us assume then that $a \rightarrow b \notin \chi$. Set $\alpha_0 = [a)$, $\beta_0 = \{b' : \exists x \in \chi, a \circ x \leq b'\}$. It is easy to see that this is the filter generated from $[a) \circ \chi$. It is clear that $R_\circ \chi \alpha_0 \beta_0$; for if $x \in \chi$ and $a \leq a'$, then $x \circ a \leq x \circ a'$. Further it is clear that $b \notin \beta$; for otherwise (residuation) for some $x \in \chi$, $x \leq a \rightarrow b$, so $a \rightarrow b \in \chi$, contrary to our assumption. We now apply the Exclusion Lemma and then the Squeeze Lemma to obtain first $\beta \supseteq \beta_0$ and then $\alpha \supseteq \alpha_0$, so that $R_\circ \alpha\chi\beta$. The argument that h preserves $\leftarrow$ is similar. ⊣

COROLLARY 5.5. *Every Peirce semi-group is isomorphic to a framed Peirce semi-group on an associative a-frame.*

Proof. We need only verify that the canonical relation $R_\circ$ is associative, i.e.,

$$\exists \chi (R_\circ \alpha\beta\chi \;\&\; R_\circ \chi\gamma\delta) \text{ iff } \exists \theta (R_\circ \alpha\theta\delta \;\&\; R_\circ \beta\gamma\theta). \tag{75}$$

We show only the direction from left-to-right, the other direction being analogous. Suppose $R_\circ \alpha\beta\chi \,\&\, R_\circ \chi\gamma\delta$. Set

$$\theta_0 = \{x : \exists b \in \beta, d \in \gamma, b \circ d \leq x\}. \tag{76}$$

This is a filter, and it is almost immediate that $R_\circ \beta\gamma\theta_0$. We must show that $R_\circ \alpha\theta_0\delta$. Suppose $a \in \alpha$, and $x \in \theta_0$. Then $\exists b \in \beta$, $c \in \gamma$, $b \circ c \leq x$. By virtue of $R_\circ \alpha\beta\chi$, $a \circ b \in \chi$, and using $R_\circ \chi\gamma\delta$, we obtain that $(a \circ b) \circ c \in \gamma$, i.e, by associativity, $a \circ (b \circ c) \in \gamma$. But since $\circ$ is isotone, we have $a \circ (b \circ c) \leq a \circ x$, and so $a \circ x \in \gamma$ as needed.

We can now apply the Squeeze Lemma to obtain a prime filter θ with $R_\circ \alpha\theta\delta$. Moreover, since the Squeeze Lemma assures us that $\theta \supseteq \theta_0$, $R_\circ \beta\gamma\theta$ as well, and we are through. ⊣

We next go about the business of accommodating an identity element. We actually analyze the proof for requirements weaker than a full left and right identity element, but we state the theorem only for this strongest case.

THEOREM 5.6. *Every Peirce monoid is isomorphic to a framed Peirce monoid on an associative assertional a-frame.*

Proof. Canonically, we define $Z = \{\zeta : e \in \zeta\}$. We show that if $\exists\zeta \in Z$, $R\zeta\alpha\beta$, then $\alpha \subseteq \beta$. Suppose $\exists\zeta \in Z$, $R\zeta\alpha\beta$ and $a \in \alpha$. Then since $e \in \zeta$, $e \circ a = a \in \beta$.

If $\alpha \subseteq \beta$, then clearly $R_\circ[e)\alpha\beta$; for if $e \leq x$ and $a \in \alpha$, then $a = e \circ a \leq x \circ a$ and so $x \circ a \in \beta$. Then by the Squeeze Lemma, $\exists\zeta \in Z$, $R_\circ\zeta\alpha\beta$.

A symmetrical argument, using $a \circ e = a$, shows that $\exists\zeta \in Z$, $R_\circ\alpha\zeta\beta$ *iff* $\alpha \subseteq \beta$. ⊣

Let us now address the question of what happens if a Peirce monoid has a unary operation $^{-1}$ on it corresponding to converse.

THEOREM 5.7. *Every pre relation algebra is isomorphic to a framed pre relation algebra on an associative assertional a-frame on which is defined an involution satisfying tagging.*

Proof. Given a prime filter α, define:

$$\breve{\alpha} = \{a : a^{-1} \in \alpha\}. \tag{77}$$

Clearly $^\cup$ is of period two, since $^{-1}$ is of period two. $Z = \{\zeta : e \in \zeta\}$. It is easy to see then that Z is closed under $^\cup$, for if $e \in \zeta$, then $e^{-1} = e \in \zeta$.

All that remains is to demonstrate that tagging holds in the canonical a-frame. Let us suppose that $R_\circ\alpha\beta\gamma$. In order to show $R_\circ\breve{\beta}\breve{\alpha}\breve{\gamma}$, assume $x \in \breve{\beta}$ and $y \in \breve{\alpha}$, i.e., $x^{-1} \in \beta$, $y^{-1} \in \alpha$. Since $R_\circ\alpha\beta\gamma$, $y^{-1} \circ x^{-1} = (x \circ y)^{-1} \in \gamma$, i.e., $x \circ y \in \breve{\gamma}$ as needed. ⊣

We next show:

THEOREM 5.8. *Every proto De Morgan monoid is isomorphic to a framed proto De Morgan monoid on an associative assertional a-frame on which is defined an involution satisfying antilogism.*

Proof. The only trick is how to define the involution given that we no longer have an operation corresponding to inverse. The idea comes right out of the canonical definition of the Routley-Meyer (*1973*) "star operator",[21] and for reasons of history as well as for making a contrast with the definition (77), we shall use the star notation in this proof (again reminding the reader that it has nothing to do with transitive closure):

$$\alpha^* = \{a : {\sim}a \notin \alpha\}. \tag{78}$$

[21] It comes even earlier from A. Białynicki-Birula and H. Rasiowa (*1957*).

It is straightforward that the operation so-defined is of period two, since $\sim$ is. We must show antilogism holds on the canonical frame. Let us suppose that $R_{\circ}\alpha\beta\gamma$, and show that $R_{\circ}\gamma^{*}\alpha\beta^{*}$. To this end, let us suppose that $c' \in \gamma^{*}, a \in \alpha$. We need that $c' \circ a \in \beta^{*}$, i.e., $\sim(c' \circ a) \notin \beta$. Let us suppose the opposite: $\sim(c' \circ a) \in \beta$. Since $\sim(c' \circ a) = a \to \sim c'$, we have $a \to \sim c' \in \beta$. Since $R_{\circ} = R_{\to}$, we have $\sim c' \in \gamma$. But this contradicts $c' \in \gamma^{*}$. The other half of antilogism follows analogously. ⊣

We next show how to represent Meyer monoids.

THEOREM 5.9. *Every Meyer monoid is isomorphic to a framed Meyer monoid on an associative assertional classical frame satisfying antilogism.*

Proof. The two new items we must address are classicality and antilogism. The first of these comes from the fact that a relation algebra is Boolean, and in the canonical representation, as in *Stone 1936,* the points become maximal filters, since in a Boolean algebra (proper) prime filters and maximal filters coincide. Since in the canonical representation of converse algebras, $\sqsubseteq$ is just $\subseteq$, clearly the relation must reduce to identity since one maximal filter cannot extend another.

We can now also show that the canonical frame satisfies antilogism. We shall depend heavily upon the "presence of absence" which Boolean complement gives us, allowing us to interchange $y \notin \chi$ with $-y \in \chi$. The point is that it is well-known that in a Boolean algebra a prime (proper) filter χ (being maximal) must contain exactly one of an element y or its Boolean complement. Thus assume $R_{\circ}\alpha\beta\gamma$. We must show $R_{\circ}\breve{\gamma}\alpha\breve{\beta}$. Let us assume that $x \in \breve{\gamma}$, $a \in \alpha$. Then $x^{-1} \in \gamma$. In order to show $x \circ a \in \breve{\beta}$, we must show $(x \circ a)^{-1} \in \beta$, which we do by assuming the opposite in order to derive a contradiction. From $(x \circ a)^{-1} \notin \beta$, it follows ("presence of absence") that $-[(x \circ a)^{-1}] \in \beta$, since

$$-[(x \circ a)^{-1}] = -[a^{-1} \circ x^{-1}] = -[a^{-1} \circ - - (x^{-1})] = a \to -(x^{-1}). \quad (79)$$

So $a \in \alpha$, $a \to -(x^{-1}) \in \beta$. But since $R_{\circ} = R_{\to}$, we have $R_{\to}\alpha\beta\gamma$, and so we infer that $-(x^{-1}) \in \gamma$, i.e., ("presence of absence") $x^{-1} \notin \gamma$. But this contradicts our assumption that $x \in \breve{\gamma}$, i.e., $x^{-1} \in \gamma$.

We must also show the other form of antilogism, but this follow similarly using $\leftarrow$. ⊣

REMARK. An enlightening way to see the proof above is that we had already previously shown in the proof of Theorem 5.9 that the canonical frame for a proto De Morgan monoid satisfies antilogism. Since a Meyer monoid is a proto De Morgan monoid, we might expect this to immediately transfer over to the canonical frame for a Meyer monoid. But we are momentarily stopped by the fact that we had done this for α^{*}, defined by (78), rather than $\breve{\alpha}$, defined by (77). But for a Meyer monoid, $\alpha^{*} = \breve{\alpha}$. Because of the " presence of absence," $\alpha^{*} = \{a : \sim a \notin \alpha\}$ can be exchanged with $\breve{\alpha} = \{a : a^{-1} \in \alpha\}$.

Let us now pull out all the stops and deal with the full set of postulates for relation algebras.

THEOREM 5.10. *Every relation algebra is isomorphic to a framed relation algebra on an associative assertional frame on which is defined an involution satisfying tagging and antilogism.*

Proof. This can be constructed by combining the proofs above representing pre-relation algebras and Meyer monoids. The chief points are that we can now show that the canonical frame satisfies both tagging (as we did for a pre relation algebra) and antilogism (as we did for a Meyer monoid), since a relation algebra is both. ⊣

6. PHILOSOPHICAL INTERPRETATION OF THE REPRESENTATION

We now discuss a possible interpretation of the above representation for relational algebras. At first blush the representation in terms of ternary frames must be admitted to be more abstract than one would want. An ideal representation of relation algebras would send the elements into relations, but Lyndon (*1950*) has shown that such a representation is impossible. It turns out that we are in one sense close to such an ideal representation, but we are off a type level. Elements are **not** carried into *relations*, but into *sets of relations.*[22] In the present section we shall explain the philosophical intuitions behind this statement, and in the next section we shall make proper mathematics of it and also relate it to the computer science notion of a "relational database".

Peter Woodruff suggested early on that the ternary relation $R\alpha\beta\gamma$ that arises in the Routley-Meyer semantics for relevance logic should be viewed as an "indexed" binary accessibility relation $R_{\beta}\alpha\gamma$.[23] An equivalent idea, with slightly different metaphysical overtones, is to view β as itself a binary relation. More accurately, we view β as something like the Fregean "object-correlate" of a binary relation. By thus not literally identifying β with a set of ordered pairs, but rather thinking of it as determining a set of ordered pairs, we avoid unnecessary set-theoretical problems about what happens when one has an otherwise "untyped" relation, as with e.g., $R\beta\beta\beta$. We thus think of the ternary relation $R\alpha\beta\gamma$ as saying that the (binary) relation β relates (is exemplified by) the pair of terms α and γ. This has definite Bradleian overtones where the relations themselves are objects needing to be related to their terms,[24] but we stop the Bradleian regress (as did G. Bergmann,

[22] It is important to notice as well that this difference of type-level distinguishes our approach from other so-called "relational semantics" such as that of, say, Orlowska (*1992*).

[23] Actually Woodruff thought of the first position α as the index, but for reasons of notational convenience we take the second position β. The paper of *Dunn and Meyer 1997* takes the first position.

[24] Cf. *Bradley 1897.*

1967[25]) at the second level, not requiring that there be a yet higher-level relation relating the "exemplification relation" R to its terms.

There is another, more "technological" interpretation of essentially the same ideas. Barwise (*1993*) distinguishes between "sites" and "channels"; roughly, the first being situations between which information is to be communicated and the second being what enables the communication. Barwise uses the notation $s_1 \overset{c}{\mapsto} s_2$ to indicate the 3-placed relation that exists when the two sites s_1 and s_2 are connected by the channel c. In the notation of Routley and Meyer this could be expressed as Rs_1cs_2. Channels hence determine binary relations between sites, but they need not be identified with those relations. Barwise leaves explicitly open the idea that channels might be sites and *vice versa*, which is of course the situation with the frames we have been examining here.[26]

For mnemonic reasons we shall often use ρ and σ for the second (relational) term of the ternary relation, and sometimes suppress R, e.g., writing $\alpha\rho\beta$ for $R\alpha\rho\beta$.

Before going on it is good to explicitly introduce modifications of the notational conventions of Routley and Meyer introduced above in (43):

$$R(\alpha\beta)\gamma\delta =_{def} \exists\chi(R\alpha\beta\chi \;\&\; R\chi\gamma\delta), \tag{80}$$

$$R\alpha(\beta\gamma)\delta =_{def} \exists\chi(R\alpha\chi\delta \;\&\; R\beta\gamma\chi). \tag{81}$$

Routley and Meyer actually write "R^2" where we write simply "R" above, but for reasons of both suggestiveness and simplicity we shall often simply let the number of terms affixed tell that there is a power.

One of the requirements imposed by Routley-Meyer is the requirement (see (43)) that these two compositions are equivalent:

$$R(\alpha\beta)\gamma\delta \textit{ iff } R\alpha(\beta\gamma)\delta \quad (R\textit{-associativity}). \tag{82}$$

Let us define the composition of two "binary relations" ρ and σ:

DEFINITION 6.1. $\alpha(R_\rho \otimes R_\sigma)\beta$ *iff* $\exists\chi(R\alpha\rho\chi \;\&\; R\chi\sigma\beta)$.

Clearly using this definition and *R-associativity*, we obtain

$$\alpha(R_\rho \otimes R_\sigma)\beta \textit{ iff } R(\alpha\rho)\sigma\beta \textit{ iff } R\alpha(\rho\sigma)\beta, \tag{83}$$

where the right-hand sides of the last two conditions are understood using the compositional notations of Routley-Meyer. One could write (83) as

[25]Our R serves the same role as his "nexus".

[26]I was working on the present representation of relation algebras roughly during the same period that Jon Barwise was working on his "sites and channels". Publication of my paper had been delayed (see footnote *). My work was not motivated by his, but rather from the Routley-Meyer semantics for relevance logic and from my "gaggle theory", although Jon Barwise and I have been discussing the connection between his ternary relation and the ternary relation of Routley-Meyer. More recently Greg Restall has noticed this relationship and has been working out some of its details.

$$(R_\rho \otimes R_\sigma)\beta \text{ iff } R_{(\rho\sigma)}\alpha\beta, \tag{84}$$

but one should be clear that "$(\rho\sigma)$" is a mere notation and does not denote an actual point.

The above expresses what one might label a "notional homomorphism". But it can be made into an actual homomorphism if we "refine" the Routley-Meyer relation into smaller bits, in effect interpreting $R\alpha\beta\gamma$ as $\alpha \bullet \beta \sqsubseteq \gamma$. Computer scientists might think of this as a "curried" version.

This refinement is implicit in the definition of the canonical frame in the completeness theorems of Routley-Meyer and explicit in K. Fine's independent semantics.[27] Our presentation is strongly influenced by Fine, but we adapt his ideas to make them fit the present context.

DEFINITION 6.2. A *refined relation algebra frame* is a structure $(U, M, \circ, \sqsubseteq, {}^\cup, Z)$, with $M, Z \subseteq U, \bullet : U^2 \to U$, $\sqsubseteq$ a partial order on U, ${}^\cup : U \to U$, all subject to the following conditions:

1. $\circ$ is required to be associative and isotonic with respect to the partial order,
2. $\forall\alpha, \beta \in U : \exists\zeta \in Z\ (\zeta \bullet \alpha \sqsubseteq \beta)$ *iff* $\alpha \sqsubseteq \beta$ *iff* $\exists\zeta \in Z\ (\alpha \bullet \zeta \sqsubseteq \beta)$ (*"ordered"*),
3. $\forall\alpha, \beta, \gamma \in U : \alpha \bullet \beta \sqsubseteq \gamma$ *only if* $\breve{\beta} \bullet \breve{\alpha} \sqsubseteq \breve{\gamma}$ (*"tagging"*),
4. $\forall\alpha, \beta, \gamma \in M : \alpha \bullet \beta \sqsubseteq \gamma$ *iff* $\breve{\gamma} \bullet \alpha \sqsubseteq \breve{\beta}$ (*"antilogism"*),
5. $\forall\zeta \in Z : \breve{\zeta} \in Z$ (*"converse closure"*),
6. $\forall\alpha \in M, \beta \in U : \alpha \sqsubseteq \beta$ *only if* $\alpha = \beta$ (*"maximality"*).

We put quotes around the labels because they do not literally label the same conditions we earlier imposed on Routley-Meyer frames, but they are close enough, given the reading of $\alpha \bullet \beta \sqsubseteq \gamma$ as $R\alpha\beta\gamma$.

In canonical refined frames, U is the set of all (proper) filters, and M is the set of (proper) prime filters, and in relation algebras these last are of course just the maximal filters. This is the meaning of (6). For two filters α and $\beta \in U$, we interpret $\alpha \bullet \beta$ as the filter generated by the set $\{a \circ b : a \in \alpha$ and $b \in \beta\}$. Immediately there is a wrinkle, because there is no reason that $\alpha \bullet \beta$ should be a prime filter, even when α and β are both prime filters. This is the reason that Routley and Meyer went to a ternary relation, $R_\circ\alpha\beta\gamma$, because γ can be a prime filter extending $\alpha \bullet \beta$. So in a canonical refined frame, U must be taken as the set of all filters, and then M is interpreted as the set of prime filters. Z and ${}^\cup$ are defined as before, with Z being the set of

[27] A. Urquhart should also be mentioned. For further history and details, cf. *Dunn 1986*, and *Anderson, Belnap, and Dunn et al. 1992*.

filters containing e, and $\breve{\alpha}$ being defined as in (77).[28] The proofs of (1), (2), (3), and (5) are basically the same as for Routley-Meyer frames. Regarding (4), the reader can similarly check the proof of Theorem 5.9 and see that the "trick" used depends upon α, β, γ being maximal.

Given a refined relation algebra frame, we can define relation algebras on it by modifying the definitions (39), (40), (41) by simply substituting $\alpha \bullet \beta \sqsubseteq \gamma$ for $R\alpha\beta\gamma$. We leave to the interested reader the task of redoing our representation of relation algebras (and the other related algebras we introduced in section 2) in terms of refined frames.

We do however want to state and prove the following propositions, which are somewhat pointless in terms of the Routley-Meyer style classical frames which we used in our first representation of relation algebras (for the simple reason that $\rho \sqsubseteq \sigma$ never happens except in the degenerate case when $\rho = \sigma$).

PROPOSITION 6.3. *For a refined relation algebra frame, for $\rho, \sigma \in U$:*

$$\rho \sqsubseteq \sigma \textit{ iff } \breve{\rho} \sqsubseteq \breve{\sigma}. \tag{85}$$

Proof. Suppose that $\rho \sqsubseteq \sigma$. Then by assertion $\exists\zeta \in Z$ such that $\rho \bullet \zeta \sqsubseteq \sigma$. But then by tagging, $\breve{\zeta} \bullet \breve{\rho} \sqsubseteq \breve{\sigma}$,and $\breve{\zeta} \in Z$ since Z is closed under $^{\cup}$. So by assertion, $\breve{\rho} \sqsubseteq \breve{\sigma}$. The converse follows since $^{\cup}$ is of period two. ⊣

REMARK. The reader can verify that in the definition of a refined frame "tagging" can be replaced equivalently by (86) and $(\alpha \circ \beta)^{\cup} = \breve{\beta} \circ \breve{\alpha}$.

PROPOSITION 6.4. *For a refined relation algebra frame, for $\rho, \sigma \in U$:*

$$\rho \sqsubseteq \sigma \textit{ iff } R_\rho \supseteq R_\sigma. \tag{86}$$

Proof. Assume that $\rho \sqsubseteq \sigma$ and that $(\chi_1, \chi_2) \in R_\sigma$, i.e., $R\chi_1\sigma\chi_2$. It follows using (38) that $R\chi_1\rho\chi_2$, i.e., $(\chi_1, \chi_2) \in R_\rho$. For the converse assume that $R_\rho \supseteq R_\sigma$. By weak reflexivity (50), we have $\exists\zeta \in Z, R\zeta\sigma\sigma$. But then $R\zeta\rho\sigma$, and so by (48), $\rho \sqsubseteq \sigma$. ⊣

REMARK. Note that the direction of inclusion on the right-hand side of (86) is the reverse of what the reader might have expected. The necessity of this is clear in the proof. But we can define an "information" order between binary relations R and S : $R \sqsubseteq S$ iff $S \subseteq R$. This again reverses the direction, so the right-hand side ends up $R_\rho \sqsubseteq R_\sigma$ as desired.[29] Upon reflection, this is more than mere notation since one relation is "stronger" (carries more information than another) just when the first is included in the other. Mother-of carries more information than does parent-of.

[28]This definition is the reason that we can have $^{\cup}$ defined upon all of U, and not just on M as does Fine. The point is that our definition (78) using a^{-1} carries filters to filters, whereas the definition (78) using $\sim$, which Fine uses, does not (although it does carry *prime* filters to *prime* filters). But our definition is available only for those algebras that have a^{-1} as an operator (here primitive for relation algebras, or defined as $-(a^{-1})$ as in a Meyer monoid).

[29]This same phenomenon occurs in the ternary semantics for the combinatory logic of Dunn and Meyer (*1997*) and is discussed there.

THEOREM 6.5. *For a refined relation algebra frame,* R *(defined as* $\alpha \bullet \beta \sqsubseteq \gamma$*) carries the composition of relations to the composition of their correlates, i.e.,*

$$(R_\rho \otimes R_\sigma) = R_{(\rho \bullet \sigma)}, \text{ i.e.,} \tag{87}$$

$$\alpha(R_\rho \otimes R_\sigma)\beta \textit{ iff } \alpha R_{(\rho \bullet \sigma)}\beta. \tag{88}$$

Proof. Given the Routley-Meyer compositional notation described above, $\alpha(R_\rho \otimes R_\sigma)\beta$ can be expressed as:

$$R(\alpha\rho)\sigma\beta. \tag{89}$$

What we are trying to establish is that

$$R(\alpha\rho)\sigma\beta \textit{ iff } R\alpha(\rho \bullet \sigma)\beta \tag{90}$$

or, using our convention of "R suppression", simply as

$$(\alpha\rho)\sigma\beta \textit{ iff } \alpha(\rho\sigma)\beta. \tag{91}$$

In other bastard notation, this might be expressed as

$$(\alpha\rho)\sigma\beta = \alpha(\rho\sigma)\beta. \tag{92}$$

Putting it this way suggests that associativity of $\bullet$ will figure in the formal proof, and this is indeed the case, once we curry R, obtaining

$$\exists\chi(\alpha \bullet \rho \sqsubseteq \chi \,\&\, \chi \bullet \sigma \sqsubseteq \beta) \textit{ iff } \alpha \bullet (\rho \bullet \sigma) \sqsubseteq \chi. \tag{93}$$

Right-to-left is easy. Set $\chi = \alpha \bullet \rho$, so $\alpha \bullet \rho \sqsubseteq \chi$ follows from reflexivity, and then use associativity to obtain $\chi \bullet \sigma \sqsubseteq \beta$. Left-to-right is barely harder. Let us assume

$$\alpha \bullet \rho \sqsubseteq \chi \,\&\, \chi \bullet \sigma \sqsubseteq \beta. \tag{94}$$

Then by isotonicity,

$$(\alpha \bullet \rho) \bullet \sigma \sqsubseteq \beta, \tag{95}$$

and so by associativity we obtain,

$$\alpha \bullet (\rho \bullet \sigma) \sqsubseteq \beta, \tag{96}$$

as desired. ⊣

Given the involution $^\cup$ with the postulates we required of it, we can now show the following.

THEOREM 6.6. *R carries the converse of a relation to the involute of its correlate, i.e.,*

$$(R_\rho)^{-1} = R_{(\breve{\rho})}. \tag{97}$$

Proof. This is similar to that for (66) above but with respect to the second position instead of the first. Thus using antilogism, tagging and period two, we obtain: $\alpha(R_\rho)^{-1}\beta$ *iff* $\beta R_\rho \alpha$ *iff* $R\beta\rho\alpha$ *iff* $R\rho\breve{\alpha}\breve{\beta}$ *iff* $R\alpha\breve{\rho}\beta$ *iff* $\alpha R_{(\breve{\rho})}\beta$. ⊣

We have demonstrated that the function which interprets ρ as the relation R_ρ is a homomorphism from a refined frame (with $\bullet, \sqsubseteq, ^\cup$), carrying any point ρ of the frame to a relation R_ρ (a set of ordered pairs), $\bullet$ to relative product, and $^\cup$ to converse. This allows us to think of the points of a frame as "concept-correlates" of relations. Note that if concept-correlates were intensional, there might be two different concept-correlates that determine the same relation. But it would be nice to be able to treat them extensionally so that two concept correlates are identical if (and of course only if) they determine the same set of ordered pairs.

In the Routley-Meyer semantics there is a favored point 0. They introduce the definition $\alpha \sqsubseteq \beta$ *iff* $R0\alpha\beta$ and show from various postulates that $\sqsubseteq$ has the properties of a pre-order (reflexive, transitivity). In our representation of relation algebras we were not able to utilize a single such favored point, but instead, introduced a set Z and the requirement (51) that the frame be assertional. Let us break down (51) into two parts:

$$\exists\zeta \in Z(R\zeta\alpha\beta) \textit{ iff } \alpha = \beta, \tag{98}$$

$$\exists\zeta \in Z(R\alpha\zeta\beta) \textit{ iff } \alpha = \beta. \tag{99}$$

The meaning of (99) is reasonably clear: each R_ζ is an identity relation on some subset of U, and taken together (take their union) they constitute the identity relation on all of U. In symbols:

THEOREM 6.7.

$$\bigcup_{\zeta\in Z} R_\zeta = I_U. \tag{100}$$

But how are we to interpret (98)? In all honesty, its sense is less than straightforward. But let us try to think of it as follows.

Let us suppose that we could add some new points ζ as "identity indicators". The job of the points ζ is to tell us when two relations α and β are identical. We might then postulate (98) as above and, in effect, extend the meaning of R (and indirectly that of α, thought of as a relation-correlate). The puzzle is: how can the set Z, which by (99) contains the concept-correlates of the various identity relations, also at the same time play this other role of as the set of "identity indicators"? But (98) says it does, so let us just accept it and follow the consequences.

Using (98) we can now show:

THEOREM 6.8 **Extensionality**. *Two relation correlates are identical just when they determine the same sets of ordered pairs, i.e.,*

$$\rho = \sigma \textit{ iff } R_\rho = R_\sigma. \tag{101}$$

Proof. Left-to-right is just logic (substitution of identicals). For right-to-left we assume $R_\rho = R_\sigma$, i.e., $\forall\alpha, \beta(R\alpha\rho\beta \textit{ iff } R\alpha\sigma\beta)$. Now for some $\zeta \in Z$,

$R\zeta\rho\rho$ holds, since $\rho = \rho$. Instantiating both α and β to that ζ we obtain $R\zeta\rho\rho$ *iff* $R\zeta\sigma\rho$, and hence $R\zeta\sigma\rho$. But then by (98), we obtain $\sigma = \rho$. ⊣

We conclude this section by summarizing the above observations in the following, appropriately using the famous lambda notation of Church.

THEOREM 6.9. *The function* $\lambda\rho(R_\rho)$ *is an isomorphism, carrying* • *into relative product,* $^\cup$ *into converse, and* Z *into a set of partial identity relations whose union is the identity relation on* U. *In the refined semantics* • *is an actual operation, whereas in the Routley-Meyer version it is notional.*

There is another way to make sense of (51) that is in fact more useful in interpreting the frame semantics in terms of relations. We keep the idea that each state ρ can be traded off for a binary relation $R_\rho = \{(\alpha, \beta) : R\alpha\rho\beta\}$, so our "concrete" frames will have relations as their points. But to say what the points are is not yet to say how the ternary accessibility relation is to be interpreted. We shall give up the "untyped" interpretation of the ternary accessibility relation $R\alpha\rho\beta$ as "ρ relates α to β", and adopt instead the more mundane "typed" interpretation: "the composition of R_α and R_ρ is included in R_β". Then (99) and (98) make equal sense with Z as a set of partial identity relations. We shall have more to say about this interpretation of the accessibility relation in the next section.

7. REPRESENTATION RESULTS USING RELATIONS

The intuitive idea of a representation is that we take some structures viewed abstractly, and show how they are each isomorphic to structures which are (relatively) concrete.[30] A paradigm of this is Stone's (*1936*) showing how a Boolean algebra can be regarded as a collection of subsets of some given set, with the Boolean operations interpreted as intersection, union, and relative complement. Compared to this paradigm, the representation results of section 5 must be acknowledged as disappointing. We have represented relation algebras using frames, which themselves are about as abstract. This is much worse than what happens in, say, intuitionistic logic, where one can give a representation of Heyting algebras in terms of binary frames. There at least one can give some meaning to the frames in terms of information states (Kripke talked of "evidential states") and the information order ("evidential states being possible relative to one another"). But "at ten yards," it is not so clear what our ternary frames have to do with any intuitive interpretation of relation algebras.

But this is only at first blush, because the reader knows that we have already made some headway in section 6 regarding a philosophical interpretation of our ternary frames. We make this precise by the following definition:

[30] Concreteness is in the eye of the beholder (and not just for motes). I am reminded of Bob Meyer's joke on this subject. A mathematician says, "Take a concrete example, say an infinite-dimensional Hilbert space."

DEFINITION 7.1. By a *concrete relation algebra frame* we shall mean a structure $(\mathsf{U}, \mathsf{R}, {}^{\cup}, \mathsf{Z})$ where U is a set of relations on some set X, $\mathsf{R}\alpha\beta\gamma$ is defined as $\gamma \subseteq \alpha \circ \beta$, ${}^{\cup}$ is just converse, and $\mathsf{Z} \subseteq \mathsf{U}$ is subject to the condition that $\bigcup \mathsf{Z} = I_X$ (the identity relation on X).

PROPOSITION 7.2. *Every "concrete relation algebra frame" is indeed a relation algebra frame.*

Proof. The needed facts can be easily found in standard sources regarding the algebra of relations[31] and/or easily verified by the reader.

THEOREM 7.3. *Every relation algebra is isomorphic to a framed relation algebra on a concrete relation algebra frame.*

The theorem is a direct consequence of the following:

LEMMA 7.4. *Every relation algebra frame $(U, R, {}^{\cup}, Z)$ is isomorphic to some concrete relation algebra frame.*

Proof. We set $\mathsf{U} = \{R_\rho : \rho \in U\}$ and $\mathsf{Z} = \{R_\zeta : \zeta \in Z\}$. Note that $X = U$. The desired isomorphism should be $\lambda\rho(R_\rho)$, which maps each $\rho \in U$ to $R_\rho \in \mathsf{U}$. We know from (6.9) that this map is 1-1, onto, and preserves ${}^{\cup}$. We must also show that the map preserves the ternary accessibility relation R, i.e.,

$$R\alpha\beta\gamma \textit{ iff } \mathsf{R}R_\alpha R_\beta R_\gamma, \text{ i.e.,} \tag{102}$$

$$R\alpha\beta\gamma \textit{ iff } R_\gamma \subseteq R_\alpha \otimes R_\beta. \tag{103}$$

For left-to-right, assume that $R\alpha\beta\gamma$ and that $(\chi_1, \chi_2) \in R_\gamma$, i.e., $R\chi_1\gamma\chi_2$. By (42) we then have that for some $\delta, R\chi_1\alpha\delta$ and $R\delta\beta\chi_2$, i.e., $(\chi_1, \delta) \in R_\alpha$ and $(\delta, \chi_2) \in R_\beta$, which is just the meaning of $(\chi_1, \chi_2) \in R_\alpha \otimes R_\beta$. For the converse assume that $R_\gamma \subseteq R_\alpha \otimes R_\beta$. By weak reflexivity (49) we have for some $\zeta \in Z$ that $R\zeta\gamma\gamma$, i.e., $(\zeta, \gamma) \in R_\gamma$. But then using the assumed inclusion, we get $(\zeta, \gamma) \in R_\alpha \otimes R_\beta$, i.e., for some δ, $R\zeta\alpha\delta$ and $R\delta\beta\gamma$. The first means that $\alpha = \delta$ and so by substitution of identicals in the second we obtain $R\alpha\beta\gamma$ as needed.

Finally we must show that the union of Z is the identity relation restricted to $X = U$. This is just (100). ⊣

REMARK. In terms of an intuitive interpretation, we might view the elements of U as relations in some relational database. These might be parent-of, sister-of, etc., and it is natural that they should be closed under relative product (so we should also have grandparent-of, aunt-of, etc.), as well as converse (so we would have child-of, grandchild-of, etc.). Because of Proposition (6.4) (and the Remark following it), the elements of M (being maximal in the information order) would be minimal in the inclusion order, i.e., maximally strong relations. Why should we require A to be closed under $\sqsubseteq$? Let us view

[31] Cf. *Tarski and Givant 1985*, and *Maddux 1991*.

A as a "query". If we want information about who is a parent of whom, this query would be at least partially answered by the stronger information about whom is a mother of whom. This account is not entirely satisfying, but there it is.[32]

8. CORRESPONDING SUBSTRUCTURAL LOGICS

Peirce monoids correspond to the (associative) Lambek Calculus permitting empty left hand sides, supplemented with conjunction and disjunction (which distribute over each other). Proto De Morgan monoids arise by the addition of De Morgan negation, and Meyer monoids can be viewed as arising from the addition of Boolean negation in addition to De Morgan negation (cf. Theorems 2.9 and 2.10). We have shown in Theorem 2.13 that relation algebras can be viewed as differing from Meyer monoids only by the addition of (29).

The question occurs as to whether there are natural ways to restrict the relations in a concrete relation algebra frame so as to satisfy laws corresponding to various structural rules of Gentzen, in particular permutation, contraction, and thinning. This would give us the ability to model "substructural relation algebras," or maybe they should be called simply "structural relation algebras," since they arise by adding laws corresponding to the Gentzen structural rules.

The corresponding frame properties are respectively R-commutativity (62), R-reflexivity (63), and

$$\alpha \circ \beta \sqsubseteq \alpha \quad (R\textit{-lower-bound}). \tag{104}$$

(Of course in the absence of commutativity one should also consider its commuted form.)

Let us first address the question of square-increasingness, where the answer is straightforwardly "yes". Corresponding to $R_\rho \bullet R_\rho \sqsubseteq R_\rho$ is the condition that $R_\rho \subseteq R_\rho \otimes R_\rho$, i.e., if $(\alpha, \beta) \in R_\rho$, then $\exists\chi[(\alpha,\chi),(\chi,\beta) \in R_\rho]$. This is a standard property of relations sometimes referred to as "pseudo-denseness." Further, if a frame satisfies square-increasingness, it is easy to see that the corresponding concrete frame will be pseudo-dense. Thus suppose $R\alpha\rho\beta$, i.e.,

$$\alpha \bullet \rho \sqsubseteq \beta. \tag{105}$$

Set $\chi = \alpha \bullet \rho$. Clearly $\alpha \bullet \rho \sqsubseteq \chi$ since $\sqsubseteq$ is reflexive. We now must show that $R\chi\rho\beta$, i.e., $(\alpha \bullet \rho) \bullet \rho = \alpha \bullet (\rho \bullet \rho) \sqsubseteq \beta$. But this follows from our hypothesis (105) using isotonicity.[33]

[32] Another possible interpretation of a concrete relation algebra frame is to treat its points as relations on states, i.e., as sets of transitions, i.e., as possible actions on states. But we shall not pursue this here.

[33] We carried out this reasoning with respect to a refined frame, but it works just as well with an unrefined frame. From $R\rho\rho\rho$ and the hypothesis $R\alpha\rho\beta$ and we obtain $R^2\alpha(\rho\rho)\beta$, i.e., by R–associativity, $R^2(\alpha\rho)\rho\beta$, i.e., $R\alpha\rho\chi$ and $R\chi\rho\beta$ for some χ.

Turning now to commutativity, I do not know of a natural property of relations that makes relative product commutative. (Symmetry does not work.) I will point out that our relations R_ρ corresponding to points on an R-commutative frame do in fact commute with each other using relative product. Thus $(\alpha, \beta) \in R_\rho \otimes R_\sigma$ iff $R\alpha\rho\chi$ and $R\chi\sigma\beta$ iff $R(\alpha\rho)\sigma\beta$ iff $R\alpha(\rho\sigma)\beta$ iff $R\alpha(\sigma\rho)\beta$ iff $R(\alpha\sigma)\rho\beta$ iff $(\alpha, \beta) \in R_\sigma \otimes R_\rho$. So a crude way to assure commutativity is just to make the global requirement that relative product is commutative on the collection of relations taken as the points in the frame. This works, but it is unsatisfying. A similar point seems to apply to the requirement that the collection of relations satisfies R-lower-bound.

It must be pointed out that it seems unlikely that adding one or more of the above "structural postulates" will give a class of frames which is characteristic for the corresponding substructural logic. It will rather be characteristic for the "relation algebra" version. The reason is that relations always satisfy

$$T \subseteq R \otimes S \text{ only if } \breve{T} \subseteq \breve{S} \otimes \breve{R} \quad \textit{(tagging)},$$

and so concrete frames always validate (56). This observation can be made to prove for example that R-commutative concrete frames do not characterize commutative Meyer monoids, or that R-commutative R-reflexive concrete frames do not characterize commutative square-increasing Meyer monoids ("Boolean De Morgan monoids").

Unfortunately this does not decisively solve the problem for the corresponding substructural logics (distributive linear logic without exponentials, **R**), since substructural logics are not usually formulated with Boolean negation and converse (Meyer actually used "star") as primitive. It was R. K. Meyer's insight that the relevance logic **R** can be so formulated with Boolean negation and star as conservative additions, and that then the usual De Morgan negation of **R** can be viewed as arising by way of the definition $\sim\varphi = -(\varphi^{-1})$).

But (56) $[(A \circ B)^{-1} = B^{-1} \circ A^{-1}]$ does not correspond to any formula in the primitive vocabulary of **R**, nor do I know of any such formula that it implies which is not also a theorem of **R**. So we are left with a nagging question.

9. FRAME MODELS OF THE λ-CALCULUS

Reflection on the above interpretation of the representation of relation algebras gives a perhaps helpful way of looking at relational frames as intuitive models of combinatory logic and hence, by the well-known translations between the two, as intuitive models of the λ-calculus. This section will be programmatic in character and not focus on the details of development, which can be found in a recent paper I have written with R. K. Meyer.[34]

[34]The paper referenced in *Dunn and Meyer 1997*.

The λ-calculus of Church can be viewed as a theory of functionality. In its original versions, it is untyped. Its terms stood for functions, and given two terms M, N one can always "apply" one to the other to obtain $M(N)$. Also given a term M, one can always form the function $\lambda x M$.

The problem in interpreting the (untyped) λ-calculus has always been how to interpret an expression such as $M(M)$, which treats M as simultaneously standing for both a function and an argument. In the early seventies, Scott (*1976*) gave two different models of the λ-calculus: the limit model D_∞ and the graph model $P\omega$. The last of these is most relevant to our present concerns, and uses the well-known correspondence of Cantor between a natural number and a pair of natural numbers. Thus a set X of natural numbers can be simultaneously regarded as a relation between natural numbers, and so we can make sense of $X(Y)$ as the X-image of Y. R. K. Meyer has a similar model, constructed by taking a set U, and closing it under pairing so as to obtain the set $U^* = U \cup U^2 \cup U^3 \ldots$. Then for $X, Y \subseteq U^*$, $X(Y) = \{\chi : \exists\gamma \in Y, <\chi, \gamma> \in X\}$.[35]

The discussion of the previous section showed how to achieve the effect of a similar type-defying interpretation: a proposition A can be simultaneously thought of as both a set of states and a set of relations between states (note well that we are up a type-level from Scott and Meyer). Where ρ is a relation, and A is a set, the ρ-image of $A = \{\chi : \exists\alpha \in A, \alpha\rho\chi)\}$. Going up a type-level, where A is a set of relations, it is natural to define a corresponding image $BA = \{\chi : \exists\rho \in B, \exists\alpha \in A, \alpha\rho\chi\}$ and to think of this as a kind of application. Now regarding ρ as simultaneously a relation and a state, we stick in the ternary relation R of exemplification, rewriting this as:

$$BA = \{\chi : \exists\rho \in B, \exists\alpha \in A, R\alpha\rho\chi\} = A \circ B.$$

It is clear that the application of one proposition to another is just our old friend fusion (but with the quantifiers in reverse order, an awkwardness which is repaired simply by taking the relation to be determined by the first position: R_α). We now naturally interpret $AA = A \circ A$. The key difference between the interpretation of $\circ$ application and as relative product has to do with associativity.

This is about enough for the present occasion, although we should say that rather than discuss the care and feeding of the λ-operator directly, Meyer and I instead exploit the translation of Curry of λ-calculus into combinatory logic, and show how the combinators can be represented as certain sets of states. This representation does not depend upon the particular choice of combinators (as long as they are "proper"), and so the representation can be applied to combinatorial bases appropriate to various substructural logics.

[35]Although Meyer, Bunder, and Powers' (*1991*) "fool's model" was not published until 1991, I remember discussing it in detail with Meyer about 1970. He was influenced in this model by some ideas L. Powers had in connection with the so-called "P-W Problem" from relevance logic.

There is the well-known "Curry-Howard Isomorphism" that establishes a connection between combinators and provable implicational formulas. The standard translation of the λI-calculus into combinatory logic uses the combinators $\mathbf{I}, \mathbf{B}, \mathbf{C}$, and $\mathbf{S}$, and these are well-known to correspond to the implicational axioms of the relevance logic **R** by way of the Curry-Howard Isomorphism. R. K. Meyer has said that it cannot be a mere accident that Church was the creator of both the λI-calculus and the Weak Calculus of Implication (the implicational fragment of which agrees with the relevance logic **R**). I agree.

REFERENCES

Allwein, G., and J. M. Dunn

1993 A Kripke semantics for linear logic, *The Journal of Symbolic Logic*, vol. 58, pp. 514-545.

Anderson, A. R., and N. D. Belnap *et al.*

1975 *Entailment: The logic of relevance and necessity, vol. I*, Princeton University Press, Princeton, New Jersey.

Anderson, A. R, N. D. Belnap, and J. M. Dunn *et al.*

1992 *Entailment: The logic of relevance and necessity, vol. II*, Princeton University Press, Princeton, New Jersey.

Barwise, J.

1993 Constraints, channels, and the flow of information, *Situation theory and its applications* (S. Peters and DIsrael, editors), vol. 3, CSLI Lecture Notes, University of Chicago Press, Chicago, pp. 3–27.

Bergmann, G.

1967 *Realism*, University of Milwaukee Press.

Białynicki-Birula, A., and H. Rasiowa

1957 On the representation of quasi-Boolean algebras, *Bulletin de l' Académie Polonaise des Sciences*, vol. 5, pp. 259–261.

Birkhoff, G.

1967 *Lattice theory*, American Mathematical Society, Providence.

Bradley, F. H.

1897 *Appearance and reality*, Oxford University Press, Oxford.

Brown, C., and D. Gurr

1995 Relations and non-commutative linear logic, *Journal of Pure and Applied Algebra*, vol. 105, pp. 117–136.

Church, A.

1951 The weak theory of implication, *Kontrolliertes Denken, Untersuchgen zum Logikkalkul und zur Logik der Einzelwissenschaften*, (A. Menne, A. Wilhelmy, and H. Angsil, editors), Kommissions-Verlag Karl Alber, Munich, pp. 22–37.

Dŏsen, K.

1992 A brief survey of frames for the peirce calculus, *Zeitschrift für Mathematische Logik und Grundlagen der Mathematik*, vol. 38, pp. 179–187.

Dunn, J. M.

1966 The algebra of intensional logics, *Doctoral Dissertation*, University of Pittsburgh, University Microfilms, Ann Arbor; portions relevant to this paper are reprinted in *Anderson, Belnap et al. 1975* as §8 and §28.2.

1982 A Relational Representation of quasi-Boolean algebras, *Notre Dame Journal of Formal Logic*, vol. 23, pp. 353–357.

1986 Relevance logic and entailment, *Handbook of philosophical logic* (D. Gabbay and F. Guenthner, editors), vol. 3, Reidel, pp. 117–224.

1991 Gaggle theory: An abstraction of Galois connections and residuation with applications to negation and various logical operations, *Logics in AI, Proceedings European workshop JELIA 1990* (J. van Eijck, editor), Lecture Notes in Computer Science 478, Springer-Verlag, Berlin.

1993a Partial-gaggles applied to substructural logics, *Substructural logic* (P. Schröder-Heister and K. Dŏsen, editors), Oxford Press, pp. 63–108.

1993b Perp and star: Two treatments of negation, *Philosophical perspectives: Philosophy of language and logic* (J. Tomberlin, editor), vol. 7, pp. 331-357.

Dunn, J. M., and R. K. Meyer

1997 Combinators and structurally free logic, *Logic Journal of the IGPL*, vol. 5, pp. 505–537.

Girard, J.-Y.

1987 Linear logic, *Theoretical Computer Science*, vol. 50, pp. 1–102.

Gyssens, M., L. V. Saxton, and D. van Gucht

1993 Tagging as an alternative to object creation, *Proceedings of the conference on query processing in object-oriented, complex-object and nested relation databases.*

Jónsson, B., and A. Tarski

1951–52 Boolean algebras with operators, *American Journal of Mathematics*, vols. 73–74, pp. 891–939, 127–162.

Kraegeloh, K.-D., and P. C. Lockemann

1976 Hierarchies of database languages: An example, *Information Systems*, vol. 1, pp. 79–90.

Lambek, J.

1958 The mathematics of sentence structure, *American Mathematical Monthly*, vol. 65, pp. 153–172.

Lyndon, R. C.

1950 The representation of relation algebras, *Annals of Mathematics*, ser. 2, vol. 51, pp. 707–729.

Maddux, R.

1991 The origin of relation algebras in the development of the axiomatization of relations, *Studia Logica*, vol. 91, pp. 421–455.

Meyer, R. K.

1979 A Boolean valued semantics for **R**, Research Paper no. 4, Logic group, Research School of Social Sciences, Australian National University, Canberra.

Meyer, R. K., M. Bunder, and L. Powers
1991 Implementing the 'fool's model' for combinatory logic, *Journal of Automated Reasoning*, vol. 7, pp. 597–630.

Meyer, R. K., and M. A. McRobbie
1982 Multisets and relevant implication, *Australasian Journal of Philosophy*, vol. 60, pp. 265–281.

Meyer, R. K., and R. Routley
1972 Algebraic analysis of entailment, *Logique et Analyse*, n.s., vol. 15, pp. 407–428.
1973 Classical relevant logics (I), *Studia Logica*, vol. 32, pp. 51–66.
1974 Classical relevant logics (II), *Studia Logica*, vol. 33, pp. 183–194.

Orlowska, E.
1992 Relational proof systems for relevant logics, *The Journal of Symbolic Logic*, vol. 57, pp. 1425–1440.

Pratt, V.
1991 Action logic and pure induction, *Logics in AI, Proceedings of the European workshop JELIA 1990* (J. van Eijck, editor), Lecture Notes in Computer Science 478, Springer-Verlag, Berlin.

Routley, R., and R. K. Meyer
1972 The semantics of entailment, II–III, *Journal of Philosophical Logic*, vol. 1, pp. 53–73, 192–208.
1973 The semantics of entailment, *Truth, syntax and modality* (H. Leblanc, editor), North-Holland, Amsterdam, pp. 199–243.

Scott, D. S.
1976 Data types as lattices, *SIAM Journal of Computing*, vol. 5, pp. 522–587.

Schröder-Heister, P., and K. Dŏsen
1993 (editors), *Substructural logic,* Oxford University Press, Oxford.

Stone, M.
1936 The theory of representations for Boolean algebras, *Transactions of the American Mathematical Society*, vol. 40, pp. 37–111.

Stone, M.
1937 Topological representations of distributive lattices and Brouwerian logics, *Časopsis pro Pěstování Matematiky a Fysiky*, vol. 67, pp. 1–25.

Tarski, A.
1941 On the calculus of relations, *The Journal of Symbolic Logic*, vol. 6, pp. 73–89.

Tarski, A., and S. Givant
1985 *A formalization of set theory without variables*, American Mathematical Society Colloquium Publications, vol. 41, Providence.

Van Benthem, J.
1991 *Language in action: Categories, lambdas and dynamic logic*, Studies in Logic and the Foundations of Mathematics, vol. 130, North-Holland, New York.

Ward, M., and R. P. Dilworth
1939 Residuated lattices, *Transactions of the American Mathematical Society*, vol. 45, pp. 335–354.

THOMAS FORSTER

CHURCH'S SET THEORY WITH A UNIVERSAL SET*

For Alonzo Church on the occasion of his ninetieth birthday.

Abstract. A detailed and fairly elementary introduction is given to the techniques used by Church to prove the consistency of his set theory with a universal set by constructing models of it from models of *ZF*. The construction is explained and some general facts about it proved.

1. INTRODUCTION

In 1974 Alonzo Church and Urs Oswald simultaneously and independently lit upon a refinement of Rieger-Bernays permutation models which enabled them to give elementary proofs of consistency for some set theories with a universal set. My own interest began much later, when I made the aquaintance of Flash Sheridan, a former student of Church's who was then writing an Oxford D.Phil. thesis on Church's work in this area. I am greatly endebted to Flash for kindling my interest in this engaging bywater, and for showing me some relevant and (even now) unpublished material.

Church was not generally known for having an interest in set theory, and one might wonder what moved him to write this late and isolated piece. An interest in set theory with a universal set was even less mainstream twenty years ago than it is now, but the consistency question for such theories is something that those whose interest in set theory was purely philosophical might well have mused about at any stage in its history, and I suspect that Church's motive was to make a small polemical point to the effect that there are consistent set theories with a universal set. This point is worth making to students even now. Against this view is that fact that there were already two papers making the same point, and we know that one of them, *Jensen 1967,* was known to Church. (The manuscript of Ward Henson's review of *Jensen 1967* in *The Journal of Symbolic Logic* archives has comments on it in Church's hand.) However it is possible that Church felt that this paper did not make the point convincingly, since the system whose consistency is proved therein is not a pure set theory, but admits *urelemente*. The other, *Grishin 1969,* (which proves the consistency of NF_3, a pure set theory with a universal set and no *urelemente*) was apparently not known to Church.[1]

---

* There is considerable overlap between this paper and the final chapter of my book on set theory with a universal set, *Forster 1995*.

[1] I am endebted to Herb Enderton for historical details about *The Journal of Symbolic*

C. Anthony Anderson and M. Zelëny (eds.), *Logic, Meaning and Computation*, 109-138.

The appropriate place to start is with a high-speed treatment of Rieger-Bernays permutation models.

2. RIEGER-BERNAYS PERMUTATION MODELS

We will start by swiftly recapitulating and updating a treatment of Rieger-Bernays models from *Forster 1995*. If $\langle V, R\rangle$ is a structure for the language of set theory, and π is any (possibly external) permutation of V, then we say $x\ R_\pi\ y$ iff $x\ R\ \pi`y$.[2] $\langle V, R_\pi\rangle$ is a *permutation model* of $\langle V, R\rangle$. We call it $V^\pi$. Alternatively we could define $\Phi^\pi$ as the result of replacing every atomic wff $x \in y$ in $\Phi$ by $x \in \pi`y$. We do not rewrite equations in this operation: $=$ is a logical constant, not a predicate letter. The result of our definitions is that $\langle V, R\rangle \models \Phi^\pi$ iff $\langle V, R_\pi\rangle \models \Phi$.

A wff ϕ is stratified iff we can find a *stratification* for it, namely a map f from its variables (after re-lettering where appropriate) to $\mathbb{N}$ such that if the atomic wff '$x = y$' occurs in ϕ then $f(`x') = f(`y')$, and if '$x \in y$' occurs in ϕ then $f(`y') = f(`x') + 1$. Variables receiving the same integer in a stratification are said to be of the same *type*. Values of f are *types*.

To discuss these topics properly we will also need $j =_{\text{df}} \lambda f \lambda x.(f``x)$. The map j commutes with group-theoretic operations: $j`(\pi\sigma) = (j`\pi)(j`\sigma)$. A (possibly external) permutation $\sigma$ of a set $X$ is *setlike* if, for all $n$, $j^n`\sigma$ is defined and is a permutation of $\mathcal{P}^n`X$ (if this last thing exists). *ZF* proves that every definable permutation is setlike. This is actually a formulation of the axiom scheme of replacement!

We shall start with a lemma and a definition, both due to Henson (*1973*). The definition arises from the need to tidy up Φ^τ. A given occurrence of a variable 'x' which occurs in 'Φ^τ' may be prefixed by 'τ' or not, depending on whether or not that particular occurrence of 'x' is after an '$\in$'. This is messy. If there were a family of rewriting rules around that we could use to replace $x \in \tau`y$ by $\sigma`x \in \gamma`y$ for various other σ and γ, then we might be able to rewrite our atomic subformulæ to such an extent that for each variable, all its occurences have the same prefix.

Why bother? Because once a formula has been coerced into this form, every time we find a quantifier Qy in it, we know that all occurrences of y within its scope have the same prefix. As long as that prefix denotes a permutation then we can simply remove the prefixes! This is because $(Qx)(\ldots\sigma`x\ldots)$ is the same as $(Qx)(\ldots x\ldots)$. If we can do this for all variables, then τ has disappeared completely from our calculations and we have an invariance result. When can we do this?

Henson's insight was as follows. Suppose we have a stratification for Φ and permutations τ_n (for all n used in the stratification) related somehow to

---

Logic.

[2] $\pi`y$ is the value given by the function π to the argument y. $\pi``y$ is the set of values given by π to arguments in y.

τ, so that, for each n,

$$x \in \tau`y \text{ iff } \tau_n`x \in \tau_{n+1}`y.$$

Then by replacing ' $x \in \tau`y$ ' by ' $\tau_n`x \in \tau_{n+1}`y$ ' whenever '$x$' has been assigned the subscript $n$, every occurrence of '$x$' in ' $\Phi^\tau$ ' will have the same prefix. Next we will want to know that $\tau_n$ is a permutation, so that in any wff in which '$x$' occurs bound—$(\forall x)(\ldots \tau_n`x \ldots)$—it can be re-lettered $(\forall x)(\ldots x \ldots)$ so that 'τ' has been eliminated from the bound variables. It is not hard to check that the definition we need to make this work is as follows

DEFINITION 2.1. $\tau_0 =$ identity, $\tau_{n+1} = (j^n`\tau)\tau_n$.

This definition is satisfactory as long as $j^n`\tau$ is always a permutation of V whenever τ is, for each n. For this we need τ to be setlike. The trick of re-lettering variables that this facilitates is of crucial importance and will be used often.

This gives us immediately a proof of the following result.

LEMMA 2.2 (Henson, *1973*). *Let Φ be stratified with free variables 'x_1', ..., 'x_n', where 'x_i' has been assigned an integer k_i in some stratification. Let τ be a setlike permutation and V any model of NF. Then*

$$(\forall \vec{x}) V \models (\Phi(\vec{x})^\tau \longleftrightarrow \Phi(\tau_{k_1}`x_1 \ldots \tau_{k_n}`x_n)).$$

In the case where Φ is closed and stratified, we infer that if τ is a setlike permutation, then

$$V \models \Phi \longleftrightarrow \Phi^\tau.$$

This was proved by Scott in *1962*. We can actually prove something with the flavor of a completeness theorem about these things: a formula is equivalent to a stratified formula iff the class of its models is closed under permutation models using setlike permutations. The proof of this is too long to be done here, but see *Forster 1995*, where it is Theorem 3.0.4.

REMARK 2.3. If $\langle V, \in \rangle \models ZF$ and τ^{-1} is a setlike permutation of V, check that $\langle V, \in_\tau \rangle \models ZF$.

Proof. The stratified axioms are no problem. The only unstratified axiom scheme is replacement. Easy enough to check for any ϕ that if $\forall x \exists! y \phi$, then $\forall x \exists! y \phi^\tau$, so that for any set X the image of X in ϕ^τ is also a set. Call it Y. But then $\tau^{-1}`Y$ is the image-of-X-under-ϕ (in the sense of V^τ). ⊣

We will also need the following observation:

LEMMA 2.4. *If f is a stratification of Φ thought of as the partial function* $\mathbb{N} \to \mathbb{N}$ *that sends the variable's* subscript *(rather than the variable itself) to the type then*

$$\Phi(x_1, \ldots, x_k) \longleftrightarrow \Phi((j^{f`1}`\sigma)`x_1, \ldots, (j^{f`k}`\sigma)`x_k).$$

Proof. It might be a good idea to have an illustration before a full proof. For example, the lemma tells us that

$$x \in y \longleftrightarrow \sigma`x \in \sigma``y.$$

The theorem is simply a more general assertion, true for the same reasons.

Now for a full proof. By definition of j we have $x \in y$ iff $\tau`x \in (j`\tau)`y$ for any $\tau$. In particular if '$x$' has been assigned type $n$ and '$y$' the type $n+1$, we invoke the case where $\tau$ is $j^n`\sigma$ to get $x \in y \longleftrightarrow (j^n`\sigma)`x \in (j^{n+1}`\sigma)`y$. By substitutivity of the biconditional we can do this simultaneously for all atomic subformulae in $\Phi(x_1, \ldots, x_k)$. Variables 'y' that were bound in '$\Phi(x_1 \ldots x_k)$' now have prefixes like '$j^n`\sigma$' in front of them but, since '$\Phi(x_1, \ldots, x_k)$' was stratified, they will be constant for each such variable '$y$'. We then use the fact that $j^n`\sigma$ is a permutation of V so that any formula $(Qy)(\ldots(j^n`\sigma)`y\ldots)$ (Q a quantifier) is equivalent to $(Qy)(\ldots y \ldots)$. ⊣

DEFINITION 2.5. A $\mathcal{P}$-embedding from $\mathcal{A}$ into $\mathcal{B}$ is a map $i : \mathcal{A} \to \mathcal{B}$ for which the power set operation is absolute. "No new members or subsets of old sets." If i is the identity we say $\mathcal{B}$ is a **$\mathcal{P}$-extension of $\mathcal{A}$.**

THEOREM 2.6. *Let $\mathcal{M} = \langle M, \in \rangle$ be a wellfounded model of ZF, and let σ be a setlike permutation of M. Let $i : \mathcal{M} \hookrightarrow \mathcal{M}^\sigma$ be recursively defined by $i`x =: \sigma^{-1}(i``x)$. Then* (i) *i is a $\mathcal{P}$-embedding and* (ii) *i is elementary for stratified formulæ.*

Proof.

(i) If $x \in_\sigma i`y$, then $x$ is a value of $i$, so $i$ is an end-extension. Suppose $(x \subseteq i`y)^\sigma$: we want x to be a value of i. $(x \subseteq i`y)^\sigma$ is just $\sigma`x \subseteq \sigma(i`y) = i``y$ so $\sigma`x$ is a set of values of i, so x is a value of i.

(ii)[3] Let $\phi(\vec{x})$ be a stratified formula whose free variables are precisely the $\vec{x}$, a tuple of length k. Assume

$$\mathcal{M}^\sigma \models \phi(i`x_1, i`x_2, \ldots i`x_k).$$

By Lemma 2.2 this is the same as

$$\mathcal{M} \models \phi(\sigma_{n_1}`(i`x_1), \sigma_{n_2}`(i`x_2), \ldots \sigma_{n_k}`(i`x_k)),$$

where, for each j, the variable x_j has been assigned the type n_j in some fixed stratification of ϕ. By the recursive definition of i, $i`x_1$ is equal to $\sigma^{-1}(i``x)$ or to $\sigma^{-1} \circ (j`\sigma)^{-1} \circ (j^2`i)`x_1$ and so on, by unravelling steps in the recursive definition of $i`x$ *ad libitum* up to n_1, which gives us $(\sigma^{-1})_{n_1}`((j^{n_1}`i)`x_1)$. So

[3] Randall Holmes has found an error in the proof of Theorem 2.6, part (ii), which must be withdrawn *pro tem*. A discussion will be published elsewhere in due course.

applying σ_{n_1} to this gives us $(j^{n_1}`i)`x$ and similarly for the other x variables, so the formula becomes:

$$\mathcal{M} \models \phi((j^{n_1}`i)`x_1, (j^{n_2}`i)`x_2, \ldots (j^{n_k}`i)`x_k).$$

ϕ is stratified, so by Lemma 2.4 this is just the same as

$$\mathcal{M} \models \phi(\vec{x}). \qquad \dashv$$

COROLLARY 2.7. *The axiom of foundation is independent of ZF.*

Proof. Use the transposition $(\Lambda, \{\Lambda\})$. (Λ is the empty set.) In the new model the old empty set has become an object equal to its own singleton. ⊣

The completeness theorem for setlike permutations and stratified formulæ is a powerful and satisfying piece of 1950's-style model theory, not unlike Birkhoff's theorem in flavor, but it is actually a nuisance. It's all very well if you have a model of a stratified theory, and want to generate lots more models of it, but it does mean that if you start with a model of a theory that does not have a universal set, then the result will not have a universal set either. (After all, the assertion that there is no universal set is stratified, and if it starts off false, will remain false.) It gives us the independence of the axiom of foundation, but not the relative consistency of a universal set. If we want a universal set we have to generalize the construction slightly.

3. CHURCH-OSWALD MODELS

Let $\langle V, \in \rangle$ be a model of *ZF* and let k (for '*k*ode') be a bijection between V and $V \times \{0, 1\}$. Next we define

$$x \in_{\mathbf{co}} y \text{ iff } \begin{cases} 1.\ k`y = \langle y', 0 \rangle \text{ and } x \in y', \textit{ or} \\ 2.\ k`y = \langle y', 1 \rangle \text{ and } x \notin y'. \end{cases}$$

This is the simplest version of the construction, and is the one in *Oswald 1976*—though Oswald considered specifically the case where V was V_ω and presented it very differently. (Oswald's model is presented more fully in section 3.1.1.)

The first thing to notice is that every set has a complement in the sense of $\in_{\mathbf{co}}$. (Perhaps the first thing of all to notice is that $\in_{\mathbf{co}}$ really is extensional!) In fact the resulting model is a model of a theory known as NF_2, which Oswald was studying at the time. The axioms of NF_2 are (i) Extensionality (ii) Complementation ($-x$ is a set, always) (iii) $x \cap y$ and (iv) existence of $\{x\}$. This axiomatization is finite but it is perhaps more convenient to replace (iv) by the scheme giving the existence, for each $n \in \mathbb{N}$ and for all n-tuples $\vec{x}$, of the set $\{x_1, x_2, \ldots x_n\}$.

We took k to be a bijection between V and $V \times \{0,1\}$, but it could have been $V\times$ anything K we like ('K' for 'Kode'), as long as $0 \in K$ and we have suitable clauses for $\in_{\mathbf{co}}$ x where the second component of $k`x$ is not 0. The idea is that in general we define $x \in_{\mathbf{co}} y$ if either (i) $\mathtt{snd}(k`y) = 0$ and $x \in \mathtt{fst}(k`y)$ or[4] (ii) various other clauses concerning (new) membership in sets $y$ such that $\mathtt{snd}(k`y) \neq 0$. The idea is that other entries in K will correspond to other operations on sets (in the case we have just seen 1 corresponds to complementation). In general there is a problem of extensionality. There is of course no difficulty in showing that x and x' satisfy extensionality as long as $\mathtt{snd}(k`x) = \mathtt{snd}(k`x') = 0$, but there are other cases to consider. The task of verifying extensionality in the new structure will be much easier if x and y such that $\mathtt{snd}(k`x) \neq \mathtt{snd}(k`y)$ are so different that the possibility of them having the same members in the new sense simply never arises. For example, in the case we have just seen, if $\mathtt{snd}(k`x)$ is 0, then $x$ has only a set (in the old sense) of members-in-the-new-sense, whereas if $\mathtt{snd}(k`x)$ is 1, it has a proper class (in the old sense) of members-in-the-new-sense. This tells us that if the theory for which we are trying to obtain a model by this construction is T, then the extensionality problem for the model is deeply related to the word problem for T. In particular if we have a good notion of normal forms for T words over a set of generators, then we will be able to take K to be (roughly) a set of such normal words. *What this reveals is that this technique is not a great deal of use for constructing models of a theory T unless T has an easy word problem.* Set theories with an easily solvable word problem are unlikely to be of interest.

For the moment we will—in the name of generality—ignore the question of what the remaining components of $k`x$ can be. We will assume that $K$ is some arbitrary collection such that $0 \in K$ and there are rules to ensure that $\in_{\mathbf{co}}$ is extensional and that when $\mathtt{snd}`k`y = 0$, then $x \in_{\mathbf{co}} y \longleftrightarrow x \in \mathtt{fst}(k`y)$. We will try to prove some general results about this construction. Let us call structures built in this way **CO structures**. ("Church-Oswald").

To summarize, there are three parts to a **CO** construction over a model $\langle V, \in \rangle$. There is a collection K of objects available to be used as second components of ordered pairs; there is a bijection k between V and $V \times K$, and finally there is a family of rules telling us how possible membership (in the new sense) of x in y depends on the second component of $k`y$. Rieger-Bernays permutation model constructions can be seen as those special cases of **CO**-constructions where K is a singleton. Accordingly the only thing one is free to play with is the bijection k.

Since we are assuming that one of the clauses in the definition of $\in_{\mathbf{co}}$ is always $\mathrm{snd}(k`x) = 0 \to (\forall y)(y \in_{\mathbf{co}} x \longleftrightarrow y \in \mathtt{fst}(k`x))$, we can make the following definition.

DEFINITION 3.1. A **low set** is a set x such that $\mathtt{snd}(k`x) = 0$.

[4] $\mathtt{fst}(x)$ and $\mathtt{snd}(x)$ are the first and second components of the ordered pair x.

This is at variance with other definitions in the literature, but it is the most useful. This is not to be confused with the notion of a **hereditarily low** set, for this will turn out to be important as well.

DEFINITION 3.2. H_{low} is the greatest fixed point for the function that sends an argument x to the set of low subsets of x. That is to say, H_{low} is the collection of sets x such that everything in the transitive closure of x is low.

(Notice that the least fixed point must consist entirely of wellfounded sets.)

DEFINITION 3.3. The Axiom scheme of **Low Comprehension** states that, for any formula $\phi(x, \vec{y})$, and for all $\vec{y}$, if the collection of all x such that $\langle V, \in_{\mathbf{co}} \rangle \models \phi(x, \vec{y})$ is a set of the original model, then $\langle V, \in_{\mathbf{co}} \rangle \models$ "$\{x : \phi(x, \vec{y})\}$ is a (low) set".

Thus it is by no means obvious that low comprehension is axiomatizable. For this reason the most profitable approach to this topic is to think of the **CO** constructions as things that give us models, rather than to attempt to be specific in saying what the axioms are of the theory whose consistency we have proved.

The following triviality is central to what is to come.

THEOREM 3.4. *All* **CO** *structures satisfy low comprehension.*

Proof. Let $\phi(x, \vec{y})$ satisfy the antecedent, and consider the class of all x such that $\langle V, \in_{\mathbf{co}} \rangle \models \phi(x, \vec{y})$ which is a set of the original model, X, say. Then $\{x : \phi(x, \vec{y})\}$ in the sense of the new model is simply $k^{(-1)}\text{`}\langle X, 0 \rangle$, and is of course low. ⊣

In particular we have an axiom of **pairing**.

In a typical **CO** construction there will be plenty of new sets containing all low sets: V for one. However,

PROPOSITION 3.5. *No new set containing all low sets can be low.*

Proof. Suppose there was a low set containing all low sets. Then, by low comprehension, the collection of all low sets is low. That is to say, there is an x such that $(\forall y)(y \in_{\mathbf{co}} x \longleftrightarrow \mathtt{snd}(k\text{`}y) = 0)$. But there is certainly a proper class of $y$ such that $\mathtt{snd}\text{`}(k\text{`}y) = 0$, so this x has a proper class of members and is therefore not low. ⊣

Attempts to reconstruct the usual paradoxes in this new context give rise to demonstrations that certain sets are not low. Take Russell's paradox for example. The new model cannot contain the set of all things that are not members of themselves. If the collection of x such that $\neg(x \in_{\mathbf{co}} x)$ were a set of the original model, then the Russell class would be a set of the new model. So there is a proper class of x such that $\neg(x \in_{\mathbf{co}} x)$.

COROLLARY 3.6. *Every surjective image of a low set is low and every subset of a low set is low.*

Proof. The first part follows from replacement in the original model and the second from comprehension. ⊣

The first follows from the second by AC but a proof without AC is preferable since we will not otherwise be making any use of AC in this development.

COROLLARY 3.7. *Every low set has a power set, which is also low.*

Proof. When x is low, $\mathcal{P}`x$ (in the sense of $\in_{\mathrm{co}}$) must be

$$k^{(-1)}`\langle k^{(-1)}``(\mathcal{P}`(\mathtt{fst}(k`x)) \times \{0\}), 0\rangle. \quad ⊣$$

COROLLARY 3.8. *If x is a low set of low sets, then it has a sumset, which is low.*

Proof. The object we need to play the role of $\bigcup x$ is

$$k^{-1}\langle \bigcup (\mathtt{fst} \circ k)``(\mathtt{fst} \circ k)`x, 0\rangle.$$

This depends only on the availability of the axiom of sumset in the model we start with, and does not tell us about stronger, less restricted forms of the axiom. ⊣

This seems to be an argument for setting up this theory with low comprehension in the way I have done it rather than with low replacement in the way Church originally did. Low replacement in the form "The image of a wellfounded set in a function is a set" certainly implies existence of power set for low sets and sumset, but we seem to need AC to deduce sumset for low sets of low sets. This is very messy.

3.1. Applications of the technique

3.1.1 Models of NF_2

Oswald did not set up his first illustration with the apparatus of K and k as here. Instead he defined a binary relation on $\mathbb{N}$ as follows:

$n\ E\ m$ iff either

1. m is even and the nth bit of the binary expansion of $m/2$ is 1; *or*
2. m is odd and the nth bit of the binary expansion of $(m-1)/2$ is 0.

This obviously derives from the old trick (due to Ackermann) of defining $n\ E\ m$ $(n, m \in \mathbb{N})$ iff the nth bit of the binary expansion of m is 1.

It might be a good idea to concentrate briefly on the three salient features of the simplest possible case.

$K = \{0, 1\}$, and the other clause gives us complements;
The $\in$ relation of the original model is wellfounded;
The rank of $\mathtt{fst}(k`x)$ is no greater than the rank of x.

When these conditions are met we can say a lot.

DEFINITION 3.9. An antimorphism is a permutation π of the universe satisfying $(\forall xy)(x \in y \longleftrightarrow \pi`x \notin \pi`y)$.

REMARK 3.10. Under the three assumptions above the new model admits an antimorphism.

Proof. Declare the following recursive definition

$$\sigma`x =: k^{-1}`\langle \sigma``(\mathtt{fst}(k`x)), (1 - \mathtt{snd}`(k`x))\rangle.$$

By considering putative counterexamples of minimal rank we can show that this is everywhere defined.

Suppose $\sigma`y \in_{\mathbf{co}} \sigma`x$. This is

$$\sigma`y \in_{\mathbf{co}} k^{-1}`\langle \sigma``(\mathtt{fst}(k`x)), (1 - \mathtt{snd}`(k`x))\rangle.$$

Now either $\mathtt{snd}`(k`x) = 0$ or $\mathtt{snd}`(k`x) = 1$.

0. If $\mathtt{snd}`(k`x) = 0$, the displayed formula becomes

$$\sigma`y \in_{\mathbf{co}} k^{-1}`\langle \sigma``(\mathtt{fst}(k`x)), 1\rangle,$$

which is

$$\sigma`y \notin \sigma``(\mathtt{fst}(k`x)),$$

which is $y \notin (\mathtt{fst}(k`x))$. But if $\mathtt{snd}`(k`x) = 0$, this becomes $y \notin_{\mathbf{co}} x$.

1. On the other hand if $\mathtt{snd}`(k`x) = 1$, then

$$\sigma`y \in_{\mathbf{co}} k^{-1}`\langle \sigma``(\mathtt{fst}(k`x)), (1 - \mathtt{snd}`(k`x))\rangle$$

is

$$\sigma`y \in_{\mathbf{co}} k^{-1}`\langle \sigma``(\mathtt{fst}(k`x)), 0\rangle.$$

This is $\sigma`y \in \sigma``(\mathtt{fst}(k`x))$, which of course simplifies to $y \in \mathtt{fst}(k`x)$, and (since $\mathtt{snd}`(k`x) = 1$) this becomes $y \notin_{\mathbf{co}} x$.

So either way we have $\sigma`y \in_{\mathbf{co}} \sigma`x \longleftrightarrow y \notin_{\mathbf{co}} x$. ⊣

There is another nice result we get as a reward for making these assumptions. Consider the game G_x, played as follows, by two players, I and II. I picks $x_1 \in x$, II picks $x_2 \in x_1$, I picks $x_3 \in x_2 \ldots$, with the first player unable to move (you cannot pick a member of the empty set!) losing.

Clearly the axiom of foundation implies that G_x is always determinate in the sense of admitting a winning strategy for one player or the other. This is **$\in$-determinacy**. The axiom of foundation is obviously implicated because if $x = \{x\}$, G_x is clearly not determinate. The converse is not true, however, and we can prove the following.

REMARK 3.11. Under the three assumptions above, the new model obeys $\in$-determinacy.

Proof. G_x is a win for player I if there is a $y \in x$ such that G_y is a win for player II. Similarly G_x is a win for player II iff for every $y \in x$, G_y is a win for player I.

First we prove that there are x of arbitrarily high rank such that x is hereditarily low and G_x is a win for player II. (And player I, but we do not need that.) This is because, under the assumptions given, H_{low} is an isomorphic copy of the original model and there are clearly x of arbitrarily high rank with those properties.

Next we prove by induction on rank of x that (in the sense of the new model) G_x is determinate (admits a winning strategy for I or II). Suppose this is not true, and let x be a counterexample of minimal rank. Then $k`x = \langle y, 1\rangle$ or $k`x = \langle y, 0\rangle$, for some y. If $k`x = \langle y, 0\rangle$ for some $y$, then by induction hypothesis $G_z$ is determinate for every $z \in_{\text{co}} x$ and $G_x$ must be determinate as well. If $k`x = \langle y, 1\rangle$, then there is some ordinal α (namely rank of $y + 1$) such that, for all z, if z is an element of the original model of rank at least α, then $z \notin y$, so (since $k`x = \langle y, 1\rangle$) $z \in_{\text{co}} x$. Now we have just proved that at least some of these z give rise to G_z that are wins for player II. So there is at least one $z \in_{\text{co}} x$ such that II has a winning strategy for G_z. But then I has a winning strategy for G_x which he launches by picking z. ⊣

3.1.2 An elementary example

We are going to consider two models, more complicated than the original Oswald model, in which in addition to complements for all x, we also have $B`x$ for all $x$. $B`x$ is $\{y : x \in y\}$.[5] The theory that has boolean axioms ($\cup$, $\cap$ and $-$), singleton and B is *NFO*. (The 'O' is intended to suggest **O**pen. This theory is more naturally axiomatized as Extensionality plus existence of $\{x : \phi(x, \vec{y})\}$ where ϕ is stratified and quantifier-free. It is a relatively straightforward exercise to check that these axiomatizations are equivalent.) The second model will satisfy all the boolean axioms (and so is a model of *NFO*), but we start with the first, which does not.

[5] I use this notation because it was Boffa who first impressed on me the importance of this operation. In fact the first use of this operation was by Quine, and Whitehead suggested to him that $\{y : x \in y\}$ should be called the *essence* of x.

Consider $\mathcal{S}$ = the set of reduced words in the semigroup-with-unity with two generators c and b, and the equation $\mathtt{c}^2 = 1_{\mathcal{S}}$. 'c' is intended to recall "Complement" and 'b' to recall "B'x". We now let k be a bijection between V and $V \times \mathcal{S}$. We define $\in_{\mathbf{co}}$ by recursion by cases:

DEFINITION 3.12.

1. If $\mathtt{snd}(k`x) = \mathtt{c}w$ for some word $w \in \mathcal{S}$, then

$$y \in_{\mathbf{co}} x \text{ iff } y \notin_{\mathbf{co}} k^{(-1)}\langle \mathtt{fst}(k`x), w\rangle.$$

2. If $\mathtt{snd}(k`x) = \mathtt{b}w$ for some word $w \in \mathcal{S}$ then

$$y \in_{\mathbf{co}} x \text{ iff } k^{(-1)`}\langle \mathtt{fst}(k`x), w\rangle \in_{\mathbf{co}} y.$$

3. If $\mathtt{snd}(k`x)$ is $1_{\mathcal{S}}$, then $y \in_{\mathbf{co}} x$ iff $y \in \mathtt{fst}(k`x)$.

PROPOSITION 3.13. *The $\in_{\mathbf{co}}$ of Definition 3.12 is extensional.*

Proof. The proof is an inductive proof (by cases) on the length of words in $\mathcal{S}$. The situation we are contemplating is two distinct x_1 and x_2 which have the same $\in_{\mathbf{co}}$-members. Naturally we will be interested in the second components of $k`x_1$ and $k`x_2$. There are several cases to consider:

1. The second components are both $1_{\mathcal{S}}$, the unit of the semigroup. In this case the first components must be the same, and $x_1 = x_2$ follows.

2. One of the second components is $1_{\mathcal{S}}$ and the other is not. Suppose $\mathtt{snd}(k`x_1) = 1_{\mathcal{S}}$ and $\mathtt{snd}(k`x_2)$ is $\mathtt{bx}$ or $\mathtt{cbw}$. (We cannot have two adjacent c's since the words are reduced and $\mathtt{c}^2 = 1_{\mathcal{S}}$.) Then the collection of y such that $y \in_{\mathbf{co}} x_1$ is $\mathtt{fst}(k`x_1)$, which is a set in the sense of $\langle V, \in\rangle$. In contrast, the collection of $y$ such that $y \in_{\mathbf{co}} x_2$ is either (i) the collection of $y$ such that $y \notin_{\mathbf{co}} k^{(-1)}\langle\mathtt{fst}(k`x_2),\mathtt{b(x)}\rangle$—which is to say, by an application of part 2 of Definition 3.12—the same as the collection of y such that $k^{(-1)}\langle\mathtt{fst}(k`x_2),\mathtt{x}\rangle \notin_{\mathbf{co}} y$, or (ii) is the collection of $y$ such that $k^{(-1)`}\langle\mathtt{fst}(k`x_2), w\rangle \in_{\mathbf{co}} y$. The collection in case (i) cannot be a set because for any object a we can easily find proper-class-many unordered pairs which do not have a as a member. The collection in case (ii) cannot be a set because for any object a we can easily find proper-class-many unordered pairs which *do* have a as a member.

3. $\mathtt{snd}(k`x_1) = \mathtt{c(x_1)}$ and $\mathtt{snd}(k`x_2) = \mathtt{c(x_2)}$ where the first letter of both $\mathtt{x_1}$ and $\mathtt{x_2}$ is b. Assume $(\forall y)(y \in_{\mathbf{co}} x_1 \longleftrightarrow y \in_{\mathbf{co}} x_2)$ with a view to deducing $x_1 = x_2$. We can expand this in accordance with part 2 of definition

$$(\forall y)(y \notin_{\mathbf{co}} k^{(-1)}\langle\mathtt{fst}(k`x_1),\mathtt{x_1}\rangle \longleftrightarrow y \notin_{\mathbf{co}} k^{(-1)}\langle\mathtt{fst}(k`x_2),\mathtt{x_2}\rangle),$$

which is to say

$$(\forall y)(y \in_{\mathbf{co}} k^{(-1)}\langle \mathtt{fst}(k`x_1),\mathtt{x}_1\rangle \longleftrightarrow y \in_{\mathbf{co}} k^{(-1)}\langle \mathtt{fst}(k`x_2),\mathtt{x}_2\rangle).$$

That is to say, if we have distinct x_1 and x_2 with the same members-in-the-sense-of-$\in_{\mathbf{co}}$, where the first letter of $\mathtt{snd}(k`x_1)$ is the same as the first letter of $\mathtt{snd}(k`x_2)$, namely $\mathtt{c}$, then there are distinct y_1 and y_2 (namely $k^{(-1)}\langle \mathtt{fst}(k`x_1),\mathtt{x}_1\rangle$ and $k^{(-1)}\langle \mathtt{fst}(k`x_2),\mathtt{x}_2\rangle$) satisfying $(\forall z)(z \in_{\mathbf{co}} y_1 \longleftrightarrow z \in_{\mathbf{co}} y_2)$ and $\mathtt{snd}(k`y_1)$ and $\mathtt{snd}(k`y_2)$ are shorter than $\mathtt{snd}(k`x_1)$ and $\mathtt{snd}(k`x_2)$—indeed they are terminal segments.

This clearly reduces to the case where the two words in question are $\mathtt{x}_1$ and $\mathtt{x}_2$ both beginning with $\mathtt{b}$, which we now treat.

4. The second components of $k`x_1$ and $k`x_2$ are words $\mathtt{b}(\mathtt{x}_1)$ and $\mathtt{b}(\mathtt{x}_2)$. We have $(\forall y)(y \in_{\mathbf{co}} x_1 \longleftrightarrow y \in_{\mathbf{co}} x_2)$. Expanding this by clause (2) in Definition 3.12 we obtain

$$(\forall y)(k^{(-1)}`\langle \mathtt{fst}(k`x_1),\mathtt{b}(\mathtt{x}_1)\rangle \in_{\mathbf{co}} y \longleftrightarrow k^{(-1)}`\langle \mathtt{fst}(k`x_2),\mathtt{b}(\mathtt{x}_2)\rangle \in_{\mathbf{co}} y).$$

Now as long as

$$k^{(-1)}`\langle \mathtt{fst}(k`x_1),\mathtt{b}(\mathtt{x}_1)\rangle \neq k^{(-1)}`\langle \mathtt{fst}(k`x_2),\mathtt{b}(\mathtt{x}_2)\rangle$$

we can falsify this biconditional by taking y to be the singleton of one of these two. Singletons exist by low comprehension: the singleton of x (in the sense of $\in_{\mathbf{co}}$) is just $k^{(-1)}`\langle\{x\},0\rangle$.

5. $k`x_1$ has second component $\mathtt{cbw}_1$ and $k`x_2$ has second component $\mathtt{b}(\mathtt{x}_2)$. This is really the case where one word begins with a $\mathtt{c}$ and the other begins with a $\mathtt{b}$, but since the words are all reduced, we also know that the letter following the $\mathtt{c}$ in the word beginning with a $\mathtt{c}$ must be a $\mathtt{b}$. If we have

$$(\forall y)(y \in_{\mathbf{co}} x_1 \longleftrightarrow y \in_{\mathbf{co}} x_2),$$

this becomes

$$(\forall y)(y \notin_{\mathbf{co}} k^{(-1)}\langle \mathtt{fst}(k`x_1),\mathtt{b}(\mathtt{x}_1)\rangle \longleftrightarrow k^{(-1)}`\langle \mathtt{fst}(k`x_2),\mathtt{x}_2\rangle \in_{\mathbf{co}} y).$$

This is the same as

$$(\forall y)(y \in_{\mathbf{co}} k^{(-1)}\langle \mathtt{fst}(k`x_1),\mathtt{b}(\mathtt{x}_1)\rangle \longleftrightarrow k^{(-1)}`\langle \mathtt{fst}(k`x_2),\mathtt{x}_2\rangle \notin_{\mathbf{co}} y).$$

Now $y \in_{\mathbf{co}} k^{(-1)}\langle \mathtt{fst}(k`x_1),\mathtt{b}(\mathtt{x}_1)\rangle$ iff (by clause (2) in Definition 3.12)

$$k^{(-1)}\langle \mathtt{fst}(k`x_1),\mathtt{x}_1\rangle \in_{\mathbf{co}} y$$

and substituting this for '$y \in_{co} k^{(-1)}\langle \texttt{fst}(k`x_1),\mathtt{b}(\mathtt{x}_1)\rangle$' in

$$(\forall y)(y \in_{co} k^{(-1)}\langle \texttt{fst}(k`x_1),\mathtt{b}(\mathtt{x}_1)\rangle \longleftrightarrow k^{(-1)}`\langle \texttt{fst}(k`x_2),\mathtt{w}_2\rangle \notin_{co} y)$$

we obtain

$$(\forall y)(k^{(-1)}\langle \texttt{fst}(k`x_1),\mathtt{w}_1\rangle \in_{co} y \longleftrightarrow k^{(-1)}`\langle \texttt{fst}(k`x_2),\mathtt{w}_2\rangle \notin_{co} y).$$

Now anything like $(\forall y)(a \in y \longleftrightarrow b \notin y)$ must always be false because of the existence of the empty set. ⊣

The rest is easy:

PROPOSITION 3.14. *With $\in_{co}$ as in Definition 3.12, $\langle V, \in_{co}\rangle \models (\forall x)(-x$ exists) and $(\forall x)(B`x$ exists).*

Proof. The complement of x will be $k^{(-1)}\langle \texttt{fst}(k`x),\texttt{csnd}(k`x)\rangle$, for, by clause (1) in Definition 3.12, we have

$$y \in_{co} k^{(-1)}\langle \texttt{fst}(k`x),\texttt{csnd}(k`x)\rangle \text{ iff}$$
$$y \notin_{co} k^{(-1)}\langle \texttt{fst}(k`x), \texttt{snd}(k`x)\rangle \text{ iff}$$
$$y \notin_{co} x.$$

$B`x$ will be $k^{(-1)}\langle \texttt{fst}(k`x),\texttt{bsnd}(k`x)\rangle$, for, by clause (2) in Definition 3.12,

$$y \in_{co} k^{(-1)}\langle \texttt{fst}(k`x),\texttt{bsnd}(k`x)\rangle \text{ iff}$$
$$k^{(-1)}\langle \texttt{fst}(k`x), \texttt{snd}(k`x)\rangle \in_{co} y \text{ iff}$$
$$x \in_{co} y.$$

⊣

3.1.3 $\mathcal{P}$-extending models of Zermelo to models of NFO

Before we contemplate the second construction (which gives us a model containing $x \cap y$ and $x \cup y$ for all x and y) we had better ask ourselves why we did not get it last time. After all, these axioms hold when $K = \{0, 1\}$. The point is that in that case everything is the same size as an old set or the complement of an old set. Also if x and y are both the same size as an old set or the complement of an old set, then so are $x \cap y$ and $x \cup y$, and so $\cap$ and $\cup$ do not construct anything that is not already there. Once we have $B`x$ this breaks down, and if we want $x \cup y$ and $x \cap y$ in general we have to construct them specially. This makes the construction of models of *NFO* altogether more daunting.

THEOREM 3.15. *Every model of ZF+foundation has a $\mathcal{P}$-extension that is a model of NFO + low comprehension.*

Proof. (We certainly need not restrict ourselves to models of *ZF*, for this is certainly true for well-founded models of Z and presumably weaker theories as well.)

The idea of the construction was originally that $k`x$ is to be a pair $\langle y, w\rangle$ where y is a set and w is a reduced word in some algebra with operations that correspond to the operations under which we want the universe of the new model to be closed. Although this can be made to work, the approach it gives is very much less smooth than an approach that creates a gigantic free *NFO* model over a proper class of generators where there is one generator for each set of the old model. Unfortunately this second, smoother approach is not really a **CO** construction, and so strictly does not belong here as an illustration. The excuse for putting it in here is that the construction is, in spirit, very close to the **CO** constructions that precede it and will succeed it, and its presence here will help.

In the last case the only operations in which we were interested were unary (complementation and B) so the case had a spurious simplicity. Recall that the axioms of *NFO* are (apart from extensionality) existence of $\{x\}$ and closure under B, $\cup$ and $\cap$. If this were a **CO** construction *au pied de la lettre* we would get closure under singleton free because of low comprehension, so we could explicitly forget about it. Here too we can forget about it, and will indeed do so—for the moment. Later we will see why this is all right after all. We should be able to make do with only one of $\cap$ and $\cup$, but conjunctive and disjunctive normal forms for boolean words are so useful that we will retain both.

Our language of terms has a constant term g_x for each old set x. The constants will eventually correspond to low sets of the new model. We also have function letters $\cup$, $\cap$ $-$ and B. We have to do a bit of work to find the correct notion of *reduced word* for this algebra. There is the irritating feature that we do not want to have both $a\cap b$ and $b\cap a$ but we can get round that by wellordering the alphabet and extending the order lexicographically to the words. It will turn out that we will want to augment the language by adding Δ (symmetric difference).

DEFINITION 3.16. A **restricted word** is either a constant, or is $W\Delta g$ where W is a boolean combination of Bs of restricted words, and g is a constant. If g is missing, we can speak of a **pure** restricted word.

Now we have to show that everything that we wish to construct can be denoted by a restricted word.

We can think of a word $\mathbf{w}$ as a boolean combination as a union of intersections, where the things being intersected are constants or Bs and complements of either. Consider an intersection like

$$a\cap b\cap c\cap d\ldots.$$

If even one of these is a constant (i.e., will correspond to a low set), then the whole intersection can be represented by just one (new) constant. Intersections of any number of complements of low sets can be represented as one complement of a low set. Thus the intersections are either constants, or intersections of Bs and $-B$s with the complement of a low set, which is to say, an intersection of Bs and $-B$s minus some low set.

Thus **w** can be rewritten in the form

$$[(w_1 \cap \overline{g_1}) \cup (w_2 \cap \overline{g_2}) \cup (w_3 \cap \overline{g_3})] \cup g_{n+1} \tag{1}$$

where the w_i are intersections of values of B or complements of values of B, and the g_i are low sets. We will work through this in the case where $n = 3$, so that the reader than see how to do the general case. (This is probably more helpful than a rigorous proof of the general case would be!) For the moment we are interested only in the expression inside the square bracket of formula (1):

$$(w_1 \cap \overline{g_1}) \cup (w_2 \cap \overline{g_2}) \cup (w_3 \cap \overline{g_3}).$$

This expands to an intersection of $2^3 = 8$ subformulæ as follows:

$$(w_1 \cup w_2 \cup w_3) \cap \text{ seven other terms.}$$

A typical example of these other terms is $w_1 \cup w_2 \cup \overline{g_3}$. It is typical in that it contains at least one entry of the kind $\overline{g_i}$. Now $w_1 \cup w_2 \cup \overline{g_3}$ is the same as

$$-(\overline{w_1} \cap \overline{w_2} \cap g_3).$$

This is the complement of a low set, so we can think of this as $\overline{g_{\text{novel}_1}}$ for some novel low set g_{novel_1}. This can be done to all the six remaining terms, so the expression-inside-the-square-bracket in formula (1) now looks like

$$(w_1 \cup w_2 \cup w_3) \cap \overline{g_{\text{novel}_1}} \cap \overline{g_{\text{novel}_2}} \cap \overline{g_{\text{novel}_3}} \cap \ldots \overline{g_{\text{novel}_7}}.$$

Subtracting finitely many low sets is the same as subtracting one, so the expression-inside-the-square-bracket in formula (1), reduces to

$$(w_1 \cup w_2 \cup w_3) \cap \overline{g_{\text{novel}}}\ .$$

So, in the general case, we have reduced **w** to

$$(\bigcup_{i \leq n} w_n - g_{\text{novel}}) \cup g_{n+1}.$$

This is actually

$$(\bigcup_{i \leq n} w_n) \Delta G,$$

where G is

$$((\bigcup_{i\leq n} w_n) \cap g_{\text{novel}}) \cup (g_{n+1} \cap -(\bigcup_{i\leq n} w_n)),$$

which is a low set.

Notice that we started with $\mathtt{w}$ as a boolean combination of B's and constants and have ended up with something rather simpler: $\bigcup_{i\leq n} w_n$ is a boolean combination of B's, and G can be taken to be a constant.

This tidying-up process has not turned $\mathtt{w}$ into a restricted word, but if we invoke a notion of rank (where the rank of a word is simply the maximal depth of nesting of Bs in it), then we can see that none of the manipulations involved in the foregoing increases the rank of $\mathtt{w}$, so that is, we perform these manipulations successively on words of increasing rank, every word will eventually be manipulated into a restricted word.

Every word is equivalent to a restricted word. Verifying extensionality in this model will depend on our being able to show that it is equivalent to a *unique* restricted word. (Note that it has not yet been made clear quite *how* it will depend on this uniqueness!) To do this it will be sufficient to show that if $\mathtt{w}_1$ and $\mathtt{w}_2$ are two pure restricted words, then $\mathtt{w}_1 \,\Delta\, \mathtt{w}_2$ is not low. Suppose this were not so. Then there is some word W which evaluates to a low set. Without loss of generality we can think of W as a union of lots of terms each of which of course must themselves evaluate to something low and each such term is of the form $B`w_1 \cap B`w_2 \cap \ldots B`w_n \cap -B`u_1 \cap -B`u_2 \cap \ldots -B`u_k$ (where the $\vec{u}$ and the $\vec{w}$ are all distinct), which is supposed to be a low set. (Here $B`x$ here means the set (in the sense of $\langle V, \in\rangle$) of those y such that $x \in_{\mathbf{co}} y$.) Now $\{w_1, \ldots w_n\} \cup \{x\}$ belongs to this collection as long as x is not a u or a w, and there is certainly going to be a proper class of such x, so it is clear that this collection must be a proper class (in the sense of $\langle V, \in\rangle$).

Now we have to define a new membership relation $\in_{\mathbf{co}}$. There is a bijection between the words of this new algebra and the sets of our old model. We may as well call this k as before. The elements of the model will be the words of the algebra. We will have to define $\in_{\mathbf{co}}$ by recursion on the structure of the terms.

If w is a molecular term of rank n then, in full generality, it is $w'\Delta g$ for some constant g and some boolean combination w' of $B`z_i$ for various i. We then say $x \in_{\mathbf{co}} w$ iff $(x \not\in_{\mathbf{co}} g) \longleftrightarrow$ (the obvious boolean combination of things like $z_i \in_{\mathbf{co}} x$).[6] This is a recursive definition that appeals to a rank function on *NFO* terms, where rank is the depth of nesting of Bs. What are we to make of '$x \in_{\mathbf{co}} g_y$', where g_y is a constant? This is defined to be $k^{-1}`x \in y$. This ensures that every constant corresponds to a low set and *vice versa*. As before we can define an injection $i$ from the old model into the new, by recursion. This time the recursion is $i`x = g_{(k^{(-1)} \circ i)``x}$.

[6]Where the $\vec{z}$ are of course the free variables occurring in words of rank $n-1$ in w.

We'd better verify that this does actually define a $\mathcal{P}$-extension. Suppose $y \in_{\mathbf{co}} i`x$. Then $y \in_{\mathbf{co}} g_{(k^{(-1)} \circ i)``x}$ which is to say $k^{-1}`y \in (k^{(-1)} \circ i)``x$, whence $y \in i``x$ and $y$ is a value of $i$. So the range of $i$ is transitive. Next suppose $i`y \in_{\mathbf{co}} i`x$. This is $i`y \in_{\mathbf{co}} g_{(k^{(-1)} \circ i)``x}$, which in turn is $k^{-1}`i`y \in (k^{(-1)} \circ i)``x$, which is just $y \in x$. So $i$ is an isomorphism. Finally let $y$ be a low set of values of $i$. Then $i`y$ will have to be $g_{(k^{(-1)} \circ i)``y}$, if this last thing is defined. But $(k^{(-1)} \circ i)``y$ is certainly a set, so $g_{(k^{(-1)} \circ i)``y}$ is defined and is available to be $i`y$. ⊣

One corollary of this is that any fragment of Z strong enough to execute this construction proves every $\Pi_1^{\mathcal{P}}$ theorem of *NFO*. See *Forster-Kaye 1991* for an explanation of $\Pi_1^{\mathcal{P}}$ expressions.

3.1.4 Church's Model

Church's models are all rough **CO** constructions in which every hereditarily low set has an n-cardinal. What is an n-cardinal? The 1-cardinal of x is the cardinal of x in the usual sense. And for larger n? Actually it does not matter a great deal what happens for larger n, since whatever we decide to mean by it it can be made to work. Sheridan is developing a construction that accomodates a version different from the one here (with the effect that the singleton function, considered as a set of Wiener-Kuratowski ordered pairs, is a union of finitely many of these cardinals). One natural version of the idea of n-cardinal is quite well-developed in *NF* studies. Recall that j is the operation on permutations defined by $(j`\pi)`x = \pi``x$. Then we say x is n-equivalent to y if there is some permutation π of V so that $(j^n`\pi)`x = y$. See *Forster 1995* for more on this sort of n-cardinal. These ideas are anticipated in the eminently readable *Tarski 1986*.

In the version used here, two sets have the same $n+1$-cardinal iff there is a bijection between them such that the elements paired by the bijection have the same n-cardinal. We start off with the case $n = 1$ for simplicity's sake.

As with the example of the last section, this is not a strict **CO** construction. We need a bijection (k, as ever) between V and a set of codes for objects. We will use the notation '$\overline{\overline{x}} = \overline{\overline{y}}$' to mean that x and y are the same size. We will see very soon how these fake terms ('$\overline{\overline{x}}$' etc.) can be treated as genuine denoting terms.

DEFINITION 3.17. The things that are are values of k are either:

1. ordered pairs $\langle x, i\rangle$, where x is an arbitrary set and i is 0 or 1 (this will provide low sets and complements of low sets as usual), or

2. ordered pairs $\langle i, \kappa\rangle$, where κ is a cardinal (other than 0) and i is either I or II. I and II are two unspecified distinct objects that were not in

the original ground model. The idea is that these objects are to be cardinals (and complements of cardinals) in the new model.

Now we say $y \in_{co} x$ iff

1. $\mathtt{snd}(k`x) = 0$ and $y \in \mathtt{fst}(k`x)$, *or*
2. $\mathtt{snd}(k`x) = 1$ and $y \notin \mathtt{fst}(k`x)$, *or*
3. $\mathtt{fst}(k`x) = \mathrm{I}$ and $\mathtt{snd}(k`y) = 0$ (so y is low) and $\overline{\overline{\mathtt{fst}(k`y)}} = \mathtt{snd}(k`x)$, *or*
4. $\mathtt{fst}(k`x) = \mathrm{II}$ and $\mathtt{snd}(k`y) \neq 0$ (so y is low) or $\overline{\overline{\mathtt{fst}(k`y)}} \neq \mathtt{snd}(k`x)$.

Clauses 1 and 2 make sure that every low set has a complement. Notice that nothing has been said about what cardinal numbers *are*. Notice also that this does not matter! All we need is that there should be a definable class C and a definable relation `belongs-to` between sets and members of C satisfying

$$(\forall x)(\exists! y \in C)(x \ \mathtt{belongs\text{-}to}\ y),$$

$$(\forall x \forall y)(\overline{\overline{x}} = \overline{\overline{y}} \longleftrightarrow (\forall z \in C)(x \ \mathtt{belongs\text{-}to}\ z \longleftrightarrow y \ \mathtt{belongs\text{-}to}\ z)).$$

The term $\overline{\overline{x}}$ can then be taken to denote the appropriate member of C. We do not need the axiom of choice to define cardinal numbers since as long as we have foundation (which we are assuming here) we can use Scott cardinals. The Scott cardinal of x is the set of all things the same size as x that are of minimal rank with this property.

Clause 3 will ensure that every low set has a cardinal in the new model (in the strong sense that for every low set x, the collection of all sets that have the same cardinal as x is a set of the new model). We stipulate that cardinals used do not include 0. We do this for two reasons: (i) to keep 0 free to signal low sets as usual, also (ii) because the extension of the cardinal number 0 (the set of all empty sets) is a set by low comprehension anyway, and we do not wish it to make difficulties for ourselves with extensionality by manufacturing it twice. Clause 4 ensures that the complement of every such cardinal is a set.

The usual apparatus of low comprehension can now be taken for granted. It should by now be clear that this model is a model of complementation. It is the existence of cardinals that we had better spend a bit of time verifying.

PROPOSITION 3.18. *The clauses of Definition 3.17 give a model in which every low set x has a cardinal:* $\{y : \overline{\overline{y}} = \overline{\overline{x}}\}$.

Proof. Notice that the cardinals that we have created by this means are demonstrably neither low nor are the complements of low sets, which makes life much easier. Let x be any low set. Consider the ordered pair $\langle \mathrm{I}, \mathtt{fst}(k`x)\rangle$. We will check that $k^{(-1)}`\langle \mathrm{I}, \mathtt{fst}(k`x)\rangle$ is the cardinal of $x$ (in the sense that it is the set of all things the same size as $x$) in the new model. By clause 3 we have $y \in_{\mathbf{co}} k^{(-1)}`\langle \mathrm{I}, \mathtt{fst}(k`x)\rangle$ iff $y$ is low and $\mathtt{fst}(k`y)=\mathtt{fst}(k`x)$. Since x and y are both low, this is the same as saying that the set (in the old sense) of things $\in_{\mathbf{co}}$ y is the same size (in the old sense) as the set of things $\in_{\mathbf{co}}$ x, so there is a bijection between these two (old) sets. This bijection is an (old) set of (old) ordered pairs. By low comprehension (Theorem 3.4) the corresponding (new) set of (new) ordered pairs is also a set, so x and y are of the same size in the new sense as well. The other direction is easy. Therefore $k^{(-1)}`\langle \mathrm{I}, \mathtt{fst}(k`x)\rangle$ is indeed the cardinal of x in the new model. Correspondingly $k^{(-1)}`\langle \mathrm{II}, \mathtt{fst}(k`x)\rangle$ is the complement of that cardinal, which we have to have if complementation is to be true. ⊣

We have to do a little bit of work to see how to generalize this correctly to the case $n = 2$, the model where every low set of low sets has a 2-cardinal. What Church actually claims is that for each n his construction gives us a model where every well-founded set has an n-cardinal. I prefer the statement in terms of low sets, low sets of $\ldots^n$ low sets.

For the case $n = 2$ we have to add two more clauses 5 and 6 to Definition 3.17 in the same style. We will need two more novel constants in the style of I and II, which we may as well write 'III' and 'IV'. Objects x such that $\mathtt{fst}(k`x) = \mathrm{III}$ will be 2-cardinals and objects $x$ such that $\mathtt{fst}(k`x) = \mathrm{IV}$ will be complements of 2-cardinals. We will need the notation '$\mathtt{2\text{-}card}`x$' for the 2-cardinal of x, and we will use lower case Greek letters to range over 2-cardinals as over cardinals. Then there are to be two further kinds of ordered pairs in the range of k: pairs whose first components are III and pairs whose first components are IV. In both cases the second components are 2-cardinals. We will need the two following new clauses in the definition of $y \in_{\mathbf{co}} x$.

DEFINITION 3.19.

5. $\mathtt{fst}(k`x) = \mathrm{III}$ and y is a low set of low sets and
$$\mathtt{2\text{-}card}`\{\mathtt{fst}(k`z) : z \in \mathtt{fst}(k`y)\} = \mathtt{snd}(k`x).$$
6. $\mathtt{fst}(k`x) = \mathrm{IV}$ and (y is not a low set of low sets or
$$\mathtt{2\text{-}card}`\{\mathtt{fst}(k`z) : z \in \mathtt{fst}(k`y)\} \neq \mathtt{snd}(k`x)).$$

Clause 6, of course, ensures that 2-cardinals, too, have complements. The details will be omitted.

PROPOSITION 3.20. *The membership relation of Definition 3.19 gives a model in which each low set of low sets has a 2-cardinal.*

Proof. Let x be a low set of low sets. Then the 2-cardinal (in the sense of $\in_{\mathbf{co}}$) of x will be $k^{(-1)}$‘$\langle$III, $\texttt{2-card}$‘$\{\texttt{fst}(k‘z) : z \in \texttt{fst}(k‘x)\}\rangle$. We had better check this. Suppose y is a low set of low sets. Then

$$y \in_{\mathbf{co}} k^{(-1)}‘\langle \mathrm{III}, \texttt{2-card}‘\{\texttt{fst}(k‘z) : z \in \texttt{fst}(k‘x)\}\rangle$$

iff

$$\texttt{2-card}‘\{\texttt{fst}(k‘z) : z \in \texttt{fst}(k‘x)\} = \texttt{2-card}‘\{\texttt{fst}(k‘z) : z \in \texttt{fst}(k‘y)\}.$$

What we actually want is for x and y to have the same 2-cardinal in the new sense. As before, if there is an (old) bijection between $\texttt{fst}(k‘x)$ and $\texttt{fst}(k‘y)$, there will be a new bijection between x and y by low comprehension. And the same goes not only for x and y, but for each $x' \in_{\mathbf{co}} x$ and $y' \in_{\mathbf{co}} y$ that are paired by the bijection: if there is an (old) bijection between $\texttt{fst}(k‘x')$ and $\texttt{fst}(k‘y')$, there will be a new bijection between x' and y' by low comprehension as desired. ⊣

It should now be clear how to tinker with this construction to add simultaneously for all $n \in \mathbb{N}$, the assertion that every (low set of)^n low sets has an n-cardinal. It is perhaps worth noting that it does not seem to be necessary to argue that, for each $n \in \mathbb{N}$, we can do this for all $m < n$ and then use compactness.

3.2. Wellfounded sets in CO-structures

The roots that this technique has in Rieger-Bernays permutation models still have fruit to bear, as witness the following two theorems.

THEOREM 3.21. *For a given choice of K and V and rules, all* **CO** *structures are permutation models of each other.*

Proof. Suppose we have two **CO** structures over a model $\langle V, \in\rangle$ with the same K but different coding functions k and k' respectively. We wish to find $\sigma \in \mathit{Symm}(V)$ such that the first model thinks that $x \in y$ iff the second thinks $x \in \sigma‘y$. σ must be $(k')^{-1} \circ k$. ⊣

We also have the following:

PROPOSITION 3.22. *Every permutation model* $\langle V, \in_{\mathbf{co}}\rangle^\sigma$ *of* $\langle V, \in_{\mathbf{co}}\rangle$ *is obtained from it by replacing k by some k′ with a corresponding new membership relation* $\in_{\mathbf{co}'}$. *If the permutation is* σ, *then the new* k' *is* $\sigma^{-1}k$.

Proof.

$$\langle V, \in_{\mathbf{co}}\rangle^\sigma \models x \in y \text{ iff } \begin{cases} \langle V, \in_{\mathbf{co}}\rangle \models x \in \sigma‘y, \\ \langle V, \in\rangle \models x \in_{\mathbf{co}} \sigma‘y, \\ \langle V, \in\rangle \models x \in_{\mathbf{co}'} y, \\ \langle V, \in_{\mathbf{co}'}\rangle \models x \in y. \end{cases}$$

⊣

THEOREM 3.23.

(1) H_{low} *is always isomorphic to a permutation model of the original universe.*

(2) *Whatever* K *we started with, for any permutation* σ *of the old universe we can find a coding function* k *so that* $\langle H_{\text{low}}, \in_{\mathbf{co}}\rangle \simeq \langle V, \in_\sigma\rangle$.

Proof.
(1) There will be a bijection $\pi : V \longleftrightarrow H_{\text{low}}$. We seek a σ so that

$$(\forall xy)(x \in \sigma`y. \longleftrightarrow .\pi`x \in_{\mathbf{co}} \pi`y).$$

What is $\sigma`y$? Clearly it has to be $\{x : \pi`x \in_{\mathbf{co}} \pi`y\}$. This is a set, since $\pi`y$ is low. We must check that this definition gives us a σ that is 1-1 and onto. It is certainly 1-1 by extensionality of $\in_{\mathbf{co}}$. Is it onto? Given z we must find a y so that $z = \{x : \pi`x \in_{\mathbf{co}} \pi`y\}$. This y must be $\pi^{-1}`k^{-1}`\langle \pi``z, 0\rangle$.

(2) We know—however we choose k—that H_{low} is a proper class whose complement is a proper class, so let π be a bijection between V and such a class, and let us fasten on that class to be H_{low} and resolve to cook up k so that it actually *is* the H_{low} of the new model. Dugald Macpherson has used the word "moiety" for things that are both infinite and coinfinite: we will borrow it here to describe proper classes whose complements are proper classes. Let σ be a permutation of V. We want to cook up k so that $\langle H_{\text{low}}, \in_{\mathbf{co}}\rangle \simeq \langle V, \in_\sigma\rangle$. We want

$$(\forall xy)(x \in \sigma`y. \longleftrightarrow .\pi`x \in_{\mathbf{co}} \pi`y).$$

The right hand side is $\pi`x \in \mathtt{fst}(k`(\pi`y))$ (and $\mathtt{snd}(k`\pi`y) = 0$ since $\pi`y$ is a low set). Now $\mathtt{fst}(k`\pi`y) = \pi``\sigma`y$, so we want $k`\pi`y = \langle \pi``\sigma`y, 0\rangle$. It is true that this only tells us what k should do to values of π, but since the range of π is a moiety and the range of $k \circ \pi$ is also a moiety, there will be no problem extending this to a bijection between V and all the ordered pairs we need. ⊣

Analogously with the embedding $i : V \hookrightarrow V^\sigma$ defined in the theorem model into the new model $\langle V, \in_{\mathbf{co}}\rangle$ defined by recursion on $\in$:

DEFINITION 3.24. $i`x =: k^{(-1)}`\langle i``x, 0\rangle$.

Like the i of Theorem 2.6, this embedding is a $\mathcal{P}$-embedding.

THEOREM 3.25. *If* $\langle V, \in\rangle$ *is well-founded, then* i *is defined and is a* $\mathcal{P}$*-embedding from* $\langle V, \in\rangle$ *into* $\langle V, \in_{\mathbf{co}}\rangle$.

Proof. First we prove by $\in$-induction that i is defined on all sets.

We must next check that i is an isomorphism, so we want $i`x \in_{\mathbf{co}} i`y \longleftrightarrow x \in y$. Since the second component of $k`(i`y)$ is 0, $i`x \in_{\mathbf{co}} i`y$ iff $i`x \in \mathtt{fst}(k`(i`y)) = i``y$ iff $x \in y$.

Next we show that the range of i is transitive$^{(\in_{co})}$. Suppose y is in the range of i, and $y = i`z$. So $x \in_{co} i`z = k^{(-1)}`\langle i``z, 0\rangle$ iff $x \in i``z$, so x would also be in the range of i.

Finally we must check that any subset of something in the range of i is likewise in the range of i. Suppose x is in the range of i, so that $k`x = \langle i``z, 0\rangle$ for some $z$. Suppose also that $(\forall w)(w \in_{co} y \rightarrow w \in_{co} x)$. Consider the set (in the sense of the original model) of those things that are $\in_{co}$ $y$. This is indeed a set of the original model, since it is a subset of $i``z$. If it is $i``u$, then $k`y$ must be $\langle i``u, 0\rangle$, so y is in the range of i. ⊣

(Essentially, this proof is in *Church 1974* though he prefers to allege that the well-founded sets form a model of *ZF* and does not have the concept of a $\mathcal{P}$-embedding.)

The difference here from Theorem 2.6 is that this time we cannot expect i to be elementary for stratified formulæ. (It's a good question which formulæ it is elementary for, actually!) However we can at least derive the following:

COROLLARY 3.26. *The $\mathcal{P}$-embedding $i : \langle V, \in\rangle \rightarrow \langle H_{\text{low}}, \in_{co}\rangle$ is elementary for stratified formulæ.*

Proof. Since H_{low} is isomorphic to a permutation model of V (by Theorem 3.23) we can use Theorem 2.6. ⊣

Does this mean that we can take the well-founded sets of the new model to be the range of i? No, because we can arrange for the range of i to be a set of the new model, and it will be in some sense well-founded.

PROPOSITION 3.27. *The range of the canonical embedding can be a set of the new model.*

Proof. We use a slight modification of Oswald's original construction from section 3.1.1: if x is a finite subset of H, then $\Sigma_{n \in x} 2^{n+1} \in H$. Then say $x \in_{co} y$ iff *either*: y is $2n$ and the xth bit of n is 1, *or* y is $2n+3$ and the xth bit of n is 0, *or* $y = 1$ and $x \in H$.

To complete the proof we think of $\mathbb{N}$ as a copy of V_ω by associating with it the Ackermann relation alluded to above. The canonical embedding i is then defined on everything in V_ω which is a set of the new model. ⊣

However, although it is possible for the range of i to be a set, it is not always a set.

PROPOSITION 3.28. *If the three conditions of section 3.1.1 are met, the range of i is not a set.*

Proof. We want to show that if X satisfies $(\forall y \in_{co} X)$(y is in the range of i), then X is in the range of i too. The case $\mathtt{snd}(k`X) = 0$ we have already considered. There remains the case $\mathtt{snd}(k`X = 1)$.

We will show that this case cannot occur. If it did, $x \in_{\mathbf{co}} X$ iff $x \notin$ $\mathtt{fst}(k`X)$. So $X$ would be so large that there is only a set of things that are not members$^{(\in_{\mathbf{co}})}$ of it, and all the things that *are* would be values of $i$. But the collection of values of $i$ is not so big that its complement is a set, since if $x \in$ range $i$, $\mathtt{snd}(k`x) = 0$, and there is a proper class of x such that $\mathtt{snd}(k`x) = 1$. Therefore $\mathtt{snd}(k`X) \neq 1$. ⊣

The possibility of adding new well-founded sets in the new model in this *ad hoc* way restricts the things we can say in general about well-founded sets, at least if we take well-founded sets to be those defined in the natural way by means of the inductive definition or as regular sets (Definition 3.29).

The inductive definition is the obvious one: if we think of well-founded sets as those over which we can do $\in$-induction then we are led to the inductive definition:

$$WF(x) \longleftrightarrow (\forall y)((\forall z)(z \subseteq y \to z \in y) \to x \in y). \tag{2}$$

The trouble with this definition is that if we have very little comprehension there may be so few y such that $(\forall z)(z \subseteq y \to z \in y)$ that lots of sets might turn out to be well-founded that should not. In fact, in *ZF* there are no such sets at all and this definition is completely useless! We talk of **regular** sets instead.

DEFINITION 3.29. x is regular iff $(\forall y)(x \in y \to (\exists z \in y)(z \cap y = \emptyset))$.

If we have an axiom of complementation, we can prove:

PROPOSITION 3.30. *A set is well-founded in the inductive sense iff it is regular.*

Proof. Suppose $(\forall y)((\forall z)(z \subseteq y \to z \in y) \to x \in y)$. Substitute $-y$ for y, getting $(\forall y)((\forall z)(z \cap y = \emptyset \to z \notin y) \to x \notin y)$. Contrapose, getting $(\forall y)(x \in y \to \neg(\forall z)(z \cap y = \emptyset \to z \notin y))$. This is $(\forall y)(x \in y \to (\exists z)(z \cap y = \emptyset \wedge z \in y))$, which says that x is regular. ⊣

The inductive definition enables us to prove that every well-founded set has ϕ as long as the extension of ϕ is a set and any set of things-which-are-ϕ is itself ϕ. This is a lot less applicable than it might seem because of the extreme unlikelihood that the extension of ϕ should be a set. We can prove that if $x \in x_1 \in x_2 \ldots x_n \in x$ then $-\{x, x_1 \ldots x_n\}$ is a set y such that $(\forall z)(z \subseteq y \to z \in y)$, and so no well-founded set is a member^n of itself, but one would expect to be able to prove a lot more.

There is also a principle of induction for regular sets. This is standard in modern treatments of set theory, and we do it as follows. Suppose we know $(\forall x)((\forall y \in x)(\phi(y)) \to \phi(x))$, and suppose there is a regular set z such that $\neg\phi(z)$. Let Y be a transitive set containing z. Let $X = \{y \in Y : \neg\phi(y)\}$. Then $z \in X$ and z is regular, so X must be disjoint from one of its members. But this contradicts the induction hypothesis, so $\phi(z)$.

How does this fare in the Church-Oswald context? If we know that any set of things-which-are-ϕ is itself ϕ, and that for any regular z there is a transitive set Y containing z such that $\{y \in Y : \neg\phi(y)\}$ exists, then we can prove that every regular set is ϕ. Can we do this for all regular z and all ϕ? If we could show that for every regular z there is a low transitive Y with $z \in Y$, then we can do the rest by low comprehension. We could do this if we knew that every well-founded set is in the range of i.

H_{low}, the hereditarily low sets, might capture better the idea of sets well-founded in the new model. Here we have to remember that foundation fails and that therefore there are several notions of hereditarily low. In particular there may be hereditarily low sets that are not well-founded, for example Quine atoms, which are sets equal to their own singletons. Though this is easy to arrange, it is also easy to avoid. We can arrange that every set that is hereditarily low (in the sense that everything in its transitive closure is low) is also in the range of i. Suppose $\langle x_n : n \in \mathbb{N}\rangle$ is a descending ω-sequence under $\in_{\mathbf{co}}$ and that x_0 is hereditarily low. Then for all n, $\mathtt{snd}(k`x_n)$ is 0, so $x_{n+1} \in_{\mathbf{co}} x_n$ is just $x_{n+1} \in \mathtt{fst}(k`x_n)$, so x_{n+1} is a set of lower rank than $\mathtt{fst}(k`x_n)$. If we set up $k$ so that for all $y$ the rank of $\mathtt{fst}(k`y)$ is no greater than the rank of y, then there can be no such infinite descending sequence in the new model and all hereditarily low sets are well-founded. In particular this will happen whenever the three conditions of section 3.1.1 are met.

What this says is that we can arrange that the only sets that are illfounded are large, or at least have something large in their transitive closure. There is a similar problem in *NF*: must self-membered sets be large? When I asked this question in the first edition of *Forster 1995*, Maurice Boffa came up with a permutation model which partly answered my question. The permutation he used (and which he and Pétry reported in *1993*) uses a device a bit like a rank function and it admits a great deal of refinement. It is now possible using results of Körner (*1994*) to show that it is consistent relative to *NF* that $\in$ restricted to finite sets is well-founded. This is done by Rieger-Bernays permutation models. It sounds a weaker result than arranging for $\in$ restricted to low sets to be well-founded, but bear in mind that in the present case all we have to do is prevent the appearance of new ill-founded small sets in the new model, whereas in the *NF* case we not only have to do that but we have to kill off any small ill-founded sets in the model we started with, and—being a model of *NF*—it may have had lots.

Nice though it is to know that the greatest fixed point for "set of all low subsets of" and the least fixed point can be the same, what really matters is that it is the least fixed point we want for our concept of well-founded in the new model. A treatment now follows.

DEFINITION 3.31. Let us say x is well-founded* if $(\forall X)((\forall y)((low(y) \wedge y \subseteq X) \rightarrow y \in X) \rightarrow x \in X)$.

In other words the collection of well-founded* sets is the least fixed point corresponding to the greatest fixed point H_{low}.

We will justify a principle of unrestricted $\in$-induction for well-founded* sets.

First we check (i) that the collection of well-founded* sets is transitive and (ii) that everything in it is low.

(i) If x is well-founded*, and $(\forall y)((low(y) \wedge y \subseteq X) \to y \in X)$, then $x \in X$. It will suffice to show that $x \subseteq X$ as well.

Suppose $x \not\subseteq X$. We will show that $(\forall y)(low(y) \wedge (y \subseteq (X - \{x\})) \to y \in (X - \{x\}))$, whence $x \in (X - \{x\})$ (since x is well-founded*). This is impossible.

Suppose $y \subseteq (X - \{x\})$ and is low. Then $y \subseteq X$ and $y \in X$. To deduce $y \in (X - \{x\})$ it will suffice to show $y \neq x$, which would follow from $x \not\subseteq (X - \{x\})$. But we have assumed that $x \not\subseteq X$, so *a fortiori* $x \not\subseteq (X - \{x\})$.

(ii) Suppose x is well-founded* but not low, and that $(\forall y)((low(y) \wedge y \subseteq X) \to y \in X)$. Let y be a low subset of $(X - \{x\})$. Then y is a low subset of X and is therefore a member of X. We want it to be a member of $X - \{x\}$ so we want $y \neq x$. But y is low and x is not. Therefore $(\forall y)((low(y) \wedge y \subseteq X) \to y \in (X - \{x\}))$ and $x \in (X - \{x\})$, since x is well-founded*.

Next we need to know that $TC(x)$ exists if x is well-founded*. We prove by induction on $\mathbb{N}$ that if x is well-founded*, then $\bigcup^n x$ exists and is low. Low comprehension tells us that a low set of low sets has a low sumset. The induction is good because $\mathbb{N}$ is a low set and so there is enough comprehension for full induction over it. So the collection $\{\bigcup^n x : n \in \mathbb{N}\}$ is a low set of low sets and its sumset—$TC(x)$—is a low set.

Next by relativising the proof of proposition 3.30, we show that x is well-founded* iff $(\forall y)(x \in y \to (\exists z \in y)(x \cap y = \Lambda \wedge low(z)))$. ($\Lambda$ is the empty set.)

Finally, we prove $\in$-induction as follows. Suppose we know $(\forall x)((\forall y \in x)(\phi(y)) \to \phi(x))$, and suppose there is a well-founded* set z such that $\neg\phi(z)$. Let $X = \{y \in TC(\{z\}) : \neg\phi(y)\}$ which exists by low comprehension. $z \in X$ and z is well-founded*, so X must be disjoint from one of its members. But this contradicts the induction hypothesis, so $\phi(z)$. ⊣

We can prove this unrestricted scheme of $\in$-induction scheme for well-founded* sets despite being unable to prove it for sets well-founded in the inductive sense we began with because there may well be large well-founded sets about which we know nothing.

Armed with this scheme of $\in$-induction for well-founded* sets we can prove that every well-founded* set is in the range of i.

Perhaps the moral of this discussion is that we should restrict ourselves to coding functions k which add no new well-founded sets. In those circumstances "low" is definable in the new language as "same size as a well-founded set" as Church originally intended, and the whole enterprise becomes axiomatizable. It is true that we lose some generality by making this restriction, but we sacrificed generality at the outset of this section by not considering models of *ZF* without foundation. Neither of these seems a harsh sacrifice.

4. OPEN PROBLEMS

4.1. The Axioms of Sumset and Power set

At present the only form of the axiom of sumset we can prove for this construction is that a low set of low sets has a sumset (which will be low). Can we tweak this construction to get a less restricted axiom of sumset, dropping the second or perhaps even the first occurence of 'low' in the above? Similarly the only form of the axiom of power set that this construction apparently gives us is power sets of low sets. With neither of these two axioms is there a standard paradox obviously skulking in the wings waiting to cause trouble should we adopt unrestricted forms of them. It's natural to see if we can do better than this. Mitchell's set theory allows power set, but has other disadvantages. For example, $x \cup y$ and $x \cap y$ do not exist in general.

The question is also discussed in Sheridan's thesis.

4.2. Natural strengthenings of Theorem 3.15

This is the most general and the most important of the open problems in this area. Such extensions would be theorems of the form "Every well-founded model of T_1 is the well-founded part of a model of T_2". It is perhaps in this connection that Church's remark in *1974* seems most apposite: "One source of added axions to be studied for their consistency with the basic axioms is ... Quine set theory. ... An interesting possibility ... is a synthesis or partial synthesis of *ZF* and Quine set theory." Let *KF* be the subsystem of Zermelo set theory obtained by dropping the axiom of infinity and restricting the assonderung (comprehension) scheme to stratified Δ_0 formulæ. A natural conjecture for *NF* to make in this context would be that:

CONJECTURE 4.1. *If NF is consistent, then every (well-founded) model of KF is the well-founded part of a model of NF.*

Another possibility is:

CONJECTURE 4.2. *Every model of KF has a $\mathcal{P}$-extension that is a model of NF+ low replacement.*

There is a basic difficulty in the path of anyone trying to prove anything like Conjectures 4.1 and 4.2. The **CO** construction is a method which—on being presented with (i) a robust method of constructing term models for a trivial subsystem T of *NF*, and (ii) a model $\mathcal{M} \models ZF$—outputs a model of

T which has $\mathcal{M}$ as an initial segment of its well-founded part. To make this work, we have to have the method of generating term models for T *in the hand*, as it were. It is not enough to know that T is consistent. (The system *NF*∀ of Forster (*1987*) has a canonical term model so we might be able to use Church-Oswald constructions to show that every (well-founded) model of *KF* is the well-founded part of a model of *NF*∀.) *A fortiori* we do not have a method of proving conditionals of the kind: if $T \subseteq NF$ is consistent, and $\mathcal{M}$ is a model of *ZF*, then there is $\mathcal{M}' \models T$ of which $\mathcal{M}$ is the well-founded part. There are plenty of interesting assertions of this kind, and it would be very nice to know which of them were true.

4.3. Axiomatizability

We should axiomatize the theories that **CO** constructions give us: low comprehension is obviously important but not obviously axiomatizable. We could do this if we could find in the language of $\in_{\mathbf{co}}$ and $=$ a predicate which meant "low". We can do this if the model we started with satisifes the three condition of section 3.1.1, but it just may be possible to say something even without these extra assumptions.

4.4. Constructive **CO** *constructions*

Another question is: is there a sensible constructive treatment of **CO** constructions? The problem is the very classical nature of **CO** constructions: $\mathtt{snd}(k`x)$ is always equal to 0, or to something else. This seems to mean that if we want to execute intuitionistic **CO** constructions, we will have to assume *tertium non datur* for atomics. It would be quite illuminating to develop **CO** constructions inside intuitionistic *ZF* and see what happens to the transfinite recursive versions of the negative interpretation. It is probably quite hard. Indeed, as unpublished work of Dzierzgowski has shown, it is extraordinarily diffcult to get nontrivial models for intuitionistic versions even of systems as straightforward as NF_2.

4.5. Schröder-Bernstein

The point has been made that in Church's theory and its kin we have no comprehension to speak of for big sets.[7] This means that even apparent banalities can turn out to be hard to prove. For example can we prove Schröder-Bernstein for big sets in Church's set theory? There are various proofs of the Schröder-Bernstein theorem. There is a slick one suitable for use with computer science students and others who have been exposed to the Tarski-Knaster fixpoint theorem for complete lattices. Suppose $f : A \hookrightarrow B$

[7] *Big* sets are things like complements of singletons. They are not *large* in the way that, say, measurable cardinals are large.

and $g : B \hookrightarrow A$. Then $\lambda A'.(A - g``(B - f``(A'))$ is a continuous function on the complete lattice $\mathcal{P}`A$ and must have a fixpoint. However this proof depends on the power set axiom, which is not available in Church's theory. There is a much more low-tech proof *not* using power set for which the outlook is slightly more hopeful. It goes like this.

Consider the two sequences defined by a mutual recursion:

$$b_0 = B - f``A; \qquad b_{n+1} = f``a_n,$$

$$a_0 = A - g``B; \qquad a_{n+1} = g``b_n.$$

Set
$$A' = \bigcup_{n \in \mathbb{N}} a_n.$$

The bijection we want is $f|A' \cup g^{-1}|(A - A')$.

There seem to be two major hurdles. (i) Are all the b_n and the a_n sets? If so then $\{a_n : n \in \mathbb{N}\}$ is a set by low comprehension. (ii) If $\{a_n : n \in \mathbb{N}\}$ is a set, is its sumset a set? If A and B are low, then, by low comprehension, all the b_n and the a_n are low sets, so $\{a_n : n \in \mathbb{N}\}$ is a low set of low sets and its sumset is a low set, and there seems no reason to expect that the bijection $f|A' \cup g^{-1}|(A - A')$ will not also be a low set. However if A and B are merely sets, not low sets, there does not seem to be any reason to expect this proof to work. There may be some other way of proving Schröder-Bernstein but there is no reason to expect that.

One special case can be disposed of easily. In models satisfying the three conditions of section 3.1.1, Schröder-Bernstein will be true, for the following very unsatisfactory reason. In such models every set is either low or co-low. Schröder-Bernstein will certainly hold for the low sets, and no low set will be the same size as a co-low set. If A and B are two co-low sets then the model cannot contain any injections from A into B or B into A since any such injection would be neither low nor co-low (it will be a *moiety*) and must be absent from the model. So Schröder-Bernstein will be vacuously true for non-low sets. (Indeed, because of the absence of moieties, no non-low set can even be the same size as itself!) This argument cannot be used in the case of Church's model, because there are plenty of moieties there.

4.6. Mitchell's set theory

Emerson Mitchell's Ph.D. Thesis (*1976*) contains a **CO**-like construction. His system has power set for all sets (not just low sets) but does not have closure under (binary) $\cup$ and $\cap$.

4.7. Extensional quotients?

I close by raising an obvious but completely unexplored possibility. The enduring difficulty with these constructions is extensionality. It's easy to show that x and y have the same members (in the sense of $\in_{\mathbf{co}}$) as long as $\mathtt{snd}(k`x) = \mathtt{snd}(k`y)$, but if $\mathtt{snd}(k`x) \neq \mathtt{snd}(k`y)$, we have our work cut out, and it tends to be doable only for theories that have an easily solvable word problem. One thing one could naturally do is start by being much more reckless in one's choice of K and membership conditions for non-low sets, and pick up the pieces later by taking an extensional quotient of the result. It's hard to see what might be preserved in a development like this, but that means that there may be many consistency proofs waiting to be revealed by such an extension of the method. My present feeling is that the possibility of developing CO techniques along these lines is their most exciting feature. There is no obvious reason why this should not hold the key that will one day unlock the consistency question of *NF*.

REFERENCES

Boffa, M., and Pétry, A.

1993 On self-membered sets in Quine's set theory *NF*, *Logique et Analyse*, vol. 141–142, pp. 59–60.

Church, A.

1974 Set theory with a universal set, *Proceedings of the Tarski symposium*, Proceedings of Symposia in Pure Mathematics (L. Henkin, editor), vol. XXV, Providence, pp. 297–308; also in *International Logic Review*, vol. 15, pp. 11–23.

Forster, T. E.

1987 Term models for weak set theories with a universal set, *The Journal of Symbolic Logic*, vol. 52, pp. 374–387.

1995 *Set theory with a universal set: Exploring an untyped universe*, second edition, Oxford Logic Guides, no. 31, Clarendon Press, Oxford.

Forster, T. E., and R. Kaye

1991 End-extensions preserving power set, *The Journal of Symbolic Logic*, vol. 56, pp. 323–328.[8]

Grishin, V. N.

1969 Consistency of a fragment of Quine's *NF* system, *Soviet Mathematics Doklady*, vol. 10, pp. 1387–1390.

Henson, C. W.

1973b Permutation methods applied to *NF*, *The Journal of Symbolic Logic*, vol. 38, pp. 69–76.

Jensen, R. B.

1969 On the consistency of a slight(?) modification of Quine's *NF*, *Synthese*, vol. 19, pp. 250–263.

[8] Errata, p. 327: Line 11 should read 'and $a \in M$ such that $M \models \overline{\overline{\pi}}`a = \overline{\overline{\mathcal{P}}}`a$'. Line 13: the expression following '$M \models$' should be '$\overline{\overline{\pi}}`a = \overline{\overline{\mathcal{P}}}`a$'. Line 26: '(not just $\pi`a = \mathcal{P}`a$)' should read '(not just $\overline{\overline{\pi}}`a = \overline{\overline{\mathcal{P}}}`a$)'. Line 28: '$\pi`a$' should read '$\overline{\overline{\mathcal{P}`\pi}}`a$'.

Körner, F.

1994 Cofinal indiscernibles and some applications to New Foundations, *Mathematical Logic Quarterly*, vol. 40, pp. 347–356.

Mitchell, E.

1976 A model of set theory with a universal set, *Ph.D. thesis*, Madison, Wisconsin.

Oswald, U.

1976 Fragmente von "New Foundations" und typentheorie, *Ph.D. thesis*, ETH, Zürich, 46 pp.

Scott, D. S.

1962 Quine's individuals, *Logic, Methodology and Philosophy of Science* (E. Nagel, editor), Stanford University Press, pp. 111–115.

Tarski, A.

1986 What are logical notions?, *History and Philosophy of Logic*, vol. 7, pp. 143–154.

ROBIN O. GANDY

AXIOMS OF INFINITY IN CHURCH'S TYPE THEORY

Abstract. In his 1940 paper Church gave an elegant formulation of the simple theory of function-types. Higher order arithmetic is represented in it almost without artifice; the only artificial device used is a particular formulation of the axiom of infinity. We discuss the significance of this axiom, and possible alternatives.

1.

Church's system has two basic types: the type o of propositions and the type ι of individuals. If α and β are type symbols then $(\alpha\beta)$ is the symbol for the type $(\beta \to \alpha)$ of functions from type β into type α. Terms (with types indicated by subscripts) are formed from variables and constants by application and λ-abstraction. Brackets may be omitted from both type symbols and terms with association to the left. We adopt the convention that type subscripts may be omitted from constants introduced by definition or by hypothesis, when this can be done without ambiguity, and from all occurrences of a bound variable except its binding occurence.

Church's rules I–VI and axioms (1)–(6) cover the rules of λ-conversion, classical propositional calculus and quantification theory at all types. Equality, at all types, is defined by:

$$A_\alpha = B_\alpha =_{\text{df}} (f_{o\alpha})(fA_\alpha \supset fB_\alpha).$$

This system, together with Church's axiom:

(7) $(\exists x_\iota, y_\iota)(x \neq y)$

we shall call (C).

Further possible axioms are the axioms of descriptions:

9^α $(f_{o\alpha})((\exists! x_\alpha)(fx) \supset f(\iota_{\alpha(o\alpha)}f))$

and of extensionality

$10^{\alpha\beta}$ $(f_{\alpha\beta}, g_{\alpha\beta})((x_\alpha)(fx = gx) \supset f = g)$;

we shall also count

10° $(p_o, q_o)(p \equiv q \supset p = q)$

as an axiom of extensionality. We use (C), (CD), (CDE) for the corresponding systems.

C. Anthony Anderson and M. Zelëny (eds.), Logic, Meaning and Computation, 139-147.

2.

Church introduces the natural numbers as iterators over the type. Let ι' stand for $\iota\iota(\iota\iota)$. Then we define:

$$\begin{aligned} 0_{\iota'} &=_{\text{df}} \lambda f_{\iota\iota} x_\iota . x, \\ S_{\iota'\iota'} &=_{\text{df}} \lambda n_{\iota'} f_{\iota\iota} x_\iota . f(nfx), \\ N_{o\iota'} &=_{\text{df}} \lambda n_{\iota'} . (f_{o\iota'})[f0 \,\&\, (m_{\iota'})(fm \supset f(Sm)) . \supset fn]. \end{aligned}$$

These definitions may also be made over an arbitrary type α (with α' for $\alpha\alpha(\alpha\alpha)$). We shall omit the type subscripts from $O_{\iota'}, S_{\iota'\iota'}, N_{o\iota'}$ and adopt the convention that the letters m, n, p, q (without subscripts) stand for variables of type ι' restricted to range only over $N_{o\iota'}$.

A consequence of axiom (7) is

$$(m)(0 \neq Sm).$$

Church's axiom (8) is

$$(m, n)(Sm = Sn \supset m = n).$$

Thus (in (C)) the numbers in type ι' satisfy all of Peano's axioms, and hence all theorems of classical higher-order arithmetic can be proved in (CDE)+(8).

Unlike most recursive functions, + and $\times$ can be defined without using descriptions:

$$\begin{aligned} m + n &=_{\text{df}} \lambda f_{\iota\iota} x_\iota . mf(nfx); \\ m \times n &=_{\text{df}} \lambda f_{\iota\iota} x_\iota . m(\lambda y_\iota . nfy)x. \end{aligned}$$

3.

We will write $(P3)^\alpha$ for the result of replacing ι by α in (8); it is Peano's third postulate for the numbers in type α'. In *Newman and Turing 1942* it was shown that if ι occurs in α then $(P3)^\alpha$ is provable in (CDE)+(8). They introduced the notion 'n belongs to its own posterity' defined by

$$\text{BOP}_{o\iota'} =_{\text{df}} \lambda n. (\exists m)(n = S(m+n)) \tag{3.1}$$

and they proved, in (C);

$$\sim(8) \equiv \exists n.\text{BOP}n \tag{3.2}$$

(and the corresponding theorem with ι replaced by α).

And, assuming $(\exists n)\text{BOP}n$, they prove, in effect,

$$(\exists! p)(\exists! q)[\sim \text{BOP}p \,\&\, S(Sp+q) = Sp]. \tag{3.3}$$

Plainly if p and q satisfy the formula within the [ ] for (3.3), then BOP(Sp) and in an intuitive sense (which will be formalized in the next section) Sp+q is the 'largest' possible number. We need a name for it. When we are assuming $\sim$(8) (or $(\exists n)$BOPn) we assume that M satisfies

$$N_{o\iota'}M \ \& \ (\exists p)(\sim \text{BOP}p \ \& \ SM = Sp). \tag{3.4}$$

In fact, M is unique; either it is its own successor or its successor 'precedes' it.

4.

We are going to prove that (8) is equivalent to 'there are not finitely many individuals'. We need a definition of finite which can be applied even when we assume (8) is false. To construct this we need to work in (CD). Church showed that in this system, definition by primitive recursion is permissible (as, also, is the use of conditional definitions). Of course if we are assuming that M satisfies (3.4), then the recursion equation cannot be used to define the value of a function at SM, but it will work at other arguments. Also when using induction, the inductive step is

$$(n)(n \neq M \ \& \ A(n) \,.\supset A(Sn)).$$

4.1.

In (CD) we define a function $D_{\iota'\iota'\iota'}$ by recursion as follows:

(i) $Dx_{\iota'}y_{\iota'} = 0$ if $\sim Nx_{\iota'}$ or $\sim Ny_{\iota'}$.

(ii) $Dm0 = 0$.

(iii) $Dm(Sn) = Dmn$ if $n \neq M$ and $m \neq n$,
$= 1$ if $n \neq M$ and $m = n$.

And we write $m < n$ for $Dmn = 1$.

Then $<$ is a linear ordering of the numbers, with M as the greatest. Specifically in (CD) the following can be (successively) inferred from (3.4) using induction on n where necessary.

4.2.

(1) $n \neq M \supset n < Sn$.

(2) $n \neq M \supset .\, m < Sn \equiv m \leq n$.

(3) $p < m \supset .\, m < n \supset p < n$.

(4) $n \not< n$.

(5) $n \neq M \supset n \neq Sn$.

(6) $m \neq M \supset . Sm < n \equiv (m < n \,\&\, n \neq Sm)$.

(7) $m < n \vee m = n \vee n < m$.

(8) $n \leq M$.

(9) $SM \leq M$.

4.3. Remarks

(1) Rather than using descriptions it would be possible simply to postulate the existence of a function satisfying 4.1. (2) If one defines $m \leq n$ by $(\exists p)(n = m + p)$, then BOPm & BOPn . $\supset m \leq n$, so this definition is not suitable. (3) If we assume (8) instead of ~(8), then 4.1 and 4.2 are to be modified by omitting all clauses which involve M.

4.4.

Now we write

$$Seg_n =_{\text{df}} \lambda x_{\iota'} . Nx \,\&\, x \leq n.$$

We shall not be concerned with the empty set, so we use $\leq$ rather than $<$; Seg_n has, intuitively, $n+1$ members, and is a paradigm for sets of cardinal $n+1$. Since we have not assumed axiom 10° our use of the term 'set' here, and in what follows should be treated as a convenient abuse of language.

4.5.

Now we write

$$\text{INJ}^\alpha(f_{\alpha\beta} : X_{o\beta}) =_{\text{df}} (x_\beta, y_\beta)[(X_{o\beta}x \,\&\, X_{o\beta}y \,\&\, f_{\alpha\beta}x = f_{\alpha\beta}y) \supset x = y].$$

This says that f is an injection from the domain given by $X_{o\beta}$ into the type α. Similarly we can define $\text{SUR}^\alpha(f_{\alpha\beta} : X_{o\beta})$ to say that $f_{\alpha\beta}$ maps the domain $X_{o\beta}$ onto type α, and $\text{BIJ}^\alpha(f_{\alpha\beta} : X_{o\beta})$ for bijection, and we define

$$\text{INT}^\alpha(f_{\alpha\beta} : X_{o\beta}) =_{\text{df}} \text{INJ}^\alpha(f_{\alpha\beta} : X_{o\beta}) \,\&\, {\sim}\text{SUR}^\alpha(f_{\alpha\beta} : X_{o\beta})$$

to express that $f_{\alpha\beta}$ is an injection from $X_{o\beta}$ which is *strictly* into type α.

4.6.

Finally we write

$$\text{FIN} =_{\text{df}} (\exists n)(\exists f_{\iota\iota'})(\text{BIJ}^\iota(f : Seg_n))$$

to express that there are finitely many individuals. It is a strong assertion, as it implies that there are many functions in type $\iota\iota$.

5.

Now we prove

THEOREM 5.1. $\sim$FIN $\supset$ (8) *is provable in* (CD).

We take $\sim$FIN as a hypothesis, H say.

LEMMA 5.2. *Let* $A(n)$ *be* $(\exists f_{\iota\iota'})(\mathrm{INT}^{\iota}(f : Seg_n))$. *Then* $(n)A(n)$ *can be inferred from* H *in* (CD).

Proof. $A(0)$ is an immediate consequence of axiom (7). And

$$A(n) \supset (\exists g_{\iota\iota'})(\mathrm{INJ}^{\iota}(g : Seg_n))$$

is easily proved; hence, using H we have

$$A(n) \supset A(Sn). \qquad \dashv$$

Note that this proof, unlike a proof of $(\exists f_{\iota\iota'})(\mathrm{INJ}^{\iota}(f : N_{o\iota'}))$, does not require an axiom of choice.

Now to prove the theorem we assume $\sim$(8) and that M satisfies (3.4) and seek a contradiction.

By the Lemma we may suppose that $F_{\iota\iota'}$ satisfies

$$\mathrm{INT}^{\iota}(F : Seg_M)$$

and there is a z_ι not in the range of F. Define $G_{\iota\iota'}$ by

$$\begin{aligned} Gx_\iota = {} & F(S(\mu m)(m < M \mathbin{\&} x = Fm)) \\ & \text{if } (\exists m)(m < M \mathbin{\&} x = Fm), \\ & = z \text{ ow}. \end{aligned}$$

(In (CD) the μ-operator is definable.) Then, since $(\exists p)(p < M \mathbin{\&} M = Sp)$ follows from (3.4), by induction we prove

$$mG(F0) = Gm \text{ for } m \leq M.$$

But then $SMG(F0) = G(FM) = z$. Therefore $SMG(F0) \neq mG(F0)$, and $SM \neq m$, for $m \leq M$. Thus $SM \not\leq M$ which contradicts 4.2(1).

So, on eliminating the various hypotheses we have a proof of $\sim$FIN $\supset$ (8) in (CD). $\dashv$

To prove the converse of Theorem 5.1 we need an axiom of extensionality.

THEOREM 5.3. *In* (CDE) *one can prove*

$$\text{FIN} \supset \sim (8).$$

Proof. We take FIN and (8) as hypotheses, and seek a contradiction. Since all ordinary mathematical arguments can be formalized in (CDE)+(8) we argue informally (and write f^m in place of mf).

Suppose there are $N+1$ individuals. Then for any x_ι and any $f_{\iota\iota}$

$$x, f^1 x, \ldots, f^{N+1} x$$

cannot all be distinct. Hence for some $0 \leq m \leq N, 1 \leq n \leq N+1$.

$$f^m x = f^{m+n} x.$$

So if $n \mid q$ then

$$f^N x = f^{N+q} x.$$

Let P be the LCM of $1, 2, \ldots, N+1$. Then for all f, x we have

$$f^N x = f^{N+P} x.$$

Then, using the axiom of extensionality, we have

$$N = N + P. \tag{*}$$

Thus BOPN, which by (3.2), contradicts (8). ⊣

5.1. *Remarks*

(1) The use of extensionality is unavoidable. One can make a model for (CD)+FIN in which all the terms

$$N, SN, S(SN), \ldots,$$

are distinct, and so BOP(N) fails and (8) holds.

(2) The numbers in (*) are the best possible. In (CD) one can prove that if $n < N$ or $p < P$ then

$$n \neq n + p.$$

(3) In particular, by considering a cyclic permutation of the $N+1$ individuals one can prove $N+1 \neq N$ and hence that if M satisfies (3.4), then

$$M > N.$$

5.2.

Church also considers the proposition, called 30^ι, which states that the function $T_{\iota'\iota''}$ defined on $N_{o\iota''}$ by

$$Tm_{\iota''} = m_{\iota''}SO$$

is injective: there are not more numbers in $N_{o\iota''}$ than there are in $N_{o\iota'}$. He proves in (C)

$$(8) \supset 30^\iota.$$

In a footnote he credits Turing with a proof of the converse in (CD). We can use remark (3), lifted up to $N_{o\iota''}$ to get such a proof. Suppose $\sim$(8) and hence that there is a largest number M as in sections 3 and 4.

Define

$$m'_{\iota''} = (\mu n_{\iota''})(Tn = m).$$

Then T gives a bijection between $Seg_{M'}$ and $N_{o\iota'}$. Now we replace the type of individuals by $N_{o\iota'}$ and consider numbers in $N_{o\iota''}$. Let $Q_{\iota''}$ be the largest number: i.e. it satisfies (3.4) with ι replaced by ι''. Lifting remark (3) to ι'' we see that

$$Q > M'.$$

But then T cannot be an injection; thus

$$\sim(8) \supset \sim 30^\iota$$

is a theorem of (CD). Whether it is also a theorem of (C) remains open (cf. 5.4 below).

5.3.

It is not hard to see that the use of descriptions in the proof of Theorem 5.1 is essential. For one can construct a model for (CE) which has three individuals $\kappa_0, \kappa_1, \kappa_2$ but in which the only functions in type $\iota\iota$ are the identity and the constant functions. Hence in this model $S1 = 1$. One can define $<$ by $\lambda mn.\, m = 0 \;\&\; n = 1$, and then the results of §4 hold, with $M = 1$. Even if, as one may, one adds an injection from Seg_1 into type ι, it is plain that FIN and (8), and hence $\sim$FIN $\supset$ (8) are all false in this model.

5.4.

In fact I have discovered a general method for constructing models (like that in Theorem 5.3) of (C). Let (C^+) be obtained from (C) by adding constants for individuals, for equality between individuals, and if required, for (extensional) functions in various types, together with axioms governing the added constants. Then one can show that the (normal forms of the) closed terms of

(C$^+$) form a model for (C$^+$E). A proof of this theorem will appear in *Gandy 1995.* It can be used to construct a model in which $(P3)^{\iota'}$ holds but (8) fails. I have tried (using a small number of individuals, and selected functions) to make models in which $(30)^{\iota}$ is true but (8) is false. I had hoped to make good use of Schwichtenberg's characterization (in *his 1976*) of the λ-definable functions from $N_{o\iota'}$ to $N_{o\iota'}$, but the presence of added constants complicates matters.[1]

6. CONCLUDING REMARKS

(1) It is well known that without the axiom of choice (for type ι) one cannot infer from ~FIN that type ι is Dedekind infinite. Hence Theorem 5.1 fails if one replaces FIN by 'Dedekind finite'.

(2) From a conceptual (and from a model-theoretic) point of view, ~FIN is a more natural axiom of infinity than (8). But Turing, who had done a lot of formal work on Church's system, was surprised when in 1953, I showed him the informal proof that they are equivalent. The emphasis, both at Princeton and Göttingen, on formal proofs and metamathematics tended to discourage thought about the intuitive meaning of axioms.

(3) Church's paper shows how arithmetic can be founded using the idea that the natural numbers are iterators, rather than finite cardinals or ordinals. If, like Church, one seeks to avoid the use of descriptions and extensionality whenever possible, then (8) is the preferred axiom of infinity. But the proofs in *Gandy 1956* can be easily modified to show that (CDE) is consistent relative to (C). So, provided one is ready to put up with the formal details of sections 3 and 4 (which are necessary for the definition of FIN), one can take ~FIN as the axiom of infinity.

REFERENCES

Church, A.

1940 A formulation of the simple theory of types, *The Journal of Symbolic Logic*, vol. 5, pp. 56–68.

Gandy, R. O.

1956 On the axiom of extensionality, *The Journal of Symbolic Logic*, vol. 21, pp. 36–48.

1995 A closed term model for Church's type theory, in preparation.

Newman, M. H. A., and A. M. Turing

1942 A formal theorem in Church's theory of types, *The Journal of Symbolic Logic*, vol. 7, pp. 28–33.

[1]Since writing the above, I have managed to show that $30^{\iota} \supset (8)$ is a theorem of (CE), but is false in a non-extensional model of (C).

Schwichtenberg, H.
1976 Definierbare funktionen im λ-kalcül mit typen, *Archiv für mathematische Logik und Grundlagenforschung,* vol. 17, pp. 113–114.

EDWARD L. KEENAN

LOGICAL OBJECTS

INTRODUCTION

This paper began, and begins, as a reflection on the following claim by Putnam (*1981,* page 217): "*Theorem*: Let $\mathcal{L}$ be a language with predicates F_1, $F_2, \ldots, F_k$ (not necessarily monadic). Let $\mathcal{I}$ be an interpretation, in the sense of an assignment of an intension to every predicate of $\mathcal{L}$. Then if $\mathcal{I}$ is non-trivial in the sense that at least one predicate has an extension which is neither empty nor universal in at least one possible world, there exists a second interpretation $\mathcal{J}$ which disagrees with $\mathcal{I}$, but which makes the same sentences true in every possible world as $\mathcal{I}$ does. ... If at least one predicate, say, F_u, has an extension R_{uj} which is neither empty nor all of $\mathcal{U}_j$, select a permutation P_j of $\mathcal{U}_j$ such that $P_j(R_{uj}) \neq R_{uj}$. ..."

Lewis (*1984*) accepts some form of Putnam's claim, and Lakoff (*1987,* chapter 15: "Putnam's Theorem ... The Inconsistency of Model Theoretic Semantics") uses it to argue against the applicability of model theoretic semantics to the study of natural language.

But we show, in section 1, that Putnam's claim, as stated, is false. We reformulate the claim, proving that *permutation invariance* (PIness) rather than non-triviality is what binds reference (extensions of predicates) to truth. In section 2 we argue, contra Etchemendy (*1990*), that invariance under automorphisms (AIness, a conceptual generalization of PIness) characterizes the denotations of "logical" expressions, including but not limited to classical logical constants. In section 3 we defend our main thesis: that the logical objects of a given type in a given ontology are the AI ones. Our analysis derives without circularity (Theorem 22) that being "logical" on our thesis is itself a logical property. Section 4 concludes with an open problem.

Previous work, originating with Klein (*1872*) and including *Weyl 1949, Mostowski 1957, Silva 1945, van Fraassen 1990, Keenan 1991, Keenan and Stabler 1991*; *Keenan and Stabler 1995* identifies the *structural* properties of objects in a given domain with those invariant under the structure preserving permutations of that domain. That the *logical* properties of objects may be identified with those invariant under all permutations was, to my knowledge, first proposed in a lecture by Tarski in 1966, edited and published as *Tarski 1986* by J. Corcoran. Tarski there explicitly builds on Klein's Erlanger Program (*1872*; see *Klein 1892* and *van Fraassen 1990,* page 266) in

C. Anthony Anderson and M. Zelëny (eds.), Logic, Meaning and Computation, 149-180.

which geometries of various sorts are distinguished according as the objects of study are invariant under one or another class T of transformations (= permutations). E.g., "Euclidean objects" are ones invariant under translations, rotations, etc. Then the weaker the invariance conditions the transformations must satisfy the larger the equivalence classes of objects. And Tarski makes the natural extrapolation: the most general objects, the "logical" ones, are those invariant under all transformations.[1] It is also this association of generality with PIness under the heading "symmetry" in the physical science literature that lies behind van Fraassen's use of the role of symmetry in characterizing scientific laws.

The idea that "PIness" = "logicality" is developed most extensively in *van Benthem 1989a.* Plotkin *1980,* Keenan and Stavi *1986,* van Benthem *1991,* Marshall and Chuaqui *1991,* and Sher *1991* also support this idea. We differ only slightly from these in emphasizing the type and structure dependency of "logical" objects. This enables us to explicate the sense in which, e.g., set membership both is, and is not, a logical relation. We draw on natural language studies to exhibit a variety of novel logical objects in high types (*van Benthem 1983, 1991*; *Westerståhl 1985*; *Keenan and Moss 1985*; *Keenan and Stavi 1986*).

Notation. $[A \rightarrow B]$ is the set of functions from A into B. If $f \in [A \rightarrow B]$ and $\mathrm{K} \subseteq A$, $f(\mathrm{K})$ is $\{f(x) \mid x \in \mathrm{K}\}$; for $s \in A^n$, $f(s)$ is $\langle f(s_1), \ldots, f(s_n)\rangle$. The power set of A is denoted $\mathcal{P}_A$ or $\mathcal{P}(A)$. Languages $\mathcal{L}$ are assumed first order relational (without zero place relation symbols) unless indicated otherwise. *Models* for such $\mathcal{L}$s are pairs $\mathrm{M} = (\mathrm{E_M}, \mathcal{I})$, $\mathrm{E_M}$ a non-empty set, the *universe* of M (usually denoted E), and $\mathcal{I}$ a function mapping the relation (function) symbols of $\mathcal{L}$ to relations (functions) over E. As usual, $\mathcal{I}$ lifts recursively to an *interpretation* $\mathcal{I}_\mathrm{M}$ of $\mathcal{L}$ relative to assignments of values to the variables. For ϕ a sentence write $\mathrm{M} \models \phi$ for $\mathcal{I}_\mathrm{M}(\phi) = 1$, the latter meaning that $\mathcal{I}_\mathrm{M}(\phi)(g) = 1$, for all assignments g. $\mathsf{Th}[\mathrm{M}]$, the theory of M, $=_{\mathrm{df}} \{\phi \in \mathsf{Sent}(\mathcal{L}) \mid \mathrm{M} \models \phi\}$. Models M, M' are *elementarily equivalent*, $\mathrm{M} \cong \mathrm{M}'$, if $\mathsf{Th}[\mathrm{M}] = \mathsf{Th}[\mathrm{M}']$. For $\mathrm{M} = (\mathrm{E}, \mathcal{I})$ and α a bijection with domain E, $\alpha\mathrm{M} =_{\mathrm{df}} (\alpha\mathrm{E}, \alpha \circ \mathcal{I})$ is a model. For M, M' models, M' is isomorphic ($\simeq$) to M iff $\mathrm{M}' = \alpha\mathrm{M}$, some bijection $\alpha : \mathrm{E_M} \rightarrow \mathrm{E_{M'}}$. And of course

(1) $\mathrm{M} \simeq \mathrm{M}' \Rightarrow \mathrm{M} \cong \mathrm{M}'$ (so $\mathrm{M} \cong \alpha\mathrm{M}$, all bijections α with domain $\mathrm{E_M}$).

---

[1]The notion of invariance developed here differs in two respects from that in *Tarski 1986*: (1) Tarski does not treat Truth and Falsity, or truth functions (AND, OR, ...) built on them, as objects to be preserved by permutations. He just considers a set theoretic hierarchy built up from the universe and notes that none of its elements are PI, only two subsets are, only four binary relations are. At higher levels relations like ***subset*** and ***disjoint*** will be PI. (2) Tarski's paper is informal. He does not explicitly define a type hierarchy and identify the PI objects as those fixed by all permutations. He even remarks that while the logical properties of classes are the numerical ones, the idea "is quite difficult to formulate in a precise way" (page 151).

1. BINDING REFERENCE AND TRUTH

To evaluate Putnam's claim let $\mathcal{L}_P$ be of signature $\{P\}$, P a P_1 (one place predicate symbol). Then, in a possible world w with universe $\mathcal{U}_w$, if $\mathcal{I}$ interprets P as non-empty and non-universal, so that $\emptyset \subset \mathcal{I}(P) \subset \mathcal{U}_w$, let π be a permutation of $\mathcal{U}_w$ interchanging some $x \in \mathcal{I}(P)$ with some $y \notin \mathcal{I}(P)$ and fixing everything else. As per (1) then, $(\mathcal{U}_w, \mathcal{I}) \cong (\mathcal{U}_w, \pi \circ \mathcal{I})$, but $\mathcal{I} \neq \pi \circ \mathcal{I}$ since $x \in \mathcal{I}(P)$ and $x \notin (\pi \circ \mathcal{I})(P)$, in conformity with Putnam's claim.

But Putnam's claim fails when P is a P_n, $n \geq 2$. Let $\mathcal{U}_w$ be the doubleton $\{a, b\}$ and let $\mathcal{I}(P) = \{\langle a, b\rangle, \langle b, a\rangle\}$. $\mathcal{I}(P)$ is not $\emptyset$ nor universal $= \{a, b\} \times \{a, b\}$, but all models $(\{a, b\}, \mathcal{J})$ with $\mathcal{J} \neq \mathcal{I}$ disagree on some sentence of $\mathcal{L}_P$.

Proof. (a), (b) and (c) below are false in $\mathrm{M} = (\mathcal{U}_w, \mathcal{I})$:

(a) $\exists x(xPx)$ (b) $\forall x \forall y \neg(xPy)$ (c) $\exists x \exists y(xPy \,\&\, \neg yPx)$

Let $(\{a, b\}, \mathcal{J})$ be a model of $\mathcal{L}$ with $\mathcal{J}(P) \neq \mathcal{I}(P)$. If $\langle c, c\rangle \in \mathcal{J}(P)$ for some $c \in \mathcal{U}_w$ then (a) above is true in $\mathcal{J}$. So we may assume that for all $c \in \mathrm{E}$, $\langle c, c\rangle \notin \mathcal{J}(P)$. If no pair $\langle c, d\rangle$ of distinct elements of $\mathcal{U}_w$ is in $\mathcal{J}(P)$ then $\mathcal{J}(P)$ is empty and (b) above is true. So at least one of $\langle a, b\rangle$, $\langle b, a\rangle$ is in $\mathcal{J}(P)$. If both are then $\mathcal{J} = \mathcal{I}$ contrary to assumption. So just one is, and in each case (c) is true. This exhausts the cases. ⊣

Lemma 2 reformulates Putnam's claim in terms of *permutation invariance* (PIness) not non-triviality (the two notions coinciding on P_1s). And Theorem 4 tells us that PIness is equivalent to covariation of truth and reference (holding the universe constant). Theorems 7 and 12 partially generalize Theorem 4 to show that for sufficiently rich $\mathcal{L}$s, PIness guarantees a kind of expressive completeness: such $\mathcal{L}$s describe their models up to isomorphism. First,

DEFINITION 1.

(i) An n-ary relation R over a universe E is *permutation invariant* (PI) iff $\pi R = R$, for all permutations π of E. (Recall: $\pi(R) = \{\pi(d) \mid d \in R\}$). For $R \in [\mathrm{E}^m \to \mathrm{E}]$ this just says that $R(\pi d) = \pi(R(d))$, all $d \in \mathrm{E}^m$.

(ii) A *model* (E,I) is *PI* iff for all $P \in \mathsf{Dom}(\mathcal{I})$, $\mathcal{I}(P)$ is PI.

LEMMA 1. *For* $R \subseteq \mathrm{E}^n$, R *is not* PI iff $\exists s \in \mathrm{E}^n\, \exists \pi \in \mathrm{PERM(E)}$, $s \in R$ *and* $\pi s \notin R$.

Proof.

⇐ If $s \in R$ then $\pi s \in \pi R \neq R$ since $\pi s \notin R$; so R is not PI.

⇒ Let R such that for $\delta \in \mathrm{PERM(E)}$, $\delta R \neq R$.

Case 1: $\exists t \in R - \delta R$. Then the righthand side above holds choosing $s = t$ and $\pi = \delta^{-1}$.

Case 2: $\exists t \in \delta R - R$. Then $\delta^{-1}t \in \delta^{-1}\delta R = R$ and $t \notin R$ so the righthand side above holds choosing $\pi = \delta$ and $s = \delta^{-1}t$. ⊣

LEMMA 2. *Let* $\mathrm{M} = (\mathrm{E}, \mathcal{I})$ *a model for* $\mathcal{L}$ *and* P *a relation symbol of* $\mathcal{L}$ *with* $\mathcal{I}(P)$ *not* PI. *Then* $\mathcal{L}$ *has a model* $\mathrm{M}' = (\mathrm{E}, \mathcal{J}) \cong \mathrm{M}$ *such that* $\mathcal{J}(P) \neq \mathcal{I}(P)$.

Proof. Given $\mathrm{M} = (\mathrm{E}, \mathcal{I})$ with $\mathcal{I}(P)$ not PI. By Lemma 2 there is a tuple s and a permutation π of E such that $s \in \mathcal{I}(P)$ and $\pi(s) \notin \mathcal{I}(P)$. Then $\pi\mathrm{M} = (\mathrm{E}, \pi \circ \mathcal{I}) \simeq \mathrm{M}$ so $\mathsf{Th}[\pi\mathrm{M}] = \mathsf{Th}[\mathrm{M}]$. But $\mathcal{I} \neq \pi \circ \mathcal{I}$ since $\pi(s) \notin \mathcal{I}(P)$ but $\pi(s) \in (\pi \circ \mathcal{I})(P) = \pi(\mathcal{I}(P))$ since $s \in \mathcal{I}(P)$. ⊣

COROLLARY 3. *Every non-*PI *model has an elementarily equivalent model with the same universe which disagrees with it on some predicate symbol.*

THEOREM 4 **Stability of Reference**. *Let* $\mathcal{L}$ *have equality. Then for all models* M *for* $\mathcal{L}$,

$$\mathrm{M} \textit{ is } \mathrm{PI} \Leftrightarrow \forall \mathrm{M}' \textit{ with } \mathrm{E}_{\mathrm{M}'} = \mathrm{E}_{\mathrm{M}},\ \mathrm{M}' \cong \mathrm{M} \leftrightarrow \mathrm{M}' = \mathrm{M}.$$

Proof.

⇒ Let $\mathrm{M} = (\mathrm{E}, \mathcal{I})$ be PI. Let $\mathrm{M}' = (\mathrm{E}, \mathcal{J})$. The ← direction of the consequent is trivial. Assume $\mathrm{M}' \neq \mathrm{M}$, so $\mathcal{J} \neq \mathcal{I}$. We show $\mathrm{M}' \not\cong \mathrm{M}$. Let P be of arity n with $\mathcal{J}(P) \neq \mathcal{I}(P)$. Let $t \in (\mathcal{I}(P) - \mathcal{J}(P)) \cup (\mathcal{J}(P) - \mathcal{I}(P))$, let x be an n-tuple $x_1, \ldots, x_n$ of distinct variables and let $\mathsf{Dist}(x)$ be the formula $\text{AND}\{(x_i = x_j) \mid t_i = t_j, 1 \leq i \neq j \leq n\}$ & $\text{AND}\{(x_i \neq x_j) \mid t_i \neq t_j, 1 \leq i \neq j \leq n\}$, where for K a finite set of formulas, $\text{AND}(\mathrm{K}) = \forall x(x = x)$ if $\mathrm{K} = \varnothing$, and otherwise AND(K) is the conjunction of the formulas in K. Consider $\phi = \forall x(\mathsf{Dist}(x) \rightarrow Px)$. If $t \in \mathcal{I}(P) - \mathcal{J}(P)$, then $\mathrm{M} \models \phi$ and $\mathrm{M}' \not\models \phi$; if $t \in \mathcal{J}(P) - \mathcal{I}(P)$, then $\mathrm{M}' \models \phi$ and $\mathrm{M} \not\models \phi$.

⇐ Follows immediately from Lemma 2. ⊣

Theorem 4 tells us that it is precisely PIness that guarantees that changing extensions of predicates changes the truth of sentences. Does Theorem 4 generalize, replacing identity of universes with sameness of cardinality and equality of models with isomorphism? In fact only half the generalization obtains. We first establish the useful:

LEMMA 5. *A model* M *is* PI iff *for all* $\pi \in \text{PERM}(\mathrm{E}), \pi\mathrm{M} = \mathrm{M}$.

Proof.

⇒ Let M be PI and $\pi \in \text{PERM}(\mathrm{E})$. Then for all $P \in \mathsf{Dom}(\mathcal{I})$, $\mathcal{I}(P) = \pi(\mathcal{I}(P)) = (\pi \circ \mathcal{I})(P)$. Since P was arbitrary $\mathcal{I} = \pi \circ \mathcal{I}$, so $\pi\mathrm{M} = (\pi\mathrm{E}, \pi \circ \mathcal{I}) = (\mathrm{E}, \mathcal{I}) = \mathrm{M}$.

⇐ Let $P \in \mathsf{Dom}(\mathcal{I})$. For $\pi \in \text{PERM}(\mathrm{E})$, $\pi(\mathcal{I}(P)) = (\pi \circ \mathcal{I})(P) = \mathcal{I}(P)$, so $\mathcal{I}(P)$ is PI. ⊣

LEMMA 6. *Let* M *be a model for* $\mathcal{L}$. *Then* [a] $\Rightarrow$ [b] *and if* $\mathcal{L}$ *has equality,* [b] $\Rightarrow$ [a].

[a] *For all models* M', $M \cong M'$ $\Rightarrow$ *for all bijections* $\pi : E_M \to E_{M'}$, $M' = \pi M$,

[b] M *is* PI.

Proof.

[a] $\Rightarrow$ [b]. Assume [a] and let $M = (E, \mathcal{I})$, let $\delta \in \text{PERM}(E)$. Show $\delta M = M$, whence M is PI by Lemma 5. Since $M \simeq \delta M$, then $M \cong \delta M$, so for all bijections $\pi : E \to E_{\delta M} = \delta E = E$, $\delta M = \pi M$. Choosing $\pi = \mathrm{id}_E$ yields $\delta M = M$.

[b] $\Rightarrow$ [a]. Let $\mathcal{L}$ have $=$, let $M = (E, \mathcal{I})$ be PI and let $M' = (E', \mathcal{I}')$ with $M \cong M'$. If $|E| \neq |E'|$, then [a] holds vacuously. So let π be a bijection: $E \to E'$. Show $\pi M = M'$. Assume, leading to a contradiction, that equality fails. Then $\pi \circ \mathcal{I} \neq \mathcal{I}'$, so for some $P \in \mathsf{Dom}(\mathcal{I})$, $\mathcal{I}'(P) \neq (\pi \circ \mathcal{I})(P)$. Now if there is a $t \in (\pi \circ \mathcal{I})(P) - \mathcal{I}'(P)$, then, where $\mathsf{Dist}(x)$ is formed as in Theorem 4, $\phi = \forall x(\mathsf{Dist}(x) \to Px)$ is true in πM and false in M'. But $\pi M \simeq M$ so $\pi M \cong M$ so ϕ is true in M, contradicting $M \cong M'$. And if there is a $t \in \mathcal{I}'(P) - (\pi \circ \mathcal{I})(P)$, then ϕ is false in πM and true in M'. And again since $\mathsf{Th}[\pi M] = \mathsf{Th}[M]$, ϕ is false in M, contradicting that $\mathsf{Th}[M] = \mathsf{Th}[M']$. This exhausts the cases proving the lemma. ⊣

THEOREM 7. *Let* M *be a model of a relational language* $\mathcal{L}$ *with equality. Then*

$$M \text{ is PI} \Rightarrow \forall M' \text{ with } |E_{M'}| = |E_M|,\ M' \cong M \text{ iff } M' \simeq M.$$

Proof. Given M PI and M' with $|E_{M'}| = |E_M|$. Let $\pi : E_{M'} \to E_M$ be a bijection. Assume $M' \cong M$. By Lemma 6, $\pi M = M'$ and, (1), $M \simeq \pi M$. ⊣

FACT 1. The converse to Theorem 7 is false.

Proof. Let $\mathcal{L}$ have signature $\{P\}$, P of arity 2. Let $M = (\{a, b\}, \mathcal{I})$ with $\mathcal{I}(P) = \{\langle a, b\rangle\}$. Then for all M' with $|E_{M'}| = 2, \mathsf{Th}[M'] = \mathsf{Th}[M] \Rightarrow M' \simeq M$. In general (*Enderton 1972,* page 96, example 17) finite models which are elementarily equivalent are isomorphic. Here is a demonstration for this case. Note that $\mathsf{Th}[M]$ includes:

$$\text{(a) } \neg \exists x Pxx \qquad \text{(b) } \exists xy Pxy \quad \text{and} \quad \text{(c) } \forall x, y(Pxy \to \neg Pyx).$$

Let $M' = (\{\alpha, \beta\}, \mathcal{I}')$ with $\alpha \neq \beta$. Assume $\mathsf{Th}[M'] = \mathsf{Th}[M]$. By (a) neither $\langle \alpha, \alpha\rangle$ nor $\langle \beta, \beta\rangle$ is in $\mathcal{I}'(P)$. By (b) either $\langle \alpha, \beta\rangle$ or $\langle \beta, \alpha\rangle$ is in $\mathcal{I}'(P)$ and, by (c), not both. So $\mathcal{I}'(P) = \{\langle \alpha, \beta\rangle\}$ or $\mathcal{I}'(P) = \{\langle \beta, \alpha\rangle\}$. In the first case the map sending a to α and b to β is an isomorphism; in the second the map sending a to β and b to α is. So $M' \simeq M$. But $\mathcal{I}(P)$ is not PI. The π interchanging a and b fails to fix $\{\langle a, b\rangle\}$. ⊣

FACT 2. In Theorem 7 the condition that $|\mathrm{E_M}| = |\mathrm{E_{M'}}|$ is necessary.

Proof. Let $\mathcal{L}_=$ be the zero-signature first order $\mathcal{L}$ with equality. For any E, $(\mathrm{E}, \mathcal{I})$ is a model of $\mathcal{L}_=$, since $\mathcal{I}(=)$ is required to be $\{\langle a, a\rangle \mid a \in \mathrm{E}\}$. Now let M and M′ be models of $\mathcal{L}_=$, where $\mathrm{E_M} = \mathbb{N}$, the natural numbers, and $\mathrm{E_{M'}} = \mathcal{P}(\mathbb{N})$. By Theorem 9, $\mathsf{Th}[\mathrm{M}] = \mathsf{Th}[\mathrm{M'}]$; but $\mathrm{M} \neq \mathrm{M'}$ since their universes differ in cardinality. ⊣

LEMMA 8. *Let* M, M′ *be infinite models for* $\mathcal{L}_=$. *Then for all formulas* ϕ *in* $\mathcal{L}_=$, *for all assignments* $g : \mathrm{VAR} \to \mathrm{E_M}$ *and all assignments* $g' : \mathrm{VAR} \to \mathrm{E_{M'}}$

if $g(x) = g(y)$ iff $g'(x) = g'(y)$, *for all* $x, y \in \mathrm{FREEVAR}(\phi)$,
then $\mathrm{M} \models \phi[g]$ iff $\mathrm{M'} \models \phi[g']$.

Proof. By recursion on formula complexity (Appendix). ⊣

THEOREM 9. *All infinite models of* $\mathcal{L}_=$ *have the same theory.*

Proof. Let $\phi \in \mathsf{Sent}(\mathcal{L}_=)$, let M, M′ infinite models of $\mathcal{L}_=$. Clearly all $g : \mathrm{VAR} \to \mathrm{E_M}$ and $g' : \mathrm{VAR} \to \mathrm{E_{M'}}$ identify the same variables free in ϕ. So for all g, g', $\mathrm{M} \models \phi[g]$ iff $\mathrm{M'} \models \phi[g']$, so $\mathrm{M} \models \phi$ iff $\mathrm{M'} \models \phi$. Thus $\mathsf{Th}[\mathrm{M}] = \mathsf{Th}[\mathrm{M'}]$. ⊣

N.B.: It is *only* the inability to state the cardinality of the universe which prevents the more general statement that for PI models, sameness of theories and model isomorphism coincide. This latter statement is naturally understood as a kind of expressive completeness requirement: languages whose models satisfy it can describe themselves up to isomorphism (Theorem 12).

DEFINITION 2 **Cardinality Quantifiers**. Let $\mathcal{L}$ be a first order language. We expand $\mathcal{L}$ to $\mathcal{L}(\mathrm{CARD})$ by adding the cardinality quantifiers Q_λ for each cardinal λ, as follows:

syntax: For each cardinal λ, Q_λ is a *quantifier symbol* (as are ∀ and ∃). For all quantifier symbols Q, all variables x, and all formulas ϕ, $\mathrm{Q}x\phi$ is a formula.

semantics: $\mathrm{M} \models \mathrm{Q}_\lambda x\phi[g]$ iff $|\{a \in \mathrm{E_M} \mid \mathrm{M} \models \phi[g[x/a]]\}| = \lambda$.

In expanding $\mathcal{L}$ to $\mathcal{L}(\mathrm{CARD})$ we add new clauses to the definition of satisfaction but we do not add new symbols to be interpreted. Hence the models $\mathrm{M} = (\mathrm{E}, \mathcal{I})$ for $\mathcal{L}(\mathrm{CARD})$ are exactly those for $\mathcal{L}$; the definition of PIness for models is unchanged. We need:

LEMMA 10. PI*ness is preserved by model isomorphism.*

Proof. Let $\mathrm{M} = (\mathrm{E}, \mathcal{I})$ be PI, let $\delta : \mathrm{M} \simeq \mathrm{M}' = (\mathrm{E}', \mathcal{I}')$. Show M′ is PI. Let $\pi \in \text{PERM}(\mathrm{E}')$. Show $\pi(\mathcal{I}'(P)) = \mathcal{I}'(P)$, for all $P \in \mathsf{Dom}(\mathcal{I}')$. Note that $\mathcal{I}'(P) = (\delta \circ \mathcal{I})(P) = \delta(\mathcal{I}(P))$. And since $\delta^{-1} \circ \pi \circ \delta \in \text{PERM}(\mathrm{E})$ and M is PI, $(\delta^{-1} \circ \pi \circ \delta)(\mathcal{I}(P)) = \mathcal{I}(P)$, so $\delta((\delta^{-1} \circ \pi \circ \delta)(\mathcal{I}(P))) = \delta(\mathcal{I}(P)) = \mathcal{I}'(P)$. But $\delta((\delta^{-1} \circ \pi \circ \delta)(\mathcal{I}(P))) = (\pi \circ \delta)(\mathcal{I}(P)) = \pi(\mathcal{I}'(P))$, whence $\pi(\mathcal{I}'(P)) = \mathcal{I}'(P)$, so M′ is PI. ⊣

LEMMA 11. *Let* $\pi : \mathrm{M} \simeq \mathrm{M}'$. *Show* $\forall \phi \in \mathcal{L}(\text{CARD})$, $\forall g : \text{VAR} \to \mathrm{E}_{\mathrm{M}}$,

$$\mathrm{M} \models \phi[g] \text{ iff } \mathrm{M}' \models \phi[\pi \circ g].$$

Proof. Set $S = \{\phi \in \mathcal{L}(\text{CARD}) \mid \forall g : \text{VAR} \to \mathrm{E}_{\mathrm{M}}, \mathrm{M} \models \phi[g] \text{ iff } \mathrm{M}' \models \phi[\pi \circ g]\}$. See *Enderton 1972* (pages 91–92) for the proof that S includes the atomic formulas and is closed under conjunction, negation and universal quantification. We show that S is closed under cardinal quantification.

Let $\phi \in S$. Show for all g, all cardinals λ that $\mathrm{M} \models \mathrm{Q}_\lambda x\phi[g]$ iff $\mathrm{M}' \models \mathrm{Q}_\lambda x\phi[\pi \circ g]$.

Observe that $\forall g : \text{VAR} \to \mathrm{E}_{\mathrm{M}}, \forall x \in \text{VAR}, \forall a \in \mathrm{E}_{\mathrm{M}}$, $\pi \circ g[x/a] = (\pi \circ g)[x/\pi a]$:

$$\text{(a) } (\pi \circ g[x/a])(y) = \begin{cases} \pi(g[x/a](y)) = \pi(a) & \text{if } y = x, \\ \pi(g(y)) & \text{if } y \neq x, \end{cases}$$

$$\text{(b) } (\pi \circ g)[x/\pi a](y) = \begin{cases} \pi a & \text{if } y = x, \\ (\pi \circ g)(y) = \pi(g(y)) & \text{if } y \neq x. \end{cases}$$

And we infer: $\mathrm{M} \models \phi[g[x/a]]$ iff $\mathrm{M}' \models \phi[\pi \circ g[x/a]]$ Induction Hypothesis.
iff $\mathrm{M}' \models \phi[(\pi \circ g)[x/\pi a]]$ the observation.

Thus the map $\pi \restriction \{a \in \mathrm{E}_{\mathrm{M}} \mid \mathrm{M} \models \phi[g[x/a]]\} \to \{\pi a \in \pi(\mathrm{E}_{\mathrm{M}}) = \mathrm{E}_{\mathrm{M}'} \mid \mathrm{M}' \models \phi[(\pi \circ g)[x/\pi a]]\}$ is onto (and one-to-one, since π is one-to-one). Thus

$$|\{a \in \mathrm{E}_{\mathrm{M}} \mid \mathrm{M} \models \phi[g[x/a]]\}| = |\{\pi a \in \mathrm{E}_{\mathrm{M}'} \mid \mathrm{M}' \models \phi[(\pi \circ g)[x/\pi a]]\}|.$$

Whence for all cardinals λ, $\mathrm{M} \models \mathrm{Q}_\lambda x\phi[g]$ iff $\mathrm{M}' \models \mathrm{Q}_\lambda x\phi[\pi \circ g]$, completing the proof. ⊣

THEOREM 12 **Expressive Completeness**. *For all* $\mathcal{L}$ *with* =, *for all models* M *for* $\mathcal{L}(\text{CARD})$,

$$\mathrm{M} \textit{ is } \mathrm{PI} \Rightarrow \textit{for all models } \mathrm{M}' \textit{ for } \mathcal{L}(\text{CARD}),\ \mathrm{M} \cong \mathrm{M}' \text{ iff } \mathrm{M} \simeq \mathrm{M}'.$$

Proof. Given $\mathcal{L}$ with =, let M and M′ be models of $\mathcal{L}(\text{CARD})$ with M PI.

⇒ Let $\mathsf{Th}[\mathrm{M}] = \mathsf{Th}[\mathrm{M}']$. For $\lambda = |\mathrm{E}_{\mathrm{M}}|$, $\mathrm{M}' \models \mathrm{Q}_\lambda(x = x)$ since $\mathrm{M} \models \mathrm{Q}_\lambda(x = x)$, so $|\mathrm{E}_{\mathrm{M}'}| = |\mathrm{E}_{\mathrm{M}}|$. Let $\pi : \mathrm{E}_{\mathrm{M}} \simeq \mathrm{E}_{\mathrm{M}'}$. By Lemma 11, $\mathrm{M}' = \pi\mathrm{M}$, so $\mathrm{M}' \simeq \mathrm{M}$ by Lemma 10.

$\Leftarrow$ For M, M′ isomorphic models of $\mathcal{L}$(CARD) and $\phi \in$ Sent($\mathcal{L}$(CARD)), Lemma 11 implies that M $\models \phi$ iff M′ $\models \phi$, whence Th[M] = Th[M′], completing the proof. ⊣

Theorem 12 entails that elementary equivalence of models of $\mathcal{L}$(CARD) preserves PIness. But we have a stronger result, whose proof incorporates some observations used later:

THEOREM 13. *For $\mathcal{L}$ with =, M and M′ models of $\mathcal{L}$, if M is PI & M $\cong$ M′ then M′ is PI.*

Proof sketch. As noted earlier, the PI n-ary relations form a complete atomic subalgebra of $\mathcal{P}(\mathrm{E}^n)$, denoted PI($\mathrm{E}^n$).

13.1 For each $s \in \mathrm{E}^n$, set $[s] =_{\mathrm{df}} \{t \in \mathrm{E}^n \mid t_i = t_j \text{ iff } s_i = s_j, \text{ for all } 1 \leq i, j \leq n\}$. Then $[s]$ is non-empty ($s \in [s]$) and PI, and no proper non-empty subset of $[s]$ is PI, so $[s]$ is an atom of PI(E^n). Whence for R an n-ary PI relation on E, $R = \bigcup_{s \in R}[s]$.

13.2 For $s \in \mathrm{E}^n$ let $s_\equiv$ be that equivalence relation on $\{1, \ldots, n\}$ given by: $i\ s_\equiv\ j$ iff $s_i = s_j$. Let $F(s)$ be the formula (AND$\{(x_i = x_j) \mid i\ s_\equiv\ j\}$ & AND$\{(x_i \neq x_j) \mid \neg i\ s_\equiv\ j\}$), where AND(K) is as defined earlier for K a finite set of formulas. Also we write OR(K) for $\exists x(x \neq x)$ if K = $\varnothing$; otherwise OR(K) is the disjunction of the elements of K.

13.3 For M = (E, $\mathcal{I}$) a model of $\mathcal{L}$ with = and P a P_n, one proves that

$$\mathcal{I}_{\mathrm{M}}(P) \text{ is PI iff } \mathrm{M} \models \forall x_1, \ldots, x_n(Px_1, \ldots, x_n \leftrightarrow \mathrm{OR}\{F(s) \mid s \in \mathcal{I}_{\mathrm{M}}(P)\}).$$

Write 'P' is *PI* for the formula to the right of '$\models$' above. It says that 'P' denotes $\varnothing$ or the non-trivial union of the $[s]$ for the tuples s in its extension. That is, it says "'P' is PI".

13.4 Now, let $\mathcal{L}$ have =, let M and M′ be models with M $\cong$ M′ and M PI. Then for each predicate symbol P (including '='), 'P' is *PI* is true in M and thus in M′, so M′ is PI. ⊣

2. LOGICAL OBJECTS AND LOGICAL CONSTANTS

We have seen that PI relations play a special role with respect to the truth of sentences. Here we extend this observation to a characterization of "logical objects", objects we intend to include as denotations of classical logical constants as well as an infinitely enriched class drawn from natural language. So we cannot ad hocly define the logical objects to be the denotations of a listed set of constants. Rather we need a criterion for including new objects among the "logical" ones. So contra Etchemendy (*1990*), we seek a property common to denotations of logical expressions. And we claim that a certain generalization of PIness is that property.

First, just as the "equals" predicate is a standard logical one and for each E, its denotation $\mathsf{ID}_\mathrm{E} = \{\langle a, a\rangle \mid a \in \mathrm{E}\}$ is PI, so, when interpreted directly, the denotations of other standard logical expressions will be "logical objects". For example, interpreting the standard quantifiers directly (*Westerståhl 1985*) we think of $\exists x\phi$ as $\exists\lambda x\phi$, where $\lambda x\phi$ is interpreted as a set (or its characteristic function) and $\exists$ is interpreted as a function from sets to truth values, viz., that function mapping a subset A of the universe to *True* iff A is not empty. Similarly $\forall$ maps A to *True* iff A is the universe. (Note the universe dependence of $\forall$.) These functions must count as logical objects, though we must say something about their PIness.

Similarly we want to say that the truth functions denoted by classical *and, or,* and *not,* are logical objects, though naive application of PIness here appears to yield a wrong result: the table for AND contains the line $\langle\langle \mathrm{T}, \mathrm{F}\rangle, \mathrm{F}\rangle$. But under a permutation of $\{\mathrm{T}, \mathrm{F}\}$ which interchanges T and F the resulting table contains $\langle\langle \mathrm{F}, \mathrm{T}\rangle, \mathrm{T}\rangle$ which is not a line in the table for *and,* so the denotation of *and* is (obviously) not invariant under all permutations of $\{\mathrm{T}, \mathrm{F}\}$.

Now in distinction to the universe of a model, the set of truth values is *structured.* It has two elements so that Ss (sentences) and their negations can be interpreted differently and its boolean structure is given by a $\leq$ relation. One of the two elements is "least" in that it bears the $\leq$ relation to both elements; it is the one assigned to a disjunction of Ss just in case each disjunct is assigned that value. The other element is "greatest" in that it bears the $\leq$ only to itself; it is the value assigned to a conjunction of Ss iff each conjunct is assigned that value. Thus the permutations π of truth values we want are just the boolean *automorphisms,* those that satisfy $x \leq y$ iff $\pi x \leq \pi y$ (which is to say $\pi(\leq) = \leq$, that is, π *fixes* $\leq$). We recall that all two element boolean lattices are isomorphic and that the only automorphism of such a lattice is the identity function.

In general the automorphisms of a structure $(A, R_1, R_2, \ldots)$ are the permutations of the domain A which fix the relations R_i. The automorphisms of a structure (like the universe of a model) with no relations (functions) given are simply the permutations of the domain. Note that being structured, like the truth values, is typical of the primitives we need to model natural language. For example to represent tense marking (present, past, ...) we might include as a primitive a set T of points of time (or intervals), equipped with a precedence (or overlap) relation. To account for intensional operators we might include a set W of possible worlds, equipped with an accessibility relation. Doubtless more is needed. And the permutations we need are just the automorphisms. This guarantees that the properties of models determined by the primitives hold as well of their images under an automorphism, and thus those images are also models for the language. And the notion of PIness we require of objects to be "logical" is *invariance under automorphisms,* AIness.

Before formally defining AIness we consider some of the novel expressions whose denotations we wish to count among the logical objects. These include ones which are not first order definable and ones of high logical type. So we must consider automorphisms of structures other than the primitive ones we build our models on.

First, there are infinitely many expressions of the same category—Determiner (Det) in linguistic parlance—as *some* and *all*. (See *Keenan and Westerståhl 1994* and references cited there for an overview of recent work in this area; note in particular the collections *van Benthem and ter Meulen 1985*, *Gärdenfors 1987* and *van der Does and van Eijck 1996*; *Keenan and Moss 1985* and *Keenan and Stavi 1986* are probably the most extensive empirical studies.)

(1) a. *Not more than fifty* students here major in math.
    b. *Most* undergraduates study some Linguistics but *less than two percent* major in it.
    c. *All but two* of the students signed up for advanced Ulithian this quarter.
    d. *Both* students answered question six correctly but *neither* got an A on the exam.

Many of these Dets are first order: *no, both, neither* and those of the form *more/fewer than n, exactly n, between n and m, all but n, the n*. But many are not: *just finitely many, all but finitely many, infinitely many,* as well as (*Barwise and Cooper 1981*) the properly proportional Dets like *most, more/ less than half, exactly ten percent, less than a third* (even if the universe is required to be finite). But all share the logical character of the classical first order quantifiers.

Now, given an *ontology* Ω, namely a set E_Ω of entities and a boolean set 2_Ω of truth values (subscripts usually omitted), we treat P_1s like *major in math, get an A on the exam*, etc. as denoting properties of entities, represented as functions from E into 2. So P_1s, like common nouns (*student, undergraduate*, etc.), denote in $[E \to 2]$. NPs (Noun Phrases) like *some student, John and some student*, etc. combine with P_1s to form *S*s (P_0s) and denote in $[[E \to 2] \to 2]$, the set of *generalized quantifiers over* E. So we think of a sentence like *Every student laughed* as denoting the truth value assigned to the denotation of *laugh* by the function denoted by *every student*. And as Dets like *every* combine with nouns to form NPs we can treat them as denoting in $[[E \to 2] \to [[E \to 2] \to 2]]$.

More systematically now let us represent denotation sets relative to a choice of ontology (E, 2) in terms of *types*. The notion of type used in linguistic semantics derives from Church (*1940*) but replaces his "ι, o" notation with Montague's (*1973*) more mnemonic "e, t"—"e" for entity, "t" for truth value. We extend it to include product types:

DEFINITION 3. *type* is the least set containing the *initial types* 'e' and 't' and satisfying (a) and (b):

(a) $\alpha, \beta \in$ *type* $\Rightarrow (\alpha, \beta) \in$ *type* (functional types),

(b) $\alpha_1, \ldots, \alpha_k \in$ *type* $\Rightarrow (\alpha_1 \bullet \ldots \bullet \alpha_k) \in$ *type* (product types).

Type indexes denotation sets relative to an ontology $\Omega = (\mathrm{E}, 2)$. An expression is of type τ if, for each ontology Ω, it is interpreted as an element of $\mathsf{Den}_\Omega \tau$, defined by:

(2) a. $\mathsf{Den}_\Omega = \mathrm{E}$,
b. $\mathsf{Den}_\Omega t = \{0_2, 1_2\}$,
c. $\mathsf{Den}_\Omega(\alpha, \beta) = [\mathsf{Den}_\Omega(\alpha) \to \mathsf{Den}_\Omega(\beta)]$,
d. $\mathsf{Den}_\Omega(\alpha_1 \bullet \ldots \bullet \alpha_k) = \mathsf{Den}_\Omega \alpha_1 \times \ldots \times \mathsf{Den}_\Omega \alpha_k$.

DEFINITION 4. For each ontology Ω, the Ω *objects* are just those in the union of the $\mathsf{Den}_\Omega \tau$.

Our concern here is to characterize those Ω objects appropriately called *logical*. But we anticipate. First, certain of the $\mathsf{Den}_\Omega \tau$ are always boolean, as defined by:

DEFINITION 5. *bool-type* $=_{\mathrm{df}}$ the least subset of *type* satisfying:

a. $t \in$ *bool-type,*
b. If $\alpha \in$ *type* and $\beta \in$ *bool-type* then $(\alpha, \beta) \in$ *bool-type*, and
c. If $\alpha_1, \ldots, \alpha_k \in$ *bool-type* then $(\alpha_1 \bullet \ldots \bullet \alpha_k)$ *bool-type.*

For τ boolean, $\mathsf{Den}_\Omega \tau$ is a complete atomic boolean lattice, given by:

(3) a. For $x, y \in \mathsf{Den}_\Omega t$, $x \leq y$ iff $x = 0_2$ or $y = 1_2$,
b. For $f, g \in \mathsf{Den}_\Omega(\alpha, \beta)$, $f \leq g$ iff for all $a \in \mathsf{Den}_\Omega(\alpha)$, $f(a) \leq g(a)$,
c. For $f, g \in \mathsf{Den}_\Omega(\alpha_1 \bullet \ldots \bullet \alpha_k)$, $f \leq g$ iff for each $1 \leq i \leq k$, $f_i \leq g_i$.

And we are now in a position to define *automorphism invariance*:

DEFINITION 6. A map π with domain type is an automorphism of an ontology $\Omega = (\mathrm{E}, 2)$ iff

1. For all $\tau \in \{e, t\}$, $\pi(\tau)$ is an automorphism of $\mathsf{Den}_\Omega \tau$,
2. $\pi(\alpha, \beta)$ is that map sending each f in $\mathsf{Den}_\Omega(\alpha, \beta)$ to $\{\langle \pi(\alpha)(a), \pi(\beta)(b) \rangle \mid \langle a, b \rangle \in f \rangle\}$, and
3. $\pi(\alpha_1 \bullet \ldots \bullet \alpha_k)$ sends each k-tuple a in $\mathsf{Den}_\Omega(\alpha_1 \bullet \ldots \bullet \alpha_k)$ to $\langle \pi(\alpha_1)(a_1), \ldots, \pi(\alpha_k)(a_k) \rangle$.

FACT 3. For all automorphisms π of Ω, all types τ, $\pi(\tau)$ is a permutation of $\mathsf{Den}_\Omega(\tau)$. If $\tau \in$ bool-type then $\pi(\tau)$ is a boolean automorphism of $\mathsf{Den}_\Omega(\tau)$. Moreover, every permutation of E extends uniquely to an automorphism of $\Omega = (\mathrm{E}, 2)$.

DEFINITION 7.

(i) For all ontologies Ω and types τ, a $d \in \mathsf{Den}_\Omega\tau$ is *automorphism invariant* (AI) iff $\pi(\tau)(d) = d$, all automorphisms π of Ω. Write $\mathrm{AI}_\Omega\tau$ for the AI elements of $\mathsf{Den}_\Omega\tau$.

(ii) For d a closed expression of $\mathcal{L}$, d is *AI* iff for all models M for L, $\mathcal{I}_{\mathrm{M}}(d)$ is AI.[2]

Notation: We usually write $\pi(d)$ for $\pi(\tau)(d)$. And for π some Ω-automorphism we write π^{-1} for that map sending each $\tau \in$ *type* to $(\pi(\tau))^{-1}$, the inverse of the permutation $\pi(\tau)$. $\pi \circ \pi^{-1}$ is understood similarly and is thus the identity automorphism.

THEOREM 14. *Let $\mathcal{L}$ have '$=$' of type $((e \cdot e), t)$, '$\exists$' and '$\forall$' of type $((e,t),t)$,* and *and* or *of type $((t \cdot t), t)$ and* not *of type (t,t), all interpreted as indicated earlier. Then each of these expressions is* AI.

Below we present many theorems of this sort omitting proofs for reasons of space. We illustrate one case here to give the reader a feel for the algebraic character of these proofs. Observe first that for π an automorphism, (1) $a \in A$ iff $\pi(a) \in \pi(A)$, and (2) $(\pi F)(\pi a) = \pi b$ says the same as $\langle \pi a, \pi b\rangle \in \pi F$. Now we show that, given E, $F : \mathcal{P}(\mathrm{E}) \to 2$ is AI, where $F(\mathrm{A}) = 1$ iff $\mathrm{A} \neq \varnothing$. Let π an arbitrary automorphism. We show that $\pi F = F$:

$(\pi F)(A) = 1$	iff $(\pi F)(\pi(\pi^{-1}A)) = 1$	π, π^{-1} are inverses
	iff $\pi(F(\pi^{-1}(A)) = 1$	observation (2)
	iff $F(\pi^{-1}(A)) = 1$	π is the identity on truth values
	iff $\pi^{-1}(A) \neq \varnothing$	def F
	iff $A \neq \varnothing$	observation (1); π^{-1} an automorphism
	iff $F(A) = 1$	def F

Of course the truth functions denoted by *and*, *or*, and *not* are not the only AI ones. All truth functions are AI (*van Benthem 1989a*). Let us define the pure *t-types* to be the closure of $\{t\}$ under the formation of functional and product types. Then,

THEOREM 15. *For all ontologies* (E, 2), *all $\tau \in$ t-type, each $d \in \mathsf{Den}_{(\mathrm{E},2)}\tau$ is* AI.

And as well, the Dets discussed above, of type $((e,t),((e,t),t))$, are AI. Here are some sample definitions, assuming an ontology with universe E and A, B arbitrary subsets of E.

[2] We need not require that d have no free variables; but then the definition must mention $\mathcal{I}_{\mathrm{M}}(d)(g)$, for all assignments g.

(4) a. $\text{ALL}(A)(B) = 1$ iff $A - B = \emptyset$.
b. $\text{NO}(A)(B) = 1$ iff $A \cap B = \emptyset$.
c. $\text{SOME}(A)(B) = 1$ iff $A \cap B \neq \emptyset$.
d. $\text{MOST}(A)(B) = 1$ iff $|A \cap B| > |A - B|$.
e. $(\text{ALL BUT TWO})(A)(B) = 1$ iff $|A - B| = 2$.
f. $(\text{EXACTLY TEN})(A)(B) = 1$ iff $|A \cap B| = 10$.
g. $(\text{THE TWO})(A)(B) = 1$ iff $|A| = 2$ and $A - B = \emptyset$.

(4a) says in effect that *All poets daydream* is true iff the poets less the daydreamers is the empty set. (5) below illustrates the type structure and semantic interpretation we assume:

(5)

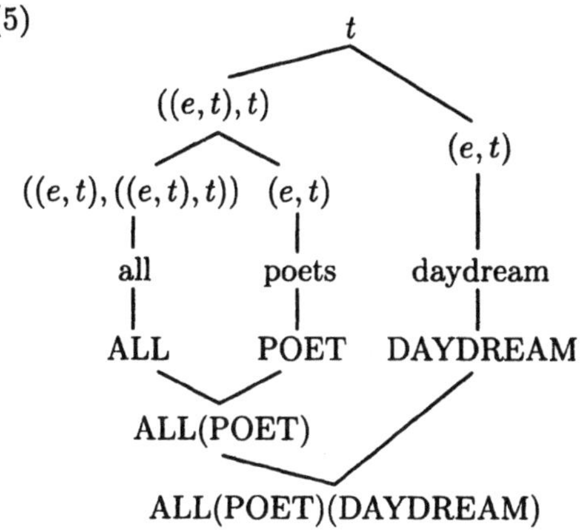

THEOREM 16. *The* Dets *given in* (4) *are* AI.

Observe that not all expressions which share the syntactic distribution of the Dets above do present a "logical" character. Contrast:

(6) a. *All / John's* cats are black,
b. *Most / More male than female* students read the Times.

In (6a) the interpretation of *John's* depends both on which individual *John* denotes and on what objects he possesses. Many non-AI functions can interpret *John's* (*Keenan and Stavi 1986*). In (6b) *more male than female* admits of non-AI interpretations, as it depends on which objects are male and which female, a contingent matter.

We return now to "logical" expressions in natural language. They share with the Dets in (4) the property of being AI. Consider first the comparative quantifiers in (7) (*Keenan and Moss 1985, Beghelli 1992*). Their type structure is represented in (8), using 'p' ("property") for (e, t):

(7) a. *More/Fewer* students *than* teachers signed the petition,
b. *Exactly as many* students *as* teachers signed the petition,
c. *More than twice* (*n times*) *as many* students *as* teachers signed,
d. *The same number of* students *as* teachers signed the petition.

(8)

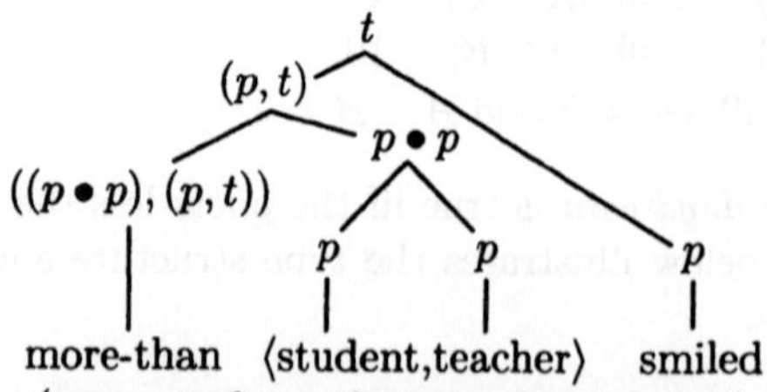

(more students than teachers smiled)

The interpretation of these cardinal comparatives is straightforward: MORE *As* THAN *Bs* have *C* iff $|A \cap C| > |B \cap C|$. Quite generally cardinal comparatives are not first order definable: *more-than* defines *most* (over a universe E) by MOST$(A)(B)$ iff (MORE $A \cap B$ THAN $A - B$)(E). That is, most *As* are *Bs* iff more *As* who are *Bs* than *As* who are not *Bs* exist. So if MORE-THAN were first order definable MOST would be, but by Barwise and Cooper (*1981*) it is not.

The examples given so far suggest that logical expressions of high type depend on cardinalities. But this is not in general the case (See *Keenan 1987*, page 92; *van Benthem 1989b*):

(9) a. *Different* people like *different* things,
b. *No two* students read *the same* papers,
c. *Each* student answered *the same* questions (on the exam).

Keenan (*1992*) shows that we do not obtain a correct semantic analysis of these sentences if for example *different things, ..., the same questions* are interpreted as generalized quantifiers. A semantically correct analysis interprets *different-different* (*no two-the same*, etc.) as functions mapping pairs of properties to functions from binary relations into 2. So writing r for $(e, (e, t))$, *different-different* has type $((p \bullet p), (r, t))$. As illustrated in (10a) and defined in (10b) this function is AI, and its values depend primarily on identity of objects, not cardinality.

(10) a. *Different people like different things* is true iff at least two people like something, and for any two such people, the things one likes are not exactly the same as those the second likes.

b. (DIFF-DIFF)$(A, B)(R) = 1$ iff $|\mathsf{Dom}(R \cap (A \times B))| \geq 2$ and $\forall a \neq a' \in \mathsf{Dom}(R \cap (A \times B))$

$$\{y \in B \mid (a, y) \in R \cap (A \times B)\} \neq \{y \in B \mid (a', y) \in R \cap (A \times B)\}.$$

An example which moves away from the Determiner class of expressions are reflexives, (11a), and reciprocals, (11b):

(11) a. Every senator admires *himself*,
b. The candidates criticized *each other*.

himself and *each other* contrast with non-AI Noun Phrases (NPs) like *the President* in *Every senator admires the President* and *the reporters* in *The candidates criticized the reporters*. We treat *himself* as denoting that function SELF from binary relations (e.g., ADMIRE) to properties given by: $\text{SELF}(R) = \{a \in \text{E} \mid \langle a, a\rangle \in R\}$. Clearly SELF is AI. And we interpret *each other* as that AI map EO from binary relations to maps from sets to truth values given (on first approximation) by: $\mathsf{EO}(R)(A) = 1$ iff $\forall a \neq a' \in A, aRa'$.

A last, non-quantificational, example: Arguably the P_1 *is opened* in (12b) is derived by a logical operation, Passive, from the P_2 *opens* mapping binary relations to sets: $\text{PASS}(R) = \{b \in \text{E} \mid \exists a \in \text{E}\ aRb\} = \mathsf{Ran}(R)$. Obviously PASS is AI. In (12c) the role of the preposition *by* is to say that its object, *John*, is the "logical subject" of the P_2 *opens*. It combines with NPs to yield binary relation modifiers, and so has type $((p, t), (r, r))$.

(12) a. John opens the door every morning at nine,
b. The door is opened every morning at nine,
c. The door is opened by John every morning at nine.

Keenan (*1980*) defines: $\text{BY}(z)(R)(y)(x) = 1$ iff $\langle x, y\rangle \in R$ and $x = z$. Analyzing (12c) as (the door)(Pass(by John)(open))) yields correct truth conditions. And clearly $\text{BY} \in \text{AI}_\Omega((p, t), (r, r))$.

2.1. Logical constants

We have found a property, that of being AI, which classical logical constants have in common. And we have shown that a variety of novel expressions which are pretheoretically "logical" also have this property. This would seem to refute Etchemendy (*1990*, chapter 15 "The Myth of the Logical Constant", page 128) who claims that there is no property of expressions whose interpretations we hold fixed that accounts for their role in establishing logical truths and valid inferences. But denoting AI elements in their type is a good candidate. In fact:

THEOREM 17 (Keenan and Stavi, *1986*). *Constantly interpreted expressions denote* AI *objects.*

We extend the theorem to type theoretic $\mathcal{L}$s, rich enough to include logical expressions of the sort discussed above. And we define *constantly interpreted* for these $\mathcal{L}$s.

Syntax. Type theoretic $\mathcal{L}$s with equality have, for each type τ, countably many (possibly zero) constants Con_τ and denumerably many variables with

$\mathsf{Var}_\tau \cap \mathsf{Var}_{\tau'} = \emptyset$ if $\tau \neq \tau'$. If d and d' are expressions of the same type, then $(d = d')$ is an expression of type t. Expressions of type (α, β) concatenate with ones of type α to form ones of type β. If $d_1, \ldots, d_k$ are expressions of types $\tau_1, \ldots, \tau_k$ respectively, then $\langle d_1, \ldots, d_k \rangle$ is an expression of type $(\tau_1 \bullet \ldots \bullet \tau_k)$. Boolean compounding, universal (existential) quantification apply just to expressions of boolean types.

Semantics. $\mathrm{M} = (\mathrm{E_M}, 2_\mathrm{M}, \mathcal{I})$ is a model for a type theoretic $\mathcal{L}$ iff $(\mathrm{E_M}, 2_\mathrm{M})$ is an ontology and $\mathcal{I}$ is a function mapping each $c \in \mathsf{Con}_\tau$ into $\mathrm{D}_{(\mathrm{E},2)}(\tau)$, all $\tau \in$ *type* (subscripts omitted). An *assignment* s maps each $x \in \mathsf{Var}_\tau$ into $\mathrm{D}_{(\mathrm{E},2)}\tau$. An *interpretation* $\mathcal{I}_\mathrm{M}$ *of* $\mathcal{L}$ *relative to* M maps $\mathcal{L}$-expressions to functions from assignments into the $\mathrm{D}_{(\mathrm{E},2)}\tau$ such that: For $c \in \mathsf{Con}_\tau$, $\mathcal{I}_\mathrm{M}(c)(s) = \mathcal{I}(c)$. For $v \in \mathsf{Var}_\tau, \mathcal{I}_\mathrm{M}(v)(s) = s(v)$. For d of type (α, β) and b of type α, $\mathcal{I}_\mathrm{M}(d^\frown b)(s) =_{\mathrm{df}} \mathcal{I}_\mathrm{M}(d)(s)(\mathcal{I}_\mathrm{M}(b)(s))$. $\mathcal{I}_\mathrm{M}\langle d_1, \ldots, d_k\rangle(s) = \langle \mathcal{I}_\mathrm{M}(d_1)(s), \ldots, \mathcal{I}_\mathrm{M}(d_k)(s)\rangle$. $\mathcal{I}_\mathrm{M}(d = d')(s) = 1$ iff $\mathcal{I}_\mathrm{M}(d)(s) = \mathcal{I}_\mathrm{M}(d')(s)$. Boolean compounds are interpreted pointwise on the assignments, and for each $x \in \mathsf{Var}_\tau$ and each expression ϕ of boolean type, $\mathcal{I}_\mathrm{M}(\exists x\phi)(s) = \bigvee\{\mathcal{I}_\mathrm{M}(\phi)[s/a] \mid a \in \mathrm{D}_{(\mathrm{E},2)}(\tau)\}$. (Each boolean $\mathrm{D}_{(\mathrm{E},2)}\tau$ is complete.) Note: $\mathcal{L}$-models are closed under isomorphism: if $\mathrm{M} = (\mathrm{E}, 2, \mathcal{I})$ is a model and π an automorphism of E and of 2, then $\pi\mathrm{M} =_{\mathrm{df}} (\pi\mathrm{E}, \pi 2, \pi \circ \mathcal{I})$ is an $\mathcal{L}$-model.

For d an expression of $\mathcal{L}$ and M a model M, when $\mathcal{I}_\mathrm{M}(d)$ is constant on the assignments we write $\mathcal{I}_\mathrm{M}(d)$ for $\mathcal{I}_\mathrm{M}(d)(s)$, s any assignment. If d is *closed* (= no free variables), $\mathcal{I}_\mathrm{M}(d)$ is constant on the assignments. In what follows we need Theorem 20, the appropriate generalization of $\mathrm{M} \simeq \mathrm{M}' \Rightarrow \mathrm{M} \cong \mathrm{M}'$; the proof uses the technical Lemmas 18 and 19. '$\mathcal{L}$' ranges over type theoretic languages.

LEMMA 18. *Given* $\mathrm{M} = (\mathrm{E}, I)$, π *a bijection with domain* E, s *an assignment*, $x \in \mathsf{Var}_\tau$, $a \in \mathrm{D}_\Omega\tau$ *and writing* πs *for* $\pi \circ s$, *we see that* $\pi(s[x/a]) = (\pi s)(x/\pi a)$. *Moreover,* $A = B$, *where*

$$A = \{\mathcal{I}_{\pi\mathrm{M}}(d)((\pi s)[x/a])) \mid a \in \mathrm{D}_\Omega\tau\} \text{ and}$$
$$B = \{\mathcal{I}_{\pi\mathrm{M}}(d)((\pi s)(x/\pi a)) \mid a \in \mathrm{D}_\Omega\tau\}.$$

Proof.

(1) $A \subseteq B$. Let $\delta \in A$. Then for some $a \in \mathrm{D}_\Omega\tau$, $\delta = \mathcal{I}_{\pi\mathrm{M}}(d)(\pi s)[x/a]$. Since $\pi \restriction \mathrm{D}_\Omega\tau$ is onto, $a = \pi a'$ for some $a' \in \mathrm{D}_\Omega\tau$, so $\delta = \mathcal{I}_{\pi\mathrm{M}}(d)(\pi s)[x/\pi a']$, so $\delta \in B$.

(2) $B \subseteq A$. Let $\delta \in B$. Then for some $a \in \mathrm{D}_\Omega\tau$, $\delta = \mathcal{I}_{\pi\mathrm{M}}(d)((\pi s)[x/\pi a])$. Hence for some $b \in \mathrm{D}_\Omega\tau$, $\delta = \mathcal{I}_{\pi\mathrm{M}}(d)((\pi s)[x/b])$, so $\delta \in A$. ⊣

LEMMA 19. *For* $\mathrm{M} = (\mathrm{E}, \mathcal{I})$ *a model for* $\mathcal{L}$, π *a bijection with domain* E, *and* s *an assignment,*

$$\pi(\mathcal{I}_\mathrm{M}(d)(s)) = \mathcal{I}_{\pi\mathrm{M}}(d)(\pi s), \qquad \textit{for all expressions } d \textit{ of } \mathcal{L}.$$

We give the proof in full in the Appendix.

THEOREM 20. *For type theoretic $\mathcal{L}$s, Model Isomorphism $\Rightarrow$ Elementary Equivalence.*

Proof. Let $\pi : \mathrm{M} \simeq \mathrm{M}'$, $\phi \in \mathsf{Sent}(\mathcal{L})$, s any M-assignment. Then,

$\mathcal{I}_{\mathrm{M}}\phi = \mathcal{I}_{\mathrm{M}}\phi(s) = \pi(\mathcal{I}_{\mathrm{M}}\phi(s))$, π fixes truth values, $= \mathcal{I}_{\mathrm{M}'}\phi(\pi s)$, Lemma 19, $= \mathcal{I}_{\mathrm{M}'}\phi$ since ϕ is closed. ⊣

DEFINITION 8. An expression d of $\mathcal{L}$ is *constant* iff for all models M, $\mathcal{I}_{\mathrm{M}}d$ is constant on the assignments and for all models M′, all bijections $\pi : \mathrm{E}_{\mathrm{M}} \to \mathrm{E}_{\mathrm{M}'}$, $\pi(\mathcal{I}_{\mathrm{M}}(d)) = \mathcal{I}_{\mathrm{M}'}(d)$.

THEOREM 21. *A closed expression d of type τ is constant*
$$\Rightarrow \forall \mathrm{M} = (\mathrm{E}, 2, \mathcal{I}), \mathcal{I}_{\mathrm{M}}(d) \in \mathrm{AI}_{(\mathrm{E},2)}\tau.$$

Proof. Given $\mathrm{M} = (\mathrm{E}, 2, \mathcal{I})$, let $\pi \in \mathrm{PERM}(\mathrm{E}_{\mathrm{M}})$. Show $\mathcal{I}_{\mathrm{M}}(d) = \pi(\mathcal{I}_{\mathrm{M}}(d))$. Let s an arbitrary assignment. Then $\pi(\mathcal{I}_{\mathrm{M}}(d)) = \pi(\mathcal{I}_{\mathrm{M}}(d)(s)) = \mathcal{I}_{\pi\mathrm{M}}(d)(\pi s)$, Lemma 18, $= \mathcal{I}_{\mathrm{M}}(d)(\pi s)$ since $\mathrm{E}_{\mathrm{M}} = \mathrm{E}_{\pi\mathrm{M}}$ and d is locally constant, $= \mathcal{I}_{\mathrm{M}}(d)(s)$ since $\mathcal{I}_{\mathrm{M}}d$ is constant on the assignments, $= \mathcal{I}_{\mathrm{M}}(d)$. ⊣

THESIS 1. The logical constants in a type theoretic $\mathcal{L}$ are the constants as defined in (8).

Logical constants like '=', 'and', 'not', and '∀' are in fact constant. Some might want to require that logical constants be syntactically simple, though this does not seem essential and would have the unpleasant consequence that a defined simple expression could be logically constant even though the complex defining statement was not. (E.g., let *Taut* abbreviate $\forall x, y(x = y \text{ or } x \neq y)$). Also Westerståhl (*1985*), van Benthem (*1989a*), Keenan and Stavi (*1986*) and Sher (*1991*) note that Definition 8 allows as logical constants expressions defined by AI conditions, like the hypothetical *lottsa* below, where for all models $\mathrm{M} = (\mathrm{E}, \mathcal{I})$,

$$(13)\quad \mathcal{I}(\text{lottsa}) = \begin{cases} \text{ALL} & \text{if E is finite,} \\ \text{INFINITELY MANY} & \text{if E is infinite.} \end{cases}$$

We stress that Theorem 21 and Thesis 1 assume closure under isomorphism for the class of models of the $\mathcal{L}$s in question. Here for example is an attempt to construct a constant whose denotation is not AI: Let *blik* be an expression of type (e, t) and suppose that the definition of interpretation requires that for $\mathrm{M} = (\mathrm{E}, 2, \mathcal{I}), \mathcal{I}(\text{blik}) = \{3\}$ if $3 \in \mathrm{E}$, otherwise $\mathcal{I}(\text{blik}) = \mathrm{E}$. *Blik* is constantly interpreted but $\mathcal{I}(\text{blik})$ is not AI when $\mathrm{E} = \{3, 4\}$. But this condition violates closure under isomorphism. If π is a bijection: $\{3, 4\} \to \{5, 6\}$ then $\pi\mathrm{M} = (\{5, 6\}, \pi 2, \pi \circ \mathcal{I})$ is not a model of $\mathcal{L}$ since $(\pi \circ \mathcal{I})(\text{blik}) = \{\pi 3\}$ is a unit set, but the definition of interpretation requires it to be $\{5, 6\}$.

Note that $\mathcal{L}$s may present "logical" expressions which are not constant. Candidates are Dets like *several* and *a few* ($\neq$ *many* and *few*). *Several* seems vague. Just how many sparrows on a clotheline count as several? At least two, and perhaps on some occasions that suffices. One account of this vagueness (*Keenan and Stavi 1986*) would merely require of models of English that $\mathcal{I}$(several) $\in$ {AT LEAST n | $2 \leq n$}. Then we would have models with the same universe in which *several* was interpreted differently, and so not constantly interpreted, but its denotation would always be AI in its type. We turn now to the central thesis of this article:

3. THESIS 2

THESIS 2. For all ontologies Ω and all types τ, the *logical* elements of $\text{Den}_\Omega\tau$ are the AI ones.

3.1.

Empirical support for Thesis 2 is given by Theorem 21: given an ontology Ω, $\text{AI}_\Omega\tau$ provides the denotations of logical constants of type τ. And this observation generalizes. The constants considered so far are ones that, given an ontology Ω, uniquely determine an element of some $\text{Den}_\Omega\tau$. But many expressions, including ones already considered, determine objects in multiple types, where, in each type, the object is AI. Here are some examples:

(14) *and, or, neither...nor* ... combine in English with expressions in most boolean types to form expressions in that type:

VerbPhrases: John both laughed and cried; neither laughed nor cried.

Transitive VPs: John both hugged and kissed / neither hugged nor kissed his cat.

Adjectives: a smart and industrious / neither smart nor industrious student.

Prepositions: John lives either in or near / neither in nor near New York City.

FACT 4. For all Ω, all boolean τ, AND_τ of type $((\tau \bullet \tau), \tau)$ which maps each (a, b) to $a \wedge_\tau b$, the greatest lower bound of $\{a, b\}$ in $\text{D}_\Omega\tau$, is in $\text{AI}_\Omega((\tau \bullet \tau), \tau)$.

And needless to say, all booleanly definable constants, functions, and relations determine objects which are logical in the appropriate type. Here are some additional AI functors:

(15) a. **Typed Equality:** For all Ω, all τ, $=_\tau$ of type $((\tau \bullet \tau), t)$ given by: $=_\tau (a, b) = 1_2$ iff $a = b$, is in $\mathrm{AI}_\Omega((\tau \bullet \tau), t)$

b. **Typed Membership:** For all Ω, all τ, $\in_\tau$, of type $((\tau \bullet (\tau, t)), t)$ which sends each (a, g) in its domain to $g(a)$ is in $\mathrm{AI}_\Omega((\tau \bullet (\tau, t)), t)$, as is $\notin_\tau$, defined in the obvious way.

Note that $\in_\tau$ is the special case of $\text{APPLY}_{\tau,\sigma}$ with $\sigma = t$:

c. **Typed Application:** For all Ω, all types τ, σ, $\text{APPLY}_{\tau,\sigma}$ is that element of $\mathrm{D}_\Omega((\tau \bullet (\tau, \sigma)), \sigma)$ mapping each (a, g) in its domain to $g(a)$. It is in $\mathrm{AI}_\Omega((\tau \bullet (\tau, \sigma)), \sigma)$.

(15b,c) characterize the sense in which set membership is a "logical" notion. Note that in a $\mathcal{L}$ in which '$\in$' is a P_2 of type $((e \cdot e), t)$ interpreted as set membership, its denotation is not in general $\mathrm{AI}_\Omega((e \cdot e), t)$. The only AI binary relations over E are $\emptyset$, $\mathrm{E} \times \mathrm{E}$, $\mathsf{ID}_\mathrm{E} = \{\langle a, a\rangle \mid a \in \mathrm{E}\}$, and $\neg\,\mathsf{ID}_\mathrm{E} = \{\langle a, b\rangle \mid a, b \in \mathrm{E} \;\&\; a \neq b\}$; see Theorem 28. Set membership fails to coincide with any of these relations except when no x in E is a member of any y in E.

Other "transcendental" (*van Benthem 1989a*) functors like Composition and Substitution, so useful in linguistic description, are also easily seen to be AI in the appropriate type.

d. $\text{COMP}_{\sigma,\tau,\upsilon}$ in $\mathrm{D}_\Omega(((\sigma, \tau) \bullet (\tau, \upsilon)), (\sigma, \upsilon))$ mapping $\langle f, g\rangle$ to $g \circ f$, is in $\mathrm{AI}_\Omega(((\sigma, \tau)(\tau, \upsilon))(\sigma, \upsilon))$

e. $\text{SUB}_{\sigma,\tau,\upsilon}$, that element S of $\mathrm{D}_\Omega(\sigma, ((\sigma, (\upsilon, \tau)), (\sigma, \upsilon)))$ such that $S(a)(f)(g) = f(a)(g(a))$, is in $\mathrm{AI}_\Omega((\sigma, ((\sigma, \tau), (\sigma, \upsilon)))$.

Finally, observe that the property of being "logical" is itself "logical". That is, without circularity, the property of being AI is itself AI:

THEOREM 22 **"Being logical is logical"**. *For all ontologies* $\Omega = (\mathrm{E}, 2)$ *4 and all types* τ, *that element* AI_τ *of* $\mathsf{Den}_\Omega(\tau, t)$ *which maps each* d *in* $\mathsf{Den}_\Omega\tau$ *to* 1_2 *iff* $d \in \mathrm{AI}_\Omega\tau$, *is in* $\mathrm{AI}_\Omega(\tau, t)$.

Proof. We sketch the direct proof by cases to confirm the absence of paradox. Let Ω be an arbitrary ontology, τ an arbitrary type. Either τ is 'e', 't', a functional type or a product type. We show that in each case the function AI_τ is AI (that is, is an element of $\mathrm{AI}_\Omega(\tau, t)$). In what follows we often write "d is AI in type τ" for $d \in \mathrm{AI}_\Omega\tau$. Also we use repeatedly that an object is AI in a type iff its image under any automorphism is AI in that type.

$\tau = t$. Each element (omitting subscripts) 0,1 of $\mathsf{Den}_\Omega t$ is AI, so AI_t maps each to 1. We must show that $\pi(\mathrm{AI}_t) = \mathrm{AI}_t$, for each Ω automorphism π. Now $\pi(\mathrm{AI}_t)(x) = \pi(\mathrm{AI}_t(\pi^{-1}x)) = \pi(1) = 1$, so $\pi(\mathrm{AI}_t) = \mathrm{AI}_t$.

$\tau = e$. If $\mathsf{Den}_\Omega e$ has just one element, b, then b is AI so $\mathrm{AI}_e(b) = 1$. And $\pi(\mathrm{AI}_e)(b) = \pi(\mathrm{AI}_e(\pi^{-1}(b)) = \pi(\mathrm{AI}_e(b)) = \pi(1) = 1$, so $\pi(\mathrm{AI}_e) = \mathrm{AI}_e$. If

$\text{Den}_\Omega e$ has more than one element, then for each $x \in \text{Den}_\Omega e$ x is not AI, so $\text{AI}_e(x) = 0$ and $\pi(\text{AI}_e)(x) = \pi(\text{AI}_e(\pi^{-1}(x))) = \pi(0) = 0$, so $\pi(\text{AI}_e) = \text{AI}_e$.

τ is a functional type (a,b). We must show that $\text{AI}_{(a,b)} = \pi(\text{AI}_{(a,b)})$. Assume first that $g \in \text{Den}_\Omega a$ is AI, so $\text{AI}_{(a,b)}(g) = 1$. Then $\pi(\text{AI}_{(a,b)})(g) = \pi(\text{AI}_{(a,b)})(\pi g) = \pi(\text{AI}_{(a,b)}(g)) = \pi(1) = 1$, so the two functions take the same values at AI arguments. Now assume g is not AI. So $\text{AI}_{(a,b)}(g) = 0$. And $\pi(\text{AI}_{(a,b)})(g) = \pi(\text{AI}_{(a,b)}(\pi^{-1}g)) = \pi(0) = 0$. Thus $\text{AI}_{(a,b)}$ and $\pi(\text{AI}_{(a,b)})$ take the same values at all g and so are the same function.

τ is a product type $a = a_1 \cdot \ldots \cdot a_n$. Let $x \in \text{Den}_\Omega a$. Observe that x is AI in its product type iff each x_i is AI in its type a_i. If x is AI then $\text{AI}_a(x) = 1$. And $\pi(\text{AI}_a)(x) = \pi(\text{AI}_a(\pi^{-1}(x))) = \pi(\text{AI}_a(x)) = \pi(1) = 1$, so equality holds in this case. If x is not AI then $\text{AI}_a(x) = 0$ and $\pi(\text{AI}_a)(x) = \pi(\text{AI}_a(\pi^{-1}(x))) = \pi(0) = 0$, completing the proof. ⊣

3.2. Plausibility

Granted that the technically defined $\text{AI}_\Omega \tau$ provide denotations of logical expressions, why should we expect this be so? We offer answers from two perspectives.

1. Regarding our type theoretical primitives, the elements of 2 are given as logical: 1_2 is uniquely the value a conjunction of sentences has iff each conjunct has that value, and 0_2 is uniquely the (distinct) value assigned to the negation of a sentence with value 1_2. The permutations we study do not change these logical objects. So to some extent we are asking what other objects must not change as a result of holding the truth values constant? These are ones whose "logicality" follows from the logicality of the truth values.

E, however, is randomly chosen; its elements do not in general have any logical character. But the *permutations* of E do. They are, in effect, the functions which fully respect the boolean structure of properties of elements of E, that is, they are the boolean automorphisms of P_{E}. Writing ATOM(P_{E}) for the set of unit sets in P_{E}, observe:

FACT 5.

(a) If δ is a boolean automorphism of P_{E}, then $\delta \upharpoonright \text{ATOM}(P_{\text{E}})$ is a permutation of ATOM(P_{E}). So δ determines $\delta' \in$ PERM(E), where $\delta'(x) = y$ iff $\delta\{x\} = \{y\}$.

(b) Conversely each permutation π of E uniquely determines a boolean automorphism π^* of P_{E} by: $\pi^*(\text{K}) = \{\pi(k) \mid k \in \text{K}\}$.

To say that a permutation π of E preserves the boolean structure of P_{E} is to say informally that π preserves the "and, or, and not" structure of P_{E}. E.g., π(LAUGH OR CRY) $= \pi$(LAUGH) $\cup\ \pi$(CRY); π(NEITHER LAUGH

NOR CRY) = $\neg(\pi(\text{LAUGH}) \cup \pi(\text{CRY}))$. π also preserves the universal and existential quantificational structure of P_{E}. Note that the set denoted by *praise every boy* is the intersection of the sets denoted by *praise x* taken over all boys x. And $\pi(\text{PRAISE EVERY BOY}) = \pi(\bigcap_{X\in\text{BOY}} \text{PRAISE } X) = \bigcap_{X\in\text{BOY}} \pi(\text{PRAISE } X)$. Similarly $\pi(\text{PRAISE SOME BOY}) = \pi(\bigcup_{X\in\text{BOY}} \text{PRAISE } X) = \bigcup_{X\in\text{BOY}} \pi(\text{PRAISE } X)$.

Finally, as a bijection, a permutation π also respects distinctness of individuals: π does not map distinct individuals to the same one, and π does not omit any individuals. This guarantees that π maps the identity relation to itself. Thus,

(16) ***AI objects are ones that remain invariant under substitutions which don't change truth, boolean structure, or identity.***

That AI objects have a logical character then is not surprising.

2. But the intuition behind using AIness to characterize logicality is not that permutations fix properties we pre-theoretically consider logical—most of these observations are theorems, i.e., derived. Rather, our intuition concerns the *generality-specificity* dimension: purely logical relations cannot discriminate one object from another but must treat them uniformally and thus must remain unchanged under systematic substitutions of individuals. One way this uniform treatment shows up is in definability properties. For example, consider definability of n-ary relations over the universe of models for first order $\mathcal{L}$s. Given $\mathcal{L}$ and a model $\text{M} = (\text{E}, 2, \mathcal{I})$ for $\mathcal{L}$, we say that $R \supseteq \text{E}^k$ is $\mathcal{L}$-definable (in M) iff for some $\mathcal{L}$-formula ϕ with exactly the k distinct $x_1, \ldots, x_k$ free, $R = \{t \in \text{E}^k \mid \text{M} \models \phi[t]\}$, where $\phi[t]$ is $\phi[g]$, $g : \text{VAR} \rightarrow \text{E}$ such that $g(x_i) = t_i$, all $1 \leq i \leq k$. Then, recalling that $\mathcal{L}_=$ is the zero signature first order $\mathcal{L}$ with equality,

THEOREM 23. *An n-ary relation R over* E *is* AI *iff R is definable in $\mathcal{L}_=$.*

Proof. Appendix. ⊣

Note that in distinction to definability in higher types (*Läuchli 1970*; *Plotkin 1980*; *van Benthem 1989a,* page 330) Theorem 23 does not require that the universe be finite. No matter how large E is, the AI n-ary relations over E are finite in number (Theorem 28), so the number of such relations of all arities never exceeds the number of expressions in $\mathcal{L}_=$.

COROLLARY 24. *For $R \subseteq \text{E}^n$, R is* AI *iff for all first order $\mathcal{L}$ with =, R is definable in all $\mathcal{L}$-models* M *with* $\text{E}_\text{M} = \text{E}$.

Proof. The proof consists in observing that R is preserved under language and model expansions, and any first order $\mathcal{L}$ with = is an expansion of $\mathcal{L}_=$. ⊣

Clearly any relation R definable in $\mathcal{L}_=$ is "logical", as all we can use to define R is equality, boolean operators, and universal and existential quantification. The fact that all AI n-ary relations over an arbitrary universe are definable in $\mathcal{L}_=$ then says that we cannot limit the "logical" relations to a proper subset of the AI ones unless we are prepared to accept that non-logical objects can be defined in terms of logical ones (equality) by logical operations. And a much more general result given by van Benthem (*1989a*) is:

(17) All closed terms in the theory of types define PI objects in their type. Further, in a type hierarchy built from a finite E, every PI item in any type is definable by some closed type-theoretic term of that type.

(Type-theoretic terms here differ from expressions in the type-theoretic $\mathcal{L}$s discussed earlier in that (1) they lack constants of arbitrary type, and (2) they may be built by lambda abstraction.) Our last result here concerns *uniform definability*. First,

DEFINITION 9. Let $\tau = \tau_1 \bullet \ldots \bullet \tau_k$ be a product type and R be a functor of type (τ, t), i.e., R maps each $\Omega = (\mathrm{E}, 2)$ into $\mathrm{Den}_\Omega(\tau, t)$. (So $R(\Omega) \subseteq \mathrm{D}_\Omega\tau_1 \times \ldots \times \mathrm{D}_\Omega\tau_k$.) Then for all languages $\mathcal{L}$, R is *uniformly definable* in $\mathcal{L}$ iff for some $v = \langle v_1, \ldots, v_k\rangle$ of k distinct variables of types $\tau_1, \ldots, \tau_k$ respectively, for some $\mathcal{L}$-formula ϕ free for v and for all models $\mathrm{M} = (\mathrm{E}, 2, \mathcal{I})$ for $\mathcal{L}$, $\mathrm{M} \models \phi[s]$ iff $\langle sv_1, \ldots, sv_k\rangle \in R(\Omega)$.

THEOREM 25. *$\forall\mathcal{L}$, $\forall$functors R of type (τ, t), R is uniformly definable in $\mathcal{L} \Rightarrow R(\Omega)$ is AI, all Ω.*

Proof. Let R be uniformally definable in $\mathcal{L}$ by ϕ. Let $\Omega = (\mathrm{E}, 2)$ arbitrary, π an automorphism of Ω. Show $R(\Omega) = \pi(R(\Omega))$.

$\Rightarrow$ Let $u \in R(\Omega)$. By uniform definability $\exists\mathrm{M}$, $\mathrm{M} \models \phi[u]$. So $\pi^{-1}\mathrm{M} \models \phi[\pi^{-1}u]$ by Lemma 19 since π^{-1} is an automorphism, so $\pi^{-1}u \in R(\Omega)$, whence $u \in \pi(R(\Omega))$.

$\Leftarrow$ Let $u \in \pi(R(\Omega))$. So $u = \pi t$ for some $t \in R(\Omega)$. Thus $\exists\mathrm{M}'$, $\mathrm{M}' \models \phi[t]$. So $\pi\mathrm{M}' \models \phi[\pi t]$ by Lemma 19, so by uniform definability $\pi t = u \in R(\pi\Omega) = R(\Omega)$ since π is an automorphism. ⊣

These results hold because AIness is preserved by the operations in the definitions:

THEOREM 26. *For $\Omega = (\mathrm{E}, 2)$ given,*

(1) *The type forming operations preserve* AI*ness: if A and B are sets of* AI *objects then all elements of $[A \to B]$, $A \times B$ and of course $\mathcal{P}(A) \simeq [A \to 2]$ are* AI.

(2) *For each boolean τ, $\mathrm{AI}_\Omega\tau$ is a (complete, atomic) boolean subalgebra of* $\mathrm{Den}_\Omega\tau$. *In other words, boolean functions preserve* AI*ness.*

(3) *Generalized converses: If R is a* AI *k-ary relation and* γ *is a permutation of* $\{1,\ldots,k\}$, *then* $\gamma R =_{\mathrm{df}} \{\langle a_{\gamma(1)},\ldots,a_{\gamma(k)}\rangle \mid \langle a_1,\ldots,a_k\rangle R\}$ *is a* AI *k-ary relation.*

(4) *Cylindrification (existential quantification): If R is a k+1-ary* AI *relation, then for all* $1 \le i \le k$, $C_i(R) =_{\mathrm{df}} \{\langle a_1,\ldots,a_{i-1},a_{i+1},\ldots,a_{k+1}\rangle \mid \exists a \in A_i\langle a_1,\ldots,a_{i-1},a,a_{i+1},\ldots,a_{k+1}\rangle \in R\}$ *is* AI.

(5) *The value of a* AI *function at a* AI *argument is itself* AI. *As a special case, composition of* AI *relations (functions) yields a* AI *relation (function).*

3.3 Additional conditions? Granting the logical objects are AI ones, are we ready to grant that all AI objects (of a given type, in a given ontology) are "logical"? Our remarks above on definability suggest that we do, pending plausible additional conditions. Here are two suggestions, the first of which is only apparent, the second of which rules out too much.

3.3.1 The notion of AIness generalizes naturally to invariance under isomorphisms (ISOM) and trivially objects which are ISOM are also AI. In fact equality holds, so the apparently stronger requirement does not in fact eliminate any candidates. We sketch the result.

First, a map π with domain *type* is an *isomorphism* from an ontology $\Omega = (\mathrm{E}, 2)$ to an ontology $\Omega' = (\mathrm{E}', 2')$ iff for all initial types τ, $\pi(\tau)$ is an isomorphism: $\mathsf{Den}_\Omega\tau \simeq \mathsf{Den}_{\Omega'}\tau$ and the value of π at functional and product types lifts just as in the case of automorphisms. An element d of $\mathsf{Den}_\Omega\tau$ is said to be *isomorphism invariant* iff for all ontologies Ω', all isomorphisms π, δ from Ω to Ω', $\pi(d) = \delta(d)$. And we prove:

THEOREM 27. *For all* $\Omega = (\mathrm{E}, 2)$, *all* $\tau \in$ *type, all* $d \in \mathrm{D}_\Omega\tau$, $d \in \mathrm{ISOM}_\Omega\tau \Leftrightarrow d \in \mathrm{AI}_\Omega\tau$.

Proof. Only $\Leftarrow$ is non-obvious. The proof is by recursion on *type*.

$$\text{Set } \mathrm{K} = \{\tau \in \textit{type} \mid \text{for all } \Omega = (\mathrm{E},2), \text{ all } d \in \mathrm{D}_\Omega\tau,\ d \in \mathrm{AI}_\Omega\tau \Rightarrow d \in \mathrm{ISOM}_\Omega\tau\}.$$

i. Obviously $t \in \mathrm{K}$ since for any $\Omega' = (\mathrm{E}', 2')$ there is only one isomorphism: $2 \to 2'$. And $e \in \mathrm{K}$ since if $\mathrm{AI}_\Omega e$ is non-empty, then E has just one element, whence $|\mathrm{E}'| = 1$ so again there is only one bijection from E to E$'$.

ii. Let $\alpha, \beta \in \mathrm{K}$, show $(\alpha, \beta) \in \mathrm{K}$. Let $d \in \mathrm{D}_\Omega(\alpha,\beta)$, let π, δ be isomorphisms from Ω to Ω'. Then $\pi_{(\alpha,\beta)}(d)(\pi_\alpha(a)) = \pi_\beta(d(a))$, by def ISOM, $= \delta_\beta(d(a))$ by the IH, $= \delta_{(\alpha,\beta)}(d)(\delta_\alpha(a)) = \delta_{(\alpha,\beta)}(d)(\pi_\alpha(a))$ by IH, so $\pi_{(\alpha,\beta)}(d) = \delta_{(\alpha,\beta)}(d)$, so $(\alpha,\beta) \in \mathrm{K}$.

iii. Closure under products is straightforward. ⊣

3.2 Universe independence? One might have hoped for a more "absolute" notion of logical object, one that did not depend on the choice of universe or type. Note that on the view presented here,

FACT 6. For each ontology Ω and type τ, $\mathsf{Den}_\Omega\tau$ is in $\mathrm{AI}_\Omega(\tau, t)$. It is in fact its boolean unit.

So each element in the type hierarchy built on an $(\mathrm{E}, 2)$ is AI in a certain type relative to that ontology. But it doesn't make sense to claim that, e.g., $\{3,4\}$ is "absolutely" PI. It is in $\mathrm{AI}_{(\mathrm{E},2)}(e,t)$ when $\mathrm{E} = \{3,4\}$ but not otherwise. Similarly $\{\langle 2,2\rangle, \langle 3,3\rangle\}$ is AI over the universe $\mathrm{E} = \{2,3\}$ but fails to be AI over any other universe, say $\{1,2,3\}$ or $\{1,2\}$. More generally, for $\Omega = (\mathrm{E}, 2)$ and $\Omega' = (\mathrm{E}', 2')$ with E, E' disjoint and $2, 2'$ disjoint, we have that for all τ, $\mathrm{D}_\Omega\tau$ and $\mathrm{D}_{\mathrm{E}'}\tau$ are disjoint. Thus the attempt to characterize invariants independent of ontology seems bootless. Nor can we omit mention of type. An object d in some $\mathrm{AI}_\Omega\tau$ fails to be in $\mathrm{AI}_{\Omega'}(e)$, where the universe of Ω' includes $\{d, \langle d, 0\rangle, \langle d, 1\rangle\}$. And we have thus shown:

FACT 7. For d in any $\mathrm{AI}_\Omega\tau$ there is an ontology Ω' and type τ' such that $d \in \mathsf{Den}_\Omega\tau - \mathrm{AI}_\Omega\tau$.

4. AN OPEN PROBLEM

Van Benthem (*1989a*) notes as open the problem of characterizing the AI elements of the type hierarchy built from an arbitrary ontology $(\mathrm{E}, 2)$, an obvious desideratum given the close association between AIness and logicality. We do not have a general answer, but we shall answer two basic cases, and make a few remarks concerning some others. (We tend to write PI instead of AI when dealing with types not involving 't'.)

THEOREM 28. *The* PI *n-ary relations over a universe* E *with at least n elements correspond one-to-one to the sets of equivalence relations over an n-element set.*

Proof sketch. $\mathcal{P}(\mathrm{E}^n)$ is a complete, atomic boolean algebra. Direct calculation shows that PIness is preserved under complements and arbitrary intersections, so the PI n-ary relations form a complete and thus atomic subalgebra of $\mathcal{P}(\mathrm{E}^n)$. To construct its atoms let T be any equivalence relation over $\{1, \ldots, n\}$. Set $R_T = \{s \in \mathrm{E}^n \mid \forall 1 \leq i, j \leq n, s_i = s_j \text{ iff } iTj\}$. Then R_T is AI, non-empty (since $|\mathrm{E}| \geq n$), and no non-empty proper subset of R_T is AI. So R_T is an atom. Clearly different T's give rise to different R'_Ts, and if R is AI and non-empty, then for each $s \in R$, $R_s =_{\mathrm{df}} \{t \in \mathrm{E}^n \mid t_i = t_j \text{ iff } s_i = s_j$, all $1 \leq i, j \leq n\}$ is an atom which is a subset of R. So the R'_Ts are all the atoms, and since each AI R is the union of its atoms, the AI R's correspond one-to-one to the sets of atoms. Hence $|\{R \in \mathcal{P}(\mathrm{E}^n) \mid R \text{ is AI}\}| = 2^{\mathsf{EQ}(n)}$, where $\mathsf{EQ}(n)$ is the number of equivalence relations over an n element set. ⊣

Westerståhl (*1985*) also gives the figure $2^{\mathsf{EQ}(n)}$ and provides a recursive means of calculating $\mathsf{EQ}(n)$. Our interest in the figure is that the number of AI n-ary relations over E does not grow with the size of (sufficiently large) E but is bounded as a function of the arity of the relations. The same property obtains for the PI functions from E^n into E:

THEOREM 29. *If* $|\mathrm{E}| \geq n+2$,
then $|\{f \in [\mathrm{E}^n \to \mathrm{E}] \mid f \textit{ is } \mathrm{PI}\}| = n \cdot \prod_{1\leq k\leq n-1}(n-k)^{\mathrm{C}[n;k+1]}$.

Proof sketch. The PI maps from E^n into E are the projection functions which can choose different coordinates at non-isomorphic tuples, e.g., $\langle a,b,c\rangle$ and $\langle a,b,a\rangle$. If $f \in [\mathrm{E}^n \to \mathrm{E}]$ is PI, then $\forall\sigma \in \mathrm{E}^n f(\sigma) = \sigma_i$ for some $1 \leq i \leq n$, otherwise permuting $f(\sigma)$ with some element not $f(\sigma)$ or any coordinate of σ fails to preserve f. So f restricted to the set of n-tuples with no coordinates identified is one of the n projection functions. For $k > 0$, $\mathrm{C}[n;k+1] =_{\mathrm{df}} n!/(k+1)!\cdot(n-(k+1))!$ is the number of ways of identifying k+1 coordinates of an n-tuple. For each such way the number of different coordinates of the resulting tuple and hence the number of possible values a PI function can take, is $(n-k)$. It takes its values independently on tuples formed by different ways of identifying k+1 coordinates. ⊣

So (for E sufficiently large) no $x \in \mathrm{E} \simeq [\mathrm{E}^0 \to \mathrm{E}]$ is PI (so there can be no logically constant proper nouns); only the identity map in $[\mathrm{E} \to \mathrm{E}]$ is PI, and only the two projection functions in $[\mathrm{E}\times\mathrm{E} \to \mathrm{E}]$ are PI. There are $3\cdot 2^3 = 24$ PI maps in $[\mathrm{E}^3 \to \mathrm{E}]$ and $4\cdot 3^6\cdot 2^4 = 46,656$ PI maps in $[\mathrm{E}^4 \to \mathrm{E}]$.

Richer e types seem to behave somewhat erratically. For $|\mathrm{E}| > 1$, there are no PI maps in $[[\mathrm{E} \to \mathrm{E}] \to \mathrm{E}]$: Let h be in that set and let π move $h(id)$. Then $(\pi h)(id) = (\pi h)(\pi id) = \pi(h(id)) \neq h(id)$, so h is not PI. This observation is a special case of:

THEOREM 30 (Van Benthem, *1989a*). *For τ in the closure of $\{e,t\}$ under functional types,* $\mathsf{Den}_\Omega\tau$ *has no* PI *elements iff* $\tau = (\sigma_1,(\sigma_2,\ldots(\sigma_n,e)\ldots))$ *where each* $\mathsf{Den}_\Omega\sigma_i$ *has* PI *elements.*

By contrast in $[[\mathrm{E} \to \mathrm{E}] \to [\mathrm{E} \to \mathrm{E}]]$ the number of PI maps grows exponentially in $|\mathrm{E}|$, whence for infinite E the number outstrips the number of expressions in the language. Note that different subsets K of $\{0,\ldots,|\mathrm{E}|\}$ in (18) give rise to different f_{K}'s, where

$$(18)\quad f_{\mathrm{K}}(g)(a) =_{\mathrm{df}} \begin{cases} g(a) \text{ if } |\{a \mid ga = a\}| \in \mathrm{K}, \\ a \text{ otherwise.} \end{cases}$$

Similarly, the growth of AI generalized quantifiers is exponential in the size of the universe:

$$(19)\quad |\mathrm{AI}[\mathcal{P}(\mathrm{E}) \to 2]| = 2^{|\mathrm{E}|+1}.$$

But we still lack a general means of computing cardinalities of AI sets in a type hierarchy.

APPENDIX

LEMMA 8. *Let* M, M′ *be infinite models for* $\mathcal{L}_=$. *Then for all formulas* ϕ *in* $\mathcal{L}_=$, *all* $g : \text{VAR} \to E_M$ *and all* $g' : \text{VAR} \to E_{M'}$

$$\text{if } g(x) = g(y) \text{ iff } g'(x) = g'(y), \text{ for all } x, y \in \text{FREEVAR}(\phi),$$
$$\text{then } M \models \phi[g] \text{ iff } M' \models \phi[g'].$$

Proof. By recursion on the formulas of $\mathcal{L}_=$. Let M, M′ be infinite models for $\mathcal{L}_=$.

Set $S = \{\phi \in \mathsf{Fmla}(\mathcal{L}_=) \mid$ for all $g : \text{VAR} \to E_M$ and all $g' : \text{VAR} \to E_{M'}$ if $g(x) = g(y)$ iff $g'(x) = g'(y)$, for all $x, y \in \text{FREEVAR}(\phi)$, then $M \models \phi[g]$ iff $M' \models \phi[g']\}$.

(i) All atomic formulas ϕ are in S. ϕ is $(x = y)$ for some $x, y \in \text{VAR}$. Let g, g' be as in the lemma with $g(x) = g(y)$ iff $g'(x) = g'(y)$. Then $M \models (x = y)[g]$ iff $g(x) = g(y)$, iff $g'(x) = g'(y)$, iff $M' \models (x = y)[g']$, so $(x = y) \in S$.

(ii) S is closed under conjunction and negation.

(a) Let $\phi, \psi \in S$. Show $(\phi \,\&\, \psi) \in S$. Let g, g' be such that $g(x) = g(y)$ iff $g'(x) = g'(y)$, for all x, y free in $(\phi \,\&\, \psi)$. Then for all x, y free in ϕ, $g(x) = g(y)$ iff $g'(x) = g'(y)$ since if, e.g., g identified free variables in ϕ that g' distinguished, then g would have identified them in $(\phi \,\&\, \psi)$ but g' would distinguish them, contrary to hypothesis. Then

$M \models (\phi \,\&\, \psi)[g]$	iff $M \models \phi[g]$ and $M \models \psi[g]$	def interpretation,
	iff $M' \models \phi[g']$ and $M \models \psi[g']$	IH + the observation,
	iff $M' \models (\phi \,\&\, \psi)[g']$	def int.

So $(\phi \,\&\, \psi) \in S$.

(b) Let $\phi \in S$. Show that $\neg\phi \in S$. Note that $\text{FREEVAR}(\neg\phi) = \text{FREEVAR}(\phi)$. So

$M \models \neg\phi[g]$	iff $M \not\models \phi[g]$	def int,
	iff $M' \not\models \phi[g']$	IH + note,
	iff $M' \models \neg\phi[g']$	def int.

So $\neg\phi \in S$.

(iii) S is closed under existential quantification.

Let $\phi \in S$. Show for all variables z, $\exists z\phi \in S$. Let z be arbitrary, let g, g' as in the lemma distinguish the same variables in $\exists z\phi$. Show $M \models \exists z\phi[g]$ iff $M' \models \exists z\phi[g']$.

Left to right. Let M $\models \exists z\phi[g]$. Then M $\models \phi[g_z]$, where g_z is a z-variant of g.

case 1: $g_z(z) = g(x)$ for some $x \in$ FREEVAR($\exists z\phi$). Let g'_z be that z-variant of g' such that $g'_z(z) = g'(x)$. Then for all x, y free in ϕ, $g_z(x) = g_z(y)$ iff $g'_z(x) = g'_z(y)$. So by IH, M$' \models \phi[g'_z]$, whence M$' \models \exists z\phi[g']$.

case 2: $g_z(z) \neq g(x)$ for any x free in $\exists z\phi$. Then find a z-variant g'_z of g' such that $g'_z(z) \neq g'(x)$, for any x free in ϕ. Such a g'_z exists since FREEVAR(ϕ) is finite and $E_{M'}$ is infinite, so there is an object not in the range of $g'_z \restriction$ FREEVAR(ϕ). And again for all $x, y \in$ FREEVAR(ϕ), $g_z(x) = g_z(y)$ iff $g'_z(x) = g'_z(y)$. So by IH, M$' \models \phi[g'_z]$, whence M$' \models \exists z\phi[g']$.

This exhausts the cases. The right to left direction is done analogously. ⊣

THEOREM 20. *Model Isomorphism ⇒ Elementary Equivalence for Type Theoretic Languages.*

For M = (E, $\mathcal{I}$) a model for $\mathcal{L}$, π a bijection with domain E, s any M-assignment and d an expression of any type in $\mathcal{L}$,

$$\pi(\mathcal{I}_M(d)(s)) = \mathcal{I}_{\pi M}(d)(\pi s).$$

Set $S = \{d \in \mathcal{L} \mid \text{for all assignments } s,\ \pi(\mathcal{I}_M(d)(s)) = \mathcal{I}_{\pi M}(d)(\pi s)\}$. Then,

1. For all τ, $\mathsf{Con}_\tau \subseteq S$. Let $c \in \mathsf{Con}_\tau$. Then,

$\pi(\mathcal{I}_M(c)(s)) = \pi(\mathcal{I}(c))$, def int,
and $\mathcal{I}_{\pi M}(c)(\pi s) = (\pi \circ \mathcal{I})(c)$, def int, $= \pi(\mathcal{I}(c))$.

2. For all τ, $\mathsf{Var}_\tau \subseteq S$. Let $x \in \mathsf{Var}_\tau$. Then,

$\pi(\mathcal{I}_M(v)(s)) = \pi(s(v))$, def int,
and $\mathcal{I}_{\pi M}(v)(\pi \circ s) = (\pi \circ s)(v) = \pi(s(v))$.

3. For $d, d' \in S$, d, d' both of type τ. Show $(d = d') \in S$.

$\pi(\mathcal{I}_M(d = d')(s)) = 1$	
iff $\mathcal{I}_M(d = d')(s) = 1$	π fixes each $x \in 2$,
iff $\mathcal{I}_M(d)(s) = \mathcal{I}_M(d')(s)$	def interpretation,
iff $\pi(\mathcal{I}_M(d)(s)) = \pi(\mathcal{I}_M(d')(s))$	π is injective,
iff $\mathcal{I}_{\pi M}(d)(\pi s) = \mathcal{I}_{\pi M}(d')(\pi s)$	Induction Hypothesis,
iff $\mathcal{I}_{\pi M}(d = d')(\pi s) = 1$	def interpretation.

4. For $d_1, \ldots, d_k \in S$, show $d = \langle d_1, \ldots, d_k \rangle \in S$.

$\pi(\mathcal{I}_{\mathrm{M}}(d)(s)$

$= \pi(\langle \mathcal{I}_{\mathrm{M}}(d_1)(s), \ldots, \mathcal{I}_{\mathrm{M}}(d_k)(s)\rangle)$	def interpretation,
$= \langle \pi(\mathcal{I}_{\mathrm{M}}(d_1)(s)), \ldots, \pi(\mathcal{I}_{\mathrm{M}}(d_k)(s))\rangle$	def π on tuples,
$= \langle \mathcal{I}_{\pi\mathrm{M}}(d_1)(\pi s)), \ldots, \mathcal{I}_{\pi\mathrm{M}}(d_k)(\pi \circ s))\rangle$	Induction Hypothesis,
$= \mathcal{I}_{\pi\mathrm{M}}(\langle d_1, \ldots, d_k\rangle)(\pi \circ s)$	def interpretation.

5. For $d, b \in S$, d of type (α, β) and b of type β, show $d^\frown b \in S$.

$\pi(\mathcal{I}_{\mathrm{M}}(d^\frown b)(s))$

$= \pi(\mathcal{I}_{\mathrm{M}}(d)(s)(\mathcal{I}_{\mathrm{M}}(b)(s))$	def interpretation,
$= \pi(\mathcal{I}_{\mathrm{M}}(d)(s)(\pi(\mathcal{I}_{\mathrm{M}}(b)(s)))$	def π on function spaces,
$= \mathcal{I}_{\pi\mathrm{M}}(d)(\pi s)(\mathcal{I}_{\pi\mathrm{M}}(b)(\pi \circ s))$	Induction Hypothesis,
$= \mathcal{I}_{\pi\mathrm{M}}(d^\frown b)(\pi \circ s)$	def interpretation.

6. For $d, d' \in S$, d, d' of boolean type τ, show $(d \mathbin{\&} d') \in S$ and $\neg d \in S$.

$\pi(\mathcal{I}_{\mathrm{M}}(d \mathbin{\&} d')(s))$

$= \pi(\mathcal{I}_{\mathrm{M}}(d)(s) \wedge \mathcal{I}_{\mathrm{M}}(d')(s))$	def interpretation,
$= \pi(\mathcal{I}_{\mathrm{M}}(d)(s)) \wedge \pi(\mathcal{I}_{\mathrm{M}}(d')(s))$	π a boolean automorphism,
$= \mathcal{I}_{\pi\mathrm{M}}(d)(\pi \circ s)) \wedge \mathcal{I}_{\pi\mathrm{M}}(d')(\pi \circ s))$	Induction Hypothesis,
$= \mathcal{I}_{\pi\mathrm{M}}(d \mathbin{\&} d')(\pi \circ s)$	def interpretation.

The proof that S is closed under negation follows the same pattern.

7. For $x \in \mathsf{Var}_\tau$ and ϕ of boolean type σ, assume $\phi \in S$. Show $\exists x\phi \in S$.

$\pi(\mathcal{I}_{\mathrm{M}}(\exists x\phi)(s))$

$= \pi(\bigvee\{\mathcal{I}_{\mathrm{M}}(\phi)(s[x/a] \mid a \in \mathrm{D}_\Omega\tau\})$	def interpretation,
$= \bigvee\{\pi(\mathcal{I}_{\mathrm{M}}(\phi)(s[x/a] \mid a \in \mathrm{D}_\Omega\tau)\}$	$\pi \restriction \mathrm{D}_\Omega\tau$ a boolean automorphism,
$= \bigvee\{\mathcal{I}_{\pi\mathrm{M}}(\phi)(\pi(s[x/a])) \mid a \in \mathrm{D}_\Omega\tau\}$	Induction Hypothesis,
$= \bigvee\{\mathcal{I}_{\pi\mathrm{M}}(\phi)(\pi \circ s)[x/\pi a])) \mid a \in \mathrm{D}_\Omega\tau\}$	Lemma 18,
$= \bigvee\{I_{\pi\mathrm{M}}(\phi)(\pi \circ s)[x/a])) \mid a \in \mathrm{D}_\Omega\tau\}$	Lemma 18,
$= \mathcal{I}_{\pi\mathrm{M}}(\exists x\phi)(\pi \circ s)$	def interpretation.

⊣

THEOREM 23. $\Rightarrow$ *The $\mathcal{L}_=$-definable $n \geq 1$-ary relations over* E *are just the* $R_{\phi,v} =_{\mathrm{df}} \{t \in \mathrm{E}^k \mid \mathrm{M} \models \phi[t]\}$, *any* M *with* $\mathrm{E_M} = \mathrm{E}$, $v = \langle v_1, \ldots, v_k\rangle$. *We show that each $R_{\phi,v}$ is* AI, *ϕ free for $\{v_i \mid 1 \leq i \leq k\}$. Set $S = \{\phi \in \mathcal{L}_= \mid R_{\phi,v}$ is* AI, *all v such that ϕ is free for $v\}$.*

Proof.

a. all atomic formulas ϕ are in S.

case 1: ϕ is $(x = y)$ for x, y distinct. Let v be a tuple of distinct variables such that for some i, j $v_i = x$, $v_j = y$. $R_{\phi,v} = \{s \in \mathrm{E}^k \mid \mathrm{M} \models (x = y)[s]\} = \{s \in \mathrm{E}^k \mid s_i = s_j\}$, which is AI.

case 2: ϕ is $(x = x)$. Let v be a tuple of distinct variables with $v_j = x$. Then $R_{\phi,v} = \{s \in \mathrm{E}^k \mid \mathrm{M} \models (x = x)[s_j]\} = \mathrm{E}^k$, again AI.

b. S is closed under conjunction and negation

1. Let $\phi, \psi \in S$. Let $(\phi \,\&\, \psi)$ be free in v, so ϕ, ψ are also free in v. Then

$$R_{\phi\&\psi,v} = \{s \in \mathrm{E}^k \mid \mathrm{M} \models (\phi \,\&\, \psi)[s]\} = \{s \in \mathrm{E}^k \mid \mathrm{M} \models (\phi)[s]\} \cap \{s \in \mathrm{E}^k \mid \mathrm{M} \models (\psi)[s]\},$$
$$= R_{\phi,v} \cap R_{\psi,v},$$ which is AI since intersection preserves AIness.

2. Let $\phi \in S$. Let $\neg\phi$, and hence also ϕ, be free in v.

$R_{\neg\phi,v} = \{s \in \mathrm{E}^k \mid \mathrm{M} \not\models \phi[s]\} = -R_{\phi,v}$ is again AI since complement preserves AIness.

c. S is closed under existential quantification.

Let $\phi \in S$, let $\exists z\phi$ be free in v. We may assume z is free in ϕ, otherwise $R_{\exists z\phi,v} = R_{\phi,v}$, which is AI. Then $R_{\exists z\phi,v} = \{s \in \mathrm{E}^k \mid \mathrm{M} \models \exists z\phi[s]\} = \{s \in \mathrm{E}^k \mid \exists b\mathrm{M} \models \phi[v_i/s_i, z/b]\}$. If $R_{\exists z\phi,v} = \varnothing$, then it is AI. Set $v' = \langle v_1, \ldots, v_k, z\rangle$. Let $u \in R_{\exists z\phi,v}$, so $\exists b$, $\mathrm{M} \models \phi[s/b]$ and $\langle u, b\rangle \in R_{\phi,v'}$. Let $\pi \in \mathrm{PERM(E)}$. So $\langle \pi u, \pi b\rangle \in R_{\phi,v'}$ by the IH, so $\pi u \in R_{\exists z\phi,v}$, whence $R_{\exists z\phi,v}$ is AI.

$\Leftarrow$ Let $R \subseteq \mathrm{E}^n$ be AI. Show that R is definable in $\mathcal{L}_=$.

i. If $R = \varnothing$, then $R = R_{\phi,v}$ where ϕ is $(x \neq x)$ and $v = \langle x\rangle$.

ii. Let $R \neq \varnothing$. Let $s \in R$. Recall $R_s = \{t \in \mathrm{E}^n \mid t_i = t_j \text{ iff } s_i = s_j, \text{ all } 1 \leq i, j \leq n\}$.

Then fixing the sequence $v = \langle x_1, \ldots, x_n\rangle$ of distinct variables, R_s is defined by the conjunction of $\mathrm{AND}\{(x_i = x_j) \mid s_i = s_j\}$ with $\mathrm{AND}\{(x_i \neq x_j) \mid s_i \neq s_j\}$. Since $R = \bigcup_{s \in R} R_s$, R is defined by the disjunction of the finitely many sentences used to define each R_s. ⊣

ACKNOWLEDGEMENTS

For constructive discussion of specific issues raised in this paper I would like to thank John Corcoran, David Kaplan, Jim Lambek, Michael Makkai, Tony Martin, Uwe Moennich, Yiannis Moschovakis, Ed Stabler, and Gonzalo

Reyes. Specific thanks to: Michael Barr for substantive help with issues concerning definability in languages without equality; to Gila Sher for constructive discussion of the early theorems, most notably a significant simplication of my original proof of Theorem 4; and to Michael Zelëny for drawing my attention to Putnam's work and reminding me of Lakoff's discussion.

REFERENCES

Barwise, J., and R. Cooper

1981 Generalized quantifiers and natural language, *Linguistics and Philosophy*, vol. 4, pp. 159–219.

Beghelli, F.

1992 Comparative quantifiers, *Proceedings of the VIII Amsterdam colloquium on logic, language, and information* (P. Dekker et al., editors), ITLC, Amsterdam.

Church, A.

1940 A formulation of the simple theory of types, *The Journal of Symbolic Logic*, vol. 5, pp. 56–68.

Enderton, H. B.

1972 *A mathematical introduction to logic*, Academic Press, New York.

Etchemendy, J.

1990 *The concept of logical consequence*, Harvard University Press, Cambridge, Massachusetts.

Gärdenfors, P.

1987 *Generalized quantifiers. Linguistic and logical approaches*, Reidel, Dordrecht.

Keenan, E. L.

1980 Passive is phrasal (not sentential or lexical), *Lexical grammar* (T. Hoekstra, H. van der Hulst, and M. Moortgat, editors), Foris.

1987 Unreducible n-ary quantifiers in natural language, in *Gärdenfors 1987*.

1991 Anaphora invariants and language universals, *Proceedings of the West Coast Conference of Formal Linguistics* (D. Bates, editor), vol. X, Stanford Linguistics Association, Stanford University.

1992 Beyond the Frege boundary, *Linguistics and Philosophy*, vol. 15, pp. 199–221.

Keenan, E. L., and L. Faltz

1985 *Boolean semantics for natural language*, Reidel, Dordrecht.

Keenan, E. L., and L. Moss

1985 Generalized quantifiers and the expressive power of natural language, in *van Benthem and ter Meulen 1985*.

Keenan, E. L., and E. P. Stabler

1991 Language invariants, *Proceedings of the eighth Amsterdam colloquium* (P. Dekker and M. Stokhof, editors), Institute for Logic, Language, and Computation, University of Amsterdam, Rodopi, Atlanta, pp. 309–329.

1995 There is more than one language, *Proceedings of langues et grammaire I* (G. Tsoulas and L. Nash, editors).

Keenan, E. L., and J. Stavi

1986 A semantic characterization of natural language determiners, *Linguistics and Philosophy*, vol. 9, pp. 253–326.

Keenan, E. L., and D. Westerståhl

1994 Generalized quantifiers in linguistics and logic, *The handbook of logic and linguistics* (J. van Benthem and A. ter Meulen, editors), Elsevier, ch. X, pp. 837–893.

Klein, F. A.

1872 Comparative review of recent researches in geometry (M. W. Haskell, translator), *Bulletin of the New York Mathematical Society*, vol. II, 1892-93, pp. 215–249.

Lakoff, G.

1987 *Women, fire and dangerous things,* University of Chicago Press, Chicago.

Läuchli, H.

1970 An abstract notion of realizability for which intuitionistic predicate calculus is complete, *Intuitionism and proof theory,* North-Holland, Amsterdam.

Lewis, D.

1984 Putnam's paradox, *Australasian Journal of Philosophy,* vol. 62, no. 3.

Marshall, M. V. and R. Chuaqui

1991 Sentences of type theory: The only sentences preserved under isomorphisms, *The Journal of Symbolic Logic*, vol. 56, no. 3, pp. 932–948.

Montague, R.

1973 The proper treatment of quantification in ordinary English, *Approaches to natural language* (J. Hintikka *et al.*, editors), Reidel; reprinted in *Formal philosophy* (R. H. Thomason, editor), Yale University Press, New Haven.

Mostowski, A.

1957 A generalization of quantifiers, *Fundamenta Mathematicae*, vol. 44, pp. 12–36.

Plotkin, G.

1980 Lambda-definability in the full type hierarchy, *To: H. B. Curry: Essays on combinatory logic, lambda calculus and formalism,* Academic Press, New York.

Putnam, H.

1981 *Reason, truth and history,* Cambridge University Press, Cambridge, England.

Sher, G.

1991 *The bounds of logic,* MIT Press, Cambridge, Massachusett.

Silva, J. S.

1945 On automorphisms of arbitrary mathematical systems (A. J. Franco De Oliveira, translator), *History and Philosophy of Logic*, vol. 6, 1985, pp. 91–116.

Tarski, A.

1986 What are logical notions? (J. Corcoran, editor), from a 1966 lecture, *History and Philosophy of Logic*, vol. 7, pp. 143–154.

Van Benthem, J.

1983 Determiners and logic, *Linguistics and Philosophy*, vol. 6.4, pp. 447–478.

1989a Logical constants across varying types, *Notre Dame Journal of Formal Logic*, vol. 30, no. 3, pp. 315–342.

1989b Polyadic quantifiers, *Linguistics and Philosophy*, vol. 12, pp. 437–464.

1991 *Language in action*, North-Holland, Amsterdam.

Van Benthem, J., and A. ter Meulen

1985 (editors), *Generalized quantifiers*, Foris, Dordrecht.

Van der Does, J., and J. van Eijck

1996 *Quantifiers, logic, and language*, CSLI Lecture Notes 54, CSLI Publications, Stanford.

Van Fraassen, B.

1989 *Laws and symmetry*, Oxford University Press, Oxford; reprinted in 1990.

Westerståhl, D.

1985 Logical constants in quantifier languages, *Linguistics and Philosophy*, vol. 8.4, pp. 387–413.

Weyl, H.

1949 *Philosophy of mathematics and natural science*, Princeton University Press, Princeton, New Jersey; reprinted by Athenaum, 1963.

SAUNDERS MAC LANE

THE LAMBDA CALCULUS AND ADJOINT FUNCTORS

Abstract. The well known lambda calculus was first formulated by Alonzo Church (*1932*). It was originally intended as a new foundation of mathematics; but soon the remarkable connection between lambda definability and recursive functions was developed. Only much later was the connection with rewrite systems noted, while the remarkable influence of the calculus in computer languages was also noted later. This note is to point out that the Galois connections and this lambda calculus are perhaps the first appearances of an explicit pair of adjoint functors. These functors in general were not found until the work of Daniel Kan in 1958.[1]

1. LAMBDA CALCULUS AND ADJOINTS

In the typed lambda calculus consider functions $f(x, y)$ where x and y are variables of types X and Y, and with values of type W. Then $\lambda x f(x, y)$ denotes the function g of y whose values are functions of x given by $(g(y))(x) = f(x, y)$. This defines lambda. Or, consider a category with objects X, Y, and W with product objects $X \times Y$. Then the exponential set of all functions on X to W is written W^X. It is determined by the fact that there is a "natural" bijection (one-to-one correspondence) between functions

$$f : (X \times Y) \to W$$

and functions

$$g : Y \to W^X.$$

One says that the operation "product with X" is left adjoint to the operation, "raise to the power of X"; such operations are called "functors". This bijection is often expressed by the natural isomorphism

$$\mathrm{Hom}(X \times Y, W) = \mathrm{Hom}(Y, W^X),$$

with the left adjoint on the left, and where $\mathrm{Hom}(Y, W)$ is the conventional notation for the set of all morphisms, i.e., all functions in the category from Y to W. A category with products and such exponentials is called a *Cartesian closed category.*

[1]This and other references may be found in *Mac Lane and Moerdijk 1993.* Also many fine further comments on the lambda calculus are presented in an article by Henk Barendregt, "The impact of the lambda calculus in logic and computer science", *Bulletin of Symbolic Logic,* vol. 3, no. 2 (1997), pp. 155–215.

C. Anthony Anderson and M. Zelëny (eds.), *Logic, Meaning and Computation,* 181-184.

In this presentation the two notions of lambda and exponential adjunctions are almost identical. This can be stated more precisely, as in the book by Lambek and Scott, as a suitable equivalence between the category of typed lambda calculii and that of Cartesian closed categories with a natural numbers object. It took a long time for this connection to be clearly formulated.

The notion of an adjoint covers many more cases than that of product and exponent. Given categories B and A, a functor $F : B \to A$ is said to be left adjoint to a functor $G : A \to B$ when there is a bijection

$$\mathrm{Hom}(Fb, a) = \mathrm{Hom}(b, Ga)$$

which is "natural" for all b in B and a in A. For example, let B be the category of sets and A that of groups, while G sends each group to its underlying set and F turns each set into the free group generated by that set. The bijection then expresses the fact that a given map of the generators b into this underlying set of a uniquely determines a group homomorphism from the free group to the group a.[2]

2. THE CATEGORY OF SETS

In his initial work on the lambda calculus, Church considered this calculus as a formulation of basic properties of functions, so his considerations applied to what would now be called the category of sets. The presence there of products and exponentials is an essential part of the Lawvere proposal to use the elementary theory of the category of sets as a foundation of mathematics—a beautiful idea which is neglected by many people enmeshed in the rococo properties of the higher set theory. However in the 1930's, the astounding connections of lambdas to recursive functions and computability, as developed by Church and his students Kleene, Rosser, and Turing, were properly in the center of interest. A functorial point of view became possible only after the development of algebraic topology and its representation of geometrical maps by algebraic ones, that is, with functors—just as the real understanding of forcing depends on the geometrical notion of sheaf (see Mac Lane and Moerdijk).

The discovery of adjoint functors also came in stages: in the adjunction above, set $a = Fb$ to get the bijection

$$\mathrm{Hom}(Fb, Fb) = \mathrm{Hom}(b, GFb).$$

Then the identity map on the left (every category has identities, in keeping with ancient wisdom) must correspond to some map $\eta : b \to GFb$; in the case of groups this is the evident map which sends each group generator to itself,

[2]One might recall the philosophical origins of the terminology: "Category" was purloined from Aristotle or Kant, at your choice, while "functor" was taken from a book by Rudolf Carnap.

as seen in the underlying set. The adjunction then takes the form that each map g of sets determines a unique map f of groups so that the triangular diagram d below "commutes".

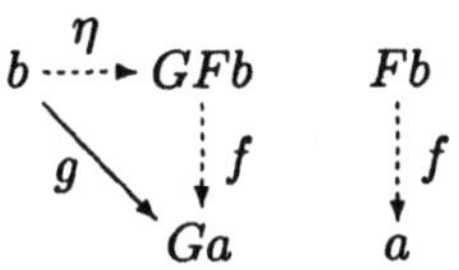

One commonly says that this is the universal way of embedding the set b in a group a. If follows from this that the free group is determined only up to an isomorphism. This reveals a basic aspect of mathematics: its well known objects are not unique, but only so up to isomorphism. Despite von Neuman, he did not have *the* ordinal numbers, but only one of many possible isomorphic models—and the back of my hand to the familiar definition of *the* ordered pair.

3. QUANTIFIERS AS AGENTS

There are other striking connections with logic. The connective "And" is just a product, so is properly defined by an adjunction, as is "Or" and "Not". More surprising is the fact that quantifiers are really adjoints. For consider a propositional function $H(x, y)$ as if it were a subset of the product set $X \times Y$, so that $(\forall y)H(x, y)$ and $(\exists y)H(x, y)$ are subsets of X. Write p for the projection

$$p : X \times Y \to X.$$

For any subset $S \subset X \times Y$, write p^*S for its "inverse image" in $X \times Y$; that is, for the set of all pairs (x, y) with $x \in S$. Then for any set $T \subset X$, one has the inclusions

$$p^*T \subset H(x, y) \text{ iff } T \subset (\forall y)X(x, y),$$

$$H(x, y) \subset p^*T \text{ iff } (\exists y)H(x, y) \subset T.$$

Now read inclusion as an arrow; these equivalences then state that the existential quantifier is left adjoint and the universal one right adjoint to inverse image. This striking idea, due to Lawvere, finally replaces those previously clumsy attempts to make quantifiers algebraic, as in cylindric algebra or polyadic logic.

These examples go to exemplify the main thesis: Logic is a branch of mathematics, and not vice versa.[3]

[3]Often it is not noted that the recognition that logic could be practiced in departments of mathematics was possible only because of Hilbert, Gödel, and in this country Alonzo Church.

Before Church, there was no hope. In 1928, as an undergraduate at Yale, I asked my pro-

REFERENCES

Church, A.

1936 Mathematical logic, Department of Mathematics, Princeton University, mimeographed notes (113 pages); reviewed in *The Journal of Symbolic Logic*, vol. 2, p. 39.

1941 *The calculi of lambda conversion,* Princeton University Press, Princeton, New Jersey.

Lambek, J., and P. J. Scott

1986 *Introduction to higher order categorical logic,* Cambridge University Press, Cambridge, England.

Lawvere, F. W.

1970 Equality in Hyperdoctrines and comprehension scheme as an adjoint functor, *Applications of categorical algebra*, American Mathematical Society, Providence.

Mac Lane, S.

1938 Review of "Logical syntax of language" by Rudolf Carnap, *Bulletin of the American Mathematical Society*, vol. 44, pp. 171–176.

Mac Lane, S., and I. Moerdijk

1993 *Sheaves in geometry and logic, A first introduction to topos theory,* Springer-Verlag, New York.

---

fessor of mathematics to direct my reading of Principia Mathematica. He responded, "Oh no, read Hausdorff instead." In 1930, as a graduate student of mathematics at Chicago, I found no one to direct my intended thesis in logic, In 1933, when I presented my (Gottingen) thesis in logic to the annual meeting of the American Mathematical Society, the leading mathematician at Yale, Oystein Ore, followed my 10 minute talk on logic with a ten minute denunciation: Algebra, not logic! Little did he know the future.

American help was on the way. Oswald Veblen, of Princeton University mathematics, in his retiring address as President of the American Mathematical Society, said:

> "The conclusion seems inescapable that formal logic has to be taken over by the mathematicians. The fact is that there does not exist an adequate logic at the present time, and unless the mathematicians create one, no one else is likely to do so."

Unlike many words in a retiring address, this statement led to results. Veblen himself had (at least) two PhD. students in logic: A. A. Bennett, who was subsequently influential in forming the Association for Symbolic Logic, and Church, in 1927. The striking and penetrating results of Church himself and his early American students Kleene and Rosser together made logic an appropriate subject for mathematical departments.

GERALD J. MASSEY

ATOMIC BOOLEAN ALGEBRAS AND CLASSICAL PROPOSITIONAL LOGIC

1. BOOLEAN ALGEBRA AND CLASSICAL PROPOSITIONAL LOGIC

From the time of Boole and De Morgan, algebraic methods have been used to study logics. In this century, the customary vehicle for the application of algebra to logic has been the so-called *natural* or *Lindenbaum algebra* of the logical system being investigated.[1] In the case of a formal axiomatization (logistic system) S of classical propositional logic (hereafter PL), the Lindenbaum algebra $\boldsymbol{A}$ happens to be a *Boolean Algebra.* (The elements of the Lindenbaum algebra $\boldsymbol{A}$ are the equivalence classes $|A|$ of the wffs of S with respect to the relation $A \leftrightarrow B$, where $A \leftrightarrow B$ if and only if there is an S-proof of B from A and of A from B, and the complement, meet, and join operations of $\boldsymbol{A}$ are defined thus: $-|A| = |\sim A|$, $|A| \bigwedge |B| = |A \mathbin{\&} B|$, and $|A| \bigvee |B| = |A \vee B|$, where the tilde, ampersand, and wedge are either primitive or defined connectives of S.)

As an illustration of the use and power of algebraic methods, consider the derivation of the completeness theorem for a (formal) axiomatization of PL from *Stone's Representation Theorem,* which states that every Boolean algebra is isomorphic to the set algebra of a field of sets. (In a set algebra, the algebraic operations are set complement [relative to the union of the elements of the algebra], set intersection, and set union.) This derivation employs two additional and easily proved results, namely:

LEMMA 1.1. *If Φ is the set algebra of a field of sets and x is an element of Φ other than the unit, then there is a 2-valued homomorphism h of Φ such that $h(x) = 0$. (A 2-valued homomorphism of a Boolean algebra is a structure-preserving mapping of the algebra into the 2-element Boolean algebra.)*

LEMMA 1.2. *Every 2-valued homomorphism h of the Lindenbaum algebra of an axiomatization S of PL determines a classical valuation V on the wffs of S, where $V(A) =$ t or f according as $h(|A|) = 1$ or 0.*

(By a *classical valuation* on the wffs of S is meant a mapping from the wffs of S onto the truth values (t, f) that conforms to the classical truth tables for the connectives of S.)

[1]For the algebraic terminology used in this paper and for a lucid presentation of the application of algebra to logic, see *Rasiowa and Sikorski 1970.*

C. Anthony Anderson and M. Zelëny (eds.), *Logic, Meaning and Computation*, 185-189.

The derivation of the completeness theorem for an axiomatization S of PL from Stone's Representation Theorem goes thus:

Proof. Let A be any non-theorem of S. Then, by the construction of the Lindenbaum algebra for S, $|A| \neq 1$. So, by Stone's Representation Theorem together with Lemma 1.1, there is a 2-valued homomorphism h of the Lindenbaum algebra such that $h(|A|) = 0$. Hence by Lemma 1.2, there is a classical valuation V such that $V(A) = $ f. So, A is not a tautology. Therefore, by contraposition, if A is a tautology, then A is a theorem of S.[2] ⊣

2. THE ALLEGED IRRELEVANCE OF ATOMIC BOOLEAN ALGEBRAS

Stone's Representation Theorem is difficult to prove, a claim supported by the fact that it was not proved until 1936. In brief, what makes Stone's Representation Theorem hard to prove is the non-existence in an arbitrary Boolean algebra of a class X of elements that obey the following three laws (where $x \leq y$, which is read as x *is included in* y, is defined as $x \wedge y = x$):

(2.1) If $x \in X$, then $(y)(x \wedge y = 0$ or $x \wedge y = x)$.

(2.2) $(y)[$if $y \neq 0$, then $(\exists x)(x \in X$ and $x \leq y)]$.

(2.3) If $x \in X$, then $(y)(z)[x \leq (y \vee z)$ iff $x \leq y$ or $x \leq z]$.

If a Boolean algebra contains a set X of elements that satisfy these three laws, one can readily construct an isomorphic set algebra thus: For any element z of the given algebra, let $f(z) = \{x : x \in X$ and $x \leq z\}$. Then, let the elements of the set algebra be the f-images of the elements of the given algebra. Finally, let f map the complement, meet, and join operations of the given algebra with set complementation (relative to the f-image of the unit of the algebra), intersection, and union respectively. It is easy to establish that f is an isomorphism from the given algebra onto the set algebra.

Certain simple but nevertheless important Boolean algebras—the *atomic Boolean algebras*—boast a class X of elements that satisfy the laws (2.1), (2.2), (2.3), namely, the set of atoms of the algebra. (An *atom* a is an element that strictly includes only the zero element, i.e., if $y < a$, then $y = 0$. A Boolean algebra is *atomic* if and only if every non-zero element includes an atom.[3]) It is easy to show, by the line of reasoning sketched in the preceding paragraph, that every atomic Boolean algebra is isomorphic to a sub-algebra of the powerset algebra of its atoms (*Representation Theorem for Atomic Boolean Algebras*). One simply establishes that the function that maps each element of the given Boolean algebra the set of the element's atoms (the

[2] This derivation of the completeness theorem from Stone's Representation Theorem follows closely that of Bell and Slomson (*1971*).

[3] See *Mendelson 1970* for proofs of representation theorems for both atomic and finite Boolean Algebras.

atoms included in the element) is an isomorphism from the given atomic algebra to a sub-algebra of the powerset algebra of the set of atoms of the given algebra. (Such a sub-algebra is, of course, itself a field of sets.)

If the Lindenbaum algebra of an axiomatization of *PL* were atomic, one could derive the completeness theorem for it from the Representation Theorem for Atomic Boolean Algebras—in conjunction with Lemmas 1.1 and 1.2—by the same reasoning used to derive it from Stone's Representation Theorem in section 1 above. But none of the Lindenbaum algebras of axiomatizations of *PL* are atomic; rather, these algebras are altogether *atomless.* This elementary observation has caused mathematicians and logicians to conclude that the Representation Theorem for Atomic Boolean Algebras is not germane to the investigation of classical propositional logic. The principal burden of this paper is to debunk this widely-shared belief.

3. CLASSICAL PROPOSITIONAL LOGIC AS AN ENSEMBLE OF FRAGMENTS

An axiomatization S of *PL* will be said to be *regular* if and only if, for each theorem T of S, there is a proof of T in S that contains no propositional variables other than those that occur in T. (This terminology is motivated by the conviction that there is something intuitively *irregular* about axiomatizations of *PL* that are not regular in the sense just defined.) It is important to note that axiomatizations of *PL* that employ axiom schemes and structural inference rules (rules given by means of rule schemes, e.g., *modus ponens*) are regular. To see that this is so, note that simultaneous substitution of the propositional variables of a theorem T for propositional variables not in T preserves the property of being a proof in any axiomatization of *PL* that uses axiom schemes and structural inference rules. Hence, we do not forfeit any generality when in the remainder of this paper we confine our attention to regular axiomatizations of classical propositional logic.

By a *finite-variable fragment* of a regular axiomatization S of *PL* is meant the axiom system that results when one restricts the primitive basis of S to a finite subset of its propositional variables. Notice that nothing goes on, so to speak, in S unless it also goes on in some of its finite-variable fragments, and conversely. In particular, a finite sequence of wffs qualifies as a proof of a wff A in S if and only if it also qualifies as a proof of A in any or all of the finite-variable fragments of S, the vocabularies of which contain every propositional variable that occurs anywhere in the given sequence of wffs of S. Hence, A is a theorem of S if and only if A is a theorem of the finite-variable fragment determined by the propositional variables of A.

Note that a finite-variable fragment F of S contains only finitely many equivalence classes $|A|$ of wffs. The Lindenbaum algebra of F, therefore, will be a finite Boolean algebra. All finite Boolean algebras are atomic, so the Lindenbaum algebra of F will be an atomic Boolean algebra, (One can prove a Representation Theorem for Finite Boolean algebras that is slightly stronger than the one for atomic Boolean algebras, to wit: every finite Boolean algebra is isomorphic to the powerset algebra of the set of its atoms.) The derivation in section 1 of the completeness theorem for axiomatizations of *PL* from Stone's Representation Theorem made use of the lemmas (1.1) and (1.2). Lemma 1.2 needs to be amended slightly before it can be applied to the Lindenbaum algebras of the finite-variable fragments of a regular axiomatization *S* of *PL* because the 2-valued homomorphisms of these algebras determine classical valuations only on the wffs of their respective finite-variable fragments, and not on all the wffs of *S*. Here, then, is the required amended version of Lemma 1.2:

LEMMA 1.2*. *Every 2-valued homomorphism h of the Lindenbaum algebra of a finite-variable fragment F of a regular axiomatization of PL determines a classical valuation V on the wffs of F, where* $V(\mathrm{A}) = \mathrm{t}$ *or* f *according as* $h(|A|) = 1$ *or* 0.

By a *tautology* of a regular axiomatization *S* of *PL* is meant a wff that has the value truth in all classical valuations on the wffs of *S*. Note, then, that a wff *A* of *S* is a tautology if and only if *A* has the value truth in all classical valuations on the wffs of any finite-variable fragment of *S* that contains *A*. Therefore, by making use of Lemmas 1.1 and 1.2*, for each finite-variable fragment *F* of *S* we can derive the following completeness theorem from the Representation Theorem for Atomic Boolean Algebras: *If a wff A of F is a tautology, then A is a theorem of F.* The completeness theorem for *S* itself follows as a corollary from these completeness theorems for its finite-variable fragments.

Why should one bother to relate a regular axiomatization *S* of *PL* to the *ensemble* of its finite-variable fragments rather than deal with *S* as a single unified system? There is a significant algebraic advantage in so doing, for *the Lindenbaum algebras of the finite-variable fragments are atomic Boolean algebras*; whereas, the Lindenbaum algebra of *S* is atomless. (It is the infinity of the propositional variables of *S* that makes its Lindenbaum algebra non-atomic, indeed atomless.) Consequently, by invoking only the Representation Theorem for Atomic Boolean Algebras, we can derive a completeness theorem for each finite-variable fragment of *S* from Lemmas 1.1 and 1.2*. Then, simply by relating *S* to the ensemble of its finite-variable fragments, we obtain the completeness theorem for *S* as a corollary of the Representation Theorem for Atomic Boolean Algebras.

The method used in this paper to bring the Representation Theorem for Atomic Boolean Algebras to bear upon classical propositional logic—namely, to relate or reduce a regular axiomatization of the logic to the ensemble of its finite-variable fragments—can be used to study many other propositional logics. For example, one can investigate the modal propositional logic S5 by relating or reducing some regular axiomatization of it to the ensemble of its finite-variable fragments. Similarly, one can study modal propositional logic

S4, or intuitionistic propositional logic, by relating regular axiomatizations of these logics to the ensembles of their respective finite-variable fragments.

An axiomatization S of propositional logic will be said to be *infinitistic* if and only if S contains an infinite set Δ of pairwise non-interderivable wffs (i.e., no two wffs in Δ can be proved from each other). S will be said to be *intrinsically infinitistic* if and only if it contains an infinite set of pairwise non-interderivable wffs Δ such that only finitely many propositional variables occur in the members of Δ. Finally, an infinitistic axiomatization of propositional logic will be said to be *extrinsically infinitistic* if and only if it is not intrinsically infinitistic. For example, axiomatizations of classical propositional logic and of S5 modal propositional logic are extrinsically infinitistic; only the presence of infinitely many propositional variables in their respective vocabularies makes these systems infinitistic. On the other hand, regular axiomatizations of S4 modal propositional logic are intrinsically infinitistic; even the finite-variable fragments of these axiomatizations are (intrinsically) infinitistic.

The advantage of investigating an extrinsically infinitistic regular axiomatization of some propositional logic by relating or reducing it to the ensemble of its finite-variable fragments is obvious; for, unlike the parent system, none of its finite-variable fragments is infinitistic. But the method of relating or reducing a regular axiomatic propositional system to the ensemble of its finite-variable fragments is advantageous even when applied to intrinsically infinitistic propositional systems such as regular axiomatizations of S4 modal propositional logic. Even though the finite-variable fragments of a regular axiomatization of S4 will themselves be intrinsically infinitistic, they are nonetheless more tractable than the parent system itself.

REFERENCES

Bell, J., and A. Slomson

1971 *Models and ultraproducts: An introduction,* North-Holland, Amsterdam, pp. 42–43.

Mendelson, E.

1970 *Boolean algebras and switching circuits,* Schaum's Outline Series, McGraw-Hill, New York, pp. 135–137.

Rasiowa, H., and R. Sikorski

1970 *The mathematics of metamathematics,* third edition, Polish Scientific Publishers, Warsaw.

ROBERT K. MEYER

IMPROVED DECISION PROCEDURES FOR PURE RELEVANT LOGIC

Dedicated to Professor Alonzo Church, in his 91st year, and to Professor Saul Kripke, who provided the impetus for these results.

Abstract. We study Church's "weak implicational calculus" of 1951, which is the pure implicational part $\mathbf{R}_{\rightarrow}$ of the relevant logic **R**. We investigate and develop, again following Church, the effect of adding propositional quantifiers to $\mathbf{R}_{\rightarrow}$; this allows also the specification of a minimal, a De Morgan and finally a Boolean negation. The bulk of the paper deals with finite model properties for various fragments of **R**, including $\mathbf{R}_{\rightarrow}$. A modification of an argument due to Saul Kripke is the chief tool in this project, yielding an Infinite Division Prinicple (IDP) for "Church monoids" that facilitates our finitization. Thus systems intermediate between $\mathbf{R}_{\rightarrow}$ and **LR** (non-distributive **R**) have the finite model property and are hence model-theoretically decidable. A concluding section shows how to define various logical particles using Church's propositional quantifiers, which turns what are conservative extension results at the quantifier-free level into axiomatic extensions when quantifiers are present.

1. INTRODUCTION TO RELEVANT IMPLICATION

Among the many contributions of Alonzo Church to Logic, his invention of relevant implication is by no means the least.[1] His principal paper on the subject is *Church 1951a*; so far as I know, he has never returned to it. As Church deals with relevant implication, it has two levels: a fundamental one, just based on $\rightarrow$ and propositional variables; and an advanced one, in which propositional quantifiers (based on $\forall$) are also introduced. Following Church's notational conventions,[2] we axiomatize $\mathbf{R}_{\rightarrow}$, choosing the following

[1] In fact, Church (*1951a*) called it the "*weak* implicational calculus", while in *Church 1951* he again speaks of "weak" implication. Also he used $\supset$, which I reserve here for a *material* implication. Church's invention of this pure $\rightarrow$ system (afterwards called $\mathbf{R}_{\rightarrow}$ by Anderson and Belnap, and **BCIW** by combinatory logic fans) was mildly anticipated by Moh Shaw-Kwei (*1950*), and more significantly anticipated in *Orlov 1928*; also see *Došen 1993*. Adapting a terminological suggestion of Bacon, $\mathbf{R}_{\rightarrow}$ was extended to the system **R** of relevant implication. See *Anderson and Belnap 1975*.

[2] That is, we associate equal connectives to the left, using dots (sparingly) as parentheses to reverse the direction of association. We extend Church's conventions by ranking binary connectives, including additional primitives or defined ones, in order of increasing scope: &, o, $\vee$, +, $\oplus$, $\supset$, $\rightarrow$, $\leftrightarrow$. Of these, o, $\oplus$, and $\leftrightarrow$ are the "relevant" analogues of 'and', 'or', and 'co-implies' introduced in *Church 1951a*; + (fission) is in *Anderson and Belnap 1975* and is a DeMorgan dual of o (fusion), while &, $\vee$, and $\supset$ are "truth-functional" particles for **R** in the sense of *Anderson and Belnap 1975*.

C. Anthony Anderson and M. Zelëny (eds.), *Logic, Meaning and Computation*, 191-217.

axiom schemes and rule:[3]

AxI. $A \to A$
AxB. $B \to C \to .\ A \to B \to .\ A \to C$
AxC. $(A \to .\ B \to C) \to .\ B \to .\ A \to C$
AxW. $(A \to .\ A \to B) \to .\ A \to B$

Rule $\to$E. $A \to B \Rightarrow .\ A \Rightarrow B$

Church viewed $\mathbf{R}_{\to}$ as the result of weakening the intuitionist implicational calculus $\mathbf{J}_{\to}$ by trading in the (paradoxical) axiom

AxK. $A \to .\ B \to A$

for AxI. Thus $\mathbf{R}_{\to}$ bears the same sort of relation to $\mathbf{J}_{\to}$ as the $\lambda\mathbf{I}$ calculi of Church (*1941*) bears to the preferred $\lambda\mathbf{K}$ calculus of Curry and Feys (*1958*).[4]

2. PROPOSITIONAL QUANTIFIERS

We extend $\mathbf{R}_{\to}$ to $\mathbf{R}^{\forall p}_{\to}$ with the following definitions and additional axiom schemes and rule:[5]

Do. $A \circ B =_{df} \forall q((A \to .\ B \to q) \to q)$, q does not occur in A or B
D$\leftrightarrow$. $A \leftrightarrow B =_{df} (A \to B) \circ (B \to A)$
D$\oplus$. $A \oplus B =_{df} (B \to A \to .\ A \to B \to B)$
D$\exists$. $\exists p A(p) =_{df} \forall q(\forall p(A(p) \to q) \to q)$, q does not occur in $A(p)$
DF. $\mathbf{F} =_{df} \forall pp$
DT. $\mathbf{T} =_{df} \mathbf{F} \to \mathbf{F}$
Dt. $\mathbf{t} =_{df} \forall p(p \to p)$

Ax$\forall$E. $\forall p A(p) \to A(B)$
Ax$\to\forall$. $\forall p(A \to B(p)) \to .\ A \to \forall p B(p)$, p not free in A
Rule $\forall$I. $A \Rightarrow \forall pA$

The definitions above are largely taken from Church, who has himself adapted them from Russell. Among them, Church's analogue o of conjunction

---

[3] Church (*1951a*) stated particular axioms, building substitution into his notion of deducibility. We dispense with an explicit substitution rule by stating axioms schematically, following von Neumann. Rules are formulated with $\Rightarrow$ as a metalogical 'if'. Thus, read $\to$E below as "If $A \to B$ is a theorem then if A is a theorem then B is a theorem."

[4] Or, as Curry once put it, "Church never did much like the combinator K." Incidentally, Church (*1951a*) notes that others have previously suggested doing without AxK.

[5] $\mathbf{R}^{\forall p}_{\to}$ is what Church (*1951a*) calls the "weak theory of implication", adapting Russell's use of the "theory of implication" to characterize the classical theory based on material implication and universal propositional quantification (near enough). In stating quantificational axiom schemes for $\mathbf{R}^{\forall p}_{\to}$, we adopt the convention that $A(p)$ stands for a formula in which p may occur free; where B is any formula, $A(B)$ shall then denote the result of *proper* substitution of B for p in $A(p)$, rewriting bound variables as necessary to avoid confusion of bound variables.

has played a central role in subsequent developments, largely on account of Dunn's algebraic work contributed to *Anderson and Belnap 1975.* Similarly, Church's co-implication $\leftrightarrow$ comes to the same thing in **R** as the $\leftrightarrow$ defined in *Anderson and Belnap 1975.* But Church's disjunction, for which I use $\oplus$, has (more or less) vanished without trace. Church also has various thoughts about negation, none of which have really caught on (yet, anyway); we look at these below.

What has caught on, however, is Church's form of the deduction theorem. What it amounts to, in outline, is that if there is a deduction of B from $A_1, \ldots, A_n$ in $\mathbf{R}_{\rightarrow}$, then either there is already a deduction of B from $A_1, \ldots, A_{n-1}$, or else there is a deduction of $A_n \rightarrow B$ from $A_1, \ldots, A_{n-1}$. The usual formulation of the deduction theorem involves only the second clause; but it makes essential use of AxK to provide a "fictitious" use of A_n in case there were no "real" ones. Church's deduction theorem for $\mathbf{R}_{\rightarrow}$is more vertebrate; only those $A_{i,j}$ actually used (by the lights of $\mathbf{R}_{\rightarrow}$) show up in the theorem $A_{i,1} \rightarrow . \ldots A_{i,m} \rightarrow B$ that *records* the deduction. In the limiting case, where B itself is a theorem, there may (or may not) be a way to use one or more of the A_i in the deduction of B; still, the first clause of Church's version of the deduction theorem lets us off the hook here, since the theoremhood of B is sufficient for (what Anderson and Belnap, *1975*, calls) its Official deducibility. Church (*1951a*) shows how to extend Church's deduction theorem to all of $\mathbf{R}^{\forall\mathrm{p}}_{\rightarrow}$.

Among the properties of our just defined connectives o and $\leftrightarrow$, are the $\leftrightarrow$ elimination theorems $(A \leftrightarrow B) \rightarrow .\ A \rightarrow B, (A \leftrightarrow B) \rightarrow .\ B \rightarrow A$, and the introduction theorem $A \rightarrow B \rightarrow .\ B \rightarrow A \rightarrow .\ A \leftrightarrow B$. Even nicer is the following:

Residuation Theorem. $(A \rightarrow .\ B \rightarrow C) \leftrightarrow A \circ B \rightarrow C$.

3. NEGATION

Church (*1951a*) tries out various definitions of negation. The most straightforward is a *minimal* negation, in the style of Johansson (*1936*), obtained by simply adding a constant f to the language of $\mathbf{R}_{\rightarrow}$, together with

D$\sim$. $\sim A =_{\mathrm{df}} A \rightarrow \mathrm{f}$.

We call this system $\mathbf{MR}^{\mathrm{f}}_{\rightarrow}$. Since f, in this case, is simply playing the role of a propositional variable, it lacks those properties of classical negation which intuitionist logic lacks (like full double negation and contraposition principles). But, to get those properties, it is sufficient to formulate a system $\mathbf{R}^{\mathrm{f}}_{\rightarrow}$ with f primitive, D$\sim$, and the single additional axiom scheme

Ax$\sim\sim$E. $\sim\sim A \rightarrow A$.

While Ax$\sim\sim$E would induce collapse to classical logic if applied to intuitionist (or even minimal) logic, it produces a conservative extension of $\mathbf{R}_{\rightarrow}$.

And it permits *another* definition of the *relevant* analogue of conjunction in $\mathbf{R}^{\mathrm{f}}_{\rightarrow}$, namely the *fusion* o, as follows:[6]

Do. $A \circ B =_{\mathrm{df}} \sim(A \rightarrow \sim B)$.

Given Do, we can even define an equivalence connective by D$\leftrightarrow$ above. And we also may define

Dt. $\mathrm{t} =_{\mathrm{df}} \sim\mathrm{f}$.

As we have seen above, o and $\leftrightarrow$ were, to all intents and purposes, already present in *Church 1951a*. We have also the promised DeMorgan dual + of o in $\mathbf{R}^{\mathrm{f}}_{\rightarrow}$.

D+. $A + B =_{\mathrm{df}} \sim A \rightarrow B$.

$\mathbf{R}^{\mathrm{f}}_{\rightarrow}$ is an interesting alternative, not considered by Church but built into **R** in *Anderson and Belnap 1975* as an adaptation of *Ackermann 1956*. Making double negation explicit suffices to give $\sim$ its usual truth-functional properties (more or less). There is moreover a simple Gentzen system, due to Kripke, which yields (pretty trivially) that $\mathbf{R}^{\mathrm{f}}_{\rightarrow}$ is a conservative extension of $\mathbf{R}_{\rightarrow}$. Thus, unlike intuitionist logic, $\mathbf{R}_{\rightarrow}$ lives quite comfortably with a standard truth-functional view of negation; on the other hand, it resists nonsense like $A \rightarrow . \sim A \rightarrow B$.[7]

4. LATTICE EXTENSIONS OF $\mathbf{R}_{\rightarrow}$

We may also, following Anderson and Belnap (*1975*, and eventually Ackermann), introduce *lattice connectives* & and $\vee$, subject to the following additional axiom schemes and rule:

Ax&E. $A \,\&\, B \rightarrow A$ and $A \,\&\, B \rightarrow B$
Ax$\rightarrow$&I. $(A \rightarrow B) \,\&\, (A \rightarrow C) \rightarrow . \, A \rightarrow B \,\&\, C$
Ax$\rightarrow\vee$E. $(A \rightarrow C) \,\&\, (B \rightarrow C) \rightarrow . \, A \vee B \rightarrow C$
Ax$\vee$I. $A \rightarrow A \vee B$ and $B \rightarrow A \vee B$

Rule &I. A and $B \Rightarrow A \,\&\, B$

[6]In fact, our two versions of Do become equivalent in the presence of Ax$\sim\sim$E above. Otherwise, it seems reasonable to take the previous Do (from *Church 1951a*) as the more fundamental. Another approach is simply to take fusion o as an additional primitive, subject to $\rightarrow$ axioms corresponding to the two halves of the residuation theorem scheme above. On all such schemes, residuation still holds. So do the elimination and introduction properties of $\leftrightarrow$.

[7]Church (*1951a*) does consider $\mathrm{f} \rightarrow \sim A$ as a possible additional axiom scheme, which Church used to prove various Glivenko-style theorems. I think that he got carried away with trying to make $\mathbf{MR}_{\rightarrow}{}^{\mathrm{f}}$ as much like Johansson's minimal logic as possible. Note that $\mathrm{f} \rightarrow \sim A$ does become a theorem scheme if one defines f as $\forall p \forall q(p \rightarrow . \, q \rightarrow p)$, which Church (*1951a*) also considers. But, while AxK may be bad, it hardly seems *sufficiently* bad to qualify as *das Falsche*.

In the sequel, the extension of $\mathbf{R}_{\rightarrow}$ obtained by adding just & and its governing axioms and rules (Ax&E, Ax → &I, and Rule &I) will have a distinguished place. We call this system $\mathbf{R}_{\rightarrow\&}$, and we note that it is a conservative extension of $\mathbf{R}_{\rightarrow}$.[8] We may also consider it a system $\mathbf{MR}^{\mathrm{f}}_{\rightarrow\&}$, since any propositional variable irrelevant to a context may play the role of a *minimal* f. But the systems split if we add the Ax∼∼E form of double negation, naturally enough. In the presence of double negation, ∨ becomes definable in the usual DeMorgan way from & and ∼, and its lattice axioms Ax→∨E and Ax∨I become provable; but only the Ax∨I axioms are provable in $\mathbf{MR}^{\mathrm{f}}_{\rightarrow\&}$.

Instead, let us take ∨ as an additional primitive, subject to the lattice axioms and rule above. This produces variants of the system **LR** (for "lattice R").[9] We then have systems as follows:

$\mathbf{MLR}_{\rightarrow\&\vee} = \mathbf{R}_{\rightarrow\&} + \vee\text{-axioms}$[10]
$\mathbf{MLR}^{\mathrm{f}}_{\rightarrow\&\vee} = \mathbf{MLR}_{\rightarrow\&\vee} + \mathrm{D}\sim$ (minimal negation)
$\mathbf{LR}_{+} = \mathbf{MLR}_{\rightarrow\&\vee}$
$\mathbf{LR} = \mathbf{MLR}^{\mathrm{f}}_{\rightarrow\&\vee} + \mathrm{Ax}\sim\sim\mathrm{E} = \mathbf{LR}_{+} + \mathrm{D}\sim + \mathrm{Ax}\sim\sim\mathrm{E}$

Note that (here anyway) we follow Church in sticking to *inferential* definitions of negation.

Finally, to get the full system **R**, Anderson and Belnap (*1975*) add the distribution principle

AxDist. $A \,\&\, (B \vee C) \rightarrow A \,\&\, B \vee A \,\&\, C.$

But **R**, being undecidable, will only concern us peripherally in the present essay.

5. THE COMBINATORICS OF RELEVANT DECIDABILITY

Kripke (*1959*) notes (by a Gentzen argument) that $\mathbf{R}_{\rightarrow}$ is decidable. Kripke's argument was adapted to all of **LR** in *Meyer 1966.*[11] Unfortunately, the most that one gets out of Kripke is *raw* decidability; the sense that one has obtained the sort of *control* over a system that decidability proofs often yield is sadly missing.[12]

[8]Moreover, **R** is a conservative extension of $\mathbf{R}_{\rightarrow\&}$.

[9]Mechanized in *Thistlewaite, McRobbie, and Meyer 1988.*

[10]We may call this system simply $\mathbf{MLR}_{+}$.

[11]See *Thistlewaite, McRobbie, and Meyer 1988* for the smoothest Gentzen-type formulation of **LR**.

[12]One ought not, however, read too much into this lack of control. Even truth-functional logic is no better than NP-complete, which tends to leave complexity theorists tearing their hair. But relevantists are lucky to have any hair left to tear. In the first place, Urquhart (*1984*) showed that **R** (and its close cousins, like **E**) are *undecidable.* More distressing (for $\mathbf{R}_{\rightarrow}$ fans) is the fact that Urquhart (*1990*) placed frightening bounds on the complexity of *any* decision procedure for $\mathbf{R}_{\rightarrow}$ and relevant decidable extensions (like **LR**).

Still, the decision question even for $\mathbf{R}_{\rightarrow}$ remained open for some years. Kripke solved it with a *very nice* combinatorial insight.[13] We shall express this insight, in the first place, in terms of the *divisibility* ordering | on the set $\mathbb{N}^+$ of positive integers.[14]

We may write the non-zero members s of $\mathbb{N}^+$ in the form

$$s = 2^{f(1)} \cdot \ldots \cdot p_i^{f(i)} \cdot \ldots \cdot p_{n(s)}^{f(n(s))},$$

where p_i is the ith prime; $f : \mathbb{N}^+ \rightarrow \mathbb{N}^+$ is a function giving the exponent on p_i in prime decomposition of s, and $n : \mathbb{N}^+ \rightarrow \mathbb{N}^+$ yields the *last* relevant prime in the decomposition of s.[15]

Let now S be any subset of $\mathbb{N}^+$. We shall call S *finitely grounded* iff there is some prime p such that, for all s in S, every prime p_i with *positive* exponent in the above prime decomposition is such that $p_i \leq p$. This means, in our notation above, that $p \geq n(s)$ for *all* $s \in S$. Alternatively, where $\mathbb{N}_k$ is the set of all positive integers built out of the first k primes, S is finitely grounded iff there exists a k such that $S \subseteq \mathbb{N}_k$.

I shall shortly prove my favorite equivalent of Kripke's lemma, expressed as a fact of number theory. Again, let S be any set of positive integers. We call $s \in S$ an *infinite divisor* for S iff there exists an infinite subset S' of S such that, for all s' in S', we have $s \mid s'$. That is, an infinite divisor for S divides infinitely many other members of S. And now we show the following:

THEOREM 5.1 Infinite Division Principle (IDP). *Let S be any infinite, finitely grounded subset of $\mathbb{N}^+$. Then there exists an infinite divisor s for S.*

Proof. We prove IDP for all $\mathbb{N}_k$, by induction on k. The first applicable case is $k = 1$, in which case all members of S are multiples of a single prime p (which, as we have set things up, is 2). Let s be the *least* member of S. It will, evidently, divide all the others. So $S' = S$ will do, with s as the infinite divisor. The inductive case, which is more fun, we leave to the reader to look up or work out, ending the proof of the IDP. ⊣

DEFINITION 5.2. Let us call a sequence $\langle n_i \rangle_{1<i<\omega}$ of natural numbers **redundant** provided that, there exist positive integers h and j, $h < j$, such that $n_h \mid n_j$. Otherwise $\langle n_i \rangle$ is **irredundant**. If there exists a k in $\mathbb{N}^+$ such that all members of this sequence belong to $\mathbb{N}_k$, then we call $\langle n_i \rangle$ **finitely grounded.**[16]

[13]It was all the nicer because Kripke was still in high school when he had it!

[14]We might as well express it as in *Meyer and Ono 1994* in terms of the *product* ordering on direct powers of $\mathbb{N}$—which, in view of the fundamental theorem of arithmetic, comes to the same thing in cases of interest here.

[15]In a word, we are simply building in here the fundamental theorem of arithmetic, taking advantage of the correspondence between positive integers and the multisets of their prime factors.

[16]I.e., according to definitions above, if the set of all the n_i is finitely grounded.

We then have, adapted to the present context,

LEMMA 5.3 Kripke's Lemma. *Every finitely grounded irredundant sequence of positive integers is finite.*[17]

Proof. Let $\langle n_i \rangle$ be a finitely grounded sequence of positive integers. To prove the lemma, assume $\langle n_i \rangle$ infinite; we show that, for some $h < j$, $n_h | n_j$; whence, contraposing, if $\langle n_i \rangle$ is irredundant, it must also be finite. Let S now be the set of all the n_i. There is a k such that $S \subseteq \mathbb{N}_k$, since $\langle n_i \rangle$ is finitely grounded. If S is finite, then some n_h is repeated infinitely often in the sequence, which is then surely redundant. So we may assume S infinite. But then, by the IDP, there is a member n_h of S which is an infinite divisor for S, for h a positive integer. I.e., there is an *infinite* subset S' of S such that n_h divides *every* member of S'. Evidently, infinitely many of these properly *succeed* n_h in the sequence ordering; letting j be the least positive integer $> h$ such that $n_j \in S'$, we have $n_h \,|\, n_j$, ending the proof of (the present version of) Kripke's lemma. ⊣

There are a number of interesting principles which either imply or are equivalent to Kripke's lemma.[18] A final one that I shall consider here goes back to Dickson (*1913*). Let $S \subseteq \mathbb{N}^+$, and let $S_{\min} \subseteq S$ be those members of S that are *minimal* on the divisibility ordering. Dickson showed that:

LEMMA 5.4 Dickson's Lemma. *If $S \subseteq \mathbb{N}_k$, then $S_{\min}$ is finite.*

Proof. Again we apply the IDP. Suppose that $S_{\min}$ is infinite, for *reductio*. Then, by the IDP, it has an infinite divisor. But the *minimal* members of a set, by definition, are totally unordered, which yields a contradiction. This ends the proof of Dickson's lemma. ⊣

This is enough combinatorics. Now we need some model theory.

6. OPERATIONAL SEMANTICS FOR $\mathbf{R}_{\rightarrow\&}$

There have been a number of semantical analyses, dating from the early '70's, of relevant logics. I shall concentrate here on those that have targeted **R** and its subsystems and shall stick to a style of semantical analysis that, in a wider perspective, I more or less do not prefer. Still, it works for present purposes.

DEFINITION 6.1. A **commutative monoid** (henceforth, **c-monoid**) is a structure

$$\mathrm{M} = \langle \mathrm{M}, \circ, 1 \rangle$$

[17]For something closer to Kripke's original lemma, towards which we are working here, see *Anderson and Belnap 1975,* page 139. This is based on a letter from Kripke to Anderson and Belnap in 1959.

[18]For a survey of some of these principles, see Riche's ANU thesis (*1991*).

where M is a set, 1 is an (identity) element in M, and o is a commutative, associative binary operation on M.

We shall often use simple juxtaposition for 'o'.

DEFINITION 6.2. A **partially ordered c-monoid** (henceforth, **c-po-monoid**) is a structure $\langle M, \leq \rangle$, where M is a c-monoid and $\leq$ partially orders M; i.e., $\leq$ is a reflexive, transitive, antisymmetric relation on M; moreover, it has the monotonicity property; i.e., for all a, b, c in M,

$$\text{(m) If } a \leq b \text{ then } ca \leq cb.$$

(The other monotonicity property, "$a \leq b$ implies $ac \leq bc$," is immediate by commutativity of o.)

The idea of using c-monoids to give a semantics for $\mathbf{R}_{\rightarrow\&}$ goes back to Urquhart and to Routley, independently; also independently, but a little later, the ideas were extended to all of **R** (and other relevant logics) by Fine.[19]

DEFINITION 6.3. Let L be the set of formulas of $\mathbf{R}_{\rightarrow\&}$, let M be a c-monoid, and let 2 be the set {T,F} of truth-values. By an **interpretation** we mean a function $\mathcal{I} : L \times M \rightarrow 2$ subject to the following conditions, for all $A, B \in L$, $x \in M$:

(T&) $\mathcal{I}(A \,\&\, B, x) = T$ iff $\mathcal{I}(A, x) = \mathcal{I}(B, x) = T$;
(T→) $\mathcal{I}(A \rightarrow B, x) = T$ iff,
for all $a \in M$ if $\mathcal{I}(A, a) = T$, then $\mathcal{I}(B, xa) = T$.

DEFINITION 6.4. A formula A is **verified** on $\mathcal{I}$ just in case $\mathcal{I}(A, 1) = T$. And the *idea* of the modelling is that A should be a theorem of $\mathbf{R}_{\rightarrow\&}$ just in case A is verified on all interpretations in all c-monoids.

Unfortunately, this idea does not work out for $\mathbf{R}_{\rightarrow\&}$. Since it does work out for the corresponding fragment of Girard's linear logic, the problem is with AxW, which separates the systems. To dodge this problem, we define:

DEFINITION 6.5. A **Church monoid** is a c-po-monoid, subject for all a in M to Dunn's square-increasing postulate

(Sq) $a \leq a \circ a$.

[19]Urquhart (*1972*) provided a *semilattice* semantics for $\mathbf{R}_{\rightarrow\&}$. This was Routley's original idea as well, though he did drop hints toward the *relational* semantics that Routley and Meyer (*1973*) eventually developed. Fine's *1974* sets out his *operational-relational* ideas—which, with extra (and somewhat cumbersome) machinery, provides a complete semantics for **R** and other relevant logics. See *Routley, Plumwood, Meyer, and Brady 1982* (if you can find it) for a general setting out of Routley's and my ideas; what I think (which differs from just about everyone on *some* salient points) is covered in the papers (many with Routley) listed in Wolf's bibliography (*Anderson, Belnap, and Dunn 1992*).

And:

DEFINITION 6.6. A **hereditary interpretation** $\mathcal{I}$ (henceforth, **h-interpretation**) in a Church monoid is required to satisfy

(H) If $a \leq b$, then if $\mathcal{I}(A, b) = \mathrm{T}$, then $\mathcal{I}(A, a) = \mathrm{T}$,

for all $A \in \mathrm{L}$, $a, b \in \mathrm{M}$.

A simple induction suffices to show that it is enough to impose (H) on propositional variables to get it for all formulas of L.

DEFINITION 6.7. A formula A is **h-valid** in $\langle \mathrm{M}, \leq \rangle$ just in case A is verified on all h-interpretations therein. And A is **h-valid**, *simpliciter*, just in case A is h-valid in all Church monoids.

For a snappy completeness proof for $\mathbf{R}_{\rightarrow\&}$, we appeal first to a canonical model in the Lindenbaum algebra of (all of) $\mathbf{R}$, identifying provably equivalent formulas of $\mathbf{R}$ and defining o and t by Do and Dt of section 3.[20] Taking $\mathbf{R}$ as synonymous with its Lindenbaum algebra (which will do for the present) we define:

DEFINITION 6.8. $\mathcal{C}$ is the **canonical interpretation** of $\mathbf{R}_{\rightarrow\&}$ in $\mathbf{R}$ iff

$$\mathcal{C}(B, A) = \mathrm{T} \text{ iff } A \rightarrow B \text{ is a theorem of } \mathbf{R},$$

for all formulas B in the language of $\mathbf{R}_{\rightarrow\&}$ and A in the language of $\mathbf{R}$.

So parsed, $\mathbf{R}$ is a DeMorgan monoid (in the terminology of Anderson and Belnap, *1975*), and a fortiori a Church monoid (in present terms). Moreover, it is trivial that $\mathcal{C}$ satisfies the hereditary condition (H) and the truth-conditions (T$\rightarrow$) and (T&) above, by deductive properties of $\mathbf{R}$. Accordingly we have:

THEOREM 6.9 Canonical Completeness Theorem for $\mathbf{R}_{\rightarrow\&}$. *$A$ is a theorem of $\mathbf{R}_{\rightarrow\&}$ iff A is verified in $\mathbf{R}$ on the canonical h-interpretation $\mathcal{C}$.*

Proof. Little needs to be said. In view of the $\mathbf{R}$-theorem $A \leftrightarrow \mathrm{t} \rightarrow A$, exactly the theorems of $\mathbf{R}_{\rightarrow\&}$ will be true at 1 on $\mathcal{C}$. We have already noted that $\mathcal{C}$ is an h-interpretation. ⊣

THEOREM 6.10 First Completeness Theorem for $\mathbf{R}_{\rightarrow\&}$. *$A$ is a theorem of $\mathbf{R}_{\rightarrow\&}$ iff A is h-valid.*

Proof. For "only if", argue by deductive induction that $\mathcal{I}(A, 1) = \mathrm{T}$, for an arbitrary h-interpretation $\mathcal{I}$ in an arbitrary Church monoid M. If A is a non-theorem, it is canonically refuted as above. ⊣

[20] This is as in Dunn's (*1966*) thesis. The monoid 1 is identified with (the set of formulas provably equivalent to) t, while $\leq$ is the provable entailment relation in $\mathbf{R}$. We depend on the fact that $\mathbf{R}$ is a conservative extension of $\mathbf{R}_{\rightarrow\&}$, which I have proved elsewhere.

COROLLARY 6.11. *A is a theorem of* $\mathbf{R}_{\rightarrow\&}$ *iff A is h-valid in all countable Church monoids.*

Proof. Observe that, formulated in a countable vocabulary, the Lindenbaum algebra **R** has countably many elements. ⊣

THEOREM 6.12 Second Completeness Theorem for $\mathbf{R}_{\rightarrow\&}$. *A is a theorem of* $\mathbf{R}_{\rightarrow\&}$ *iff A is h-valid in all* finitely generated *Church monoids.*

Proof. Only right-to-left requires thought. So assume that A is a non-theorem of $\mathbf{R}_{\rightarrow\&}$. We find a finitely generated Church monoid $\langle \mathrm{M}, \leq \rangle$ and an interpretation $\mathcal{I}$ therein such that $\mathcal{I}(A, 1) = \mathrm{F}$. Let $\mathsf{Sub}(A)$ be the collection of subformulas of A, and let $\mathsf{Sub}(\mathbf{R})$ be the sub-monoid of the Lindenbaum algebra **R** generated by (equivalence classes of) members of $\mathsf{Sub}(A)$ and t, closing under fusion o.[21] We then use the canonical interpretation $\mathcal{C}$ to *impose* an interpretation $\mathcal{I}$ in $\mathsf{Sub}(\mathbf{R})$, by setting $\mathcal{I}(p, x) = \mathcal{C}(p, x)$ for all propositional variables p and elements x of $\mathsf{Sub}(\mathbf{R})$. Imposing T→ and T&, $\mathcal{I}$ is then defined on all formulas of $\mathbf{R}_{\rightarrow\&}$. Moreover, since (H) continues to hold on propositional variables (since $\mathcal{C}$ is an h-interpretation), it will hold on all formulas. We can conclude the proof by showing that, for all subformulas B of A and for all elements x of $\mathsf{Sub}(\mathbf{R})$, $\mathcal{I}(B, x) = \mathcal{C}(B, x)$. We show this by structural induction. The only interesting case occurs when B is of the form $D \rightarrow E$, when $\mathcal{C}(B, x) = \mathrm{F}$. This means that, in **R**, x does not entail $D \rightarrow E$. But then, by residuation, $x \circ D$ does not entail E; i.e., in bastard symbols, $\mathcal{C}(E, xD) = \mathrm{F}$, while of course $\mathcal{C}(D, D) = \mathrm{T}$. But then, since D is a subformula of A (because B is), we have on inductive hypothesis the same counterexample under $\mathcal{I}$; whence, by T→, $\mathcal{I}(B, x) = \mathrm{F}$ as well. Other cases being trivial, we have in particular that $\mathcal{I}(A, 1) = \mathrm{F}$. So each non-theorem of $\mathbf{R}_{\rightarrow\&}$ is refuted in some finitely generated $\mathsf{Sub}(\mathbf{R})$. ⊣

It is not immediately obvious how to extend the theorem just proved to the Urquhart-Routley *semilattice* semantics set out in *Anderson, Belnap, and Dunn 1992.* (If it were, we should already have a finite model property for $\mathbf{R}_{\rightarrow\&}$, since finitely generated semilattices are finite.) This aspect will be considered below.

7. RELEVANT DIVISION AND FREE COMMUTATIVE MONOIDS

We shall not, for the time being, pursue semilattices. Our emphasis will be on Church monoids and what it takes to finitize them. We pause for the following disconcerting observation. Little children, we have heard, are indoctrinated with the idea that 2 divides 12. While this is OK for some

[21] We also stipulate that $\leq$ shall *agree* in $\mathsf{Sub}(\mathbf{R})$ and **R**. The trick of closing under o but not under other connectives of **R** was suggested to me by Segerberg, when I discussed these problems with him in 1972.

purposes, like being promoted to third grade, it runs immediately into a violation of the Church *use* criterion. For 12 has a factor, namely 3, which is not *used* in performing the division. But we could not object, on such grounds, to "6 divides 12". For every prime factor of 12 (specifically, 2 and 3) is also a prime factor of 6. Let us invent some notation for this situation. We use, as before, $m \mid n$ for *ordinary* divisibility. But we shall write $m \,/\, n$ to mean that m *relevantly divides* n. Specifically, for all positive integers m and n, we have

$$\text{(D/)} \quad m \,/\, n \ \text{ iff } \ \text{(a) } m \mid n \ \text{ and } \ \text{(b) } \sigma(m) = \sigma(n),$$

where $\sigma : \mathbb{N}^+ \to 2^{\text{Primes}}$ is a function that assigns to each positive integer the *set* of its prime factors. Note that, satisfyingly, we have not only 6 / 12 but also 111 / 999. But, despite (a), both 2 / 12 and 37 / 999 *fail*.

We next observe that all of the refutation built into our first semantical completeness theorem can already be done in $\mathbb{N}^+$, ordered by /. Indeed, it can be done in the $\mathbb{N}_k$. Let us call a formula A *numerically valid* iff A is verified on every hereditary interpretation I in every $\langle \mathbb{N}_k, / \rangle$.[22] We then have:

THEOREM 7.1 Numerical Completeness Theorem for $\mathbf{R}_{\to\&}$. *A is a theorem of $\mathbf{R}_{\to\&}$ iff A is numerically valid.*

Proof. From left to right, the theorem is a consequence of our first completeness theorem. For the rest, assume that A is not a theorem. We shall find an $\mathbb{N}_k$ and a hereditary interpretation $\mathcal{I}$ therein such that $\mathcal{I}(A, 1) = \text{F}$.

By the second completeness theorem, there is a finitely generated Church monoid $\langle \text{M}, \leq \rangle$ and a hereditary interpretation $\mathcal{J}$ therein such that $\mathcal{J}(A, 1) = \text{F}$ in M. Every finitely generated c-monoid is an image of the *free* c-monoid $\mathbb{N}_k$ in k generators, for some k in $\mathbb{N}^+$. Moreover, relevant divisibility / is the *smallest* relation under which $\mathbb{N}_k$ becomes a Church monoid.

So, where G is a set of k generators for M, let $h : \mathbb{N}_k \to \text{M}$ be the homomorphism determined by any surjection from the first k primes onto G. Note, in particular, that if $m \,/\, n$, then $h(m) \leq h(n)$. And let us now define $\mathcal{J}$ by setting, for all propositional variables p and numbers m in $\mathbb{N}_k$, $\mathcal{J}(p, m) = \text{T}$ iff $\mathcal{I}(p, h(m)) = \text{T}$. Note that this satisfies (H) on propositional variables; for if $m \,/\, n$, then certainly $h(m) \leq h(n)$. We complete the specification of J by imposing the truth-conditions T$\to$ and T&. This makes $\mathcal{J}$, certainly, an h-interpretation in the Church monoid $\mathbb{N}_k$, ordered by /.

We conclude the proof by showing, for all m in $\mathbb{N}_k$ and formulas D of $\mathbf{R}_{\to\&}$, $\mathcal{J}(D, m) = \mathcal{I}(D, h(m))$. We show this by structural induction on D. It is true by fiat if D is a propositional variable, and it follows quickly if D is $B \& C$, on

[22]We make / explicit here only to make the point that the $\mathbb{N}_k$ are Church monoids, by an easy verification. But, since / is definable in terms of monoid operations, we essentially get it for free.

inductive hypothesis. So suppose D is $B \to C$. Then $\mathcal{J}(D,m) = \mathrm{F}$ iff there exists an n in $\mathbb{N}_k$ such that $\mathcal{J}(B,n) = \mathrm{T}$ and $\mathcal{J}(C,mn) = \mathrm{F}$, iff (on inductive hypothesis) $\mathcal{I}(B,h(n)) = \mathrm{T}$ and $\mathcal{I}(C,h(mn)) = \mathrm{F} = \mathcal{I}(B,h(m) \circ h(n))$. Since h is an epimorphism, this argument goes smoothly in both directions; its upshot is that $\mathcal{J}(A,1) = \mathcal{I}(A,h(1)) = \mathcal{I}(A,1) = \mathrm{F}$, transmuting our refutation of A in finitely generated M into its refutation in one of the $\mathbb{N}_k$. ⊣

8. FINITIZING FREE COMMUTATIVE MONOIDS

Primitive peoples, not having been indoctrinated in the sophistry that the natural numbers *fail* to run out, are said to have counted $1, 2, \ldots,$ HEAP, where HEAP was (from their viewpoint) the greatest natural number. It is time to resuscitate this primordial intuition of our species.[23] Our aim is to refute every non-theorem of $\mathbf{R}_{\to\&}$ in a finite model. But even the $\mathbb{N}_k$, our preferred models to this point, do not run out. We wish to use Kripke's combinatorial insights, jammed in a few pages back, to shrink them until they are finite. And for this we shall require a value for HEAP.

At any rate, we already have a value for k. Our strategy in the second semantical completeness proof above was to build a refuting model for a non-theorem A out of Sub(A), throwing in 1 and closing under o. Thus k, the cardinality of Sub(A), is an upper bound for the size of our generating set.[24] Note too that our manipulations of the last section, trading in M for one of the $\mathbb{N}_k$, did not increase the value of k.

Still, unless we can use A to induce a value for HEAP, after which iterated multiplication will just produce HEAP again, we have naught but cold comfort. But Kripke's lemma and its cousins give us a way to do just this. In the first place, recall our three principles from section 5—the IDP, Kripke's lemma, and Dickson's lemma. We note immediately that all three principles continue to go through if all talk of *ordinary divisibility*, $|$, is replaced with talk of *relevant divisibility*, $/$.[25] For example, let us call:

DEFINITION 8.1. $s \in S \subseteq \mathbb{N}_k$ is a **relevant infinite divisor** for S just in case there is an infinite subset S' of S such that $s \,/\, s'$ for *every* s' in S'.

Then we have:

THEOREM 8.2 Relevant IDP. *Let S be an infinite subset of an $\mathbb{N}_k$. Then there exists a relevant infinite divisor for S.*

[23] Being a Sixist, I have always thought that HEAP = 7; i.e., there are just six *objects* in the world, with an extra one if one throws in the metaphysical subject. ("Monism is right in what it intends," as Wittgenstein taught; but it is off by 5, at least.) But Slaney, author of the Most Magnificent Result (the MMR) ever achieved in logic, showed in *Slaney 1985,* that there are *exactly* 3088 Ackermann constants in **R**.

[24] Alas, it too often induces an *enormous* upper bound.

[25] So putting the matter is a little unfair to Kripke—since, to all intents and purposes, his lemma dealt with relevant divisibility from the start.

Proof. All of the members of $\mathbb{N}_k$, and hence of S, are built out of the same *finite* set $P_k = p_1, \ldots, p_k$ of primes. There are only finitely many subsets of P_k. Accordingly, since S is infinite, there is an infinite subset S^* of S and a particular subset P of P_k such that all members of S^* are built out of members of P. By the ordinary IDP, S^* has an infinite divisor s, which will be a *relevant* infinite divisor for S^*. Accordingly, s will also be a relevant infinite divisor for S. ⊣

COROLLARY 8.3. *Dickson's and Kripke's lemmas hold in relevant form.*

Proof. From the *relevant* IDP, as above (*mutatis mutandis*). ⊣

Now we can get back to the business of finding HEAP.

DEFINITION 8.4. Let A be a non-theorem of $\mathbf{R}_{\rightarrow\&}$. Fix k and $\mathcal{J}$ by the numerical completeness theorem so that $\mathcal{J}(A,1) = \mathrm{F}$ in $\mathbb{N}_k$. Call an element n in $\mathbb{N}_k$ ***A*-critical** just in case, for some subformula B of A, we have both

(i) $\mathcal{J}(B,n) = \mathrm{F}$, and

(ii) For all m such that $m \,/\, n$, $\mathcal{J}(B,m) = \mathrm{T}$.

Since the hereditary condition (H) assures that Truth is propagated *downwards* under relevant division, it similarly assures that Falsehood is propagated *upwards*. So the A-critical elements of $\mathbb{N}_k$ are numbers at which some subformula of A (including, perhaps, A itself) is *minimally false*. This sets things up for a relevant application of Dickson's lemma.

FACT 8.5 Dickson's Fact. Let $\mathcal{J}$ be an h-interpretation in $\mathbb{N}_k$ such that $\mathcal{J}(A,1) = \mathrm{F}$. There are *finitely many* A-critical elements of $\mathbb{N}_k$.

Proof. By contradiction. Suppose there are infinitely many A-critical n in $\mathbb{N}_k$. Since the number of subformulas of A is finite, there is some particular subformula B of A which has properties (i) and (ii) immediately above, for infinitely many n. But all those n are *minimal*, under /, with respect to $\mathcal{J}(B,n) = \mathrm{F}$. By (relevant) Dickson, this cannot be; any totally unordered subset of $\mathbb{N}_k$ must be finite. ⊣

We now use Dickson's fact to find HEAP. Let $\mathbb{N}_k$, $\mathcal{J}$, and A be as in the fact, and let S be the set of A-critical elements of $\mathbb{N}_k$. Since S is finite, there is some highest exponent $m-1$ which occurs among the prime powers which characterize its members. We define a new c-monoid $\mathbb{N}_k^m$ whose elements shall be just those members of $\mathbb{N}_k$ which, in prime power decomposition, shall have no exponent greater than m on any prime factor.[26] Thus, for example, $\mathbb{N}_2^3 = \{1, 2, 3, 4, 6, 8, 9, 12, 18, 24, 27, 36, 54, 72, 108, 216\}$.

[26] A concern for efficiency here would suggest that we keep track of exponents on *particular* prime powers individually. After all, if no A-critical element has exponent greater than 2 on the prime 3, it is otiose to allow these exponents to rise to 10 just because some other A-critical element has divisor 512. But we are chasing a finite model property here; it is notationally simpler, and no impediment to the development, if we allow exponents on *any* prime to rise to the maximal one.

Having set HEAP $= m$, we define / on $\mathbb{N}_k^m$ so that it agrees with / on $\mathbb{N}_k$. But, since some products of members of $\mathbb{N}_k^m$ threaten to "go off the edge", we *redefine* o on prime powers so that $p^j \circ p^h = p^{j+h}$ if $j + h \leq m$, and $p^j \circ p^h = p^m$ otherwise; on the obvious epimorphism from $\mathbb{N}_k$ to $\mathbb{N}_k^m$, o is then defined in the latter for all elements. We set out, just for fun, the o and / tables for $\mathbb{N}_2^2$.

o	1	2	3	4	6	9	12	18	36
1	1	2	3	4	6	9	12	18	36
2	2	4	6	4	12	18	12	36	36
3	3	6	9	12	18	9	36	18	36
4	4	4	12	4	12	36	12	36	36
6	6	12	18	12	36	18	36	36	36
9	9	18	9	36	18	9	36	18	36
12	12	12	36	12	36	36	36	36	36
18	18	36	18	36	36	18	36	36	36
36	36	36	36	36	36	36	36	36	36

/	1	2	3	4	6	9	12	18	36
1	T	F	F	F	F	F	F	F	F
2	F	T	F	T	F	F	F	F	F
3	F	F	T	F	F	T	F	F	F
4	F	F	F	T	F	F	F	F	F
6	F	F	F	F	T	F	T	T	T
9	F	F	F	F	F	T	F	F	F
12	F	F	F	F	F	F	T	F	T
18	F	F	F	F	F	F	F	T	T
36	F	F	F	F	F	F	F	F	T

While it is overkill, let us set about refuting the mingle axiom $p \rightarrow . \, p \rightarrow p$ in $\mathbb{N}_2^2$. (This is also a refutation of AxK.) Set $\mathcal{I}(p,2) = \mathrm{T}$, $\mathcal{I}(p,3) = \mathrm{T}$, $\mathcal{I}(p,6) = \mathrm{F}$. Then, by T→, $\mathcal{I}(p \rightarrow p, 2) = \mathrm{F}$, since for $j = 3$ we have $\mathcal{I}(p,j) = \mathrm{T}$ and $\mathcal{I}(p,2j) = \mathrm{F}$. But then $\mathcal{I}(p \rightarrow . \, p \rightarrow p, 1) = \mathrm{F}$, since it has a true antecedent and a false consequent on $\mathcal{I}$ at 2.

What we have done for mingle we can do quite generally. For we have the:

LEMMA 8.6 Finitizing Lemma for $\mathbf{R}_{\rightarrow \&}$. *Let $\mathbb{N}_k$ be a finitely generated submonoid of $\mathbb{N}^+$, and let $\mathcal{J}$ be an h-interpretation (under /) and A be a formula of $\mathbf{R}_{\rightarrow}$ such that $\mathcal{J}(A,1) = \mathrm{F}$. Then there exists a natural number m and an h-interpretation $\mathcal{I}$ such that $\mathcal{I}(A,1) = \mathrm{F}$ in $\mathbb{N}_k^m$.*

Proof. We fix $m - 1$ by Dickson's fact, and define $\mathbb{N}_k^m$ accordingly. For each propositional variable p and element x of $\mathbb{N}_k^m$, set $\mathcal{I}(p,x) = \mathcal{J}(p,x)$. Complete the definition of $\mathcal{I}$ by imposing the truth-conditions T→ and T&. Clearly $\mathcal{I}$ is hereditary, because it agrees with the h-interpretation $\mathcal{J}$ on propositional variables, while T→ and T& preserve (H). We finish the proof by showing, for all subformulas B of A and for all x in $\mathbb{N}_k^m$, that $\mathcal{I}(B,x) = \mathcal{J}(B,x)$.

We proceed by structural induction. The condition is imposed on propositional variables. It is evident on inductive hypothesis when B is of the form $C \,\&\, D$. So suppose B is $C \rightarrow D$. Assume first that $\mathcal{I}(B,x) = \mathrm{F}$. Then there is a c in $\mathbb{N}_k^m$ such that $\mathcal{I}(C,c) = \mathrm{T}$ and $\mathcal{I}(D, x \circ c) = \mathrm{F}$, where o is defined in the $\mathbb{N}_k^m$ way. But then, on inductive hypothesis, $\mathcal{J}(C,c) = \mathrm{T}$ and $\mathcal{J}(D, x \circ c) = \mathrm{F}$. Of course $x \circ c$ may not be the *true* product xc. Never mind. Falsehood is propagated *upward* under /, and $x \circ c / xc$. So $\mathcal{J}(C \rightarrow D, x) = \mathrm{F}$ as well.

We finish the argument by assuming $\mathcal{J}(C \to D, x) = \mathrm{F}$ and showing the same for $\mathcal{I}$, for x in $\mathbb{N}_k^m$ and $C \to D$ a subformula of A. There is now a c in $\mathbb{N}_k$ such that $\mathcal{I}(C, c) = \mathrm{T}$ and $\mathcal{I}(D, xc) = \mathrm{F}$ in $\mathbb{N}_k$. If $xc = x \circ c$, we have on inductive hypothesis nothing to prove; for we shall have, on inductive hypothesis, the same counterexample in $\mathbb{N}_k^m$. Otherwise, we had better reduce exponents until we *put* both c and $x \circ c$ in $\mathbb{N}_k^m$. We can reduce exponents freely on c; for, since Truth propagates *down* under / on $\mathcal{J}$ in $\mathbb{N}_k$, we shall have on inductive hypothesis $\mathcal{I}(C, c') = \mathrm{T}$ for any $c' \in \mathbb{N}_k^m$ such that $c' \,/\, c$. And we can reduce exponents *a little bit* on $x \circ c$; for we chose m so that A-critical elements have in $\mathbb{N}_k$ exponents at most $m - 1$ on any prime. Since D is a subformula of A, which is false on $\mathcal{J}$ at xc, there is a y such that, in prime decomposition, every exponent i on any prime p satisfies $i < m$; moreover, $y \,/\, xc$, and $\mathcal{J}(D, y) = \mathrm{F}$. Fix c' by reducing any exponents greater then m to m. Then $c' \in \mathbb{N}_k^m$ and $\mathcal{I}(C, c') = \mathrm{T}$. Moreover $y \,/\, x \circ c'$ in $\mathbb{N}_k^m$, whence, since Falsehood propagates *up*, $\mathcal{J}(D, y) = \mathcal{I}(D, y) = \mathcal{I}(D, x \circ c') = \mathrm{F}$. So we still have a counterexample to $C \to D$ at x under $\mathcal{I}$. This completes the inductive argument, whence $\mathcal{I}(A, 1) = \mathcal{J}(A, 1) = \mathrm{F}$. So A is refuted in a finite model. ⊣

We sum all this up in the following:

THEOREM 8.7 Finite Model Theorem for $\mathbf{R}_{\to\&}$. *The following conditions are equivalent:*

(1) *A is a theorem of* $\mathbf{R}_{\to\&}$.

(2) *A is h-valid.*

(3) *A is h-valid in all the* $\mathbb{N}_k$.

(4) *A is h-valid in all finite Church monoids.*

(5) *A is h-valid in all the* $\mathbb{N}_k^m$.

Proof. by previous theorems, lemmas, and facts. ⊣

COROLLARY 8.8. $\mathbf{R}_{\to\&}$ *is decidable.*

We knew the corollary anyway, of course, by Kripke. Indeed, so far we are behind Kripke (though we shall rectify that shortly), since his methods yield decidability for all of **LR**. Meanwhile, the decidability that we have is less than completely satisfying. Usually, finite model properties place some sort of bound (usually horrible, but that's another story) on a test for theoremhood. Here, however, we have *naked* decidability. We set someone, perhaps Belnap, the task of enumerating proofs in $\mathbf{R}_{\to\&}$. We set someone else, maybe Dunn, the task of enumerating refutations in the $\mathbb{N}_k^m$. Dunn's job is eased because he can bound k, for a given A, just by counting up its subformulas. He may curse us nonetheless, since we have given him no recipe to bound m. All he knows (or we know) is that, if A is refutable, there exists an m that will do the job. In brief, this is the "Don't call us, we'll call you" method; if A is provable, Dunn must still wait for Belnap (or Thistlewaite's KRIPKE program, or its refinement by Riche) to pick up the phone, before he knows that the job is done.

9. EXTENDING THE FINITE MODEL PROPERTY TO LR

The results of the previous section apply immediately to $\mathbf{R}_{\rightarrow}$ (since it is an exact subsystem of $\mathbf{R}_{\rightarrow\&}$) and to $\mathbf{MR}^{\mathrm{f}}_{\rightarrow\&}$ (since it is just $\mathbf{R}_{\rightarrow\&}$, with f identified syntactically with a variable irrelevant to context).[27] A little trickier is $\mathbf{R}^{\mathrm{f}}_{\rightarrow}$, for which ~~E is explicit. But it's not much trickier, since $\mathbf{R}^{\mathrm{f}}_{\rightarrow}$ exactly translates into $\mathbf{MR}^{\mathrm{f}}_{\rightarrow}$ with the translation of A being $p_1 \rightarrow \mathrm{f} \rightarrow \mathrm{f} \rightarrow p_1 \rightarrow . \ldots \rightarrow . \; p_n \rightarrow \mathrm{f} \rightarrow \mathrm{f} \rightarrow p_n \rightarrow A$, where $p_1, \ldots, p_n$ is some selection from among the propositional variables of A.[28] So our finite models for $\mathbf{R}_{\rightarrow}$ will go on refuting (translations of) non-theorems of $\mathbf{R}^{\mathrm{f}}_{\rightarrow}$.

When it comes to adding ∨, the situation becomes trickier. For the translation that embeds $\mathbf{R}^{\mathrm{f}}_{\rightarrow}$ in $\mathbf{MR}^{\mathrm{f}}_{\rightarrow}$ relies on the fact that $\sim\sim B \rightarrow B \rightarrow . \sim\sim(A \rightarrow B) \rightarrow . \; A \rightarrow B$ is already a theorem of $\mathbf{MR}^{\mathrm{f}}_{\rightarrow}$. A similar trick works in the presence of &; but the induction breaks down on ∨. But if one trick won't work another will—a Glivenko-style translation, which we borrow in principle from Church (*1951a*).

DEFINITION 9.1. For each formula A of **LR**, we define its ***-translation A^*** recursively as follows:

(i) $\mathrm{f}^* = \mathrm{f}$

(ii) $p^* = \sim\sim p$, where $\sim$ is defined inferentially by D~

(iii) $(B \rightarrow C)^* = B^* \rightarrow C^*$

(iv) $(B \& C)^* = B^* \& C^*$

(v) $(B \vee C)^* = \sim(\sim B^* \& \sim C^*)$

Evidently, A^* is always in the vocabulary of $\mathbf{MR}^{\mathrm{f}}_{\rightarrow\&}$.

LEMMA 9.2 ~~E Lemma. *For all formulas A of* **LR**, $\sim\sim A^* \rightarrow A^*$ *is a theorem of* $\mathbf{MR}^{\mathrm{f}}_{\rightarrow\&}$.

[27] The $\mathbf{R}_{\rightarrow}$ result was proved in *Meyer and Ono 1994*, together with a similar one for **BCK**. Indeed, the **BCK** result was the excuse for *Meyer and Ono 1994*. For I realized, in thinking about it, that the present methods (which had long lain fallow) would apply to **BCK** as well.

[28] If A contains no propositional variables, it translates simply as itself. As another triumph for Sixism, there are exactly 6 such non-equivalent formulas in $\mathbf{MR}^{\mathrm{f}}_{\rightarrow}$.

Proof. By structural induction on A. The most interesting case (since we were quick about it before) is when A is of the form $B \,\&\, C$. On inductive hypothesis, $\sim\sim B^* \rightarrow B^*$ and $\sim\sim C^* \rightarrow C^*$ are theorems of $\mathbf{MR}^{\mathrm{f}}_{\rightarrow\&}$. Elementary properties of & and $\rightarrow$ yield $\sim\sim B^* \,\&\, \sim\sim C^* \rightarrow B^* \,\&\, C^*$ as another theorem. But, suffixing twice in Ax&E, we have $\sim\sim(B^* \,\&\, C^*) \rightarrow \sim\sim B^*$ on *minimal* properties of $\sim$. The same holds for C^*, whence adjoining and applying Ax→&I, we have $\sim\sim(B^* \,\&\, C^*) \rightarrow \sim\sim B^* \,\&\, \sim\sim C^*$. Transitivity of provable $\rightarrow$ in $\mathbf{MR}^{\mathrm{f}}_{\rightarrow\&}$ then yields $\sim\sim A^* \rightarrow A^*$ in this case. ⊣

THEOREM 9.3 **LR** Containment Theorem. ***LR** is an exact subsystem of* $\mathbf{MR}^{\mathrm{f}}_{\rightarrow\&}$ *under * translation.*

Proof. We must show that A is a theorem of **LR** iff A^* is a theorem of $\mathbf{MR}^{\mathrm{f}}_{\rightarrow\&}$. From right to left this is easy, since if A^* is provable *without* $\sim\sim$E it is certainly provable *with* it; moreover $A^* \leftrightarrow A$ is a theorem scheme of **LR**. For the converse, proceed by deductive induction. Since the *rules* of **LR** and $\mathbf{MR}^{\mathrm{f}}_{\rightarrow\&}$ are the same, it suffices to show that when A is an axiom of **LR**, then A^* is a *theorem* of $\mathbf{MR}^{\mathrm{f}}_{\rightarrow\&}$. Even the axioms don't require much checking; the only additional one is $\sim\sim$E, which our $\sim\sim$E lemma suffices to handle on * translation. But we had better look in particular at the $\vee$ axioms, since this is the particle left out of the $\mathbf{MR}^{\mathrm{f}}_{\rightarrow\&}$ vocabulary. As for Ax∨I, half of it will suffice; anyway, by Ax&E we have $\sim B^* \,\&\, \sim C^* \rightarrow \sim B^*$. But then, minimally, we get $\sim\sim B^* \rightarrow \sim(\sim B^* \,\&\, \sim C^*)$ by suffixing. But $B^* \rightarrow \sim\sim B^*$ is minimally OK. (There is nothing wrong with $\sim\sim$I even at the $\mathbf{MR}^{\mathrm{f}}_{\rightarrow}$ level.) So, by transitivity, $B^* \rightarrow (B \vee C)^*$. Finally, we look at Ax→∨E. We need to show $(B^* \rightarrow D^*) \,\&\, (C^* \rightarrow D^*) \rightarrow .\ \sim(\sim B^* \,\&\, \sim C^*) \rightarrow D^*$ in $\mathbf{MR}^{\mathrm{f}}_{\rightarrow\&}$. At any rate, by Ax&I we have $(\sim D^* \rightarrow \sim B^*) \,\&\, (\sim D^* \rightarrow \sim C^*) \rightarrow .\ \sim D^* \rightarrow \sim B^* \,\&\, \sim C^*$. Making replacements using minimally acceptable forms of contraposition, we get $(B^* \rightarrow D^*) \,\&\, (C^* \rightarrow D^*) \rightarrow .\ \sim(\sim B^* \,\&\, \sim C^*) \rightarrow \sim\sim D^*$. The $\sim\sim$E lemma and replacement theorems allow us to remove the $\sim\sim$ on the ultimate consequent, ending the verification of the last potentially troubling axiom on * translation. ⊣

One is disturbed over the failure of AxDist to enter into these pleasant reflections. As an axiom relating & and $\vee$, there is much to be said for AxDist.[29] But as a principle relating & and minimal $\sim$, it isn't very appetizing.

THEOREM 9.4 Finite Model Theorem for **LR**. *The following conditions are equivalent:*

(i) *A is a theorem of* **LR**.

(ii) *A^* is h-valid.*

(iii) *A^* is h-valid in all the* N^m_k.

[29]Much of this *is said* by Belnap (*1993*).

Proof. By the **LR** containment theorem just proved, and the finite model theorem for $\mathbf{R}_{\rightarrow\&}$ above. ⊣

Well, that wasn't much of a finite model theorem.[30] Still, it suffices for yet another proof of the decidability of **LR**.

10. THE FINITE MODEL PROPERTY FOR THE SEMILATTICE SEMANTICS

It might be expected, from our results so far, that it is easy to get a finite model property for the semilattice semantics that Urquhart contributed to *Anderson, Belnap, and Dunn 1992,* and which Routley discovered independently.[31] For we quickly found above, for a given non-theorem A, a *finitely generated* commutative monoid $\mathbb{N}_k$ that refutes A. If we could pull the same trick with semilattices, the finite model property would be a piece of cake. For finitely generated semilattices are, after all, finite.

As the reader probably expects by now, things are not necessarily so simple. On Urquhart's original formal intuitions, the semilattice elements are (or might as well be) sets[32] of *relevance numerals*, corresponding to the *tags* on formulas that appear in the adaptation in *Anderson and Belnap 1975* of Fitch-style natural deduction to $\mathbf{R}_{\rightarrow}$ and other relevant logics. So thinking does yield a completeness proof for the semilattice semantics for $\mathbf{R}_{\rightarrow}$, which Urquhart contributed to *Anderson, Belnap, and Dunn 1992.* In our terminology so far:

DEFINITION 10.1. An **Urquhart semilattice** may be parsed as a c-monoid $\mathrm{M} = \langle \mathrm{M}, \circ, 1\rangle$, where M is a sets of sets closed under finite union ∪, fusion ∘ is ∪, and the unit 1 is the null set ∅, which belongs to M. We may at the same time consider an Urquhart semilattice to be a Church monoid $\langle \mathrm{M}, =\rangle$, since idempotence of ∪ assures that the square-increasing condition will be met.

Similarly, on this construal, every interpretation satisfying T→ (and T&, if present) will be an h-interpretation. Accordingly, in discussing Urquhart semilattices we can drop all talk of heredity, taking the notions of *h-interpretation* and *interpretation* to be synonymous. Because of this synonymy, we call *valid* the h-valid formulas in an Urquhart semilattice M. And a formula valid in all Urquhart semilattices is just *Urquhart valid.*

[30] I had, I think, a better way of dealing with **LR**, once upon a time. I even gave a seminar to the ANU logic group on the topic, after Kripke had suggested to me in correspondence that of course the methods of the present paper worked for all of **LR**.

[31] Conscience requires the admission of my profound debt to Urquhart, in all that I have thought semantically about relevant logics. It was hardly appropriate to say so in my joint work with Routley, since he was not so indebted. But I never thought independently about anything like T→ above. And since that, through many a *metamorphosis*, is for my money the *key* to such relevant semantical insight as there has been, it is pleasant to record my gratitude to the fellow who slipped me the key.

[32] Urquhart motivates these as *pieces of information.*

I think, as I reported with McRobbie (*Meyer and McRobbie 1982*), that the *data type* most naturally associated with systems of relevant implication is that of *multisets.* All of our talk of $\mathbb{N}^+$, the product $\cdot$, and divisibility relations $|$ and $/$, is just a way of recalling what we have known about multisets for a long time. But it is *possible* (even if less than completely *natural*) to *represent* multisets as sets. Both the success and (again, as I see it) the ultimate failure of both the Fitch-style natural deduction systems and the semilattice semantics for **R** and its fragments rest on such representation. At any rate, what we shall do here will press to the fore a set-theoretic representation of multisets as sets, transforming the c-monoid semantics into an Urquhart semantics, which we can then finitize along lines laid down above.

It is trivial that all theorems of $\mathbf{R}_{\rightarrow\&}$ are Urquhart valid, by our first semantical completeness theorem. And we need not redo Urquhart by proving the converse; our interest, rather, is in *finitizing* him. To this end we concentrate on the $\mathbb{N}_k^m$. With each prime p in $\mathbb{N}_k$, we associate m *copies* $p_1, \ldots, p_m$. We let the *Urquhart universe* $\mathcal{U}_k^m$ consist of all these copies of primes; and we let $\mathcal{P}_k^m$ be the power set of $\mathcal{U}_k^m$. We define a *representation function* r from $\mathcal{P}_k^m$ to $\mathbb{N}_k^m$ as follows:

(i) $r(\varnothing) = 1$.

(ii) Where p_i is a copy of the prime p, $r(p_i) = p$.

(iii) Suppose $S \subseteq \mathcal{U}_k^m$, $\mathrm{T} \subseteq \mathcal{U}_k^m$, $S \cap \mathrm{T} = \varnothing$. Then $r(S \cup \mathrm{T}) = r(S) \circ r(\mathrm{T})$ in $\mathbb{N}_k^m$.

Since every member S of $\mathcal{P}_k^m$ is expressible as a disjoint union of (unit sets of) copies of primes $\in \mathbb{N}_k$, and since the c-monoid properties of $\mathbb{N}_k$ make it indifferent which representation we choose, r is well-defined as a function from $\mathcal{P}_k^m$ onto $\mathbb{N}_k^m$. Moreover, since we make exactly m copies of each prime in $\mathcal{U}_k^m$, images under r stay in a natural way *within* $\mathbb{N}_k^m$.

Now suppose that we have built an $\mathbb{N}_k^m$ for the refutation of a particular non-theorem A of $\mathbf{R}_{\rightarrow\&}$. We shall now prove a Xerox[33] lemma that shows that we can do the same work in $\mathcal{P}_k^m$.

LEMMA 10.2 Xerox Lemma. *Let $\mathcal{I}$ be an h-interpretation of $\mathbf{R}_{\rightarrow\&}$ in $\mathbb{N}_k^m$, for some positive integers k, m. And let A be a formula of $\mathbf{R}_{\rightarrow\&}$ such that $\mathcal{I}(A, 1) = \mathrm{F}$, where $m-1$ is the highest exponent that occurs on any A-critical elements of $\mathbb{N}_k^m$ under $\mathcal{I}$. There is then an interpretation $\mathcal{I}'$ in the associated Urquhart semilattice such that $\mathcal{I}'(A, 1) = \mathrm{F}$.*

Proof. That there are k, m, and $\mathcal{I}$ that meet the conditions of the lemma for any non-theorem A is part of the content of our finite model theorem for $\mathbf{R}_{\rightarrow\&}$ in section 8. Where r is the representation function defined by (i)–(iii)

[33]Langford suggests that use of this trade name witnesses how far behind the times I am. Maybe so. He would suggest, I guess, that I call it the "photocopy lemma".

just above, we define $\mathcal{I}'$ by setting, for each propositional variable q in A and subset S of $\mathcal{U}_k^m$, $\mathcal{I}'(q,S) = \mathcal{I}(q, r(S))$. Imposing T $\rightarrow$ and T& to complete the definition of $\mathcal{I}'$ on all of $\mathcal{P}_k^m$, we show for each subformula B of A and for all S in $\mathcal{P}_k^m$ that $\mathcal{I}'(B,S) = \mathcal{I}(B, r(S))$, by structural induction on B. This is laid down by fiat for propositional variables, and it is readily apparent on inductive hypothesis when B is C & D, just applying T& in both cases.

So the interesting case is $B = C \rightarrow D$. Suppose first that $\mathcal{I}'(B,S) = \mathrm{F}$. Then there is a counterexample—a $V \subseteq \mathcal{U}_k^m$ such that $\mathcal{I}'(C,V) = \mathrm{T}$ and $\mathcal{I}'(D, S \cup V) = \mathrm{F}$. On inductive hypothesis, $\mathcal{I}(C, r(V)) = \mathrm{T}$ and $\mathcal{I}(D, r(S \cup V)) = \mathrm{F}$ as well. We wish now to *massage* this counterexample so that it is also one on the c-monoid semantics for $\mathbb{N}_k^m$ as well.

How could it *fail*, one wonders, to remain a counterexample? It could fail, alas, if S and V are not disjoint, whence $r(S \cup V)$ is *not* the $r(S) \circ r(V)$ that we require for a bona fide counterexample under T→ in $\mathbb{N}_k^m$ on $\mathcal{I}$. But this is no real problem; if S and V have copies of some particular prime p *in common*, all this means is that $r(S \cup V) \,/\, r(S) \circ r(V)$. Since Falsehood is propagated up under relevant divisibility we shall still have $\mathcal{I}(D, r(S) \circ r(V))$, falsifying $C \rightarrow D$ at $r(S)$ on $\mathcal{I}$ as required.

Finally, we turn to the case where $\mathcal{I}(C \rightarrow D, r(S)) = \mathrm{F}$. We need to show that $\mathcal{I}'(C \rightarrow D, S) = \mathrm{F}$ as well, on inductive hypothesis. At any rate, there is a $c \in \mathbb{N}_k^m$ such that $\mathcal{I}(C,c) = \mathrm{T}$ and $\mathcal{I}(D, r(S) \circ c) = \mathrm{F}$, by T→. We need to convert this to a refutation in $\mathcal{P}_k^m$. There is at least some set $V \subseteq \mathcal{U}_k^m$ (and probably many) such that $r(V) = c$. We want to choose V so that it is as disjoint from S *as possible*. If there is a V such that $S \cap V = \varnothing$ and $r(V) = c$, we shall have on inductive hypothesis the same counterexample in $\mathcal{P}_k^m$ that we already have in $\mathbb{N}_k^m$, and we are done.

But we may not be done. For it is conceivable that there is no V disjoint from S such that $r(V) = c$. This could happen, for example, if all m copies of some prime p already occur in S, and $p \mid c$. However, we chose m so that all A-critical elements of $\mathbb{N}_k^m$ would have exponent *at most* $m-1$ in prime decomposition, on every prime p. So let S' be like S, except for the removal of one copy of every prime p which divides c, when m copies occur already in S. And let c' be the product (without iteration) of the primes that divide c. Since Truth propagates *down* under /, and since c' / c, we have $\mathcal{I}(C,c') = \mathrm{T}$ as well. And since $r(S') \circ c' = r(S) \circ c$ in $\mathbb{N}_k^m$, we still have $\mathcal{I}(D, r(S') \circ c') = \mathrm{F}$ as well. This suffices to refute $C \rightarrow D$ at $r(S')$ in $\mathbb{N}_k^m$. (Well, we knew that $C \rightarrow D$ could not be *minimally false* at $r(S)$.) And choosing now V' so that it contains one copy of each prime divisor of c', disjoint from the copies in S', we have on inductive hypothesis that $\mathcal{I}'(D, S' \cup V') = \mathrm{F}$. But $S' \cup V' = S \cup V'$, whence refutation of $C \rightarrow D$ on $\mathcal{I}'$ at S is completed with the observation, on inductive hypothesis, that $\mathcal{I}'(C, V') = \mathrm{T}$. ⊣

We can now state the following:

THEOREM 10.3 Finite Model Theorem for the Semilattice Semantics. *The following are equivalent:*

(1) *A is a theorem of* $\mathbf{R}_{\to\&}$.

(2) *A is Urquhart valid.*

(3) *A is valid in all finite semilattices.*

Proof. By the finite model theorem for $\mathbf{R}_{\to\&}$ and the Xerox lemma. ⊣

We can even use previous tricks to get **LR** under the tent.

COROLLARY 10.4 Finite Semilattice Corollary for **LR**. *Let A be a formula of* **LR**, *and let* A^* *be the result of translating A into* $\mathbf{R}_{\to\&}$ *as above. Then the following conditions are equivalent:*

(1) *A is a theorem of* **LR**.

(2) A^* *is a theorem of* $\mathbf{R}_{\to\&}$.

(3) A^* *is valid in all finite semilattices.*

Proof. All immediate from previous theorems. ⊣

11. EXTENDING THE THEORY OF PROPOSITIONAL QUANTIFICATION

We left propositional quantifiers back in section 2, with the observation that Church had provided them as original equipment in his theory of relevant implication. In the climate of the time, when any use of the word *proposition* tended to expose one to the *wrath of Quine*, this was a daring thing to do.[34] It is time to delve a little more into what propositional quantifiers can do for us, in a theory of relevant implication.

A conspicuous use of propositional quantifiers, already plumbed by Church (*1951a*), lies in their utility for defining things. We have seen how already in $\mathbf{R}^{\forall p}_{\to}$ that we can define fusion o, equivalence ↔, existential quantification ∃, and so forth. Moreover, in so defining them, we give them the properties that they *ought* to have. Clearly this kind of thing can be carried considerably further. How *much* further is a question that we shall look at now, borrowing results from the draft paper *Meyer 1978*.[35]

Church contented himself with *analogues* of the truth-functional connectives. But could we, following Ackermann, Anderson, and Belnap, define something closer to the real things? I do not see how, although I have made some attempts. First, there is negation. I am content with the inferential definition of negation via D∼ and a constant f. But I am *not* content with the

[34] I used to avoid *propositions* myself, until Belnap convinced me at last that I was being silly.

[35] Anderson, Belnap, and Dunn *1992* pass on some of my observations on page 68f.

failure of $\sim\sim$E and associated principles. Solution (at least for my money): Just add $\sim\sim$E as a new axiom scheme. The result, without quantifiers, is $\mathbf{R}^{\mathrm{f}}_{\rightarrow}$. Let us look now at this system *with* quantifiers. Call it $\mathbf{R}^{\mathrm{f}\forall\mathrm{p}}_{\rightarrow}$, and subject it to both the quantificational principles and definitions of section 2 and the constant f and the $\sim\sim$E axiom scheme of section 3.

We note immediately that $\mathbf{R}^{\mathrm{f}\forall\mathrm{p}}_{\rightarrow}$ has a conspicuous theorem that $\mathbf{R}^{\forall\mathrm{p}}_{\rightarrow}$ lacks. For universally and then existentally generalizing on the $\sim\sim$E axiom, we get

(1) $\exists q \forall p(p \rightarrow q \rightarrow q \rightarrow p)$.

The closest that we can get to (1) in $\mathbf{R}^{\forall\mathrm{p}}_{\rightarrow}$ itself is

(2) $\forall p \exists q(p \rightarrow q \rightarrow q \rightarrow p)$,

which is a long way from the same thing. So, contradicting the expectations of conservative extension that we have come to expect in systems based on $\mathbf{R}_{\rightarrow}$, we have a failure here.

Perhaps, though, we can do something with that failure. It occurs to one that we can introduce something like a Hilbert (indefinite description) $\in$-operator according to a rubric like

$$\text{D}\in.\ \in q(A(q)) =_{\text{df}} \exists q(q \circ A(q)).$$

Let us see how the D$\in$ trick works on f. Define, in $\mathbf{R}^{\mathrm{f}\forall\mathrm{p}}_{\rightarrow}$,

$$\text{Df. } \mathrm{f} = \in q \forall p(p \rightarrow q \rightarrow q \rightarrow p),$$

then we can prove the *equivalence* of the f just defined to the primitive constant of $\mathbf{R}^{\mathrm{f}\forall\mathrm{p}}_{\rightarrow}$. So we do not need that constant at all. (Before continuing, we *do need* the $\sim\sim$E axiom; without it, back in $\mathbf{R}^{\mathrm{f}\forall\mathrm{p}}_{\rightarrow}$, the definition is still legal, but $\sim\sim$E remains, unfortunately, a non-theorem.)

It is time also, in $\mathbf{R}^{\mathrm{f}\forall\mathrm{p}}_{\rightarrow}$, to clear up our alternative definitions of o (and hence of $\leftrightarrow$). We need to show, in that system, that $\sim(A \rightarrow \sim B)$ and $\forall q((A \rightarrow .\ B \rightarrow q) \rightarrow q)$ are in fact equivalent. From right to left, simply instantiate q to $\sim(A \rightarrow \sim B)$; after this, right-to-left goes through in $\mathbf{R}^{\mathrm{f}}_{\rightarrow}$. From left to right, it will suffice to show, by generalization and confinement properties of $\forall$, that $\sim(A \rightarrow \sim B)$ entails in $\mathbf{R}^{\mathrm{f}}_{\rightarrow}$ the formula $(A \rightarrow .\ B \rightarrow q) \rightarrow q$. Minor exercise, left to reader.

In addition to defining constants, it is also interesting to experiment with the direct definition of connectives, via a rubric like our $\in$ operator. Here is $\sim$ in $\mathbf{R}^{\mathrm{f}\forall\mathrm{p}}_{\rightarrow}$.

$$\text{D}\sim.\ \sim A =_{\text{df}} \exists q((A \rightarrow q) \circ \forall p(p \rightarrow q \rightarrow q \rightarrow p)).$$

Again the definition corresponds to a provable equivalence. From left to right, $\sim A$ does imply $A \rightarrow \mathrm{f}$, to which we may fuse the $\mathbf{R}^{\mathrm{f}}_{\rightarrow}$ theorem $\forall p(p \rightarrow \mathrm{f} \rightarrow \mathrm{f} \rightarrow p)$ in the consequent; existentially generalizing on f completes the left to right argument. In the other direction, it suffices (by properties of $\exists$) simply to show that $(A \to q) \circ \forall p(p \to q \to q \to p)$ entails $\sim A$ in $\mathbf{R}^{\mathrm{f}\forall\mathrm{p}}_{\to}$. The trick here is to instantiate the second "fusejunct" to $\sim q$; in view of the $\mathbf{R}^{\mathrm{f}}_{\to}$ theoremhood of $\sim q \to q \to q$; this "fusejunct" will entail $\sim q$, after which a step of "*modus tollens*" gets us $\sim A$. That will do.

What happens if one throws in the lattice connectives? Still more interesting things. For example, form $\mathbf{R}^{\forall\mathrm{p}}_{\to\&}$ by adding to $\mathbf{R}_{\to\&}$ the quantificational machinery of section 2. It then turns out that lattice $\vee$ is already definable, by

$$\text{D}\vee.\ A \vee B =_{\text{df}} \forall q((A \to q) \,\&\, (B \to q) \to q).$$

It is then an easy exercise to use Ax&E and Ax $\to$ &I to prove the dual principles Ax$\vee$I and Ax $\to$ $\vee$E in $\mathbf{R}^{\forall\mathrm{p}}_{\to\&}$. So, while $\vee$ is independent of the other positive particles at the sentential level, this is no longer the case in the presence of $\forall$. But AxDist, after all, remains independent. Not all lattices are distributive.

This merely scratches the surface of what can be done. Especially interesting is what happens when we pass to all of **R**, adding not only AxDist but, at the quantifier level, the accompanying "infinite distribution" principles

$$\text{Ax}\forall\vee.\ \forall p(A \vee B(p)) \to A \vee \forall p B(p), \text{ if } p \text{ is not free in } A,$$

$$\text{Ax}\,\&\,\exists.\ A \,\&\, \exists p B(p) \to \exists p(A \,\&\, B(p)), \text{ if } p \text{ is not free in } A.$$

The resulting system is the $\mathbf{R}^{\forall\mathrm{p}}$ of Anderson, Belnap, and Dunn (*1992*). Indeed, one can go further. One gets a conservative extension **CR** of **R** by adding a new, explicitly Boolean negation $\neg$ to the DeMorgan $\sim$ which comes in **R** as original equipment.[36] Sufficient to govern $\neg$ are the (paradoxical) axioms:

$$\text{AxAbsurd.}\ A \,\&\, \neg A \to B,$$

$$\text{AxTrivial.}\ A \to B \vee \neg B.$$

Our rubric above that produced something like a Hilbert $\in$-operator appealed crucially to o. We can do something similar in $\mathbf{CR}^{\forall\mathrm{p}}$ using &, via

$$\text{D}\neg.\ \neg \text{A} =_{\text{df}} \exists p(p \,\&\, (A \,\&\, p \to \text{F}) \,\&\, (\text{T} \to A \vee p)),$$

where p does not occur in A.

First, the left side of D$\neg$ implies the right side in **CR** when we drop the initial $\exists p$, replacing all the subsequent occurrences of p with $\neg A$; existentially

[36] While the extension is conservative in the original vocabulary of Anderson-Belnap **R**, built up from variables under &, $\vee$, $\sim$, and $\to$, it is ***not*** conservative as we are formulating things here, with primitive Ackermann constant f. To the contrary, **CR** (of which my most recent study is ***Meyer 1979***) counteracts the Slaney explosion of Ackermann constants by reducing the non-equivalent ones to 8—almost, but not quite, enough to gladden the Sixist heart.

generalizing completes the argument in $\mathbf{CR}^{\forall p}$. To go from right to left, it suffices to show that $B = p \,\&\, (A \,\&\, p \rightarrow \mathrm{F}) \,\&\, (\mathrm{T} \rightarrow A \vee p)$ entails $\neg A$ in **CR**. Well, certainly $\neg A \,\&\, B$ entails $\neg A$, by &E. And $A \,\&\, B$ has (near enough) $A \& p \& (A \& p \rightarrow \mathrm{F})$ as a conjunct, which entails F, which entails anything—in particular, $\neg A$. So, by Ax$\rightarrow\vee$E and AxDist, we get $(A \vee \neg A) \& B \rightarrow \neg A$. But, appealing to AxTrivial, this is equivalent to B itself. So B does entail A by the lights of **CR**, whence (slapping an existential quantifier on the antecedent, which is OK because p does not occur in A) we complete the verification that $\neg A$ is $\mathbf{CR}^{\forall p}$ equivalent to its suggested *definiens.*[37] Note again that what is at the quantifier-free level a ***linguistic*** extension becomes, with quantifiers, an *axiomatic* extension. For $D\neg$ is, of course, already available at the $\mathbf{R}^{\forall p}$ level as a *definition.* But, like our definition of f in $\mathbf{R}^{\forall p}_{\rightarrow}$, our defined Boolean $\neg A$ only has Boolean properties when we make them explicit through specific axioms.

12. CONCLUSION

I began this paper with a note of appreciation for Alonzo Church (who, alas, has died since it was written). Let me conclude on the same note. Ideas that he threw off more or less in passing have become, for many of us, the work of years. He built better than he knew. The Church "use criterion", in particular, is susceptible of almost indefinite variation. It need not imply AxW, if one is fussy about reusing premisses; that way lies linear logic, and much else. Nor need the use criterion even imply that fusion is commutative, or associative; that way lie ***minimal*** relevant and other substructural logics, in the style, e.g., of Lambek. In another direction, one may expand the $\mathbf{R}_{\rightarrow}$ insights so that fusion becomes idempotent, as in the Dunn-McCall system **RM** of Anderson and Belnap (*1975*). But, however modified, mutilated, or extended, $\mathbf{R}_{\rightarrow}$ and $\mathbf{R}^{\forall p}_{\rightarrow}$ lie at the very center of a great deal of most valuable research in logic. Thanks, Professor Church.

ACKNOWLEDGEMENTS

In addition to Church, I am also seriously in debt to a number of other scholars. I have mentioned Kripke, who first thought up the tricks that here produce finite model properties. Indeed, I would have preferred a joint paper with Kripke, which I proposed to him on several occasions. His consistent response was that he would think about it. I am also much indebted to Urquhart, who was my colleague at the University of Toronto when I produced the first draft of this paper; I had many stimulating conversations with

[37] While we used AxTrivial in the ***verification*** that $D\neg$ defines Boolean negation in **CR**, we may chop its stand-in $\mathrm{T} \rightarrow A \vee p$ from the suggested definiens. I.e., the following also works ***in*** $\mathbf{CR}^{\forall p}$:

$$\mathrm{D}\!\sim\!. \ \sim\! A =_{\mathrm{df}} \exists p\, (p \,\&\, (A \,\&\, p \rightarrow \mathrm{F})).$$

him, especially about the semilattice semantics, at that time. I am *always* indebted to Dunn; the job of finding, for each non-theorem A of $\mathbf{R}_{\rightarrow}$, a finite DeMorgan monoid, *a la* Dunn, that would refute A was my first aim. (There is always one; just tweak the $\mathbb{N}_k^m$ as in *Routley and Meyer 1973* to find it.) This material is presented at some length in *Anderson, Belnap, and Dunn 1992* and in *Routley, Plumwood, Meyer, and Brady 1982.* Past and present members of the Logic Group of the Australian National University—the late Richard Sylvan, Slaney, Martin, McRobbie, Meglicki, Restall, and many others—have been a constant source of fruitful ideas. (This group, which began life in the Philosophy department of the Research School of Social Science, has now mainly migrated to the Automated Reasoning Project in the Research School of Information Science and Engineering.) Jacques Riche, a Ph.D. graduate of the ARP, related the IDP and its cousins to many well-known mathematical principles in *Riche 1991.* I project with him (and with Kripke as well, this time, I hope) a retrospective "39 years of Kripke's lemma" to study these principles. Though I was not so explicitly influenced by him, Kit Fine—also a sometime visitor to the Logic group—will recognize the coincidence between semantical ideas set out here and ones that he had quite independently. Near enough, h-validity in c-monoids is Fine's criterion for theoremhood in $\mathbf{R}_{\rightarrow\&}$. Naturally I am also indebted to Anderson and particularly to Belnap, who not only introduced me to $\mathbf{R}$ (starting from $\mathbf{R}_{\rightarrow}$) but who suggested the decision question for $\mathbf{R}$ (toward which I once hoped that this paper would contribute) as a good problem. Finally, I am indebted to Professor Hiroakira Ono, who bugged me into turning this argument on its head to produce a finite model property for **BCK**. In *Meyer and Ono 1994*, $\mathbf{R}_{\rightarrow}$ was almost an afterthought. So it is pleasant to return the argument to its feet, extending it as indicated. Thanks to all.

REFERENCES

Ackermann, W.

1956 Begründung einer strengen Implikation, *The Journal of Symbolic Logic*, vol. 21, pp. 113–128.

Anderson, A. R., and N. D. Belnap, Jr.

1975 *Entailment, vol. I,* Princeton University Press, Princeton, New Jersey.

Anderson, A. R., N. D. Belnap, Jr., and J. M. Dunn

1992 *Entailment, vol. II,* Princeton University Press, Princeton, New Jersey.

Belnap, N. D.

1993 Life in the undistributed middle, *Substructural logics* (P. Schroeder-Heister and K. Došen, editors), Oxford University Press, Oxford, pp. 31–41.

Church, A.

1941 *The calculi of lambda-conversion,* Princeton University Press, Princeton, New Jersey.

1951 The weak positive implicational propositional calculus (abstract), *The Journal of Symbolic Logic*, vol. 16, p. 238.

1951a The weak theory of implication, *Kontrolliertes Denken, Untersuchungen zum Logikkalkül und zur Logik der Einzelwissenschaften,* (A. Menne, A. Wilhelmy, and H. Angsil, editors), Kommissions-Verlag Karl Alber, Munich, pp. 22–37.

Curry, H. B., and R. Feys

1958 *Combinatory logic*, vol. 1, North-Holland, Amsterdam.

Dickson, L. E.

1913 Finiteness of the odd perfect primitive abundant numbers with n distinct prime factors, *American Journal of Mathematics*, vol. 35, pp. 413–422.

Došen, K.

1993 A historical introduction to substructural logics, *Substructural logics* (P. Schroeder-Heister and K. Došen, editors), Oxford University Press, Oxford, pp. 1–30.

Dunn, J. M.

1966 The algebra of intensional logics, *Doctoral dissertation*, University of Pittsburgh, University Microfilms, Ann Arbor; see *Anderson and Belnap 1975*, pp. 352–369, contributed by Dunn.

Fine, K.

1974 Models for entailment, *The Journal of Philosophical Logic*, vol. 3, pp. 347–372; an updated version appears in *Anderson, Belnap, and Dunn 1992*, pp. 208–231.

Johansson, I.

1936 Der Minimalkalkül, einer reduzierter intuitionistischer Formalismus, *Compositio Mathematica*, vol. 4, pp. 119–136.

Kripke, S.

1959 The problem of entailment (abstract), *The Journal of Symbolic Logic*, vol. 24, p. 324.

Meyer, R. K.

1966 Topics in modal and many-valued logic, *Doctoral dissertation*, University of Pittsburgh, University Microfilms, Ann Arbor.

1978 The relevant theory of propositions is undecidable, unpublished manuscript.

1979 A Boolean-valued semantics for **R**, research paper no. 4, Research School of Social Sciences, Logic Group, Australian National University.

Meyer, R. K., and M. A. McRobbie

1982 Multisets and relevant implication, *Australasian Journal of Philosophy*, vol. 60, pp. 265–281.

Meyer, R. K., and H. Ono

1994 The finite model property for BCK and BCIW, *Studia Logica*, vol. 53, pp. 107–118.

Orlov, I. E.

1928 The calculus of compatibility of propositions (Russian), *Matematicheskii Sbornik*, vol. 35, pp. 263–286.

Riche, J.

1991 Decidability, complexity and automated reasoning in relevant logic, *Doctoral dissertation*, Australian National University, Canberra.

Routley, R., and R. K. Meyer

1973 The semantics of entailment (I), *Truth, syntax, modality* (H. Leblanc, editor), North-Holland, Amsterdam, pp. 199–243.

Routley, R., V. Plumwood, R. K. Meyer, and R. T. Brady

1982 *Relevant logics and their rivals*, Part I, Ridgeview, Atascadero, California.

Shaw-Kwei, M.

1950 The deduction theorems and two new logical systems, *Methodos*, vol. 2, pp. 56–75.

Slaney, J. K.

1985 3088 varieties. A solution to the Ackermann constant problem, *The Journal of Symbolic Logic*, vol. 52, pp. 487–501.

Thistlewaite, P. B., M. A. McRobbie, and R. K. Meyer

1988 *Automated theorem-proving in non-classical logics*, Wiley, New York.

Urquhart, A.

1972 Semantics for relevant logics, *The Journal of Symbolic Logic*, vol. 37. pp. 159–169; an updated version appears in *Anderson, Belnap, and Dunn 1992*, pp. 142–155.

1984 The undecidabilty of entailment and relevant implication, *The Journal of Symbolic Logic*, vol. 49, pp. 1059–1073; reprinted with additional material in *Anderson, Belnap, and Dunn 1992*, pp. 348–375.

1990 The complexity of decision procedures in relevance logic, *Truth or consequences: Essays in honor of Nuel Belnap* (J. M. Dunn and A. Gupta, editors), Kluwer, London, pp. 61–76.

STEWART SHAPIRO

THE "TRIUMPH" OF FIRST-ORDER LANGUAGES*

If "all" and "there exists" are applied to variable propositional functions, the question arises: what is the totality of all propositional functions?

Skolem 1928

It is true that the ... notion of consequence ... presupposes a certain absolute notion of ALL propositional functions ... But this is presupposed also in classical mathematics, especially classical analysis.

Church 1956, page 326n

1. INTRODUCTION

Is second-order logic the answer? As the saying goes, it depends on the question. There are several important metaphysical and epistemic issues in the vicinity: the nature and status of classes, and their place in logic, and the relationship between logic and mathematics. We also have to address significant details of logic: variables, connectives, and quantifiers.

A few ahistorical remarks on "the question" might be in order. Here is its form:

What is the correct or best [language or logic] in which to [   ]?

The most important item is the blank, which concerns the goals or purposes for which a language/logic is chosen. Without attempting to be exhaustive, the possibilities have included:

(1) to develop calculi that (more or less) accurately describe (correct) inference patterns of mathematics—each calculus to describe inference in one or more areas;

---

* ***Originally published in Foundations without Foundationalism: A Case for Second-order Logic by Stewart Shapiro (1991). Reproduced in this new revised form by permission of Oxford University Press.***

I owe a large debt to Gregory Moore, both for suggestions and encouragement to pursue this project. I would also like to thank the participants of the Ohio University Conference on Inference (October 1986), the Ohio State Logic Colloquium (January 1987), the Center for Philosophy of Science at the University of Pittsburgh (April 1987), and the Tel Aviv University Philosophy Colloquium (December 1987). This paper also benefited from John Corcoran, Michael Detlefsen, Matthew Foreman, Harvey Friedman, Harold Hodes, Ulrich Majer, David McCarty, Barbara Scholz, Robert Turnbull, and two anonymous referees.

C. Anthony Anderson and M. Zelëny (eds.), Logic, Meaning and Computation, 219-259.

(2) to codify the underlying logic of *all* rational or scientific discourse—to capture the most general features of reasoning competence (logicism);

(3) to provide a framework in which all (or most) of mathematics can be (re)formulated—a foundation;

(4) to formalize a particular branch of mathematics, such as arithmetic or geometry, which can involve (4a) codifying the true or knowable propositions of that branch (Hilbert), or (4b) describing the structure (or structures) studied by that branch, or (4c) both;

(5) to enhance human reason, by providing clear models;

(6) to establish a canonical language of science (logical positivism).

Some of these goals are associated with different philosophical programs. The second presupposes, or postulates, that *there is* an underlying logic for all rational or scientific discourse, presumably a single logic. The third postulates that there is (or can be) a foundation for all of mathematics. The fourth seems to involve a variety of realism toward the branches in question, in that it assumes that there is a collection of truths or knowable propositions to codify and/or some structures to describe. On the other hand, a substantial anti-metaphysical agenda underlies the positivist program, item (6).

Clearly, the different goals suggest different kinds of logical systems, and they involve different criteria of success. For example, if the development of a calculus is taken to be part of the goal of logic, as in (1), then an effectively enumerable (deductive) consequence relation is requisite. Indeed, it seems to be part of the notion of "calculus" that the correctness of deductions should be mechanically checkable. Less trivially, if one is concerned with *codifying correct inference*, either in a given area (4a) or in all areas at once (2), then one would require the consequence relation to match human inference competence in appropriate ways. If one further takes inference competence to be effective in some sense, then the deductive consequence relation should correspond and be effective.[1] If, however, one holds a more grandiose view of human inference competence, then an effectively enumerable consequence relation is not requisite, and may not be desired. As we shall see, Zermelo held a view like this. Gödel did as well, but apparently he did not accept the goal of describing inference competence. In cases where the goal is the *description* of a mathematical structure (4b), one would also not insist on an effective (semantic) consequence relation.[2]

[1] For a Fregean account of the relevance of effectiveness for goal (2), see *Wagner 1987*. In *1983*, I argue that problems like these were central factors in the mathematical characterization of the notion of effectiveness more than three decades later. See also *Gandy 1988*.

[2] Veblen (*1904*) seemed aware of the fact that different goals suggest different logical systems. The purpose of his axiomatization of geometry was to *describe* Euclidean space.

The other item in the question form, "language or logic" is only a matter of terminology, but clarity can forestall confusion. There are, I believe, deeper issues that lie nearby. It is now common to define a "logic" to be a logical vocabulary, a set of operators and variable sorts. Such vocabulary is combined with non-logical terminology to determine a language. Although this is a useful conception for many purposes (see the papers in *Barwise and Feferman 1985*), it is unnatural for historical analysis. The problem is that the definition presupposes that the boundary between "logical" and "non-logical" terminology has already been drawn. There is a sense in which the disputes between advocates of first-order logic and advocates of higher-order logic concern just this boundary. In particular, the status of certain quantifiers and the predication, or membership, relation is in question. The higher-order camp takes these items to be logical, while the first-order camp takes them to be non-logical. It would thus be imprudent to decide this issue in advance.

With some hindsight, let us define a *logic* to consist of a formal language, or class of formal languages, together with a deductive system and/or a semantics. First, a language is a collection of formulas. By itself, it is neither first-order nor higher-order. As Gilmore (*1957*) has pointed out, languages with "predicate variables" can be interpreted as many-sorted first-order systems. In effect, this is the insight behind the so-called "non-standard" semantics for higher-order languages developed by Henkin (*1950*) (for which common deductive systems are complete; see *Shapiro 1991*, chapters 3-4). That is to say, a first-order model-theoretic semantics is available for virtually all of the (finitary) languages under consideration here, and this first-order semantics is sound and complete for common deductive systems. In such cases, the "predication" relation (or the membership relation) between objects and properties (or sets) is taken to be *non-logical*.

The terms "second-order" or "higher-order" refer to interpreted, or partially interpreted, formal languages that contain variables ranging over *all* collections, or all properties, or all relations, or all propositional functions, of whatever is in the range of the ordinary, or first-order, variables. At least in part, this is a matter of the "meaning" of some terms and, thus, "semantics" broadly construed. For our purposes, it is a matter of interpretation.

If a language with predicate (or relation) variables is presented with an

---

An explicit aim was "to show that there is ***essentially only one*** class of which ... the axioms are valid ..." He added that once such a "categorical" characterization is accomplished, "any further axiom would have to be considered redundant", to which a footnote is added "... even were it not deducible from the [other] axioms by a finite number of syllogisms." Such a "further axiom" is redundant only to the description of the structure, not to the codification of the true principles thereof. Compare Veblen's orientation to that of Hilbert (*1900a*, problem 2): "When we are investigating the foundations of a science, we must set up a system of axioms which contains an exact and complete description of the ... ideas of that science ... no statement within the realm of the science ... is held to be true unless it can be deduced from the axioms by means of a finite number of logical steps."

explicitly articulated model-theoretic semantics, the situation is, or can be, straightforward. We only need to see if the *stated* range of the predicate variables (in each model) is the entire powerset of the domain, or includes *all* of its properties (or relations, etc.). If so, the logic is higher-order. Of course, we can *still* wonder whether the use of the word "all" in the meta-language really means *all*, and for this we can look to the meta-meta-theory (if there is one). But perhaps this is perverse. More substantially, one can wonder whether the author has the same notion of "property" that we do. It may be that not every subclass (in the present sense) of the domain is the extension of a "property". Some authors, for example, are not disposed to countenance properties that are not definable, or are definable only impredicatively.

Of course, model theory is not much help in historical analysis, being a relatively late development (see *Etchemendy 1988*). Typically, we are often left without an articulated semantics at the very beginning. In such cases, if predicate (or relation) variables are introduced, we must figure out what they are "intended" to range over. One possibility, of course, is that they aren't intended to range over anything. The author might be developing a "meaningless" formal language, perhaps with a deductive system. In this case, the logic is neither first-order nor higher-order. Another possibility is that the language is to be regarded as at least partially interpreted, but the interpretation is left informal, or intuitive. Then it is a matter of exegesis. Some authors clearly intend their predicate variables to range over "all" properties, but, as noted, they may not mean by "property" what we mean by "property". As shall be shown, in almost all the cases under study here before, say, 1930, even if a standard higher-order interpretation distorts the author's intentions, a Henkin interpretation is a much greater distortion.

Before the historical roots can be traced, I must make a number of distinctions, all of which concern the development of logic since the latter half of the last century. We must attend to differing views on the scope of logic, and various conceptions of class, variable, and quantifier.

Three traditions can be distinguished in the history of modern logic (provided one does not insist on sharp boundaries). One of them originates with Boole and includes, among others, Peirce, Venn, and Schröder. This *algebraic school* focused on relationships, or analogues, between canons of inference and algebraic operations like addition and multiplication. A primary aim was to develop *calculi* common to the reasoning in different areas of mathematics. Boole's seminal work (*1847, 1854*), for example, was a formal calculus that could be interpreted among propositions, classes and probabilities. As the name suggests, the orientation is that of abstract algebra, along the lines of group theory or field theory. One begins with one or more systems of related operations and abstracts a common structure. One set of axioms is then formulated which is satisfied by each system and which illuminates all of them. In some of the cases at hand, the axioms are dubbed "laws of thought".

The second tradition dates back at least to Euclid, and, for present purposes, includes Dedekind, Peano, Hilbert, and the postulate theorists like Veblen and Huntington. The aim of this *mathematical school* is the axiomatization of particular branches of mathematics. Euclid, Hilbert (*1899*), and Veblen (*1904*) developed geometry; Dedekind (*1888*) and Peano (*1889*) developed arithmetic; and Hilbert (*1900*) and Huntington (*1902, 1905*) developed analysis. An axiomatization produced by the algebraic school is intended to apply to several systems at once, while those of the mathematical school are primarily intended to apply to a single system. It should be noted that in at least some cases, this is a difference of emphasis rather than a difference of principle. The mathematicians occasionally considered alternate interpretations of axiom systems, typically when establishing independence results. Some members of the mathematical school did not indicate interest in the study of logic *per se*. They are included here because of their role in the history of logic.

Our third tradition, called the *logicist school*, includes Frege, Russell, and, perhaps, the early Wittgenstein. Their aim was to codify the underlying logic of *all* rational, scientific discourse. For them, logic is *not* the result of abstractions from the reasoning in various disciplines, or in various systems. Logic concerns the most general features of actual (precise) discourse, features independent of subject matter. Frege's writings contain numerous comparisons between his work and that of the algebraists. In *1883*, for example, he wrote:

> I did not wish to represent an abstract logic by formulas, but to express a content ... in more exact and clear fashion...

The difference seems to be that the algebraists (and the mathematicians) were concerned with certain systems, or algebras, and with reasoning *relative* to these systems, while the logicists were attempting to codify *unrelativized* reasoning, reasoning that is applicable in every context.[3]

Goldfarb (*1979*) and van Heijenoort (*1967a*) argue that the thesis that logic does not involve (a process of) abstraction and the idea of a universal language applicable in all contexts separates the logicists not only from the mathematicians and algebraists, but also from contemporary model theory. To use present terminology, logicist systems have no non-logical terminology. Today, typical treatments do not refer to a particular subject matter. Rather, schematic letters are used for nonlogical terminology, such as predicates, relations, and individual constants. These letters get different interpretations in the various models of the language. In this respect, the orientation resembles

[3]Using contemporary techniques, relativized reasoning can be accommodated in a logicist system. One simply restricts the quantifiers in some formulas to particular structures. When an algebraist or mathematician says "for all x ..." in reference to a domain d, a logicist could write "for all x, if x is in d, then ..." From the other perspective, the notion of unrelativized reasoning was foreign to the algebraic approach. Something similar might be obtained if, conceivably, one could invoke a notion of "all systems whatsoever" and abstract "laws of thought" from it, or if the entire universe could be taken as a single system. But this is hindsight.

that of the algebraists. Instead of schematic letters, logicist systems have variables, which have associated quantifiers. The whole notion of "model" or "interpretation" is foreign to the logicist project. There is nothing to "interpret" in various domains.[4]

One important distinction concerns the intended *ranges* of variables and quantifiers. There are three cases. The algebraic school of logicians took variables to range over an *unspecified*, but *fixed* domain. As in algebra, this domain is the result of an abstraction from more specific contexts.[5] For the mathematical school, the variables of each system are intended to range over a *fixed and specified* domain, such as the natural numbers or Euclidean space. The logicists Frege, Russell, and the early Wittgenstein, took their systems to represent the underlying logic of all rational discourse. For them, (first-order) variables range over *all objects whatsoever*.

All of the schools developed quantifiers, but they were understood differently. Frege and Peano, for example, took free variables and quantifiers in the contemporary sense except, perhaps, that they allowed (at least) second-order terminology. The algebraic school, on the other hand, took quantified propositions to be extended conjunctions and disjunctions. This led to the consideration of infinitely long expressions, discussion of "convergence", etc. Notice that a *domain* is needed to expand quantifiers into conjunctions and disjunctions, and, thus, once again the algebraic school is relying on a fixed domain.

The final distinction (for now) concerns the notion of class and its role in logic, together with the status of the concomitant membership relation. There is a tradition dating at least to Boole that takes classes (or intensional counterparts like properties or propositional functions) to be under the purview of logic. Gödel acknowledged this in his paper on Russell (*Gödel 1944*), writing that mathematical logic deals with "classes [and] relations ... instead of numbers, geometric forms, etc." As noted, the algebraic school took the subsets of a fixed domain to be under study by logic—this was one of the stated interpretations of their systems. There is some hindsight here, however, since they did not always distinguish between the membership relation and the subset relation, and sometimes they did not distinguish the singleton from its element. Frege also took the study of concepts and their extensions to be within logic (see for example, *Frege 1884*), as did Russell before his no-class period. What is crucial here is that the focus is on the subsets of a *fixed* universe, or domain. That is, the context of the theory

[4]As noted, it is not claimed that this taxonomy is exclusive or comprehensive. Indeed, members of the different schools drew from each other. Moreover, the division is limited to the period under study here. Later logicians cannot be classified along these lines, and contemporary logic has elements of all three. There is, of course, no need to force earlier logicians, such as Aristotle, the Stoics, and the Scholastics, into this classification.

[5]In the axioms for group theory, for example, the (object-language) variables range over a fixed, but unspecified group. The purpose is to make statements applicable to all groups.

determines, or presupposes, a universe of discourse—a range of the (first-order) variables. A set is a subdomain of this universe. Call this the *logical conception* of set. Thus, in arithmetic, a logical set is a collection of natural numbers; in geometry, a logical set is a collection of points, etc. To reiterate, there are no logical sets *simpliciter*, only logical sets within a given theory, although, of course, the context can be left unspecified (see *Shapiro 1991*, section 1.3).

On the other hand, current axiomatic set theory, as it developed from the work of Cantor, Zermelo, Skolem, Fraenkel, etc., fits the mold of the mathematical school. The axiomatization refers to a binary relation on a domain consisting of a (possibly empty) collection of urelements, sets of those, sets of sets of those, etc., with the "set of" relation possibly iterated into the transfinite. Call this the *iterative conception* of set. An iterative set is sometimes called a "well-founded set".

In short, then, an iterative set is a member of the set-theoretic hierarchy, the domain of a single theory, axiomatic set theory. The corresponding membership relation is *non-logical* (like the predication relation in non-standard semantics). In contrast, the logical conception of set always refers to a domain fixed by context. A logical set is a subclass of the universe of discourse. Its membership relation is a part of the logical systems of the algebraic and logicist schools, as well as second-order logic and type theory. Part of the issue concerning higher-order systems is the extent to which the latter are "legitimate" parts of logic.

Confusion between the two concepts of set can arise in set theory, since there the "fixed domain" (for the logical conception) is itself the set-theoretic hierarchy. An iterative set is a *member* of that hierarchy and, in this context, a logical set is a *collection* of iterative sets—what is today called a "class". Thus, within set theory, the entire hierarchy, the collection of all iterative sets, is a subdomain of itself. So the entire hierarchy is a logical set. With this terminology, the argument yielding Russell's paradox shows that in set theory, there are logical sets (classes) that are not iterative sets, or in other words, there are collections of iterative sets that are not members of the set-theoretic hierarchy. In particular, the entire hierarchy is not an iterative set. The custom today is that, when discussing set theory, the word "set" is used only for "iterative set" and, in this context, the word "class" is used for "logical set". A "proper class" is a logical set that is not an iterative set.

This suggests that Russell's paradox should trouble only those theorists using both a logical notion of set *and* an all-inclusive domain, one that might *include* all of its subdomains as members. To some extent, history confirms this "expectation". For the most part, only the logicists were affected, for they took variables to range over all objects whatsoever, which at least prima facie includes all collections of objects. Although Frege clearly separated concepts from objects, he came to associate each concept with an object—its extension. The assertion of the existence of extensions, and the axiom that two concepts have the same extension only if they are co-extensive led to the paradox (*Frege 1893*). Russell himself avoided this (and other) antinomies by separating the all inclusive domain into individuals, classes (or properties) of individuals, classes of classes of individuals, etc.

For present purposes, the interesting cases are those authors who did *not* feel threatened by the paradox. Zermelo independently discovered Russell's paradox a full year before Russell (see *Rang and Thomas 1981*), but did not publish it, and continued his work in set theory. Whether or not Zermelo held the "iterative conception" of set at this time, he apparently did not believe that set theory, as he construed it, was undermined.[6] In the aforementioned paper on Russell, Gödel (*1944*) also says that the iterative concept of set was never affected by the paradoxes. For the most part, the algebraic school was also unaffected, since they only considered the subclasses of a *fixed* domain, and did not envision a universal domain—one that contains all of its subdomains as members. In *1890* (page 245), Schröder remarked that one cannot, without contradiction, let the fixed domain consist of all objects. The contrast with the logicists is striking.[7]

It should be mentioned that this relatively neat division of treatments of set and class requires some hindsight. Around the turn of the century, Hilbert was more or less in the mathematical school (with some influence from the algebraists), and his correspondence with Frege indicates just how far he was from the logicists (see *Resnik 1980*, chapter 3). As far as I know, he never considered a universal domain. Thus, according to the thesis at hand, Hilbert should not have been bothered by Russell's paradox. Yet, after learning of the paradox, he took a cautious view of sets. In *1905* (page 176), he wrote:

> ... the conceptions ... of logic, conceived in the traditional sense, do not measure up to the rigorous demands that set theory makes ... Yet if we observe attentively, we realize that in the traditional treatment of ... logic, certain fundamental notions are used, such as the notion of set.

The proposed solution, of course, was to develop logic and mathematics, including set theory, simultaneously, and then to provide a finitary consistency proof of the whole system. This was not to be.

[6]See *Moore 1978*. Zermelo (*1908a*) does mention Russell's paradox in the opening paragraph, noting that "the very existence of [set theory] seems to be threatened". However, by careful analysis of the historical context, Moore shows that the operative word here is "seems", for Zermelo took the threat to be more apparent than real (see also *Zermelo 1908*, section 20). For example, the stated plan is to "start with set theory as it is historically given". This program would be a rather poor one if set theory "as it is historically given" is not fundamentally sound. In Zermelo's mind, the culprit is an unrestricted principle that every "collection ... of ... well-distinguished objects" is a set. This seems close to the way I formulated the paradox above—not every logical set is a member of the hierarchy.

[7]In the relevant passage, Schröder criticizes an earlier suggestion by Boole that one interpretation of "class 1" is the universal class, the class which contains everything conceivable. Schröder's argument is rather obscure, and includes the conflation of the element and subset relations. Frege (*1895*) had little trouble disposing of Schröder's argument.

2. NARRATIVE

In the more direct sketches of the relevant histories, there are three major items, two of which overlap considerably. The first is the emergence of first-order *languages* (and semantics) as the standard in formal logic. The second is the development of *set theory* and the emergence of first-order ZFC.[8] The third is the contemporary controversy over the status of second-order logic. The first two are presented here, the third is deferred to the next section. Much of the historical material in this section is based on the work of Gregory Moore (*1980, 1988*), and on several suggestions he made to me. Of course, I maintain responsibility for any errors.

Once disagreements over these issues began in earnest, Skolem and Gödel were the main proponents of first-order languages. For different reasons, each held that only first-order axiomatizations are legitimate. The higher-order language "opposition" was championed by Zermelo, Hilbert, and Bernays.[9] There is an interesting absence of correlation between the line-up across this battle line and that over more traditional philosophical matters. Gödel and Bernays were Platonists (each in his own way), whereas Skolem and Hilbert embraced various versions of formalism, relativism, and finitism.

2.1. The development of first-order logic

Prior to Löwenheim's theorem (*1915*), no one had discussed first-order logic as even a distinguished part of logic. In various ways, authors from all schools introduced variables ranging over both objects and properties (or propositional functions, or classes). As noted, one stated interpretation of the systems of Peirce and Schröder was the subclasses of a fixed domain. Each proposition (with a free variable) determines a class. They also introduced second-order variables ranging over proposition-subclasses, but did not carry this very far. By contrast, Frege (*1879, 1884*) introduced and brilliantly exploited second-order variables ranging over concepts (but, again, extensions of concepts were regarded as objects and thus in the range of the first-order variables). The notions of identity, equipollence, and the ancestral were all defined in second-order language, the latter playing a central role in the formulation of the natural numbers. Frege did not separate the first-order part of the system, and in a sense he couldn't, since the language contains no predicate and relation symbols other than higher-order variables. A variation on this theme is Russell, who went on to develop ramified type theory, with different variable sorts for individuals, propositional functions, propositional

[8] The words "emergence" and "standard" in this paragraph (as well as the word "triumph" in the title) are not meant to endorse a thesis that the first-order theories ought to be standard. As noted in the final section of this paper, my own sympathies are on the higher-order side. See *Shapiro 1991* for a further articulation of higher-order logic.

[9] John Corcoran informs me that Tarski eventually rejected higher-order logic, even though much of his early work concerns such systems.

functions of propositional functions, etc.

For Russell, at least, the properties or propositional functions are intensional. Not much turns on this, however, since a notion of "identity between properties" plays no significant role. Indeed, such a notion was not formulated. In Frege's case, the identity relation only applies to objects, the items in the range of the first-order variables. The closest thing to "identity" that Frege applies to concepts is coextensiveness, which, of course, is the criterion of identity among extensional items.[10]

It does not take a major exegetical study to show that the logicist systems are at least close to contemporary (standard) higher-order logics, and it would be a major distortion to think of logicist systems as first-order (e.g., to impose a non-standard semantics). First, as noted above, the systems were not intended to be uninterpreted formal calculi. Second, and more important, in contemporary terms, the systems do not have any non-logical terminology, much less a non-logical predication relation. Every item has a fixed meaning, every variable a fixed range. It is true that Russell (at least) may not have envisioned the *range* of property variables in contemporary higher-order systems, because his propositional functions were predicative, but even this difference is not clear in light of the axiom of reducibility. In any case, the difference is due to Russell's views on the nature of propositional functions, not on the extent of the variable-ranges. In contrast with non-standard semantics, he did not regard the ranges as indeterminate—the variables range over *all* propositional functions of appropriate type and level.

Peano also introduced variables ranging over "classes" of natural numbers, and similar practices were adopted by other members of the mathematical school. The axiomatizations of geometry by Hilbert (*1902*) and Veblen (*1904*), and the axiomatizations of analysis by Huntington (*1902, 1905*), all contain second-order terminology, or something equivalent, essentially. Veblen and Huntington employ the more usual second-order axioms of continuity, essentially due to Dedekind. For Hilbert, it was the extremal axiom that there are to be no proper extensions of the system.[11] It must be noted, however, that the consequence relation was not discussed. Some of these authors had very little to say about logic at all. But the second-order variables were clearly intended to range over all classes or properties of the relevant domain. If, through an act of hindsight, one restricts these systems to first-order languages, Hilbert's extremal axiom is not satisfiable and, of course, the categoricity theorems proved by Veblen and Huntington are false.[12]

[10]It is consistent with Frege's view that coextensiveness is a *congruence* relation among concepts. That is, if $\forall x(Yx \equiv Xx)$ and $\Phi(X)$, then $\Phi(Y)$ (provided that Y is free for X in $\Phi(X)$). I am grateful to Harold Hodes for pointing this out.

[11]As Moore (*1988*) notes, this "completeness axiom" did not appear in the original German edition (*1899*), but was added in the French (*1902*) and English translations.

[12]It follows from the so-called upward and downward Löwenheim-Skolem theorems that no first-order theory with an infinite model is categorical. Indeed, for any such theory T and any infinite cardinal κ, there is a model of T whose domain has cardinality κ.

Our attempt to characterize the emergence of first-order languages and logic is complicated by the use or discussion of infinitary formulas. This dates back (at least) to the interpretation of the quantifiers in the algebraic school, and continued through the period under study here. For example, Hilbert adopted such an interpretation in *1905* (but later came to reject infinitary systems), and in the discussion surrounding *Principia Mathematica*, Lewis (*1918*) and Ramsey (*1925*) both introduced infinitary formulas. With hindsight, such systems provide a hierarchy of logics that are "intermediate" in some sense between first-order and higher-order.

The algebraic school, as well as Whitehead and Russell (*1910*), often discussed the logic of propositions, the logic of classes, and the logic of relations, but these did not correspond to the present distinction between first-order logic and higher-order logics. Concerning the algebraists, there is the complication of infinitary formulas, and the conflation of syntax and semantics left an unclarity concerning what the restricted systems were like.

The separation of first-order systems, or *sub*-systems, is due to Löwenheim and Hilbert. I'll begin with the latter.

In a series of lectures in 1917, Hilbert applied the axiomatic method to geometry *and logic* (see *Moore 1988*). This is easily recognized as an extension of the methodology of the mathematical school. He explicitly developed primitive symbols and axioms for first-order logic, which he named "the functional calculus". Hilbert noted that this system is sufficient to codify deductions *within* various branches of mathematics; but he added that it is not adequate when the "foundations of mathematical theories themselves [become] an object of investigation". For "foundational study", an "extended calculus", containing variables ranging over properties, was developed. This extended calculus is, in effect, a version of ramified type theory (with the axiom of reducibility). Hilbert's logical systems did not appear in print, however, until *Principles of mathematical logic*, by Hilbert and Ackermann, was published (*1928*). Eventually, Hilbert came to accept the simple theory of types, an ω-order logic. The important point is that in both places Hilbert regarded first-order logic as a distinct *sub-system* of all of logic. It was dubbed the "*restricted* functional calculus". He suggested that set theory—and even arithmetic—is incapable of an adequate treatment in the restricted system. Hilbert's point seems to be that the first-order systems cannot adequately formulate many propositions and concepts necessary for meta-mathematical

---

Thus, if the categoricity statements proved by Veblen and Huntington are applied to first-order theories, they are false. As noted, Hilbert's extremal axiom states that there are to be no proper extensions of the systems. But if the theory is first-order and if it has an infinite model (which it does), then there is no maximal model. Every infinite model of any first-order theory has a proper extension. To put this historical point in Quinean terms, suppose we were engaged in "translating" the systems of these authors into current mathematical dialect. If it is insisted that the "output" of the translation be formulated in the corresponding contemporary first-order systems, then we must attribute some rather elementary mistakes to the authors.

study of various theories. Examples include the principle of induction for formulas, the membership relation, number, and cardinality (see *Shapiro 1991*, chapter 5). He showed how all of these notions have straightforward formulations in the full, higher-order system.

Hilbert's co-workers, such as Bernays (*1918, 1928*), Schönfinkel (*1924*) (and *Bernays and Schönfinkel 1928*), Ackermann (*1924*), and von Neumann (*1927*), all used essentially the same system. In *1918,* Bernays formulated and proved the completeness of propositional logic. Within the Hilbert program, this result was the first precise solution to a completeness problem for a *proper part* of logic. Similar problems for other sub-systems of logic became crucial components of the program, and represented stages toward what would ultimately be a consistency proof for all of mathematics. In this spirit, the problem concerning the completeness of first-order logic was proposed in *Hilbert 1929* (and *Hilbert and Ackermann 1928*).[13] This, of course, was established by Gödel (*1930*) (who also adopted the phrase "restricted functional calculus" to denote first-order logic).

In contrast with Hilbert, Löwenheim was firmly placed in the algebraic school. He names Schröder as the source of his interest in logic, and all of Löwenheim's work concerns variants of Schröder's systems (see *Thiel 1977*, page 237, and *Moore 1988*). The celebrated theorem, appearing in *1915*, was no exception. Löwenheim first made a distinction between a "relational expression" and an "individual expression". The latter had to be of finite length and could only contain variables over individuals. This is close to Hilbert's distinction between first-order and higher-order formulas, but in keeping with the algebraic school, Löwenheim thought of these as expandable into infinitely long strings of connectives and quantifiers. On this matter, however, the crucial difference between Löwenheim and his predecessors is that the only infinitary formulas envisioned here were those equivalent (in the system) to finitary formulas. This variation on the algebraic theme permitted Löwenheim to carefully distinguish the equivalent of the first-order part of his system. His version of the Löwenheim-Skolem theorem concerned that part, but his *proof* relied heavily on the full system. That is, the proof involved both higher-order and infinitary formulas. Löwenheim also showed that the result does not hold for the full (higher-order) system.

It was Skolem who showed that Löwenheim's theorem could be proved *in* a first-order (meta-)language (by using Skolem functions). He extended the theorem somewhat, applying it to countably infinite sets of formulas and to some infinitary languages (*Skolem 1920*). By *1922,* however, he ceased mention of the extensions of the theorem to infinitary languages, and he

[13]By this time, the distinction between proof theory and semantics was getting clear. Some of the problems posed in *Hilbert 1929* concern the deductive completeness of (second-order) arithmetic. After noting that the Peano postulates are categorical, Hilbert asked for a proof that if a sentence is shown to be consistent with arithmetic, then its negation cannot be shown consistent with arithmetic.

began to urge that first-order logic is the proper basis for set-theory and, indeed, for all of mathematics. I say "urge" here, not "argue". As far as I can determine, neither Skolem nor Gödel gave detailed *reasons* for their insistence on first-order logic. Nevertheless, their assertions are not merely directives that first-order logic is the most interesting or fruitful area for research. They are claims that higher-order variables do not belong in logic.

It was at this time that Skolem began discussing the "Skolem paradox", which concerns the interpretation of axiomatic systems. It is to the axiomatization of set theory that I now turn.

2.2. Axiomatization of set theory

In this part of our story, the crucial item is the axiom of separation:

$$\forall x \exists y \forall z (z \in y \equiv z \in x \,\&\, Pz).$$

The issue concerns the status of the variable P. In first-order ZFC, the axiom is a scheme, each instance of which is obtained by replacing the letter P with a first-order formula (not containing y free). In second-order treatments, the letter P is a predicate variable, regarded as either free or as bound by a prenex universal quantifier (see *Shapiro 1991,* chapter 5).

Here we begin with Zermelo. Earlier set theorists, like Cantor, were not particularly concerned with axiomatization and underlying logic. Late in 1904, Zermelo proved the well-ordering theorem (*Zermelo 1904*) and, almost immediately, the proof drew intense resistance (see *Moore 1982*). Virtually every aspect of Zermelo's treatment was attacked. During 1907, in response to his critics, Zermelo composed his first axiomatization of set theory (*Zermelo 1908* and *1908a*). Within the tradition of Hilbert's methods, and what I have called the "mathematical school", the goal was to elucidate (and defend) the assumptions behind the proof of the well-ordering theorem.

The version of the axiom of separation that appeared in Zermelo's first formulation asserted that for every propositional function $P(z)$, if $P(z)$ is *definit* for a set S, then there exists a set containing exactly those elements of S for which $P(z)$ is true. Present concern, of course, is with the opening universal quantifier and with the word "*definit*". At the time, Zermelo only noted that a given $P(z)$ is *definit* for S if the membership relation on the domain and the "universally valid laws of logic" determine whether P holds for each element of S. The only "laws of logic" that are even hinted at here are the principles of bivalence and excluded middle. Perhaps there is little point in speculating on this, but as Moore (*1980,* page 109) notes, it is reasonably clear that Zermelo did not intend a restriction to those properties definable in first-order logic. First, the axiomatization occurred almost a decade before Löwenheim and Hilbert separated first-order logic as a subsystem of logic; second, the axiom of separation was presented as a single sentence with a variable ranging over propositional functions; and, third, Zermelo later attacked proposals to limit set theory to a first-order language.

Like the previous proof of the well-ordering theorem, this axiomatization attracted criticism, but now the critics had something more tangible (or, shall we say, more *definit*) to attack. Much of this was directed at the axiom of choice, but separation was not ignored. For example, Russell, writing to Jourdain in 1908, called it "so vague as to be useless" (see *Grattan-Guinness 1977,* page 109); and Poincaré (*1909*) objected that a property that is *definit* might not be well-defined.[14]

In attempting to improve the axiom of separation, Hermann Weyl (*1910*) proposed that a property is *definit* iff it is either of the form $x \in y$ or $x = y$, or is obtained from such propositions by finitely many uses of negation, conjunction, disjunction, existential quantification, and substitution of a constant for a variable. He did not indicate what sorts of existential quantification are permitted here. Again, recall that first-order languages had yet to be published as separate systems. *Weyl 1917* contains a similar proposal, but there he explicitly stated that variables ranging over properties are not countenanced. He went on to explicitly *reject* higher-order variables as unsuitable for logic, probably the first person to do so in print. However, one should not infer that Weyl (*1917*) explicitly *accepted* classical first-order logic; sometime after this, Weyl adopted the intuitionist perspective in mathematics (see *Moore 1988*).

Weyl (*1910*) noted that his formulation of *definit* seems to presuppose the notion of *natural number*, as evidenced by the phrase "finitely many uses of". This raises a foundational issue concerning whether sets or numbers are more fundamental, the question being which of these are to be formulated in terms of the other. If set theory is to serve as a foundation for all of mathematics, it would seem that one ought to be able to develop it independently of arithmetic, and *then* formulate the natural numbers within set theory (as, for example, *Dedekind 1888*).[15] After several attempts to explicate *definit* without using the notion of number, Weyl came to reject Zermelo's (and

[14] See also *Schoenflies 1911*. Poincaré's attack was based in part on his rejection of the actual infinite and of impredicative definition (see *Goldfarb 1988*).

[15] Harold Hodes suggested that, conceivably, one could attempt a bootstrap construction. First formulate a rudimentary set theory without separation and define "the natural numbers" to be, say, the finite von Neumann ordinals. Then, with this as meta-theory, one might employ the newly-defined notion of natural number to formulate a set theory containing a principle of separation, perhaps along the lines of *Weyl 1910*. However, it is not clear that the arithmetic formulated in the rudimentary set theory is sufficient, since the usual way to derive a general induction scheme within set theory uses separation. Moreover, the elementary set theory does not appear to be capable of coding some collections necessary to do meta-theory (e.g., the collection of well-formed formulas). If an induction principle were explicitly stated and added as an axiom scheme, it would seem that one is taking the natural numbers as fundamental after all. In any case, this sort of thing was not attempted by anyone during the period under study here. I would suggest that one reason that such a set theory was not conceived is that the principle of separation was (and is) regarded as essential to the notion of set.

Dedekind's (*1888*)) set-theoretic formulation of the natural numbers.

Fraenkel and Skolem, who began writing on the subject in the early 1920's, enjoyed considerable influence on the development of set theory. Although the systems they developed were remarkably similar, their attitudes toward logic, set theory, and foundations of mathematics were quite different.

Fraenkel (*1922, 1922a, 1925*) was rather skeptical of the whole enterprise of mathematical logic. He attacked Zermelo's notion of *definit* just because it relied on the imprecise basis of "general logic". Presumably, this is a reference to Zermelo's invoking the "universally valid laws of logic" in the formulation of *definit* property. Fraenkel thus seemed to suggest that a sufficiently precise notion was not characterized. To improve it, he introduced an alternate: $f(x)$ is *definitorisch* if it can be obtained by the finite iteration of the operation of powerset, union, and unordered pair. Notice that, like Weyl's version, this formulation contains a reference to "finitely many iterations". With hindsight, it seems that Fraenkel took the theory of natural numbers to be more secure than set theory.

In *1922*, Skolem formulated a revision of *definit* property similar to (but independently of) that given by Weyl. Like Weyl's later *1917*, Skolem explicitly restricted the existential quantification to first-order variables, and he was the first to formulate separation as a scheme—one instance for each *definit* property *definable in the first-order language.*[16] Skolem's proposal was, in short, to treat the membership relation as non-logical, and, more radically, to formulate set theory in a first-order language. In short, Skolem was moving away both from the algebraic view of quantifiers as conjunctions and disjunctions, and from the algebraic and logicist view of classes as logical entities.

In the same paper (*1922*), Skolem began the attack on the absoluteness of set-theoretic notions, arguing that there is no single intended interpretation to set-theoretic concepts. The Löwenheim-Skolem theorem indicates that like any first-order axiomatization, set theory has a countable model—a model within the natural numbers (if it has a model at all). This, of course, is despite the fact that within set theory, one can prove that, say, the real numbers are uncountable. He concluded from this, the so-called "Skolem paradox", that set-theoretic notions are "unavoidably relative". For example, one cannot claim that a given domain D is uncountable simpliciter, but only relative to a given model (containing D) of set theory. For any such D, one cannot rule out the possibility that D may be countable relative to a (richer)

[16]Skolem and Fraenkel (independently) also realized the need for an axiom of replacement. Skolem's version was an axiom scheme formulated in terms of his (first-order) notion of*definit* property. This, of course, is contemporary practice. Fraenkel first gave a rather imprecise account: "If M is a set, and M' is obtained by replacing each member of M with some object, then M' is a set." Thus the name "replacement". The principle was soon modified in terms of Fraenkel's notion of *definitorisch* function, but then it was too weak to serve its purpose. Von Neumann (*1928*) demonstrated that this version of the axiom of replacement could be proved in Zermelo's original system.

model, one that contains a function from the natural numbers onto D. Skolem held that this relativity also applies to the Dedekind notions of "finite" and "simply infinite system".[17] Given the modern trend of regarding virtually all mathematical notions as set-theoretic, Skolem's conclusions would entail the relativity of just about everything. Skolem eventually held such a view, but at the time he seemed content to reject the set-theoretic foundation (see *Benacerraf 1985*). Skolem (of *1922*) seemed to accept Weyl's conclusion (and Fraenkel's apparent presupposition) that there is no gain in formulating intuitive mathematical notions, like that of natural number, in set-theoretic terms. Zermelo did not agree.

Questions concerning categoricity were often raised in this context. From contemporary perspectives, the discussion can be read as if it were full of confusion over the distinction between the semantics of first-order languages and the semantics of higher-order languages, confusion over the range of logic and the range of categoricity, and/or confusion over the applicability of the Löwenheim-Skolem theorems. Those whose minds concerning these matters are settled might understand it that way. The historical assertions that conform to one's views are to be praised as clear while those that are incompatible are dubbed "confused". On the other hand, most of the remarks by the historical figures have counterparts in the current debate over second-order logic. Once presuppositions are noted, some of the pronouncements are at least relatively clear and, in some cases, remarkably clear, even by modern lights.

With hindsight, there are several sources of fundamental disagreement. One concerns the distinction between syntax and semantics and, in particular, the distinction between proof theory and model theory. A closely related matter concerns the status of the second-order variables, involving what I above call the "logical notion of set". Recall that on this conception, a fixed domain is presupposed, or determined by context, and the variables range over its subdomains (or its properties and relations). On one level, the question of whether this notion is a legitimate part of logic is a matter of deciding where to draw a border: Is the (logical) notion of set in the jurisdiction of logic, or does it belong to mathematics proper? The deeper issue concerns the extent of our "intuitive" understanding of the second-order variables and the extent to which they can be *used* in formulating the syntax and semantics of formal languages. The higher-order camp seems to hold that the terminology in question is already clear and does not stand in need of "foundation" (see my *1991*). Indeed, the claim is that this very terminology

[17]For Dedekind, a ***simply infinite system*** is a countably infinite set with a successor operation. Dedekind defined a set S to be ***finite*** if there is no one-to-one correspondence between S and a proper subset of S. Notice the variable ranging over "one-to-one correspondences". At least conceptually, this differs from the usual definitions of finitude which refer to (or presuppose) the natural numbers. Here we have a another choice between a set-theoretic and a number-theoretic foundation.

is at the heart of the foundational enterprise. Any successful attempt at formulating an axiomatic foundation of second-order logic would itself have terminology involving the logical notion of set (perhaps in presenting the semantics) and would thus beg any questions.[18] The first-order advocates, on the other hand, hold that (what I call the) "logical" notion of set is not clear enough for use in foundational systems, at least not as it stands. To be used at all, the relevant notions must first be formulated as part of an axiomatic theory. In short, the notion of set must be regarded as non-logical. And any assertions about sets must follow from the axioms. This, of course, is the model of the mathematical school of logic. With hindsight, the best candidate for such an axiomatic formulation is the study of the iterative hierarchy. Thus, the first-order camp rejects my distinction between the iterative conception and the logical conception of set. There are only iterative sets, and the terminology is non-logical.

Skolem (*1922*) launched, in effect, a two-fold attack against the categoricity of set theory. One aspect was the general relativistic thesis that *no* theory with an infinite model can be categorical. This, of course, *follows* from the Löwenheim-Skolem theorems and *presupposes* that all theories are first-order.[19] Skolem, of course, embraced this presupposition. The other critique was specific to set theory (and would apply to the second-order versions). Skolem pointed out that even when the axiom of replacement is added, Zermelo's system is consistent with the existence of various non-well-founded sets and various collections of urelements. For example, given any domain D that satisfies the axioms, if D contains urelements, then the subdomain consisting of those elements of D whose transitive closures do not contain urelements also satisfies the axioms. Conversely, it is always possible to add new urelements to D.

One way to attack the more general relativistic thesis would be to reject the presupposition that all languages are first-order. The latter aspect of Skolem's critique seems to call for new axioms, some of which were suggested by Skolem himself, trying to be helpful. Both approaches were forthcoming, some in response to Skolem, others independently.

Fraenkel (*1921*) attempted to formulate a categorical version of set theory with an "axiom of restriction", which was similar, but opposite to Hilbert's axiom of completeness in geometry. Fraenkel's "axiom" asserted that the only sets to be considered are those whose existence follows from the (other) axioms. In short, no model of the theory is to have a proper subset that is itself a model of the (other) axioms.[20] This rules out models with non-well-

[18]One can give a set-theoretic definition of second-order logical consequence in contemporary set theory, which is first-order. But this pushes the question back one level, to the issue of whether first-order set theory is adequate. See my *1991*, sections 5.4 and 9.3.

[19]As far as I can determine, the first explicit observation that first-order theories like arithmetic are not categorical did not occur until *Skolem 1933, 1934*, but it follows from the incompleteness of arithmetic and the completeness of first-order logic.

[20]Fraenkel's axiom of restriction might be called a "minimal principle", since it rules out

founded sets and models with urelements (as well as models with inaccessible cardinals). Fraenkel seemed to believe that the intersection of all domains satisfying the rest of the axioms would be a model of the axiom of restriction.

By standards of modern logic, the axiom of restriction and Hilbert's axiom of completeness conflate object language and meta-language, model-theoretic semantics in particular. The axioms of a theory are to refer to the relevant subject matter, sets in this case. It is only in the meta-language that one can refer to what follows from the axioms, or to its models. But perhaps it is not proper to impose such standards on historical figures.[21] A deeper "problem" with Fraenkel's axiom of restriction is similar to the "problems" with second-order languages. It *presupposes* familiarity with sets or, as von Neumann (*1925*) put it, the axiom relies on "naive" set theory. If the purpose of the axiomatization is to *clarify* the notion of (iterative) set, and to do so while avoiding "naive" set theory (and the antinomies), then the axiom of restriction is circular. It might be noted, however, that those who hold that the concept of set is intuitively clear—and does not stand in (further) need of "foundation"—could argue that the circularity is not vicious, no more so than the *use* of connectives and quantifiers in a meta-language to explicate the corresponding items in the object language. On such a view, the purpose of axiomatization is not to provide a "foundation" as such, but to codify the reasoning in a theory, or to characterize its models.

Von Neumann went on to suggest that at least some of the problems with the axiom of restriction can be overcome by precisely formulating the relevant model-theoretic notions in set theory. After all, set theory deals with the basic building blocks of model theory. Von Neumann's own axiomatization, which includes separate variables for sets and classes, is suited for this. However, the notion of a "model" of *this* theory requires a "higher" set theory, one with the sets and (proper) classes of the original theory as *elements*, i.e., as sets. Of course, the higher theory would have its own (super)classes.[22] Working in such a higher set theory, one can *use* the relations of subset and subclass to formulate the notion of a class satisfying various statements, and,

the existence of proper substructures that are models of the other axioms. In contrast, Hilbert's axiom of completeness is maximal, since it rules out proper superstructures that are models of the other axioms. For an extensive treatment of "extremal axioms" like Fraenkel's axiom of restriction and Hilbert's axiom of completeness, see *Carnap and Bachmann 1936*. On the axiom of restriction, see *Fraenkel, Bar Hillel, and Levy 1973*, pages 113–119.

[21] If one allows the conflation of object language and semantics, then Fraenkel's system does provide a categorical characterization. The only models of *first-order* ZFC that have no proper submodels are those isomorphic to the minimal constructible model—a countable set. The only standard models of *second-order* ZFC that have no proper submodels are those that are isomorphic to the first inaccessible rank. See *Shapiro 1991*, chapter 5.

[22] A "model" of von Neumann set theory would be a collection of sets (i.e., a class) together with the collection of its subclasses. But in the original theory, (proper) classes are not to be elements of collections. So, in the original theory, one cannot discuss (class) models of set theory. For this, we require a more encompassing theory and system.

thus, the notion of a class being a *model* of ordinary set theory. In this context, the axiom of restriction has a precise meaning—one can state in the higher-theory that a given class has no non-trivial subclasses that are models of ordinary set theory. Von Neumann observed that the intersection (in the higher system) of all models of ordinary set theory might turn out to be itself a model of set theory. If so, then it also satisfies the axiom of restriction, and no other model in the postulated system does. So far, so good. He then argued, however, that even if this process succeeds in fixing a model, the resulting theory—ordinary set theory together with the axiom of restriction—may still not be categorical. In some cases, perhaps, a single model has been characterized, but only *relative to* the choice of the higher set theory and the system satisfying it. If one had begun the construction with an even larger (or still "higher") system, there may be a different collection of models and, thus, a different intersection and so a different model of the axiom of restriction.

The last section of *von Neumann 1925* is a general discussion of categoricity. He notes that it is easy to *prove* the categoricity of theories like Euclidean geometry, but added, parenthetically, that to accept the arguments, one must "disregard the fact that the axioms of geometry ... depend on those of set theory." After all, the (second-order) axiom of continuity refers to the *subsets* of the plane and Hilbert's axiom of completeness refers to models of the axioms, just like Fraenkel's axiom of restriction. The point is that the categoricity proofs are based on higher-order, or set-theoretic premises and, thus, are relative to a particular system of set theory. Von Neumann then reiterated the aforementioned conclusion that set theory, as then formulated, is *not* categorical, and he argued that considerations like the Löwenheim-Skolem theorems suggest the depressing conclusion that

> ... no categorical axiomatization of set theory seems to exist at all; for probably no axiomatization will be able to avoid the difficulties connected with the axiom of restriction and the "higher systems". And since there is no system for mathematics, geometry, and so forth, that does not presuppose set theory, there probably cannot be any categorically axiomatized infinite system at all.

He went on to suggest that the same analysis might apply to the axiomatic treatment of basic notions of cardinality, since even these are formulated with variables which occur "with reference to the entire [set-theoretic] system". Like Skolem (*1922*), the Dedekind notion of finitude is given as an example. Von Neumann concludes that if this Skolemite relativity is sustained, then, of the notion of finitude,

> ... nothing but the shell of its formal characterization would ... remain ... It is difficult to say whether this would militate more strongly ... against its intuitive character, ... or its formulation as given by set theory.

We have seen this dilemma before, in the guise of whether sets or numbers are more fundamental. Von Neumann is ambivalent about whether, say, finitude "needs" a (higher-order or) set-theoretic foundation. If it does, he sees no way to regard the notion as unequivocal.

It might be noted in passing that by this time, Hilbert had repudiated the algebraic school's understanding of quantified formulas as infinitary. He also came to accept in part the intuitionistic—and Skolemite—claim that infinite sets cannot be treated uncritically like finite ones (see, for example, *Hilbert 1923*). Nevertheless, he continued to hold that the *use* of class-terminology is essential to the foundational enterprise (i.e., to the Hilbert program). One cannot characterize the inter-relations between theories without it. The proposed solution to this dilemma, of course, was finitary proof theory, in which the logic and the mathematics are to be developed simultaneously.

Zermelo was beginning to move in the opposite direction. In a lecture of 1929, he accepted a version of the restricted "logicist" thesis that mathematics is "a systematization of the provable and, as such, an applied logic", but he came to reject the view that proofs must be of finite length.[23] Mathematics was dubbed "the logic of the infinite".

The same year, Zermelo returned to his axiomatization of set theory (*Zermelo 1929*). In discussing his earlier work, and its critics, he indicated that the explication of the crucial notion of *definit* depended on the logical resources available:[24]

> At the time there did not exist a generally recognized "mathematical logic" to which I could appeal, any more than it does today, where every ... researcher has his own system. (*Zermelo 1929*, page 340)

Zermelo rejected Fraenkel's proposal concerning the notion of *definit* since it relied on a *construction*. Such "procedures" contradicted what he took to be the essence of the axiomatic method. Zermelo apparently took an axiomatization to be a *characterization* of its subject, not a construction of it. In this spirit, he proposed to axiomatize the notion of *definit* property. The result was an essentially *second-order* formulation. For present purposes, the crucial clause is that if $P(g)$ is *definit* for each propositional function g (with only individual variables), then so are $\forall f P(f)$ and $\exists f P(f)$. He then noted that this characterization does *not* rely on the notion of natural number. This remark seems to be aimed at authors like Skolem and Weyl who asserted that set theory is less fundamental than arithmetic. For Zermelo, set theory is sufficient to serve as a "foundation"—as a theory in which to (re)formulate other mathematical theories, including arithmetic.

---

[23]Zermelo's lecture is published in *Moore 1980*, section 10.2.

[24]The irony in Zermelo's remark is that today there is a (more or less) generally accepted mathematical logic, classical first-order predicate calculus, but he would be reluctant to appeal to it, because he thought it too weak. An appeal to first-order logic would play into Skolem's hands.

Skolem (*1930*) was quick to respond in print, pointing out that his own characterization of *definit* (in *1922*) is quite similar to Zermelo's new one. The difference, of course, is that Skolem did not allow the clause concerning second-order variables—variables ranging over propositional functions. He took this move to be obscure and, possibly, contradictory. He asked whether Zermelo intended to characterize "propositional functions" by composing even more axioms. Of course, if Zermelo did this, and the axioms were first-order, the Löwenheim-Skolem theorem would apply.[25] Skolem then repeated his thesis of set-theoretic relativity. In this context, the claim is that an unequivocal notion of *definit* property, as Zermelo surely intended it, *cannot* be characterized by axioms. This, it seems, is the closest that Skolem came to arguing for his insistence on first-order languages. Perhaps the underlying thesis is that any notion that is not cast in a first-order language is too vague or obscure to be useful in foundational studies.

In his rejoinder, Zermelo (*1930*) held his ground. A second-order version of ZF set theory was formulated.[26] In describing the theory, Zermelo presented the cumulative hierarchy in virtually its present form. He then proved all models of his system are characterized up to isomorphism by two numbers, chosen independently of each other. One is the number of urelements (a cardinal) and the other is the structure of the inaccessible cardinals (an ordinal). This is correct for current second-order ZFC, but, of course, not for current first-order ZFC.

The dialogue ended with this standoff/regress. In an expository article during this period (*1928*), Skolem made a few remarks on second-order logic. After developing and explaining a language with first-order variables and quantifiers, he showed how variables ranging over "propositional functions" could be introduced; and he raised the possibility of quantifiers over these second-order variables. I repeat the passage at the top of this paper.

> If "all" and "there exists" are applied to variable propositional functions, the question arises: what is the totality of all propositional functions?

The latter question, of course, is the sort of thing Zermelo did not think had to be raised. Once a domain is fixed, the range of the locution "all subsets" (or "all propositional functions") is determined. After all, similar questions are not raised about the first-order variables and the other logical

[25]In fact, around this time, Hilbert and Ackermann (*1928*) did publish an axiomatic treatment of propositional functions, an ω-order logic.

[26]This version contained the axiom of replacement and the axiom of foundation. The axiom of choice was not included, however, but it seems to have been assumed as a general logical principle. He did not develop the meta-theory, but it might be noted that in current second-order ZF, a (global) principle of choice can be proved if it is assumed in the meta-theory. In discussing the models of the theory, Zermelo also seemed to accept a thesis to the effect that the collection of inaccessible cardinals is not bounded.

terminology. Moreover, from this perspective, the only way to answer such questions is to *use* the corresponding terminology in the meta-language.

But for Skolem the question is legitimate and pressing. He asserted (in *1928*) that only two conceptions for the range of the higher-order variables are "scientifically tenable". The first is, in effect, ramified type theory, in which propositional functions are associated with formulas of an expanding language. The second formulation is to introduce the notion of propositional function axiomatically:

> The axioms will then become first-order propositions, since the ... "propositional functions" ... will assume the role of individuals. The relation between arguments and functions will then appear ... as primitive.

This is a foreshadowing of non-standard semantics. Of course, these "first-order propositions" would be subject to the Löwenheim-Skolem theorems and would thus have a variety of "interpretations". Second-order logic would itself suffer from the "relativity" of set-theoretic notions.

In sum, the clash between Zermelo and Skolem was rather fundamental. Skolem took the second-order terminology to be obscure and in need of further axiomatization, which would have to be first-order. That would allow various "interpretations" of the terminology—the result being a confirmation of relativism. Zermelo took the higher-order terminology to be clear and not to stand in need of further "foundation". Such terminology is an essential part of the very framework for doing foundations.

Of course, neither propositional functions nor sets are the first controversial entities to be studied by mathematicians. Negative, irrational, and complex numbers come readily to mind. In such cases, there are, to speak (very) roughly, three different stances that have been taken by proponents. Here, however, there are troubling circularities at every turn. The first orientation is simply to *postulate* the existence of the entities. If any axioms are given, they are taken to *describe* the postulated entities. One insists on some autonomy between the axioms and the entities themselves, and, thus, alternate interpretations of the axioms are irrelevant. It need hardly be mentioned that postulation, by itself, is not going to convince the wary. It begs the question if anything does. Often, postulation is accompanied by arguments concerning the usefulness of the entities, or the fruitfulness of the resulting theory. Much of Zermelo's and Cantor's writings fit this mold. However, the fruitfulness would apply to *any* structure that satisfies the axioms. Thus, the second possibility is *implicit definition.* One first gives axioms and then holds that the defined entities "are" or "can be" anything that satisfies them. In arithmetic, for example, Dedekind (*1888*) defines a "simply infinite system" to be any collection of objects that has an operation with certain properties. He then gave a categoricity proof and defined "the natural numbers" to be one such system. Skolem's second "scientifically tenable" attitude toward propositional functions, the axiomatic treatment, is an implicit definition. The problem here is that the very issue at hand concerns just which systems satisfy which axioms. Sets, properties, propositional functions, or the like, are central items in that very enterprise. We can't do model theory without something like set theory. If the *axioms* used in the definitions are taken to be first-order, as Skolem explicitly intended, then many (non-isomorphic) systems are so defined, and there is no way to distinguish a preferred interpretation. This is relativism. If the axioms are (standard) second-order, as Zermelo intended, there is no relativism, but the implicit definitions are circular. One "defines" the entities in question by using those same entities. The third outlook toward problematic entities is *construction*. One shows how the objects in question can be taken as combinations of less problematic entities. Examples include the Dedekind construction of real numbers as sets of rationals, and the "definition" of complex numbers as pairs of reals. Skolem's first conception of propositional functions along the lines of ramified type theory is of this form. A series of languages is constructed and propositional functions are associated with formulas thereof. From the perspective of the advocates of higher-order logic, this construction is inadequate. Only countably many propositional functions are constructed this way. More importantly, virtually every construction appeals to the intuitive notion of set and, as Skolem acknowledged, natural number. Again, these are the "problematic" entities under study.[27] As we have seen, Zermelo explicitly rejected "construction" as incompatible with axiomatic method.

2.3. Gödel

Influenced by the methodology of the Hilbert school, but himself no finitist, Gödel established the completeness of first-order logic in his 1929 dissertation (*Gödel 1930*) and the incompleteness of arithmetic a year later (*Gödel 1931*). The former, of course, only applies to the logic of first-order languages while the latter concerns virtually every recursive axiomatization of arithmetic (and implies the incompleteness of second-order logic). Gödel soon joined Skolem as a proponent of first-order axiomatizations.

In a lecture of 1931—after he became aware of the incompleteness theorem—Zermelo developed a radically new perspective on the relationship between mathematics and logic (*Zermelo 1931*, see *Moore 1980*). He deplored what he called "Skolemism, the doctrine that *every* mathematical theory ... is satisfiable in a countable model", and he rejected the relativism of

[27] Skolem (*1928*) states that both of his "scientifically tenable" characterizations (of propositional functions) make essential use of the notion of natural number. He then concedes that we cannot characterize *this* notion without using propositional functions (in the induction axiom). He concludes, "The attempt to base the notions of logic upon those of arithmetic, or vice versa, seems to me to be mistaken. The foundations of both must be laid simultaneously and in an interrelated way."

set theory. This is certainly not a case of a stubborn refusal to accept the Löwenheim-Skolem theorems. In effect, Zermelo rejected the thesis that all mathematical theories should be formulated in first-order languages. He derided the view as the "finitistic prejudice":

> ... the true subject matter of mathematics is *not*, as many would have it, 'combinations of signs', but *conceptually-ideal* relations between elements of a conceptually determined *infinite manifold*. Thus our system of signs is always an *incomplete* device...

Zermelo went on to propose a massive infinitary logic. In accordance with the algebraic school, quantified formulas were to be replaced by infinite conjunctions and disjunctions, but he went on to allow for each ordinal α, formulas of length α. In this context, Zermelo defined "proof" in terms of logical consequence, and was careful to formulate the relation of implication as well-founded. He took it as crucial that every true sentence be "provable", and proposed that Gödel's incompleteness theorem relied on an overly restrictive conception of quantification and proof. In fact, Zermelo argued that Gödel's reasoning shows that any finitistic notion of "proof" is inadequate. In short, compactness is an unacceptable property of logic.

Over a period of six weeks, Zermelo wrote to Gödel, Gödel responded, and Zermelo wrote once more.[28] After an initial confusion on Zermelo's part was cleared up, the two seemed to be at cross purposes. Gödel insisted on a finitistic and first-order meta-language, from which the essential incompletability of arithmetic could be established. Zermelo suggested "new methods of proof", along the lines of his infinitary system. He was clear on the source of the problem:

> ... a 'finitistically restricted' proof-schema *does not* suffice to 'decide' the propositions of an uncountable mathematical system ... For what 'proof' is cannot be 'proved', but must be *presupposed* or *assumed* in some form ... What does one understand by a 'proof'? In complete generality one understands thereby *a system of propositions such that from the assumption of the premises the validity of the conclusion can reasonably be asserted*. And still the question remains, what is 'reasonable'? In any case, *not merely* (and this you [Gödel] have shown) the propositions of a finitistic schema ... from the beginning, I have been working on the basis of a *more general* schema, ...
>
> (translated in *Moore 1980*, page 128)

Gödel's proof is Zermelo's *reductio ad absurdum* against finitary, first-order languages.

It is ironic that, according to Wang (*1987*), Gödel accepted Zermelo's conclusions, or at least something close to them (see also Gödel's correspondence reported in *Wang 1974*). However, Gödel held that this part of logic,

[28]The last two letters are published in *Grattan-Guinness 1979*. The first is reprinted (in Zermelo's hand) and translated in *Dawson 1985*.

the study of deductive systems, should focus on the *formal*. Thus, for example, proofs within object-language systems should be mechanically checkable. This rules against the second-order consequence relation and the consequence relation enumerated by Zermelo. But Gödel also held that the human ability to understand and work with mathematical concepts goes beyond the mechanical. As far as I know, Gödel never objected to higher-order Euclidean geometry, or even higher-order set theory, as a viable branch of mathematics. We use the theories, work in them, and understand what we do. It is only that these theories are not *formal*, and so not in the purview of logic.

Zermelo published a few articles developing his infinitary logic, but did not carry it very far. In 1935 he was fired for failing to salute Hitler and his proposal was not pursued by anyone else until, perhaps, the revival of infinitary languages in the 1950's.

In a conference in 1938, Skolem returned to his discussion of the Löwenheim-Skolem theorems and extended his relativism to mathematics as a whole (*Skolem 1941*). At the same conference, Bernays still resisted both the relativism and the restriction to first-order languages. He suggested that the limitations of the Löwenheim-Skolem theorems are due to the "restrictiveness" of the formalism and proposed that this reveals "a certain inadequacy of the method under discussion ... for making axiomatizations precise" (*Gonseth 1941*).

Gödel, of course, did not accept the relativism, but he did insist on a first-order language. This combination prevailed. From this point, the explicit controversy over higher-order languages subsided, and most logicians began to accept the Skolem-Gödel proposal that only first-order languages are appropriate for their work.[29] I would suggest that this consensus was not based on a philosophy of foundational studies. It was more of a research program, suggesting that first-order model theory is the best place to focus intellectual attention. Perhaps one can say that first-order logic became a Kuhnian paradigm. Given the apparent lack of explicit arguments from the major figures in the first-order camp, one can only speculate on the reasons.[30]

[29] For example, the vast majority of textbooks in logic written after 1940 hardly mention higher-order terminology. Two (very) notable exceptions are *Church 1956* and *Boolos and Jeffrey 1980*. The debate over *relativism* continued throughout Skolem's career. See, for example, the exchange between Tarski and Skolem in *Skolem 1958* (reprinted in part in *Moore 1988*).

[30] Surely, Gödel's completeness theorem makes first-order systems attractive for many purposes. The theorem shows, in effect, that the proof theory and the semantics of first-order systems are a perfect match. This entails that there is but one consequence relation, one sort of "independence" statement, etc., and it allows one to (automatically) transfer results about the semantics of a system to its proof theory, and vice versa. Thus, one can reliably study semantics in order to shed light on provability. The emphasis on proof theory (and the formulation of the completeness problem) is traced to the Hilbert program, which of course predates Gödel's work. By contrast, the second-order consequence relation cannot be axiomatized. So it may be that first-order languages were adopted in part because the resulting proof theory is more fruitful, more conceptually clear, or more elegant.

In any case, the range of the locution "all subclasses" was (and is) widely regarded as sufficiently problematic (or insecure) to suggest that the corresponding terminology be non-logical. This amounts to an insistence that only the resources of first-order languages and semantics are reliable enough to serve logic. I don't know the extent to which this ambivalence toward class terminology is the cause or the effect of the "triumph" of first-order languages.

To close this section, it is to be noted that higher-order logic never quite died out. For example, interest in type theory remained alive, perhaps prompting Skolem (*1961*) to reiterate his support for its interpretation as a many-sorted first-order system (referring to *Gilmore 1957*, see also *Hazen 1983*). *Church 1951* is a straightforward higher-order system, and, by way of transition to the next section, *Church 1956* contains a discussion of second-order logic.

3. THE PRESENT STATE

This final section consists of a few remarks on the contemporary controversy concerning second-order languages. From one perspective, the discussion continues the border dispute over what I call the "logical" conception of set. In those terms, however, the question does not seem to amount to much. The best resolution, perhaps, is simply to hold that there is no sharp border between logic and mathematics. The analogy, of course, is with political boundaries but, unlike those, our lines can be drawn and moved at will (unless emotions of sovereignty run high). As we have seen, the deeper issues concern the epistemic and semantic status of the higher-order terminology, especially the ranges of the variables. Advocates of first-order logic hold that second-order terminology is not sufficiently clear, or is otherwise inappropriate to play a role in logic. In some cases, the underlying thesis is that notions used in logic should not have any possible obscurity or avoidable uncertainty. Accordingly, concepts used in the formulation and study of correct reasoning should be self-evident. A less encompassing thesis may simply be that terminology for *sets* is too obscure or uncertain to serve logic. The set-theoretic antinomies are still used to defend this claim.

The central opponent of higher-order systems is W. V. O. Quine. A full elaboration of the attack on second-order logic occurs in his *1986*, but his critique dates back (at least) to his *1941*. In the earlier paper, Quine argues that the intensional character of the attributes in *Principia Mathematica* make them too obscure for foundational work, or for any scientifically respectable study. He proposes that attributes be supplanted by sets. In most contemporary higher-order systems, properties and classes are extensional (see *Shapiro 1991*). In this respect, they resemble sets, and so the development of higher-order logic has followed Quine's advice. He argues, however, that in invoking set-like entities, we have crossed a significant border, out of logic and into mathematics. His thesis is that so-called "second-order logic" is "set theory in disguise". He points out that some substantial set-theoretic theses can be formulated in second-order logic. There is, for example, a second-order sentence that is a "logical truth" if and only if the continuum hypothesis is true (see Shapiro *1991*, chapter 5).

It should be noted that Quine is not a critic of set theory (unlike, say, Skolem's early *1922*). He was (and remains) an advocate of the theory, and has contributed to its development. The claim here is that set-terminology does not belong to *logic*. To state the obvious, Quine (both *1941* and *1986*) follows Skolem's later work in taking set terminology and membership to be non-logical. The idea seems to be that logic, properly so-called, should be topic neutral and should not presuppose anything about any subject matter. There are no "logical objects" or, in other words, logic should not have an ontology:[31]

> Set theory's staggering existential assumptions are... hidden... in the tacit shift from schematic predicate letter to quantifiable set variable.
> (*Quine 1986*, page 68)

Notice the divergence between this view of logic and that held by both the algebraic school and the logicists, all of whom took sets, or classes (or properties, or attributes) to be under the purview of logic, as did many of the mathematicians. As we have seen, historically the "shift" went in the direction opposite to that suggested by Quine. It went from the quantifiable variables of higher-order logic to the schematic, non-logical predicate letters of first-order logic.

It is curious that another prominent Quinean theme is the lack of a sharp border between respectable academic disciplines. He speaks often of our scientific enterprise as a "seamless web". Along these lines, Quine shows that mathematical concepts, involving a "commitment" to mathematical ontology, are essential to all of the sciences. Why should logic, especially the logic *of mathematics* be different? Why is it important to decide where logic leaves off and mathematics properly begins (see *Shapiro 1991*, chapters 5 and 8)?

The border skirmish between logic and set theory was joined by Tharp (*1975*). Noting that first-order logic is compact and complete, whereas second-order logic is neither, he argued that other things equal, first-order logic ought to be preferred.

Arguments like this are sometimes bolstered by a theorem, proved in *Lindström 1969*, that first-order logic is the only logic that is both compact and has the (downward) Löwenheim-Skolem property.[32] It is interesting to speculate on why these properties (and completeness) are regarded as desirable.

[31] A similar attack on second-order logic was recently made by Hartry Field (*1984*), who suggested that logic should not have substantial existence theorems. Field embraces a Kantian thesis that there should be no *a priori*, or purely logical, arguments for the existence of anything (e.g., *Critique of pure reason* B622–623 on "ontological arguments").

[32] Lindström's theorem concerns semantics. Let L be an extension of first-order logic

The completeness theorem indicates a perfect match between proof theory and semantics (see note 30). It follows that the first-order semantic consequence is axiomatizable, and, thus, the calculi developed in the proof theory are the best possible. The "cost", however, lies in the expressive resources of the languages (through the semantics). First-order languages are inadequate to the task of *describing intended* or *standard* models of various theories. Wang (*1974*, page 154) agrees:

> When we are interested in set theory or classical analysis, the Löwenheim-Skolem theorem is usually taken as a sort of defect (often thought to be inevitable) of the first-order logic. Therefore, what is established (by Lindström's theorems) is not that first-order logic is the only possible logic but rather that it is the only possible logic when we in a sense deny reality to the concept of uncountable...

So does Barwise and Feferman (*1985*, page 5):

> As logicians, we do our subject a disservice by convincing others that logic is first-order and then convincing them that almost none of the concepts of modern mathematics can really be captured in first-order logic.

Somewhat consistently, Skolem insisted (in his later work) that, in effect, most mathematical theories *do not have* unique or unequivocal "standard" interpretations (even up to isomorphism). Such is relativism. But, as far as I know, Tharp did not embrace relativism, nor do most contemporary advocates of first-order languages (see also *Myhill 1951*). Quine, for his part, has articulated and defended a quite general relativism in ontology, but not on Skolemite grounds. In fact, he has criticized the Skolem paradox in several writings. An extended treatment of the connections between Quinean ontological relativity, Skolemite relativism, and Quine's views on first-order logic would go beyond the scope of this paper.

Meanwhile, a few voices continued the tradition of Hilbert, Zermelo, and Bernays, arguing that first-order languages do not capture crucial aspects of pre-formal mathematical practice. A clear and forceful example is the last chapter of *Church 1956,* devoted to second-order logic. Either explicitly or through exercises, Church shows how many mathematical theories have natural and straightforward formulations in higher-order systems, and that first-order counterparts are awkward and inadequate. He points out that due to incompleteness, second-order semantic notions, like consequence,

---

that has the same class of interpretations. Then L has the *downward Löwenheim-Skolem property* if and only if for each sentence Φ of L, if Φ has a model with an infinite domain, then Φ has a model with (at most) a countable domain; L is *compact* if and only if for each set S of sentences of L, if each finite subset of S has a model, then S itself has a model. Lindström's theorem is that if L is compact and has the downward Löwenheim-Skolem property, then for every sentence Φ of L there is a first-order sentence Φ' such that Φ and Φ' have exactly the same models, i.e., $\models \Phi \equiv \Phi'$. See *Shapiro 1991*, section 6.5.

satisfaction, and validity, must be distinguished from their proof-theoretic counterparts of deductive consequence, consistency, and theoremhood. The important semantic ones are not axiomatizable. He concedes in a footnote (page 326n) that this raises doubts in some minds. Church agrees that there would be cogent objections to a logic whose very grammar is not effective. There would be no mechanical way to determine if a given string is a well-formed-formula. But objections to a non-effective consequence relation are on a different level:

> Objections may indeed be made to this new point of view, on the basis of the sort of *absolutism* it presupposes ... But it should be pointed out that this ... is already inherent in classical mathematics generally, and it is not made more acute or more doubtful, but only more conspicuous, by its application to theoretical syntax [i.e., logic]. For our definition of the consequences of a system of postulates ... can be seen to be not essentially different from [those] required for the ... treatment of classical mathematics ... It is true that the noneffective notion of consequence, as we have introduced it ... presupposes a certain absolute notion of ALL propositional functions of individuals. But this is presupposed also in classical mathematics, especially classical analysis, and objections against it lead to such modifications ... as intuitionism ...

Except for Weyl, none of the aforementioned critics of higher-order logic favor intuitionism. All accept classical mathematics as it stands.[33]

Another widely read item was Kreisel's "Informal rigour and completeness proofs" (*1967*). The attack is directed at first-order versions of theories, like arithmetic and analysis (and set theory), which have schemes with infinitely many instances in place of second-order axioms. It is argued that such theories allow only unnatural epistemologies and are not true to the informal meaning—the "informal rigor"—of actual mathematical practice (see my *1991*, section 5.3).

One of Kreisel's arguments can be seen as an elaboration of Hilbert's thesis that first-order theories are inadequate to the task of illuminating the *interactions* between mathematical theories. The use of schemes depends on the *language* of the theory, and the precise set of non-logical terms chosen. Thus, for example, instances of the first-order induction scheme only contain the terminology of one particular language of arithmetic, say one with the successor function alone. The introduction of new terminology (such as symbols for addition and multiplication, or variables ranging over sets of numbers)

[33]Church uses the word "Platonism" to describe the attitude shared by the employment of the second-order consequence relation and the pursuit of classical mathematics generally (referring to *Bernays 1935*). This can be misleading. The matters at hand are methodological, concerning things like non-effective relations, the law of excluded middle, and impredicative definition. The issues surrounding Platonistic epistemology are not raised, nor, for that matter, are traditional matters of ontology.

yield new instances of the *axioms themselves*, and, thus, constitute a "new" theory. A more natural outlook is to see these as the addition of new terms to an old theory. Similarly, first-order arithmetic and first-order analysis are presented as isolated theories which are independent of both each other and the rest of mathematics. This is not in accord with the practice of viewing such structures as interrelated, a practice manifest in the common technique of "embedding" or "modeling" one structure in another. Kreisel (*1967*, page 166) wrote:

> ... very often the mathematical properties of a domain D become only graspable when one embeds D in a larger domain D'. Examples: (1) D integers, D' complex plane; use of analytic number theory. (2) D integers, D' p-adic numbers; use of p-adic analysis. (3) D surface of a sphere, D' 3-dimensional space; use of 3-dimensional geometry. Non-standard analysis [also applies] here...

To follow the first of these examples, when one realizes that the complex plane (or for that matter, the set-theoretic hierarchy) contains the natural numbers—or an "isomorphic copy" of the natural numbers—then one can use complex analysis (or set theory) to shed light on the natural numbers. That is, since isomorphic structures have the same set of truths, a theorem of complex analysis that refers only to the "natural numbers" of the complex plane is *true of* the natural numbers. It is well known that techniques like this yield results that are not provable in the original theories. From the first-order perspective, however, there does not seem to be any reason to regard the new theorems as results of arithmetic or as truths *about* the natural numbers. Why think, for example, that the "natural numbers of complex analysis" really are isomorphic to the natural numbers? Indeed, the completeness theorem indicates that there are models of the original theory in which the new theorems are false. In short, "the natural numbers" has an informal interpretation that outruns what is captured in first-order arithmetic, and this informal interpretation plays a central role in actual mathematical practice. Along these lines, Montague (*1965*) proposed that the informal, but much used, notion of "standard model" of arithmetic, analysis, and even set theory is explicated by the notion of "model of the (respective) second-order theory".

A second observation that Kreisel makes is that each first-order theory is tied too closely to the terminology in use. For example, first-order arithmetic with the successor symbol alone is a rather simple theory. It is complete and its models are easily characterized. First-order arithmetic with addition and multiplication symbols is another matter entirely, as evidenced by the incompleteness theorems. Consequently, addition and multiplication are not definable in the former theory. It might be added that addition and multiplication have straightforward definitions in second-order arithmetic with successor alone. Dedekind's (*1888*) own characterization, which explicitly includes definitions of addition and multiplication, is straightforwardly second-order.

John Corcoran (*1973*) and George Boolos (*1975*) engaged in studies of aspects of mathematical practice that resist first-order treatments. Boolos went on to challenge Quine's claim that the ontological presuppositions of second-order logic are excessive. He pointed out that an interpretation of a second-order theory is exactly the same as an interpretation of a first-order theory—a domain of discourse together with interpretations of the non-logical terminology. The difference, of course, is that once a domain is fixed, second-order theories have variables ranging over its subdomains (and relations). That is, in conformity with what I call the "logical conception" of set, the variables range over the subclasses of a fixed domain. In two recent articles (*1984, 1985*), Boolos proposed that second-order quantifiers be regarded as "plural quantifiers" are in ordinary language, and has developed a semantics along these lines. The claim is that, understood this way, second-order variables and quantifiers introduce *no* ontology beyond that of a corresponding first-order theory. This interpretation is employed by a number of philosophers interested in ontology. Even with traditional interpretation, however, the "increase" in ontology relative to a first-order theory is not severe. The first-order theories have variables ranging over a domain; the second-order theories also have variables ranging over all subdomains (and relations). There is no "powerset" operator that gets iterated twice, let alone into the transfinite (unless, of course, such an operator is part of the nonlogical terminology). The "staggering existential assumptions" that Quine mentions apply to the iterative conception of set, not to logic.

My own contributions (*1985, 1990, 1991*), on the higher-order side, elaborate and extend Church's note that notions directly represented in second-order languages are implicit in mathematical practice and that the epistemic and semantic presuppositions of second-order logic are no greater than those of classical mathematics. I propose that categorical characterizations are necessary to capture the informal communication of mathematics (see also *Corcoran 1980*), both the manifestly successful description of certain structures (like that of the natural numbers or the real numbers) and certain concepts like finitude, well-foundedness, and minimal closure, that resist first-order treatment. I also develop and extend the Hilbert/Kreisel view concerning the significance of higher-order languages in capturing the relationships between mathematical theories. In short, higher-order logic provides the best explication of classical mathematics.[34]

This is not to say that the consequence relation of second-order logic is completely understood, or intuitively evident, nor is it *as* evident as first-order consequence. There are, to be sure, deep philosophical and mathe-

---

[34] *Shapiro 1991* (chapters 3 and 4) also contains a detailed presentation of the languages, semantics, and deductive systems of higher-order logic.

matical problems associated with second-order logic. For example, there are problems concerning reference and problems concerning how the language can be learned. But these are present in classical mathematics as well, and most of the proponents of first-order logic are not willing to give that up, or even to modify it or take it at other than face value. To echo Church, the problems are more perspicuous in the case of higher-order logic, but not more troublesome.[35] The thesis is that the language *and semantics* used to formulate various mathematical theories should be in line with the (presumed) descriptive and communicative power of the language of informal mathematics.

The contemporary arguments in favor of higher-order logic are not successful against a thorough relativism like that of Skolem (and perhaps von Neumann). These views, in effect, deny outright that these structures and concepts are unequivocal. Accordingly, the failure of the categoricity of first-order theories accurately reflects the philosophical situation. On the other hand, at least some advocates of first-order languages, especially among mathematicians, reject Skolemite relativism and, thus, presumably agree that we do succeed in the description (up to isomorphism) and communication of at least some infinite structures. The attitude is that *informal* mathematics (i.e., informal semantics) is sufficient to describe and communicate the structures and concepts in question, but formal semantics must fail where informal practice succeeds. Myhill (*1951*) develops this theme explicitly.

On the other hand, Skolemite relativism is not dead, among philosophers. A common response is that the (semantic) "benefits" of second-order languages are the results of the (naive) set theory hidden in the meta-language. This is an elaboration of von Neumann's (*1925*) remark that the categoricity of, say, Euclidean geometry, is won by temporarily ignoring problems with the underlying set theory. Some writers, such as Weston (*1976*), claim that the non-categoricity that pervades first-order systems reappears in the *set-theoretic* semantics of higher-order languages. These objectors propose that with infinite sets, the powerset operation is not unique in an absolute sense—different models of set theory have different powerset operations. If this is correct, then the very semantics for second-order languages is itself not unique. In fact, there would be a wealth of such semantics—one for each "powerset". Notice that these considerations are a form of skepticism, and it is well known that skepticism is difficult to refute, especially on its own terms.

Another suggestion is that it would be more natural, and more revealing,

[35]Completeness is not a cure-all for the sorts of philosophical problems indicated here. First-order logic is not immune, nor is propositional logic. One group of issues concerns the extent to which explicit or implicit, formal or informal, rules for using a language determine what it "means". Present attention is focused on the consequence relation. Carnap (*1943*) shows in great detail that even the eminently complete deductive system for *propositional* logic is consistent with different "interpretations".

to formulate various mathematical theories within (first-order) set theory. This re-raises the issue concerning the border between mathematics and its logic, and issues concerning the understanding of set theory. First-order set theory has a wealth of non-standard models.

No doubt, the consequence relation of second-order logic is not a suitable ingredient in a strict rationalist or foundationalist program (see *Shapiro 1991*, chapter 2). First-order logic probably isn't either, because the most illuminating formulation of its consequence relation is set-theoretic (as, for example, the completeness theorem requires the axiom of infinity), but perhaps it is closer. One who takes logic to be the last outpost of foundationalism, or rationalism, might be inclined toward first-order logic. However, Quine and his followers are thoroughly anti-foundationalist, even concerning logic (see, for example, *Quine 1960*).[36]

To sum up, the period between, say, the 1930's and the present has produced a detailed elaboration of the various positions and, consequently, not as much discussion is obviously at cross-purposes. Nevertheless, the present stand-offs are not all that different from their historical counterparts. The higher-order advocates claim that what I call the "logical conception" of set—variables ranging over the subclasses of a fixed domain—is sufficiently clear for use in logic and, moreover, the use of such terminology is essential to at least part of the foundational enterprise. Advocates of first-order logic reject the use of set terminology in logic, regarding it as problematic or as not sufficiently topic-neutral. The alternative, once again, is that membership be treated as non-logical and that the ranges of the set-variables be regarded by the underlying semantics as indeterminate. The task of characterizing structures and the "higher-order" concepts (like finitude and well-foundedness) is passed to *first-order* set theory (pretty much as Skolem formulated it). The latter, of course, has a steep ontology (as conceived informally), but it is associated with the mathematical theory of sets, not with logic. Of course, first-order set theory is itself subject to the Löwenheim-Skolem theorems and thus has a wealth of non-standard models. The "first-order" options are

[36]See *Wagner 1987* for a lucid elaboration of rationalism in logic. The historical and philosophical connections between "foundations of mathematics" and "foundationalism" are worth pursuing, but it would take us too far afield here (see *Shapiro 1991*, chapter 2). In *1950*, Skolem sketched several possible "goals" of foundational research in mathematics. The "logicistic" desire is "... to obtain a way of reasoning which is logically correct so that it is clear and certain in advance that contradictions will never occur, and what we prove are truths in some sense." This, of course, is foundationalism. A more modest desire is "... to have a foundation which makes it possible to develop present day mathematics, and which is consistent so far as is known yet." This is called the "opportunistic" outlook. Skolem states that it has the "unpleasant" feature that we are never certain when the foundational work is complete: "We are not only adding new floors at the top of our building, but from time to time it may be necessary to make changes at the basis." Skolem's third "desire" is the "Hilbert program", a valiant attempt at "giving up the logicistic standpoint and not being content with the opportunistic one". Skolem's preference for foundationalism is evident, but he was well aware of its difficulties. The contrast with Quine is striking.

either to embrace relativism and deny that there is a single range to the locution "all subclasses" (after a domain is fixed) or to insist that formal semantics must ignore the inherent success of mathematical understanding, practice, and communication. In short, the foundational "goals" for which higher-order terminology is invoked are regarded as unattainable in formal semantics, either because they are not attainable at all—there is nothing unequivocal to describe—or because communication is attained "only informally".

The stand-off is particularly frustrating because there seems to be no common standpoint from which to evaluate or even discuss the matter. What one side regards as an essential, integral part of the whole foundational enterprise—including the enterprise of "logic choice"—the other side rejects as obscure or problematic.

Stewart Shapiro
Ohio State University
Department of Philosophy
350 University Hall
230 North Oval Mall
Columbus, Ohio 43210 USA
Email: shapiro+@osu.edu

REFERENCES

Ackermann, W.

1924 Begründung des 'Tertium non datur' mittels der Hilbertschen Theorie der Widerspruchsfreiheit, *Mathematische Annalen*, vol. 93, pp. 1–36.

Barwise, J., and S. Feferman

1985 *Model-theoretic logics,* Springer-Verlag, New York.

Benacerraf, P.

1985 Skolem and the skeptic, *Proceedings of the Aristotelian Society, Supplementary Volume 59*, pp. 85–115.

Bernays, P.

1918 *Beiträge zur axiomatischen Behandlung des Logik-Kalküls*, Habilitationsschrift, Göttingen.

1928 Die Philosophie der Mathematik und die Hilbertsche Beweistheorie, *Blätter für Deutsche Philosophie*, vol. 4, pp. 326–367.

1935 Platonism in mathematics, *Philosophy of mathematics* (P. Benacerraf and H. Putnam, editors), second edition, Cambridge University Press, Cambridge, England, 1983, pp. 258–271.

Bernays, P., and M. Schönfinkel

1928 Zum Entscheidungsproblem der mathematischen Logic, *Mathematische Annalen*, vol. 99, pp. 342–372.

Boole, G.
1847 *The mathematical analysis of logic, being an essay toward a calculus of reasoning,* London.
1854 *An investigation of the laws of thought on which are founded the mathematical theories of logic and probabilities,* London.

Boolos, G.
1975 On second-order logic, *The Journal of Philosophy,* vol. 72, pp. 509–527.
1984 To be is to be a value of a variable (or to be some values of some variables), *The Journal of Philosophy,* vol. 81, pp. 430–449.
1985 Nominalist Platonism, *The Philosophical Review,* vol. 94, pp. 327–344.

Boolos, G., and R. Jeffrey
1980 *Computability and logic,* second edition, Cambridge University Press, Cambridge, England.

Carnap, R.
1943 *Formalization of logic,* Harvard University Press, Cambridge, Massachusetts.

Carnap, R., and F. Bachmann
1936 Über Extremalaxiome, *Erkenntnis,* vol. 6, pp. 166–188; translated by H. G. Bohnert, *History and Philosophy of Logic,* vol. 2, 1981, pp. 67–85.

Church, A.
1951 A formulation of the logic of sense and denotation, *Structure, method and meaning* (P. Henle *et al.*, editors), Liberal Arts Press, New York, pp. 3–24.
1956 *Introduction to mathematical logic, vol. 1,* Princeton University Press, Princeton, New Jersey.

Corcoran, J.
1973 Gaps between logical theory and mathematical practice, *The methodological unity of science* (M. Bunge, editor), Reidel, Dordrecht, pp. 23–50.
1980 Categoricity, *History and Philosophy of Logic,* vol. 1, pp. 187–207.

Dawson, J.
1985 Completing the Gödel-Zermelo correspondence, *Historia Mathematica,* vol. 12, pp. 66–70.

Dedekind, R.
1888 The nature and meaning of numbers, *Essays on the theory of numbers* (W. W. Berman, editor), Dover, New York, 1963, pp. 31–115.

Etchemendy, J.
1988 Tarski on truth and logical consequence, *The Journal of Symbolic Logic,* vol. 53, pp. 51–79.

Field, H.
1984 Is mathematical knowledge just logical knowledge?, *The Philosophical Review,* vol. 93, pp. 509–552.

Fraenkel, A.
1921 Über die Zermelosche Begründung der Mengenlehre, *Jahresbericht der Deutschen Mathematiker-Vereinigung,* vol. 30, second section, pp. 97–98.
1922 Zu den Grundlagen der Mengenlehre, *Jahresbericht der Deutschen Mathematiker-Vereinigung,* vol. 31, second section, pp. 101–102.

1922a Der Begriff 'definit' und die Unabhängigkeit des Auswahlaxioms, *Sitzungsberichte der Preussichen Akademie der Wissenschaften, Physikalisch-mathematische Klasse*, pp. 253–257; translated in *van Heijenoort 1967*, pp. 284–289.

1925 Untersuchungen Über die Grundlagen der Mengenlehre, *Mathematische Zeitschrift*, vol. 22, pp. 250–273.

Fraenkel, A., Y. Bar-Hillel, and A. Levy

1973 *Foundations of set theory*, second revised edition, North Holland, Amsterdam.

Frege, G.

1879 *Begriffsschrift, eine der arithmetischen nachgebildete Formelsprache des reinen Denkens*, Louis Nebert, Halle; translated in *van Heijenoort 1967*, pp. 1–82.

1883 Über den Zweck der Begriffsschrift, *Sitzungsberichte der Jenaischen Gesellschaft für Medicin und Naturwissenschaft*, vol. 16, pp. 1–10.

1884 *Die Grundlagen der Arithmetik*, Koebner, Breslau; translated by J. Austin, second edition, Harper, New York, 1960.

1893 *Grundgesetze der Arithmetik 1*, Olms, Hildescheim.

1895 Review of Schröder's *Vorlesungen Über die Algebra der Logik*, *Archiv für systematische Philosophie*, vol. 1, pp. 433–456; translated in *Translations from the philosophical writings of Gottlob Frege* (P. Geach and M. Black, editors), Blackwell, Oxford, pp. 86–106.

Gandy, R.

1988 The confluence of ideas in 1936, *The universal Turing machine* (R. Herken, editor), Oxford University Press, New York, pp. 55–111.

Gilmore, P.

1957 The monadic theory of types in the lower-predicate calculus, *Summaries of talks presented at the Summer Institute of Symbolic Logic at Cornell*, Institute for Defense Analysis.

Gödel, K.

1930 Die Vollständigkeit der Axiome des logischen Funktionenkalkuls, *Montatshefte für Mathematik und Physik*, vol. 37, pp. 349–360; translated in *van Heijenoort 1967*, pp. 582–591.

1931 Über formal unentscheidbare Sätze der Principia Mathematica und verwandter Systeme I, *Montatshefte für Mathematik und Physik*, vol. 38, pp. 173–198; translated in *van Heijenoort 1967*, pp. 596–616.

1944 Russell's mathematical logic, *Philosophy of mathematics* (P. Benacerraf and H. Putnam, editors), second edition, Cambridge University Press, Cambridge, England, 1983, pp. 447–469.

Goldfarb, W.

1979 Logic in the twenties: The nature of the quantifier, *The Journal of Symbolic Logic*, vol. 44, pp. 351–368.

1988 Poincaré against the logicists, *History and philosophy of modern mathematics* (W. Aspray and P. Kitcher, editors), Minnesota Studies in the Philosophy of Science, vol. 11, University of Minnesota Press, pp. 61–81.

Gonseth, F.

1941 *Les entretiens de Zurich, 6–9 décembre 1938,* Leeman, Zurich.

Grattan-Guinness, I.

1977 *Dear Russell — Dear Jourdain,* Columbia University Press, New York.

1979 In memoriam Kurt Gödel: His 1931 correspondence with Zermelo on his incompletability theorem, *Historia Mathematica,* vol. 6, pp. 294–304.

Hazen, A.

1983 Predicative logics, *Handbook of philosophical logic 1* (D. Gabbay and F. Guenthner, editors), Reidel, Dordrecht, pp. 331–407.

Henkin, L.

1950 Completeness in the theory of types, *The Journal of Symbolic Logic,* vol. 15, pp. 81–91.

Hilbert, D.

1899 *Grundlagen der Geometrie,* Teubner, Leipzig; *Foundations of geometry,* (E. Townsend, translator), Open Court, La Salle, Illinois, 1959.

1900 Über den Zahlbegriff, *Jahresbericht der Deutschen Mathematiker-Vereinigung,* vol. 8, pp. 180–194.

1900a Mathematische Problem, *Bulletin of the American Mathematical Society,* vol. 8 (1902), pp. 437–479.

1902 *Les principes fondamentaux de la géometrie,* Gauthier-Villars, Paris; French translation of *Hilbert 1899.*

1905 Über der Grundlagen der Logik und der Arithmetik, *Verhandlungen des dritten internationalen Mathematiker-Kongresses in Heidelberg vom 8. bis 13 August 1904,* Teubner, Leipzig; translated in *van Heijenoort 1967,* pp. 129–138.

1923 Die logischen Grundlagen der Mathematik, *Mathematische Annalen,* vol. 88, pp. 151–165.

1929 Probleme der Grundlegung der Mathematik, *Mathematische Annalen,* vol. 102, pp. 1–9.

Hilbert, D, and W. Ackermann.

1928 *Grundzüge der theoretischen Logik,* Springer, Berlin.

Huntington, E.

1902 A complete set of postulates for the theory of absolute continuous magnitude, *Transactions of the American Mathematical Society,* vol. 3, pp. 264–279.

1905 A complete set of postulates for ordinary complex algebra, *Transactions of the American Mathematical Society,* vol. 6, pp. 209–229.

Kreisel, G.

1967 Informal rigour and completeness proofs, *Problems in the philosophy of mathematics* (I. Lakatos, editor), North-Holland, Amsterdam, pp. 138–186.

Lewis, C. I.

1918 *A survey of symbolic logic,* University of California Press, Berkeley.

Lindström, P.

1969 On extensions of elementary logic, *Theoria,* vol. 35, pp. 1–11.

Löwenheim, L.

1915 Über Möglichkeiten im Relativkalkül, *Mathematische Annalen*, vol. 76, pp. 447–479; translated in *van Heijenoort 1967*, pp. 228–251.

Montague, R.

1965 Set theory and higher-order logic, *Formal systems and recursive functions* (J. Crossley and M. Dummett, editors), North-Holland, Amsterdam, pp. 131–148.

Moore, G.

1978 The origins of Zermelo's axiomatization of set theory, *Journal of Philosophical Logic*, vol. 7, pp. 307–329.

1980 Beyond first-order logic, the historical interplay between logic and set theory, *History and Philosophy of Logic*, vol. 1, pp. 95–137.

1982 *Zermelo's axiom of choice: Its origins, development, and influence*, Springer-Verlag, New York.

1988 The emergence of first-order logic, *History and philosophy of modern mathematics* (W. Aspray and P. Kitcher, editors), Minnesota Studies in the Philosophy of Science, vol. 11, University of Minnesota Press, pp. 95–135.

Myhill, J.

1951 On the ontological significance of the Löwenheim-Skolem theorem, *Academic freedom, logic and religion* (M. White editor), American Philosophical Society, Philadelphia, pp. 57-70; also in *Contemporary readings in logical theory* (I. Copi and J. Gould, editors), Macmillan, New York, 1967, pp. 40–54.

Peano, G.

1889 *Arithmetices principia, Nova methodo exposita,* Turin; translated in *van Heijenoort 1967*, pp. 85–97.

Poincaré, H.

1909 La Logique de l'infini, *Revue de métaphysique et morale*, vol. 17, pp. 461–482.

Quine, W. V. O.

1941 Whitehead and the rise of modern logic, *The philosophy of Alfred North Whitehead* (P. A. Schilpp, editor), Tudor Publishing Company, New York, pp. 127–163.

1960 *Word and object,* The MIT Press.

1986 *Philosophy of logic*, second edition, Prentice-Hall, Englewood Cliffs.

Ramsey, F.

1925 The foundations of mathematics, *Proceedings of the London mathematical society,* series 2, *25*, pp. 338–384.

Rang, B., and W. Thomas

1981 Zermelo's discovery of the 'Russell paradox', *Historia Mathematica*, vol. 8, pp. 15–22.

Resnik, M.

1980 *Frege and the philosophy of mathematics,* Cornell University Press, Ithaca.

Schoenflies, A.

1911 Über die Stellung der Definition in der Axiomatik, *Jahresbericht der Deutsche Mathematiker-Vereinigung*, vol. 20, pp. 222–255.

Schönfinkel, M.

1924 Über die Bausteine der mathematischen Logik, *Mathematische Annalen*, vol. 92, pp. 305–316; translated in *van Heijenoort 1967*, pp. 355–366.

Schröder, E.

1890 *Vorlesungen über die Algebra der Logik, vol. 1*, Teubner, Leipzig.

Shapiro, S.

1983 Remarks on the development of computability, *History and Philosophy of Logic*, vol. 4, pp. 203–220.

1985 Second-order languages and mathematical practice, *The Journal of Symbolic Logic*, vol. 50, pp. 714-742.

1990 Second-order logic, foundations, and rules, *The Journal of Philosophy*, vol. 87, pp. 234-261.

1991 *Foundations without foundationalism: A case for second-order logic*, Oxford University Press, Oxford.

Skolem, T.

1920 Logisch-kombinatorische Untersuchungen über die Erfüllbarkeit oder Beweisbarkeit mathematischer Sätze nebst einem Theoreme über dichte Mengen, *Videnskapsselskapets skrifter I. Matematisk-naturvidenskabelig klasse*, no. 4; section 1 translated in *van Heijenoort 1967*, pp. 252–263.

1922 Einige Bemerkungen zur axiomatischen Begründung der Mengenlehre, *Matematikerkongressen i Helsingfors den 4-7 Juli 1922*, Akademiska Bokhandeln, Helsinki, pp. 217–232; translated in *van Heijenoort 1967*, pp. 291–301.

1928 Über die mathematische Logik, *Norsk matematisk tidsskrift*, vol. 10, pp. 125–142; translated in *van Heijenoort 1967*, pp. 508–524.

1930 Einige Bemerkungen zu der Abhandlung von E. Zermelo: 'Über die Definitheit in der Axiomatik', *Fundamenta Mathematicae*, vol. 15, pp. 337–341.

1933 Über die Unmöglichkeit einer vollständigen Charakterisuerung der Zahlenreihe mittels eines endlichen Axiomsystems, *Norsk matematisk forenings skrifter*, series 2, no. 10, pp. 73–82.

1934 Über die Nicht-charakterisierbarkeit der Zahlenreihe mittels endlich oder abzählbar unendlich vieler Aussagen mit ausschliesslich Zahlenvariablen, *Fundamenta Mathematicae*, vol. 23, pp. 150–161.

1941 Sur la porté du théormè de Löwenheim-Skolem, in Gonseth *1941*, pp. 25–52.

1950 Some remarks on the foundation of set theory, *Proceedings of the international congress of mathematicians (Cambridge, Massachusetts)*, American Mathematical Society, Providence, 1952, pp. 695–704.

1958 Une relativisation des notions mathématiques fondamentales, *Colloques internationaux du Centre National de la Recherche Scientifique*, Paris, pp. 13–18.

1961 Interpretation of mathematical theories in the first-order predicate calculus, *Essays on the foundations of mathematics, Dedicated to A. A. Fraenkel* (Y. Bar-Hillel *et al.*, editors), Magnes Press, Jerusalem, pp. 218–225.

Tharp, L.

1975 Which logic is the right logic?, *Synthese*, vol. 31, pp. 1–31.

Thiel, C.

1977 Leopold Löwenheim: Life, work, and early influence, *Logic colloquium 76* (R. Gandy and M. Hyland, editors), North-Holland, Amsterdam, pp. 235–252.

Van Heijenoort, J.

1967 *From Frege to Gödel*, Harvard University Press, Cambridge, Massachusetts.

1967a Logic as calculus and logic as language, *Synthese*, vol. 17, pp. 324–330.

Veblen, O.

1904 A system of axioms for geometry, *Transactions of the American Mathematical Society*, vol. 5, pp. 343–384.

Von Neumann, J.

1925 Eine Axiomatisierung der Mengenlehre, *Journal für die reine und angewandte Mathematik*, vol. 154, pp. 219–244; translated in *van Heijenoort 1967*, pp. 393–413.

1927 Zur Hilbertschen Beweistheorie, *Mathematische Zeitschrift*, vol. 26, pp. 1–46.

1928 Über die Definition durch transfinite Induktion und verwandte Fragen der allgemeinen Mengenlehre, *Mathematische Annalen*, vol. 99, pp. 373–391.

Wagner, S.

1987 The rationalist conception of logic, *Notre Dame Journal of Formal Logic*, vol. 28, pp. 3–35.

Wang, H.

1974 *From mathematics to philosophy*, Routledge and Kegan Paul, London.

1987 *Reflections on Kurt Gödel*, MIT Press.

Weston, T.

1976 Kreisel, the continuum hypothesis and second-order set theory, *Journal of Philosophical Logic*, vol. 5, pp. 281–298.

Weyl, H.

1910 Über die Definitionen der mathematischen Grundbegriffe, *Mathematisch-naturwissenschaftliche Blätter*, vol. 7, pp. 93–95, 109–113.

1917 *Das Kontinuum*, Veit, Leipzig.

Whitehead, A. N., and B. Russell

1910 *Principia Mathematica*, vol. 1, Cambridge University Press, Cambridge, England.

Zermelo, E.

1904 Beweis, dass jede Menge wohlgeordnet werden kann, *Mathematische Annalen*, vol. 59, pp. 514–516; translated in *van Heijenoort 1967*, pp. 139–141.

1908 Neuer Beweis für die Möglichkeit einer Wohlordnung, *Mathematische Annalen*, vol. 65, pp. 107–128; translated in *van Heijenoort 1967*, pp. 183–198.

1908a Untersuchungen über die Grundlagen der Mengenlehre. I, *Mathematische Annalen*, vol. 65, pp. 261–281; translated in *van Heijenoort 1967*, pp. 199–215.

1929 Über den Begriff der Defintheit in der Axiomatik, *Fundamenta Mathematicae*, vol. 14, pp. 339–344.

1930 Über Grenzzahlen und Mengenbereiche: Neue Untersuchungen über die Grundlagen der Mengenlehre, *Fundamenta Mathematicae*, vol. 16, pp. 29–47.

1931 Über stufen der Quantifikation und die Logik des Unendlichen, *Jahresbericht der Deutsche Mathematische Verein*, vol. 31, pp. 85–88.

RAYMOND SMULLYAN

EQUIVALENCE RELATIONS AND GROUPS

Abstract. Our purpose is to show how the logic of relations can be uitilized in the study of group theory. There are some striking similarities in certain theorems in group theory and certain results about equivalence relations, and we show how the former can be derived as consequences of the latter. This transition is accomplished by means of certain ismorphism theorems, proved in considerable generality in section 2, and applied to groups in section 3. In section 1 we give several miscellaneous theorems on equivalence relations, which later turn out to have their analogues in the theory of groups.

0. PRELIMINARIES ON RELATIONS

We let R be a binary relation, $\mathsf{Dom}(R)$ its domain, $\mathsf{Ran}(R)$ its range, and S its field $\big(\mathsf{Dom}(R) \cup \mathsf{Ran}(R)\big)$, or as we say, R is *on* the set S. As usual, for any two relations R_1 and R_2 on S, by their *relative product* R_1R_2 (sometimes in the literature written $R_1 \mid R_2$) is meant the set of all ordered pairs $\langle x, y\rangle$ of elements of S such that xRz and zRy both hold for at least one element z of S. Thus $x(R_1R_2)y$ iff[1] $(\exists z)(xRz \wedge zRy)$. By R^2 we mean RR. For any $x \in S$, by xR we mean the set of all y such that xRy. Thus $y \in xR$ iff xRy. For any subset A of S, by AR we shall mean the set of all y such that aRy holds for at least one $a \in A$. Thus $y \in AR$ iff $(\exists a)(a \in A \wedge aRy)$. In particular, for any singleton $\{x\}$, where $x \in S$, $\{x\}R$ is xR. By the *inverse* R^{-1} of R is meant the set of all ordered pairs $\langle x, y\rangle$ such that yRx. Thus $xR^{-1}y$ iff yRx.

The following properties are all trivial (R, R_1, R_2 are relations on S, A and B are subsets of S, and x is any element of S).

P$_1$ $(A \subseteq B) \supset (AR \subseteq BR)$. Thus also, $(x \in B) \supset (xR \subseteq BR)$.

P$_2$ $A(R_1R_2) = (AR_1)R_2$. Thus also, $x(R_1R_2) = (xR_1)R_2$.

P$_3$ $(A \cup B)R = AR \cup BR$.

P$_4$ $A(R_1 \cup R_2) = AR_1 \cup AR_2$.

P$_5$ $(R_1 \subseteq R_2) \supset (RR_1 \subseteq RR_2)$.

P$_6$ $R(R_1 \cup R_2) = RR_1 \cup RR_2$.

P$_7$ $R_1(R_2R_3) = (R_1R_2)R_3$. (Multiplication of relations is associative.)

P$_8$ $(R_1R_2)^{-1} = R_2^{-1}R_1^{-1}$.

[1]We use "iff" to abbreviate "if and only if".

C. Anthony Anderson and M. Zelëny (eds.), Logic, Meaning and Computation, 261-271.

$\mathbf{P}_9$ $(R^{-1})^{-1} = R$.

$\mathbf{P}_{10}$ $(R_1 \cup R_2)^{-1} = R_1^{-1} \cup R_2^{-1}$.

$\mathbf{P}_{11}$ $(R_1 \cap R_2)^{-1} = R_1^{-1} \cap R_2^{-1}$.

R is called *symmetric* if $R \subseteq R^{-1}$—which means that $xRy \supset yRx$, for all x and y.

$\mathbf{P}_{12}$ R is symmetric iff $R = R^{-1}$.

1. EQUIVALENCE RELATIONS

R is called *transitive* if $R^2 \subseteq R$—which means $(\forall x)(\forall y)(\forall z)\big((xRy \wedge yRz) \supset xRz\big)$.

R will be called *triangular* if $R^2 \subseteq R^{-1}$—which means $(\forall x)(\forall y)(\forall z)\big((xRy \wedge yRz) \supset zRx\big)$.

By I_S we shall mean the restriction of the identity relation I to $S \times S$—i.e., I_S is the set of all ordered pairs $\langle x, x \rangle$ such that $x \in S$.

We are assuming that every element of S is either in the domain of R or the range of R. We call R *reflexive* if $I_S \subseteq R$—which means that xRx holds for every $x \in S$. The following is well known.

PROPOSITION 1.1. *If R is transitive and symmetric, then R is reflexive.*

Proof. Suppose R is transitive and symmetric. If x is in the domain of R, then xRy holds for some $y \in S$, hence yRx (since R is symmetric), hence xRx (since $xRy \wedge yRx$ and R is transitive). Now suppose x is in the range of R. Then yRx for some y in S, hence xRy (by symmetry), hence again xRx (by transitivity). Thus xRx holds for every $x \in S$. ⊣

R is called an *equivalence relation* if R is both symmetric and transitive. Thus Proposition 1.1 is that every equivalence relation is reflexive. (Very often, reflexivity is taken as part of the *definition* of equivalence relation.)

In the elegant notation of relative products, an equivalence relation is a relation R such that $R^2 \subseteq R \subseteq R^{-1}$. (This should be well remembered, as it will frequently crop up.)

PROPOSITION 1.2. *R is an equivalence relation if and only if $R^2 = R = R^{-1}$.*

Proof. Suppose $R^2 \subseteq R \subseteq R^{-1}$. Then $R = R^{-1}$ (by $\mathbf{P}_{12}$), so it only remains to show that $R \subseteq R^2$. Well, suppose xRy. Also yRy (by Proposition 1.1), hence xR^2y. Thus $xRy \supset xR^2y$ for all x, y in S, hence $R \subseteq R^2$. Thus if R is an equivalence relation, then $R^2 = R = R^{-1}$. (The converse is trivial.) ⊣

PROPOSITION 1.3. *R is an equivalence relation if and only if $I_S \subseteq R$ and $R^2 \subseteq R^{-1}$—in other words, an equivalence relation is a relation that is both reflexive and triangular.*

Proof. If R is an equivalence relation, then R is reflexive (Proposition 1.1) and the transitivity and symmetry obviously imply triangularity. Conversely, suppose that R is reflexive and triangular. To show symmetry, suppose xRy. Also yRy (since R is reflexive). Hence yRx (by triangularity), and so R is symmetric. As for transitivity, this is obviously implied by triangularity and symmetry. ⊣

PROPOSITION 1.4. *A necessary and sufficient condition that R be an equivalance relation is that for every x and y in the field of R, xRy iff $xR = yR$*

Proof.

(a) *Necessity*: Suppose R is an equivalence relation. Let x and y be any element of the field of R. We will show: (1) $xRy \supset (xR = yR)$; (2) $(xR = yR) \supset xRy$.

To show (1), suppose xRy. Then $y \in xR$. Hence $yR \subseteq (xR)R$ (by $\mathbf{P}_1$). Hence $yR \subseteq x(RR)$ (since $(xR)R = x(RR)$ by $\mathbf{P}_2$), but $RR = R$ (by Proposition 1.3), hence $yR \subseteq xR$. Also, since xRy, then yRx (by symmetry), so by similar reasoning, $yR \subseteq xR$. Thus $xR = yR$.

To show (2), suppose $xR = yR$. Now by Proposition 1.1, R is reflexive, hence $y \in yR$ (i.e., yRy), hence $y \in xR$, and so xRy.

(b) *Sufficiency*: Suppose that for all x and y in the field of R, xRy iff $(xR = yR)$.

To show symmetry, suppose xRy. Then $xR = yR$ (by hypothesis). Hence $yR = xR$, and so yRx (by hypothesis). Thus R is symmetric.

As for transitivity, suppose xRy and yRz. Then $xR = yR$ and $yR = zR$ (by hypothesis), hence $xR = zR$, and so again by hypothesis, xRz. Thus R is transitive. ⊣

DEFINITION 1.5 **Partitions.** A collection Σ of sets is called a *partition* if for any two elements A and B of Σ, either $A = B$ or $A \cap B$ is empty.

For any relation R we let Σ_R be the collection of all sets xR, where x is in the domain of R. It is well known that if R is an equivalence relation, then Σ_R is a partition, and this fact is a simple corollary of Proposition 1.4 as follows. Suppose R is an equivalence relation and that x and y are elements of its domain such that xR and yR have a common element z. Then xRz and yRz, hence also zRy (since R is symmetric), hence xRy (since R is transitive), hence $xR = yR$ by Proposition 1.4.

It is not conversely true that if Σ_R is a partition, then R is necessarily an equivalance relation, but we have the following.

PROPOSITION 1.6. *If Σ_R is a partition and R is reflexive, then R is an equivalence relation.*

Proof. Assume the hypothesis. Let x and y by any elements of the field of R. We will show that xRy iff $xR = yR$, and the conclusion will then follow by Proposition 1.4.

(a) Suppose xRy. Then $y \in xR$. Also $y \in yR$ (since R is reflexive). Hence $xR = yR$ (since Σ_R is a partition).

(b) Conversely, suppose $xR = yR$. Since $y \in yR$ then $y \in xR$, hence xRy.⊣

We thus have the following.

PROPOSITION 1.7.

(1) *If R is an equivalence relation, then Σ_R is a partition.*

(2) *If R is reflexive, then R is an equivalence relation if and only if Σ_R is a partition.*

Here are some further facts about equivalence relations.

PROPOSITION 1.8. *R is an equivalence relation iff $RR^{-1} = R$.*

Proof. If R is an equivalence relation, then $R = R^{-1}$ (Proposition 1.2), hence $RR^{-1} = RR$, but $RR = R$ (Proposition 1.2), so $RR^{-1} = R$.

Conversely, suppose $RR^{-1} = R$. To show that R is symmetric, suppose xRy. Then $x(RR^{-1})y$, hence $xRz \wedge zR^{-1}y$ for some z. Hence xRz and yRz, so yRz and $zR^{-1}x$, hence $y(RR^{-1})x$, hence yRx. Thus R is symmetric.

Since R is symmetric, $R = R^{-1}$ (by $\mathbf{P}_{12}$), and since $RR^{-1} = R$, then $RR = R$, hence R is transitive. ⊣

PROPOSITION 1.9. *For any two equivalance relations E_1 and E_2 on S, a necessary and sufficient condition that E_1E_2 be an equivalence relation is that $E_1E_2 = E_2E_1$.*

Proof. Suppose E_1 and E_2 are both equivalence relations on S.

(a) *Necessity*: If E_1E_2 is an equivalence relation, then $E_1E_2 = (E_1E_2)^{-1}$ (by Proposition 1.2) $= E_2^{-1}E_1^{-1}$ ($\mathbf{P}_8$) $= E_2E_1$ (by $\mathbf{P}_{12}$, since E_1 and E_2 are symmetric).

(b) *Sufficiency*: Suppose $E_1E_2 = E_2E_1$. To show that E_1E_2 is transitive, $(E_1E_2)(E_1E_2) = E_1(E_2(E_1E_2))$ (by $\mathbf{P}_7$) $= E_1((E_2E_1)E_2)$ (since $E_2(E_1E_2) = (E_2E_1)E_2$ by $\mathbf{P}_7$) $= E_1((E_1E_2)E_2)$ (since $E_2E_1 = E_1E_2$) $= (E_1E_1)(E_2E_2)$ (by two applications of $\mathbf{P}_7$) $= E_1E_2$ (since $E_1^2 = E_1$ and $E_2^2 = E_2$). Thus $(E_1E_2)^2 = E_1E_2$.

As for symmetry, $(E_1E_2)^{-1} = E_2^{-1}E_1^{-1}$ ($\mathbf{P}_8$) $= E_1E_1$ (since $= E_2^{-1} = E_2$ and $E_1^{-1} = E_1$, by Proposition 1.2). ⊣

2. OPERATIONS AND THEIR INNER RELATIONS

In what follows, S will be a set fixed for the discussion and O will be a function that assigns to each ordered pair $\langle x, y\rangle$ of elements of S, an element denoted xy (not necessarily in S). A subset H of S will be called *closed* (under O) if for every x and y in H, xy is in H. We call O *closed* if S is closed under O. For any subsets H and K of S, by HK is meant the set of all elements hk, where $h \in H$ and $k \in K$. Also for any element $x \in S$, by xH we mean $\{x\}H$, which is the set of all elements xH, where $h \in H$. The sets xH (where x ranges over S) are called the (left) *cosets* of H. By H^2 we mean HH. We note that H is closed iff $H^2 \subseteq H$.

The following properties are trivial (H and K are subsets of S and x is any element of S).

$\mathbf{P}_{13}$ $(H \subseteq K) \supset (xH \subseteq xK)$.

$\mathbf{P}_{14}$ $x(H \cup K) = xH \cup xK$.

DEFINITION 2.1 **Associativity**. We shall define this in a way that does not presuppose closure. We shall call O *associative* if for all x, y and z in S, *if* xy and yz are both in S, *then* $x(yz) = (xy)z$.

The following is again trivial.

$\mathbf{P}_{15}$ If O is both closed and associative, then for all subsets H, K and L of S, $H(KL) = (HK)L$. Also, for any $x \in S$, $x(KL) = (xK)L$.

DEFINITION 2.2 **Cancellation Laws**. We shall call O *left concellative* if it obeys the left cancellation law: $(xy = xz) \supset (y = z)$ (for all x, y, z in S). We call O *right concellative* if for all x, y, z in S, $(xz = yz) \supset (x = y)$.

PROPOSITION 2.3. *The following three conditions are equivalent.*

(1) *For all subsets H, K of S and every $x \in S$, $H \subseteq K$ iff $xH \subseteq xK$.*

(2) *For all subsets H, K of S and every $x \in S$, $H = K$ iff $xH = xK$.*

(3) *O is left cancellative.*

Proof. Obviously (1) implies (2). Also (3) is that special case of (2) in which H and K are singletons. And so it remains to show that (3) implies (1).

Suppose O is left cancellative. Of course $(H \subseteq K) \supset (xH \subseteq xK)$ ($\mathbf{P}_{13}$), and so it remains to show the converse. So suppose $xH \subseteq xK$. We must show $H \subseteq K$. Let h be any element of H. Then $xh \in xH$, and so $xh \in xK$, hence $xh = xk$ for some $k \in K$, hence $h = k$ (by left cancellation), hence $h \in K$. Thus for all $h \in H$, h is also in K, and so $H \subseteq K$. ⊣

DEFINITION 2.4 **Right Soluble**. We next define O to be *right soluble* if for all a and b in S, there is some $x \in S$, such that $ax = b$—equivalently, if for every a and b in S, $b \in aS$—or again equivalently that for every element $a \in S$, $S \subseteq aS$.

(The definition that O is *left* soluble—for every a and b in S, the equation $xa = b$ has a solution in S—will play less of a role.)

We come now to the central notion.

DEFINITION 2.5 **Inner Relations.** For any subset H of S, we define $[H]$ to be the set of all ordered pairs $\langle x, y\rangle$ such that $y \in xH$. Thus $x[H]y$ iff $y \in xH$. Also $x[H]$ is xH. For a single element h of S, by $[h]$ we mean $[\{h\}]$, and thus $x[h]y$ iff $y \in x\{h\}$ iff $y = xh$. Thus also $x[h] = xh$. The relations $[H]$, where H ranges over all subsets of S, will be called the *inner relations* of O.

Inner relations will play a key role in just about all that follows.

The following is an easy consequence of $\mathbf{P}_{14}$.

$\mathbf{P}_{16}$ For any subsets H, K of S, $[H \cup K] = [H] \cup [K]$.

DEFINITION 2.6 **Isomorphisms.** We now let π be the mapping that assigns to each subset H of S the inner relation $[H]$. Thus $\pi(H) = [H]$.

PROPOSITION 2.7. *If O is left concellative, then π is a 1-1 correspondence.*

Proof. Suppose O is left cancellative. Suppose $[H] = [K]$. We are to show that $H = K$. Well, let x by any element of S. Then $x[H] = x[K]$, which means $xH = xK$, and hence $H = K$ by Proposition 2.3. ⊣

PROPOSITION 2.8. *If O is closed, then the following two conditions are equivalent:*

(1) *For any subsets H and K of S, $[H][K] = [HK]$.*

(2) *O is associative.*

Proof. Suppose O is closed (S is closed under O).

(a) Suppose O is associative. Then for any subsets H, K of S and any element $x \in S$, $x([H][K]) = (x[H])[K]$ (by $\mathbf{P}_2$) $= (xH)K = x(HK)$ (by $\mathbf{P}_{15}$) $= x[HK]$. Since for all $x \in S$, $x([H][K]) = x[HK]$, then $[H][K] = [HK]$.

(b) Conversely, suppose (1) holds. Then in particular, for any elements y and z of S, $[y][z] = [yz]$, and so for any x in S, $x(yz) = x[yz] = x([y][z]) = (x[y])[z]$ (by $\mathbf{P}_2$) $= (xy)z$, and so O is associative. ⊣

We shall say that π is an *ismorphism with respect to multiplication* if π is 1-1 and if for all subsets H and K of S, $[H][K] = [HK]$. From Proposition 2.7 and 2.8 we have the following.

THEOREM 2.9. *A sufficient condition that π be an isomorphism with respect to multiplication is that O is closed, associative and left cancellative. Thus if O has these three properties, then for any subsets H and K of S:*

(1) $[H] = [K]$ *iff* $H = K$,

(2) $[H][K] = [HK]$.

Proof. For each $x \in S$, the inner relation $[x]$ is single valued (since for any $y \in S$, $y[x]$ consists of the single element yx). Thus if H consists of only one element, then $[H]$ is single valued. The converse also holds if O is left cancellative, for suppose that O is left cancellative and that $[H]$ is single valued. Then for any $x \in S$, xH contains only one element, so if h and k are both elements of H, then xh and xk are both in xH, and so $xh = xk$, hence $h = k$ by left concellation. ⊣

Let us call an inner relation *singulary* if it is $[x]$, for some element x. We now see that every singulary relation is single valued, and if O is left cancellative, then the only single valued inner relations are the singulary ones.

We shall say that S is *isomorphic* (under π) to the set of singulary inner relations if for each x and y in S,

(1) $[x] = [y]$ iff $x = y$;

(2) $[xy] = [x][y]$.

By Theorem 2.9 we have the following.

COROLLARY 2.10. *If O is closed, associative and left cancellative, then S is isomorphic to the set of singulary inner relations of O.*

DEFINITION 2.11 **Permutations**. If O is right concellative, then each singulary inner relation $[x]$ is not only single valued, but is 1-1, because if $a[x]b$ and $c[x]b$ both hold, then $b = ax$ and $b = cx$, hence $ax = bx$, hence $a = b$ by right cancellation. If furthermore S is finite, then $[x]$ is a *permutation.*

Thus by Corollary 2.10 we have the following.

THEOREM 2.12. *If O is closed, associative and both left and right cancellative, then S is isomorphic to a set of 1-1 relations.*[2]

DEFINITION 2.13 **Boolean Isomorphism**. For any subset H of S, by H' we shall mean the complement of H relative to S (the set of all elements of S not in H). For any relation R on S, by R' we shall mean the complement of R relative to $S \times S$—i.e., the set of all ordered pairs $\langle x, y\rangle$ of elements of S such that xRy *doesn't* hold. Thus for any x and y in S, $xR'y$ iff $\neg(xRy)$.

[2]From this easily follows Cayley's Theorem that every group is isomorphic to a transformation group, and if the group is finite, to a permutation group—cf. application $\mathbf{A}_3$ in section 3.

We know that for any subsets H and K of S, $[H \cup K] = [H] \cup [K]$ ($\mathbf{P}_{16}$). If also π is 1-1, then we might call π an isomorphism with respect to union. If also $[H'] = [H]'$ holds for every subset H of S, then we will call π a *Boolean isomorphism.* And so if π is 1-1 and $[H'] = [H]'$ holds for all subsets H of S, then π is a Boolean isomorphism (by $\mathbf{P}_{16}$). (If π is a Boolean isomorphism, then of course also $[H \cap K] = [H] \cap [K]$.)

We now wish to show that a necessary and sufficient condition for π to be a Boolean isomorphism is that O is both left cancellative and right soluble. This will follow from the next two propositions.

PROPOSITION 2.14. *O is left cancellative iff for all subsets H of S, $[H'] \subseteq [H]'$.*

PROPOSITION 2.15. *O is right soluble iff for all subsets H of S, $[H]' \subseteq [H']$.*

Proof of Proposition 2.14. Suppose O is left cancellative. We show that $x[H']y \supset x[H]'y$ for all x and y in S, and hence that $[H'] \subseteq [H]'$, so suppose $x[H']y$. Then $y \in xH'$, which means that $y = xk$ for some k outside H. Hence by left cancellation, $y \neq xh$ for any $h \in H$ (for otherwise we would have $xh = xk$ and hence $h = k$, hence k would be in H). Thus $y \notin xH$, so $x[H]y$ doesn't hold, hence $x[H]'y$. This proves that if O is left cancellative, then $x[H']y \supset x[H]'y$ for all x and y in S, and hence that $[H'] \subseteq [H]'$.

Conversely, suppose that $[H'] \subseteq [H]'$ for all subsets of H of S. Then for any element $a \in S$, $[\{a\}'] \subseteq [\{a\}]'$, hence $[\{a\}] \subseteq [\{a\}']'$, hence for any $x \in S$, $x[\{a\}] \subseteq x[\{a\}']'$, and so for any element y, $y \in x[\{a\}] \supset y \notin x[\{a\}']$. Now, $xa \in x[\{a\}]$, and so $xa \notin x[\{a\}']$. This means that $xa \neq xb$ for any $b \in \{a\}'$, and thus $xa \neq xb$ for any $b \neq a$, or equivalently, $xa = xb$ implies $a = b$, and so O is left cancellative. ⊣

Before proving Proposition 2.15, let us note that O is right solvable iff $[S] = S \times S$, because O is right soluble iff for all x and y in S, $y \in xS$, or equivalently, for every x and y in S, $x[S]y$, which means $S \times S \subseteq [S]$ (and of course $[S] \subseteq S \times S$).

Proof of Proposition 2.15. We shall show, in fact, that for any subset H of S, $[H]' \subseteq [H']$ iff O is right soluble. Well, $[H]' \subseteq [H']$ iff $[H] \cup [H'] = S \times S$. But $[H] \cup [H'] = [H \cup H']$ (by ($\mathbf{P}_{16}$) $= [S]$, and so $[H]' \subseteq [H']$ iff $[S] = S \times S$, which in turn is the case iff O is right soluble (as already noted). ⊣

REMARK 2.16. We see from the proof above that if so much as one subset H of S has the property that $[H]' \subseteq [H']$, then every subset H of S has that property.

It follows from the last two propositions that if O is both left cancellative and right soluble then for every subset H of S, $[H'] = [H]'$, and also that

π is 1-1 (by Proposition 2.7), and hence that π is a Boolean isomorphism. Conversely if π is a Boolean isomorphism, then $[H'] = [H]'$ for every subset H of S, hence O must be both left cancellative and right soluble. And so we have the following:

THEOREM 2.17. *π is a Boolean isomorphism iff O is both left cancellative and right soluble.*

3. EQUIVALENCE RELATIONS AND GROUPS

An element i of S is called an *identity* (relative to O) if $ix = xi = x$ for every x in S. As is well known, S can have at most one identity (since if i and j are both identities, then $ij = j$ and also $ij = i$, hence $i = j$). An element b of S is called an *inverse* of an element a of S if ab and ba are both identities (and hence also both equal). As is well known, if O is closed and associative, then an element a of S cannot have more than one inverse (because if b and c are both inverses of a, then $b = b(ac) = (bc)c = c$). If an element x has an inverse, this inverse (which is unique) is denoted x^{-1}, and for any subset H of S, by H^{-1} is meant the set of inverses of all the elements of H. We shall call H *symmetric* if $H \subseteq H^{-1}$. Then H is symmetric iff every element of H has an inverse in H. We shall say that π is an *isomorphism with respect to inversion* if π is 1-1 and for every subset H of S, $[H]^{-1} = [H^{-1}]$.

Now, a non-empty subset G of S is called a *group* (with respect to O) if O is associative and if $G^2 \subseteq G \subseteq G^{-1}$ (i.e., G is closed and symmetric). These conditions automatically imply that G has an identity, since G, being non-empty, has an element x, and x then has an inverse x^{-1} in G, and then by closure, xx^{-1}, which is the identity, is in G. A subset H of G is called a *subgroup* of G if H is also a group (with respect to O).

In what follows, G will be assumed to be a group with respect to O and that $G \times G$ is the domain of O. As in section 2, π will be the mapping that assigns to each subset H of G the inner relation $[H]$.

Since G is a group, then, as is well known, O is left and right cancellative (since $(xa = xb) \supset (x^{-1}(xa) = x^{-1}(xb)) \supset ((x^{-1}x)a = (x^{-1}x)b) \supset (a = b)$, and similarly for right cancellation) and both left and right soluble (since $ax = b$ has the solution $x = a^{-1}b$, and $xa = b$ has the solution $x = ba^{-1}$), and so by earlier results π is a Boolean isomorphism and an isomorphism with respect to multiplication. We now show the following.

THEOREM 3.1. *π is an isomorphism with respect to inversion.*

Proof. We are to show that for any subset H of G, $[H^{-1}] = [H]^{-1}$. We first show that $[H]^{-1} \subseteq [H^{-1}]$. Well, suppose $x[H]^{-1}y$. Then $y[H]x$, so $x = yh$ for some $h \in H$. Then $xh^{-1} = (yh)h^{-1} = y(hh^{-1}) = y$, so $y = xh^{-1}$. Therefore $y \in xH^{-1}$ (since $h^{-1} \in H^{-1}$), hence $x[H^{-1}]y$. Thus for all x and y in G, $x[H]^{-1}y \supset x[H^{-1}]y$, and so $[H]^{-1} \subset [H^{-1}]$.

To show that $[H^{-1}] \subseteq [H]^{-1}$, suppose $x[H^{-1}]y$. Then $y \in xH^{-1}$, so $y = xh^{-1}$ for some $h \in H$, hence $yh = (xh^{-1})h = x(h^{-1}h) = x$, hence $yh = x$, so $x \in yH$, hence $y[H]x$, and so $x[H]^{-1}y$. Thus $x[H^{-1}]y \supset x[H]^{-1}y$ for all x and y in G, so $[H^{-1}] \subseteq [H]^{-1}$. ⊣

By Theorem 2.9, 2.17 and 3.1 we now have the following.

THEOREM 3.2 **Main Isomorphism Theorem**. *Suppose that G is a group with respect to O and π is the mapping that assigns to each subset H of G the inner relation $[H]$. Then π is a Boolean isomorphism and an isomorphism with respect to multiplication and inversion. Thus for any subsets H and K of G:*

(1) $[H] = [K]$ *iff* $H = K$,
(2) $[H][K] = [HK]$,
(3) $[H'] = [H]'$,
(4) $[H \cup K] = [H] \cup [K]$ *and* $[H \cap K] = [H] \cap [K]$,
(5) $H \subseteq K$ *iff* $[H] \subseteq [K]$,
(6) $[H^{-1}] = [H]^{-1}$.

4. APPLICATIONS

Many results in group theory—standard and otherwise—are corollaries of our isomorphism theorems. Here are some applications.

$\mathbf{A}_1$ For any subset H of G, H is a group iff $H^2 \subseteq H \subseteq H^{-1}$, but by the main isomorphism theorem, $H^2 \subseteq H \subseteq H^{-1}$ iff $[H]^2 \subseteq [H] \subseteq [H]^{-1}$, which is the condition that $[H]$ is an equivalence relation! We thus have a neat proof of the known result that a subset H of G is a subgroup of G iff $[H]$ is an equivalence relation. More specifically we can say for any subset H of G:

(1) H is closed ($H^2 \subseteq H$) iff $[H]$ is transitive ($[H]^2 \subseteq [H]$),
(2) H is symmetric ($H^{-1} \subseteq H$) iff $[H]$ is symmetric ($[H]^{-1} \subseteq [H]$);
(3) We also note that for any non-empty subset H of G, $i \in H$ iff H is reflexive (i being the identity of G).

To see that (3) holds, suppose $i \in H$. Then for any x in G, $xi \in xH$, hence $x \in xH$, hence $x[H]x$, and so $[H]$ is reflexive. Conversely, if $[H]$ is reflexive, then $i[H]i$, hence $i = ih$ for some $h \in H$, but $ih = h$, so $i \in H$.

$\mathbf{A}_2$ For any subset H of G, we let Σ_H be the set of all left cosets xH of H. Actually Σ_H is $\Sigma_{[H]}$, as defined in section 1. From the fact that Σ_R is a partition if R is an equivalance relation (Proposition 1.7, (1)) it follows by $\mathbf{A}_1$ that if H is a subgroup of G, then Σ_H is a partition (a well known fact). From this in turn follows Lagrange's Theorem that if G is finite, then the number of elements of any subgroup H of G must divide the number of elements of G (since each coset xH of H has the same number of elements as H, by left concellation).

A$_3$ Since in a group we have closure, associativity, and both cancellation laws, then, as remarked earlier, Cayley's Theorem that every group is isomorphic to a transformation group is a consequence of Theorem 2.12.

Actually, Theorem 2.9 is a substantial generalization of Cayley's Theorem, which is to the effect that the set of singletons $\{x\}$ is multiplicatively isomorphic to the set of corresponding singulary inner relations $[x]$, whereas Theorem 2.9 tells us that the set of *all* subsets H is multiplicatively isomorphic to the set of *all* inner relations $[H]$.

For the next few applications we will simply state the result and alongside it, in parentheses, the theorem on the logic of relations from which it follows by virtue of our isomorphism theorems.

A$_4$ If Σ_H is a partition and $i \in H$, then H is a subgroup of G. (If Σ_R is a partition and R is reflexive, then R is an equivalence relation, by Proposition 1.6.)

A$_5$ H is a subgroup of G iff $i \in A$ and $H^2 \subseteq H^{-1}$. (R is an equivalence relation iff R is reflexive and $R^2 \subseteq R^{-1}$, by Proposition 1.3.)

A$_6$ H is a subgroup of G iff for all x and y in G, ($y \in xH$ iff $xH = yH$). (R is an equivalence relation iff for all x and y in G, (xRy iff $xR = yR$, by Proposition 1.4.)

A$_7$ H is a subgroup of G iff $HH^{-1} = H$. (R is an equivalence relation iff $RR^{-1} = R$, by Proposition 1.8.)

A$_8$ For any two subgroups H and K of G, HK is a group iff $HK = KH$. (For any two equivalence relations E_1 and E_2, a necessary and sufficient condition that E_1E_2 be an equivalence relation is that $E_1E_2 = E_2E_1$, by Proposition 1.9.)

We have now seen some parallels between the theory of equivalence relations and the theory of groups. This approach might well be pushed further—e.g., in the study of normal subgroups, or perhaps more complex structures such as rings and fields.

PART II

COMPUTATION

HENK BARENDREGT

DISCRIMINATING CODED LAMBDA TERMS

Abstract. A *coding* for a (type-free) lambda term M is a lambda term $\ulcorner M \urcorner$ in normal form such that M (and its parts) can be reconstructed from $\ulcorner M \urcorner$ in a lambda definable way. Kleene (*1936*) defined a coding $\ulcorner M \urcorner^K$ and a *self-interpreter* $\mathbf{E}^K \in \Lambda^o$ such that:

$$\forall M \in \Lambda^o \ \mathbf{E}^K \ulcorner M \urcorner^K = M. \tag{1}$$

(Kleene did it for the $\lambda\mathbf{I}$-calculus, but the result is valid also for the $\lambda\mathbf{K}$-calculus.) In this style one can construct a *discriminator* $\Delta^K \in \Lambda^o$ such that:

$$\forall M, N \in \Lambda \ \ \Delta^K \ulcorner M \urcorner^K \ulcorner N \urcorner^K = \begin{cases} \texttt{true} & (\equiv \lambda xy.x) \quad \text{if } M \equiv N, \\ \texttt{false} & (\equiv \lambda xy.y) \quad \text{else.} \end{cases} \tag{2}$$

The terms $\mathbf{E}^K$ and Δ^K are complicated. $\mathbf{E}^K$ appears also in *Church 1941* under the name `form`. They depend on the lambda definability of functions on the integers dealing with coded syntactic properties. Inspired by a construction of P. de Bruin (see *Barendregt 1991*) a different coding $\ulcorner M \urcorner$ and an efficient self-interpreter $\mathbf{E} \in \Lambda^o$ was constructed by Mogensen (*1992*) such that even

$$\forall M \in \Lambda \ \mathbf{E} \ulcorner M \urcorner = M. \tag{3}$$

This construction does not represent syntax via an encoding as numbers, but directly as lambda terms. This results in a much less complex $\mathbf{E}$. Mogensen's construction was simplified even further in *Böhm et al. 1994*. In this paper we construct a simple discriminator $\Delta \in \Lambda^o$ such that:

$$\forall M, N \in \Lambda^o \ \ \Delta \ulcorner M \urcorner \ulcorner N \urcorner = \begin{cases} \texttt{true} & \text{if } M \equiv_\alpha N, \\ \texttt{false} & \text{else.} \end{cases} \tag{4}$$

Note that in (1) and (4) the statement is only about closed lambda terms, while that in (2) and (3) is about all lambda terms. Moreover in (2) syntactic equality on terms is considered literary, while in (4) we can deal in an easy way with the better notion of equality modulo a change of names for bound variables. It will become clear why this is so.

This paper first appeared in *From Universal Morphisms to Megabytes—a Baayen Space Odyssey*, printed at CWI, Amsterdam.

1. INTRODUCTION

The most important notations for the type-free lambda calculus will be given here. Background can be found in *Barendregt 1984*.

DEFINITION 1.1. Variables and terms of the lambda calculus are defined by the following abstract syntax.

$$\begin{aligned} \texttt{var} &= a \mid \texttt{var}' \\ \texttt{term} &= \texttt{var} \mid \texttt{term term} \mid \lambda\, \texttt{var term} \end{aligned}$$

C. Anthony Anderson and M. Zelëny (eds.), Logic, Meaning and Computation, 275-285.

NOTATION.

(i) $M, N, \ldots, P, Q, \ldots$ range over λ-terms. The letters $x, y, z, \ldots$ range over variables. Note that the variables are $\{a, a', a'', \ldots, a^{(n)}, \ldots\}$.

(ii) Λ is the set of lambda terms. $\mathrm{FV}(M)$ is the set of free variables of M. The set of closed terms is $\Lambda^{\circ} = \{M \in \Lambda \mid \mathrm{FV}(M) = \emptyset\}$.

(iii) The relation $\equiv$ denotes syntactic equality; the relation $\equiv_\alpha$ denotes syntactic equality up to a change of names of the bound variables. For example

$$\lambda x.x \equiv_\alpha \lambda y.y \not\equiv \lambda x.x.$$

(iv) The relation $=$ denotes β-convertibility, axiomatized by

$$(\lambda x.M)N = M[x := N].$$

Here $[x := n]$ denotes substitution of N for the free occurrences of x. E.g.,

$$(x(\lambda x.x))[x := a] \equiv a(\lambda x.x).$$

(v) $\mathbb{N}$ is the set of natural numbers. For $n \in \mathbb{N}$ the terms $\mathbf{c}_n \equiv \lambda f x.f^n x$, where $f^0 x \equiv x$ and $f^{n+1}x \equiv f(f^n x)$, denote the so called Church numerals. Note that the $\mathbf{c}_n$ are distinct normal forms; hence

$$\mathbf{c}_n = \mathbf{c}_m \Rightarrow n = m$$

by the Church-Rosser theorem.

A lambda term can be seen as an executable: the redexes want to be evaluated. In this sense a normal form is not executable anymore. For a lambda term M its *code* $\ulcorner M \urcorner$ is a normal form such that M is reconstructible from M. Kleene (*1936*) defined a code $\ulcorner M \urcorner^K$ essentially as follows.

DEFINITION 1.2.

(i) By induction on the structure of M we define $\#M$.

$$\begin{aligned} \#(a^{(n)}) &= \; < 0, n >, \\ \#(PQ) &= \; < 1, < \#(P), \#(Q) >>, \\ \#(\lambda x.P) &= \; < 2, < \#(x), \#(P) >> . \end{aligned}$$

Here $< -, - >$ denotes a recursive pairing function on $\mathbb{N}$ with the recursive projections $(-)_0, (-)_1$:

$$(< n_0, n_1 >)_i = n_i.$$

(ii) The map $\ulcorner - \urcorner^K : \Lambda \to \Lambda$ is defined by

$$\ulcorner M \urcorner^K = \mathbf{c}_{\#M}.$$

Note that for all $M \in \Lambda$ the term $\ulcorner M \urcorner^K$ is in normal form. Moreover,

$$\ulcorner M \urcorner^K = \ulcorner N \urcorner^K \Rightarrow M \equiv N.$$

PROPOSITION 1.3. *There is no lambda term Q such that for all $M \in \Lambda^{(o)}$ one has*

$$QM = \ulcorner M \urcorner^K.$$

Proof. Suppose Q exists. Then for $\mathbf{I} \equiv \lambda x.x$ one has

$$Q(\mathbf{II}) = \ulcorner \mathbf{II} \urcorner^K = \mathbf{c}_{\#(\mathbf{II})} = \mathbf{c}_{<1,<\#\mathbf{I},\#\mathbf{I}>>}.$$

But also

$$Q(\mathbf{II}) = Q\mathbf{I} = \ulcorner \mathbf{I} \urcorner^K = \mathbf{c}_{\#\mathbf{I}} = \mathbf{c}_{<2,<\#(x),\#(x)>>}.$$

Hence $< 1, < \#(\mathbf{I}), \#(\mathbf{I}) >>=< 2, < \#(x), \#(x) >>$, a contradiction. ⊣

In spite of this fact that the 'quote' Q does not exists, the inverse 'evaluation' $\mathbf{E}$ can be constructed.

THEOREM 1.4 (Kleene, *1936*). *There exists an $\mathbf{E}^K \in \Lambda^\circ$ such that for all $M \in \Lambda^\circ$ one has*

$$\mathbf{E}^K \ulcorner M \urcorner^K = M.$$

Proof. See *Kleene 1936* or *Barendregt 1984*, Theorem 8.1.16. ⊣

The self-interpreter $\mathbf{E}$ can work only for closed terms M (or terms having at most a fixed finite set of free variables). The reason is that if

$$\mathbf{E}^K \ulcorner M \urcorner^K = M,$$

then

$$\mathrm{FV}(M) \subseteq \mathrm{FV}(\mathbf{E}^K \ulcorner M \urcorner^K) = \mathrm{FV}(\mathbf{E}^K).$$

Therefore if $\mathbf{E}^K$ is closed, then the M have to be closed as well. This causes one difficulty in the construction of $\mathbf{E}^K$. The closed terms do not form a context-free language. Kleene solved this problem by constructing $\mathbf{E}$ first for the set of combinatory terms $\mathcal{C}^\circ$ built from the basis $\{\mathbf{K}, \mathbf{S}\}$ using application only; then the real self-interpreter can be obtained by translations between Λ° and $\mathcal{C}^\circ$.

A different construction of a self-interpreter was given by a former student of mine, using ideas from denotational semantics. The equations for the self-interpreter are those of the semantic interpretation and the F plays the role of a valuation (environment).

THEOREM 1.5 (P. de Bruin). *There exists an $\mathbf{E}_0 \in \Lambda^\circ$ such that for all $M \in \Lambda$ and all $F \in \Lambda$ one has*

$$\mathbf{E}_0 \ulcorner M \urcorner F = M[x_1, \ldots, x_n := F\ulcorner x_1 \urcorner, \ldots, F\ulcorner x_n \urcorner] \tag{5}$$

(simultaneous substitution), where $\{x_1, \ldots, x_n\} = \mathrm{FV}(M)$.

Proof. By the representability of computable functions and the fixedpoint theorem there is a term $\mathsf{E}_0 \in \Lambda^o$ such that

$$\begin{aligned} \mathsf{E}_0 \ulcorner x \urcorner^K F &= F \ulcorner x \urcorner^K, \\ \mathsf{E}_0 \ulcorner PQ \urcorner^K F &= (\mathsf{E}_0 \ulcorner P \urcorner^K F)(\mathsf{E}_0 \ulcorner Q \urcorner^K F), \\ \mathsf{E}_0 \ulcorner \lambda x.P \urcorner^K F &= \lambda x.(\mathsf{E}_0 \ulcorner P \urcorner^K F_{[\ulcorner x \urcorner \mapsto x]}), \end{aligned}$$

where $F_{[\ulcorner x \urcorner \mapsto x]} = F'_x$ with

$$\begin{aligned} F'_x \ulcorner x \urcorner &= x, \\ F'_x \ulcorner y \urcorner &= F \ulcorner y \urcorner, \text{ if } y \not\equiv x. \end{aligned}$$

Note that F'_x can be written as GFx, with G closed. By induction on the structure of $M \in \Lambda$ one can show that the statement holds. ⊣

COROLLARY 1.6. *There exists an* $\mathsf{E}^{dB} \in \Lambda^o$ *such that for all* $M \in \Lambda^o$ *one has*

$$\mathsf{E} \ulcorner M \urcorner^K = M.$$

Proof (P. de Bruin). We can take

$$\mathsf{E}^{dB} \equiv \lambda m.\mathsf{E}_0 m \mathsf{I}.$$

Indeed, for closed terms M it follows from (5) that

$$\mathsf{E}^{dB} \ulcorner M \urcorner = \mathsf{E}_0 \ulcorner M \urcorner \mathsf{I} = M.$$ ⊣

2. REPRESENTING DATA TYPES

After seeing the method of P. de Bruin, an improved version of it was given by Mogensen (*1992*) by representing data types directly (i.e., not using the natural numbers) in the lambda calculus as done in, e.g., *Böhm and Berarducci 1985*. This approach was improved later by Böhm *et al.* (*1994*) by constructing a new representation of data types into the type-free lambda calculus. This new representation will be treated in a slightly modified form in this section.

DEFINITION 2.1. Write

$$\begin{aligned} \langle M_1, \ldots, M_n \rangle &= \lambda z.zM_1 \ldots M_n, \\ \mathsf{U}_i^n &= \lambda x_1 \ldots x_n.x_i, \\ \texttt{true} &= \mathsf{U}_1^2, \\ \texttt{false} &= \mathsf{U}_2^2. \end{aligned}$$

Note that

$$\begin{aligned} \langle M_1, \ldots, M_n \rangle \mathbf{U}_i^n &= M_i, \\ \mathtt{true}\, PQ &= P, \\ \mathtt{false}\, PQ &= Q. \end{aligned}$$

In particular we have $\langle M \rangle = \lambda z.zM$ and $\langle\ \rangle = \lambda x.x = \mathbf{I}$. Now we define the notion of lists inspired by the language LISP (*McCarthy et al. 1961*).

DEFINITION 2.2.

(i) Write

$$\begin{aligned} \mathtt{nil} &= \langle\ \rangle, \\ \mathtt{cons} &= \lambda xy.\langle x, y \rangle, \\ \mathtt{car} &= \langle \mathbf{U}_1^2 \rangle, \\ \mathtt{cdr} &= \langle \mathbf{U}_2^2 \rangle, \\ \mathtt{null?} &= \langle \mathbf{U}_3^3, \mathbf{U}_1^2, \mathtt{false}, \mathtt{true} \rangle. \end{aligned}$$

(ii) Define

$$\begin{aligned} [\ ] &= \langle\ \rangle, \\ [M_1, \ldots, M_{n+1}] &= \mathtt{cons}\, M_1 [M_2, \ldots, M_{n+1}]. \end{aligned}$$

So for example

$$[M_1, M_2, M_3] = \langle M_1, \langle M_2, \langle M_3, \langle\ \rangle \rangle \rangle \rangle.$$

(In *Barendregt 1984* this term is written as $[M_1, M_2, M_3, \mathbf{I}]$. At the time of writing that book we did not yet see the usefulness of terminating a list with a special constructor.) Note that

$$\begin{aligned} \mathtt{car}\,(\mathtt{cons}\, PQ) &= P, \\ \mathtt{cdr}\,(\mathtt{cons}\, PQ) &= Q, \\ \mathtt{null?}\,\mathtt{nil} &= \mathtt{true}, \\ \mathtt{null?}(\mathtt{cons}\, \mathtt{PQ}) &= \mathtt{false}. \end{aligned}$$

PROPOSITION 2.3. *There exist lambda definable functions* $(\ )_i$ *such that for* $1 \le i \le n$ *one has*

$$([M_1, \ldots, M_n])_i = M_i.$$

Proof. Take

$$\begin{aligned} (l)_1 &= \mathtt{car}\, l, \\ (l)_{i+1} &= (\mathtt{cdr}\, l)_i. \end{aligned}$$

⊣

DEFINITION 2.4. An *(algebraic) signature* s consists of a number $n \in \mathbb{N}$ (thought of as the list of symbols $[f_1, \ldots, f_n]$) together with a list of numbers $[s_1, \ldots, s_n]$ (thought of as the *arity* of the respective f_i's). We write $s = [s_1, \ldots, s_n]$.

For example a field has signature $s = [2, 2, 1, 1, 0, 0]$ (thought of as the arities of the function symbols $[+, \times, -, ^{-1}, 0, 1]$; so $f_1 = +, f_2 = \times$, et cetera).

DEFINITION 2.5. If s is a signature then $\mathtt{term}_s$, the set of *terms of signature* s, is defined as follows.

$$\begin{array}{rcl} x \in \mathtt{var} & \Rightarrow & x \in \mathtt{term}_s, \\ t_1, \ldots, t_{s_i} \in \mathtt{term}_s & \Rightarrow & f_i(t_1, \ldots, t_{s_i}) \in \mathtt{term}_s. \end{array}$$

For example, in the signature of fields the term $f_1(f_1(x, f_3(f_2(y, f_4(z))), f_6)$ is usually written as $x - yz^{-1} + 1$.

DEFINITION 2.6. Let $s = [s_1, \ldots, s_n]$ be a signature.

(i) A *lambda interpretation* of s is a list of 'constructors' $C_1, \ldots, C_n \in \Lambda$.

(ii) Let $C_1, \ldots, C_n$ be a lambda interpretation of s. Then we define a map

$$T : \mathtt{term}_s \rightarrow \Lambda$$

as follows.

$$\begin{array}{rcl} T_x & = & x, \\ T_{f_i(t_1, \ldots, t_{s_i})} & = & C_i[T_{t_1}, \ldots, T_{t_{s_i}}], \end{array}$$

where $[T_{t_1}, \ldots, T_{t_{s_i}}]$ is the list operation on lambda terms defined in 2.2.

Example. The signature of *binary trees* is $[0, 2]$. The term $t = f_2(f_2(f_1, f_1), f_1)$ denotes a simple tree and $t' = f_2(f_1, f_2(f_1, f_1))$ its miror image. Can we find a lambda interpretation for this signature in such a way that mirroring becomes lambda definable, i.e., for some $F \in \Lambda^o$ one has $FT_t = T_{t'}$? The following result, due to Böhm *et al.* (*1994*), will affirm this. We present the result in a modified form that will be useful for §4.

THEOREM 2.7. *For every algebraic signature $s = [s_1, \ldots, s_n]$ there exists a lambda interpretation $C_1, \ldots, C_n$ such that the following hold.*

(i) $\forall A_1 \ldots A_n \, \exists F$

$$FT_{f_i(t_1, \ldots, t_{s_i})} = A_i[T_{t_1}, \ldots, T_{t_{s_i}}]F, \qquad 1 \le i \le n. \tag{6}$$

(ii) *The $C_1, \ldots, C_n$ only depend on n, not on the $[s_1, \ldots, s_n]$. In* (6) *we can take $F \equiv \langle\langle A_1, \ldots, A_n \rangle\rangle$.*

Proof. Define $C_{f_i} \equiv \lambda le.e\mathbf{U}_i^n l\langle e\rangle$.

(i) Given $A_1, \ldots, A_n$, we see whether $F \equiv \langle\langle A_1, \ldots, A_n\rangle\rangle$ works. Indeed,

$$\begin{aligned} FT_{f_i(t_1,\ldots,t_{s_i})} &= \langle\langle A_1, \ldots, A_n\rangle\rangle(C_{f_i}[T_{t_1}, \ldots, T_{t_{s_i}}]) \\ &= C_{f_i}[T_{t_1}, \ldots, T_{t_{s_i}}]\langle A_1, \ldots, A_n\rangle \\ &= \langle A_1, \ldots, A_n\rangle \mathbf{U}_i^n [T_{t_1}, \ldots, T_{t_{s_i}}]\langle\langle A_1, \ldots, A_n\rangle\rangle \\ &= A_i[T_{t_1}, \ldots, T_{t_{s_i}}]F. \end{aligned}$$

(ii) By the construction. ⊣

COROLLARY 2.8. *Let $s = [s_1, \ldots, s_n]$ be an algebraic signature. Let $C_1, \ldots, C_n$ be the lambda interpretation of s constructed in Theorem 2.7. Then for all $B_1 \ldots B_n$ there exists an F such that*

$$F(C_{f_i}[x_1, \ldots, x_{s_i}]) = B_i x_1 \ldots x_{s_i} F, \qquad 1 \leq i \leq n. \tag{7}$$

Proof. Let $B_1, \ldots, B_n$ be given. Define $A_i = \lambda l.B_i(l)_1 \ldots (l)_{s_i}$. Then

$$A_i[x_1, \ldots, x_{s_i}] = B_i x_1, \ldots, x_{s_i}.$$

Then the F for the A_i found in Theorem 2.7 is the F satisfying (7). ⊣

Now we can program the function that 'mirrors' trees. In the signature [0,2] for binary trees let

$$\begin{aligned} \texttt{leaf} &= T_{f_1} &&= C_{f_1}\langle\,\rangle, \\ \texttt{tree} &= \lambda ab.T_{f_2(a,b)} &&= \lambda ab.C_{f_2}\langle a, b\rangle. \end{aligned}$$

By Corollary 2.8 there exists an F such that

$$\begin{aligned} F\texttt{leaf} &= \texttt{leaf}, \\ F(\texttt{tree}\, a\, b) &= \texttt{tree}(fb)(fa). \end{aligned}$$

This F has the mirror effect. E.g., $F(f_2(f_2(f_1, f_1), f_1)) = f_2(f_1, f_2(f_1, f_1))$.

3. A SIMPLE SELF-INTERPRETER

In *Mogensen 1992* a simple coding and self-interpreter for lambda terms is defined, using the fact that data types (term algebras of a signature s) have a lambda interpretation. The method was simplified by Böhm et al. (*1994*) by making use of their lambda representation of algebraic signatures given in §2.

DEFINITION 3.1. Let s be the signature $[1, 2, 1]$. Define

$$\begin{aligned} \texttt{const} &= C_{f_1} &&\equiv \lambda le.e\mathbf{U}_1^3 l\langle e\rangle, \\ \texttt{app} &= C_{f_2} &&\equiv \lambda le.e\mathbf{U}_2^3 l\langle e\rangle, \\ \texttt{abs} &= C_{f_3} &&\equiv \lambda le.e\mathbf{U}_3^3 l\langle e\rangle. \end{aligned}$$

DEFINITION 3.2. For $M \in \Lambda$ define $\ulcorner M \urcorner$ as follows.

$$\begin{aligned}\ulcorner x \urcorner &= \mathtt{const}\,[x],\\ \ulcorner PQ \urcorner &= \mathsf{app}\,[\ulcorner P \urcorner, \ulcorner Q \urcorner],\\ \ulcorner \lambda x.P \urcorner &= \mathsf{abs}\,[\lambda x.\ulcorner P \urcorner].\end{aligned}$$

Note that $\mathrm{FV}(\ulcorner M \urcorner) = \mathrm{FV}(M)$.

THEOREM 3.3 (Mogensen, *1992*). *There exists an* $\mathbf{E} \in \Lambda^{\circ}$ *such that*

$$\forall M \in \Lambda\ \mathbf{E}\ulcorner M \urcorner = M.$$

Proof (Böhm *et al., 1994*). By Corollary 2.8 there exists a term $\mathbf{E} \in \Lambda^{\circ}$ such that

$$\begin{aligned}\mathbf{E}(\,\mathtt{const}\,[p]) &= p,\\ \mathbf{E}(\,\mathsf{app}\,[p,q]) &= (\mathbf{E}p)(\mathbf{E}q),\\ \mathbf{E}(\,\mathsf{abs}\,[p]) &= \lambda x.\mathbf{E}(px).\end{aligned}$$

Then

$$\begin{aligned}\mathbf{E}\ulcorner x \urcorner &= x,\\ \mathbf{E}\ulcorner PQ \urcorner &= (\mathbf{E}\ulcorner P \urcorner)(\mathbf{E}\ulcorner Q \urcorner),\\ \mathbf{E}\ulcorner \lambda x.P \urcorner &= \lambda x.\mathbf{E}\ulcorner P \urcorner.\end{aligned}$$

Now the results follows by induction on the structure of M. ⊣

Using the constructions in §2 the self-interpreter becomes

$$\mathbf{E} \equiv \langle\langle \lambda lf.(l)_1, \lambda lf.f(l)_1(f(l)_2), \lambda lfx.f((l)_1 x)\rangle\rangle.$$

The construction in *Böhm et al. 1994* is simpler. They take

$$\begin{aligned}\mathtt{const} &= \lambda xe.e\mathbf{U}^3_1 xe,\\ \mathsf{app} &= \lambda xye.e\mathbf{U}^3_2 xye,\\ \mathsf{abs} &= \lambda xe.e\mathbf{U}^3_3 xe.\end{aligned}$$

The resulting self-interpreter then becomes $\mathbf{E}^B = \langle\langle \mathbf{K}, \mathbf{S}, \mathbf{C}\rangle\rangle$. Here $\mathbf{K} \equiv \lambda xy.x$, $\mathbf{S} \equiv \lambda xyz.xz(yz)$ and $\mathbf{C} \equiv \lambda xyz.x(zy)$. For reasons of uniformity we have given the definition of const, app and abs as in 3.1. This will be useful in §4.

4. A SIMPLE DISCRIMINATOR

In this section we will construct a simple term discriminating between coded closed lambda terms. The discrimination is even modulo α-conversion. For open terms discrimination is possible only for the coding $\ulcorner\ \urcorner^K$ of Kleene.

LEMMA 4.1.

(i) *There exists a term* $\delta_{\mathbb{N}} \in \Lambda^{\circ}$ *such that*

$$\delta_{\mathbb{N}} \mathbf{c}_n \mathbf{c}_m = \begin{cases} \mathtt{true} & \textit{if } n = m, \\ \mathtt{false} & \textit{else.} \end{cases}$$

(ii) *There exists a term* **and** $\in \Lambda^{\circ}$ *such that*

$$\begin{aligned} \mathbf{and}\,\mathtt{true}\,\mathtt{true} &= \mathtt{true}, \\ \mathbf{and}\,\mathtt{true}\,\mathtt{false} &= \mathtt{false}, \\ \mathbf{and}\,\mathtt{false}\,\mathtt{true} &= \mathtt{false}, \\ \mathbf{and}\,\mathtt{false}\,\mathtt{false} &= \mathtt{false}. \end{aligned}$$

Proof.

(i) By the representability of the recursive functions.

(ii) Take **and** $= \lambda ab.a\,\mathtt{true}\,b$. ⊣

PROPOSITION 4.2. *There exists a term* $\delta \in \Lambda^{\circ}$ *such that (writing* δ_n *for* $\delta\mathbf{c}_n$*) one has*

$$\begin{aligned} \delta_n \ulcorner x \urcorner \ulcorner x' \urcorner &= \delta_{\mathbb{N}} xy, \\ \delta_n \ulcorner x \urcorner \ulcorner P'Q' \urcorner &= \mathtt{false}, \\ \delta_n \ulcorner x \urcorner \ulcorner \lambda x'.P' \urcorner &= \mathtt{false}, \\[1ex] \delta_n \ulcorner PQ \urcorner \ulcorner x' \urcorner &= \mathtt{false}, \\ \delta_n \ulcorner PQ \urcorner \ulcorner P'Q' \urcorner &= \mathbf{and}\,(\delta_n \ulcorner P \urcorner \ulcorner P' \urcorner)(\delta_n \ulcorner Q \urcorner \ulcorner Q' \urcorner), \\ \delta_n \ulcorner PQ \urcorner \ulcorner \lambda x'.P' \urcorner &= \mathtt{false}, \\[1ex] \delta_n \ulcorner \lambda x.P \urcorner \ulcorner x' \urcorner &= \mathtt{false}, \\ \delta_n \ulcorner \lambda x.P \urcorner \ulcorner P'Q' \urcorner &= \mathtt{false}, \\ \delta_n \ulcorner \lambda x.P \urcorner \ulcorner \lambda x'.P' \urcorner &= \delta_{n+1}(\ulcorner P \urcorner [x := \mathbf{c}_n])(\ulcorner P' \urcorner [x' := \mathbf{c}_n]). \end{aligned}$$

Proof. We introduce the following ad hoc notation.

(i) Let $A_1, \ldots, A_n \in \Lambda$. Then we write

$$\lambda\vec{x}![A_1, \ldots, A_n] \equiv \langle\langle \lambda\vec{x}.A_1, \ldots, \lambda\vec{x}.A_n \rangle\rangle.$$

(ii) If $B_i \equiv [A_{i1}, \ldots, A_{in}]$, then we write

$$\lambda\vec{x}!![B_1, \ldots, B_n] \equiv \langle\langle \lambda\vec{x}!B_1, \ldots, \lambda\vec{x}!B_n \rangle\rangle.$$

(iii) Let for $1 \le i \le n, 1 \le j \le n$ be given $A_{ij} \in \Lambda$. Then

$$\begin{aligned} [A_{ij}] \equiv\ & [[A_{11}, \ldots, A_{1n}], \\ & [A_{21}, \ldots, A_{2n}], \\ & \ldots \\ & [A_{n1}, \ldots, A_{nn}]]. \end{aligned}$$

If $n = 3$ we may write $[A_{ij}]$ as

$$\begin{bmatrix} A_{11} & A_{12} & A_{13} \\ A_{21} & A_{22} & A_{23} \\ A_{31} & A_{32} & A_{33} \end{bmatrix}.$$

Now define $\delta \equiv \lambda ntt'$.

$$\left(\lambda t d! \lambda t' d' n!! \left[\begin{array}{lll} \delta_{\mathbb{N}}(t)_1(t')_1 & \texttt{false} & \texttt{false} \\ \texttt{false} & \text{and}\,(d(t)_1(t')_1 n)(d(t)_2(t')_2 n) & \texttt{false} \\ \texttt{false} & \texttt{false} & d(tn)(t'n)(\mathbf{S}^+ n) \end{array}\right]\right) tt'n,$$

where $\mathbf{S}^+$ lambda defines the successor function. Then by Theorem 2.7 this δ satisfies the specification. ⊣

PROPOSITION 4.3. *For all $M, M' \in \Lambda$ such that $\mathrm{FV}(MM') \subseteq \{x_1, \ldots, x_n\}$ and for substitutions $* = [x_1 := \mathbf{c}_{k_1}] \ldots [x_n := \mathbf{c}_{k_n}]$ with $k_i \neq k_j$ (for $1 \leq i < j \leq n$) one has for $p > k_i$ (for all $1 \leq i \leq n$) that*

$$\delta_p \ulcorner M \urcorner \ulcorner M' \urcorner * = \begin{cases} \texttt{true} & \text{if } M \equiv_\alpha M', \\ \texttt{false} & \text{else.} \end{cases}$$

Proof. By induction on the structure of M, in each case making distinctions according to the structure of M'. We treat four instructive cases.

Case $M \equiv x, M' \equiv x'$. Then

$$\delta_p \ulcorner M \urcorner \ulcorner M' \urcorner * = \delta_{\mathbb{N}} \mathbf{c}_{k_i} \mathbf{c}_{k_{i'}},$$

where $x \equiv x_i, x' \equiv x_{i'}$. This is true or false depending on whether $x \equiv x'$ (so $i = i'$) or $x \not\equiv x'$ (so $i \neq i'$).

Case $M \equiv x, M' \equiv P'Q'$. Then

$$\delta_p \ulcorner M \urcorner \ulcorner M' \urcorner * = \texttt{false}.$$

Case $M \equiv PQ, M' \equiv P'Q'$. Then

$$\begin{array}{rcl} \delta_p \ulcorner M \urcorner \ulcorner M' \urcorner * & = & \text{and}\,(\delta_p \ulcorner P \urcorner \ulcorner P' \urcorner *)(\delta_p \ulcorner Q \urcorner \ulcorner Q' \urcorner *) \\ & =_{IH} & \text{and}\,(\texttt{true}/\texttt{false})(\texttt{true}/\texttt{false}) \\ & = & \texttt{true}/\texttt{false}, \end{array}$$

as it should ($=$ `true` only if $PQ \equiv P'Q'$, i.e., if both $P \equiv P'$ and $Q \equiv Q'$).

Case $M \equiv \lambda x.P, M' \equiv \lambda x'.P'$. Then

$$\begin{array}{rcl} \delta_p \ulcorner M \urcorner \ulcorner M' \urcorner * & = & \delta_{p+1}(\ulcorner P \urcorner [x := \mathbf{c}_p])(\ulcorner P' \urcorner [x' := \mathbf{c}_p]) * \\ & = & \delta_{p+1} \ulcorner P \urcorner \ulcorner P'[x' := x] \urcorner [x := \mathbf{c}_p] * \\ & = & \delta_{p+1} \ulcorner P \urcorner \ulcorner P'[x' := x] \urcorner *', \end{array}$$

with $*' = *[x := \mathbf{c}_p]$ being an admissible substitution. So

$$\delta_p \ulcorner M \urcorner \ulcorner M' \urcorner * =_{IH} \begin{cases} \texttt{true} & \text{if } P \equiv_\alpha P'[x' := x], \\ \texttt{false} & \text{else.} \end{cases}$$

Now $M \equiv_\alpha M'$ iff $\lambda x.P \equiv_\alpha \lambda x'.P'$ ($\equiv_\alpha \lambda x.P'[x' := x]$) iff $P \equiv_\alpha P'[x' := x]$. Hence we are done. ⊣

COROLLARY 4.4. *Write $\Delta \equiv \delta_0$. Then $\Delta \in \Lambda^\circ$ and for all $M, M' \in \Lambda^\circ$ one has*

$$\Delta \ulcorner M \urcorner \ulcorner M' \urcorner = \begin{cases} \texttt{true} & \textit{if } M \equiv_\alpha M', \\ \texttt{false} & \textit{else.} \end{cases}$$

Proof. Immediate from the proposition. ⊣

Note that this corollary cannot hold for arbitrary $M, M' \in \Lambda$. For example, it is impossible to discriminate $\ulcorner x \urcorner$ and $\ulcorner x' \urcorner$. Indeed take $x \not\equiv x'$ and make a substitution $[x' := x]$:

$$\Delta \ulcorner x \urcorner \ulcorner x' \urcorner = \texttt{false} \Rightarrow \Delta \ulcorner x \urcorner \ulcorner x \urcorner = \texttt{false},$$

a contradiction.

REFERENCES

Barendregt, H. P.

1984 *The lambda calculus, its syntax and semantics*, revised edition, Studies in Logic and the Foundations of Mathematics, North-Holland, Amsterdam.

1991 Theoretical pearls: Self-interpretation in lambda calculus, *Journal of Functional Programming*, vol. 1(2), pp. 229–233.

Böhm, C., and A. Berarducci

1985 Automatic synthesis of typed λ-programs on term algebras, *Theoretical Computer Science*, vol. 39, pp. 135–154.

Böhm, C., A. Piperno, and S. Guerrini

1994 λ-definitions of function(al)s by normal forms, *ESOP '94* (D. Sannella, editor), LNCS 788, Springer, Berlin, pp. 135–149

Church, A.

1941 *The calculi of lambda conversion*, Princeton University Press, Princeton; reprinted by Kraus Reprint Corporation, New York, 1965.

Kleene, S. C.

1936 λ-definability and recursiveness, *Duke Mathematical Journal*, vol. 2, pp. 340–353.

McCarthy, J., P. W. Adams, D. J. Edwards, T. P. Hart, and M. I. Levin

1962 *LISP 1.5 Programmer's manual*, MIT Press, Cambridge, Massachusetts.

Mogensen, T. Æ.

1992 Efficient self-interpretation in lambda calculus, *Journal of Functional Programming*, vol. 2(3), pp. 345–364.

KLAUS GRUE*

λ-CALCULUS AS A FOUNDATION FOR MATHEMATICS

Abstract. Church introduced the λ-calculus in the beginning of the thirties as a foundation of mathematics, and map theory, from around 1992, fulfilled that primary aim.

The present paper presents a new version of map theory whose axioms are simpler and better motivated than those of the original version from 1992. The paper focuses on the semantics of map theory and explains this semantics on the basis of κ-Scott domains.

The new version sheds some light on the difference between Russell's and Burali-Forti's paradoxes, and also sheds some light on why it is consistent to allow non-well-founded sets in a ZF-style system.

1. INTRODUCTION

Church introduced the λ-calculus in the beginning of the thirties (*Church 1932, 1933, 1941*) as a foundation of mathematics, and map theory, from around 1992, fulfilled that primary aim.

In the meanwhile, λ-terms have been shown very useful for expressing semantics in computer science, but there has been no natural choice of a theory for reasoning about these λ-terms.

With the lack of such a natural choice, computer science has turned to syntactic methods in which one reasons about the structure and conversion of λ-terms rather than their meaning.

Mathematics, however, has had set theory which allows reasoning about sets rather than syntax. Set theory offers the luxury of referential transparency, in which every term has a meaning and every term implicitly denotes that meaning.

Map theory resembles set theory in that it assigns meaning to λ-terms and treats λ-terms in a referentially transparent fashion. Map theory also resembles set theory when comparing metamathematical power: For every consistent set theory Z, there is a consistent map theory M more powerful than Z and vice versa. The deepest difference between set and map theories shows up in the treatment of infinite looping: Russell's sentence $\{x \mid x \notin x\} \in \{x \mid x \notin x\}$ is just a term that takes an infinitely long time to compute, but set theory deals with this sentence by forbidding it rather than taking its value seriously. Map theory, on the contrary, assigns meaning to the

* My thanks are due to the referee from whose report I copied a paragraph to the abstract.

C. Anthony Anderson and M. Zelëny (eds.), *Logic, Meaning and Computation*, 287-311.

corresponding term $(\lambda x.\neg(xx))(\lambda x.\neg(xx))$.

Map theory is not a marriage of convenience between λ-calculus and set theory (cf. *Feferman 1978*). Rather, map theory is a theory based entirely on λ-calculus in which, among other conceps, set membership, logical connectives, and quantification over all sets are definable, and in which all axioms and inference rules of ZFC are provable without resort to any syntactical considerations.

Map theory has the potential to serve as a foundation of both mathematics (due to its power that is equivalent to that of set theory) and of computer science (due to its treatment of λ-terms). Set and map theory are both inherently difficult to learn due to their level of abstraction, but set theory has the advantage of a century of pedagogical engineering that has made it easier to approach. In contrast, the version of map theory from 1992 appears as a somewhat random collection of axioms that accidentally have the power of set theory and accidentally describe λ-terms.

Since 1992 it has turned out, however, that map theory is a natural choice of a theory of λ-calculus. More specifically, it has turned out that every sufficiently large so-called κ-Scott domain $\mathcal{D}$ for which $\mathcal{D} \cong [\mathcal{D} \to \mathcal{D}] \oplus_{\perp} \mathbf{1}$ where $\mathbf{1}$ is a one-element set, contains a model for map theory (*Berline and Grue 1997*). When Dana Scott invented what is now known as Scott domains, he was fully aware that the notion could be generalised to κ-Scott domains, but did not publish this finding as he saw no application of them at that time. Hence, in some sense, a λ-based foundation of mathematics has been around for a long time without anybody recognising it.

The present paper presents a new version of map theory which will be referred to as *MTC* (Map Theory with Classical maps); the earlier version from 1992 will be referred to as *MTW* (Map Theory with Well-founded maps). The axioms of MTC are simpler and better motivated than those of MTW, and the step from MTW to MTC is intended as a step in the direction of a theory that is easier to learn and teach. The paper focuses on the semantics of MTC and explains this semantics on the basis of κ-Scott domains.

1.1. Differences from Church's approach

It is a pity that Church did not find a theory like map theory right away since it could have saved a lot of work in computer science. There are, however, three good reasons why that did not happen, and these reasons are stated in the next three sections.

1.2. Inclusion of non-functions

The first reason is that Church's theory may be seen as a theory about functions only, and as such, is a theory about only one concept. Classical logic is built around the distinction between truth and falsehood, i.e., the semantic distinction between two concepts; and the semantic nature of classical logic and set theory stems from this distinction. λ-calculus, on the other hand, deals with functions only, and in a theory with only one concept, it is impossible to make a semantic distinction. For that reason, λ-calculus can merely deal with the provability or non-provability of the equivalence of λ-terms, which is syntactic in nature.

In map theory, this problem is solved by insisting that the universe of map theory must contain at least one non-function. Just one non-function is enough to make a distinction, namely a distinction between functions and non-functions. In map theory, the minimalistic approach has been taken to include only one non-function.

Having both functions and non-functions in map theory allows it to represent truth and falsehood, and the convention has been chosen to let functions represent falsehood and let non-functions represent truth. This convention has been chosen very carefully on the basis of what makes definitions inside map theory easiest to read, but this issue will not be treated here.

Since there is only one non-function in map theory, it is convenient to introduce a name for that non-function, and since the non-function represents truth, the name T has been chosen.

In addition to functions and non-functions, the universe of map theory contains an element which is neither a function, nor a non-function. That element is denoted $\perp$ and represents infinite looping. This element violates the axiom of Tertium Non Datur, which in this context says that every object is either a function or a non-function. Nevertheless map theory is still classical in nature because it has another, similar axiom called Quartum Non Datur which says that any map is either a function or T or $\perp$, with no fourth possibility.

The inclusion of the non-function T in map theory is a trivial step, but it is a step that is very important for the semantic nature of the theory.

1.3. Set abstraction versus λ-abstraction

Set theory has set abstraction $\{x \mid p(x)\}$, and λ-calculus has λ-abstraction $\lambda x.p(x)$. It is tempting to identify the two kinds of abstraction and try to represent the class $\{x \mid p(x)\}$ by $\lambda x.p(x)$, i.e., to represent classes by their characteristic functions. This approach, however, has not succeeded, and set abstraction and λ-abstraction seem to be two completely different kinds of abstraction.

In map theory, a function g does not represent the class $\{x \mid g(x) = \mathsf{T}\}$. Rather, g represents the class $\{g(x) \mid x \in S\}$ where S is a fixed class of maps. In MTW, S is the class W of well-founded maps, and in MTC, S is the class C of classical maps. $\{g(x) \mid x \in S\}$ contains at least one element, which allows representation of the empty set; the non-function T is taken to represent that set.

With this encoding, all sets of ZFC can be represented by well-founded maps in MTW and by classical maps in MTC. And, conversely, all well-founded maps in MTW and all classical maps in MTC represent sets. Classes may also be represented, but they are represented by maps that are not well-founded/classical.

In conclusion, the non-identification of set abstraction and λ-abstraction has been an important point in turning λ-calculus into a foundation.

1.4. Selection of well-behaved maps

The third problem in turning λ-calculus into a foundation of mathematics was to find a class S of maps that was sufficiently well-behaved to represent the sets of ZFC. In MTW the class W of well-founded maps was chosen and in MTC, the class C of classical maps was chosen.

Insisting that $\{g(x) \mid x \in S\}$ should be a set of ZFC for all $g \in S$ puts many restrictions on S, and insisting that all sets of ZFC should be representable by an element of S puts strong requirements on the size and richness of S. Nevertheless, these restrictions by no means determine S uniquely, and finding a natural S is no trivial task.

1.5. Relation between MTW and MTC

One advantage of MTC over MTW is that eleven complicated axioms and inference rules that describe well-foundedness in MTW have been replaced by a single definition of classicality in MTC.

Another advantage is that MTC contains some inference rules that were missing in MTW. These are rules Y, M, and E in Appendix 2.5. In particular, rule Y says that the fixed point operator generates a minimal fixed point. It is interesting that the theorem of transfinite induction is provable in MTC from the minimality of fixed points combined with the recursive definition of classicality used in MTC. The details are worked out in *Grue 1996*.

A third advantage is the distinction between discontinuous and continuous occurrences of variables introduced in section 3.1 which allows a treatment of equations as terms. The syntax of MTW specifies the syntax of terms and well-formed formulas. MTC is simpler in that it does not distinguish between terms and well-formed formulas. This may turn out to be convenient in computer assisted proof systems for MTC because it allows a representation of both theorems and inference rules as terms. The notion of discontinuous occurrences permits the inference rule $a = \mathsf{T} \vdash a$, which is also valid in the systems of Feferman (*1984*) and Flagg and Myhill (*1989*). The notion of discontinuous occurrences also permits the opposite rule $a \vdash a = T$. The distinction between discontinuous and continuous occurrences of variables seems to be a new way to deal with the anomalies of equality.

MTW and MTC may be compared both on a syntactic and a semantic basis. A syntactic comparison is made in *Grue 1996* with the following result: If the classicality predicate of MTC is used to simulate the well-foundedness predicate in MTW, then all axioms and inference rules of MTW, except Axiom Well-2 in *Grue 1992,* are provable in MTC. Axiom Well-2 in MTW is disprovable in MTC, which is just a consequence of the slight difference between well-foundedness and classicality. In *Grue 1992,* Axiom Well-2 is only used to prove Lemma C-K and Lemma C-P, both of which are provable in MTC. Hence, all theorems of MTW proved in *Grue 1992* are also provable in MTC.

Among other results, Grue (*1992*) proves all axioms and inference rules of ZFC in MTW. Combining these observations we have that all axioms and inference rules of ZFC are provable in MTC. This shows that MTC is adequate as a foundation of mathematics (provided ZFC is considered adequate).

MTW and MTC may also be compared semantically. This paper introduces the semantics of MTC by means of a κ-Scott domain $\mathcal{D} = (D, \leq)$ where D is the universe of all maps and $\leq$ is a partial order on all maps. The κ-Scott domain used has the property that it models both MTW and MTC, so it makes sense to compare MTW and MTC in this particular model. A class A of maps will be said to be "coherent" if any two elements of A have an upper bound in $\mathcal{D}$. The correspondence between the classes W and C of well-founded and classical maps, respectively, may now be formulated as follows:

- Any well-founded map is classical.
- Any non-empty, coherent class of well-founded maps has a greatest lower bound, and that greatest lower bound is classical.
- Any classical map is the lower bound of a coherent class of well-founded maps.

Hence, the class C of classical maps can be seen as the closure of W under greatest lower bounds of non-empty, coherent sets. An important difference between W and C is:

- For all well-founded maps g except T there exists a well-founded map h such that $h < g$.
- For all classical maps g there exists a minimal classical map h such that $h \leq g$ (h is a minimal classical map if $h' \leq h \Rightarrow h' = h$ for all classical maps h').

1.6. The structure and contents of the paper

Section 2 describes the semantics of MTC based on the κ-denotational framework developed in *Berline and Grue 1997*. Section 3 gives a rather quick tour through the syntax, axioms and inference rules of MTC. Section 4 concludes by remarks on Russell's and Burali-Forti's paradoxes and non-well-founded sets. Appendix 1 outlines a model of MTC based on the model of MTW in *Berline and Grue 1997*. Only the definition of the model is stated. The satisfaction of the axioms and inference rules remains to be proved. Appendix 2 summarises MTC.

A more detailed description of the individual axioms and inference rules and explanations of how they are used may be found in *Grue 1996*. Note that the system in *Grue 1996* contains an inconsistent axiom as pointed out by Chantal Berline. *Grue 1996* plus errata may be obtained from www.diku.dk/~grue.

2. THE SEMANTICS OF MTC

2.1. Maps over finite sets

In the following, words in italics and mathematical concepts in boxes occur in the index. Figure 1 shows a *map* over the set $I = \{1, 2, 3\}$.

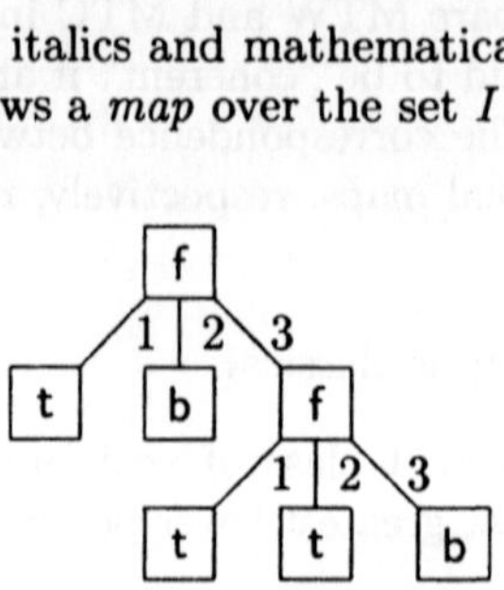

Figure 1: A map over $\{1, 2, 3\}$

In general, a map over I is a tree where each node is labelled by $\boxed{\mathsf{t}}$, $\boxed{\mathsf{f}}$, or $\boxed{\mathsf{b}}$, each edge is labelled by an element of I, each node labelled f has one downward edge for each element of I, and each node labelled t or b has no downward edges. Maps may be infinitely deep. As an example, Figure 2 shows a map over $\{1, 2\}$.

If I is a set, if $x \in I$, and if g is a map over I, then we define *g applied to* x, denoted $\boxed{g'x}$, to be the subtree of g attached to the edge labelled x that extends downwards from the root of g. As an example, if g is the map in Figure 1 then Figures 3, 4, and 5 show $g'1$, $g'2$, and $g'3$, respectively.

Application is left associative so that, e.g., $g'3'1$ means $(g'3)'1$. If g is the map in Figure 1 then $g'3'1$ is the map in Figure 3. If h is the map in Figure

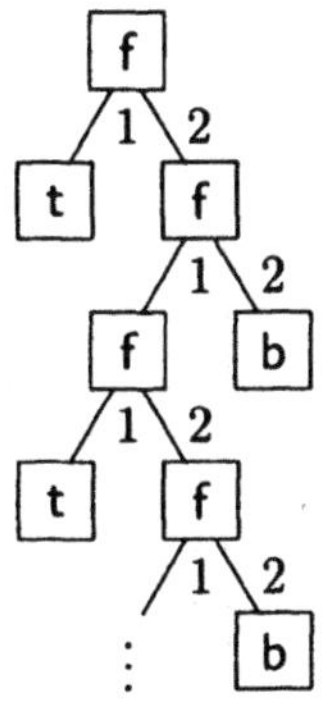

Figure 2: A map over $\{1,2\}$

Figure 3: The map T

2 then $h'2'1 = h$.

The trees in Figure 3 and 4 will be denoted $\boxed{\mathsf{T}}$ and $\boxed{\bot}$, respectively. No edges extend downward from the roots of T and $\bot$, so the definition of $g'x$ does not make sense for $g = \mathsf{T}$ and $g = \bot$. To make $g'x$ defined for all $x \in I$ and all maps g over I, we more or less arbitrarily define:

$$\begin{aligned} \mathsf{T}'x &= \mathsf{T}, \\ \bot'x &= \bot. \end{aligned}$$

2.2. Modelling of maps

Let t, f, and b be three distinct objects. For all maps g over I we define $\boxed{r(g)}$ to be the label of the root of g. Hence, $r(\mathsf{T}) = \mathsf{t}$, $r(\bot) = \mathsf{b}$, and $r(g) = \mathsf{f}$ for all maps g over I except T and $\bot$.

Let $\boxed{I^{<\omega}}$ denote the set of finite lists $\langle x_1, \ldots, x_n \rangle$ of elements of I. For all maps g over I and all $\overline{x} = \langle x_1, \ldots, x_n \rangle \in I^{<\omega}$, let $g[\overline{x}]$ denote $g'x'_1 \cdots x'_n$. As an example, $h[\langle 2,1,2,1,2,1 \rangle] = h$ where h is the map in Figure 2. If $\overline{x}$ is the empty tuple $\langle\rangle$, then $g[\overline{x}]$ denotes g itself.

If g is a map over I and if $x \in I^{<\omega}$, then $r(g[\overline{x}])$ will be referred to as the label *indexed* by $\overline{x}$. As an example, if g is the map in Figure 1, then the labels indexed by $\langle 1 \rangle$, $\langle 3 \rangle$, and $\langle 3,3 \rangle$ are t, f, and b, respectively. The label

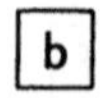

Figure 4: The map $\bot$

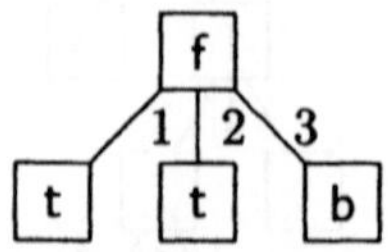

Figure 5: A map over $\{1, 2, 3\}$

indexed by $\langle 3, 3, 2\rangle$ is also b as shown by the following:

$$\begin{aligned} g[\langle 3,3,2\rangle] &= g'3'3'2, \\ &= (g'3'3)'2, \\ &= \bot'2, \\ &= \bot. \end{aligned}$$

Hence, computation of $r(g[\langle 3,3,2\rangle])$ depends on the convention that $\bot'x = \bot$.

In general, if $r(g[\langle x_1, \cdots, x_m\rangle]) = \mathsf{b}$ then $r(g[\langle x_1, \cdots, x_m, y_1, \cdots, y_n\rangle]) = \mathsf{b}$. Now for all $\overline{x}, \overline{y} \in I^{<\omega}$ let $\boxed{\overline{x} \cdot \overline{y}}$ denote the concatenation of the tuples x and y. We have

$$\begin{aligned} r(g[\overline{x}]) = \mathsf{b} &\Rightarrow r(g[\overline{x} \cdot \overline{y}]) = \mathsf{b}, \\ r(g[\overline{x}]) = \mathsf{t} &\Rightarrow r(g[\overline{x} \cdot \overline{y}]) = \mathsf{t}. \end{aligned}$$

These two statements may be combined into one:

$$r(g[\overline{x}]) \neq \mathsf{f} \Rightarrow r(g[\overline{x} \cdot \overline{y}]) = r(g[\overline{x}]).$$

Now let $\boxed{L} = \{\mathsf{t}, \mathsf{f}, \mathsf{b}\}$. For all functions u, let $\boxed{\operatorname{dom} u}$ and $\boxed{\operatorname{rng} u}$ denote the domain and range of u, respectively. For all sets G and H, let $\boxed{G \to H}$ denote the set of functions u for which $\operatorname{dom} u = G$ and $\operatorname{rng} u \subseteq H$.

If g is a map over I and if $\overline{x} \in I^{<\omega}$, then $r(g[\overline{x}]) \in L$. Two maps u and v over I are considered equal if $r(g[\overline{x}]) = r(h[\overline{x}])$ for all $\overline{x} \in I^{<\omega}$. Hence, a map g may be modelled by the function $\hat{g} \in I^{<\omega} \to L$ for which $\hat{g}(\overline{x}) = r(g[\overline{x}])$ for all $x \in I^{<\omega}$. From now on, maps are modelled this way which motivates a definition of the set M_I of maps over I by

$$\boxed{M_I} = \{g \in I^{<\omega} \to L \mid \forall \overline{x}, \overline{y} \in I^{<\omega} : (g(\overline{x}) \neq \mathsf{f} \Rightarrow g(\overline{x} \cdot \overline{y}) = g(\overline{x})).$$

2.3. Partially ordered sets

A *p.o* $\mathcal{D}$ is a *partially ordered set* $(D, \leq)$. As an example, $\boxed{\mathcal{L}} = (L, \leq_L)$ is a p.o where

$$\boxed{p \leq_L q} \Leftrightarrow p = \mathsf{b} \vee p = q.$$

For all non-empty sets I, $\boxed{\mathcal{M}_I} = (M_I, \leq)$ is a p.o where

$$\boxed{g \leq h} \Leftrightarrow \forall x \in I^{<\omega} : g(x) \leq_L h(x).$$

For all p.o's $\mathcal{D} = (D, \leq)$, define the p.o $\boxed{\mathcal{D}^{<\omega}}$ by $\mathcal{D}^{<\omega} = (D^{<\omega}, \boxed{\leq^*})$ where

$$\langle x_1, \ldots, x_m \rangle \leq^* \langle y_1, \ldots, y_n \rangle \Leftrightarrow m = n \wedge x_1 \leq y_1 \wedge \cdots \wedge x_m \leq y_m.$$

For all p.o's $\mathcal{D} = (D, \leq)$ and $\mathcal{E} = (E, \leq)$, let $x \in \mathcal{D}$, $A \subseteq \mathcal{D}$ and $g \in \mathcal{D} \to \mathcal{E}$ be shorthand for $x \in D$, $A \subseteq D$, and $g \in D \to E$, respectively.

For all non-empty sets I, define $\boxed{\bot_I}, \boxed{\mathsf{T}_I}, \boxed{\mathsf{F}_I} \in \mathcal{M}_I$ by

$$\begin{aligned}
&\bot_I(\langle x_1, \ldots, x_n \rangle) = \mathsf{b},\\
&\mathsf{T}_I(\langle x_1, \ldots, x_n \rangle) = \mathsf{t},\\
&\mathsf{F}_I(\langle\rangle) = \mathsf{f},\\
&\mathsf{F}_I(\langle u, x_1, \ldots, x_n \rangle) = \mathsf{t}
\end{aligned}$$

for all $n \geq 0$ and all $u, x_1, \ldots, x_n \in I$. $\bot_I$ is the unique bottom element of $\mathcal{M}_I$. T_I and F_I are two among many maximal elements.

2.4. κ-continuity

For all sets A and κ, A is said to be *κ-small* if A has cardinality strictly less than κ. From now on let κ be an infinite set.

An element x of a p.o $\mathcal{D} = (D, \leq)$ is said to be an *upper bound* of $A \subseteq \mathcal{D}$ if $\forall y \in A : y \leq x$. A subset $H \subseteq \mathcal{D}$ is said to be a *κ-chain* if all κ-small subsets of H have an upper bound in H. In particular, the empty set must have an upper bound in H, so any κ-chain is non-empty. A p.o $\mathcal{D} = (D, \leq)$ is a *κ-cpo* if every κ-chain $H \subseteq \mathcal{D}$ has a supremum $\boxed{\sup H}$ in $\mathcal{D}$. As an example, for all non-empty sets I, $\mathcal{M}_I$ and $(\mathcal{M}_I)^{<\omega}$ are κ-cpo's. $\mathcal{M}_I$ is an example of a κ-cpo with a bottom element and $(\mathcal{M}_I)^{<\omega}$ is an example of one without.

For all κ-cpo's $\mathcal{D}$ and $\mathcal{E}$, a function $g \in \mathcal{D} \to \mathcal{E}$ is said to be *κ-continuous* if

$$g(\sup H) = \sup \{g(x) \mid x \in H\}$$

for all κ-chains $H \subseteq \mathcal{D}$. Let $\boxed{[\mathcal{D} \to \mathcal{E}]_\kappa}$ denote the κ-cpo of κ-continuous $g \in \mathcal{D} \to \mathcal{E}$, ordered by pointwise ordering. For all $\mathcal{D} = (D, \leq)$, define

$$\begin{aligned}\boxed{M^\kappa_\mathcal{D}} &= \{g{\in}M_D \mid g \in [\mathcal{D}^{<\omega} \to \mathcal{L}]_\kappa\},\\ \boxed{\mathcal{M}^\kappa_\mathcal{D}} &= (M^\kappa_\mathcal{D}, \leq).\end{aligned}$$

We shall refer to elements of $\mathcal{M}^\kappa_\mathcal{D}$ as *κ-continuous maps over $\mathcal{D}$.*

2.5. Maps over **R**, *and maps over maps over* **R**

At this point a few examples may be illustrative. Let $\mathbf{R}$ be the set of real numbers, let $\mathbf{Z} \subseteq \mathbf{R}$ be the set of integers, and let $\mathcal{G} = (G, \leq) = \mathcal{M}_\mathbf{R}$ be the p.o of maps over $\mathbf{R}$. We have that $\mathbf{Z}$ is $\mathbf{R}$-small since $\mathbf{Z}$ has cardinality strictly less than $\mathbf{R}$.

For all $A \subseteq \mathbf{R}$ define the characteristic map $\chi_A \in \mathcal{G}$ by

$$\begin{aligned}\chi_A(\langle\rangle) &= \mathsf{f},\\ \chi_A(\langle u, x_1, \ldots, x_n\rangle) &= \begin{cases}\mathsf{t} & \text{if } u \in A,\\ \mathsf{b} & \text{otherwise.}\end{cases}\end{aligned}$$

As an example of a non-trivial $\mathbf{R}$-chain in $\mathcal{G}^{<\omega}$, we have

$$H = \{\langle\chi_A\rangle \mid A \subseteq \mathbf{R} \wedge A \text{ is } \mathbf{R}\text{-small}\}.$$

The supremum $\sup H$ of this chain is $\langle\chi_\mathbf{R}\rangle$.

Now consider the set M_G of maps over $G = M_\mathbf{R}$, i.e., of maps over maps over $\mathbf{R}$. For all $B \subseteq \mathbf{R}$, define $\forall_B \in M_G$ by

$$\begin{aligned}\forall_B(\langle\rangle) &= \mathsf{f},\\ \forall_B(\langle u, x_1, \ldots, x_n\rangle) &= \begin{cases}\mathsf{b} & \text{if } \exists v{\in}B : u(\langle v\rangle) = \mathsf{b},\\ \mathsf{t} & \text{if } \forall v{\in}B : u(\langle v\rangle) = \mathsf{t},\\ \mathsf{F}_G(x_1, \ldots, x_n) & \text{otherwise.}\end{cases}\end{aligned}$$

$\forall_B$ satisfies

$$\forall_B{}'u = \begin{cases}\bot_G & \text{if } \exists v \in B : u'v = \bot_\mathbf{R},\\ \mathsf{T}_G & \text{if } \forall v \in B : u'v = \mathsf{T}_\mathbf{R},\\ \mathsf{F}_G & \text{otherwise.}\end{cases}$$

If we take T, F, and $\bot$ to represent truth, falsehood, and undefined, respectively, then $\forall_B$ represents a universal quantifier that quantifies over B. The quantifier is strict in the sense that $\forall_B g$ is undefined if gx is undefined for some $x \in B$.

We have $\forall_\mathbf{R}(\sup H) = \mathsf{t}$ and $\sup\{\forall_\mathbf{R}(x) \mid x \in H\} = \mathsf{b}$ which shows that $\forall_\mathbf{R}$ is $\mathbf{R}$-discontinuous. Nevertheless, it is straightforward to prove that $\forall_\mathbf{Z}$

is **R**-continuous (the proof is a somewhat lengthy proof by cases, but the point of the proof is that for all **R**-chains H' there exists a function $h'' \in H'$ such that $\forall n{\in}\mathbf{Z}\colon h''(n) = h'(n)$ where $h' = \sup H'$). In general, $\forall_B$ is **R**-continuous if and only if B is **R**-small. Even more generally, quantification over a set B is κ-continuous if and only if B is κ-small.

2.6. κ-premodels

We have now seen maps over $\{1,2,3\}$, maps over **R**, and **R**-continuous maps over maps over **R**. The maps of MTC are κ-continuous maps over maps of MTC. In other words, the domain $\mathcal{D}$ of all maps of MTC satisfies $\mathcal{D} \cong \mathcal{M}^{\kappa}_{\mathcal{D}}$.

If $\mathcal{D}$ is a *κ-Scott domain* (*Berline and Grue 1997*), if σ is a *strongly inaccessible ordinal* (*Chang and Keisler 1973*), if κ is a *regular cardinal* (*Chang and Keisler 1973*) greater than σ, and if $\mathcal{D} \cong \mathcal{M}^{\kappa}_{\mathcal{D}}$ then it follows from Berline and Grue (*1997*) that $\mathcal{D}$ can be expanded into a model of MTW (to see that, one needs to prove $\mathcal{D} \cong \mathcal{M}^{\kappa}_{\mathcal{D}} \Rightarrow \mathcal{D} \cong [\mathcal{D} \to \mathcal{D}]_\kappa \oplus_\perp \{\mathsf{T}_D\}$, then let $A \in \mathcal{D} \to [\mathcal{D} \to \mathcal{D}]_\kappa \oplus_\perp \{\mathsf{T}_D\}$ be an order isomorphism, let λ be the inverse of A and apply Theorem 7.2.1 of *Berline and Grue 1997* to $(\mathcal{D}, A, \lambda)$).

A model of MTC needs to satisfy more than this, partly because of the classical maps, partly because MTC contains inference rules that were missing in MTW.

From now on, assume that there exist strongly inaccessible ordinals, assume that $\boxed{\sigma}$ is the least strongly inaccessible ordinal, and assume that $\boxed{\kappa}$ is a regular cardinal greater than σ.

A *κ-premodel* P of MTC is a structure $(\mathcal{D}, \mathsf{a}, C, q)$ that satisfies the five properties below plus one more property which is stated in section 2.7.

- $\mathcal{D} = (D, \leq)$ is a κ-Scott domain (and, in particular, a κ-cpo).
- $\mathsf{a} \in [\mathcal{D} \to \mathcal{M}^{\kappa}_{\mathcal{D}}]_\kappa$ is an isomorphism.
- $C \subseteq \mathcal{D}$ is a κ-small set of so-called κ-compact elements (cf. *Berline and Grue 1997* and Appendix 1.1).
- q is a choice function over $\mathcal{D}$, i.e., $q(A) \in A$ for $A \subseteq \mathcal{D}$, $A \neq \emptyset$.
- $q(\emptyset) \in C$.

κ-premodels are introduced here to present the intuitions behind MTC. The construction of a κ-premodel is outlined later. The detailed development and the verification of axioms and inference rules of MTC remains to be done. Now define

$\boxed{\bar{\mathsf{a}}}$ is the inverse of a,
$\boxed{\mathsf{T}} = \bar{\mathsf{a}}(\mathsf{T}_D)$,
$\boxed{\perp} = \bar{\mathsf{a}}(\perp_D)$,
$\boxed{\mathsf{F}} = \bar{\mathsf{a}}(\mathsf{F}_D)$.

For all $g, x \in \mathcal{D}$ let $\boxed{g'x}$ be the unique element of $\mathcal{D}$ for which

$$\mathsf{a}(g'x)(\langle y_1, \ldots, y_n \rangle) = \mathsf{a}(g)(\langle x, y_1, \ldots, y_n \rangle).$$

This defines the notion of applying a map g to an argument x, which was first mentioned in section 2.1. This concludes a circle: informal considerations about application of maps led to a representation of maps which was refined into a model of maps which allows a definition of application.

2.7. Classical maps

For all $S \subseteq D$ and $x, y \in D$, define $\boxed{x =_S y} \Leftrightarrow \forall z \in S^{<\omega} : \mathsf{a}(x)(z) = \mathsf{a}(y)(z)$. For all $x, y \in D^{<\omega}$, define $\boxed{x =^*_S y}$ by

$$\langle x_1, \cdots, x_m \rangle =^*_S y \langle y_1, \cdots, y_n \rangle \Leftrightarrow m = n \wedge x_1 = y_1 \wedge \cdots \wedge x_m = y_m.$$

Let $\boxed{\mathcal{P}_\sigma(A)}$ denote the set of σ-small subsets of A. Let C' be the least subset of D which satisfies

$$\begin{aligned} g \in C' \;\Leftrightarrow\; & \forall x \in C : g'x \in C' \wedge \\ & \exists V \in \mathcal{P}_\sigma(C') \forall x, y \in C^{<\omega} : (x =^*_V y \Rightarrow \mathsf{a}(g)(x) = \mathsf{a}(g)(y)). \end{aligned}$$

The existence of such a C' is easy to verify. For all $x \in \mathcal{D}$ and $A \subseteq \mathcal{D}$, let $\boxed{\uparrow x} = \{y \in \mathcal{D} \mid x \leq y\}$ and $\uparrow A = \bigcup \{\uparrow x \mid x \in A\}$. A κ-premodel P of MTC is a structure $(\mathcal{D}, \mathsf{a}, C, q)$ that satisfies the five properties in section 2.6 plus the one below:

- $C' = \uparrow C$.

3. PRESENTATION OF MTC

3.1. Syntax

The syntax $\mathcal{V}$ of variables and $\mathcal{T}$ of terms of MTC reads:

$$\begin{aligned} \mathcal{V} \;&::=\; x_1 \mid x_2 \mid \cdots, \\ \mathcal{T} \;&::=\; \mathcal{V} \mid \mathsf{T} \mid \mathcal{T}\mathcal{T} \mid \lambda \mathcal{V}.\mathcal{T} \mid \mathsf{P} \mid \varepsilon \mid \bar{\exists} \mid \epsilon(\mathcal{T}) \mid \mathcal{T} = \mathcal{T}. \end{aligned}$$

The construct $\mathcal{T}\mathcal{T}$ has higher priority than $\lambda\mathcal{V}.\mathcal{T}$, which in turn has higher priority than $\mathcal{T} = \mathcal{T}$, so that, e.g., $\lambda x_1 . x_1 x_1 = x_2$ means $(\lambda x_1 . (x_1 x_1)) = x_2$.

An occurrence of a variable v in a term t is said to be *discontinuous* if v occurs free in a subterm t' of t which has one of the forms $\epsilon(t'')$ or $t'' = t'''$. Occurrences that are not discontinuous are said to be *continuous*. As an example, the second occurrence of x_1 in $\mathsf{P} x_1 x_2 (x_1 = \lambda x_1 . x_1)$ is discontinuous, whereas the other occurrences of variables are continuous.

The following purely syntactical restriction is put on terms of MTC: Discontinuous occurrences of variables are not allowed to be bound. As an example, $\lambda x_1.(x_1 = \mathsf{T})$ is not a well-formed term.

3.2. Proofs

A proof in MTC is a sequence of terms in which each term is either an instance of an axiom scheme or follows from previous terms in the sequence by an inference rule. All axioms and inference rules of MTC are listed in Appendix 2.5.

The interpretation of a proof is that it proves the last term in the proof, i.e., it proves that the last term in the proof equals T for all values of free variables.

3.3. Truth, equality and application

The term T in section 3.1 denotes the value T defined in section 2.6. The term $a = b$ equals T when a equals b, and $a = b$ equals F otherwise. The value of $a = b$ may depend on the values of free variables in a and b. As an example, $x_1 = \mathsf{T}$ equals T when x_1 equals T and equals F otherwise. The inference rules of truth and equality read:

T	Transitivity	$a = b, a = c \vdash b = c,$
SA	Substitutivity	$a = c, b = d \vdash ab = cd,$
$S\lambda$	Substitutivity	$a = b \vdash \lambda x.a = \lambda x.b,$
$S\epsilon$	Substitutivity	$a = b \vdash \epsilon(a) = \epsilon(b),$
$=\vdash$	Equality	$a = \mathsf{T} \vdash a,$
$\vdash=$	Equality	$a \vdash a = \mathsf{T}.$

In the inference rules, a, b, c, d, g, and h denote arbitrary terms; whereas x, y, and z denote arbitrary, distinct variables.

The inference rule $a = \mathsf{T} \vdash a$ is inspired by Feferman (*1984*), Flagg and Myhill (*1989*). The opposite rule $a \vdash a = \mathsf{T}$ is made possible by the notion of discontinuous occurrences of variables, which seems to be a new way of dealing with the discontinuity of equality.

For all terms a and b, $\boxed{ab}$ denotes a applied to b, i.e., ab denotes $a'b$. Since T applied to anything equals T by convention, we have the axiom

$$AT \quad \text{Application} \quad \mathsf{T}a = \mathsf{T}.$$

As an example of a proof, we have

1	AT	$\mathsf{T}a = \mathsf{T},$
2	AT	$\mathsf{T}a = \mathsf{T},$
3	$1, 2, T$	$\mathsf{T} = \mathsf{T},$
4	$3, =\vdash$	$\mathsf{T}.$

3.4. Abstraction

For all variables x and terms $\mathcal{A}$ of ZFC, let $x \mapsto \mathcal{A}$ be shorthand for $\{(x, \mathcal{A}) \mid x \in \mathcal{D}\}$, i.e., let $x \mapsto \mathcal{A}$ be the unique function g for which $\operatorname{dom} g = \mathcal{D}$ and $g(x) = \mathcal{A}$. For all $g \in [\mathcal{D} \to \mathcal{D}]_\kappa$, let $\lambda(g)$ be the unique element of $\mathcal{D}$ for which

$$\begin{aligned} \lambda(g)(\langle\rangle) &= \mathsf{f}, \\ \lambda(g)(\langle u, x_1, \ldots, x_n\rangle) &= g(u)(\langle x_1, \ldots, x_n\rangle). \end{aligned}$$

Finally, define

$$\lambda x.\mathcal{A} = \lambda(x \mapsto \mathcal{A})$$

whenever $(x \mapsto \mathcal{A}) \in [\mathcal{D} \to \mathcal{D}]_\kappa$, i.e., whenever $x \mapsto \mathcal{A}$ is a κ-continuous function from maps to maps.

If x is a variable and $\mathcal{A}$ is a term of MTC, then $(x \mapsto \mathcal{A}) \in \mathcal{D} \to \mathcal{D}$. If furthermore, there are no discontinuous free occurrences of x in $\mathcal{A}$, then $x \mapsto \mathcal{A}$ is κ-continuous, so that $(x \mapsto \mathcal{A}) \in [\mathcal{D} \to \mathcal{D}]_\kappa$ and $(\lambda x.\mathcal{A}) \in \mathcal{D}$. This gives an interpretation of $\lambda x.\mathcal{A}$ whenever $\lambda x.\mathcal{A}$ is a well-formed term. We may now formulate two more axioms, namely:

$A\lambda$	Application	$(\lambda x.a)b = \langle a \mid x{:=}b\rangle$ if b is free for x in a,
R	Renaming	$\lambda x.\langle a \mid y{:=}x\rangle = \lambda y.\langle a \mid x{:=}y\rangle$ if x is free for y in a and vice versa.

Here $\boxed{\langle a \mid x{:=}b\rangle}$ is the term that arises when replacing all free x in a by b. See *Mendelson 1987* for a definition of *free for*. Axiom R allows renaming of bound variables and Axiom $A\lambda$ expresses that two terms are equal if they are β-equivalent. (Note, however, that terms may be equal in map theory without being β-equivalent. This holds even for terms that contain only variables, abstraction, and application. The notion of equality in map theory is the semantic notion introduced in section 2.2, and this semantic notion is not fully captured by β-equivalence.)

Having abstraction and application allows a definition of the fixed point operator Y and the bottom element $\bot$:

$$\begin{aligned} \boxed{\mathsf{Y}} &= \lambda f.(\lambda x.f(xx))\lambda x.f(xx), \\ \boxed{\bot} &= \mathsf{Y}\lambda x.x. \end{aligned}$$

Having the bottom element, we may state one more axiom and one more inference rule:

$A\bot$	Application	$\bot a = \bot$,
QND	Quartum Non Datur	$a\mathsf{T} = b\mathsf{T}, a\bot = b\bot, a\lambda y.xy = b\lambda y.xy \vdash ax = bx$.

Axiom $A\bot$ expresses that $\bot$ applied to anything yields $\bot$. Inference rule QND expresses that the root of any map is t, f, or b; there is no fourth possibility. QND together with the map P described in section 3.5 allows a development of classical propositional calculus (cf. *Grue 1992, Berline and Grue 1997,* and *Grue 1996*). In the present paper, only the intended meaning of axioms and inference rules will be stated. The details of why they express the intended meaning and how they are used is stated in *Grue 1996.*

3.5. Selection

The map P is best described by the following three axioms:

PT	Selection	$\mathsf{P}ab\mathsf{T} = a,$
Pλ	Selection	$\mathsf{P}ab\lambda x.c = b,$
P⊥	Selection	$\mathsf{P}ab\bot = \bot.$

The map P allows one to define many auxiliary concepts such as logical connectives (cf. Appendix 2.4) and a development of classical propositional calculus on the basis of QND.

For $a, b \in \mathcal{D} = (D, \leq)$, the constructs introduced so far allow a definition $a \preceq b$ such that $a \preceq b$ equals T if $a \leq b$ and equals F otherwise. The definition of $a \preceq b$ is stated in Appendix 2.4 and reads

$$a {\preceq} b \doteq (a = a{\downarrow}b)$$

where $a{\downarrow}b$ is defined on basis of P in Appendix 2.4. The definition of $\preceq$ uses $=$ which is a discontinuous construct in the sense that free occurrences of variables in $a = b$ are discontinuous occurrences. $\preceq$ inherits the discontinuity from $=$ so that free occurrences of variables in $a \preceq b$ are discontinuous, and no λ is allowed to bind a variable occurrence that is free in a subterm of the form $a \preceq b$.

We may now state three further inference rules and one more axiom:

Y	Minimality	$ga \preceq a \vdash \mathsf{Y}g \preceq a,$
M	Monotonicity	$b \preceq c \vdash ab \preceq ac,$
E	Extensionality	$\wr gxy = \wr hxy, gxyz = gab, hxyz = hab \vdash gxy = hxy$ if x, y, and z are not free in g and h,
!=	Equality	$!(a = b).$

Rule Y states that $\mathsf{Y}g$ is minimal among all fixed points of g (this does not hold in all κ-premodels, but it holds in the particular one outlined in Appendix 1). Rule M states that all maps a are monotonic in the sense that $b \leq c$ implies $ab \leq ac$. Rule E states that two maps g and h are equal if $\mathsf{a}(g)(\overline{x}) = \mathsf{a}(h)(\overline{x})$ for all $\overline{x} \in \mathcal{D}^{<\omega}$ (see *Grue 1996* for details on the interpretation and applications of rule E). Axiom != states that $a = b$ is either true or false.

3.6. Simple existential quantification

The quantifier $\bar{\exists}$ is a particularly primitive quantifier. $\bar{\exists}g = \mathsf{T}$ if $gx = \mathsf{T}$ for some $x \in \mathcal{D}$ and $\bar{\exists}g = \perp$ otherwise. Contrary to $\exists$, which is defined in section 2.4, $\bar{\exists}g$ cannot be false. $\epsilon(g)$ is an $x \in \mathcal{D}$ such that $gx = \mathsf{T}$ if such an x exists. More formally, define $E \in \mathcal{D} \rightarrow \mathcal{D}$ by

$$E(g) = \begin{cases} \mathsf{T} & \text{if } \exists x \in \mathcal{D}\colon gx = \mathsf{T}, \\ \perp & \text{otherwise.} \end{cases}$$

Then define

$$\begin{array}{rcl} \bar{\exists} & = & \lambda(E), \\ \epsilon(g) & = & q(\{x \in \mathcal{D} \mid gx = \mathsf{T}\}). \end{array}$$

Here we use the choice function q from the premodel $(\mathcal{D}, \mathrm{a}, C, q)$. The requirement $q(\emptyset) \in C \subseteq \mathcal{D}$ ensures that $\epsilon(g)$ is a map even when $gx \neq \mathsf{T}$ for all maps x. The following axioms describe $\bar{\exists}$:

$$\begin{array}{lll} \rightarrow\bar{\exists} & \text{Existence} & ab \rightarrow \bar{\exists}a, \\ \bar{\exists}\rightarrow & \text{Existence} & \bar{\exists}a \rightarrow a\epsilon(a), \\ ?\bar{\exists} & \text{Existence} & \bar{\exists}a = ?\bar{\exists}a. \end{array}$$

Axiom $\rightarrow\bar{\exists}$ expresses that if ab is true for some particular b, then $\bar{\exists}a$ is true. Axiom $\bar{\exists}\rightarrow$ expresses that if ax is true for some x then, in particular, it is true for $x = \epsilon(a)$. Axiom $?\bar{\exists}$ expresses that $\bar{\exists}a$ equals either T or $\perp$. See Appendix 2.4 for definitions of $a \rightarrow b$ and $?a$.

3.7. Hilbert's ε-operator

Define

$$\begin{array}{rcl} Q(g) & = & \begin{cases} \perp & \text{if } \exists x \in C\colon gx = \perp, \\ q(\{x \in C \mid gx = \mathsf{T}\}) & \text{otherwise,} \end{cases} \\ \varepsilon & = & \lambda(Q). \end{array}$$

The map ε is Hilbert's ε-operator (*Hilbert 1939*). The map ε allows definitions of many auxiliary concepts such as the universal and existential quantifiers (cf. Appendix 2.4) and development of first order predicate calculus. Furthermore, C is rich enough to allow all sets of ZFC to be modelled by classical maps which allows definition of set membership and proofs of all axioms and inference rules of ZFC inside MTC. The details may be found in *Grue 1992 and 1996*.

The constructs introduced so far allows the definition of the map ℓ (this definition is stated in Appendix 2.4). Now let

$$C' = \{x \in \mathcal{D} \mid \ell x = \mathsf{T}\}$$

(this C' is actually the same C' as the one defined in section 2.7). The axioms that describe ε read:

$Q1$	Quantification	$\ell a \wedge \forall b \rightarrow ba,$
$Q2$	Quantification	$\varepsilon x\colon a = \varepsilon x\colon \ell x \wedge a,$
$Q3$	Quantification	$\ell(\varepsilon x\colon a) = \forall x\colon !a,$
$Q4$	Quantification	$!\forall x\colon a = \forall x\colon !a.$

Axiom Q1 states $C' \subseteq C$, and Axiom Q2 states $C \subseteq C'$, so that $C = C'$ as in section 2.7. Axiom Q2, furthermore, expresses Ackermann's axiom (*Felgner 1976*, page 244). The definition of ε entails that $\varepsilon g \in C$ if $\forall x{\in}C\colon gx \neq \bot$ and $\varepsilon g = \bot$ if $\exists x{\in}C\colon gx = \bot$. The former is expressed directly by Axiom Q3, and the latter is expressed indirectly by Axiom Q4.

4. CONCLUSION

The most immediate translation of Russell's paradoxical sentence $\{x \mid x \notin x\} \in \{x \mid x \notin x\}$ into MTC is the term $(\lambda x.\neg(xx))(\lambda x.\neg(xx))$ whose value is $\bot$. Hence, Russell's paradox is essentially avoided by having a third truth value $\bot$ which is the value of terms that make a computer loop indefinitely. An important point in MTC is that this third truth value can be introduced without loosing the classical, semantic flavor of the theory and without resort to intuitionistic logic.

The above translation of Russell's paradox to MTC translates set abstraction to λ-abstraction, which gives insight into how Russell's paradox is avoided. As noted in section 1.3, set abstraction is not the same as λ-abstraction, so one may also look at how Russell's paradox is avoided when modelling set abstraction as in *Grue 1992*. This turns out to be trivial, however, since that modelling ensures that all sets are well-founded so that all sets x satisfy $x \notin x$. Furthermore, that modelling ensures that classes that contain all sets are not sets themselves.

A map x in MTC is classical if $\ell x = \mathsf{T}$ where

$$\ell \doteq \lambda f.f \begin{cases} \mathsf{T}, \\ (\forall x\colon \ell(fx)) \wedge \bar{\exists} S\colon \ell S \wedge \forall x \forall y\colon x \sim^2_S y \Rightarrow fx \sim fy. \end{cases}$$

The precise structure and meaning of this definition is not important here. The important observation is that the definition of ℓ is recursive in that ℓ occurs on the right hand side of $\doteq$ (recursive definitions are shorthand for definitions that use the fixed point operator Y explicitly).

The two occurrences of ℓ in the definition combined with the minimality of fixed points implies that classical maps are well-founded in two distinct ways. The first occurrence of ℓ makes classical maps well-founded in the sense that for all classical maps $g, x_1, x_2, \ldots$ there exists an n such that $gx_1 \cdots x_n = \mathsf{T}$. This kind of well-foundedness corresponds to the well-foundedness of sets expressed by the axiom of foundation in ZFC (no infinitely descending $\in$-chains).

The second occurrence of ℓ makes classical maps well-founded in a much more subtle sense. The closest analogue in ZFC to this well-foundedness is the limitation of size present in ZFC. However, maps contain more structure than sets, and the second kind of well-foundedness is more a limitation of complexity than just a simple limitation of size. In any case, it is this second kind of well-foundedness that avoids the Burali-Forti's paradox.

The first kind of well-foundedness does not avoid paradoxes. If the first kind of well-foundedness is abandoned, if the second kind is kept, and if the representation of sets used in *Grue 1992* is used, then the representable sets become those of Aczel's (*1980*) AFA set theory (the one in which the equation $X = \{X\}$ has exactly one solution). Hence, it is consistent to allow infinite descending $\in$-chains because the paradoxes are avoided by the second kind of well-foundedness.

As noted by Aczel, all known theories about non-well-foundedness share the peculiarity that they start by constructing a well-founded universe and then proceed to the non-well-founded. It is an interesting topic for further work to try to formulate a version of map theory which does not share this peculiarity. Such a theory could be a theory about non-well-founded classical maps, i.e., maps that satisfy the second kind of well-foundedness without satisfying the first.

APPENDIX

A. OUTLINE OF A MODEL OF MTC

A.1. The κ-denotational semantics

The following exposition follows *Berline and Grue 1997* except that the empty set is κ-small and κ-cpos need not have a bottom element.

A p.o $\mathcal{D}$ is a partially ordered set, $(D, \leq)$. We use $x \in \mathcal{D}$ and $x \subseteq \mathcal{D}$ as shorthand for $x \in D$ and $x \subseteq D$, respectively. $\boxed{\downarrow u}$ means $\{v \mid v \leq u\}$. $A \subseteq \mathcal{D}$ is κ-*chain* if every κ-small $B \subseteq A$ is bounded by an element of A. $\mathcal{D}$ is a κ-*cpo* if every κ-chain A has a sup (i.e., least upper bound). $\mathcal{D}$ is a κ-*ccpo* if, moreover, every bounded A has a sup (hence, as in *Berline and Grue 1997,* every κ-ccpo has a bottom). A κ-*compact* element of a κ-cpo $\mathcal{D}$ is an element $u \in \mathcal{D}$ such that, for all κ-chains A, $u \leq \sup(A)$ implies $u \leq v$ for some $v \in A$. $\boxed{\mathcal{D}_c}$ is the set of κ-compact elements and $\boxed{\downarrow_c u} = \downarrow u \cap \mathcal{D}_c$. A κ-ccpo $\mathcal{D}$ is a κ-*Scott domain* iff for all $u \in \mathcal{D}$, $u = \sup \downarrow_c u$. A *prime* element of a κ-ccpo $\mathcal{D}$ is an element u such that, for all bounded A, $u \leq \sup A$ implies $u \leq v$ for some $v \in A$. $\boxed{\mathcal{D}_p}$ is the set of prime elements of $\mathcal{D}$, and $\boxed{\downarrow_p u} = \downarrow u \cap \mathcal{D}_p$. A κ-Scott domain is κ-*prime algebraic* if for all $u \in \mathcal{D}$, $u = \sup \downarrow_p u$.

A.2. Outline of a model

A model of MTC is outlined in the following. Only the definition of the model will be stated. The satisfaction of the axioms and inference rules remains to be proved.

The model construction has many similarities with the construction in section 8 of *Berline and Grue 1997*. The construction below differs from that in *Berline and Grue 1997* in the following ways: (1) The domain of the model is constructed as coherent, complete, initial segments of κ-compact elements rather than coherent, initial segments of κ-prime elements, and the PCS of κ-prime elements is not constructed at all below. (2) The model construction is based on two fixed point constructions rather than one. The first fixed point generates the σ-compact maps of MTW which is a set large enough to contain the well-founded maps of MTW. Then the well-founded maps are turned into classical maps by discarding information, and then a second fixed point construction is used to generate the κ-compact maps of MTC. Finally, the κ-premodel of MTC is constructed from coherent, initial segments of κ-compact elements.

To simplify the exposition, assume that κ is strongly inaccessible, i.e. assume σ is the least strongly inaccessible ordinal and assume κ is a strongly inaccessible ordinal greater than σ. The assumption that κ is inaccessible rather than just regular is merely a luxury.

For all relations $\preceq$, define $\boxed{\preceq^*}$ by

$$\langle x_1, \ldots, x_m \rangle \preceq^* \langle y_1, \ldots, y_n \rangle \Leftrightarrow m = n \wedge x_1 \preceq y_1 \wedge \cdots \wedge x_m \preceq y_m.$$

An *applicative structure* $\mathcal{D}$ is a pair (D, a) where D is a set and $a \in D \to M_D$. For applicative structures $\mathcal{D} = (D, a)$, define $\boxed{\leq_{\mathcal{D}}}$ and $\boxed{M_{\mathcal{D}}}$ by

$$\begin{aligned}
&x \leq_{\mathcal{D}} y \Leftrightarrow a(x) \leq a(y),\\
&M_{\mathcal{D}} = \{g{\in}M_D \mid \forall x, y{\in}D^{<\omega}\colon (x \leq^*_{\mathcal{D}} y \Rightarrow a(g)(x) \leq_L a(g)(y))\}.
\end{aligned}$$

An applicative structure $\mathcal{D} = (D, a)$ is said to be *monotonic* if $\perp_D \in \operatorname{rng} a \subseteq M_{\mathcal{D}}$. For all monotonic applicative structures $\mathcal{D} = (D, a)$, define

$$\begin{array}{ll}
 & D_{\mathcal{D}} = D \cup M_{\mathcal{D}} \\
a^+_{\mathcal{D}} \in D_{\mathcal{D}} \to M_{\mathcal{D}} & a^+_{\mathcal{D}}(x) = \begin{cases} a(x) & \text{if } x \in D \\ x & \text{otherwise} \end{cases} \\
 & g \preceq_{\mathcal{D}} h \Leftrightarrow a^+_{\mathcal{D}}(g) \leq a^+_{\mathcal{D}}(h) \\
a_{\mathcal{D}} \in D_{\mathcal{D}} \to M_{D_{\mathcal{D}}} & a_{\mathcal{D}}(g)(x) = \sup\{a^+_{\mathcal{D}}(g)(y) \mid y \in D^{<\omega} \wedge y \preceq^*_{\mathcal{D}} x\} \\
 & \mathcal{D}^+ = (D_{\mathcal{D}}, a_{\mathcal{D}}).
\end{array}$$

To ensure the soundness of the definition, we make the arbitrary convention that $\sup A = \mathsf{b}$ when $A \subseteq L$ has no supremum.

For applicative structures $\mathcal{D} = (D, a)$ and $\mathcal{D}' = (D', a')$, define $\boxed{\mathcal{D} \sqsubseteq \mathcal{D}'} \Leftrightarrow D \subseteq D' \wedge \forall g{\in}D \forall x{\in}D^{<\omega}: a(g)(x) = a'(g)(x)$. We have $\mathcal{D} \sqsubseteq \mathcal{D}^+$. Now define $\boxed{\mathcal{D}^\alpha}$ for all ordinals α by transfinite induction as follows:

$$\begin{array}{lcl} \mathcal{D}^0 & = & \mathcal{D}, \\ \mathcal{D}^{\alpha+1} & = & (\mathcal{D}^\alpha)^+, \\ \mathcal{D}^\delta & = & \sqcup_{\alpha\in\delta}\mathcal{D}^\alpha \quad \text{for limit ordinals } \delta. \end{array}$$

Above, $\sqcup_{\alpha\in\delta}$ has the obvious interpretaion, i.e., if $\mathcal{D}^\alpha = (D_\alpha, a_\alpha)$ then $D_\delta = \cup_{\alpha\in\delta} D_\alpha$ and $a_\delta(g)(x) = a_\alpha(g)(x)$ whenever $g{\in}D_\alpha$ and $x{\in}D_\alpha^{<\omega}$.

Let $\boxed{\perp}$ be an arbitrary element of the universe of ZFC which is not a function; let $\boxed{\mathcal{D}_1} = (D_1, a_1)$ where $D_1 = \{\perp\}$ and where $a_1 \in D_1 \to (D_1^{<\omega} \to L)$ is defined by $a_1(g)(x) = \mathsf{b}$. Furthermore, let $\boxed{\mathcal{D}_2} = (D_2, a_2) = \mathcal{D}_1^\sigma$. D_2 is essentially the set of σ-compact elements of MTW.

Let $\boxed{S^\infty}$ denote the set of infinite lists of elements of S. For all $x = (x_1, \ldots)$ and $n \in \omega$, let $\boxed{(x|n)}$ denote the tuple $\langle x_1, \cdots, x_n\rangle$.

Define $\boxed{G^\circ} = \{g{\in}\mathcal{D}_2 \mid \forall x{\in}D_2^\infty \exists n{\in}\omega: a_2(g)(x|n) = \mathsf{t}\}$, and let $\boxed{\Phi}$ be the least subset of $\mathcal{D}_2$ for which $G^{\circ\circ} \subseteq \Phi$ whenever $G \subseteq \Phi$ is σ-small. This Φ is essentially the set of well-founded maps introduced in *Berline and Grue 1997* and *Grue 1992*.

Now let $\boxed{\mathcal{D}_3} = (D_3, a_3)$ where $D_3 = \Phi \cup \{\perp\}$ and where $a_3 \in D_3 \to (D_3^{<\omega} \to L)$ is defined by $a_3(g)(x) = a_2(g)(x)$. $\mathcal{D}_3$ is Φ from which some information is discarded, namely, the information about the applicative behavior of well-founded maps when applied to non-well-founded maps. This makes $\mathcal{D}_3$ useful as a model of the classical maps.

Now let $\boxed{\mathcal{D}_4} = (D_4, a_4) = \mathcal{D}_3^\kappa$. D_4 is essentially the set of κ-compact elements of MTC. Transfinite induction up to κ gives the κ-compact elements because κ is assumed inaccessible. More care would be needed in the definition of $\mathcal{D}^+$ and $\mathcal{D}^\alpha$ if κ was not assumed inaccessible.

$A \subseteq D_4$ is an initial segment if A is non-empty and $x \leq_{D_4} y \wedge y \in A \Rightarrow x \in A$. A is coherent if any two elements of A have an upper bound in D_4. A is complete if any subset of A that has an upper bound in $\mathcal{D}$ also has an upper bound in A. Let D_5 be the set of coherent, complete, initial segments, and define the relation $\leq_5$ on D_5 by $x \leq_5 y \Leftrightarrow x \subseteq y$. Define $a_5 \in D_5 \to (D_5^{<\omega} \to L)$ by $a_5(g)(\langle x_1, \ldots, x_n\rangle) = \sup\{a_4(h)(\langle y_1, \ldots, y_n\rangle) \mid h \in g \wedge y_1 \in x_1 \wedge \cdots \wedge y_n \in x_n\}$. Define $C_5 = \{\downarrow x \mid x \in D_3 \setminus \{\perp\}\}$ and let $q \in \mathcal{P}(D_5) \to D_5$ satisfy $q(A) \in A$ when A is non-empty and $q(A) \in C_5$ when A is empty.

$P = ((D_5, \leq_5), a_5, C_5, q_5)$ is claimed to be a κ-premodel that satisfies all axioms and inference rules of MTC.

B. SUMMARY OF MTC

B.1. Syntax

$$\begin{array}{lcl} \mathcal{V} & ::= & x_1 \mid x_2 \mid \cdots \\ \mathcal{T} & ::= & \mathcal{V} \mid \mathsf{T} \mid \mathcal{T}\mathcal{T} \mid \lambda\mathcal{V}.\mathcal{T} \mid \mathsf{if}(\mathcal{T},\mathcal{T},\mathcal{T}) \mid \varepsilon \mid \bar{\exists} \mid \epsilon(\mathcal{T}) \mid \mathcal{T} = \mathcal{T} \end{array}$$

B.2. Priority

The priority is as follows. Functional application fx has highest priority and appears at the top of the table. Operators on the same line have the same priority.

$$\begin{array}{l} fx \\ x, y \\ x \downarrow y \quad \downarrow x \\ x \sim y \quad x \in y \quad x \in\in y \quad x \sim^2_S y \\ \neg x \quad \wr x \quad !x \quad ?x \\ x \wedge y \quad x \mathbin{\tilde{\wedge}} y \\ x \vee y \quad x \mathbin{\tilde{\vee}} y \\ x \Rightarrow y \quad x \mathbin{\tilde{\Rightarrow}} y \quad x \Leftrightarrow y \\ x \begin{cases} y \\ z \end{cases} \\ \lambda x . y \\ x = y \quad x \preceq y \\ x \rightarrow y \end{array}$$

B.3. Associativity

fx, $x \wedge y$, $x \mathbin{\tilde{\wedge}} y$, $x \vee y$ and $x \mathbin{\tilde{\vee}} y$ are left associative so that, e.g., fxy means $(fx)y$.

$x \sim y$, $x \in y$, $x \in\in y$, $x \sim^2_S y$, $x \Rightarrow y$, $x \mathbin{\tilde{\Rightarrow}} y$, $x \Leftrightarrow y$, $x = y$, $x \preceq y$, and $x \rightarrow y$ are "and"-associative so that, e.g., $x \sim y \sim z$ means $(x \sim y) \wedge (y \sim z)$ and $x \Rightarrow y \Rightarrow z$ means $(x \Rightarrow y) \wedge (y \Rightarrow z)$.

B.4. Definitions used in axioms

$$\begin{array}{ll} \mathsf{F} & \doteq \lambda x . \mathsf{T} \\ x \begin{cases} a \\ b \end{cases} & \doteq \mathsf{P}abx \\ \mathsf{Y} & \doteq \lambda f . (\lambda x . f(xx))(\lambda x . f(xx)) \\ \bot & \doteq \mathsf{Y}\lambda x . x \end{array}$$

$$\neg x \doteq x\begin{cases} \mathsf{F} \\ \mathsf{T} \end{cases}$$

$$\wr x \doteq x\begin{cases} \mathsf{T} \\ \mathsf{F} \end{cases}$$

$$!x \doteq x\begin{cases} \mathsf{T} \\ \mathsf{T} \end{cases}$$

$$?x \doteq x\begin{cases} \mathsf{T} \\ \bot \end{cases}$$

$$x \mathrel{\tilde{\wedge}} y \doteq x\begin{cases} y \\ \mathsf{F} \end{cases}$$

$$x \mathrel{\tilde{\Rightarrow}} y \doteq x\begin{cases} y \\ \mathsf{T} \end{cases}$$

$$x \wedge y \doteq x\begin{cases} y\begin{cases} \mathsf{T} \\ \mathsf{F} \end{cases} \\ y\begin{cases} \mathsf{F} \\ \mathsf{F} \end{cases} \end{cases}$$

$$x \Rightarrow y \doteq x\begin{cases} y\begin{cases} \mathsf{T} \\ \mathsf{F} \end{cases} \\ y\begin{cases} \mathsf{T} \\ \mathsf{T} \end{cases} \end{cases}$$

$$\begin{aligned}
x \Leftrightarrow y &\doteq (x \Rightarrow y) \wedge (y \Rightarrow x) \\
x \rightarrow y &\doteq x \mathrel{\tilde{\wedge}} y = x \mathrel{\tilde{\wedge}} \mathsf{T} \\
\exists &\doteq \lambda f.\wr f(\varepsilon f) \\
\exists x\colon a &\doteq \exists \lambda x.a \\
\forall x\colon a &\doteq \neg \forall x\colon \neg a \\
\forall &\doteq \lambda f.\forall x\colon fx
\end{aligned}$$

$$f \sim g \doteq f\begin{cases} g\begin{cases} \mathsf{T} \\ \mathsf{F} \end{cases} \\ g\begin{cases} \mathsf{F} \\ \forall \lambda x.fx \sim gx \end{cases} \end{cases}$$

$$x \in y \doteq y\begin{cases} \mathsf{F} \\ \exists z\colon xz \sim y \end{cases}$$

$$\begin{aligned}
x \in\in y &\doteq \exists z\colon x \in z \wedge z \in y \\
\forall x{\in\in}a\colon b &\doteq (\lambda y.\forall x\colon x \in\in y \mathrel{\tilde{\Rightarrow}} b)a
\end{aligned}$$

$$f \sim_S^2 g \quad \doteq f \begin{cases} g \begin{cases} \mathsf{T} \\ \mathsf{F} \end{cases} \\ g \begin{cases} \mathsf{F} \\ \forall x {\in\in} S \colon fx \sim_S^2 gx \end{cases} \end{cases}$$

$$\ell \quad \doteq \lambda f.f \begin{cases} \mathsf{T} \\ (\forall x \colon \ell(fx)) \wedge \bar{\exists} S \colon \ell S \wedge \forall x \forall y \colon x \sim_S^2 y \Rightarrow fx \sim fy \end{cases}$$

$$x \downarrow y \quad \doteq x \begin{cases} y \begin{cases} \mathsf{T} \\ \bot \end{cases} \\ y \begin{cases} \bot \\ \lambda z.xz \downarrow yz \end{cases} \end{cases}$$

$$x \preceq y \quad \doteq x = x \downarrow y$$

B.5. Axioms and inference rules

Below, a, b, c, d, f, and g denote arbitrary terms whereas x, y, and z denote arbitrary, distinct variables.

T	Transitivity	$a = b, a = c \vdash b = c$
SA	Substitutivity	$a = c, b = d \vdash ab = cd$
$S\lambda$	Substitutivity	$a = b \vdash \lambda x.a = \lambda x.b$
$S\epsilon$	Substitutivity	$a = b \vdash \epsilon(a) = \epsilon(b)$
PT	Selection	$\mathsf{P}ab\mathsf{T} = a$
Pλ	Selection	$\mathsf{P}ab\lambda x.c = b$
P⊥	Selection	$\mathsf{P}ab\bot = \bot$
$A\mathsf{T}$	Application	$\mathsf{T}a = \mathsf{T}$
$A\lambda$	Application	$(\lambda x.a)b = \langle a \mid x{:=}b \rangle$
		if b is free for x in a.
$A\bot$	Application	$\bot a = \bot$
R	Renaming	$\lambda x.\langle a \mid y{:=}x \rangle = \lambda y.\langle a \mid x{:=}y \rangle$
		if x is free for y in a and vice versa.
QND	Quartum Non Datur	$a\mathsf{T} = b\mathsf{T}, a\bot = b\bot, a\lambda y.xy = b\lambda y.xy$
		$\vdash ax = bx$
$Q1$	Quantification	$\ell a \wedge \forall b \rightarrow ba$
$Q2$	Quantification	$\varepsilon x \colon a = \varepsilon x \colon \ell x \wedge a$
$Q3$	Quantification	$\ell(\varepsilon x \colon a) = \forall x \colon !a$
$Q4$	Quantification	$!\forall x \colon a = \forall x \colon !a$
Y	Minimality	$ga \preceq a \vdash \mathsf{Y}g \preceq a$
M	Monotonicity	$b \preceq c \vdash ab \preceq ac$
E	Extensionality	$\wr gxy = \wr hxy, gxyz = gab, hxyz = hab$
		$\vdash gxy = hxy$
		if x, y, and z are not free in g and h.

$\to\bar{\exists}$	Existence	$ab \to \bar{\exists}a$
$\bar{\exists}\to$	Existence	$\bar{\exists}a \to a\epsilon(a)$
$?\bar{\exists}$	Existence	$\bar{\exists}a = ?\bar{\exists}a$
$=\vdash$	Equality	$a = \mathsf{T} \vdash a$
$\vdash=$	Equality	$a \vdash a = \mathsf{T}$
$!=$	Equality	$!(a = b)$

REFERENCES

Aczel, P.

1980 Frege structures and the notions of proposition, truth and set, *The Kleene symposium,* (J. Barwise, H. J. Keisler, and K. Kunen, editors), North-Holland, Amsterdam, pp. 31–59.

Berline, C., and K. Grue

1997 A κ-denotational semantics for Map Theory in ZFC+SI, *Theoretical Computer Scienc*, vol. 179, no. 1–2 (June), pp. 137–202.

Chang, C. C., and K. J. Keisler

1973 *Model theory,* Studies in Logic and the Foundations of Mathematics, vol. 73, North-Holland, Amsterdam.

Church, A.

1933 A set of postulates for the foundations of logic I, *Annals of Mathematics,* vol. 33, pp. 346–366.

1934 A set of postulates for the foundations of logic II, *Annals of Mathematics,* vol. 34, pp. 839–864.

1941 *The calculi of lambda conversion,* Princeton University Press, Princeton, New Jersey.

Feferman, S.

1978 Recursion theory and set theory, a marriage of convenience, *Generalized recursion theory II,* Proceedings of the 1977 Oslo Symposium (J. E. Fenstad, R. O. Gandy, and G. E. Sacks, editors), North-Holland, Amsterdam, pp. 55–98.

1984 Toward useful type-free theories I, *The Journal of Symbolic Logic,* vol. 49, pp. 75–111.

Felgner, U.

1976 Choice functions on sets and classes, *Sets and classes: On the works by Paul Bernays,* North-Holland, Amsterdam, pp. 217–255.

Flagg, R. C., and J. Myhill

1989 A type-free system extending ZFC, *Annals of Pure and Applied Logic,* vol. 43, pp. 79–97.

Grue, K.

1992 Map theory, *Theoretical Computer Science,* vol. 102, no. 1 (July), pp. 1–133.

1996 *Stable map theory,* Department of Computer Science, University of Copenhagen, DIKU, Universitetsparken 1, DK–2100 Copenhagen, Denmark", DIKU Report, no. 96/10 (April); available from //www.diku.dk /~grue.

Hilbert, D., and P. Bernays
1939 *Grundlagen der Mathematic*, vol. 2, Springer-Verlag, Berlin.

Mendelson, E.
1987 *Introduction to mathematical logic*, third edition, Wadsworth and Brooks, Monterey, California.

Scott, D.
1973 Models for various type-free λ-calculi, *Proceedings of the IVth international congress for logic* (Suppes *et al.*, editors), Studies in Logic and The Foundation of Mathematics, Methodology and Philosophy of Science, North-Holland, Amsterdam, vol. 74, pp. 157–187.

DANIEL LEIVANT*

PEANO'S LAMBDA CALCULUS: THE FUNCTIONAL ABSTRACTION IMPLICIT IN ARITHMETIC

Dedicated to the memory of Alonzo Church.

Abstract. We define a fragment $\mathbf{2\lambda}$(Peano) of Girard's second order λ-calculus $\mathbf{2\lambda}$, in which arguments of type application are required to be "Peano types", namely types generated from type variables and types of the form $\forall\vec{R}.\tau$ (τ quantifier free of rank ≤ 2) using $\rightarrow$ and substitution. We show that the provably recursive functions of first order arithmetic are precisely the numeric functions that are λ-definable in $\mathbf{2\lambda}$(Peano).

One of Church's great achievements was the definition of the λ-calculus and his proof, with Kleene, that all computable functions are λ-definable. This suggests that significant classes of computable functions might arise from imposing on the λ-calculus typing disciplines, thereby identifying the functional abstraction inherent in such classes. However, the connections between natural type disciplines and natural classes of computable functions have turned out to be somewhat tricky. The most rudimentary type discipline, the simply typed lambda calculus $\mathbf{1\lambda}$, yields only the extended polynomials as definable functions (*Schwichtenberg 1976, Statman 1979*), a rather odd class that does not include even the predecessor function. If different λ-representations are allowed for input and output, then the exponentiation and predecessor functions become definable, but subtraction remains outside the class (*Statman 1979*).[1]

A powerful extension of $\mathbf{1\lambda}$ is Girard's second order type system, $\mathbf{2\lambda}$ (*Girard 1971, 1972*). Within that calculus Girard considered a polymorphic variant of Church's numerals: $\bar{n} =_{\text{df}} \Lambda t.\, \lambda s^{t\rightarrow t}, z^{t}.\, s^{[n]}(z)$, which was later rediscovered independently by Reynolds (*1974*), Fortune (*1979*), and O'Donnell (*1979*). From Girard (*1971, 1972*) it follows that the functions definable in

* Research partially supported by NSF grant CCR-9309824. The author is sincerely grateful to Thierry Coquand, Jean-Yves Marion, and Norman Danner for useful comments about earlier versions of this paper.

[1]These anomalies are repaired when the λ-calculus is defined on top of a free algebra with constants for zero, successor, predecessor, and discriminator. One obtains then the numeric functions computable in linear space, by modifying the characterization of poly-time in *Leivant 1993* to an algebra with a single successor. (This duality linear space / polynomial time is the same as in *Handley 1993* and *Leivant 1994*.)

C. Anthony Anderson and M. Zelëny (eds.), Logic, Meaning and Computation, 313-329.

2λ modulo these numerals are precisely the provably recursive functions of second order arithmetic. However, this elegant correspondence characterizes a rather vast class of computable functions.

One natural approach to restricting **2λ** so as to characterize tamer classes of functions is type stratification. We showed in *Leivant 1991* that the functions definable in the variant of **2λ** where types are finitely stratified form precisely Grzegorczyk's class $\mathcal{E}_4$ (i.e. the functions defined by one recurrence from elementary functions). Type stratification can be extended into transfinite levels, yielding characterizations of the primitive recursive and the provable recursive functions of arithmetic. While those results (which we may prove elsewhere) are useful in corroborating well-known correspondences between function classes and ordinals, they do not provide insight into the functional nature of these function classes.[2]

The system **2λ** can also be confined by syntactic restrictions on permissible type arguments. This is most natural, because it corresponds (via a Schönfinkel-Curry-Howard style "formula-as-type" homomorphism, see *Leivant 1990*) to restrictions on the comprehension principle, which have been fundamental in proof theoretic studies. Some preliminary results in this direction were noted already in *Fortune 1979, O'Donnell 1979,* and *Leivant 1981,* but these fell short of characterizing classes of functions. Here we exhibit such a characterization for the class of provably recursive functions of Peano's Arithmetic, **PA**. This class of functions has also a natural computational definition, as the class of functions generated via primitive recursion in all simple types (*Gödel 1958*), which is why we denote it by $\mathbf{Rec}^{\omega}$.

Our main result provides two characterizations of $\mathbf{Rec}^{\omega}$ by definability in subsystems of **2λ**. One is minimalist: type arguments are restricted to the types needed to permit λ-definability for primitive recursion in all simple types. The other characterization allows a broader class of type arguments, which we dub "Peano types". These may contain type parameters in the scope of ∀, but are still closed under (type-correct) substitution of Peano types for type variables. This formalism is the strongest one that we can handle when showing that all λ-definable numeric functions are provably recursive functions of **PA**. The proof of the latter result uses a formalization of Girard's strong normalization proof for **2λ**: when all types present are Peano types, the formalization can be carried out in second order arithmetic with comprehension and induction restricted to arithmetical formulas, a theory conservative over **PA**. Our formalization uses the well-known observation that formulas of the form $\forall\vec{R}.\exists\vec{x}.\,\varphi$ (φ primitive recursive) are equivalent to existential formulas.[3] A complete statement of our main result is given as the Main Theorem at the end of §2 below.

[2] A related approach, using infinite type conjunctions, is explored in *Leivant 1990a*.

[3] Such formulas are known as *strict* Π^1_1 (*Barwise 1969, 1975*), or *computational* (*Leivant 1989*).

1. THE SECOND ORDER LAMBDA CALCULUS $\mathbf{2\lambda}$

1.1. Peano Types

We refer to Girard's second order λ-calculus, $\mathbf{2\lambda}$, in its original, ontological (i.e. Church-style), formulation.[4] The *types of* $\mathbf{2\lambda}$ (*2-types* for short) are generated inductively from a countable collection of type variables by the closure conditions: if τ and σ are types, then so is $\tau \to \sigma$; and if τ is a type and R a type variable, then $\forall R.\,\tau$ is a type. Parentheses are used in concrete syntax, and $\to$ is assumed to associate to the right, e.g. $\tau \to \sigma \to \rho$ stands for $\tau \to (\sigma \to \rho)$. We write $\tau_1, \ldots, \tau_r \to \sigma$ for $\tau_1 \to \tau_2 \to \cdots \to \tau_r \to \sigma$, and $\tau^r \to \sigma$ if all types τ_i are the same type τ. We denote by $\{\sigma/R\}$ the syntactic operation of simultaneously substituting σ for every free occurrence of the variable R in the argument, modulo standard variable renaming to protect intended scoping.

A type is *functional* if its main construct is $\to$, and *universal* if it is $\forall$. A type is *first order* if it has no occurrence of $\forall$. We use $Q, Q_i, R, R_i, S \ldots$ as syntactic variables for formal type variables. A type $\forall R_1 \cdots \forall R_r.\tau$ will be abbreviated by $\forall R_1 \ldots R_r.\tau$ (or even $\forall \vec{R}.\tau$, if $R_1 \ldots R_r$ are evident or irrelevant). If $\tau = \forall R_1 \ldots R_k.\sigma$, where $k > 0$ and σ is not universal, then σ is the *matrix* of τ.

Given a set of types T we write $*T$ and $\forall T$ for the types generated from T using only $\to$ and only $\forall$, respectively. Thus $\sigma \in \forall T$ iff σ is of the form $\forall \vec{R}.\tau$ with $\tau \in T$. More restrictively, $\forall^1 T$ will denote the set of types of the form $\forall R.\tau$, with $\tau \in T$.

If τ is a first order type, then its *rank*, $\underline{\text{rnk}}(\tau)$, is the count of negative nesting of $\to$ in τ; that is: $\underline{\text{rnk}}(R) = 0$ for type variables R, and $\underline{\text{rnk}}(\tau \to \sigma) = \max(1+\underline{\text{rnk}}(\tau), \underline{\text{rnk}}(\sigma))$. For each $k > 0$ we let $\underline{k}$ denote the set of first order types of rank $\leq k$.

The *Peano types* are generated inductively by the following clauses.

(1) Every type variable is a Peano type.

(2) If τ and σ are Peano types, then so is $\tau \to \sigma$.

(3) If τ is a first order type of rank ≤ 2, then $\forall \vec{R}.\tau$ is a Peano type.

(4) If τ and σ are Peano types, then so is $\{\sigma/R\}\tau$ (correct substitution).

For example, all types in $*\forall\underline{2}$ are Peano, as are $((R \to R) \to R) \to R$ and $\forall R.((\tau \to R), (R \to \sigma) \to R)$, for any $\tau, \sigma \in *\forall\underline{2}$ in which R is not free. On the other hand, $\forall R.(((R \to R) \to R) \to R)$ and $\forall R.((\forall Q.(R \to Q)) \to R)$ are not Peano (for the latter this follows from Lemma 4 below.)

[4]For further details about $\mathbf{2\lambda}$ see e.g., *Girard 1971, Reynolds 1974, MacQueen 1982,* and *Fortune, Leivant, and O'Donnell 1983.*

1.2. A Closed Definition of Peano Types

We give an alternative, closed, definition of Peano types. This definition is an aside, and will not be used elsewhere in the paper, except for trivial uses of Theorem 1. We refer initially to types satisfying the closed definition as cPeano, and proceed to prove that cPeano and Peano types are the same.

We say that a 2-type τ is *separated* if it has no nested $\forall$ blocks with a variable bound by the outer one occurring free in the inner one. For example, $\forall R.\ (\forall S.(S \to R) \to R)$ is not separated, but $\forall R, S.(S \to R) \to R$ is. For a type $\tau \in *T$ define its *rank over* T, $\underline{\text{rnk}}_T(\tau)$, to be the count of negative nesting of $\to$ in generating τ from T; that is: $\underline{\text{rnk}}_T(\tau) = 0$ if $\tau \in T$, and $\underline{\text{rnk}}_T(\tau \to \sigma) = \underline{\max}(1+\underline{\text{rnk}}_T(\tau), \underline{\text{rnk}}_T(\sigma))$ when $\tau \to \sigma$ is not in T. For $k = 0, 1 \ldots$ we write $\underline{k}(T)$ for the set of types in $*T$ of rank $\leq k$ over T. For a tuple $\vec{R} = \langle R_1 \ldots R_k \rangle$ of type variables let $\alpha[\vec{R}]$ be the set of types with no R_i free, and let $\alpha^*[\vec{R}] =_{\text{df}} *(\{R_1 \ldots R_k\} \cup \alpha[\vec{R}])$. In particular, if τ is separated, then for every subtype $\forall \vec{R}.\sigma$ of τ we must have $\sigma \in \alpha^*[\vec{R}]$. Call a type τ *cPeano* if it is separated and for every universal subtype $\forall \vec{R}.\sigma$ of τ the matrix σ is of rank ≤ 2 over $\alpha^*[\vec{R}]$.

LEMMA 1. *A subtype of a cPeano type is cPeano.*

Proof. We need only prove that the immediate subtypes of a cPeano type τ are cPeano. This is trivial if τ is a functional type, or if τ is $\forall R.\sigma$ with σ the matrix. If τ is $\forall R. \forall S_1 \ldots S_k.\sigma$ $(k > 0)$, it suffices to observe that the rank of σ over $\alpha^*[S_1 \ldots S_k]$ is at most the rank of σ over $\alpha^*[R, S_1 \ldots S_k]$, since $\alpha[\vec{S}, R] \subseteq \alpha[\vec{S}]$. Since τ is cPeano, the latter is ≤ 2, and so the former is also ≤ 2. ⊣

LEMMA 2. *If no variable among $\vec{R}$ is free in σ, then the rank of $\{\sigma/S\}\tau$ over $\alpha^*[\vec{R}]$ is the same as the rank of τ over $\alpha^*[\vec{R}]$.*

Proof. Straightforward structural induction on τ. ⊣

LEMMA 3. *If τ and σ are cPeano types, then so is $\{\sigma/S\}\tau$.*

Proof. By structural induction on τ. The cases where τ is S, or S is not free in τ, are trivial. If τ is $\rho \to \xi$, then $\{\sigma/S\}\rho$ and $\{\sigma/S\}\xi$ are cPeano, by induction assumption; from the definition of cPeano types it follows then that $\{\sigma/S\}\tau = \{\sigma/S\}\rho \to \{\sigma/S\}\xi$ is cPeano. Finally, suppose that τ is $\forall \vec{R}.\rho$ with ρ the matrix. Then $\{\sigma/S\}\tau$ is $\forall \vec{R}.\{\sigma/S\}\rho$. By induction assumption $\{\sigma/S\}\rho$ is a cPeano type. Since τ is cPeano, the rank of ρ over $\alpha^*[\vec{R}]$ is ≤ 2, whence, by Lemma 2, the rank of $\{\sigma/S\}\rho$ over $\alpha^*[\vec{R}]$ is also ≤ 2. Therefore $\{\sigma/S\}\tau = \forall \vec{R}.\{\sigma/S\}\rho$ is cPeano. ⊣

LEMMA 4. *A 2-type is cPeano iff it is Peano.*

Proof. The forward implication is proved by structural induction on types. Type variables are all Peano by definition. If $\tau \equiv \sigma \rightarrow \rho$ is cPeano, then so are σ and ρ, by Lemma 1. By induction assumption both types are then Peano, which implies that τ is Peano. Finally, suppose that $\tau \equiv \forall\vec{R}.\sigma$ is cPeano where σ is the matrix. Since the rank of σ over $\alpha^*[\vec{R}]$ is ≤ 2, we have $\sigma \equiv \{\vec{\xi}/\vec{Q}\}\sigma_0$ for some types $\vec{\xi} \in \alpha^*[\vec{R}]$ and fresh type variables $\vec{Q}$, and where σ_0 is a first order type of rank ≤ 2. Since $\vec{\xi}$ are all subtypes of τ, they are all cPeano, by Lemma 1, and by induction assumption they are therefore Peano. Since no variable among $\vec{R}$ appears free in $\vec{\xi}$, we also have $\tau \equiv \{\vec{\xi}/\vec{R}\}\,(\forall\vec{R}.\sigma_0)$, which is Peano by definition.

The backward implication of the lemma is proved by induction on the derivation of a type as a Peano type. All type variables are cPeano by definition. If $\tau \equiv \sigma \rightarrow \rho$ is Peano, where σ and ρ are Peano, then σ and ρ are cPeano, by induction assumption. From the definition of cPeano types it immediately follows that τ is cPeano as well. If $\tau \equiv \forall\vec{R}.\sigma$, where σ is a first order type of rank ≤ 2, then τ is cPeano by definition. Finally, suppose $\tau \equiv \{\rho/R\}\sigma$, where ρ and σ are Peano; by induction assumption ρ and σ are cPeano, so τ is cPeano by Lemma 3. ⊣

1.3. Terms

The *terms* of **2λ** are generated from typed variables using object and type abstraction and application, as follows. We write $E : \tau$, or E^τ, for "E is of type τ".

(1) For each 2-type τ there is a countable collection of *variables of type* τ, which are defined to be terms of type τ. We use $x^\tau, y^\tau, x_1^\tau \ldots$ as syntactic variables ranging over formal variables;

(2) If $E : \sigma$, then $\lambda x^\tau.\, E : \tau \rightarrow \sigma$;

(3) If $E : \tau \rightarrow \sigma$ and $F : \tau$, then $EF : \sigma$.

(4) If $E : \tau$, then $\Lambda R.E : \forall R.\tau$, provided R is not free in the type of free object variables in E.

(5) If $E : \forall R.\tau$ and σ is a 2-type, then $E\sigma : \{\sigma/R\}\tau$ (modulo usual variable renaming).

As usual, we use parentheses in concrete syntax, and associate application to the left (i.e. EFG stands for $(EF)G$). If E and x are of the same type, we write $\{E/x\}$ for the syntactic operation of substituting E for every free occurrence of x in the argument, modulo variable renaming in the argument to preserve intended scoping. The notion of *subterm* is defined as usual, where we take each term to be a subterm of itself.

A term $(\lambda x^\tau.E)F$ *β-contracts* to $\{F/x\}E$, and $(\Lambda R.E)\tau$ *type-contracts* to $\{\tau/R\}E$. For each contraction the source is called the *redex* and the target the *reduct*. We say that E *β-reduces* (respectively, *t-reduces*) to F, and write $E \Rightarrow^1_\beta F$ (respectively, $E \Rightarrow^1_t F$) if F arises from E by replacing a β-redex (type-redex) subterm by the corresponding reduct. The relation $=_{\beta t}$ is the symmetric, reflexive and transitive closure of $(\Rightarrow^1_\beta) \cup (\Rightarrow^1_t)$. E is *normal* if none of its subterms is a redex, and *strongly normalizable* if every sequence of reductions starting with E terminates.

Peano's λ-Calculus, **2λ**(Peano), is the sub-calculus of **2λ** where every type argument is Peano.[5] More generally, if T is a set of types, we write **2λ**(T) for the set of terms where every type argument is in T. In addition, we say that a term E is *confined* to T if the type of every subterm of E is in T.[6]

LEMMA 5. *Suppose that E is a normal term of* **2λ**(Peano), *where the types of E and of every free (object) variable are all Peano. Then E is confined to Peano types.*

Proof. By structural induction on E. If E is a variable, then its type is Peano by assumption.

Suppose that $E \equiv \lambda x^\tau.F^\sigma : \tau \rightarrow \sigma$. By assumption $\tau \rightarrow \sigma$ is Peano, and so both τ and σ are Peano. Thus F satisfies the Lemma's conditions, and by induction assumption the type of every subterm of F is Peano, and so the same is true of E.

Suppose $E \equiv \Lambda S_1 \ldots S_r.F^\tau : \forall \vec{S}.\tau$. By assumption $\forall \vec{S}.\tau$ is Peano, and so τ is Peano, by Lemma 1. Thus the present lemma's conditions hold for F, and the type of every subterm of F is Peano. The remaining subterms of E are of the form $\Lambda S_i \ldots S_r.F$, whose type, $\forall S_i \ldots S_r.\tau$, is a subtype of $\forall \vec{S}.\tau$, and is therefore Peano, by Lemma 1.

Finally, suppose that E is an application. Since E is normal, it must be of the form $x^\sigma A_1 \cdots A_r$, where x is a variable and $A_1, \ldots, A_r$ consist of types and terms. By the Lemma's condition σ is Peano. It is easy to prove, by induction on i, that the type of every term $E_i =_{\mathrm{df}} x A_1 \cdots A_i$ $(i \leq r)$ is Peano, and so the lemma applies to each such E_i. In particular, $E \equiv E_r$ is confined to Peano. ⊣

[5] The phrase "calculus" is justified by the fact that **2λ**(Peano) is closed under β-t-reductions. This is trivial for β-reductions, because these do not generate new type arguments. A type-reduction mapping a term $(\Lambda R.E)\tau$ to $\{\tau/R\}E$ might change a type argument σ within E into $\{\tau/R\}\sigma$. However, since the input term is in **2λ**(Peano), both τ and σ are Peano, and therefore the new type argument $\{\tau/R\}\sigma$ is also Peano.

[6] If E is confined to Peano, then E is in **2λ**(Peano), except possibly if there is a vacuous type application in E; for instance, $x^{\forall R.S \rightarrow S}\tau$ is confined to Peano for any type τ. Such anomalies may be eliminated, without loss of generality, by requiring that vacuous type applications have a type variable as argument.

2. RECURRENCE SCHEMAS

Recall that the primitive recursive functions are generated from the zero-constant, successor, and projection functions by composition and the schema of primitive recursion,

$$\begin{aligned} f(0,\vec{x}) &= g(\vec{x}) \\ f(\mathsf{s}n,\vec{x}) &= h(n,\vec{x},f(n,\vec{x})) \end{aligned} \tag{1}$$

Gödel (*1958*) considered a generalization of primitive recursion to all simple types.[7] As usual, the *simple types* are generated inductively from a base type o by the binary type constructor $\rightarrow$, i.e. if τ and σ are simple types then so is $\tau\rightarrow\sigma$. We adopt the same notational convention for simple types as for quantifier free types of $\mathbf{2\lambda}$. Note that every simple type τ is of the form $\tau_1,\ldots,\tau_r\rightarrow o$ for some vector of types $\pi(\tau) =_{\mathrm{df}} (\tau_1\ldots\tau_r)$.

Primitive recursion in simple types is often stated as a reading of schema (1) above with the output of f and the parameters $\vec{x}$ permitted to have any simple types. However, the use of higher type permits a streamlining of this definition. Say that a function $f : o\rightarrow\tau$ is *defined by closed primitive recursion* from functions $g:\tau$ and $h : o,\tau\rightarrow\tau$ if

$$f(0) = g \qquad f(\mathsf{s}n) = h(n)(f(n)) \tag{2}$$

Schema (1) is easily reducible to (2): If $f(\mathsf{s}n,\vec{x}) = h(n,\vec{x},f(n,\vec{x}))$ then $f(\mathsf{s}n) = h'(n)(f(n))$, where $h'(a)(b)(\vec{x}) =_{\mathrm{df}} h(a,\vec{x},b(\vec{x}))$.

Our proofs below will be simplified by the use of an alternative definition of primitive recursion in simple types. We say that functions $f_1 : o\rightarrow\tau_1$, ..., $f_k : o\rightarrow\tau_k$ are defined by *simultaneous recurrence* from functions $g_1:\tau_1,\ldots,g_k:\tau_k$ and $h_1 : (\tau_1\ldots\tau_k\rightarrow\tau_1)$, ..., $h_k : (\tau_1\ldots\tau_k\rightarrow\tau_k)$, if

$$\begin{aligned} f_i(0) &= g_i \\ f_i(\mathsf{s}n) &= h_i(f_1(n))\cdots(f_k(n)) \\ &\quad (i = 1\ldots k) \end{aligned} \tag{3}$$

LEMMA 6. *Primitive recursion is reducible to simultaneous recurrence.*

Proof. Suppose that $f : \tau\rightarrow o$ is defined by (2). Define $f_1 : o\rightarrow o$ and $f_2 : o\rightarrow\tau$ by the following simultaneous recurrence.

$$\begin{array}{rclcrcl} f_1(\mathbf{0}) &=& \mathbf{0} & \qquad & f_2(\mathbf{0}) &=& g \\ f_1(\mathsf{s}n) &=& \mathsf{s}f_1(n) & & f_2(\mathsf{s}n) &=& h(f_1(n))(f_2(n)) \end{array}$$

Then, by induction on n, $f_1(n) = n$ and $f_2(n) = f(n)$. ⊣

[7]The generalization of primitive recursion to second order functionals seems to have been noticed first by Hilbert (*1925*), where Ackermann's function is defined by primitive recursion over type $o\rightarrow o$.

In fact, an even simpler definition of simultaneous recurrence suffices. Say that an instance of the schema (3) of simultaneous recurrence is *type-uniform* if the types $\tau_1 \dots \tau_k$ are one and the same type τ.

LEMMA 7. *The schema of simultaneous recurrence is reducible to type-uniform simultaneous recurrence.*

Proof. Without loss of generality, consider the case of two simultaneously defined functions. Suppose that $f_i : o \to \tau_i$ $(i = 1, 2)$ are defined by

$$\begin{aligned} f_1(\mathbf{0}) &= g_1 & f_2(\mathbf{0}) &= g_2 \\ f_1(\mathbf{s}n) &= h_1(f_1(n))(f_2(n)) & f_2(\mathbf{s}n) &= h_2(f_1(n))(f_2(n)) \end{aligned}$$

Let $\tau =_{\mathrm{df}} \pi(\tau_1), \pi(\tau_2) \to o$. Define the functions $U_i : \tau_i \to \tau$ and $D_i : \tau \to \tau_i$ $(i = 1, 2)$ by

$$\begin{aligned} U_i &= \lambda a^{\tau_i}\, \lambda \vec{x}_1^{\pi(\tau_1)} \vec{x}_2^{\pi(\tau_2)}.\, a(\vec{x}_i^{\pi(\tau_i)}) \\ D_1 &= \lambda z^{\tau}\, \lambda \vec{x}_i^{\pi(\tau_1)}.\, z(\vec{x}_1)(\vec{\mathbf{e}}^{\pi(\tau_2)}) \\ D_2 &= \lambda z^{\tau}\, \lambda \vec{x}_2^{\pi(\tau_2)}.\, z(\vec{\mathbf{e}}^{\pi(\tau_1)})(\vec{x}_2) \end{aligned}$$

where, for a type σ, $\mathbf{e}^{\sigma}$ is some canonical function of type σ, e.g. $\lambda \vec{x}^{\pi(\sigma)}.\mathbf{0}$. Then $f_i = D_i(f_i')$, where f_1', f_2' are defined by the type-uniform simultaneous recurrence

$$\begin{aligned} f_i'(\mathbf{0}) &= U_i(g_i) \\ f_i'(\mathbf{s}n) &= h_i'(f_1'(n))(f_2'(n)) \\ \text{with} \qquad h_i' &=_{\mathrm{df}} \lambda y_1^{\tau} y_2^{\tau}.\, U_i(\, h_i(D_1(y_1))(D_2(y_2))\,) \end{aligned}$$

⊣

Let $\mathbf{Rec}^{\omega}$ be the system of functions in all simple types generated from the constant-zero and successor functions using typed λ-abstraction and application (and hence projections and composition), and the schema of simultaneous recurrence in all simple types.

We are now ready to state precisely our main result.

THEOREM 8 **Main Theorem.** *Let f be a numeric function. The following conditions are equivalent.*

(1) *f is a provably recursive function of* **PA**.

(2) *f is definable by (closed) primitive recursion in simple types.*

(3) $f \in \mathbf{Rec}^{\omega}$.

(4) *f is λ-definable in $\mathbf{2\lambda}(*\forall^1\underline{2})$.*

(5) *f is λ-definable in Peano's λ-Calculus, $\mathbf{2\lambda}$(Peano).*

The implication (1) $\Rightarrow$ (2) is Gödel's (*1958*) "Dialectica" Theorem. (2) $\Rightarrow$ (3) is Lemma 6 above. We shall prove (3) $\Rightarrow$ (4) as Lemma 9 below. (4) implies (5) trivially. Finally, we prove the implication (5) $\Rightarrow$ (1) as Theorem 19 below.

3. LAMBDA DEFINABILITY OF SIMULTANEOUS RECURRENCE IN SIMPLE TYPE

The canonical representation of natural numbers in $\mathbf{2\lambda}$ is by type-abstracted Church numerals: $\bar{n} =_{\mathrm{df}} \Lambda R.\lambda s^{R\to R}, z^R.\, s^{[n]}(z)$, where the superscript $[n]$ stands for n-fold iteration. We denote by ν the type $\forall R.((R\to R), R \to R)$ of these numerals. For a simple type τ we write $\tau[\nu]$ for the 2-type $\{\nu/o\}\tau$ that results from replacing the base type o in τ by ν. Note that for every simple type τ the type $\tau[\nu]$ is in $*\forall^1\underline{2}$.

A function $f : \mathbb{N}^r \to \mathbb{N}$ is *defined in* $\mathbf{2\lambda}$ by a term F, if $F : \nu^r \to \nu$, and $F\bar{n}_1 \cdots \bar{n}_r =_{\beta,t} \overline{f(\vec{n})}$ for all $\vec{n} = (n_1 \ldots n_r) \in \mathbb{N}^r$. More generally, the notion of definability for numeric functionals f of simple type τ is given by structural recurrence on τ: If $\tau \equiv \tau_1, \ldots, \tau_r \to o$, then f is *defined* by a term $F : \tau[\nu]$ if for all terms $G_1 \ldots G_r$, defining over $\mathbb{N}$ functionals $g_1 \ldots g_r$ of types $\tau_1 \ldots \tau_r$, respectively, we have $FG_1 \ldots G_r =_{\beta t} \overline{fg_1 \cdots g_r}$.

Given terms $E_1, \ldots, E_k$, all of type τ, let

$$\langle E_1 \ldots E_k\rangle =_{\mathrm{df}} \lambda x^{\tau^k \to \tau}.\, xE_1 \cdots E_k,$$

and let $\langle k\tau\rangle$ stand for that term's type, namely $(\tau^k \to \tau)\to\tau$. For $i = 1 \ldots k$ let

$$\underline{i}_{k\tau} =_{\mathrm{df}} \lambda q^{\langle k\tau\rangle}.\, q(\lambda u_1^\tau \ldots u_k^\tau.\, u_i).$$

Thus $\underline{i}_{k\tau}\langle E_1 \ldots E_k\rangle =_\beta E_i$.

LEMMA 9. *Every function* $f \in \mathbf{Rec}^\omega$ *is definable in* $\mathbf{2\lambda}(*\forall^1\underline{2})$.

Proof. By Lemma 7 a function $f \in \mathbf{Rec}^\omega$ is definable from zero and successor using abstraction, application, and type-uniform simultaneous recurrence. We proceed by induction on the length of such definitions. The constant zero function is defined by $\lambda a^\nu.\bar{0}$, and the successor function by $\lambda a^\nu.\,\Lambda R.\lambda s^{R\to R} z^R.\, s(aRsz)$. Closure of the definable functions under λ-abstraction and application is trivial.

Suppose that $f_1 \ldots f_k : o \to \tau$ are defined by type-uniform simultaneous recurrence over τ, as in (3) above. By induction assumption the functions $g_1 \ldots g_k$, $h_1 \ldots h_k$ are defined by some terms $G_1 \ldots G_k$, $H_1 \ldots H_k$, respectively. Then f_i is defined by

$$
\begin{aligned}
&F_i &&=_{\mathrm{df}} && \lambda a^\nu.\, \underline{i}_{k\tau}\,(a\,\langle k\tau[\nu]\rangle\, \mathcal{H}\, \mathcal{G}),\\
\text{where}\quad &\mathcal{H} &&=_{\mathrm{df}} && \lambda q^{\langle k\tau[\nu]\rangle}.\langle H_1(\underline{1}_{k\tau}q)\cdots(\underline{k}_{k\tau}q),\ \ldots\ , H_k(\underline{1}_{k\tau}q)\cdots(\underline{k}_{k\tau}q)\rangle,\\
\text{and}\quad &\mathcal{G} &&=_{\mathrm{df}} && \langle G_1, \ldots, G_k\rangle.
\end{aligned}
$$

Note that the type argument $\langle k\tau[\nu]\rangle$ is in $*\forall^1\underline{2}$, so each F_i is in $\mathbf{2\lambda}(*\forall^1\underline{2})$. ⊣

4. COMPUTABILITY OF DEFINABLE FUNCTIONS

4.1. Normalization of Terms in $\mathbf{2\lambda}$(Peano)

We show next that every function definable in $\mathbf{2\lambda}$(Peano) is a provably recursive function of **PA**. The salient observation is that the Tait-Girard's proof of strong normalization for $\mathbf{2\lambda}$ (*Girard 1972*), when restricted to terms confined to Peano types, can be formalized in second order arithmetic with comprehension and induction restricted to arithmetical formulas. Since the latter theory is conservative over **PA**, this will imply that all functions definable in $\mathbf{2\lambda}$(Peano) are provably recursive in **PA**.

While many expositions of Girard's method have been given since its inception, we outline here an exposition tailored to the application in hand. Consider a language LAMBDA with object variables $M, \ldots$ intended to range over untyped λ-terms, and set (i.e. unary relation) variables intended to range over sets of such terms. For set variables we re-use the type variables of $\mathbf{2\lambda}$. The vocabulary of LAMBDA has identifiers for the basic syntactic operations over λ-terms, as follows. The constant identifiers of LAMBDA are the (untyped) λ-variables. The function identifiers are a binary $*$ (used in infix), and for each λ-variable x a unary $\underline{\text{lmb}}_x$ and a binary $\underline{\text{sbt}}_x$. These are intended to denote application, λ-abstraction, and (correct) substitution, respectively; that is, if M and N have the value E and F, then the values of $M * N$, $\underline{\text{lmb}}_x(M)$ and $\underline{\text{sbt}}_x(N, M)$ are EF, $\lambda x.E$ and $\{F/x\}\tilde{E}$, respectively, where $\tilde{E}$ arise from E by canonically renaming variables that are free in F. Also, LAMBDA has a unary relation identifier $\underline{\text{Var}}$, intended to hold exactly of the λ-variables.

We construe 2-types as conveying semantic properties of untyped λ-terms (as in *Leivant 1986*). To start, let $\underline{\text{Frame}}\,[B, S]$ be the conjunction of the following properties of sets B and S. We shall eventually set S to be the set $\mathbf{S}$ of strongly normalizable terms, and B the set $\mathbf{VS}^*$ of terms of the form $zE_1 \cdots E_k$ $(k \geq 0)$, with z a variable and $E_i \in \mathbf{S}$. However, we need B and S to be parameterized for now.

(1) $\{\lambda\text{-variables}\} \subseteq B \subseteq S$:

$$\forall M.\, (\underline{\text{Var}}\,(M) \rightarrow B(M)) \;\wedge\; (B(M) \rightarrow S(M)).$$

(2) If $Ez \in S$ (z a variable) then $E \in S$:

$$\forall M, N.\; S(M * N) \wedge \underline{\text{Var}}\,(N) \;\rightarrow\; S(M).$$

(3) $BS \subseteq B$:

$$\forall M, N.\; B(M) \wedge S(N) \rightarrow B(M * N).$$

Let $\underline{\text{Adeq}}\,[R, B, S]$ be the conjunction of the following two conditions.

(1) $B \subseteq R \subseteq S$.

(2) If $\{F/x\}E \in R$ with $F \in S$, then $(\lambda x.E)F \in R$:

$$\forall M, N.\ R(\underline{\text{sbt}}_x(M, N)) \wedge S(N) \rightarrow R(\underline{\text{lmb}}_x(M) * N).$$

(It suffices to take the formula above for just one λ-variable x; other cases follow modulo α-conversion.)

$\underline{\text{Adeq}}\,[R, B, S]$ is intended to state that R is "adequate as a type" within the frame B, S.

For each 2-type τ we define a formula φ^τ of LAMBDA, with a single free variable.

$$\begin{aligned} \varphi^R[M] &= R(M) \\ \varphi^{\sigma\to\rho}[M] &= \forall Y(\varphi^\sigma[Y] \rightarrow \varphi^\rho[M * Y]) \\ \varphi^{\forall R.\sigma}[M] &= \forall R.\, \underline{\text{Adeq}}\,[R, B, S] \rightarrow \varphi^\sigma[M] \end{aligned}$$

Note that the bound set variables in φ^τ are the type variables bound in τ, and the free set variables are the free type variables in τ, plus perhaps B and S.

LEMMA 10. *The formula $\varphi^{\{\rho/R\}\sigma}$ is syntactically identical to $\{\varphi^\rho/R\}\varphi^\sigma$.*[8]

Proof. Straightforward structural induction on τ. ⊣

LEMMA 11. *Assume $\underline{\textit{Frame}}\,[B, S]$, and let τ be a 2-type. Assuming $\underline{\textit{Adeq}}\,[R, B, S]$ for every variable R free in τ, we have $\underline{\textit{Adeq}}\,[\varphi^\tau, B, S]$.*

Proof. Structural induction on τ. ⊣

For a term E of $\mathbf{2\lambda}$ let $\underline{E}$ be the untyped form of E.[9]

LEMMA 12. *Assume $\underline{\textit{Frame}}\,[B, S]$. Suppose $E : \tau$ is a term of $\mathbf{2\lambda}$, with free variables $x_1^{\sigma_1} \ldots x_k^{\sigma_k}$. Assuming $\underline{\textit{Adeq}}\,[R, B, S]$ for every variable R free in τ, we have*

$$\forall M_1 \ldots M_k\ (\ \bigwedge_i \varphi^{\sigma_i}(M_i)\ \rightarrow\ \varphi^\tau(\underline{sbt}_{\vec{x}}(\vec{M}, \underline{E}))\ ). \tag{4}$$

where $\underline{sbt}_{x_1 \ldots x_k}(M_1 \ldots M_k, N)$ abbreviates $\underline{sbt}_{x_1}(M_1, \underline{sbt}_{x_2}(M_2, \ldots \underline{sbt}_{x_k}(M_k, N) \cdots))$.[10]

---

[8]Recall that the second order variables R are unary predicates, and φ^ρ is a formula with a single free variable. $\{\varphi^\rho/R\}$ denotes the syntactic operation of replacing each formula-occurrence of the form $R(\mathbf{t})$ in the argument by $\varphi^\rho[\mathbf{t}]$.

[9]The precise definition is by recurrence on E. Use $x_{i\tau}$ ($i = 1, 2 \ldots$, τ a 2-type) as the collection of untyped λ-variables. Let $\underline{x_i^\tau} =_{\text{df}} x_{i\tau}$, $\underline{\lambda x_i^\tau.E} =_{\text{df}} \lambda x_{i\tau}.\underline{E}$, $\underline{EF} =_{\text{df}} \underline{E}\,\underline{F}$, $\underline{\Lambda R.E} =_{\text{df}} \underline{E}$, and $\underline{E\tau} =_{\text{df}} \underline{E}$.

[10]Strictly speaking, $\underline{E}$ should be replaced by the term of LAMBDA that denotes $\underline{E}$, if (4) is to be a formula of LAMBDA.

Proof. By structural induction on E. For the case of application use Lemmas 11 and 10. (See e.g., Lemma 2.3 in *Leivant 1986* for details.) ⊣

COROLLARY 13. *For every term E of* $\mathbf{2\lambda}$, $\underline{E}$ *is strongly normalizable.*

Proof. Let E^τ be a term of $\mathbf{2\lambda}$. Without loss of generality, E has no free object or type variables. It is easy to see that $\underline{\text{Frame}}\,[\mathbf{VS}^*, \mathbf{S}]$ holds, where $\mathbf{S}$ is the set of strongly-normalizable untyped λ-terms, and $\mathbf{VS}^*$ is the set of expressions of the form $zE_1 \cdots E_k$ where z is a variable and $E_i \in \mathbf{S}$. So, by Lemma 11, $\underline{\text{Adeq}}\,[\varphi^\tau, \mathbf{VS}^*, \mathbf{S}]$, and in particular $\varphi^\tau \subseteq \mathbf{S}$. By Lemma 12 we have $\varphi^\tau(\underline{E})$, and so $\underline{E} \in \mathbf{S}$. ⊣

COROLLARY 14. *Every term of* $\mathbf{2\lambda}$ *is strongly normalizable, i.e. every reduction sequence terminates.*

Proof. Note that if E reduces to F by a type-contraction, then $\underline{F}$ is identical to $\underline{E}$. Thus, if $\Sigma = E_0, E_1 \ldots$ is a reduction sequence in $\mathbf{2\lambda}$, then $\underline{\Sigma} =_{\text{df}} \underline{E}_0, \underline{E}_1 \ldots$ is a sequence of β-reductions and identities. Also, type reductions do not generate new redexes, and therefore every stretch of consecutive type reductions in Σ must be finite. By Corollary 13 the subsequence of $\underline{\Sigma}$ consisting of β-reductions is finite. So $\underline{\Sigma}$ is finite, and therefore Σ is finite. ⊣

4.2. Formalization of the Normalization Proof

Let us outline a formalization of the proof above within second order arithmetic, **SOA**.[11] **SOA** is a theory based on second order logic with equality, variables for numbers and sets and quantifiers over them, and comprehension for all formulas. We take the vocabulary of **SOA** to consist of the constant **0** and an identifier for each primitive recursive function. The non-logical axioms are Peano's third and fourth axioms for the successor function, defining equations for all primitive recursive functions, and the induction axiom

$$\forall X.(X(\mathbf{0}) \wedge \forall z.X(z) \rightarrow X(\mathbf{s}z)) \rightarrow \forall z.X(z).$$

Since comprehension is postulated for all formulas, it follows that all instances of induction for formulas in the language are derivable.

Fix some canonical arithmetization of the syntax of the untyped λ-calculus, with primitive recursive numeric counterparts of the LAMBDA identifiers. It follows that the predicates $\underline{\text{Frame}}$ and $\underline{\text{Adeq}}$ are expressed by first order universal formulas (with set parameters). This is in contrast to the predicates φ^τ, whose complexity in the analytical hierarchy grows with the syntactic

[11] NB: This formalization is partial, in a sense to be explained momentarily.

complexity of τ.[12] However, if only a finite collection of types is of interest, then a formalization within **SOA** becomes possible, along the following lines.[13]

Let T be a finite set of types closed under taking syntactic subtypes. Assume, without loss of generality, that the type variables are renamed so that no type variable bound in τ is free in τ or bound by another occurrence of $\forall$. When restricted to types in T, Lemma 11 amounts to verifying a finite number of cases, using metamathematical induction on the complexity of types in T, and this can be proved in **SOA**.

Since T is finite, the property stated in Lemma 12 can be defined explicitly for terms confined to T, as follows. Set

$$\begin{aligned} \varphi^T[M] &\equiv_{\mathrm{df}} \bigvee_{\tau\in T} \varphi^\tau[M] \\ \psi^T[M] &\equiv_{\mathrm{df}} (\forall \text{ sequence } N_1\ldots,N_k \text{ of } \lambda\text{-terms}) \\ &\qquad (\forall\ \lambda\text{-variables } \vec{x})\ (\ \bigwedge_i \varphi^T(N_i) \rightarrow \varphi^T(\{\vec{N}/\vec{x}\}M\,) \end{aligned}$$

$$\underline{\mathrm{Adeq}}\,[T,B,S] \equiv_{\mathrm{df}} \bigwedge_{\tau\in T} \underline{\mathrm{Adeq}}\,[\varphi^\tau,B,S].$$

Then Lemma 12, for terms confined to T, states

$$\underline{\mathrm{Adeq}}\,[T,B,S] \rightarrow \psi^T[M],$$

which is proved within **SOA** by induction. The formalization in **SOA** of the remaining of the proof leading to Corollary 14 is routine.

4.3. Peano λ-Definability Implies Provability

The formalization above of the proof of strong normalization uses the full proof theoretic power of **SOA** in essential ways. However, for terms confined to Peano types the proof is formalizable in the subtheory $\mathbf{SOA}_0$ of **SOA**, obtained by restricting comprehension to arithmetical formulas. (Note that this implies that the induction axiom can be instantiated to such formulas only.) As usual, a formula of **SOA** is *arithmetical* if it has no set quantifiers. Recall that an arithmetical formula is Σ^0_n (Π^0_n) if it is prenex with n alternating blocks of quantifiers, starting with an existential (respectively, universal) block.

[12]This is an essential feature of the proof: if there were a formula $\varphi[t,m]$ such that $\varphi[\#\tau,\#M]$ expresses $\varphi^\tau[M]$ (where $\#\tau$ and $\#M$ are canonical numeric codes for τ and M), then the normalization proof above could be formalized within **SOA**, from which it would follow (for instance by a formula-as-type homomorphism, as in *Leivant 1990*) that **SOA** proves its own consistency.

[13]One may take instead the collection of all types whose syntactic complexity is bounded by some fixed upper limit.

The following observation seems to be due to Georg Kreisel:

LEMMA 15. *Each* **SOA** *formula* φ *of the form* $\forall\vec{R}\exists\vec{x}.\varphi_0$, *with* φ_0 *quantifier free, is equivalent (in the standard model) to a* Σ^0_1 *formula. Moreover, this fact is provable in* $\mathbf{SOA}_0$.

Proof. To avoid cluttered notation, assume that $\varphi_0 \equiv \varphi_0[x, y, R, Q]$, i.e. with singleton $\vec{R}$ and $\vec{x}$, and where φ has one free number variable y and one free set variable Q. Fixing values $P, S \in \mathcal{P}(\mathbb{N})$ for R and Q, the truth value of $\varphi_0[m, n, P, S]$ depends only on a finite number of numeric values $d_1 < \cdots < d_\ell$, and therefore remains unchanged when membership in the sets P, S is modified at values $> d_\ell$. Thus, for each choice of P there is a value $d(P)$ such that $\exists x.\varphi_0[x, n, P, S]$ is determined by the values of the characteristic function χ_P only for arguments $< d(P)$. By König's Lemma, there is an upper bound d to all values $d(P)$, and therefore a bound $b = 2^d$ on binary codes of the initial segments of χ_P for all P's needed to verify the truth of $\varphi[n, S]$. Thus,

$$\varphi[n, S] \quad \leftrightarrow \quad \exists b\, \exists x\, \forall r < b.\, \varphi_0'$$

where φ_0' arises from φ_0 by replacing each subformula of the forma $R(\mathbf{t})$ by $r@\mathbf{t} = 1$, with $r@t$ denoting the $t+1$-st bit of the binary notation of r. Since φ_0' is primitive recursive, so is $\forall r < b.\, \varphi_0'$, and φ is therefore equivalent in $\mathbf{SOA}_0$ to a Σ^0_1 formula.

The formalization of the proof above in $\mathbf{SOA}_0$ is straightforward, including König's Lemma, whose informal proof is easily formalized in $\mathbf{SOA}_0$ (see e.g., *Friedman 1969*, Theorem 3). ⊣

LEMMA 16. *If* τ *is a first order 2-type of rank* $\leq k$, *then* φ^τ *is rendered in* **SOA** *by a* Π^0_k *formula.*

Proof. Straightforward induction on k. ⊣

LEMMA 17. *If* τ *is a Peano type, then* φ^τ *is rendered in* $\mathbf{SOA}_0$ *by an arithmetical formula* Φ^τ. *That is, the formulas* Φ^τ *satisfy, modulo numeric encoding and provably in* $\mathbf{SOA}_0$, *the equivalences that define the formulas* φ^τ.

Proof. By induction on the definition of Peano types.

1. For a type variable R we have $\varphi^R \equiv R$, which is trivially arithmetical.

2. If $\tau = \sigma \to \rho$, where σ and ρ are Peano, then φ^σ and φ^ρ are rendered by arithmetical formulas Φ^σ and Φ^ρ, by induction assumption, and so $\varphi^{\sigma\to\rho}$ is rendered by

$$\Phi^{\sigma\to\rho}[m] \;\equiv_{\mathrm{df}}\; \forall y.\ \text{"}y \text{ codes a } \lambda\text{-term"} \;\wedge\; \Phi^\sigma(y) \;\to\; \Phi^\rho(app(m, y))$$

where *app* is the primitive recursive function rendering the LAMBDA identifier $*$.

3. Suppose that $\tau = \forall R_1 \ldots R_k.\sigma$, where σ is a first order type of rank ≤ 2. By definition,

$$\varphi^{\tau}[M] \equiv \forall \vec{R}. \bigwedge_i \underline{\text{Adeq}}\,[R_i, B, S] \rightarrow \varphi^{\sigma}[M].$$

The formula $\underline{\text{Adeq}}$, by its definition, is rendered by a Π^0_1 formula, and by Lemma 16 Φ^{σ} is Π^0_2. Hence the formula $\underline{\text{Adeq}}\,[R, B, S] \rightarrow \varphi^{\sigma}[M]$ is rendered by some Π^0_2 formula $\forall \vec{u}\, \exists \vec{v}.\, \psi_0$. So $\varphi^{\tau}[M]$ is rendered by the formula $\forall \vec{u} \forall R\, \exists \vec{v}.\, \psi_0[m]$. By Lemma 15 $\forall R\, \exists \vec{v}.\, \psi_0[m]$ is equivalent to some Σ^0_1 formula $\psi[m]$, and so $\varphi^{\tau}[M]$ is rendered by the Π^0_2 formula

$$\Phi^{\tau}[m] \equiv_{\text{df}} \forall \vec{u}.\, \psi[m].$$

4. If $\tau = \{\rho/R\}\sigma$ (correct substitution), where σ and ρ are Peano types, then φ^{τ} is syntactically identical to $\{\varphi^{\rho}/R\}\varphi^{\sigma}$, by Lemma 10. Since both φ^{σ} and φ^{ρ} are rendered by arithmetical formulas Φ^{σ} and Φ^{ρ}, by induction assumption, it follows that φ^{τ} is rendered by the arithmetical formula $\{\Phi^{\rho}/R\}\Phi^{\sigma}$. ⊣

THEOREM 18. *Let T be a finite set of Peano types, closed under taking syntactic subtypes. The strong normalization theorem for terms confined to T is provable in* **PA**.

Proof. From Lemma 17 and the discussion in the previous section it follows that the proof of Proposition 14 for a finite set T of Peano types is formalizable in $\mathbf{SOA}_0$. Since the strong normalization theorem is expressible in **PA**, and since $\mathbf{SOA}_0$ is conservative over **PA** (see e.g., *Troelstra 1973*, §1.9.8, or *Feferman 1977*, §5.5.1), it follows that that theorem is provable in **PA**. ⊣

THEOREM 19. *Every function definable in* **2λ**(Peano) *is a provably recursive function of* **PA**.

Proof. Suppose that E is a term of **2λ**(Peano) defining a numeric function f, unary say. Since every term of **2λ** is normalizable, we may assume that E is normal. By Lemma 5 E is confined to Peano types. Thus all terms of the form $E\bar{n}$ are confined to the finite set T of Peano types, consisting of the types of subterms of E, plus the subtypes of those. By Theorem 18 the normalization theorem for all such terms is provable in **PA**, and so the numeric function $\tilde{f}$ that maps the canonical code of $E\bar{n}$ to the code of the normal form of $E\bar{n}$ is a provably recursive function of **PA**. The function f can be defined as the composition of three functions: the primitive recursive function that maps n to the code of $E\bar{n}$, $\tilde{f}$, and the primitive recursive function than maps the code of the normal form of $E\bar{n}$ to the number it represents. Hence f is provably recursive in **PA**. ⊣

REFERENCES

Barwise, J.

1969 Applications of strict-Π^1_1 predicates to infinitary logic, *The Journal of Symbolic Logic*, vol. 34, pp. 409–423.

1975 *Admissible sets and structures,* Springer-Verlag, Berlin.

Feferman, S.

1977 Theories of finite type related to mathematical practice, *Handbook of mathematical logic* (J. Barwise, editor), North-Holland, Amsterdam, pp. 913–971.

Fortune, S.

1979 Topics in computational complexity, *Ph.D. Dissertation*, Cornell University.

Fortune, S., D. Leivant, and M. O'Donnell

1983 The expressiveness of simple and second-order type structures, *Journal of the ACM* (Association for Computing Machinery), vol. 30, no. 1, pp. 151–185.

Friedman, H.

1969 König's Lemma is weak, unpublished mimeographed notes, Stanford University.

Girard, J.-Y.

1971 Une extension de l'interprétation de Gödel à l'analyse, et son application a l'élimination des coupures dans l'analyse et la théorie des types, *Proceedings of the Second Scandinavian Logic Symposium* (J. E. Fenstad, editor), North-Holland, Amsterdam, pp. 63–92.

1972 Interprétation Fonctionnelle et Elimination des Coupures de l'Arith - métique d'Ordre Supérieur, Thèse de doctorat d'état, Université Paris VII (June).

Gödel, K.

1958 Über eine bisher noch nicht benutzte Erweiterung des finiten Standpunktes, *Dialectica*, vol. 12, pp. 280–287.

Handley, W. G.

1993 Bellantoni and Cook's characterization of polynomial time functions, manuscript, Department of Pure Mathematics, University of Leeds.

Hilbert, D.

1925 Über das Unendliche, *Mathematische Annalen*, vol. 95, pp. 161–190; translated in *van Heijenoort 1967*, pp. 367–392.

Leivant, D.

1981 The complexity of argument passing in polymorphic procedures, *Proceeding of the thirteenth symposium on theory of computing,* Association for Computing Machinery, New York, pp. 38–45.

1986 Typing and computational properties of lambda expressions, *Theoretical Computer Science*, vol. 44, pp. 51–68.

1989 Descriptive characterizations of computational complexity, *Journal of Computer and System Sciences*, vol. 39, pp. 51–83.

1990 Contracting proofs to programs, *Logic and computer science* (P. Odifreddi, editor), Academic Press, New York, pp. 179–327.

1990a Discrete polymorphism, *Proceedings of the sixth ACM conference on LISP and functional programming*, pp. 288–297.

1991 Finitely stratified polymorphism, *Information and Computation*, vol. 93, pp. 93–113.

1994 Ramified recurrence and computational complexity I: Word recurrence and poly-time, *Feasible mathematics II, Perspective in computer science*, (P. Clote and J. Remmel, editors), Birkhäuser-Boston, New York, pp. 320–343.

Leivant, D., and J.-Y. Marion

1993 Lambda calculus characterizations of poly-time, *Fundamenta Informaticae*, vol. 19, pp. 167–184; Special issue: Lambda calculus and type theory (J. Tiuryn, editor).

MacQueen, D. B., and R. Sethi

1982 A semantic model of types for applicative languages, *ACM Symposium on LISP and functional programming*, pp. 243–252.

O'Donnell, M.

1979 A programming language theorem which is independent of Peano Arithmetic, *Eleventh annual ACM symposium on theory of computing*, Association for Computing Machinery.

Reynolds, J. C.

1974 Towards a theory of type structures, *Programming symposium (colloque sur la programmation, Paris)*, LNCS #19, Springer-Verlag, Berlin, pp. 408–425.

Schwichtenberg, H.

1976 Definierbare Funktionen im Lambda-Kalkul mit Typen, *Archiv mathematische Logik u. Grundlagenforschung*, vol. 17, pp. 113–114.

Statman, R.

1979 The typed λ-calculus is not elementary recursive, *Theoretical Computer Science*, vol. 9, pp. 73–81.

Troelstra, A. S.

1973 Metamathematical investigations of intuitionistic arithmetic and analysis, *Lecture Notes in Mathematics*, vol. 344, Springer-Verlag, Berlin.

Van Heijenoort, J.

1967 (editor), *From Frege to Gödel, A source book in mathematical logic, 1879–1931*, Harvard University Press, Cambridge, Massachusetts.

RALPH LOADER*

THE UNDECIDABILITY OF λ-DEFINABILITY

Abstract. In this article, we shall show that the Plotkin-Statman conjecture (*Plotkin 1973, Statman 1982*) is false. The conjecture was that, in a model of the simply typed λ-calculus with only finitely many elements at each type, definability (by a closed term of the calculus) is decidable. This conjecture had been shown to imply many things, for example, Statman (*1982,* see also *Wolfram 1993*) has shown it implies the decidability of pure higher order pattern matching (a problem that remains open at the time of writing) and is equivalent to higher order pattern matching with δ-functions. The proof of undecidability given here uses encodings of semi-Thue systems as definability problems. It had been thought that λ-definability might be characterized by invariance under logical relations, which would imply the Plotkin-Statman conjecture. We give a relatively simple counterexample to this, using our encoding of word problems.

1. PRELIMINARIES

We shall consider the simply typed λ-calculus with a single ground type o. All models M considered here will be finite (in the sense of finitely many elements at each type) and full (for any types T and U, $M_{T\to U}$ is the set of all functions from M_T to M_U). Terms of the simply typed λ-calculus shall be assumed to be in the Church form, *i.e.* all variables are typed, but we shall usually omit the types for typographical convenience.

We shall consider terms in β-normal, η-long form. β-conversion is generated by the following rules:

$$\begin{aligned}(\lambda x\,.\,s[x])(t) &\longrightarrow_\beta && s[t],\\ s\longrightarrow_\beta s' &\Rightarrow && s(t)\longrightarrow_\beta s'(t),\\ t\longrightarrow_\beta t' &\Rightarrow && s(t)\longrightarrow_\beta s(t'),\\ s\longrightarrow_\beta s' &\Rightarrow && \lambda x\,.\,s\longrightarrow_\beta \lambda x\,.\,s',\end{aligned}$$

where bound variables may have to be renamed appropriately. η-expansion is the opposite of the usual η-contraction:

$$t^{A\to B} \longrightarrow_\eta \lambda x^A\,.\,t(x), \tag{1}$$

$$s\longrightarrow_\eta s' \Rightarrow s(t)\longrightarrow_\eta s'(t), \tag{2}$$

$$t\longrightarrow_\eta t' \Rightarrow s(t)\longrightarrow_\eta s(t'),$$

* This research was supported by the Commonwealth Scholarship Commission in the United Kingdom. I would like to thank Allen Stoughton, Lincoln Wallen (my supervisor), and the referee, for useful conversations and advice.

C. Anthony Anderson and M. Zelëny (eds.), Logic, Meaning and Computation, 331-342.

$$s \longrightarrow_\eta s' \quad \Rightarrow \quad \lambda x \,.\, s \longrightarrow_\eta \lambda x \,.\, s',$$

where, in (1), x is not free in t. To retain normalization, we make the restriction that η-expansion cannot create new β-redexes; in other words, (1) may not be applied if the term t is a λ-abstraction $\lambda x \,.\, t'$, and (2) may not be applied if $s \longrightarrow_\eta s'$ is obtained as an instance of (1).

Given a simply typed λ-term t and a valuation v defined on the free variables of t, the valuation of t with respect to v will be denoted by $[\![t]\!]_v$. The subscript v may be omitted if the intended valuation is obvious from the context or if t is closed. If $t[\bar{x}]$ is a term, and $\bar{a} \in M$ have the appropriate types, then we shall write $[\![t[\bar{a}]]\!]$ or just $t[\bar{a}]$ to denote the valuation of t w.r.t. the valuation sending $\bar{x}$ to $\bar{a}$. The valuation of a term is invariant under both β-conversion and η-expansion.

A closed term t is said to define $a \in M$ if $[\![t]\!] = a$. A term $t[\bar{x}]$ is said to define b w.r.t. $\bar{a}$ if $[\![t[\bar{a}]]\!] = b$.

The **(absolute) λ-definability problem** is: Given a full and finite model M and a member $a \in M_T$ for some type T, is there a closed term that defines a?

The **relative λ-definability problem** is: Given a model M, again full and finite, $\bar{a}, b \in M$, is there a term $t[\bar{x}]$ that defines b w.r.t. $\bar{a}$?

These two problems are related as follows:

LEMMA 1. *The λ-definability problem is decidable if and only if the relative λ-definability problem is decidable.*[1]

Proof. Suppose that the absolute problem is decidable. We reduce the relative problem to this as follows: Given $a_i \in M_{T_i}$, $i = 1 \cdots n$ and $b \in M_T$, look for $f \in M_{T_1 \to \cdots \to T_n \to T}$ such that:

1. f is λ-definable by a closed term ϕ,

2. $f(a_1) \cdots (a_n) = b$.

If such a f is found, then the term $\phi(x_1) \cdots (x_n)$ defines b relative to the a_i. If a term $t[x_1 \cdots x_n]$ defines b relative to $a_1 \cdots a_n$, then 1 and 2 above are satisfied with $\phi = \lambda x_1 \cdots x_n \,.\, t$, and $f = [\![\phi]\!]$.

Further, the search for f satisfying 1 and 2 is effective (by our assumption that the absolute problem is decidable), and terminating as we can list the finitely many objects of a given type in M. Thus the relative problem is decidable if the absolute problem is.

The converse implication is trivial as the absolute problem is a special case of the relative problem. ⊣

[1] I assume a sensible coding of finite λ-models, encoding functions as Gödel numbers of their graphs. Note that this lemma applies both to the problems in a fixed model, or the problems ranging over all (full and finite) models.

The statement that the λ-definability problem is decidable is known as the Plotkin-Statman conjecture. We will show that, in fact, these problems are undecidable, by encoding word problems as relative λ-definability problems.

2. ENCODING WORD PROBLEMS

We will encode word problems (specifically, semi-Thue systems) over the alphabet $\{A, B\}$ as members of the model M with (at least) seven elements $\{A, B, L, R, Y, N, *\}$ at ground type.

Given a word W_0 and rules $C_i \longrightarrow D_i$, $i = 1 \cdots n$, where the C_i and the D_i are words, a derivation is a sequence of words $W_0, \ldots, W_n$ such that each W_{j+1} is obtained from W_j by replacing a sub-word C_i by D_i. The word W is derivable if it is the last word of some derivation.

The word problems we consider are as follows: Given an initial word W_0, and some rules $C_i \longrightarrow D_i$, $i = 1 \cdots n$, can we derive a word W?

PROPOSITION 2. *There is a word W_0 and rules $C_i \longrightarrow D_i$ such that the problem of determining if a word W can be derived is undecidable.*

Proof. Davis (*1977*) encodes a universal Turing machine as a semi-Thue system so as to give a set of rules F_i and a word W_0 such that the problem "Given a word W, can W_0 be derived from W?" is undecidable. By reversing all the rules of this problem, we get the proposition, except with a larger alphabet. It is easy to encode such a problem in the alphabet $\{A, B\}$. ⊣

We now encode words and rules as objects in M. A word W of length n will be coded by an object $\ulcorner W \urcorner$ of type $\underbrace{o \rightarrow \cdots \rightarrow o}_{n} \rightarrow o$. If words C and D have lengths m and n respectively, then the rule $C \longrightarrow D$ will be encoded by an object $\ulcorner C \longrightarrow D \urcorner$ of type $(\underbrace{o \rightarrow \cdots \rightarrow o}_{m} \rightarrow o) \rightarrow (\underbrace{o \rightarrow \cdots \rightarrow o}_{n} \rightarrow o)$.

The encoding of a word W is defined by the following clauses:

W1. If W has $c \in \{A, B\}$ in the ith position ($i = 1 \cdots n$), then

$$\ulcorner W \urcorner \underbrace{(*) \cdots (*)}_{i-1} (c) \underbrace{(*) \cdots (*)}_{n-i} = Y,$$

W2. For $i = 1 \cdots n - 1$,

$$\ulcorner W \urcorner \underbrace{(*) \cdots (*)}_{i-1} (L)(R) \underbrace{(*) \cdots (*)}_{n-i-1} = Y,$$

W3. $\ulcorner W \urcorner (x_1) \cdots (x_n) = N$ unless stated otherwise in W1 or W2.

It may help the reader's understanding to consider $\ulcorner W \urcorner$ as answering questions about the word W, with the ith parameter corresponding to the ith letter of W. W2 may then be thought of as encoding the ordering of the letters in W; without this axiom, any permutation of W would have its encoding λ-definable from $\ulcorner W \urcorner$, which is undesirable for our purposes.

The **characteristic** $\chi S \in M_{o\to\cdots\to o\to o}$ of a set $S \subset M_o \times \cdots \times M_o$ is defined by

- $\chi S\,(x_1)\cdots(x_n) = Y$ whenever $(x_1,\ldots,x_n) \in S$,
- $\chi S\,(x_1)\cdots(x_n) = N$ whenever $(x_1,\ldots,x_n) \notin S$.

Thus W1–3 above may be stated as: The encoding of a word $c_1\cdots c_n$ is the characteristic of the union of the two sets

$$\{\,(\underbrace{*\cdots*}_{i-1},c,\underbrace{*\cdots*}_{n-i}) \mid i = 1\cdots n\}$$

and

$$\{\,(\underbrace{*\cdots*}_{i-1},L,R,\underbrace{*\cdots*}_{n-i-1}) \mid i = 1\cdots n-1\}.$$

A rule $F = C \longrightarrow D$, where C and D have length m and n, is encoded as follows:

R1. $\ulcorner F \urcorner(\chi\{(\underbrace{*,\ldots,*}_{m})\}) = \chi\{(\underbrace{*,\ldots,*}_{n})\}$,

R2. $\ulcorner F \urcorner(\chi\{(R,\underbrace{*,\ldots,*}_{m-1})\}) = \chi\{(R,\underbrace{*,\ldots,*}_{n-1})\}$,

R3. $\ulcorner F \urcorner(\chi\{(\underbrace{*,\ldots,*}_{m-1},L)\}) = \chi\{(\underbrace{*,\ldots,*}_{n-1},L)\}$,

R4. $\ulcorner F \urcorner(\ulcorner C \urcorner) = \ulcorner D \urcorner$,

R5. $\ulcorner F \urcorner(g) = \chi(\emptyset)$ in all cases not covered above.

The encoding of rules is designed so that:

LEMMA 3. *If* $W_1 = g_1\cdots g_k\,C\,h_1\cdots h_l$ *and* $W_2 = g_1\cdots g_k\,D\,h_1\cdots h_l$ *are words, and* F *is the rule* $C \longrightarrow D$, *then* $\ulcorner W_2 \urcorner$ *is given by*

$$f =_{def} \lambda\bar{x}\bar{y}\bar{z}\,.\,\ulcorner F \urcorner(\lambda\bar{u}\,.\,\ulcorner W_1 \urcorner\bar{x}\bar{u}\bar{z})(\bar{y}),$$

where $\bar{x}$, $\bar{y}$, $\bar{z}$ *and* $\bar{u}$ *are vectors of* k, n, l *and* m *distinct variables respectively.*

Proof. An easy but tedious verification. For example, by W1 and W3,

$$\lambda \bar{u} . \ulcorner W_1 \urcorner (g_1) \underbrace{(*) \cdots (*)}_{k-1} (u_1) \cdots (u_m) \underbrace{(*) \cdots (*)}_{l}$$

is the characteristic of the set $\{(*, \ldots, *)\}$ and thus by R1,

$$f(g_1) \underbrace{(*) \cdots (*)}_{k+n+l-1} = Y.$$

⊣

COROLLARY 4. *If a word W is derivable from W_0 by rules F_i, then $\ulcorner W \urcorner$ is λ-definable from $\ulcorner W_0 \urcorner$ and the $\ulcorner F_i \urcorner$.*

Proof. By an induction over the derivation of W, using the lemma. ⊣

3. FAITHFULNESS OF THE ENCODING

In this section we shall fix a word W_0 and rules $F_i = C_i \longrightarrow D_i$. We show that if $\ulcorner W \urcorner$ is λ-definable relative to $\ulcorner W_0 \urcorner$ and the $\ulcorner F_i \urcorner$, then W is derivable from W_0 by the rules F_i, and thus conclude that the Plotkin-Statman conjecture fails.

Our proofs will be by induction over terms in the following form:

DEFINITION 5. A **preword-term** is a term t of type o, containing constants among $\ulcorner W_0 \urcorner$ and $\ulcorner F_1 \urcorner \cdots \ulcorner F_n \urcorner$ such that:

- t is in β-normal, η-long form,
- t is not a variable,
- All free variables in t are of type o.

If a preword-term s has a closure $\lambda x_1 \cdots x_n . s$ that encodes a word, then $x_1 \cdots x_n$ must all occur free in s (as by W1 and W3 a function encoding a word of length n must depend on all of its n arguments) and the ordering of $x_1 \cdots x_n$ (and thus the word encoded) is unique (by W2 and W3).

In preparation for the main proof, we prove two technical lemmas about preword-terms. Although lemmas 6 and 7, and proposition 9, are stated about preword-terms, they are really used to give information about their closures, as we wish to identify which closures of preword-terms encode words. It is smoother formally, however, to prove things by induction over preword-terms.

LEMMA 6. *If s is a preword-term and v any valuation, then $[\![s]\!]_v$ is either Y or N.*

Proof. As s is in normal form and not a variable, s must be in the form $K(t_1)\cdots(t_i)$, where K is $\ulcorner W_0 \urcorner$ or one of the $\ulcorner F_i \urcorner$. That $[\![s]\!] \in \{Y, N\}$ is now obvious from W1–3 and R1–5. ⊣

LEMMA 7. *Let s be a preword-term:*

1. *If x is free in s, and v is a valuation with $v(x) \in \{A, B\}$ such that $[\![s]\!]_v = Y$, then $v(y) = *$ for all y other than x free in s.*

2. *If $[\![\lambda u_1 \cdots u_n . s]\!]_v$ encodes a word, then $v(x) = *$ for every variable x free in $\lambda\bar{u} . s$.*

Proof. We prove the lemma by induction over the preword-term s. We first derive 1 from the induction hypothesis. Consider the case when s is $\ulcorner W_0 \urcorner(t_1)\cdots(t_n)$.

- If one of the t_i is not a variable, then $[\![t_i]\!]_v \in \{Y, N\}$ by the previous lemma, so that $[\![s]\!]_v = N$ by W3, and hence there is nothing to prove.

- If all of the t_i are variables, then 1 is immediate by W1–3.

Suppose s is $\ulcorner F_i \urcorner(\lambda\bar{u} . t)(t_1)\cdots(t_n)$. For any valuation v such that $[\![s]\!]_v = Y$:

i. Each of the t_i is a variable, as else $[\![t_i]\!]_v \in \{Y, N\}$ and $[\![s]\!]_v = N$.

ii. By R5, t cannot be a variable, and thus t is a preword-term.

iii. By R5, $[\![\lambda\bar{u} . t]\!]_v \neq \chi(\emptyset)$.

We can now verify 1 for this s:

- If x is one of the variables t_j, then examining R1–5, the only way we can have $v(x) \in \{A, B\}$, $[\![s]\!]_v = Y$, is if $[\![\lambda\bar{u} . t]\!]_v = \ulcorner C_i \urcorner$ (so that by R4, $\ulcorner F_i \urcorner([\![\lambda\bar{u} . t]\!]_v) = \ulcorner D_i \urcorner$), $v(x)$ is the jth letter of the word D_i, and $v(t_k) = *$ for $k \neq j$. Now by the induction hypothesis 2 for t, $v(z) = *$ for all variables z free in $\lambda\bar{u} . t$ also.

- Suppose the variable x is free in $\lambda\bar{u} . t$, and $v(x) \in \{A, B\}$, $[\![s]\!]_v = Y$. By iii there is a valuation v' with $v'(y) = v(y)$ for all y free in $\lambda\bar{u} . t$ and $[\![t]\!]_{v'} = Y$. Applying the induction hypothesis to t and v', we see that $v(y) = v'(y) = *$ for each $y \neq x$ free in $\lambda\bar{u} . t$, and that $[\![\lambda\bar{u} . t]\!]_v = \chi\{(* \cdots *)\}$, so by R1, we must have $v(t_k) = *$ for each t_k also.

We now derive 2 from 1. If $[\![\lambda\bar{u} . s]\!]_v$ encodes a word, then either for $c = A$ or for $c = B$ we have

$$[\![\lambda\bar{u} . s]\!]_v(c)(*)\cdots(*) = Y.$$

Define a valuation v' by

- $v'(y) = v(y)$ for y free in $\lambda\bar{u} . s$,

- $v'(u_1) = c$ and $v'(u_i) = *$ for $i = 2 \cdots n$,

so that $[\![s]\!]_{v'} = [\![\lambda\bar{u} \,.\, s]\!]_v(c)(*)\cdots(*) = Y$, and thus by 1 we have $v(x) = v'(x) = *$ for any variable x free in $\lambda\bar{u} \,.\, s$. ⊣

COROLLARY 8. *If a preword-term s has a closure $\lambda\bar{u} \,.\, s$ that satisfies*

$$[\![\lambda\bar{u} \,.\, s]\!](x_1)\cdots(x_n) = Y$$

whenever

$$\ulcorner W \urcorner(x_1)\cdots(x_n) = Y$$

for some word W, then each of the variables $\bar{u}$ occurs free at most once in s.

Proof. Suppose that a variable u_i occurs more than once in s, so that we can find a preword-term $t[y, z]$ such that both y and z occur free in $t[y, z]$, and s is the term $t[u_i, u_i]$. Now by part 1 of the lemma above, $[\![t]\!]_v = N$ whenever $v(y) = v(z) \in \{A, B\}$, so that $[\![s]\!]_v = N$ whenever $v(u_i) \in \{A, B\}$. But by the hypothesis of the lemma, and W1, we must have a valuation v with $v(u_i) \in \{A, B\}$ and $v(s) = Y$, a contradiction. ⊣

PROPOSITION 9. *If a closure $\lambda\bar{u} \,.\, s$ of a preword-term s satisfies*

$$[\![\lambda\bar{u} \,.\, s]\!](x_1)\cdots(x_n) = Y$$

whenever

$$\ulcorner W \urcorner(x_1)\cdots(x_n) = Y$$

for some word W, then the word W is derivable, and $[\![\lambda\bar{u} \,.\, s]\!] = \ulcorner W \urcorner$.[2]

Proof. The proof is by induction on s. If s is $\ulcorner W_0 \urcorner(t_1)\cdots(t_n)$, then the t_i must be distinct variables by W1–3, and $\lambda\bar{u} \,.\, s$ must encode the word W_0.

If s is $\ulcorner F_i \urcorner(\lambda x_1, \ldots, x_m \,.\, t)(t_1)\cdots(t_n)$, then by R4, R5 and W1, all the t_i must be distinct variables. Then by W2 and R1-5, the closure of s that satisfies the condition of the proposition must be

$$s' = \lambda\bar{u}, t_1, \ldots, t_n, \bar{v} \,.\, s$$

for some sequences $\bar{u}$ and $\bar{v}$ of variables. Let the lengths of $\bar{u}$ and $\bar{v}$ be k and l respectively. By corollary 8, none of the t_i occurs free in $\lambda\bar{x} \,.\, t$, so that $t' = \lambda\bar{u}, \bar{x}, \bar{v} \,.\, t$ is a closed term. The induction predicate for s will follow from the induction hypothesis and lemma 3 if we show that

$$[\![t']\!](x_1)\cdots(x_n) = Y \text{ whenever } \ulcorner W' \urcorner(x_1)\cdots(x_n) = Y \tag{3}$$

[2] The reader will no doubt notice that this proposition is stronger than needed. This is a good example of saving work by choosing an induction predicate carefully — the author's original proof of what is actually needed ("if $[\![\lambda\bar{u} \,.\, s]\!] = \ulcorner W \urcorner$, then W is derivable") required several pages of tedious technical lemmas similar to lemma 7 in order to make a proof virtually identical to the one here work.

for some word W' whose sub-word at positions $k+1 \cdots k+m$ is C_i.

For each $j = 1 \cdots k$, there is $a_j \in \{A, B\}$ such that

$$s' \underbrace{(*) \cdots (*)}_{j-1}(a_j) \underbrace{(*) \cdots (*)}_{k-j+n+l} = Y,$$

and so by R1 and R5 we must have

$$t' \underbrace{(*) \cdots (*)}_{j-1}(a_j) \underbrace{(*) \cdots (*)}_{k-j+m+l} = Y \tag{4}$$

also. Similarly, for $j = 1 \cdots l$, there is $b_j \in \{A, B\}$ such that

$$s' \underbrace{(*) \cdots (*)}_{k+n+j-1}(b_j) \underbrace{(*) \cdots (*)}_{l-j} = Y,$$

so by R1 and R5 we have

$$t' \underbrace{(*) \cdots (*)}_{k+m+j-1}(b_j) \underbrace{(*) \cdots (*)}_{l-j} = Y. \tag{5}$$

We will show that (3) is satisfied when W' is the word $a_1 \cdots a_k\, C_i\, b_1 \cdots b_l$. For one of $c = A$ or $c = B$, we must have

$$s' \underbrace{(*) \cdots (*)}_{k}(c) \underbrace{(*) \cdots (*)}_{n-1+l} = Y.$$

Inspecting R1–5, this can only occur if $\lambda \bar{x} \,.\, t[\bar{u}, \bar{v} := *]$ encodes the word C_i, and c is the first letter of D_i, so that R4 applies. Thus for $j = 1 \cdots m$, we have

$$t' \underbrace{(*) \cdots (*)}_{k+j-1}(c_j) \underbrace{(*) \cdots (*)}_{m-j+l} = Y \tag{6}$$

where c_j is the jth letter of the word C_i. Similarly to the above, for $j = 1 \cdots l+m+k-1$,

$$t' \underbrace{(*) \cdots (*)}_{j-1}(L)(R) \underbrace{(*) \cdots (*)}_{l+m+k-j-1} = Y.$$

Together with (4), (5) and (6), this gives (3) with $W' = a_1 \cdots a_k\, C_i\, b_1 \cdots b_l$ as required. ⊣

THEOREM 10. *The λ-definability problem is undecidable.*

Proof. By lemma 1, it suffices to show that the relative problem is undecidable. Let W_0 and F_i be a word and set of rules such that the problem of whether or not a given word is derivable is undecidable, as given by proposition 2. As our encoding is effective, it now suffices to show that $\ulcorner W \urcorner$ is definable relative to $\ulcorner W_0 \urcorner$ and the $\ulcorner F_i \urcorner$ if and only if W is derivable from W_0 by the rules F_i. One direction is provided by corollary 4. For the other direction, suppose that $\ulcorner W \urcorner = t[\ulcorner W_0 \urcorner, \ulcorner F_1 \urcorner, \ldots, \ulcorner F_n \urcorner]$. By taking the β-normal, η-long form we may assume that t is the closure of a preword-term and hence apply the previous proposition. ⊣

The proof given is not optimal in the sense that the size of the model used may be reduced. It is however known that definability of objects of rank two or less is decidable, a decision procedure being given by logical relations, as is definability in the model with just one element at each type (this latter case is just the decidability of intuitionistic logic). In particular, the proofs given are optimal with respect to the rank of the objects concerned. Recently Jung and Tiuryn (*1993*) proved a partial decidability result by restricting the number of bound variables in a λ-term, by using modifications of the notion of logical relation.

4. THE FAILURE OF LOGICAL RELATIONS

One of the main reasons for formulating the notion of logical relations was to use them to characterize λ-definablility. For example, Plotkin (*1980*) uses a Kripke logical relation to characterize definability in infinite models.

However, as will be shown in this section, logical relations must fail to give such a characterization in finite models, as otherwise this would yield a decision procedure for definability. As our proof of undecidability is constructive, we could in principle extract an example of this failure from the proof, by diagonalizing the appropriate algorithm. Instead we shall construct a reasonably simple example of how logical relations fail to characterize relative definability.

DEFINITION 11. Given a model M of the λ-calculus, an **n-ary logical relation** is a collection of relations $R_A \subset M_A^n$ for each type A, such that $(f_1, \ldots, f_n) \in R_{A \to B}$ if and only if

$$(f_1(x_1), \ldots, f_n(x_n)) \in R_A \text{ whenever } (x_1, \ldots, x_n) \in R_B.$$

Given a Kripke frame K (*i.e.* a partial order $(K, \leq)$), an **n-ary Kripke logical relation over K** is a collection of relations $R_{A,i} \subset M_A^n$ for $i \in K$ and A a type, such that

- $R_{A,i} \subset R_{A,j}$ whenever $i \leq j$,
- $(f_1, \ldots, f_n) \in R_{A \to B,i}$ if and only if $(f_1(x_1), \ldots, f_n(x_n)) \in R_{B,j}$ whenever $i \leq j$ and $(x_1, \ldots, x_n) \in R_{A,j}$.

Note that a (Kripke) logical relation is completely determined by its behavior at ground type. We say that $a \in M_A$ is **invariant** in a (Kripke) logical relation R if $(a, \ldots, a) \in R_A$ ($(a, \ldots, a) \in R_{A,i}$ for all $i \in K$).

The following lemma is originally from Plotkin (*1973*), and is easily proved by induction over terms.

LEMMA 12. *Let R be a logical relation, t be a term and v_i $(i = 1 \cdots n)$ be valuations such that $(v_1(x), \ldots, v_n(x)) \in R$ for each variable x free in t. Then $([\![t]\!]_{v_1}, \ldots, [\![t]\!]_{v_n}) \in R$ also.*

COROLLARY 13. *If a is λ-definable (relative to $b_1 \cdots b_m$), then a is invariant in every logical relation R (such that the b_i are invariant).*

These also hold using Kripke logical relations. For classical rather than Kripke logical relations, the following is straightforward.

PROPOSITION 14. *As a predicate of a, A, M and n, "$a \in M_A$ is invariant in every n-ary Kripke logical relation" is decidable. There is a member of our seven-element model which is invariant under every Kripke logical relation, but not λ-definable.*

Proof. Given M and n, define K^1 to be

$$\left\{ R_i \;\middle|\; \begin{array}{l} K \text{ is a Kripke frame, } i \in K \text{ and } R \text{ is an} \\ n\text{-ary Kripke logical relation over } K \end{array} \right\}$$

where R_i means the collection $\{R_{B,i}\}_{B:\mathrm{Type}}$, and set $R \leq R'$ iff $R_B \subset R'_B$ for all types B. Now we can define a logical relation R^1 over K^1 by $\bar{x} \in R^1_{B,R}$ iff $\bar{x} \in R_B$. Obviously, $a \in M_A$ is invariant in R^1 if and only if a is invariant in every n-ary Kripke logical relation.

Fixing $a \in M_A$, we now reduce K^1 to a finite Kripke frame by a method similar to that of filtrations in modal logic (cf. *Chellas 1980*). Define an equivalence relation $\equiv$ on K^1 by $R \equiv R'$ iff $R_B = R'_B$ for all sub-types B of A. Let K^A be the quotient $K^1/\equiv$, and for $i, i' \in K^A$ put $i \leq i'$ iff there are $R \in i$, $R' \in i'$ such that $R \leq R'$. Now let R^A be the unique n-ary Kripke logical relation such that $R^A_{o,R/\equiv} = R_o$ for each equivalence class $(R/\equiv) \in K^A$. It is now easy to verify by induction that for any sub-type B of A, and $(R/\equiv) \in K^A$, we have $R^A_{B,R/\equiv} = R_B$ so that $a \in M_A$ is invariant in R^A iff a is invariant in R^1 iff a is invariant in every n-ary Kripke logical relation.

Further, K^A is finite, as the equivalence class of $R \in K^1$ is determined by the collection $(R_B : B$ is a sub-type of $A)$, and thus the size $|K^A|$ is bounded by $N = \prod_B |M_B|^n$ where B ranges over the sub-types of A, which is recursive in A, M and n. Thus a is invariant in every n-ary Kripke logical relation if and only if it is invariant in every n-ary Kripke logical relation over a Kripke frame of size at most N. This shows that the given predicate is recursive.

It follows that "$a \in M_A$ is invariant in every Kripke logical relation" is co-r.e. in a, A and M. λ-definability is obviously r.e., and so cannot be co-r.e. as this would contradict theorem 10. Since λ-definability implies invariance, the converse implication must fail, so that there is a member of the seven element model that is invariant but not definable. ⊣

We finish by constructing an example where relative invariance holds but relative λ-definability fails. The example is given by the encoding of a semi-Thue system in our seven element model. For $n \geq 1$, let W_n be the word consisting of n letter A's. Let F_1 be the rule $A \longrightarrow AAA$ and F_2 be the rule $AAA \longrightarrow A$, so that W_n is derivable from W_1 if and only if n is odd. However, we will see that (the encoding of) W_2 (in fact any W_k) is invariant in any logical relation such that F_1, F_2 and W_1 are.

PROPOSITION 15. *Let R be an n-ary (Kripke) logical relation.*

1. *If $k > 3n$ and $\ulcorner W_{k-1} \urcorner$ is invariant in R, then so is $\ulcorner W_k \urcorner$.*

2. *If $\ulcorner F_1 \urcorner$, $\ulcorner F_2 \urcorner$ and $\ulcorner W_1 \urcorner$ are invariant in R, then so is $\ulcorner W_2 \urcorner$. But $\ulcorner W_2 \urcorner$ is not λ-definable from $\ulcorner F_1 \urcorner$, $\ulcorner F_2 \urcorner$ and $\ulcorner W_1 \urcorner$.*

Proof. We shall assume R is a classical logical relation, although the proof works for Kripke relations also.

For the first part, suppose that $k > 3n$, and $\ulcorner W_{k-1} \urcorner$ is invariant in R. We show that $\ulcorner W_k \urcorner$ is invariant in R also. Consider x_i^j such that $(x_i^1, \ldots, x_i^n) \in R_o$ for $i = 1 \cdots k$. We must show that also

$$(\ulcorner W_k \urcorner(x_1^1) \cdots (x_k^1), \ldots, \ulcorner W_k \urcorner(x_1^n) \cdots (x_k^n)) \in R_o. \tag{7}$$

For each $j = 1 \cdots n$, we shall choose a set $C_j \subset \{1, \ldots, k\}$ with at most three members as follows:

- If there is $i_0 \in \{1, \ldots, n-2\}$ such that $x_{i_0}^j = L$, $x_{i_0+2} = R$ and $x_i^j = *$ for all other $i = 1 \cdots n$, then let $C_j = \{i_0, i_0+1, i_0+2\}$.

- If there are more than three $i \in \{1, \ldots, n\}$ such that $x_i^j \neq *$, then let C_j be the set containing the first three such i.

- Otherwise let $C_j = \{\, i \mid x_i^j \neq * \,\}$.

The C_j have been chosen so that if $i \in \{1, \ldots, k\} - C_j$, then

$$\ulcorner W_{k-1} \urcorner(x_1^j) \cdots (x_{i-1}^j)(x_{i+1}^j) \cdots (x_k^j) = \ulcorner W_k \urcorner(x_1^j) \cdots (x_k^j). \tag{8}$$

Since $k > 3n$, and the C_j have at most three members, there is $i \in \{1, \ldots, k\}$ that is not in any of the C_j. Now

$$\left(\ulcorner W_{k-1}\urcorner(x_1^j)\cdots(x_{i-1}^j)(x_{i+1}^j)\cdots(x_k^j)\right)_{1\leq j\leq k} \in R$$

as $\ulcorner W_{k-1}\urcorner$ is invariant in R, and so by (8) we see that (7) holds also.

For the second part of the proposition, suppose that $\ulcorner F_1\urcorner$, $\ulcorner F_2\urcorner$ and $\ulcorner W_1\urcorner$ are invariant in R. Take an even number $k > 3n$. Now $\ulcorner W_{k-1}\urcorner$ is derivable as $k-1$ is odd, so that by corollary 4, $\ulcorner W_{k-1}\urcorner$ is definable, and thus invariant by corollary 13. Now by the first part $\ulcorner W_k\urcorner$ is invariant also. $\ulcorner W_2\urcorner$ is derivable from $\ulcorner W_k\urcorner$ by the rule $\ulcorner F_2\urcorner$, so that similarly $\ulcorner W_2\urcorner$ is invariant. That $\ulcorner W_2\urcorner$ is not definable is obvious from proposition 9. ⊣

REFERENCES

Chellas, B. F.

1980 *Modal logic: An introduction*, Cambridge University Press, Cambridge, England.

Davis, M.

1977 Unsolvable problems, *The handbook of mathematical logic* (J. Barwise, editor), North-Holland, Amsterdam.

Jung, A., and J. Tiuryn

1993 A new characterization of lambda definability, *Typed lambda calculi and applications*, Lecture Notes in Computer Science, vol. 664, Springer-Verlag, Berlin, pp. 245–257.

Plotkin, G. D.

1973 λ-definability and logical relations, University of Edinburgh, School of Artificial Intelligence, Memorandum SAI-RM-4 (October).

1980 λ-definability in the full type hierarchy, *Combinatory logic, lambda calculus and formalism* (J. P. Seldin and J. R. Hindley, editors), Academic Press, New York.

Statman, R.

1982 Completeness, invariance, and λ-definability, *The Journal of Symbolic Logic*, vol. 47, no. 1, pp. 17–26.

Wolfram, D. A.

1993 *The clausal theory of types*, Cambridge Tracts in Theoretical Computer Science, vol. 21, Cambridge University Press, Cambridge, England.

PER MARTIN-LÖF

A CONSTRUCTION OF THE PROVABLE WELLORDERINGS OF THE THEORY OF SPECIES*

Dedicated to the memory of Alonzo Church.

1 Introduction.

1.1 It seems to be very difficult to construct a "natural" system of ordinal notations which would play the role for the theory of species that ε_0 plays for number theory and Takeuti's ordinal diagrams of finite order play for the theory of iterated inductive definitions. However if one has the more modest aim of: (1) generating certain syntactical objects to be called ordinal notations, (2) defining a recursively enumerable predecessor relation between the ordinal notations, (3) proving in the theory of species that each ordinal notation is accessible with respect to the predecessor relation, and (4) showing the unprovability in the theory of species of the universal statement that all ordinal notations are accessible, then quite a simple solution can be obtained as follows.

1.2 Consider the system of terms introduced by Howard (*1963*) which contains among other things a type symbol 2 for the type of Brouwer ordinals. Extend this system à la Girard (*1970*) by adding type variables and the associated rules of type and term formation. Let the closed terms of type 2 in the extended system serve as ordinal notations. Of two closed terms of type 2 say that one precedes the other if the Brouwer ordinal denoted by the one is intensionally a predecessor of the Brouwer ordinal denoted by the other. This predecessor relation is recursively enumerable. Prove the accessibility of each closed term of type 2 with respect to the predecessor relation in the theory of species by the method of computability. Finally, show that, given a provable wellordering of the theory of species, a closed term of type 2 can be found whose associated wellordering dominates the given one. The last goal can be achieved by means of a kind of realizability interpretation which may be of some independent interest. At least, it can be used to give an alternative proof of the fact that, as established by Girard

* ***Postscript, Nov. 1997.*** According to my memory, which agrees with that of Jean-Yves Girard, this paper was written in September 1970, just before I began working on intuitionistic type theory. It had some circulation at the time, but was never published. I have corrected the spelling of "intentional(ly)" to "intensional(ly)". Also, the style of the references has been changed by the editors so as to agree with the rest of the volume. Otherwise, this printed version agrees with the original typescript.

C. Anthony Anderson and M. Zelěny (eds.), Logic, Meaning and Computation, 343-351.

(*1970*), every provably recursive function of the theory of species is definable in his system.

2 A simplified formulation of Girard's system of terms.

2.1 Type variables: These will be denoted by $\xi, \eta, \ldots$.

2.2 Types.

2.2.1 A type variable is a type.

2.2.2 If σ and τ are types, then $\sigma \to \tau$ is a type.

2.2.3 If $\sigma(\xi)$ is a type, then so is $\Pi\xi\sigma(\xi)$.

2.3 For each type we allow ourselves to introduce as many variables $x, y, \ldots$ of that type as we please. If it is absolutely necessary, we shall indicate the type of a variable by a superscript.

2.4 Terms.

2.4.1 A variable of type σ is a term of type σ.

2.4.2 If x is a variable of type σ and $a(x)$ a term of type τ, then $\lambda x a(x)$ is a term of type $\sigma \to \tau$.

2.4.3 If $a(\xi)$ is a term of type $\sigma(\xi)$ and ξ does not occur free in the type of a variable in $a(\xi)$, then $\lambda\xi a(\xi)$ is a term of type $\Pi\xi\sigma(\xi)$.

2.4.4 If a and b are terms of type $\sigma \to \tau$ and σ, respectively, then ab is a term of type τ.

2.4.5 If a is a term of type $\Pi\xi\sigma(\xi)$ and τ is a type, then $a\tau$ is a term of type $\sigma(\tau)$.

2.5 Rules of contraction:

$$\lambda x a(x) b \ \mathrm{contr}\ a(b),$$
$$\lambda\xi a(\xi)\tau \ \mathrm{contr}\ a(\tau).$$

A term a reduces to a term b (abbreviated $a \ \mathrm{red}\ b$) if b can be obtained from a by repeated contractions of subterms.

2.6 Pairing: For any two types σ and τ put

$$\sigma \times \tau = \Pi\xi((\sigma \to \tau \to \xi) \to \xi),$$

and define the corresponding pairing and projection functions by

$$(a, b) = \lambda\xi\lambda x^{\sigma \to \tau \to \xi} xab,$$
$$Pc = c\sigma(\lambda x^{\sigma}\lambda y^{\tau} x),$$
$$Qc = c\tau(\lambda x^{\sigma}\lambda y^{\tau} y).$$

Then

$$P(a,b) \operatorname{red} a, \qquad Q(a,b) \operatorname{red} b$$

as desired.

2.7 Natural numbers: The type of the natural numbers, which we shall denote by 1, can be defined by putting

$$1 = \Pi\xi(\xi \to (\xi \to \xi) \to \xi).$$

If

$$0_1 = \lambda\xi\lambda x^{\xi}\lambda y^{\xi\to\xi}x,$$
$$S_1 = \lambda z^1\lambda\xi\lambda x^{\xi}\lambda y^{\xi\to\xi}y(z\xi xy)$$

are used to represent the natural number zero and the successor function, respectively, then we have access to ordinary recursion, because

$$0_1\tau ab \operatorname{red} a, \qquad S_1 c\tau ab \operatorname{red} b(c\tau ab).$$

In the presence of a pairing operation it is, of course, sufficient to have recursion in this simple form without a side argument.

2.8 Brouwer ordinals: In analogy with the way the natural numbers were represented, the type 2 of Brouwer ordinals, the zero element (Urelement) of type 2 and the generalized successor operation of type $(1 \to 2) \to 2$ can be defined by putting:

$$2 = \Pi\xi(\xi \to ((1 \to \xi) \to \xi) \to \xi),$$
$$0_2 = \lambda\xi\lambda x^{\xi}\lambda y^{(1\to\xi)\to\xi}x,$$
$$S_2 = \lambda z^{1\to 2}\lambda\xi\lambda x^{\xi}\lambda y^{(1\to\xi)\to\xi}y(\lambda wzw\xi xy).$$

The schema of transfinite recursion in the form

$$0_2\tau ab \operatorname{red} a, \qquad S_2 c\tau ab \operatorname{red} b(\lambda wcw\tau ab)$$

then becomes satisfied.

2.9 Although, as has just been shown, the types 1 and 2 can be defined, we shall prefer to treat them as primitive in the sequel. This makes it necessary to add the six constants:

0_1 of type 1,
S_1 of type $1 \to 1$,
R_1 of type $\Pi\xi(\xi \to (\xi \to \xi) \to 1 \to \xi)$,
0_2 of type 2,
S_2 of type $(1 \to 2) \to 2$,
R_2 of type $\Pi\xi(\xi \to ((1 \to \xi) \to \xi) \to 2 \to \xi)$;

and the four contraction rules:

$$R_1\tau ab0_1 \text{ contr } a,$$
$$R_1\tau ab(S_1c) \text{ contr } b(R_1\tau abc),$$
$$R_2\tau ab0_2 \text{ contr } a,$$
$$R_2\tau ab(S_2c) \text{ contr } b\big(\lambda w R_2\tau ab(cw)\big)$$

to the basic system.

2.9.1 A term of the form $S_1\big(\ldots S_1(S_1 0_1)\ldots\big)$ will be called a numeral.

3 Definition of the equality and predecessor relations.

3.1 The terms a and b are said to be intensionally equal and we write $a = b$ if there is a term c such that $a \operatorname{red} c$ and $b \operatorname{red} c$. The transitivity of the equality relation is guaranteed by the wellknown property of Church and Rosser, that is, that if $a \operatorname{red} b$ and $a \operatorname{red} c$ then there is a term d such that $b \operatorname{red} d$ and $c \operatorname{red} d$. The proof of this property for the system we are considering follows the usual pattern.

3.2 Let a and b be two closed terms of type 2 and n a numeral. We shall say that a is the nth predecessor of b if there is a closed term c of type $1 \rightarrow 2$ and a numeral n such that $a = cn$ and $b = S_2c$. If a is the nth predecessor of b for some n, then a is a predecessor of b. Clearly, the predecessor relation so defined is recursively enumerable.

4 The accessibility of a closed term of type 2 with respect to the predecessor relation is defined by the following generalized inductive definition.

4.1 If $a = 0_2$, then a is accessible.

4.2 If a has predecessors which are all accessible, then a is accessible.

5 Definition of computability: For the sake of simplicity, we shall define what it means for a term to be computable only if it is closed, and this we do by induction on its type.

5.1 Type 1: As in *Tait 1967*, that is, a closed term of type 1 is computable if and only if it reduces to a numeral.

5.2 Type 2.

5.2.1 0_2 is computable.

5.2.2 If ab is computable for all computable terms b of type 1, then S_2a is computable.

5.2.3 If a reduces to b and b is computable, then a is computable.

5.3 Type $\sigma \rightarrow \tau$: As in *Tait 1967*.

5.4 Type $\Pi\xi\sigma(\xi)$: As in *Girard 1970.*

6 For every closed term it can be proved in the theory of species that it is computable. Almost all of the proof can be taken from *Tait 1967* and *Girard 1970.* The only novelty is that we have to prove the computability of 0_2, S_2 and R_2.

6.1 0_2 is computable by 5.2.1.

6.2 Suppose that a is a computable term of type $1 \to 2$. By 5.3 this means that ab is computable for all computable terms b of type 1. Using 5.2.2 we can conclude that S_2a is computable. But a was arbitrary and, consequently, S_2 is computable.

6.3 In order to show that R_2 is computable it suffices to show that if τ is a type, α_τ a computability predicate of type τ, a a computable term of type τ, b a computable term of type $(1 \to \tau) \to \tau$ and c a computable term of type 2, then $R_2\tau abc$ is computable (with respect to α_τ). We do this by induction on the proof that c is computable.

6.3.1 $R_2\tau ab0_2$ is computable because it reduces to a which is computable.

6.3.2 $R_2\tau ab(S_2c)$ reduces to $b(\lambda w R_2\tau ab(cw))$ and, since b was assumed computable, it thus suffices to show that $\lambda w R_2\tau ab(cw)$ is computable, that is, that $(\lambda w R_2\tau ab(cw))d$ is computable for all computable terms d of type 1. Now, $(\lambda w R_2\tau ab(cw))d$ is indeed computable because it reduces to $R_2\tau ab(cd)$ which is computable by induction hypothesis.

6.3.3 If c reduces to d, then $R_2\tau abc$ reduces to $R_2\tau abd$. By induction hypothesis the latter term is computable and, consequently, so is the former.

7 Consequences of computability.

7.1 Comparison of the definition of accessibility and the definition of computability for closed terms of type 2 shows that they are equivalent. Thus, we have proved the accessibility of every closed term of type 2 in the theory of species.

7.2 If the definition of computability is extended so as to cover terms which are not necessarily closed (the variables functioning essentially as constants), then it can be proved that all computable terms are normalizable. Moreover, it does not require much extra labor to show that each term is computable in the extended sense and hence also normalizable. This has important consequences.

7.2.1 For any pair of terms a and b it can be decided whether or not $a = b$ by checking whether or not their normal forms are identical (neglecting differences in the naming of their bound variables).

7.2.2 If a is a closed term of type 2, then either $a = 0_2$ or $a = S_2c$ for some closed term c of type $1 \to 2$. This is seen by reducing a to normal form, because a closed normal term of type 2 must be either 0_2 or of the form S_2c.

7.2.3 If a and b are closed terms of type 2 and n is a numeral, then it can be decided whether or not a is the nth predecessor of b. Indeed, a is the nth predecessor of b if and only if the normal form of b is of the form S_2c and $a = cn$.

8 Our final goal is to show that the universal statement "all closed terms of type 2 are accessible" is unprovable in the theory of species. This will be achieved by showing that a provable wellordering of the theory of species can always be dominated by the wellordering associated with a closed term of type 2.

8.1 The language of the theory of species which we shall consider contains variables for individuals and species and the following constants.

8.1.1 Function constants: Each function constant denotes a primitive recursive function. There should at least be a constant 0 for the natural number zero, a constant s for the successor function and a binary function constant f. Think of fxy as the yth predecessor of x.

8.1.2 Predicate constants: $=$ for equality and a unary predicate constant A for accessibility.

8.1.3 Logical constants: $\to$ for implication and $\bigwedge$ for universal quantification of first and second order.

8.2 Axioms and rules of inference.

8.2.1 The defining equations of the primitive recursive functions.

8.2.2 Peano's third and fourth axioms:

$$\bigwedge x \bigwedge y(sx = sy \to x = y), \qquad \bigwedge x(0 = sx \to 0 = s0).$$

8.2.3 Equality axioms:

$$\bigwedge x(x = x), \qquad \bigwedge x \bigwedge y \bigwedge X(x = y \to Xx \to Xy).$$

8.2.4 Induction:

$$\bigwedge X(X0 \to \bigwedge x(Xx \to Xsx) \to \bigwedge xXx).$$

8.2.5 Axioms of accessibility:

$$A0, \qquad \bigwedge x(\bigwedge yAfxy \to Ax),$$
$$\bigwedge X(X0 \to \bigwedge x(\bigwedge yXfxy \to Xx) \to \bigwedge x(Ax \to Xx)).$$

8.2.6 Rules of inference for second order minimal logic:

$$\begin{array}{c} \not{F} \\ \vdots \\ \dfrac{G}{F \to G} \end{array} \qquad \dfrac{F \to G \quad F}{G}$$

$$\frac{F(x)}{\bigwedge xF(x)} \qquad \frac{\bigwedge xF(x)}{F(t)}$$

$$\frac{F(X)}{\bigwedge XF(X)} \qquad \frac{\bigwedge XF(X)}{F(T)}$$

8.3 Definition of the type associated with a formula or species term: By induction on its logical complexity,

8.3.1 The type associated with $t = u$ is 1.

8.3.2 The type associated with At is 2.

8.3.3 The type associated with Xt is ξ where ξ is a type variable uniquely associated with X.

8.3.4 If σ and τ are the types associated with F and G, respectively, then $\sigma \to \tau$ is the type associated with $F \to G$.

8.3.5 If σ is the type associated with $F(x)$, then $1 \to \sigma$ is the type associated with $\bigwedge xF(x)$.

8.3.6 If $\sigma(\xi)$ is the type associated with $F(X)$, then $\Pi\xi\sigma(\xi)$ is the type associated with $\bigwedge XF(X)$.

8.3.7 The type associated with $F(\boldsymbol{x})$ is also the type associated with $\lambda\boldsymbol{x}F(\boldsymbol{x})$.

9 Realizability predicates: Let T be a closed $n$ary species term and τ its associated type. A relation α_T between a sequence of n closed individual terms and a closed term of type τ will be called a realizability predicate of type T provided it satisfies the following two conditions.

9.1 If $\boldsymbol{t} = \boldsymbol{u}$ and $\alpha_T(\boldsymbol{t}, a)$, then $\alpha_T(\boldsymbol{u}, a)$.

9.2 If a red b and $\alpha_T(\boldsymbol{t}, b)$, then $\alpha_T(\boldsymbol{t}, a)$.

10 Definition of realizability: Let $T(\boldsymbol{X})$ be a species term which contains no free individual variables and all of whose free species variables are contained in the sequence $\boldsymbol{X} = X_1, \ldots, X_n$ and let $\boldsymbol{\alpha_T} = \alpha_{T_1}, \ldots, \alpha_{T_n}$ be a

corresponding sequence of realizability predicates. By induction on the number of logical signs in $T(\boldsymbol{X})$ we shall define a certain realizability predicate of type $T(\boldsymbol{T})$ which will be denoted $\rho_{T(\boldsymbol{X})}(\boldsymbol{\alpha_T})$. If F is a closed formula, then $\rho_F(a)$ may be read "a realizes F" or "a is a proof constant for F".

10.1 If $t = u$ and a red 0_1, then $\rho_{t=u}(\boldsymbol{\alpha_T})(a)$.

10.2 The definition of $\rho_{At}(\boldsymbol{\alpha_T})$ is by generalized induction.

10.2.1 If $t = 0$, then $\rho_{At}(\boldsymbol{\alpha_T})(0_2)$.

10.2.2 If $t \neq 0$ and $\rho_{Aftu}(\boldsymbol{\alpha_T})(au)$ for all closed individual terms u, then $\rho_{At}(\boldsymbol{\alpha_T})(S_2 a)$.

10.2.3 If a reduces to b and $\rho_{At}(\boldsymbol{\alpha_T})(b)$, then $\rho_{At}(\boldsymbol{\alpha_T})(a)$.

10.3 If $\rho_{T_i}(t, a)$, then $\rho_{X_i t}(\boldsymbol{\alpha_T})(a)$.

10.4 If $\rho_{F(\boldsymbol{X})}(\boldsymbol{\alpha_T})(b)$ implies $\rho_{G(\boldsymbol{X})}(\boldsymbol{\alpha_T})(ab)$ for all b, then $\rho_{F(\boldsymbol{X})\rightarrow G(\boldsymbol{X})}(\boldsymbol{\alpha_T})(a)$.

10.5 If $\rho_{F(t,\boldsymbol{X})}(\boldsymbol{\alpha_T})(at)$ for all closed individual terms t, then $\rho_{\bigwedge x F(x,\boldsymbol{X})}(\boldsymbol{\alpha_T})(a)$.

10.6 If $\rho_{F(X,\boldsymbol{X})}(\alpha_T, \boldsymbol{\alpha_T})(a\tau)$ for all T and α_T where τ is the type associated with T, then $\rho_{\bigwedge X F(X,\boldsymbol{X})}(\boldsymbol{\alpha_T})(a)$.

10.7 If $\rho_{F(t,\boldsymbol{X})}(\boldsymbol{\alpha_T})(a)$, then $\rho_{\lambda x F(x,\boldsymbol{X})}(\boldsymbol{\alpha_T})(t, a)$.

11 Properties of $\rho_{T(\boldsymbol{X})}(\boldsymbol{\alpha_T})$.

11.1 $\rho_{T(\boldsymbol{X})}(\boldsymbol{\alpha_T})$ is a realizability predicate of type $T(\boldsymbol{T})$. To establish this we have to verify the following two properties.

11.1.1 If $t = u$ and $\rho_{F(t,\boldsymbol{X})}(\boldsymbol{\alpha_T})(a)$, then $\rho_{F(u,\boldsymbol{X})}(\boldsymbol{\alpha_T})(a)$. Proof by induction on the number of logical signs in the formula $F(x, \boldsymbol{X})$.

11.1.2 If a red b and $\rho_{F(\boldsymbol{X})}(\boldsymbol{\alpha_T})(b)$, then $\rho_{F(\boldsymbol{X})}(\boldsymbol{\alpha_T})(a)$. Proof by induction on the number of logical signs in the formula $F(\boldsymbol{X})$.

11.2 Substitution property:

$$\rho_{F(X,\boldsymbol{X})}(\rho_{T(\boldsymbol{X})}(\boldsymbol{\alpha_T}), \boldsymbol{\alpha_T}) = \rho_{F(T(\boldsymbol{X}),\boldsymbol{X})}(\boldsymbol{\alpha_T}).$$

The verification of this property is immediate by induction on the number of logical signs in the formula $F(X, \boldsymbol{X})$.

12 Let F be a closed formula and σ its associated type. Then, from a proof of F we can find a closed term a of type σ such that $\rho_F(a)$ is provable in the theory of species.

12.1 The proof of this proceeds by induction on the length of the given derivation of F and is entirely straightforward once a stronger induction hypothesis has been made. For example, the axiom of induction is handled by means of R_1 and the axioms of accessibility by means of 0_2, S_2 and R_2, respectively. Also, the substitution property was established because it is precisely what one needs in order to handle the second order rule of $\bigwedge$-elimination.

13 We can now piece things together. Let $<$ be an arbitrary recursively enumerable binary relation. Find a primitive recursive predicate P such that

$$y < x \leftrightarrow \bigvee zPxyz.$$

Let (y, z) be a primitive recursive pairing function like $(y, z) = 2^y(2z+1) - 1$ and define a primitive recursive binary function f by putting

$$fsx(y,z) = \begin{cases} sy \text{ if } Pxyz, \\ 0 \text{ otherwise,} \end{cases}$$

and $f0(y, z) = 0$, say. Then it is easily proved in the theory of species that, for all x, x is accessible with respect to $<$ if and only if sx is accessible from 0 with respect to f. Moreover, the wellordering determined by an accessible closed individual term t with respect to $<$ is dominated by the wellordering determined by st with respect to f. Suppose now that it can be proved in the theory of species that the closed individual term t is accessible with respect to f from the initial element 0. We can then find a closed term a of type 2 such that $\rho_{At}(a)$. But $\rho_{At}(a)$ means precisely that the wellordering determined by t with respect to f is isomorphic to the wellordering determined by a. Therefore, every provable wellordering of the theory of species can be dominated by the wellordering determined by a closed term of type 2, and this is all that there remained for us to prove.

REFERENCES

Girard, J.-Y.

1970 Une extension du système de fonctionnelles récursives de Gödel et son application aux fondements de l'analyse, to appear.

Howard, W. A.

1963 Transfinite induction and transfinite recursion, *Reports of the seminar on the foundations of analysis*, sec. VI, vol. 2, Stanford.

Tait, W. W.

1967 Intensional interpretations of functionals of finite type, *The Journal of Symbolic Logic*, vol. 32, pp. 198–212.

COLIN MCLARTY

SEMANTICS FOR FIRST AND HIGHER ORDER REALIZABILITY

Abstract. First order Kleene realizability is given a semantic interpretation, including arithmetic and other types. These types extend at a stroke to full higher order intuitionistic logic. They are also useful themselves, e.g., as models for lambda calculi, for which see *Asperti and Longo 1991* and papers on PERs and polymorphism (*IEEE 1990*).

This semantics is simpler and more explicit than in *Hyland 1982,* giving the logical content of Freyd, Carboni, and Scedrov's (*1988*) assemblies. Category theory here is only an organizing device and can be skipped, except in Lemmas 11–12 verifying the higher order logic. We verify some higher order constructive recursive analysis, and prove two metatheorems by generalizing the construction to other toposes.

1. ASSEMBLIES

We enumerate the partial recursive functions and write $n.m$ for the value of the n-th function applied to m. Of course $n.m$ may be undefined. We use a surjective recursive pairing function $\langle\ ,\ \rangle$ with recursive projections, l and r, read "left" and "right". So we have $l\langle n,m\rangle = n$ and $r\langle n,m\rangle = m$, and $n = \langle ln, rn\rangle$.

An *assembly* (A) consists of a set A called the *carrier* and an infinite sequence of its subsets $(A_1, A_2, \cdots)$ whose union is all of A. Each A_n is called the n-th *caucus.* An *arrow* between assemblies, written $f : (A) \rightarrow (B)$, is a function between the carriers $f : A \rightarrow B$ with at least one *modulus*— that is a number e such that whenever $a \in A_n$ then $e.n$ is defined and $fa \in B_{e.n}$. Arrows compose, with the composite of moduli as a modulus for the composite. So we have a well defined category.

Assemblies appear in *Freyd, Carboni, and Scedrov 1988, Freyd and Scedrov 1990,* and *McLarty 1992,* and as ω-sets in *Asperti and Longo 1991.*

Let the assembly (1) have a singleton set 1 as carrier, and the same 1 for every caucus. There is an assembly $(\mathbb{N})$ with $\mathbb{N}$ the set of natural numbers and each caucus $\mathbb{N}_n$ the singleton $\{n\}$. An arrow $f : (\mathbb{N}) \rightarrow (\mathbb{N})$ must be recursive, with its own codes as moduli. Categorists note (1) is terminal and $(\mathbb{N})$ a natural number object.

Assemblies (A) and (B) have a product $(A \times B)$ whose n-th caucus is $A_{ln} \times B_{rn}$. The projections $(A \times B) \rightarrow (A)$ and $(A \times B) \rightarrow (B)$ are well defined. The exponential (B^A) has the set of arrows from (A) to (B) as carrier, and the n-th caucus contains those arrows with n as modulus. E.g.,

C. Anthony Anderson and M. Zelëny (eds.), *Logic, Meaning and Computation*, 353-363.

the carrier of $(\mathbb{N}^{\mathbb{N}})$ is the set of recursive functions, each one in those caucuses which code it. The category is cartesian closed.

A *subassembly* of (A) is a monic arrow $f : (P) \rightarrowtail (A)$, i.e., any arrow one-to-one on the carriers. This subassembly is contained in $g : (Q) \rightarrowtail (A)$ if there is an arrow $h : (P) \to (Q)$ with $g \circ h = f$. Subassemblies are equivalent if each contains the other. The restriction of (A) to a subset S of the carrier is the subassembly $(S) \rightarrowtail (A)$ with carrier S and with each caucus S_n the intersection of S with A_n. Given arrows f and g both from (A) to (B), the restriction $(E) \rightarrowtail (A)$ of (A) to those x such that $fx = gx$ is an equalizer for f and g. The category has all finite limits.

2. ASSEMBLY REALIZABILITY

The first order logic of the category of assemblies is multi-sorted, with a sort for each assembly (B).

Each variable of sort (B) is a term of sort (B). For each $b : (1) \to (B)$ we have a term b of sort (B), a constant. For any arrow $f : (A) \to (B)$ and term t of sort (A) there is a term ft of sort (B). Since a closed term of sort (B) is a constant or suitably sorted arrows applied to constants, it determines an element of the carrier B. Our notation confuses closed terms with elements when the context makes it clear.

For each subassembly $(P) \rightarrowtail (A)$ there is a predicate symbol P applicable to terms of sort (A). For each subassembly of a product, $(R) \rightarrowtail (A \times B)$, we have a two place relation symbol R, applying to pairs of terms of sort (A) and (B). Similarly for n-ary relations for any n. For any (A) the diagonal $(A) \rightarrowtail (A \times A)$ is the *equality* relation, the restriction of $(A \times A)$ to pairs $\langle a, a \rangle$. We write $=$ for its relation symbol.

We define realizability only for sentences. For any $i : (R)(A \times B \times C)$ and closed terms t_A, t_B, t_C of sorts (A), (B), and (C) respectively, n realizes the atomic sentence $Rt_At_Bt_C$ iff some r in R_n has $i(r) = \langle tA, tB, tC \rangle$. Generalization to n-ary relations is obvious.

Let S and S' be sentences and Sy any formula whose sole free variable is y of sort (B). Then n realizes:

$S \,\&\, S'$ iff ln realizes S and rn realizes S'.

$S \vee S'$ iff either: $ln = 0$ and rn realizes S or $ln = 1$ and rn realizes S'.

$S \Rightarrow S'$ iff, for any m that realizes S, $n.m$ is defined and realizes S'.

$\sim S$ iff no m realizes S.

$(\forall y)Sy$ iff, for every b in B_m, $n.m$ is defined and realizes Sb.

$(\exists y)Sy$ iff for some b in B_{ln}, rn realizes Sb.

REMARK 2.1. Just asking whether a sentence is realized by some number gives standard truth conditions, except for $\forall$. I.e., $S \Rightarrow S'$ is realized iff: if S is realized so is S'. $(\exists y.(B))Sy$ is realized iff some Sb is. Similarly for &, $\vee$, and $\sim$. Notice $(\forall y.(B)) \sim\sim Sy$ is realized iff Sb is realized for all b in B. (Constructively, throughout the "if" clauses for $\Rightarrow$ and $\forall$ read "not-not-realized".)

Let $Pxyz$ be a formula with no free variables but x, y, and z, of sorts (A), (B), and (C) respectively. We write $[xyz \mid Pxyz]$ for the *extension* of P over that list of variables. The n-th caucus $[xyz \mid Pxyz]_n$ contains those triples $\langle a, b, c\rangle$ such that $\langle a, b, c\rangle$ is in $(A \times B \times C)_{ln}$ and rn realizes $Pabc$. The inclusion makes this a subassembly of $(A \times B \times C)$.

Even if y, for example, does not occur in $Pxyz$, so b does not appear in $Pabc$, we define $[xyz \mid Pxyz]$ as a subassembly of $(A \times B \times C)$ as above. We define an extension $[x^- \mid P]$ for any list x^- of distinct variables including all those free in P. A subassembly $(P) \rightarrowtail (A)$ need not be $[x \mid Px]$ but is equivalent to it

We write "$x.(A)$" or "$x^-.(A \times B \times C)$" to show the sorts of variables.

3. THE LOGIC OF ASSEMBLIES

These extensions admit an intuitionist sequent calculus, agreeing with the first order part of topos logic (see *Bell 1988* or *McLarty 1992*). So this interpretation agrees with the categorical one.

A *sequent* is an expression $\Gamma : P$ with Γ a finite set of formulas and P a formula. We use a new formula "fa", read as "false". The extension of fa over any list of variables is the empty subassembly of the corresponding product of assemblies.

Let x^- list the variables free in $\Gamma : P$. Let $[x^- \mid \Gamma]$ be the intersection of the extensions over x^- of all formulas in Γ, or equivalently the extension of the conjunction of formulas in Γ. The sequent $\Gamma : P$ is true if $[x^- : \Gamma]$ is contained in $[x^- \mid P]$. In that case, we write

$$\Gamma \vdash P.$$

So we have $\vdash P$ iff the extension of P is the entire product it is defined over.

Calculation shows the sequents above any double line here are true iff the one below is. Here Γ is any finite set of formulas and P, P', Q, and Q' any formulas, and x any variable not free in Γ or Q

$$\frac{\Gamma : P \text{ and } \Gamma : P'}{\Gamma : P \,\&\, P'} \qquad \frac{\Gamma, Q : P \text{ and } \Gamma, Q' : P}{\Gamma, Q \vee Q' : P}$$

$$\frac{\Gamma, Q : P}{\Gamma : Q \Rightarrow P} \qquad \frac{\Gamma, Q : fa}{\Gamma : \sim Q}$$

$$\frac{\Gamma : P}{\Gamma : (\forall x)P} \qquad \frac{\Gamma, P : Q}{\Gamma, (\exists x)P : Q}$$

We use a cut rule: If $\Gamma, Q : P$ and $\Gamma : Q$ are true so is $\Gamma : P$, unless there is a free variable over the empty assembly in Q but none in $\Gamma : P$. Every sequent with a free variable over the empty assembly is true.

This logic does not prove the law of excluded middle, "$P \vee \sim P$". Any formula "P" implies "$\sim\sim P$" but the converse need not hold.

4. ON NEGATION, AND EQUIVALENCE RELATIONS

A subassembly $(C) \rightarrowtail (A)$ is *complemented*, or *decidable*, if the formula $(\forall x.(A))(Cx \vee \sim Cx)$ is true. The terminology is natural here since a subassembly of $(\mathbb{N})$ is complemented iff it is equivalent to the restriction of $(\mathbb{N})$ to some recursive set of numbers. The diagonal $(\mathbb{N}) >> (\mathbb{N} \times \mathbb{N})$ is complemented—i.e., equality is decidable for natural numbers. It is not for $(\mathbb{N}^{\mathbb{N}})$, because equality of recursive functions is not recursively decidable.

A subassembly $(C) \rightarrowtail (A)$ is *double negation closed*, or *d.n. closed* for short, iff

$$\vdash \sim\sim Cx \Rightarrow Cx. \qquad x.(A)$$

So it is d.n. closed iff it is equivalent to a restriction of (A). Every decidable subassembly is d.n. closed, as is equality on any assembly.

Call a formula in the language of assemblies *almost negative* if all its predicates and relations are d.n. closed and its only connectives are &, $\Rightarrow$, $\sim$, and $\forall$.

THEOREM 4.1. *Let P be any almost negative formula with free variables $x^-.(A \times \cdots \times K)$. Construe the predicates and relations in P as subsets and relations of the carriers $A \cdots K$. Then $[x^- \mid P]$ is equivalent to the restriction of $(A \times \cdots \times K)$ to the set theoretic extension of P over x^-.*

Proof. By the logic of assemblies, almost negative formulas are d.n. closed. And every n realizes $\sim\sim S$ if some number realizes S. Then use Remark 2.1. ⊣

An equivalence relation is a relation $(R) \rightarrowtail (A \times A)$ satisfying the usual sequents for reflexivity, symmetry, and transitivity. A quotient for (R) is defined categorically as a coequalizer for the two projections $(R) \rightarrow (A)$ with those projections as kernel pair. In terms of logic it is an arrow $q : (A) \rightarrow (Q)$ such that:

$$\vdash (Rxy \Rightarrow qx = qy) \ \& \ (qx = qy \Rightarrow Rxy), \qquad x, y.(A)$$
$$\vdash (\exists x.(A))qx = z. \qquad z.(Q)$$

THEOREM 4.2. *An equivalence relation (R) on (A) has a quotient assembly iff it is d.n. closed.*

Proof. Equality is always d.n. closed, so given the first sequent above, (R) is also. Conversely, if (R) is d.n. closed, it gives an equivalence relation on the set A with a projection $q : A \to Q$ to the set of equivalence classes. Define (Q) by making each Q_n the set of equivalence classes of members of A_n. Then $q : (A) \to (Q)$ is a quotient of (R). ⊣

REMARK 4.3. For any function $f : \mathbb{N} \to \mathbb{N}$, define an assembly (T) where each natural number m occurs only in the caucus $T\langle m, fm\rangle$. The inclusion defines a subassembly $(T) \rightarrowtail (\mathbb{N})$ whose double negation is all of $(\mathbb{N})$. And f is a well defined arrow $f : (T) \to (\mathbb{N})$. I.e., every function from $\mathbb{N}$ to $\mathbb{N}$ is an arrow from some *double negation dense* subassembly of $(\mathbb{N})$ to $(\mathbb{N})$. But $(T) \rightarrowtail (\mathbb{N})$ is equivalent to all of $(\mathbb{N})$ iff f is recursive.

5. ARITHMETIC IN ASSEMBLIES

The assembly $(\mathbb{N})$ with successor arrow s verifies the Peano axioms. By trivial realizability:

$$\vdash \sim(0 = sx), \qquad\qquad x, y.(\mathbb{N})$$
$$\vdash sx = sy \Rightarrow x = y, \qquad\qquad z.(A)$$
$$P(z,0), (\exists x)(P(z,x) \Rightarrow P(z,sx)) \vdash (\exists x)P(z,x)$$

where $(P) \rightarrowtail (A \times \mathbb{N})$ is any relation to $(\mathbb{N})$.

All the classical theory of primitive recursive functions applies in assemblies, since all the proofs in *Kleene 1952,* chapter IX, work in the logic of assemblies. Furthermore, primitive recursive definitions are almost negative. So the extension of any n-ary primitive recursive predicate in assemblies is the restriction of $(\mathbb{N} \times \cdots \times \mathbb{N})$ to that predicate's set theoretic extension. This includes all graphs of primitive recursive functions.

The classical theory of general recursive functions in *Kleene 1952,* chapter XI, works in the logic of assemblies except for the diagonal arguments. Those use excluded middle, which fails for the claim "function f is defined for argument n" in assemblies due to the undecidability of the halting problem. The classical proof that there are non-recursive functions from the natural numbers to themselves fails here, and in fact for assemblies, all such functions are recursive.

To show this, define Kleene's T-predicate $T(e,x,y)$ and output function $U(y)$ in the logic of assemblies. The usual theorems follow in that logic. So just as in the classical case, we read $T(e,x,y)$ as "e is Gödel number of a definition (or, e is a code for a partial recursive function) which when applied to argument x gives a calculation with Gödel number y", and read $U(y)$ as "the final output from calculation y".

THEOREM 5.1. *Church's Thesis is true for assemblies:*

$$\vdash (\forall f.(\mathbb{N}^{\mathbb{N}}))(\exists e)(\forall x)(\exists y)(T(e,x,y) \,\&\, U(y) = fx).$$

Proof. Since the T-predicate and U are primitive recursive they agree in extension with the classical ones. Let $c(n, m)$ be the Gödel number for the completed calculation (if there is one) obtained from definition number n applied to argument m. Church's Thesis is realized by any code for the function taking n to $\langle n, h(\mathbb{N})\rangle$ where $h(\mathbb{N})$ codes the function taking any k to $\langle c(n,k), \langle\langle n,k,c(n,k)\rangle, \langle n.k, n.k\rangle\rangle\rangle$.

This is easier to check than to read. ⊣

THEOREM 5.2. *First order Markov principle. For any relation* $(P) \rightarrowtail (\mathbb{N} \times A)$:

$$(\forall y.(\mathbb{N}))(Pyx \vee \sim Pyx), \sim\sim(\exists y)Pyx \vdash (\exists y)Pyx. \qquad x.(A)$$

Proof. It suffices to find a partial recursive ψ such that: For all a in A, if k realizes $(\forall y)(Pya \vee \sim Pya)$ and some Pra is realized, then ψk realizes $(\exists y)Pya$. Define ϕk to be the smallest i such that $l(k.i) = 0$ (so $r(k.i)$ realizes Pia). For ψk take $\langle \phi k, r(k.\phi k)\rangle$. This uses Markov's principle in the metalanguage, since constructively a realizer for $\sim\sim(\exists y)Pyx$ only implies $(\exists y)Pyx$ is not-not realized. ⊣

Since $(\mathbb{N})$ is isomorphic to $(\mathbb{N} \times \mathbb{N})$, Markov's principle extends to two or more quantifiers $\exists$ over $(\mathbb{N})$ before a decidable formula. We count such formulas as almost negative, and Theorem 4.1 still holds.

The axiom of choice is not true in assemblies, but the axiom of choice from $(\mathbb{N})$ to any assembly is.

THEOREM 5.3. *For any assemblies* (A) *and* (B) *and relation* $(P) \rightarrowtail (\mathbb{N} \times B \times A)$ *we have*:

$$(\forall y)(\exists x.(B))Pyxz \vdash (\exists f.(B^{\mathbb{N}}))(\forall y)P(y, fy, z). \qquad z.(A)$$

Proof. Given a in A, let n realize $(\forall y)(\exists x.(B))Pyxa$. I.e., for all m there exists b in $B_{l(n.m)}$ such that $r(n.m)$ realizes $Rmba$. So (assuming countable choice in our metalanguage) there is an arrow $f_a : (\mathbb{N}) \to (B)$, with the function taking m to $l(n.m)$ as modulus, such that $r(n.m)$ realizes $Rm(f_a m)a$. This easily gives a realizer for $(\exists f.(B^{\mathbb{N}}))(\forall y)P(y, fy, z)$ recursive in n. ⊣

An assembly is *modest* iff each caucus A_n has at most one member. These amount to the "strictly effective objects" of Hyland (*1982*). An arrow between them is determined by any of its moduli. They are the quotients of sub-objects of $(\mathbb{N})$ by closed equivalence relations and appear in other notations as PERs (see *IEEE 1990, Asperti and Longo 1991*).

Notice that for modest (B), the proof of choice from $(\mathbb{N})$ needs no choice principle in the metalanguage.

Ordinary arithmetized analysis defines the integers $(\mathbb{I})$ and rationals $(\mathbb{Q})$ as quotients of products of $(\mathbb{N})$ by decidable equivalence relations. The assembly

($\mathbb{Q}^{\mathbb{N}}$) of sequences of rationals, all recursive by Church's Thesis, has a sub-assembly (CS) of Cauchy sequences. Hyland (*1982*) shows that if we require a certain rate of convergence for Cauchy sequences, then their equivalence is closed. Thus, the Cauchy reals form the assembly of recursive reals. In fact, all these are modest. Hyland (*1982*) says more on this. His "canonically separated objects" amount to assemblies.

6. THE EFFECTIVE TOPOS

This construction of $\mathcal{E}\!f\!f$ was first sketched categorically in *Freyd, Carboni, and Scedrov 1988*. Further generalities are in *Freyd and Scedrov 1990* and McLarty 1992.

The objects of $\mathcal{E}\!f\!f$ are equivalence relations of assemblies. When we take $(R) \rightarrowtail (A \times A)$ as an object in $\mathcal{E}\!f\!f$ we may call it (A/R) to suggest the quotient of (A) by (R). An arrow in $\mathcal{E}\!f\!f$ from (A/R) to (B/R') is given by a relation $(F) \rightarrowtail (A \times B)$ which is equivariant for (R) and (R') in this sense:

$$Rxy, Fyz \vdash Fxz, \qquad x, y.(A)$$
$$Fxz, R'zw \vdash Fxw; \qquad z, w.(B)$$

and functional in that:

$$\begin{aligned} Fxz, Fxw &\vdash R'zw, \\ &\vdash (\exists z)Fxz. \end{aligned}$$

Relations $(F) \rightarrowtail (A \times B)$ and $(F') \rightarrowtail (A \times B)$ give the same arrow from (A/R) to (B/R') if they are equivalent sub-objects of $(A \times B)$.

Given such an $\mathcal{E}\!f\!f$ arrow (F) and another (G) from (B/R') to some (C/R''), the relation $(\exists y.(B))(Fxy \& Gyz)$ defines an arrow $(G \circ F)$. This is associative, and (R) is the identity from (A/R) to (A/R). So $\mathcal{E}\!f\!f$ is a category.

THEOREM 6.1. *The category $\mathcal{E}\!f\!f$ has all finite limits:*

(i) *The object $(1/=)$ is terminal.*

(ii) *Any (A/R) and (B/R') have a product $(A \times B/R \times R')$ with $R \times R'\langle x, z\rangle \langle y, w\rangle$ defined by $(Rxy \;\&\; R'zw)$.*

(iii) *Any arrows (F) and (G) both from (A/R) to (B/R') have an equalizer, $[x.(A) \mid (\exists y)(Fxy \& Gxy)]$ modulo the equivalence relation on it induced by R.*

Proof. The natural proofs of these for quotient sets are sound for assemblies. ⊣

Some results fit assemblies into $\mathcal{E}\!f\!f$:

LEMMA 6.2. *A relation* $(F) \rightarrowtail (A \times B)$ *satisfies:*

$$Fxy \,\&\, Fxz \vdash y = z, \qquad x.(A)$$
$$\vdash (\exists y)Fxy \qquad y, z.(B)$$

in the logic of assemblies iff there is a unique arrow of assemblies $f : (A) \to (B)$ *such that*

$$\vdash Fxy \longleftrightarrow fx = y.$$

In other words, F *is equivalent to the graph of* f.

Proof. By realizability, if the sequents are true, (F) is isomorphic to (A). Composing with the projection of (F) to (B) gives f. ⊣

So there is a full and faithful functor to $\mathcal{E}\!f\!f$ taking each assembly (A) to $(A/=)$ and each arrow to its graph. We treat assemblies and their arrows as objects and arrows of $\mathcal{E}\!f\!f$ via this functor.

LEMMA 6.3.

(i) *Each* $(R) : (A) \to (A/R)$ *is a quotient of the two projections from* (R) *to* (A)*. It is their coequalizer and they are its kernel pair.*

(ii) *If* $(F) : (A/R) \rightarrowtail (B)$ *is monic in* $\mathcal{E}\!f\!f$*, then* (A/R) *is isomorphic to an assembly.*

Proof. For (i), use the logic of assemblies. For (ii), (F) induces a relation between (A) and $[y.(B) \mid (\exists x.(A))Fxy]$ equivariant for (R) and functional in both directions, thus an isomorphism in $\mathcal{E}\!f\!f$. ⊣

For any equivalence relation $(R) \rightarrowtail (A \times A)$ it is easy to see that R-equivariance as defined by

$$Rxy, Hyz \vdash Fxz \qquad x, y.(A) \quad z.(B)$$

is equivalent to saying a relation $(H) \rightarrowtail (A \times B)$ has the same pullback along the two projections $(R \times B) \to (A \times B)$.

LEMMA 6.4. *Relations in* $\mathcal{E}\!f\!f$ *from* (A/R) *to an assembly* (B) *correspond to* R*-equivariant relations* $(H) \rightarrowtail (A \times B)$.

Proof. By Lemma 6.3(i) any relation from (A/R) to (B) pulls back to such an (H). Conversely, let (H) be R-equivariant and (R') the equivalence relation on the assembly (H) induced by the relation (R) on (A). Then (H/R') is the unique (up to equivalence) relation from (A/R) to (B) with pullback (H). ⊣

There is a kind of truth value assembly (W) with each caucus W_n the powerset of the natural numbers. For a subassembly $(M) \rightarrowtail (W)$ counting as *true*, let each M_n contain the sets of natural numbers which include n. Given any $(C) \rightarrowtail (B)$, define $f : (B) \to (W)$ with each fb the set of realizers for Cb. Then $(C) \rightarrowtail (B)$ is equivalent to the pullback of (M) along f, but not only f. We say f *weakly classifies* $(C) \rightarrowtail (B)$.

There is an equivalence relation $E(E) \rightarrowtail (W \times W)$ such that: Any $\langle f, g\rangle : (B) \to (W \times W)$ factors through (E) iff f and g weakly classify the same subassembly of (B). Each E_n contains those pairs $\langle S, S'\rangle$ of sets of numbers such that for every m in S the value $ln.m$ is defined and is in S', and for every m in S' the value $rn.m$ is defined and in S. Intuitively, n realizes $(S \Rightarrow S')\;\&\;(S' \Rightarrow S)$.

LEMMA 6.5. *Every $\mathcal{E}\!f\!f$ object (B/R') has a power object.*

Proof. Arrows $(A) \to (W^B/E^B)$ uniquely classify relations from (A) to (B). A relation from any (A/R) to (B) amounts to an arrow $(A) \to (W^B/E^B)$ coequalizing the projections $(R) \to (A)$. This induces an arrow from (A/R) to (W^B/E^B) classifying the original relation. So (B) has power object (W^B/E^B).

The equalizer $(S) \rightarrowtail (W^B)$ of the two induced arrows from (W^B) to $(W^{R'})$ has an equivalence relation (E') induced by the relation (E^B) on (W^B). And (S/E') is a power object for (B/R'). ⊣

So $\mathcal{E}\!f\!f$ is a topos and has its own higher order logic, related to the logic of assemblies by:

LEMMA 6.6. *A formula with all assembly sorts is true in the logic of $\mathcal{E}\!f\!f$ iff it is true in the logic of assemblies. An $\mathcal{E}\!f\!f$ formula Qx, with x a variable over the power object of an assembly (B), is true iff every $Q(P)$ obtained by replacing x with a relation of assemblies $(P) \rightarrowtail (A \times B)$ is true.*

Proof. Both claims follow from Lemma 6.3(ii). For the second, every $\mathcal{E}\!f\!f$ object is surjective image of an assembly, so we need only know Qs for every generalized element $s : (A) \to P(B)$ defined over an assembly, thus every subassembly $(P) \rightarrowtail (A \times B)$. ⊣

REMARK 6.7. Any (A/R) is d.n. separated in $\mathcal{E}\!f\!f$ iff it has d.n. closed equality, so iff (R) is d.n. closed in $\mathcal{E}\!f\!f$. By Lemma 6.6 that is iff (R) is d.n. closed in assemblies. And so iff (A/R) is isomorphic to an assembly. The category of sets, *Set*, is isomorphic to the full subcategory of assemblies containing those (A) with each A_n equal to the carrier A. Call these assemblies *codiscrete*. They are d.n. sheaves in the category of assemblies. Each (B) has a d.n. dense monic to the codiscrete based on the set B, and is a sheaf iff this is an isomorphism. So the d.n. sheaves in $\mathcal{E}\!f\!f$ are (up to isomorphism) the codiscretes, i.e., they are (up to functor equivalence) *Set* embedded in $\mathcal{E}\!f\!f$.

7. ARITHMETIC IN EFF

The assembly $(\mathbb{N})$ is a natural number object in $\mathcal{E}\!f\!f$. To see this, use the Peano axioms where the induction axiom has P, a variable over the power object of $(\mathbb{N})$:

$$\vdash \sim(0 = sx),$$

$$\vdash sx = sy \Rightarrow x = y,$$

$$P(0) \,\&\, (\forall x)(P(x) \Rightarrow P(sx)) \vdash (\forall x)P(x).$$

The first two deal only with an assembly and so remain true in $\mathcal{E}\!f\!f$. Lemma 6.6 proves the induction axiom from the induction scheme for assemblies. Church's Thesis also deals only with assemblies. Lemma 6.6 gives the Markov principle with P a free variable over the power object of $(\mathbb{N})$:

$$(\forall n)(Pn \vee \sim Pn), \qquad \sim\sim(\exists n)Pn \vdash (\exists n)Pn.$$

Consider choice from $(\mathbb{N})$ to any (B/R), with P a variable over the power object $P(N \times (B/R))$:

$$(\forall y.(\mathbb{N}))(\exists x.(B/R))Pyx \vdash (\exists f.((B/R)\mathbb{N}))(\forall y)P(y, fy).$$

Certainly this is true for (B/R) if it is true for (B). Then use Lemma 6.6. And again, if (B) is modest, we have choice from $(\mathbb{N})$ to any (B/R) without assuming choice in the metalanguage.

In $\mathcal{E}\!f\!f$ we can define the Dedekind real numbers by Dedekind cuts. As Hyland (*1982*) remarks, the Cauchy and Dedekind reals agree (up to isomorphism) in $\mathcal{E}\!f\!f$ since choice from $(\mathbb{N})$ to (Q) shows every Dedekind cut gives a Cauchy sequence. We have good conditions for recursive analysis in full higher order intuitionistic logic.

8. METATHEOREMS

The above reasoning is constructive, except for particulars of arithmetic, so it works in any topos. This quickly gives two standard results in our context:

THEOREM 8.1. *Realizable realizability is realizability. I.e., realizability in $\mathcal{E}\!f\!f$ of sentences with all terms of sort* $(\mathbb{N})$ *agrees with classical realizability.*

Proof. Via Gödel numbering of those sentences we regard realizability as a relation between natural numbers. Its definitions are almost negative, except for $S \vee S'$. And equality of natural numbers is decidable, so that clause is equivalent to the almost negative: If $ln = 0$ then rn realizes S; and if not $ln = 0$, then $ln = 1$ and rn realizes S'. ⊣

THEOREM 8.2. *More arithmetic is realized classically than intuitionistically; or than follows intuitionistically from the Peano axioms plus Church's Thesis.*

Proof. The second claim follows from the first as the Peano axioms and Church's thesis are intuitionistically realized. For the first we give a topos where Markov's principle fails in ordinary arithmetic. By the proof of Theorem 5.2, Markov's principle is also not realized there.

Take presheaves (Kripke models) on this poset: The set of pairs $(n,0)$ and $(n,1)$ for all natural numbers n, where each $(n,0)$ precedes $(n,1)$ and each $(m,0)$ and $(m,1)$ with $n < m$, while each $(n,1)$ is a dead-end preceding nothing. The natural numbers here are the presheaf assigning $\mathbb{N}$ at every node. Markov's principle fails for the sub-presheaf S which is empty at each $(n,0)$ and at each $(n,1)$ is the singleton $\{n\}$. ⊣

Van Oosten (*1991*) gives several other higher order realizabilities, all of which can be presented in this way, using a first order logic for variants of assemblies and then getting a topos and higher order logic by adjoining quotients. They realize different parts of arithmetic, which is much of the point. And Remark 6.7 fails for some.

REFERENCES

Asperti, A., and G. Longo
1991 *Categories, types, and structures,* MIT Press, Cambridge, Massachusett.

Bell, J. L.
1988 *Toposes and local set theories, An introduction,* Oxford.

Freyd, P., and A. Scedrov
1990 *Categories, allegories,* North-Holland, Amsterdam.

Freyd, P., Carboni, and A. Scedrov
1988 A categorical approach to realizability and polymorphic types, *Proceedings of the third ACM workshop on mathematical foundations of programming language semantics* (Morin *et al.*, editors), Lecture Notes in Computer Science, no. 298, Springer-Verlag, Berlin.

Hyland
1982 The effective topos, *The L. E. J. Brouwer centenary symposium* (A. S. Troelstra and D. van Dalen, editors), North-Holland, Amsterdam.

IEEE
1990 *Fifth annual IEEE symposium on logic in computer science,* Computer Society Press.

Kleene, S. C.
1952 *Introduction to metamathematics,* North-Holland, Amsterdam.

McLarty, C.
1992 *Elementary categories, elementary toposes,* Oxford University Press, Oxford.

Van Oosten, J.
1991 Exercises in realizability, *Dissertation,* University of Amsterdam.

JOHN C. SHEPHERDSON

LANGUAGE AND EQUALITY THEORY IN LOGIC PROGRAMMING

Abstract. The underlying language of symbols used in logic programming is usually taken to be the one consisting of the symbols occurring in the program, or the program together with the query, but sometimes infinitely many function symbols are included. As far as Prolog or SLDNF-resolution are concerned, it makes no difference since these operate entirely within the former language. But it does make a difference to the Clark completion of a program and to 'constructive negation', a recently proposed extension of negation as failure. The difference is transmitted via the equality theory. We consider the effect of the underlying language on the Clark completion, the equality theory, constructive negation and 3-valued semantics.

1. INTRODUCTION

The theory of logic programming is sometimes developed in full generality with respect to some fixed but unspecified language $\mathcal{L}$ of symbols. But often (e.g., *Lloyd 1987*) the language is taken to be $\mathcal{L}(P)$, consisting of all symbols occurring in the program P. Most people who talk about Herbrand models use this language and it is usually assumed in database applications. If one wishes to consider an arbitrary query Q, the corresponding minimal language is $\mathcal{L}(P \cup \{Q\})$ whose symbols are all those appearing in P or Q. But in his work in 3-valued semantics for logic programming, Kunen (*1988, 1989*) uses a language with infinitely many function symbols of all arities, and in their work on constructive negation and equality theory, Chan (*1988*) and Przymusinski (*1989*) assume infinitely many constant or function symbols.

As far as Prolog and SLDNF-resolution are concerned it makes no difference what language $\mathcal{L}$ is used since both of these operate within the minimal language $\mathcal{L}(P \cup \{Q\})$, but it does make a difference to the Clark completion, $\mathsf{comp}_{\mathcal{L}}(P)$, of a program P which is the most widely accepted semantics for P and to the answers given by Chan's (*1988*) constructive negation. For example, if P is

$$\begin{aligned} p &\longleftrightarrow \neg q(x) \\ q(a) &\longleftrightarrow, \end{aligned}$$

then $\mathsf{comp}_{\mathcal{L}}(P)$ consists of $CET_{\mathcal{L}}$, Clark's equality theory for $\mathcal{L}$, together with the completed definitions of p and q, viz.

$$\begin{aligned} p &\longleftrightarrow \exists x \neg q(x) \\ q(x) &\longleftrightarrow x = a, \end{aligned}$$

C. Anthony Anderson and M. Zelëny (eds.), Logic, Meaning and Computation, 365-392.

which imply

$$p \longleftrightarrow \exists x(x \neq a).$$

If $\mathcal{L}$ has only the constant a then neither p nor $\neg p$ is a consequence of $\mathsf{comp}_{\mathcal{L}}(P)$, but if $\mathcal{L}$ contains other constant or function symbols, then p is a consequence of $\mathsf{comp}_{\mathcal{L}}(P)$; also p succeeds under constructive negation. On the other hand, the Herbrand model of $\mathsf{comp}_{\mathcal{L}}(P)$ based on the language $\mathcal{L}$ of the program has only the element a, so p is false in this model.

A technical reason for using infinite languages $\mathcal{L}$ is that they simplify the treatment because $CET_{\mathcal{L}}$ is then a complete theory. But it is also claimed that they are closer to Prolog, where the user may pose queries containing symbols not in the program. However, the user presumably has in mind some model or class of models of P or of $\mathsf{comp}(P)$, and it seems absurd to suggest that these are equipped with definitions of infinitely many irrelevant functions and constants. And it is not simply a question of extending the model by giving these functions a 'free' interpretation, since as the above example shows, the addition of new terms may change the truth values of old formulas. It seems more reasonable to say that Prolog operates within the finite language $\mathcal{L}(\cup\{Q\})$ defined by the program and the query together. One peculiarity of using $\mathcal{L}(P)$ is that it causes an 'instability'. If we add to the above program the clause $r(b) \longleftarrow$, then p becomes a consequence of $\mathsf{comp}_{\mathcal{L}}(P)$ with the new $\mathcal{L}$ (and true in all Herbrand models, whereas before it was false in all Herbrand models). As Kunen says (private communication), '... we do not expect a truth value to switch from true to false upon the addition to the database of a statement in a completely disjoint language'. Using a language with infinitely many function symbols of all arities avoids this difficulty but commits one to the existence of infinitely many irrelevant and non-existent terms. In practice, one surely has in mind some particular (usually finite) language $\mathcal{L}$, possibly including symbols not explicitly occurring in the program, so the ideal development would be to state results for such a general fixed $\mathcal{L} \subseteq \mathcal{L}(P)$ as far as possible, noting where further assumptions about $\mathcal{L}$ are necessary.

In section 2 we consider the effect on $\mathsf{comp}_{\mathcal{L}}(P)$ of extending $\mathcal{L}$. If $\mathcal{L}$ is infinite (strictly: has infinitely many function or constant symbols), it has no effect (on consequences of $\mathsf{comp}_{\mathcal{L}}(P)$ exressible in $\mathcal{L}$); if $\mathcal{L}$ is finite, it has the effect of adding statements asserting the existence of finitely or infinitely many 'roots' or isolated elements, i.e., elements not in the range of any $\mathcal{L}$-function.

In section 3 we give (Theorem 3.9) a normal form for equality formulas in the theory $CET_{\mathcal{L}}$ for general $\mathcal{L}$. This generalizes the work of Chan (*1988*), Maher (*1988*), and Przymusinski (*1989*) who gave such results for the case of infinite $\mathcal{L}$ and (in *Maher 1988*) for finite $\mathcal{L}$ plus the domain closure axiom $DCA_{\mathcal{L}}$. It can be regarded as a refinement of a normal form given by Malcev (*1971*). It shows that Chan's constructive negation can be performed with a finite $\mathcal{L}$, but that the normal form of answers to queries and the algorithm for obtaining them become more complicated than with infinite $\mathcal{L}$. However, if the domain closure axiom is assumed, then a simpler normal form exists. The results on the domain closure axiom also show that Sato and Tamaki's (*1988*) replacement over the Herbrand universe based on $\mathcal{L}(P)$, of quantified inequations by predicates given by definite clause programs, extends (Theorem 3.13) to arbitrary equality formulas. Without the domain closure axiom, one can obtain (Theorem 3.14) a hierarchic normal program with 2 or 3 levels which represents a given equality formula.

In section 4 we show how the results of Kunen (*1988, 1989*) are affected by the language. His semantics based on the 3-valued analogue of the familiar T-operator corresponds to considering all 3-valued consequences of $\mathsf{comp}_{\mathcal{L}}(p)$ if $\mathcal{L}$ is infinite, and all 3-valued consequences of $\mathsf{comp}_{\mathcal{L}}(P) \cup \{DCA_{\mathcal{L}}\}$ if $\mathcal{L}$ is finite (Theorem 4.2). The 3-valued soundness of SLDNF-resolution with respect to $\mathsf{comp}_{\mathcal{L}}(P)$ and its 3-valued completeness with respect to $\mathsf{comp}_{\mathcal{L}}(P)$ for allowed programs (*Kunen 1989*) hold for all $\mathcal{L}$.

I am very grateful to the referees for making corrections and suggesting improvements.

2. HOW $\mathsf{comp}_{\mathcal{L}}(P)$ DEPENDS ON $\mathcal{L}$

NOTATION: A *language* $\mathcal{L}$ is a set of function and predicate symbols of assigned arities. Constants are treated as 0-ary function symbols. Since the predicate symbols in $\mathcal{L}$ play no role, we shall ignore them; thus when we say a language is infinite, we mean it has infinitely many function symbols (including those of arity 0).

Clark's equality theory for $\mathcal{L}$ ($CET_{\mathcal{L}}$), consists of the axioms:

(1) $f(x_1,\ldots,x_n) \neq g(y_1,\ldots,y_m)$ for all distinct f, g in $\mathcal{L}$,

(2) $f(x_1,\ldots,x_n) = f(y_1,\ldots,y_m) \longrightarrow x_1 = y_1 \wedge \ldots \wedge x_n = y_n$ for all f in $\mathcal{L}$,

(3) $x \neq t(x)$ for all terms $t(x)$ distinct from x in which x actually occurs.

We refer to this last axiom as the '*occur axiom*'.

(4) $\underline{x}$ denotes a tuple of variables $x_1,\ldots,x_n$, and $\underline{x} = \underline{y}$ stands for $x_1 = y_1 \wedge \ldots \wedge x_n = y_n$.

$\mathcal{L}$-Programs are finite sets of clauses of the form

$$A \longleftarrow \varphi$$

where A is an $\mathcal{L}$-atom and φ is an $\mathcal{L}$-formula. The *language* $\mathcal{L}(P)$ consists of the function and predicate symbols actually occurring in P.

If all the clauses of the program P which have the predicate p in their head are

$$p(\underline{t}_1) \longleftarrow \varphi, \ldots, p(\underline{t}_r) \longleftarrow \varphi_r,$$

then the *completed definition* of p is

$$p(\underline{x}) \longleftrightarrow \bigvee_{i=1}^{r} \exists \underline{y}_i (\underline{x} = \underline{t}_i \wedge \varphi_i)$$

where $\underline{x}$ are new variables and $\underline{y}_i$ are the variables in $p(\underline{t}_i) \longleftarrow \varphi_i$. The conjuction of the completed definitions of all predicates in P is denoted by $Cd(P)$. The *Clark completion* of an $\mathcal{L}$-program P with respect to $\mathcal{L}$ consists of $CET_{\mathcal{L}} \cup \{Cd(P)\}$ and is denoted by $\mathsf{comp}_{\mathcal{L}}(P)$. (We shall always interpret $=$ as equality, so the usual logical axioms of equality:

$$\begin{array}{lcl} x = x, & & \\ x = y \longrightarrow y = x, & \text{and} & \underline{x} = \underline{y} \longrightarrow (p(\underline{x}) \longleftrightarrow p(\underline{y})), \\ x = y \wedge y = z \longrightarrow x = z, & & \underline{x} = \underline{y} \longrightarrow (f(\underline{x}) = f(\underline{y})), \end{array}$$

are also assumed). For f in $\mathcal{L}$, we define $N_f(x)$—'*x is not in the range of f*'—by

$$N_f(x) =_{\text{def}} \forall y_1, \ldots, y_r (x \neq f(y_1, \ldots, y_r))$$

where r is the arity of f, and if $\mathcal{L}$ has finitely many function symbols, we define $R_{\mathcal{L}}(x)$—'*x is an $\mathcal{L}$-root*'—, by

$$R_{\mathcal{L}}(x) =_{\text{def}} \bigwedge_{f \in \mathcal{L}} N_f(x),$$

$G^k_{\mathcal{L}}$—'*there are at least k $\mathcal{L}$-roots*'—by

$$G^k_{\mathcal{L}} =_{\text{def}} \exists x_1, \ldots, x_k (R_{\mathcal{L}}(x_1) \wedge \ldots \wedge R_{\mathcal{L}}(x_k) \wedge \bigwedge_{i \neq j} x_i \neq x_j),$$

$G^\infty_{\mathcal{L}}$—'*there are infinitely many $\mathcal{L}$-roots*'—by

$$G^\infty_{\mathcal{L}} =_{\text{def}} \{G^1_{\mathcal{L}}, G^2_{\mathcal{L}}, \ldots\},$$

and $DCA_{\mathcal{L}}$—*the Domain Closure Axiom,*—by

$$DCA_{\mathcal{L}} =_{\text{def}} \neg \exists x R_{\mathcal{L}}(x).$$

This Clark completion $\mathsf{comp}_{\mathcal{L}}(P)$ is the most widely accepted semantics for a logic program P. The main justification for this is Clark's (*1978*) soundness result:

(1) *if* $P \cup \{\longleftarrow Q\}$ *has an SLDNF-refutation with answer* θ, *then* $\mathsf{comp}_{\mathcal{L}}(P) \models \forall Q\theta$,

(2) *if* $P \cup \{\longleftarrow Q\}$ *has a finitely failed SLDNF-tree, then* $\mathsf{comp}_{\mathcal{L}}(P) \models \forall \neg Q$.

This holds when $\mathcal{L} = \mathcal{L}(P \cup \{Q\})$, the language containing only the function and predicate symbols actually occurring in the program p and query Q. So it obviously holds also for any language containing this, since $\mathsf{comp}_{\mathcal{L}_1}(P) \models \mathsf{comp}_{\mathcal{L}_2}(P)$ if $\mathcal{L}_1 \supseteq \mathcal{L}_2$. However, SLDNF-resolution is usually incomplete for $\mathsf{comp}_{\mathcal{L}}(P)$, i.e., the converses of the above implications do not hold in general; there are queries which are consequences of $\mathsf{comp}_{\mathcal{L}}(P)$ which do not succeed using SLDNF-resolution. The stronger $\mathsf{comp}_{\mathcal{L}}(P)$ is, the greater is this incompleteness, so it is natural to ask how the strength of $\mathsf{comp}_{\mathcal{L}}(P)$ varies with $\mathcal{L}$. This question is answered in the next theorem.

THEOREM 2.1. *Let* P *be a program and* $\mathcal{L}_2 \supseteq \mathcal{L}_1 \supseteq \mathcal{L}(P)$.

(i) *If* $\mathcal{L}_1$ *is finite and* $\mathcal{L}_2 - \mathcal{L}_1$ *contains no function symbols of positive arity but a finite number* k *of constants, then*

$$\mathsf{comp}_{\mathcal{L}_2}(P) \text{ is a conservative extension of } \mathsf{comp}_{\mathcal{L}_1}(P) \cup \{G^k_{\mathcal{L}_1}\}.$$

(ii) *If* $\mathcal{L}_1$ *is finite and* $\mathcal{L}_2 - \mathcal{L}_1$ *contains a function symbol of positive arity, or infinitely many constants, then*

$$\mathsf{comp}_{\mathcal{L}_2}(P) \text{ is a conservative extension of } \mathsf{comp}_{\mathcal{L}_1}(P) \cup \{G^\infty_{\mathcal{L}_1}\}.$$

(iii) *If* $\mathcal{L}_1$ *is infinite, then*

$$\mathsf{comp}_{\mathcal{L}_2}(P) \text{ is a conservative extension of } \mathsf{comp}_{\mathcal{L}_1}(P).$$

Proof. Throughout, φ denotes an $\mathcal{L}_1$-sentence. Then $\mathsf{comp}_{\mathcal{L}}(P)$ is $CET_{\mathcal{L}} \cup \{Cd(P)\}$, and $\mathsf{comp}_{\mathcal{L}}(P) \models \varphi$ iff $CET_{\mathcal{L}} \models (Cd(P) \longrightarrow \varphi)$, so it is enough to prove the corresponding theorem with CET replacing $\mathsf{comp}(P)$.

(i) $CET_{\mathcal{L}_2}$ contains $CET_{\mathcal{L}_1}$, and $CET_{\mathcal{L}_2}$ implies that the k new constants of $\mathcal{L}_2 - \mathcal{L}_1$ are $\mathcal{L}_1$-roots and are distinct. So

$$CET_{\mathcal{L}_2} \vdash CET_{\mathcal{L}_1} \cup \{G^k_{\mathcal{L}_1}\}.$$

Conversely, if

$$CET_{\mathcal{L}_1} \cup \{G^k_{\mathcal{L}_1}\} \not\models \varphi,$$

then there is a model M satisfying

$$CET_{\mathcal{L}_1} \cup \{G^k_{\mathcal{L}_1}\} \cup \{\neg\varphi\}.$$

This can be extended to a model of $CET_{\mathcal{L}_2}$ without extending its domain by taking the k new constants to be k of the $\mathcal{L}_1$-roots of M and, if there are any new predicates in $\mathcal{L}_2$, defining these arbitrarily, e.g., all true. (The only axiom of $CET_{\mathcal{L}_2}$ involving the predicates is the equality axiom $\underline{x} = \underline{y} \longrightarrow (p(\underline{x}) \leftrightarrow p(\underline{y}))$, which is automatically satisfied, since we are interpreting $=$ as equality in our models. So from now on we shall ignore any new predicates.)

(ii) The case where $\mathcal{L}_2 - \mathcal{L}_1$ consists of infinitely many constants follows from (i) by compactness. If $\mathcal{L}_2 - \mathcal{L}_1$ contains a new function symbol f, then taking $g(x) = f(x, \dots, x)$, $CET_{\mathcal{L}_2}$ implies that, for any element x and any k, $g(x), g^2(x), \dots, g^k(x)$ are distinct $\mathcal{L}_1$-roots, hence implies $G^k_{\mathcal{L}_1}$. This shows that $CET_{\mathcal{L}_2}$ is an extension of $CET_{\mathcal{L}_1} \cup G^{\infty}_{\mathcal{L}_1}$.

We must show it is a conservative extension, i.e., that if

$$CET_{\mathcal{L}_2} \models \varphi, \text{ then } CET_{\mathcal{L}_1} \cup G^{\infty}_{\mathcal{L}_1} \models \varphi.$$

Suppose then that

$$CET_{\mathcal{L}_2} \models \varphi \text{ but } CET_{\mathcal{L}_1} \cup G^{\infty}_{\mathcal{L}_1} \not\models \varphi.$$

Then

$$CET_{\mathcal{L}_1} \cup G^{\infty}_{\mathcal{L}_1} \cup \{\neg\}$$

is consistent and countable, so it has a countable model M. By compactness, $CET_{\mathcal{L}_2} \models \varphi$ implies $CET_{\mathcal{L}_{2'}} \models \varphi$ where $\mathcal{L}_{2'}$ is a finite sublanguage of $\mathcal{L}_2$ which (by adjoining $\mathcal{L}_1$ if necessary) we may suppose contains $\mathcal{L}_1$. If we could extend M, without extending its domain D, or changing the interpretations of the function and predicate symbols in $\mathcal{L}_1$, to a model of $CET_{\mathcal{L}_{2'}}$ by suitably defining the new functions of $\mathcal{L}_2$ over D, then we would have left $\neg\varphi$ true and produced the desired contradiction. This is not always possible because the values of the new functions not only have to be $\mathcal{L}_1$-roots but have to satisfy the 'occur' axiom of CET which says that if $t(x)$ is a term of non-zero depth containing x, then $t(x) \neq x$. Suppose, for example, that $\mathcal{L}_1$ consists of a binary function f, and M has roots $\rho_1, \rho_2, \dots$ and elements $\alpha_0, \alpha_1, \alpha_2, \dots$ satisfying $\alpha_{i-1} = f(\rho_i, \alpha_i)$ for $i = 1, 2 \dots$ as in Figure 1. Then if $\mathcal{L}'_2$ has a

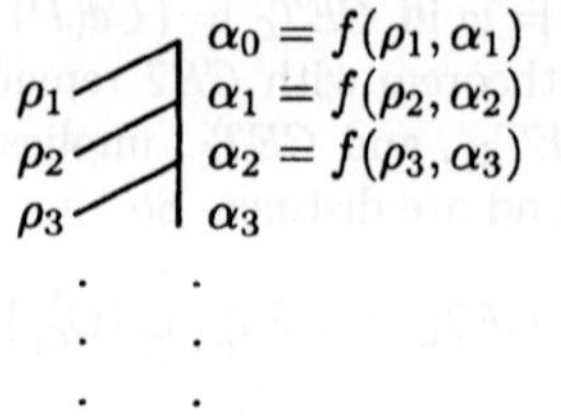

Figure 1

new unary function symbol g, then $g(\alpha_0)$ must be defined to be a root, but it cannot be ρ_1 because that would give $\rho_1 = g(f(\rho_1, \alpha_1))$ a violation of the occur axiom. Similarly, it cannot be any of $\rho_2, \rho_3, \ldots$ because that would give $\rho_2 = g(f(\rho_1, f(\rho_2, \alpha_2))), \ldots$. The problem is that all the roots are ancestors of α_0. To get around this difficulty, we use compactness again. If $CET_{\mathcal{L}_2} \models \varphi$, then for some d $CET^d_{\mathcal{L}'_2} \models \varphi$, where the superscript d indicates that the occur axiom is restricted to terms t of depth $\leq d$. We now show that M can be extended, without extending its domain, or changing the interpretations of function or predicate symbols of $\mathcal{L}_1$, to a model of $CET^d_{\mathcal{L}'_2}$.

If $g_1, \ldots, g_a$ are the functions in $\mathcal{L}'_2 - \mathcal{L}_1$, we have to define the countably many values $g_1(\alpha_1, \ldots, \alpha_{r_i}), i = 1, \ldots, a$ for $\alpha_1, \ldots, \alpha_{r_i}$ ranging over the elements of D. To satisfy $CET^d_{\mathcal{L}'_2}$ these have to be all distinct from each other, all $\mathcal{L}_1$-roots, and have to satisfy $t(x) \neq x$ for all terms t of (non-zero) depth $\leq d$. Let $v_1, v_2, \ldots$ be an enumeration of these new function values to be defined. If v_n is $g(\beta_1, \ldots, \beta_r)$ where g is one of $g_1, \ldots, g_a$ and $\beta_1, \ldots, \beta_r \in D$, then we define v_n to be the first of the $\mathcal{L}_1$-roots $\rho_1, \rho_2, \ldots$ of M which has not already been used, and which is not a $(d-1)$-ancestor of any of $\beta_1, \ldots, \beta_r$. Here a k-ancestor of β is an element α_1 of D such that there is an $\mathcal{L}'_2$-term $t(x_1, \ldots, x_m)$ of depth $\leq k$ containing x_1 and elements $\alpha_2, \ldots, \alpha_m$ of D such that at the present stage, i.e., before v_n is defined, $t(\alpha_1, \ldots, \alpha_m)$ is defined and equal to β. It is easily proved by induction on k that there are only finitely many k-ancestors of any element β.

We now show that if the $\mathcal{L}'_2$ *d-occur axiom* was satisfied before v_n was defined, then it remains satisfied afterwards. By the $\mathcal{L}'_2$ *d-occur axiom,* we mean the axiom which says that if $t(x_1, \ldots, x_m)$ is a term of non-zero depth $\leq d$ containing x_1, and if $t(\alpha_1, \ldots, \alpha_m)$ is defined, then it is not equal to α_1. Since the $\mathcal{L}'_2$ d-occur axiom is satisfied in this sense in M (since the $\mathcal{L}_1$ occur axiom is satisfied), and since eventually all functions in $\mathcal{L}'_2$ are everywhere defined, this will complete the proof. Suppose then that the definition of v_n, i.e., putting $g(\beta_1, \ldots, \beta_r) = \rho$, creates a violation of the $\mathcal{L}'_2$ d-occur axiom, i.e., there is a term $t(x_1, \ldots, x_m)$ containing x_1 and of non-zero depth $\leq d$ and elements $\alpha_1, \ldots, \alpha_m$ of D such that $t(\alpha_1, \ldots, \alpha_m)$ becomes defined and equal to α_1. Take any path from the root t to this leaf and terminate each side branch at the end of the first edge, labeling the new leaf with a new variable, as in the example of Figure 2 (ignore the third column for the time being). This new term also has the property used to define $t(x_1, \ldots, x_m)$ above, so let us take this to be $t(x_1, \ldots, x_m)$. Since the $\mathcal{L}'_2$ d-occur axiom was satisfied before $g(\beta_1, \ldots, \beta_r)$ was defined to be ρ, at least one of the functions in t must be g and occurring in a context $g(t_1(x_1, \ldots, x_m), \ldots, t_r(x_1, \ldots, x_m))$ where (after v_n is defined) $t_i(\alpha_1, \ldots, \alpha_m) = \beta_i$, for $i = 1, \ldots, r$. Take the highest such occurrence in the tree of t, and replace it by a new variable y as in the third column of Figure 2. This gives a term t' of depth $d_1 < d$ in which y occurs but x_1 does not occur, because for $x_1 = \alpha_1, \ldots, x_m = \alpha_m, y = \rho$ the value of t' is the same as the value of t, i.e., α_1, and for these values t' is defined before $g(\beta_1, \ldots, \beta_r)$ is defined, so if x_1 occurred in it, there would have been an existing violation of d-occur axiom. So we may write $t' = t'(x_2, \ldots, x_m, y)$.

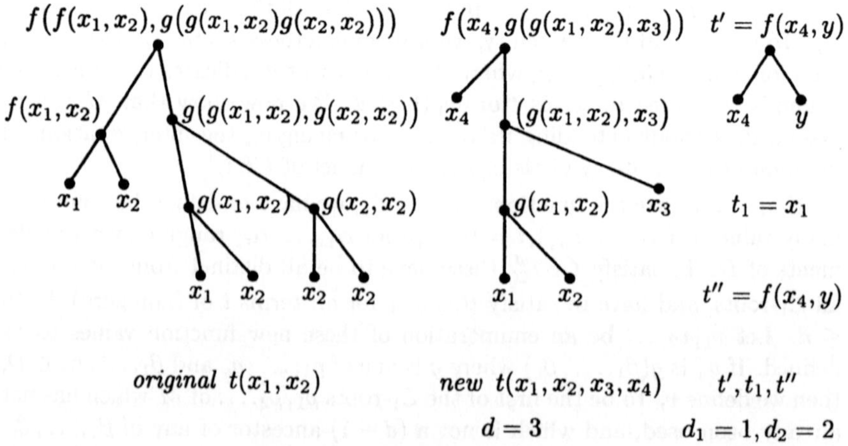

Figure 2

Because of the way we have pruned the side branches of the the t tree, d_1 is the depth in t of this highest occurrence of the new value. Now take the lowest such occurrence of the new value, at depth d_2, say. This takes the form $g\big(t_1(x_1,\ldots,x_m),\ldots,t_r(x_1,\ldots,x_m)\big)$ where $t_i(\alpha_1,\ldots,\alpha_m)=\beta_i$ for $i=1,\ldots,r$, and since it is the lowest occurrence, these values of the t_i, are defined before $g(\beta_1,\ldots,\beta_r)$ is defined. At least one of the t_i, say t_1, must contain x_1, and it has depth $\leq d-(d_2+1)$. Now

$$\beta_1=t_1(\alpha_1,\ldots,a_m)=t_1\big(t'(\alpha_2,\ldots,\alpha_m,\rho),\alpha_2,\ldots,\alpha_m\big)=t''(\alpha_2,\ldots,\alpha_m,\rho)$$

where

$$t''(x_2,\ldots,x_m,y)=t_1\big(t'(x_2,\ldots,x_m,y),x_2,\ldots,x_m\big)$$

is a term of depth $\leq d_1+d-(d_2+1)$, which is $\leq d-1$, since $d_1\leq d_2$. Also $t''(\alpha_2,\ldots,\alpha_m,\rho)$ was defined before $g(\beta_1,\ldots,\beta_r)$ was defined, and y occurs in t'', so ρ is a $(d-1)$-ancestor of β_1, contrary to the hypothesis made when it was chosen as the value of $g(\beta_1,\ldots,\beta_r)$.

(iii) If $CET_{\mathcal{L}_2}\models\varphi$, then by compactness $CET_{\mathcal{L}_2'}\models\varphi$ for some finite subset $\mathcal{L}_2'$ of $\mathcal{L}_2$. Let $\mathcal{L}_1'=\mathcal{L}_2'\cap\mathcal{L}_1$. By (i), (ii) applied to $\mathcal{L}_2'\subseteq\mathcal{L}_1'$ we have

$$CET_{\mathcal{L}_1'}\cup G^{\infty}_{\mathcal{L}_1'}\models\varphi.$$

By (ii) applied to $\mathcal{L}_1\subseteq\mathcal{L}_1'$, we obtain

$$CET_{\mathcal{L}_1}\models\varphi.$$

The converse is obvious.

3. NORMAL FORMS FOR EQUALITY FORMULA

The normal forms given here are based on those used by Chan (*1988*) in his constructive negation procedure. This is a way of extending negation as failure to non-ground negative literals by returning the negation of answers to query Q as answers to $\neg Q$. If Q has variables $x_1, \ldots, x_n$ and succeeds with answer substitution θ for the program P, we can write the answer in equational form as

$$Q \longleftarrow \exists(x_1 = x_1\theta \wedge \ldots \wedge x_n = x_n\theta),$$

where $\exists$ quantifies the variables on the RHS other than $x_1, \ldots, x_n$.

If there are a finite number of answers to the query Q and no infinite fair executions, then

$$Q \longleftrightarrow A_1 \vee \ldots \vee A_k$$

is a consequence of $\mathsf{comp}_{\mathcal{L}}(P)$; so if $\mathsf{comp}_{\mathcal{L}}(P)$ is the intended semantics, it is legitimate to return

$$\neg(A_1 \vee \ldots \vee A_n)$$

as the answer to $\neg Q$. If there are infinitely many answers to Q this procedure fails, but when it does succeed it returns as answer to a query Q an equality formula E such that

$$Q \longleftrightarrow E$$

is a consequence of $\mathsf{comp}_{\mathcal{L}}(P)$. However, the alternations of existential quantification, disjunction and negation have the effect that if the formula E is built up in the way just indicated, it can be an arbitrarily complex first order equality formula. In order to make the answer E more intelligible, Chan gives an algorithm for reducing the answer at each stage to a fairly simple normal form, to which, assuming $CET_{\mathcal{L}}$, it is equivalent. To simplify this, he appears to assume that $\mathcal{L}$ has infinitely many function symbols. (What he explicitly states is that he assume that 'a conjunction of satisfiable (universally quantified) inequalities remains satisfiable'. Now, with the meaning of 'satisfiable' usual in mathematical logic, to say that such a conjunction $C(x_1, \ldots, x_n)$ is satisfiable means that $\exists x_1 \ldots x_n C(x_1 \ldots x_n)$ is true in *some* model of $CET_{\mathcal{L}}$. In this sense the result can be shown to be true for all $\mathcal{L}$, e.g., if $\mathcal{L} = \{a, b\}$, then $\exists x(x \neq a \wedge x \neq b)$ is true in a model with elements a, b, c. What he actually does in his algorithm is to replace $\exists x_1 \ldots x_n C(x_1 \ldots x_n)$ by *true*, i.e., he wants to know that this is true in *all* models of $CET_{\mathcal{L}}$. This will be the case if $CET_{\mathcal{L}}$ is complete, which is true if $\mathcal{L}$ is infinite, but not if it is finite.)

Chan's normal form for the equality formulas returned as answers is quite simple; you only need to add some universally quantified inequations to the equations $x_1 = x_1\theta \wedge \ldots \wedge x_n = x_n\theta$ which represent the familiar substitution type answer. In fact, further restrictions can be made; these are more clearly stated in *Przymusinski 1989* (who explicitly makes the assumption that $\mathcal{L}$ is infinite). His Theorem 8.1 states that there is an algorithm for reducing any equality formula E to a *strictly normal equality formula* N_E such that

$$CET_{\mathcal{L}} \models \forall(E \longleftrightarrow N_E).$$

A *strictly normal equality formula* is a disjunction of *strictly simple equality formulas* each of which has the form

$$\exists y_1, \ldots, y_x(x_1 = t_1 \wedge \ldots \wedge x_m = t_m \wedge \forall(x'_1 \neq s_1) \wedge \ldots \wedge \forall(x'_m \neq s_m)),$$

where each t_i, s_i is either a non-variable term or one of the free variables of this formula distinct from x_i, x'_i respectively, where the $\forall$ in $\forall(x'_i \neq s_i)$ universally quantifies some (perhaps none) of the variables in s_i, the $y_1, \ldots, y_s$ are distinct from $x_1, \ldots, x_m, x'_1, \ldots, x'_m$ and each y_i occurs in at least one of the terms $t_1, \ldots, t_m$. Chan does not describe his normal form explicitly, but from his examples and his reduction algorithm, it would appear that he achieves the further restriction corresponding to the use of idempotent mgu, that the occurrence of x_i on the LHS of $x_i = t_i$ is the only occurrence of x_i in the formula. But in order to achieve this he has to admit quantified inequations of the form $\forall(y_i \neq r_i)$ as well. This is the form we generalize to arbitrary $\mathcal{L}$ in the Definition 3.7 and Theorem 3.9 below. The only difference in the normal form is that if $\mathcal{L}$ is finite, we need to add closed formulas $G^k_{\mathcal{L}}, \neg G^{k'}_{\mathcal{L}}$ asserting the existence of at least k, less than k', $\mathcal{L}$-roots. So if one is working with models for which the number of $\mathcal{L}$-roots is known, the same normal form is achievable, but our algorithm is more complicated than Chan's.

We work throughout in a finite language $\mathcal{L}$ which is supposed to contain the language $\mathcal{L}(\{E\})$ consisting of the function and predicate symbols which actually occur in the equality formula E which is to be reduced to normal form. We use the terminology of Przymusinski (*1989*) but with a different meaning; we start with a weak definition of normal equality formula which is strengthened in Definition 3.7.

DEFINITION 3.1. *A simple equality formula* has the form

$$S = \exists(E_1 \wedge \ldots \wedge E_n), \qquad n \geq 0$$

where $\exists$ quantifies some (perhaps none) of the variables in $E_1, \ldots, E_n$ and each E_i is either:

(i) an equation $t_1 = t_2$ between terms of $\mathcal{L}$, or

(ii) an inequation $t_1 \neq t_2$, or

(iii) $R_{\mathcal{L}}(v)$ for some variable v, or

(iv) $\neg G^k_{\mathcal{L}}$ for some positive integer k.

A normal equality formula is a disjunction

$$S_1 \wedge \ldots \wedge S_M, \qquad m \geq 0$$

of simple equality formulas. An empty conjunction (disjunction) is identified with true (false).

We first prove some basic reduction lemmas. From now on we assume $CET_{\mathcal{L}}$ so that when we assert a result φ (e.g., the equivalence of two formulas) we mean $CET_{\mathcal{L}} \vdash \varphi$.

LEMMA 3.2. *There is an algorithm for reducing any conjunction C of equations to equivalent solved form S':*

$$v_1 = t_1 \wedge v_2 = t_2 \wedge \ldots \wedge v_n = t_n$$

where $v_1, \ldots, v_n$ are distinct variables which do not occur in any of the terms $t_1, \ldots, t_n$.

Proof. This is just a version of the usual unification algorithm. See *Lassez, Maher, and Marriott 1988.* ⊣

LEMMA 3.3. *If all the variables $x_1, \ldots, x_n$ (and possibly other variables as well) actually occur in $t(x_1, \ldots, x_n)$, then*

$$t(x_1, \ldots, x_n) = t(y_1 \ldots, y_n) \longrightarrow x_1 = y_1 \wedge \ldots \wedge x_n = y_n.$$

Proof. By induction on the depth of t. If the depth is 0, then $t(x_1)$ is simply x_1. For the inductive step $CET_{\mathcal{L}}$ implies

$$f\big(t_1(x_1, \ldots, x_n), \ldots, t_r(x_1, \ldots, x_n)\big) = f\big(t_1(y_1, \ldots, y_n), \ldots, t_r(y_1, \ldots, y_n)\big)$$
$$\longrightarrow t_1(x_1, \ldots, x_n) = t_1(y_1, \ldots, y_n) \wedge \ldots \wedge t_r(x_1, \ldots, x_n) = t_r(y_1, \ldots, y_n),$$

so the induction hypothesis gives the result because each x_i must occur in some $t_j(x_1, \ldots, x_n)$. ⊣

LEMMA 3.4. *If x', $\underline{x} = x_1, \ldots, x_m$, $\underline{y} = y_1, \ldots, y_n$, $\underline{z} = z_1, \ldots, z_p$ are distinct variables, and $\underline{x}, \underline{y}$ are all the variables in $t(\underline{x}, \underline{y})$, then*

$$\neg \exists \underline{y}\underline{z}\big(x' = t(\underline{x}, \underline{y}) \wedge K\big) \longleftrightarrow \{\forall \underline{y}\big(x' \neq t(\underline{x}, \underline{y})\big) \vee \exists \underline{y}\big(x' = t(\underline{x}, \underline{y}) \wedge \neg \exists \underline{z} K\big)\}.$$

Proof. The LHS logically implies the RHS and the first disjunct of the RHS implies the LHS. That the second disjunct of the RHS implies the LHS follows from Lemma 3.3. ⊣

LEMMA 3.5.

$$\forall \underline{z}\big(x' \neq t(t_1, \ldots, t_r)\big) \longleftrightarrow \{\forall \underline{y}\big(x' \neq t(y_1, \ldots, y_r)\big) \vee$$
$$\exists \underline{y}\big(x' = t(y_1, \ldots, y_r) \wedge \forall \underline{z}(y_1, \neq t_1 \wedge \ldots \wedge y_r \neq t_r)\big)\},$$

where $\underline{y} = y_1, \ldots, y_r$ are variables not in the LHS, and the variables $\underline{z}$ do not occur in $t(y_1, \ldots, y_r)$.

Proof. Write the LHS in the form

$$\neg\exists \underline{y}, \underline{z}(x' = t(y_1, \ldots, y_r) \wedge y_1 = t_1 \wedge \ldots \wedge y_r = t_r)$$

and apply Lemma 3.4. ⊣

The hypothesis about $\underline{z}$ means that their only occurrences in $t(t_1, \ldots, t_r)$ are within $t_1, \ldots, t_r$.

LEMMA 3.6. *There is an algorithm for reducing disjunctions, conjunctions, and existential quantifications of normal equality formulas to equivalent normal equality formulas.*

Proof. For a disjunction there is nothing to do; for a conjunction we use the distributive law to write it as a disjunction of conjunctions of simple equality formulas, and convert the latter into simple equality formulas using $\exists y A \wedge \exists y' A' \longleftrightarrow \exists y,\ y'(A \wedge A')$ (provided y does not occur in A', and y' does not occur in A). For existential quantification we use $\exists y(A \vee B) \longleftrightarrow \exists y A \vee \exists y B$ to move the quantifier in front of the simple equality formulas, which still leaves them simple equality formulas.

By replacing $A \longrightarrow B$, $A \longleftrightarrow B$, $\forall x A$ by $\neg A \vee B$, $(\neg A \vee B) \wedge (\neg B \vee A)$ and $\neg\exists x \neg A$ respectively, we can replace E by an equivalent formula containing only the connectives $\wedge, \vee, \neg, \exists$ (it is convenient to include both $\wedge$ and $\vee$). An atomic formula $t_1 = t_2$ is a normal equality formula and Lemma 3.6 deals with $\wedge, \vee, \exists$ so it remains to show how to write the negation of a normal equality formula as a normal equality formula. Using $\neg(S_1 \vee \ldots \vee S_m) \longleftrightarrow \neg S_1 \wedge \ldots \neg S_m$, it is enough to do this for the negation $\neg S$ of a simple equality formula S. Our first step is to simplify this still further.

DEFINITION 3.7. A *strictly simple equality* formula with free variables $\underline{x} = x_1, \ldots, x_r$ has the form

$$\exists y_1, \ldots, y_s(E_1 \wedge \ldots \wedge E_n), \qquad n \geq 0$$

where

(a) each E_i is either:

(i) an equation $x_i = t$, where t is a non-variable term or is x_j for $j \neq i$; and there is at most one of these for each x_i, and calling such x_i *solved variables,* no solved variable occurs elsewhere in the formula, or

(ii) an inequation $x_i \neq t$ *or* $y_i \neq t$, or

(iii) $R_{\mathcal{L}}(x_i)$ or $R_{\mathcal{L}}(y_i)$, or

(iv) $G^k_{\mathcal{L}}$ for at most one positive integer k, or

(v) $\neg G^{k'}_{\mathcal{L}}$ for at most one positive integer k'; and

(b) each of $y_1, \ldots, y_n$ occurs in the RHS of the equations (i).

A *strictly normal equality* formula is a disjunction of strictly simple equality formulas.

Note that a strictly simple equality formula is equivalent to a simple equality formula because $G_{\mathcal{L}}^k$ is $\exists y_1, \dots, y_k \big(R_{\mathcal{L}}(y_1) \wedge \dots \wedge R_{\mathcal{L}}(y_k) \wedge \bigwedge_{i \neq j} y_i \neq y_j\big)$.

We show later, in Lemma 3.11, how to replace a simple equality formula by a strictly normal equality formula. Assuming this, we now deal with negation.

LEMMA 3.8. *There is an algorithm for replacing the negation of a strictly simple equality formula F by an equivalent strictly normal equality formula.*

Proof. The algorithm operates recursively on the number of equations in F. If this is zero, then F is a conjunction of inequations $x_i \neq t(\underline{x})$, $R_{\mathcal{L}}(x_i)$, $G_{\mathcal{L}}^k$, $\neg\, G_{\mathcal{L}}^{k'}$; and $\neg\, F$ is equivalent to a disjunction of equations $x_i = t(\underline{x})$, $\neg\, R_{\mathcal{L}}(x_i)$, $\neg\, G_{\mathcal{L}}^k$, $G_{\mathcal{L}}^{k'}$ which are all strictly simple equality formulas except $\neg\, R_{\mathcal{L}}(x_i)$, which is equivalent to the strictly normal equality formula

$$\bigvee_{f \in \mathcal{L}} \exists y_i, \dots, y_r \big(x_i = f(y_1, \dots, y_r)\big).$$

If the number of equations in F is non-zero, we write F as

$$\exists \underline{y}, \underline{z}\big(x_1 = t(\underline{x}, \underline{y}) \wedge K\big)$$

where the variables in t are all of the $\underline{y}$ and some (or none) of the $\underline{x}$ but none of the $\underline{z}$. Using Lemma 3.4, we replace $\neg\, F$ by

$$\forall \underline{y}\big(x_1 \neq t(\underline{x}, \underline{y})\big) \vee \exists \underline{y}\big(x_1 = t(\underline{x}, \underline{y}) \wedge \neg \exists \underline{z} K\big).$$

By Lemma 3.10 below, the first disjunct here can be reduced to strictly normal form, and, by induction on the number of equations, so can the second, because if $\exists \underline{u} K'$ is a strictly simple equality formula, and x_1 does not occur in K', so is

$$\exists \underline{y} \exists \underline{u}\big(x_1 = t(\underline{x}, \underline{y}) \wedge K'\big).$$

Assuming Lemmas 3.10 and 3.11 below, we can now state:

THEOREM 3.9. *There is an algorithm for replacing every equality formula by an equivalent strictly normal equality formula.*

LEMMA 3.10. *There is an algorithm which will produce an equivalent strictly normal equality formula F^* for any formula F of the form*

$$\forall \underline{z}\big(x' \neq t(\underline{x}, \underline{z})\big)$$

where $\underline{x} = x_1, \dots, x_n$, $\underline{z} = z_1, \dots, z_m$ are the distinct variables in $t(\underline{x}, \underline{z})$. Each of the strictly simple disjuncts of F^* either has no solved variables or has x' as the only solved variable.

Proof. If t is x', then $F*$ is *false*; if t is a non-variable term containing x', then F^* is *true*. Otherwise the algorithm will be defined recursively on the depth of t. If t has depth zero, i.e., is a variable, then F is either $x' \neq x_1$ which is left alone or $\forall z_1(x' \neq z_1)$, which is replaced by *false*. If t is $f(t_1, \ldots, t_r)$, we first replace F by

$$N_f(x') \vee \exists y_1 \ldots y_r(x' = f(y_1, \ldots, y_r) \wedge \forall \underline{z}(y_1 \neq t_1 \wedge \ldots \wedge y_r \neq t_r))$$

(where $y_1, \ldots, y_r$ are new variables), which is equivalent to it by Lemma 3.5. Now replace $N_f(x')$ by the strictly normal equality formula

$$R_{\mathcal{L}}(x') \vee \bigvee_{g \in \mathcal{L}, g \neq f} \exists y_1, \ldots, y_s(x' = g(y_1, \ldots, y_s))$$

to which, by $CET_{\mathcal{L}}$, F is equivalent. Now by Lemma 3.4,

$$\forall \underline{z}(y_1 \neq t_1 \wedge \ldots \wedge y_r \neq t_r)$$

is equivalent to

$$\forall \underline{z}^1(y_1 \neq t_1) \vee \exists \underline{z}^1(y_1 = t_1 \wedge \forall \underline{z}^2(y_2 \neq t_2 \wedge \ldots \wedge y_r \neq t_r))$$

where $\underline{z}^1$ are the variables in $\underline{z}$ actually occurring in t_1 and $\underline{z}^2$ the remaining variables of $\underline{z}$. Since $t_1, \ldots, t_r$ have lower depth than t, we can call the present algorithm to put $\forall \underline{z}^1(y_1 \neq t_1)$ into strictly normal form; and by an algorithm operating recursively on r, we can put the whole formula into strictly normal form, with $y_1, \ldots, y_r$ the only possible solved variables. To see this, we observe that if $\exists \underline{u} K$ is a strictly simple equality formula with $y_2, \ldots, y_r$ as the only possible solved variables (and not containing y_1) then

$$\exists \underline{z}^1 \exists \underline{u}(y_1 = t_1 \wedge K)$$

is a strictly simple equality formula with $y_1, \ldots, y_r$ as the only possible solved variables. A similar argument shows that replacing $\forall \underline{z}(y_1 \neq t_1 \vee \ldots \vee y_r \neq t_r)$ by this equivalent strictly normal equality formula in

$$\exists y_1, \ldots, y_r(x' = f(y_1, \ldots, y_r) \wedge \forall \underline{z}(y_1 \neq t_1 \vee \ldots \vee y_r \neq t_r))$$

and eliminating by substitution any of $y_1, \ldots, y_r$ which are solved, gives a formula F^* with the desired properties. (If $R_{\mathcal{L}}(y_i)$ occurs and y_i is replaced by a non-variable term, then $R_{\mathcal{L}}(y_i)$ and the whole disjunct containing it is replaced by *false*). (The formulas G^k do not occur in F^*.)

LEMMA 3.11. *There is an algorithm for replacing a simple equality formula by an equivalent strictly normal equality formula.*

Proof. We achieve the requirements (a) (i)–(iv) of Definition 3.7 in order.
(a) (i) Let the given simple equality formula S be

$$\exists y_1, \dots, y_j(E_1 \wedge \dots \wedge E_n).$$

First use Lemma 3.2 to put the equations in S into solved form $v_1 = t_1 \wedge \dots \wedge v_k = t_k$. Replace each v_i by t_i in the rest of S. If one of the v's is a y, say v_1 is y_1, then delete the equation $v_1 = t_1$ and delete $\exists y_1$ from the quantified prefix. If one of the t's is a y, say $t_1 = y_1$, then replace y_1 by v_1, delete the equation $v_1 = y_1$, and delete $\exists y_1$ from the prefix.

(ii) If there is an inequation $t \neq t'$ where neither t nor t' are variables, then put the equation $t = t'$ into solved form $v_1 = t_1 \wedge \dots \wedge v_k = t_k$ and replace S by $\bigvee_{i=1}^{k} S_i$ where S_i is obtained from S by replacing $t \neq t'$ by $v_i \neq t_i$.

(iii) If any $R_{\mathcal{L}}(t)$ with t a non-variable is now present (as a result of the substitutions of (i)), replace it and hence the whole formula by *false*.

(iv) If there is more than one $G^k_{\mathcal{L}}$, just take the one with greatest k; if there is more than one $\neg G^{k'}_{\mathcal{L}}$, take the one with least k'.

(b) Let $y_1, \dots, y_h$ be the *irrelevant variables,* i.e., those of $y_1, \dots, y_s$ which do not occur in any of the equations. If any of these, say y_1, does not have a corresponding conjunct $R_{\mathcal{L}}(y_1)$ in S, then we may simply delete $\exists y_1$ from the prefix and delete all inequations containing y_1. This is because Lemma 3.3 implies that an equation $t(y_1, y_2, \dots, y_s, x_1, \dots, x_r) = v$ has at most one solution for y_1, so the only condition y_1 has to satisfy is to be different from a finite number of elements: such a y_1 exists because $CET_{\mathcal{L}}$ implies there are infinitely many elements in the domain. At least it does provided there is at least one function symbol of non-zero arity in $\mathcal{L}$; the case where $\mathcal{L}$ consists only of constants $a_1, \dots, a_d$ needs a different treatment. In this case, we add the valid disjunction $R_{\mathcal{L}}(y_1) \vee y_1 = a_1 \vee \dots y_1 = a_d$; move the disjunction sign to the outside of the formula; and in the disjunct containing $y_1 = a_j$, replace y_1 by a_j and delete $\exists y_1$ from the prefix. If this replacement gives rise to $R_{\mathcal{L}}(a_j)$, this is replaced by *false*; if it results in an inequation $a_j \neq t$, this is left alone if t is a variable, replaced by *false* if t is a_j, and *true* if t is a non-variable term other than a_j (i.e., some a_i with $i \neq j$).

So we may assume S contains $R_{\mathcal{L}}(y_1) \wedge \dots \wedge R_{\mathcal{L}}(y_h)$. We can therefore delete any inequation $y_i \neq t$ where t is not a variable, since this is a consequence of $R_{\mathcal{L}}(y_i)$. So the inequations involving $y_1, \dots, y_h$ are of the forms $y_i \neq y_j$, $y_i \neq v$, where $i, j \leq h$, or $v \neq t(\underline{x}', \underline{y}')$, where v is a *relevant* variable, i.e., one of the variables $y_{h+1}, \dots, y_s$, $x_1, \dots, x_r$, where $\underline{x}', \underline{y}'$ denote subsets of $\underline{x}, \underline{y}$ and t is a non-variable term. The next step is to remove the last type of inequation; let us call them *indirect inequations.* We may write one of these as $v \neq t(y_1, \dots, y_a, \underline{v})$ where $y_1, \dots, y_h$ are renumbered so that $y_1, \dots, y_a$ are those actually occurring in t and $\underline{v}$ is a vector of relevant variables. By Lemma 3.5 this may be replaced by

$$\forall y'_1,\ldots,y'_a(v \neq t(y'_1,\ldots,y'_a,\underline{v})) \vee \exists y'_1,\ldots,y'_a(v = t(y'_1,\ldots,y'_a,\underline{v}) \wedge \bigvee_{s=1}^{a} (y'_j \neq y_j)).$$

We may move the disjunction signs to the outside of the whole formula so we can consider separately the replacement of $v \neq t(y_1,\ldots,y_a,\underline{v})$ by

$$\text{A: } \forall y'_1,\ldots,y'_a(v \neq t(y'_1,\ldots,y'_a,\underline{v}))$$

and by

$$\text{B: } \exists y'_1,\ldots,y'_a(v = t(y'_1,\ldots,y'_a,\underline{v}) \wedge y'_j \neq y_j).$$

In the first case we use Lemma 3.10 to replace A by a strictly normal equality formula

$$S'_1 \vee \ldots \vee S'_b$$

where the S'_i are strictly simple equality formulas. Moving the disjunction signs to the outside again we can consider separately the replacement of A by each S'_i. After changing to new variables if necessary we can move the existential quantifier prefix $\exists\underline{z}$ of S'_i to the beginning of the resulting formula. The only solved variable of S'_i can be v, so its solved form $v = t'(\underline{v},\underline{z})$ must contain all the $\underline{z}$ (and a subset $\underline{v}$ of the relevant variables). Since v occurs in an inequation of S, it is not a solved variable of S and neither are the $\underline{v}$. So v is either one of the x which is not a solved variable of S, or one of the relevant y, i.e., one which occurs in at least one of the equations of S. In both cases we substitute $t'(\underline{v},\underline{z})$, for all occurrences of v. In the first case we retain the equation $v = t'(\underline{v},\underline{z})$, making v a new solved variable; in the second case we delete this equation and delete $\exists v$ from the quantifier prefix. In both cases the variables $\underline{z}$ are still relevant, since they occur in an equation, so we have not introduced any new irrelevant variables. The replacement of v by $t'(\underline{v},\underline{z})$, if t' is not a variable, would convert $R_{\mathcal{L}}(v)$, if it were present, into $R_{\mathcal{L}}(t'(\underline{v},\underline{z}))$, which is replaced by *false*, and it converts an inequation $v \neq t_1$ into $t'(\underline{v},\underline{z}) \neq t_1$. If t_1 is a relevant variable, this can be left alone, if t_1 is an irrelevant variable y_i, it can be replaced by *false*, as a consequence of $R(y_i)$, otherwise it is reduced as in (a)(ii) above to a disjunction of inequations with variables on the LHS and the disjunction sign taken to the outside of the whole formula. Each of the resulting disjuncts contains just one of these inequations, which can only involve an irrelevant variable if t_1 does, i.e., if $v \neq t_1$ was an indirect inequation. So this does not increase the number of indirect equations and the net result is that we have got rid of one of these. When B instead of A is the replacement, the argument is similar.

So by proceeding in this way, we can get rid of all the indirect inequations. We now turn to the inequations of the form $y_i \neq v$ where v is a relevant variable and y_i an irrelevant variable. Since $R_{\mathcal{L}}(y_i)$ is already in the conjunction, we can use

$$R_{\mathcal{L}}(y_i) \wedge y_i \neq v \longleftrightarrow (R_{\mathcal{L}}(y_i) \wedge R_{\mathcal{L}}(v) \wedge y_i \neq v) \vee (R_{\mathcal{L}}(y_i) \wedge \neg R_{\mathcal{L}}(v))$$

and

$$\neg R_{\mathcal{L}}(v) \longleftrightarrow \bigvee_{f \in \mathcal{L}} \exists u_1, \ldots, u_r(v = f(u_1, \ldots, u_r))$$

to replace the formula by a disjunction of formulas in which $y_i \neq v$ is replaced by $y_i \neq v \wedge R_{\mathcal{L}}(v)$ or by $\exists u_1, \ldots, u_r\big(v = f(u_1, \ldots, u_r)\big)$. In the latter case, we replace v by $f(u_1, \ldots, u_r)$, add $v = f(u_1, \ldots, u_r)$ as a new solved equation if v is an x, or delete this equation and the existential quantifier $\exists v$ if v is a y. Inequations $v \neq t_1$ which become $f(u_1, \ldots, u_r) \neq t_1$ are dealt with as above (no new indirect equations can result).

We have now reduced the part of the formula containing the irrelevant variables to

$$\exists y_1, \ldots, y_h\big(R_{\mathcal{L}}(y_1) \wedge \ldots \wedge R_{\mathcal{L}}(y_h) \wedge R_{\mathcal{L}}(v_1) \wedge \ldots \wedge R_{\mathcal{L}}(v_k) \wedge Y \wedge V\big)$$

which we abbreviate to

$$\exists \underline{y}(R_{\mathcal{L}}(\underline{y}) \wedge R_{\mathcal{L}}(\underline{v}) \wedge Y \wedge V)$$

where Y is the conjunction of all inequations of the form $y_i \neq y_j$ between irrelevant variables and all inequalities of the form $y_i \neq v_j$ between irrelevant and relevant variables, and V is the conjunction of all inequations of the form $v_i \neq v_e$ between those relevant variables $v_1, \ldots, v_k$ which occur in such inequations $y_i \neq v_j$. The final step is to banish the irrelevant variables to contexts of the form $G_{\mathcal{L}}^k$ (i.e., $\exists y_1, \ldots, y_k\big(R_{\mathcal{L}}(y_1) \wedge \ldots \wedge R_{\mathcal{L}}(y_k) \wedge \bigwedge_{i \neq j} y_i \neq y_j\big)$). It is convenient to express this step in terms of graph-colorings. Let $\mathcal{G}$ be the graph whose vertices are $y_1, \ldots, y_h, v_1, \ldots, v_k$ which has an edge between vertices u and u' when $u \neq u'$ is in $Y \wedge V$, and $\mathcal{G}_v$ the subgraph of $\mathcal{G}$ with vertices $v_1, \ldots, v_k$. Consider each coloring C of $\mathcal{G}_v$ (two colorings which differ only in the names of the colors being considered identical). Let V_C be the conjunction which for each $i, j \leq k$ contains $v_i = v_j$ if v_i, v_j have the same color in C and contains $v_i \neq v_j$ if v_i, v_j have different colors. Let $n(C)$ be the least number of colors needed for a coloring C' of $\mathcal{G}$ extending the coloring C of $\mathcal{G}_v$. Then

$$\exists \underline{y}(R_{\mathcal{L}}(\underline{y}) \wedge R_{\mathcal{L}}(\underline{v}) \wedge Y \wedge V_C) \longleftrightarrow G_{\mathcal{L}}^{n(C)} \wedge V_C \wedge R_{\mathcal{L}}(\underline{v}).$$

This is true because a model for the LHS gives a coloring of $\mathcal{G}$ which is an extension of C by giving two vertices the same color iff they are equal in the model, so it needs at least $n(C)$ colors, i.e., has at least $n(C)$ different elements, all of them roots. Conversely, if the RHS is true in a model of $CET_{\mathcal{L}}$, then we get a model of $R_{\mathcal{L}}(\underline{y}) \wedge R_{\mathcal{L}}(\underline{v}) \wedge Y \wedge V \wedge V_C$ if we choose $y_1, \ldots, y_k$ so that they are roots and so that y_i and y_j or y_i and v_j are equal iff they have the same color in C'. This is possible because $G_{\mathcal{L}}^{n(C)}$ says there are enough roots to do this. Now

$$V \longleftrightarrow \bigvee_C V_C$$

where the disjunction is taken over all colorings C of $\mathcal{G}_v$. This is so because if C is a coloring of $\mathcal{G}_v$, then V_C implies V, and if V is true in a model, then V_C is true for the coloring C which gives v_i, v_j different colors iff $v_i \neq v_j$ is true in the model. So we obtain

$$\exists \underline{y}(R_{\mathcal{L}}(\underline{y}) \wedge R_{\mathcal{L}}(\underline{v}) \wedge Y \wedge V) \longleftrightarrow \bigvee_C (G_{\mathcal{L}}^{n(C)} \wedge V_C \wedge R_{\mathcal{L}}(\underline{v})),$$

and replacing the LHS by the RHS and taking the disjunction outside gives the required disjunction of strictly simple equality formulas, since we have now eliminated irrelevant variables. ⊣

This completes the proof of Lemma 3.11 and Theorem 3.9. We now relate our normal form of Definitions 3.1, 3.7 to those of Chan, Przymusinski, Maher, and Malcev.

The formula $R_{\mathcal{L}}(x)$ used above is a conjunction of quantified inequations of the simplest possible form

$$\forall y_1, \ldots, y_r(x \neq f(y_1, \ldots, y_r)),$$

so the type (iii) conjunct in our simple equality and strictly simple equality formula could be replaced by

'a quantified inequation $\forall(v \neq t)$ for some variable v'

or

'a quantified inequation $\forall(x_i \neq t)$ or $\forall(y_i \neq t)$',

respectively, as in Chan's normal form. Przymusinski does not allow the inequations $y_i \neq t$ and quantified inequations $\forall(y_i \neq t)$ in his strictly simple equality formula. We can also get rid of these because y_i must occur in some equation $x = t'(y_i, \underline{x}, \underline{y})$, so that when conjuncted with this, $\forall \underline{z}(y_i \neq t(\underline{x}, \underline{y}, \underline{z}))$ is equivalent to $\forall \underline{z}(x \neq t'(t(\underline{x}, \underline{y}, \underline{z}), \underline{x}, \underline{y}))$ by Lemma 3.3. After doing this, our strictly normal form gives that of Chan and Przymusinski described above, when the reduction of E is made on the hypothesis $CET_{\mathcal{L}'}$ for some infinite language $\mathcal{L}'$ containing $\mathcal{L}(\{E\})$. This is because if we carry out our reduction using $\mathcal{L} = \mathcal{L}(\{E\})$, the formulas $G_{\mathcal{L}}^k$ are true for all k, since $CET_{\mathcal{L}'}$ implies the existence of infinitely many $\mathcal{L}$-roots. Indeed, all that is necessary for this reduction to be possible is that $\mathcal{L}' - \mathcal{L}(\{E\})$ should contain one function symbol of arity greater than zero or at least k_0 constants, where k_0 is the largest k for which $G_{\mathcal{L}}^k$ or $\neg G_{\mathcal{L}}^k$ occurs in the equivalent strictly normal equality formula. In the latter case, the k_0 new constants provide the k_0 roots; in the former case, if the new function symbol has arity r and α is any element, then $g(\alpha), g^2(\alpha), \ldots, g^{k_0}(\alpha)$ provide k_0 roots, where $g(x) = f(x, \ldots, x)$. If, for example, $\mathcal{L}'$ was the language determined by a logic program and E arose from the solution of a particular query using Chan's constructive negation, it would be rather easy to check at each stage that there was a spare function in $\mathcal{L}'$. However if there was not, and one had to rely on k_0 spare constants, it would be hard to make use of this simplified reduction procedure, unless there is a way of determining k_0 without carrying out the more complicated reduction based on $\mathcal{L}(\{E\})$.

For this reduction of an equality formula E on the basis of $CET_{\mathcal{L}'}$ for an infinite language $\mathcal{L}'$, Maher (*1988,* Theorem 6) gives a normal form consisting of a boolean combination of *basic formulas.* These are of the form

$$\exists y_1, \ldots, y_s(x_1 = t_1 \wedge \ldots \wedge x_m = t_m)$$

where $y_1, \ldots, y_s$ and $x_1, \ldots, x_m$ are distinct sets of variables and $x_1, \ldots, x_m$ do not occur in $t_1, \ldots, t_m$; so they are the answer formulas corresponding to the usual answer substitutions given by SLD-resolution using idempotent mgu. This can be converted to the Chan/Przymusinski strict normal form by putting it into disjunctive normal form and reducing negations of basic formulas to strictly normal form as in Lemma 3.10 above.

Malcev (*1971,* Theorem 3) gives, for the case of finite $\mathcal{L}$, a normal form differing from the strictly normal form of Definition 3.7 in allowing $N_f(x_i)$, $N_f(y_i)$ instead of $R_{\mathcal{L}}(x_i), R_{\mathcal{L}}(y_i)$ and in not imposing condition (b), that each of $y_1, \ldots, y_s$ should occur in the RHS of an equation $x_i = t_i$, i.e., should be what we have called relevant variables. The elimination of irrelevant variables makes the Chan/Przymusinski strict normal form of an answer much more intelligible; it is the crux (Lemma 3.11(b)) of the proof of our Theorem 3.9. In two cases it is easier; when $\mathcal{L}$ is infinite, they can be removed at once, and where E is a closed formula, in which case the graph $\mathcal{G}_v$ is empty. This last case was already dealt with by Malcev (*1971,* Theorem 4).

THEOREM 3.12. *The theory $CET_{\mathcal{L}}$ is complete iff $\mathcal{L}$ contains infinitely many function symbols. If $\mathcal{L}$ contains only finitely many function symbols, a complete theory is obtained by adding to $CET_{\mathcal{L}}$ the axiom $G^k_{\mathcal{L}} \wedge \neg G^{k+1}_{\mathcal{L}}$ which says there are exactly k $\mathcal{L}$-roots. In particular, adding $DCA_{\mathcal{L}}$, which says there are no $\mathcal{L}$-roots, gives a complete theory.*

Proof. Let E be a closed $\mathcal{L}$-equality formula and $\mathcal{L}_0 = \mathcal{L}(\{E\})$. The strictly normal form of E obtained using $\mathcal{L}_0$ contains only formulas of the form $G^k_{\mathcal{L}_0}$, $\neg G^{k'}_{\mathcal{L}_0}$. If $\mathcal{L}$ contains $\mathcal{L}_0$ and has infinitely many function symbols then, as noted above in Theorem 2.1(ii), $CET_{\mathcal{L}}$ implies that all $G^k_{\mathcal{L}_0}$ are true. If $\mathcal{L}$ contains only finitely many function symbols, the strictly normal form of E obtained using $\mathcal{L}$ contains only formulas of the form $G^k_{\mathcal{L}}, \neg G^{k'}_{\mathcal{L}}$. ⊣

Note that we are counting constants as 0-ary function symbols. These completeness results are also immediate consequences of Malcev's Theorem 4. Other proofs of the completeness of $CET_{\mathcal{L}}$ for infinite $\mathcal{L}$ have been given by Kunen (*1988*) and Maher (*1988*). The latter also gives another proof of the completeness of $CET_{\mathcal{L}}$ plus the Domain Closure Axiom for finite $\mathcal{L}$ (except in the case where $\mathcal{L}$ is singular, i.e., contains only one function symbol, which was left as an open problem, but is covered by the arguments above).

The case where $\mathcal{L}$ is taken to be $\mathcal{L}(P)$ and the Domain Closure Axiom is assumed is of interest because it includes the usual Herbrand models whose domain is the set of ground terms built up from the function symbols in P. (This is studied in some detail by Malcev, *1971*.) In this case, the strictly normal form and the algorithm for obtaining it can be greatly simplified. The Domain Closure Axiom implies that $R_{\mathcal{L}(P)}(x)$ is false for all x and all $G^k_{\mathcal{L}(P)}$ are false. So the strictly simple equality formulas take the form

$$\exists y_1,\ldots,y_s(\bigwedge_{i\in A} x_i = t_i \bigwedge_{j\in B} x_j \neq r_i \bigwedge_{k\in C} y_k \neq s_k)$$

where $x_1,\ldots,x_n$ are the variables, $A, B \subseteq \{1,\ldots,n\}$, $C \subseteq \{1,\ldots,n\}$, t_i is either a non-variable term (in $x_1,\ldots,x_n,y_1,\ldots,y_s$) or is x_j for $j \neq i$, the x_i for $i \in A$ do not occur anywhere in the formula except on the LHS of $x_i = t_i$, the r_j, s_k are terms in $x_1,\ldots,x_n,y_1,\ldots,y_s$ and each of $y_1,\ldots,y_s$ occurs in one of the equations $x_i = t_i$. (This could also be derived as described above from the reduction to boolean combinations of basic formulas, given (for non-singular $\mathcal{L}$) by Maher (*1988*, Theorem 11) for the case when the Domain Closure Axiom is assumed.)

This reduction can be used to extend the result used by Sato and Tamaki (*1988*). They give a method for transforming a logic program with first order formulas in the bodies of clauses to a definite program which is in some sense equivalent to it for Herbrand models. In doing this, they need to give definite programs for representing quantified inequations $\forall \underline{y}(s(\underline{x},\underline{y}) \neq t(\underline{x},\underline{y}))$. They do this by using the fact that this is a recursive relation in $\underline{x}$ over the Herbrand universe, so there exists a definite program with a predicate $p(\underline{x})$ such that for ground terms $\underline{t}_0$, $p(\underline{t}_0)$ succeeds (fails) iff $\forall \underline{y}(s(\underline{t}_0,\underline{y}) \neq t(\underline{t}_0,\underline{y}))$ is true (false) over the Herbrand universe. But they seem unwilling to make any assumptions about what happens for non-ground terms since they insert a proviso 'it is assumed that every goal of the form $\forall y(x \neq t)$ has no free variables at the time of failure'. However, we can show that for any relation defined by an equality formula $E(x)$ there is a definite program which represents it completely for ground terms and any results it gives for non-ground terms are also correct. More precisely:

THEOREM 3.13. *If $\mathcal{L}$ is a finite language and $E(\underline{x})$ is any $\mathcal{L}$-equality formula, there exists a definite clause program P_E with predicates p_E,* equal, not_equal *such that*

(i) *for ground* $\underline{t}_0$, $p_E(\underline{t}_0)$ *succeeds in* P_E *if* $CET_{\mathcal{L}} \cup \{DCA_{\mathcal{L}}\} \vdash E(\underline{t}_0)$,

(ii) *for ground* $\underline{t}_0$, $p_E(\underline{t}_0)$ *fails in* P_E *if* $CET_{\mathcal{L}} \cup \{DCA_{\mathcal{L}}\} \vdash \neg E(\underline{t}_0)$,

(iii) *the sentences obtained by replacing* p_E, *equal*, *not_equal by* $E, =, \neq$ *in the completed definitions of* p_E, *equal*, *not_equal in* $\mathsf{comp}_{\mathcal{L}}(P_E)$ *are consequences of* $CET_{\mathcal{L}} \cup \{DCA_{\mathcal{L}}\}$,

(iv) *if* $p_E(\underline{t})$ *succeeds in* P_E *with answer* θ, *then* $CET_{\mathcal{L}} \cup \{DCA_{\mathcal{L}}\} \vdash \forall E(\underline{t}\theta)$,

(v) *if* $p_E(\underline{t})$ *fails in* P_E, *then* $CET_{\mathcal{L}} \cup \{DCA_{\mathcal{L}}\} \vdash \forall \neg E(\underline{t})$.

Proof. First note that (iv), (v) follow from (iii), the soundness of SLD-resolution with respect to P and the fact that $\mathsf{comp}(P) \vdash P$. If $\mathcal{L}$ has function symbols f_i of arity a_i for $i = 1, \ldots, l$, a suitable program P_E may be defined as follows:

$$equal(x, x) \longleftarrow$$

$$not\_equal(x, x') \longleftarrow equal(x, f_i(y_1, \ldots, y_{a_i})), equal(x', f_{i'}(y'_1, \ldots, y'_{a_{1'}})), \quad \textit{for all } i, i' = 1, \ldots, l, i \neq i'.$$

$$not\_equal(x, x') \longleftarrow equal(x, f_i(y_1, \ldots, y_{a_i})), equal(x', f_1(y'_1, \ldots, y'_{a_i})), \quad not\_equal(y_j, y'_j), \ \textit{for all } i = 1, \ldots, l \textit{ with } a_i \geq 1, \ \textit{all } j = 1, \ldots, a_i.$$

If E is the strictly simple equality formula

$$\exists y_1, \ldots, y_s(\bigwedge_{i \in A} x_i = t_i \bigwedge_{j \in B} x_j \neq r_i \bigwedge_{k \in C} y_k \neq s_k)$$

then the clause defining p_E is

$$p_E(x_1, \ldots, x_n)\theta \longleftarrow \bigwedge_{j \in B} not\_equal(x_j, r_j), \bigwedge_{k \in C} not\_equal(y_k, s_k),$$

where $\theta = \{x_i/t_i : i \in A\}$. If E is a disjunction

$$E^1 \vee E^2 \vee \ldots E^m$$

of simple equality formulas, there are m clauses for p_E obtained in this way from $E^1 \vee E^2 \vee \ldots E^m$. It is easily verified that (i), (ii), (iii), (iv) are satisfied for $=, \neq$, and finally for E. ⊣

It is not possible to get completeness for non-ground queries, i.e., to get (i) for non-ground $\underline{t}_0$, e.g., $x \neq f(x)$ is a consequence of CET but *not_equal*$(x, f(x))$ succeeds with answer θ only for θ which ground x.

Without the Domain Closure Axiom, it is not possible to deal with $\neq$ in this way and obtain a definite clause program, but by treating $x \neq y$ as $\neg(x = y)$ one can obtain a normal program which represents $E(x)$ in a stronger sense.

THEOREM 3.14. *If $E(x)$ is an $\mathcal{L}$-equality formula, there exists a normal program P_E with predicate p_E such that*

$$\mathsf{comp}_{\mathcal{L}}(P_E) \vdash \forall \underline{x}(p_E(\underline{x}) \longleftrightarrow E(\underline{x})).$$

P_E may be taken to be hierarchic with 2-levels if $\mathcal{L}$ is infinite, with 3-levels if $\mathcal{L}$ is finite.

Proof. The existence of such a P_E follows by the usual technique (*Lloyd 1987*, chapter 4), for translating first order logic programs into normal ones, starting from the fact that the program $equal(x, x) \longleftarrow$ represents $=$ in this sense. The reduction to a low level hierarchic program is obtained by using the normal forms above. If $\mathcal{L}$ is infinite, the simplest one to use is Maher's, a boolean combination of basic formulas. Write $E(\underline{x})$ in disjunctive normal form as a disjunction of basic formulas and negations of basic formulas. The disjunction is dealt with as above by taking clauses corresponding to each disjunct. If

$$E(x) \longleftrightarrow b_1 \wedge \ldots \wedge b_r \wedge \neg b'_i \wedge \ldots \wedge \neg b'_{r'},$$

where $b_1, \ldots, b_r, b'_1, \ldots, b'_{r'}$ are basic formulas, then P_E has a clause

$$p_E(\underline{x}) \longleftarrow p_{b_1}(\underline{x}), \ldots, p_{b_r}(\underline{x}), \neg p_{b_1}(\underline{x}), \ldots, \neg p_{b'_{r'}}(\underline{t})$$

and one clause for each of $p_{b_1}, \ldots, p_{b_r}, p_{b'_1}, \ldots, p_{b'_{r'}}$; if

$$b_i(\underline{x}) \longleftrightarrow \exists \underline{y}(x_1 = t_1 \wedge \ldots \wedge x_m = t_m),$$

then the clause for p_{b_i} is

$$p_{b_i}(t_1, \ldots, t_m, x_{m+1}, \ldots, x_n) \longleftarrow .$$

If we use instead the Chan normal form

$$E(\underline{x}) \longleftrightarrow \exists y_1, \ldots, y_s\Big(\bigwedge_{i \in A} x_i = t_i \bigwedge_{j \in B} \forall \underline{u}_j(x_j \neq r_j) \bigwedge_{k \in C} \forall \underline{v}_k(y_k \neq s_k)\Big),$$

Then letting $\theta = \{x_1/t_1 : i \in A\}, \varphi_j = \{x_j/r_j\}, \psi_k = \{y_k/s_k\}$, the program P_E is

$$\begin{aligned} p_E(x_1, \ldots, x_n)\theta \longleftarrow & \bigwedge_{j \in B} \neg q_j(x_1, \ldots, x_n, y_1, \ldots, y_s) \bigwedge_{k \in C} \neg r_k(x_1, \ldots, x_n, b_1, \ldots, b_s) \\ & q_j(x_1, \ldots, x_n, y_1, \ldots, y_s)\varphi_j \longleftarrow \\ & r_k(x_1, \ldots, x_n, b_1, \ldots, b_s)\psi_k \longleftarrow . \end{aligned}$$

If $\mathcal{L}$ is finite, our normal form above needs a third level to deal with the closed formulas $\neg G^k_{\mathcal{L}}$. ⊣

This shows that when Chan's constructive negation gives an answer for all queries for a program P (i.e., produces an equivalent equality formula), there is a 2-level hierarchic program which is equivalent to P in the sense that its completion is a conservative extension of comp(P). Since it is easy to show that for a hierarchic program the completed definition of each predicate is equivalent to an equality formula, it follows that every hierarchic program is in this sense equivalent to a 2 or 3-level one.

If $\mathcal{L}$ contains no function symbols of arity greater than zero, i.e., consists of constants $a_1, \ldots, a_d$, then the conjuncts in a strictly simple equality formula can be reduced to $x_i = x_j, x_i = a_j, x_i \neq x_j, x_i \neq a_j, G^k, \neg G^k$, where G^k says there are at least k elements other than $a_1, \ldots, a_d$.

4. 3-VALUED SEMANTICS

Kunen (*1988*) uses a 3-valued logic introduced by Fitting (*1985*) based on the Kleene truth tables. Here a boolean combination of the truth values ***t***, ***f***, ***u*** is ***t*** (resp. ***f***) iff all possible ways of replacing the ***u*** by ***t*** or ***f*** gives the value ***t*** (resp. ***f***) in ordinary 2-valued logic; similarly for the quantifiers, treating these as infinite conjunctions or disjunctions; thus $\exists\varphi(x)$ gets the value ***t*** if there is at least on element for which φ is ***t***, it gets value ***f*** if φ is ***f*** for all elements, and the value ***u*** otherwise. There is one exception to this, the equivalence $\longleftrightarrow$ used in forming the completed definitions of predicates in comp(P) gives $\varphi \longleftrightarrow \psi$ the value ***t*** iff φ, ψ have the same truth value and ***f*** otherwise. Equality is taken to be the 2-valued relation of identity. He uses a language $\mathcal{L}_\infty$ with infinitely many function symbols of all arities. A fundamental result which Kunen gives without proof is that SLDNF-resolution is sound for the completion of a program in 3-valued logic, i.e., that

> if $P \cup \{\longleftarrow Q\}$ has an SLDNF-refutation with answer θ, then $\text{comp}_\mathcal{L}(P) \models_3 \forall Q\theta$,
>
> if $P \cup \{\longleftarrow Q\}$ has a finitely failed SLNDF-tree, then $\text{comp}_\mathcal{L}(P) \models_3 \forall\neg Q$.

Here $X \models_3 Y$ means that Y is true in all 3-valued models of X. One would expect this to be independent of $\mathcal{L}$, i.e., true for all $\mathcal{L}$ containing $\mathcal{L}(P \cup \{Q\})$; and this follows from Theorem 5.4 of *Shepherdson 1989*; indeed it is true if the consequence relation $\models_3$ is weakened to a relation $\vdash_{3I}$ based on axioms and rules of inference which are valid both in intuitionistic and 3-valued logic. This result helps to explain why SLDNF-resolution is usually incomplete for $\text{comp}_\mathcal{L}(P)$ using classical 2-valued logic; for $P \cup \{\longleftarrow Q\}$ to have an SLDNF-resolution it is not enough that $\text{comp}_\mathcal{L}(P) \models \forall Q\theta$ in classical 2-valued logic, this must also hold using the weaker consequence relations of 3-valued or intuitionistic logic (so using either of these would give a closer fit between SLDNF-refutation and the Clark completion).

The main result of *Kunen 1988* is that 3-valued semantics based on $\mathrm{comp}_{\mathcal{L}}(P)$ is equivalent to one defined by iterating a 3-valued analogue of the familiar T_P operator of van Emden and Kowalski. At stage n, for $n = 1, 2, \ldots$, some sentences become true (are given the value $\boldsymbol{t}$) and others becoming false (are given the value $\boldsymbol{f}$), the rest have the value $\boldsymbol{u}$. At each stage, truth and falsity are assigned first for ground atoms and extended to other sentences using the Kleene truth tables and *interpreting quantifiers as ranging over the Herbrand universe, i.e., the universe of $\mathcal{L}$-terms.* This makes the notion of truth and falsity at stage n dependent on $\mathcal{L}$; when necessary, we show this dependence by the subscript $\mathcal{L}$—'true$_{\mathcal{L}}$ at stage n'. At stage 0, all ground atoms have value $\boldsymbol{u}$ and all equations $s = t$ between ground terms are true if s, t are identical, false otherwise. At stage $n+1$, a ground atom a becomes true iff for some clause $\beta \longleftarrow \varphi$ of P there is a ground substitution σ such that $\alpha = \beta\sigma$ where $\varphi\sigma$ is true at stage n, and $\mathcal{L}$ becomes false iff for *all* clauses $\beta \longleftarrow \varphi$ of P and all ground substitutions σ, if $\alpha = \beta\sigma$, then $\varphi\sigma$ is false at stage n.

Kunen's Theorem 6.3 is that for any sentence φ the following are equivalent when $\mathcal{L}$ is $\mathcal{L}_\infty$:

(1) φ is true in every 3-valued model of $\mathrm{comp}_{\mathcal{L}}(P)$.

(2) φ becomes true$_{\mathcal{L}}$ at some stage.

He proves that (1) implies (2) by constructing *one* 3-valued model N for $\mathrm{comp}_{\mathcal{L}}(P)$, such that for every sentence φ, if φ is true in N then φ becomes true at some stage n. Examination of the construction of N shows that, for any $\mathcal{L}$, it has the stronger property that φ is true in N iff φ becomes true at some stage. He omits the details of the proof that (2) implies (1) but we must examine this to see how it depends on $\mathcal{L}$. I am very grateful to him for supplying the following argument. The first lemma holds for arbitrary $\mathcal{L}$ so we omit the subscript $\mathcal{L}$.

LEMMA 4.1. *With each formula φ with free variables $x_1, \ldots, x_r$ and each $n = 0, 1, 2,$ can be associated two equality formulas $T_n(\varphi)$, $F_n(\varphi)$ such that,*

(a) *for ground terms $t_1, \ldots, t_r$,*

$$\varphi(t_1, \ldots, t_r) \text{ is true at stage } n \text{ iff } T_n(\varphi)(t_1, \ldots, t_r) \text{ is true,}$$
$$\varphi(t_1, \ldots, t_r) \text{ is false at stage } n \text{ iff } F_n(\varphi)(t_1, \ldots, t_r) \text{ is true;}$$

and

(b) *for elements $a_1, \ldots, a_r$ of any 3-valued model M of* comp(P),

$$\text{if } T_n(\varphi)(a_1, \ldots, a_r) \text{ is true, then } \varphi(a_1, \ldots, a_r) \text{ is true,}$$
$$\text{if } F_n(\varphi)(a_1, \ldots, a_r) \text{ is true, then } \varphi(a_1, \ldots, a_r) \text{ is false.}$$

Proof. $T_n(\varphi), F_n(\varphi)$ are defined by induction on n, and, for each n, by induction on the structure of φ. If φ is an equation $s = t$, then $T_n(\varphi)$ is $s = t$ and $F_n(\varphi)$ is $s \neq t$. If φ is an atom $p(\underline{s})$ and the clauses with p in their heads are

$$p(\underline{t}_1) \longleftarrow \varphi_1, \ldots, p(\underline{t}_k) \longleftarrow \varphi_k$$

and $\underline{y}_i$ are the free variables in the ith clause, then $T_0(\varphi), F_0(\varphi)$ are both $\boldsymbol{f}$ and

$$T_{n+1}(\varphi) \text{ is } \bigvee_{i=1}^{k} \exists \underline{y}_i (\underline{s} = \underline{t}_i \wedge T_n(\varphi_i))$$

$$F_{n+1}(\varphi) \text{ is } \bigwedge_{i=1}^{k} \forall \underline{y}_i (\underline{s} \neq \underline{t} \vee F_n(\varphi_i)).$$

The boolean connectives and quantifiers are dealt with in the obvious way:

$$\begin{aligned} T_n(\neg\varphi) &\text{ is } F_n(\varphi), & F_n(\neg\varphi) &\text{ is } T_n(\varphi), \\ T_n(\varphi \wedge \psi) &\text{ is } T_n(\varphi) \wedge T_n(\psi), & F_n(\varphi \wedge \psi) &\text{ is } F_n(\varphi) \vee F_n(\psi), \\ T_n(\varphi \vee \psi) &\text{ is } T_n(\varphi) \vee T_n(\psi), & F_n(\varphi \vee \psi) &\text{ is } F_n(\varphi) \wedge F_n(\psi), \\ T_n(\varphi \longrightarrow \psi) &\text{ is } T_n(\varphi) \vee F_n(\varphi), & F_n(\varphi \longrightarrow \psi) &\text{ is } T_n(\varphi) \wedge F_n(\psi), \\ T_n(\forall x\varphi) &\text{ is } \forall x T_n(\varphi), & F_n(\forall x\varphi) &\text{ is } \exists x F_n(\varphi), \\ T_n(\exists x\varphi) &\text{ is } \exists x T_n(\varphi), & F_n(\exists x\varphi) &\text{ is } \forall x F_n(\varphi). \end{aligned}$$

It is easily proved, by induction on n and on the structure of φ that $T_n(\varphi)$, $F_n(\varphi)$ defined in this way satisfy the conditions of the lemma. ⊣

THEOREM 4.2.

(a) *If φ is an $\mathcal{L}$-sentence then, if $\mathcal{L}$ is infinite, φ is true in every 3-valued model of* $\mathsf{comp}_{\mathcal{L}}(P)$ *iff φ becomes true$_{\mathcal{L}}$ at some stage,*

(b) *If $\mathcal{L}$ is finite φ is true in every 3-valued model of* $\mathsf{comp}_{\mathcal{L}}(P) \cup \{DCA_{\mathcal{L}}\}$ *iff φ becomes true$_{\mathcal{L}}$ at some stage.*

Proof.

(a) The 'only if' half follows because, if φ is true in every 3-valued model of $\mathsf{comp}_{\mathcal{L}}(P)$, it is true in Kunen's model N, hence true at some stage. If φ becomes true at stage n, then by Lemma 4.1(a), $T_n(\varphi)$ is true at stage n. Now $T_n(\varphi)$ is an equality sentence, so it is either true or false already at stage 0, and it is true at stage 0 iff it is true in the Herbrand model of $CET_{\mathcal{L}}$ whose domain consists of all $\mathcal{L}$-terms. Now 3-valued models of $CET_{\mathcal{L}}$ are also 2-valued models since we interpret $=$ as 2-valued, so by the completeness of $CET_{\mathcal{L}}$, $T_n(\varphi)$ must be true in all 3-valued models of $CET_{\mathcal{L}}$, hence in all 3-valued models of $\mathsf{comp}_{\mathcal{L}}(P)$. So, by Lemma 4.1(b) φ is true in all 3-valued models of $\mathsf{comp}(P)$.

(b) The 'only if' half follows because $DCA_{\mathcal{L}}$ is true at stage 0, hence true in Kunen's model N. If φ becomes true at stage n, then as in (a), $T_n(\varphi)$ is true in the Herbrand model of $CET_{\mathcal{L}}$ whose domain consists of all $\mathcal{L}$-terms. This model satisfies $DCA_{\mathcal{L}}$, and $\mathsf{comp}_{\mathcal{L}}(P) \cup \{DCA_{\mathcal{L}}\}$ is complete, so the argument is completed as in (a). ⊣

COROLLARY 4.3. *If φ is an $\mathcal{L}(P)$ sentence, then*

(a) *if $\mathcal{L}$-$\mathcal{L}(P)$ contains a function of positive arity, or infinitely many constants, then φ is true in every 3-valued model of $\mathsf{comp}_{\mathcal{L}(P)}(P) \cup G^{\infty}_{\mathcal{L}(P)}$ iff φ becomes true$_{\mathcal{L}}$ at some stage,*

(b) *if $\mathcal{L}$-$\mathcal{L}(P)$ contains only k constants, then φ is true in every 3-valued model of $\mathsf{comp}_{\mathcal{L}(P)}(P) \cup \{G^{k}_{\mathcal{L}(P)}, \neg G^{k+1}_{\mathcal{L}(P)}\}$ iff φ becomes true$_{\mathcal{L}}$ at some stage.*

Proof. Since the equality relation in our 3-valued models is 2-valued, Theorem 2.1 applies to 3-valued models. So if $\mathcal{L}$ is infinite, (a) follows from (a) of Theorem 4.2 and (ii) of Theorem 2.1. If $\mathcal{L}$ is finite but $\mathcal{L}$-$\mathcal{L}(P)$ contains a function of positive arity, then (a) follows from (b) of Theorem 4.2 and (ii) of Theorem 2.1 together with the fact that the extension of M defined in the proof of (ii) does, in the case where $\mathcal{L}_{2'}$ is taken to be $\mathcal{L}_2$, satisfy $DCA_{\mathcal{L}_2}$ since all $\mathcal{L}_1$-roots are used up in the construction. (To see this, take $g(x) = f(x, \dots, x)$ where f is one of the new functions. By induction on d, one shows that at no stage can ρ_i be a $d-1$ ancestor of $g^d(\rho_i)$, so unless ρ_i has already been used as a value of a new function it will be taken as the value of $g^{d+1}(\rho_i)$ when that is defined.)

Part (b) follows from (b) of Theorem 4.2 and the fact that in this case $CET_{\mathcal{L}} \cup \{DCA_{\mathcal{L}}\}$ is, by an argument like that that of Theorem 2.1 part (i), a conservative extension of $\mathsf{comp}_{\mathcal{L}(P)}(P) \cup \{G^{k}_{\mathcal{L}(P)}, \neg G^{k+1}_{\mathcal{L}(P)}\}$. ⊣

It should be noted that these results only hold for φ which are built up using the Kleene truth tables, which have the property that if the value of a truth function is ***t*** or ***f***, it does not change if one of its arguments is changed from ***u*** to ***t*** or ***f***. This is not true of the connective $\longleftrightarrow$ used in $\mathsf{comp}_{\mathcal{L}}(P)$ and indeed the sentences of $\mathsf{comp}_{\mathcal{L}}(P)$, which are obviously true in all 3-valued models of $\mathsf{comp}_{\mathcal{L}}(P)$, may not be true$_{\mathcal{L}}$ at any stage. For example, if P is $p(0), p(s(x)) \longleftarrow p(x)$, the completed definition of p is

$$\forall y(p(y)) \longleftrightarrow y = 0 \vee \exists x(y = s(x) \wedge p(x)),$$

which is false at each stage n because for $y = s^n(0)$ the LHS is ***u*** and the RHS is ***t***. In fact, the notion of truth at some stage does not make sense for sentences using non-Kleene connectives because they can switch from true at some stage to false at a later stage and vice-versa. Take, for example, the instance

$$p(s(0)) \longleftrightarrow s(0) = 0 \vee \exists x(s(0) = s(x) \wedge p(x))$$

of the completed definition of p. This is true at stage 0, false at stage 1, and true at all later stages. There is also something strange about the clauses of P itself. If these are written with the Kleene $\longrightarrow$ then they may not be true in all models of $\mathsf{comp}(P)$, e.g., in this example there are models with elements a for which $p(a)$ and $p(s(a))$ are $\boldsymbol{u}$ so that the clause $\forall x(p(x)) \longrightarrow p(s(x))$ is also $\boldsymbol{u}$. If you want P to be a 3-valued consequence of $\mathsf{comp}(P)$, you have to write the clauses of P with 2-valued implication $p \supset q$, 'if p is $\boldsymbol{t}$, then q is $\boldsymbol{t}$' (you can add 'and if q is $\boldsymbol{f}$, then p is $\boldsymbol{f}$' if you like). But then, like the sentences of $\mathsf{comp}(P)$ they are not covered by Theorem 4.2, indeed, the notion of true at some stage is not sensible.

Following Fitting, Kunen extends the notion of truth at stage n into the transfinite where it eventually stabilizes at some closure ordinal; if we call truth at this stage 'truth at ∞' then Kunen's Theorem 6.6 states that sentence φ is true in all 3-valued *Herbrand* models iff it is true at ∞. This is easily seen to be true for all languages $\mathcal{L}$. So are the completeness results of *Kunen 1989*, e.g., Theorem 4.3 which states that if P is an allowed program and Q is an allowed query, then using SLNDF-resolution from P, if $\mathsf{comp}_{\mathcal{L}}(P) \models_3 \forall Q\theta$ then Q succeeds with answer θ, and if $\mathsf{comp}_{\mathcal{L}}(P) \models_3 \forall \neg Q$, then Q finitely fails. Kunen proves this for his infinite language $\mathcal{L}$ using his Theorem 6.3; since $\mathcal{L} \supseteq \mathcal{L}_1$ implies $\mathsf{comp}_{\mathcal{L}}(P) \models_3 \mathsf{comp}_{\mathcal{L}_1}(P)$, it must be true for all $\mathcal{L}_1$.

REFERENCES

Chan, D.

1988 Constructive negation based on the completed database, *5th International conference and symposium on logic programming, Seattle,* MIT Press, Cambridge, Massachusetts, vol. 1, pp. 111–125.

Clark, K. L.

1978 Negation as failure, *Logic and data base* (M. Gallaire and J. Minker, editors), Plenum, New York, pp. 293–322.

Fitting, M. R.

1985 A Kripke-Kleene semantics for logic programs, *Journal of Logic Programming,* vol. 2, pp. 295–312.

Kunen, K.

1988 Negation in logic programming, *Journal of Logic Programming,* vol. 4, pp. 289–308.

1989 Signed data dependencies in logic programs, *Journal of Logic Programming,* vol. 7, pp. 231–245.

Lassez, J.-L., M. J. Maher, and K. Marriott

1988 Unification revisited, *Introduction to foundations of deductive databases and logic programming* (J. Minker, editor), Kaufmann, Los Altos, California, pp. 587–625.

Lloyd, J. W.

1987 *Foundations of logic programming,* second edition, Springer-Verlag, Berlin.

Maher, M. J.

1988 Complete axiomatization of the algebras of finite, infinite and rational trees, *3rd Annual symposium on logic in computer science, Edinburgh,* pp. 348–357.

Malcev, A.

1971 Axiomatizable classes of locally free algebras of various types, *Metamathematics of algebraic systems: Collected papers,* ch. 23, North-Holland, Amsterdam, pp. 262–281.

Mancarella, P., S. Martini, and D. Pedreschi

1988 Complete logic programs with domain closure axiom, *Journal of Logic Programming,* vol. 5, pp. 263–276.

Przymusinski, T. C.

1989 On constructive negation in logic programming, *Proceedings of the North American logic programming conference, Cleveland, Ohio, Addendum,* MIT Press, Cambridge, Massachusetts.

Sato, T., and H. Tamaki

1988 Deterministic transformation and deterministic synthesis, *Proceedings of the 2nd Franco-Japanese symposium on programming of future generation computers II* (K. Fuchi and M. Nivat, editors), North-Holland, Amsterdam, pp. 307–327.

Shepherdson, J. C.

1989 A sound and complete semantics for a version of negation as failure, *Theoretical Computer Science,* vol. 65, pp. 343–371.

PART III

PHILOSOPHY, MEANING, AND INTENSIONAL LOGIC

C. ANTHONY ANDERSON

ALTERNATIVE (1*): A CRITERION OF IDENTITY FOR INTENSIONAL ENTITIES

Abstract. The problem of formulating an adequate criterion of identity for propositions and other intensionalia was at one time considered the principle obstacle to the construction of an acceptable general intensional logic. The objection was urged time and again, principally by W. V. Quine, and still has its influence.[1] Alonzo Church responded directly to the challenge and attempted to incorporate various criteria of identity into fully formalized intensional logics, different implementations of his logic of sense and denotation. The logistic system which was to be based on the most important of Church's ideas, synonymous isomorphism, remains unfinished.

In this paper I attempt to clarify the point and nature of demands for criteria of identity and to evaluate various objections against Church's criterion. My conclusion is that a modification suggested by certain alleged counterexamples appears to fit the data better and so merits further study. Important parts of the overall project require the development of a detailed logistic treatment meeting current standards of rigor. This will not be attempted here, but is reserved for a technical sequel.

1. CRITERIA OF IDENTITY

To fix ideas,[2] define a (*potential*) *one-level criterion of identity for being an* F^1 to be a relation R_o satisfying:

$$(\mathrm{CI}_1)\quad F^1(z_1)\,\&\,F^1(z_2) \supset .\ z_1 = z_2 \equiv R_o(z_1, z_2).$$

As an example of this, we may take the familiar one, the *Axiom of Extensionality* of set theory. Here F^1 is being a set and '$R_o(z_1, z_2)$' means that all the members of z_1 are members of z_2, and vice-versa. In this case and analogous ones to be discussed below we shall speak of criteria of identity of this sort as *ontological* criteria of identity.

Now suppose that G^1, F^2, and R_σ satisfy:

$$(\mathrm{CI}_2)\quad G^1(x)\,\&\,G^1(y) \supset .\ F^2(x, z_1)\,\&\,F^2(y, z_2) \supset .\ z_1 = z_2 \equiv R_\sigma(x, y),$$

where F^2 is a function (on G^1's). That is, F^2 obeys the two conditions:

$$(\mathrm{E})\quad G^1(x) \supset \exists y F^2(x, y),$$

$$(\mathrm{U})\quad G^1(x) \supset .\ F^2(x, u) \supset .\ F^2(x, v) \supset u = v.$$

[1] See, for example, *Grim 1951*, page 20.

[2] And following *Williamson 1990*, pages 144–148.

C. Anthony Anderson and M. Zelëny (eds.), *Logic, Meaning and Computation*, 395-427.

Then the relation R_σ will be said to be a (*potential*) *two-level criterion of identity.* If certain other conditions are satisfied, we might be willing to say that R_σ is a (two-level) criterion of identity for the-range-of-F^2-with-its-domain-confined-to-G^1.

Many of the examples in the literature, proposed as illustrative of the general idea of a criterion of identity can be made to fit one or the other of these patterns. One much discussed two-level criterion is the case where the G^1's are predicates or open sentences (of some given language) and F^2 is the relation between these and their *extensions.* Here R_σ is the relation which holds between two open sentences when they are *true-of,* or *satisfied by,* the same objects. Since we will be concerned with this example and its analogues for intensions, we call such cases of two-level criteria of identity by the more suggestive name, "semantical criteria of identity".[3] The most pleasant case occurs when we have criteria of both sorts and if F^1 is related to F^2 and G^1 in this way:

(T) $F^1(z) \supset \exists x(G^1(x) \, . \, F^2(x,z))$ ("Totality Condition").

There are various interesting logical relationships between these things.[4] If R_σ is well understood, we may even be able to usefully define an appropriate R_o. In some cases, if other things are favorable, we can "construct" the F^1's as equivalence classes of G^1's. Here we are interested in obtaining further conditions which can guide us in our search for criteria of identity for intensions.

Consider first our paradigm of an ontological criterion, the Axiom of Extensionality just mentioned. There are various ways of choosing an appropriate companion two-level criterion, but we take our cue from Quine:

> Observe, in contrast, how well the corresponding requirement is met in the individuation of classes. I began by saying that classes are identical when their members are identical; but what we now want is a satisfactory formulation of a relation between two open sentences 'Fx' and 'Gx' which holds if and only if 'Fx' and 'Gx' determine the same class. The desired formulation is of course immediate: it is simply '$(x)(Fx \equiv Gx)$'. It does not talk of classes; it does not use class abstraction or epsilon, and it does not presuppose classes as values of

[3]The terminology of *semantical/ontological* criteria of identity is not appropriate for many of the examples discussed in the literature. John Myhill almost recognizes the distinction between semantical and ontological criteria of identity for intensions in his discussion of Church's Alternative (0):

> This [Church's criterion] does not determine the identity or diversity of those senses which are not senses of expressions, but it does suggest a series of axioms concerning the identity and diversity of senses from which various interesting consequences can be deduced. (*Myhill 1958,* page 82)

[4]Timothy Williamson (*1986*) has begun a study of logical questions about criteria of identity which should be emulated.

> variables. It is as pure as the driven snow. Classes, whatever their foibles, are the very model of individuation on this approach.
> (*Quine 1975,* page 7)

Unfortunately, there is a use-mention error here which obscures an important point. Quine merely cites a formula of first-order logic as specifying the relation between the open sentences 'Fx' and 'Gx', but the formula itself does not speak of these syntactical entities. Rather the (semantical) criterion in question must be the relation which holds between these open sentences when '$(x)(Fx \equiv Gx)$' is true. Equivalently, we may speak of the relation which holds between these open sentences when they are *true-of* or *satisfied by* the same objects. It is evidently required that truth-of (or satisfaction) either be taken as primitive in the metalanguage, or be defined in set-theoretical terms therein for the given language.[5] And there is a hidden parameter in the criterion as stated by Quine, namely the language whose open sentences are in question.

So let us take as our instance of (CI_2), the case where G^1 is *being an open formula* (*of such-and-such a language* L), F^2 is *has as extension* (in L) (or *determines-* or, perhaps, *denotes-in-L*), and R_σ is the relation which holds between x and y when they are open formulas *true-of-in-L* (or *satisfied-by-in-L*) for exactly the same objects. Thus, in semi-English:

(CI_2)^S If x and y are open sentences of language L, then if z_1 is the extension of x (in L) and z_2 is the extension of y (in L), then $z_1 = z_2$ if and only if x and y are satisfied (in L) by the same objects.

The superscript 'S' is supposed to suggest that we are ultimately interested in a theory of sets. Here, alas, if we take F^1 as *being a set* in (CI_1), then the Totality Condition with respect to G^1 and F^2 does not hold. For if L is a language of the familiar kind, there will be sets which are not the extensions of any open formulas in L. Still, this seems to be the semantical criterion of interest.[6]

The semantical principle is supposed to have a kind of epistemic priority over the corresponding ontological principle.[7] We imagine that open sentences and their properties are familiar to us. Perhaps in the general case we

[5] This observation might lead one to question the "purity" of (CI_2)^S below. I am aware that one might try to finesse the semantical concepts by means of some sort of "deflationary" tactic. But even apart from doubts about the feasibility of such projects, we are seeking a *general* criterion of identity for sets as extensions of predicates, and this seems to require the use of semantical concepts.

[6] There are interesting "intermediate" principles that might serve as criteria of identity for sets. For example, we might see something like Frege's Basic Law V as telling us that concepts have the same sets as extensions if they are "materially equivalent". This does not quite qualify as semantical or as ontological in our sense. It does rightly emphasize the epistemological priority of attributes or concepts over sets. It suffers from the fact that we currently have no detailed theory of concepts.

[7] Compare *Anderson 1980,* page 220.

might just require that the entities in question be treated by a systematic and well-understood theory. And it is supposed that we either understand directly, or by way of a theory, the relation of satisfaction or being-true-of for a given language. Extension-of-in-L is just a restriction of the idea of a set to those supposed to correspond to open sentences of the given language and is to be antecedently understood, or at least clarified by the criterion itself, perhaps with the aid of an accompanying theory.

In the present case, the epistemic priority comes out clearly when we consider the possibility of *proving* that two sets are identical. Typically this is done by showing, in a particular language, a certain biconditional—the constituent formulas of which are taken as determining as extensions the sets in question.

Now what is the desired relationship between $(\mathrm{CI}_2)^{\mathrm{S}}$ and $(\mathrm{CI}_1)^{\mathrm{S}}$? Well, it is evidently that we are to find R_o, formulated in set theoretical terms, using $\in$ if desired, so that $(\mathrm{CI}_1)^{\mathrm{S}}$ has the semantical criterion $(\mathrm{CI}_2)^{\mathrm{S}}$ as "observational consequence". Of course this will be of little assistance if we do not have a well-developed theory about sets—but we in fact do, namely, set theory with the Axiom of Extensionality functioning as ontological criterion.

Here there may arise the threat of a regress. Quine (*1975*) observes that the identity of sets is relative to the identity of their members. But in a set theory allowing individuals, or non-sets of some kind, the question may arise as to the ontological criterion of identity for these things. Somewhere the regress must halt and explanations must come to an end.

To illustrate the point, let us consider the simpler case of the extensional entities corresponding to sentences. The desired criteria of identity in that case would seem to be:

$(\mathrm{CI}_2)^{\mathrm{T}}$ If x and y are sentences of language L and if z_1 and z_2 are the truth-values determined (or denoted) by x and y, respectively, then $z_1 = z_2$ if and only if x and y are materially equivalent in L—that is, x and y are both true in L or both false in L.

and

$(\mathrm{CI}_1)^{\mathrm{T}}$ If z_1 and z_2 are truth-values, then $z_1 = z_2$ if and only if $z_1 = \mathbf{t}$ and $z_2 = \mathbf{t}$, or $z_1 = \mathbf{f}$ and $z_2 = \mathbf{f}$.

Here we have the desired relationship between the criteria—we could (and often do) even define the relation of material equivalence by reference to the truth-values.

Suppose we continue to press our Socratic question: "But what is the criterion of identity for being either **t** or **f**?" Here the only possible answer is that these two things are distinct and we can tell this in some cases. If we take **t** and **f** as being the values of something known to be true and something known to be false—say as of the propositions $\forall x(x = x)$ and $\exists x(x \neq x)$, respectively—then in some sense we know what the truth-values are and that they are distinct.

2. CRITERIA OF IDENTITY FOR INTENSIONAL ENTITIES

Presumably, the desired analogues for attributes are of these forms:

$(\mathrm{CI}_2)^{\mathrm{A}}$ If x and y are open formulas (of a language L), then if z_1 is the attribute expressed by x and z_2 is the attribute expressed by y, then $z_1 = z_2$ if and only if $R_\sigma^{\mathrm{A}}(x, y)$,

and

$(\mathrm{CI}_1)^{\mathrm{A}}$ If z_1 and z_2 are attributes, then $z_1 = z_2$ if and only if $R_o^{\mathrm{A}}(z_1, z_2)$,

where Totality may not be assumed and the relation R_σ^{A} is well-understood. Using our analogy, it is apparently permissible that our theory of intensional entities employ the relation R_o^{A} which is not necessarily definable by appeal to R_σ^{A}. But the theory of intensions using R_o^{A}, like the theory of extensions using $\in$, should meet various formal and informal criteria of adequacy. $(\mathrm{CI}_2)^{\mathrm{A}}$ is to be "pure as the driven snow"—it may not employ the initially unfamiliar theoretical primitives involved in R_o^A and the quantification involved in R_σ^{A}, if any, shall not be explicitly[8] over intensions.

The analogy with the extensional case of sets is not quite exact. The ontological criterion for sets appeals to the identity of their elements and these elements may all be (but need not be) of a new kind. The expected relation R_o^A between complex intensions may well appeal to a comparison of "constituents", some of which are themselves necessarily intensional entities. We should certainly demand that the theory give a detailed accounting of the alleged constituent relation, but the ontological criterion is not defeated simply by the fact that it involves reference to intensions.

No regress is forthcoming if we suppose that in some fundamental cases we can just tell whether two simple expressions expressing intensions are synonymous or, what is the same, that the meanings they express are identical or distinct. But of course we can, and when we do, we may suppose that the synonymy is stated by, or is a consequence of, the semantical rules of the language—the rules which codify the knowledge attributable to a competent, perhaps ideally competent, user of the language.

To my knowledge no one has offered a compelling, or even plausible, argument that if a putative kind is not accompanied by criteria of identity, then the alleged kind is unacceptable on grounds of obscurity or other defect. Until and unless a clear, careful, and complete set of conditions for what makes something an acceptable (actual, instead of potential) criterion of identity is given, there is scant hope for such an argument. But the debate will at least be advanced if we can supply what we take to be criteria of identity for intensions, closely parallel to the two principles about sets and extensions,

[8] In the promised logistic treatment, we propose to quantify over everything whatever. However, the condition on quantification is designed to avoid circularity and it does not appear that any conceptual or epistemological circularity is reintroduced by our procedure.

and subject to the (vague but suggestive) epistemological constraints. Then it should be possible for opponents of intensions to explain the respects in which the proffered principles are inferior to their set-theoretical analogues and thereby clarify the question of the need for and value of such principles.

For simplicity in the discussion to follow, we confine attention to the question of criteria of identity for ***propositions***, but the understanding is that the more general case of attributes is also in question. Thus we seek a clear relation among sentences R_σ^P and, ultimately, a theory of propositions embodying a relation R_o^P, such that the following holds for any given language L:

$(\mathrm{CI}_2)^{\mathrm{P}}$ If x and y are sentences (of language L) and if z_1 and z_2 are the propositions expressed (in L) by x and y, respectively, then $z_1 = z_2$ if and only if $R_\sigma^{\mathrm{P}}(x, y)$,

And

$(\mathrm{CI}_1)^{\mathrm{P}}$ If z_1 and z_2 are propositions, then $z_1 = z_2$ if and only if $R_o^P(z_1, z_2)$,

is to govern propositions. The latter is to yield the former as a consequence when applied to the particular propositions expressed by sentences in L. In the present paper, we devote considerable effort to the formulation of the first sort of criterion, and only briefly indicate the formal theory which will accompany the latter.

3. SYNONYMOUS ISOMORPHISM

Rudolf Carnap (*1947*) proposed that two expressions, say sentences, are synonymous if and only if they are ***intensionally isomorphic***. Consider two expressions in a single language. Assume, for simplicity, that both are written in primitive notation.[9] If the semantically relevant parts can be matched up in such a way that the parts are L-equivalent, that is, necessarily equivalent, then the expressions are intensionally isomorphic—otherwise not.[10]

So, consider a hypothetical arithmetical language utilizing both Arabic and Roman numerals. In such a language 'V > I' might be intensionally

[9]For most formalized languages that have actually been used, this will almost eliminate the distinct expressions which are intensionally isomorphic. Really, in order to have anything significant, we would consider, following Church (*1954*), especially treatments that allow "defined" notation in the object language. That is, the formalized language is supposed to have rules of definition which allow augmentation of the object language by adding new defined notation. This is in contrast to the method, now common among logicians, of regarding all definitions as being metalinquistic abbreviations. Cf. further the discussion in *Church 1956*, note 168. Natural language synonymies, for example between 'procrastinates' and 'puts things off', are probably most usefully seen as involving only the object language.

[10]Actually, if bound variables are present, the matter is a little more complicated. Carnap in fact allows that alphabetic change of bound variables preserves intensional structure. We shall have occasion to discuss the treatment of variables briefly, in note 20 below.

isomorphic to '$5 > 1$', but not intensionally isomorphic to the *L*-equivalent '$2 + 3 > 1$'.

Strangely, Carnap does not go on to construct a theory of intensional entities[11] corresponding to intensional isomorphism—as his work and that of his successors in possible worlds semantics has produced in the case of necessary equivalence as criterion of identity. But the idea is near at hand and in his critique Church evidently has such a goal in mind. So let us consider intensional isomorphism as a criterion of identity for the propositions (of a certain sort) expressed by sentences in a given language.

Notice that this is not an ontological criterion which is directly about intensions or meanings; it's about expressions in a particular language—what we have called a semantical criterion. But, as we have urged, one might strive to construct a theory about intensions directly and use as one of the conditions of adequacy for the theory that the ontological principle entail, in application to any particular case, the corresponding semantical principle.

Church (*1954*)[12] provided a counterexample to Carnap's criterion, at least if it is intended to be used in connection with a logic of belief. Church's example is a language in which 'Qn', with 'Q' a primitive predicate, expresses that n is a positive integer greater than 1 for which there are positive integers x, y, and z, with $x^n + y^n = z^n$. In the same language, 'Pn' expresses that n is less than 3. Again 'P' is a primitive predicate of the language. If Fermat's Last Theorem is true, and we now know that it is, '$\exists n(\sim Pn \,\&\, Qn)$' is necessarily equivalent to, indeed, intensionally isomorphic to '$\exists n(\sim Pn \,\&\, Pn)$'. But, Church says, surely one could believe the proposition expressed by the first without believing the explicit contradiction expressed by the second. In the light of recent developments in number theory, one might change the argument a bit to avoid unnecessary side issues: Andrew Wiles refuted the proposition expressed by the first of these sentences. Why not say about his alleged discovery—"Well, big deal, we've known that at least since Aristotle, and probably before. The proposition that $\exists n(\sim Pn \,\&\, Pn)$ is obviously false. So why all the fuss about Wiles?" Surely this would not be taken seriously. It is relevant to observe in response that we now have a proof—and certainly we owe this to Wiles. But it is still correct to say that we now know the

[11]Rather, he employs the notion of intensional isomorphism in the metalanguage and attempts to analyze belief in those terms without ever considering the postulation of propositions corresponding to this idea. Kinichi Fukui made the oddity of this, given Carnap's other work, clear to me.

Carnap's (*1946*) use of the "Looking-around Method", so called by Ignacio Angelelli (*1984*), is really just an application of the general use of criteria of identity we have described. Certainly Angelelli is right to criticize the method if it is considered a technique for "defining by abstraction", but it need not be thus characterized. Matters here are more or less analogous to the attempts in the natural sciences to find theories which will explain the data.

[12]Others have re-discovered or re-utilized Carnap's criterion, but they do not respond to Church's criticism. See, for example, *Cresswell 1985* and its review by Anderson (*1991*).

proposition commonly called "Fermat's Last Theorem", whereas before we did not.

Various moves are still possible for the Carnapian. The most immediate suggestion is that we ascend into the metalanguage and require that the expressions involved in the semantical rules for the predicates of the object language also be intensionally isomorphic. Of course if we formalize the metalanguage, it too might pass the test, even when applied to the example before us—if it uses primitive predicates synonymous with the object language predicates '*Pn*' and '*Qn*'. Then the question must be asked again about the meta-metalanguage, and so on, up the hierarchy. It is reasonable to insist that the threatened infinite regress of metalanguages must come to a halt. To make the criterion explicit, one will probably require the idea that some meanings are intrinsically simple[13] and, if the corresponding semantical criterion is to serve its purpose, simplicity must be somehow clearly present or absent in particular cases.

The more serious difficulty, it seems, is that the criterion, when applied to two simple expressions or, really, to two expressions with simple meanings, seems utterly arbitrary.[14] That two synonymous complex expressions should have a correspondence between their meaningful parts has a certain plausibility—every meaning expressed by the one has to be expressed by the other (although it may do so by way of definitions). But why think that two simple meanings, if they (or their expressions) are logically equivalent, must

[13]Frege might have objected. In his discussion of negation, he points out (*Frege 1919*, in *Geach and Black 1966*, page 125) that it is often quite difficult to say which of two sentences expresses a negative and (presumably) more complex proposition and which expresses merely some pure affirmative. Church (*1954*, note 8) also expresses some reservations about the notion of simplicity.

[14]I have urged this criticism briefly elsewhere: in *Anderson 1987*, page 160, note 14, and in *Anderson 1991*.

Donald Davidson (*1963*) also complains about the *ad hoc* character of intensional isomorphism, but he is considering mainly its merits as a solution to the Paradox of Analysis. Given the semi-Fregean nature of Carnap's solution to Frege's Puzzle, "How can $a = b$, if true, differ in meaning from $a = a$?", and its application to the failure of the substitutivity of identity in modal contexts, Davidson argues (following Church, *1946b*) that a parallel treatment is called for in this case and in application to substitution in belief sentences. We quite agree, but would at least qualify his further criticism:

> Equivalence is firmly based on sameness of extensional reference; *L*-equivalence is provided with a similarly impressive semantic ground, sameness of intensional reference. But intensional isomorphism has no such factual or theoretical justification. (*Davidson 1963*, page 342)

"Intensional reference" is just a name for a postulated binary relation involved in Carnap's two-level criterion of identity, and the theoretical basis involves a one-level principle for Carnapian intensions and a corresponding equivalence relation between them. What is really doing the work is (i) the fact that *L*-equivalence is a relation between expressions to which we have a certain epistemic access, and (ii) Carnap provides a mathematically workable theory of intensions in terms of possible worlds (for him, state-descriptions) and functions involving them. Of course, similar things are required for intensional isomorphism and its descendants.

be identical? True, clear counterexamples are hard to come by, but the idea itself seems to have no reasonable basis.[15]

Church's emended criterion, *synonymous isomorphism*, requires that in order for two expressions to be synonymous, the replacement of parts is to involve only synonymous expressions—where the synonymy and non-synonymy of simple expressions, and the synonymy of simple expressions with complex ones, via definition, is supposed to built into the language. Such things are to be just given by the semantical rules of the language.

Now some will find this last idea puzzling or, even, a cheat. We are laboring mightily to give a clear characterization of synonymy (suitable, remember, for use in a semantical criterion of identity). Church seems to be proposing that we cut the Gordian knot by just supposing that the semantical rules already tell us what is synonymous with what!

Well, the situation is not really as just described.[16] The thought is, I believe, that two things are clear about synonymy, pre-analytically conceived. One is that, in the case of constructed languages or technical extensions of natural languages, one may just *stipulate* that a simple expression shall henceforth be synonymous with a complex (or simple, if there were any reason for it) expression already present and understood in the base language. Even Quine finds this case unproblematic (*Quine 1953a*). The second thing is that synonyms may be substituted for synonyms with preservation of meaning. I hasten to add that this last is not really literally true for some of the natural languages that we know and love. Substitution fails to preserve meaning, or even denotation and truth, in quotation contexts and others. Church is supposing, I conjecture, that the failure of substitution is such cases, like the failure of the substitutivity of identity in modal and belief contexts, is to be explained in the Fregean way—one isn't really talking about the same things in those contexts. Such anomalies are to be eliminated in the more precisely formulated languages to which the semantic theory is to be directly applied, and the application to other cases will be a matter of approximation.[17]

Notice also that there is a pretense that a language is "given" by giving its semantical rules, including perhaps extensional rules about denotation. We hope that these idealizations will prove their worth in a finished theory, but this certainly cannot be firmly predicted. Here we are just emulating the natural sciences and, especially, their use of mathematics as applied to the physical world.

Church's original statement of the motivating idea behind his criterion

---

[15]It is worth noticing that our grasp of meanings and of synonymy is in fact epistemologically prior to our knowledge of logical necessity or equivalence—and not the other way around.

[16]Recall also the dicsussion at the end of section 1 about the possibility of a vicious regress.

[17]The general assumption that certain irregularities must be removed is really already present in the pretense that extensional semantical rules and principles, e.g., Tarski's Schema T, fit actual natural languages, as applied to, say, ambiguous expressions.

appears in an abstract describing his initial work on the logic of sense and denotation (*Church 1946a*) and was apparently formulated independently of Carnap. He says there that (further) axioms about the notion of sense (in one sense) are to be obtained by using as a heuristic principle that "two names are assumed to have different senses in all cases where it is not already a consequence that the senses are the same." Later clarifications led him to see this as based on the intuitions behind synonymous isomorphism—and the heuristic principle involved came to be called "Alternative (0)".[18] A similar thought is expressed by John Myhill,

> It appears however that in view of the philosophical application we intend to make of modal [better: intensional] logics, *we should never identify intensions unless we are forced to*; for the fewer identifications we make the more flexible will be our analysis of belief-sentences."
> (*Myhill 1963*, page 306. My insertion.)

This is provocative, but what does it really mean? Church does not here tell us what we are allowed to assume in order to see what consequences there are about identity of sense, and Myhill similarly gives us no guidance as to what could force us to identify intensions or senses.[19]

The answer to this question which is presupposed by Church's synonymous isomorphism is this: we assume only that the stipulations involved in definitions produce synonymy, or sameness of sense, and that synonyms may be interchanged with preservation of sense.[20]

Thus explained, Church's suggestion has an initial intuitive appeal and seems to depend only on the preservation of the essential and clear features of synonymy (slightly idealized).

---

[18]The better terminology is to call synonymous isomorphism the semantical criterion and to reserve the name "Alternative (0)" for the strategy which involves formulating an ontological criterion and accompanying theory in such a way as to entail, when applied to particular languages, the semantical criterion for that language.

[19]In fact, Church does not really adhere to the principle as stated in his further work. See the discussion in *Anderson 1998*.

[20]We have been ignoring the fact that both Carnap and Church allow that alphabetic change of bound variables, in the usual sorts of formalized languages, preserves sense. One supposes that they would say the same about natural language analogues of such things, e.g., "he" and "that man (or male)". This is an interesting exception to the intuitive motivation and illustrates that we may be "forced" to identify senses in order to preserve the appearances.

It appears that the meaning of a bound variable is not to be explained by, say, simply referring to the semantical rule which specifies (the sense which gives) its range. Rather, the meaning of such variables also involves the relations (of their occurrences) to one another in certain contexts. Those relations must be preserved when substituting another (bound) variable. In this connection, see also *Church 1956*, page 10, note 24, Quine, page 70, and the discussion of "wiring diagrams" in *Kaplan 1986*, page 244. (The Quine-Kaplan idea needs to be supplemented to take account of the meaning contributed by the kind of the variable.)

There is nevertheless an analogy with the interchange of synonymous constants which gives some support to the exception.

4. ALLEGED COUNTEREXAMPLES TO SYNONYMOUS ISOMORPHISM

A number of different sorts of counterexamples have been urged against synonymous isomorphism as criterion of synonymy. The most telling cases[21] fall into two main classes which I shall call (1) *commutation and conversion of non-symmetricals*, and (2) *commutation of symmetricals*. Under type (1) are included such cases involving verbs (or relations) as:

'a loves b' ⇔ 'b is loved by a',
'a is taller than b' ⇔ 'b is shorter than a',
'a is greater than b' ⇔ 'b is less than a',
'a precedes b' ⇔ 'b succeeds a' (or 'b follows a').

(I use single quotes instead of corner quotes, and '⇔' to signify alleged synonymy.) A plausible case involving a non-symmetrical connective, rather than a relation, is:

'Q if P' ⇔ 'P only if Q'.

It seems clear that no explicit definitions, even pretended to be involved in the semantical basis of English, will have the consequence that such things are synonymies.

It is possible to argue, with some show of plausibility, that none of these are *really* exact synonymies, but rather that they are just obvious logical equivalents which nevertheless have some (slight) difference in information content.[22]

To consider the first and easiest example, represent 'a loves b' as '$(a \,\mathrm{L}\, b)$'. If we introduce '$\lambda x(y \,\mathrm{L}\, x)\lambda y$' as a lambda-expression of a special kind[23] which allows lambda-conversion with 'λx' operating to its left and 'λy' operating to its right, the form of 'b is loved by a' might be taken to be '$b(\lambda x(y \,\mathrm{L}\, x)\lambda y)a$'. Perhaps the first lambda operator corresponds to 'is ____-ed' and the second to 'by'. So we have to perform perhaps two steps of lambda-conversion to get from the active to the passive and these logical steps, though elementary,

---

[21] I have dealt elsewhere (*Anderson 1987*) with George Bealer's proposed counterexample, 'y outweighs z' ⇔ 'the weight of y is greater than the weight of z'. Diana (now Felicia) Ackermann proposed in conversation the example (or a close relative thereof) 'Jones is believed by Jones to be bald' ⇔ 'He himself is believed by Jones to be bald'. This is of a rather different sort and pretty clearly depends on meanings which are relative to context. That is, except for the fact that 'He himself' conveys masculinity while 'Jones' does not (a defect of the example which is easily remedied), in the context these two expressions *are* synonymous and we do not here have a real counterexample. Initially, and for simplicity, we are supposing that generally such context sensitivity is eliminated from the language before the test is applied. But see the discussion of variables in note 20.

[22] See, for example, *King 1996*, pages 503–504. I do not endorse the principle P which he discusses and rejects, but nevertheless continue to be troubled by certain of the alleged counterexamples to synonymous isomorphism.

[23] David Kaplan invented this, or a similar, operator and used it in lectures at UCLA long ago.

are sufficient to change the meaning.[24]

In the second example it seems natural to see 'taller' and 'shorter' as logically complex—the modifier 'er' even has a kind of logic. This being so, the step from '*a* is taller than *b*' to '*b* is shorter than *a*' is very plausibly taken to be a real logical step.[25] Or so one might argue.

More difficult to resist are examples of commuted symmetricals:

'*a* equals *b*' ⇎ '*b* equals *a*',[26]
'*P* and *Q*' ⇎ '*Q* and *P*'.[27]

Here we have a relation or connective which is necessarily symmetric. Other examples of this sort of thing will readily come to mind.[28]

Again, the determined Synonymy Isomorphist can offer some reasoned resistance.[29] And maybe he is in the right. None of these things seems to

[24] Of course someone might maintain that lambda-conversion preserves meaning. See the discussion of Alternative (1) below.

[25] Bertrand Russell (*1903*) argues the question and rejects the identity, partly because:

> [I]f we are to hold that "*a* is greater than *b*" and "*b* is less than *a*" are the same proposition, we shall have to maintain that both *greater and* less enter into each of these propositions, which seems obviously false ... (page 228)

This might be seen as a "distinct-denotation" argument, of the sort I offer below against associative and reflexive meanings. Such arguments are not conclusive, since it may always be claimed, with more or less plausibility, that the concept is conveyed implicitly and the relevant denotation is determined thereby.

Peter Geach (*1957*) might be seen as supporting the opposite view:

> For not only is the concept of a pair of converse relations a single and indivisible mental capacity (*eadem est scientaia oppositorum*); the exercise of the concept in judgment also brings in the two relations equally and simultaneously, for to judge that *a* bears the one relation to *b* is the very same act as judging that *b* bears the converse relation to *a*. (page 52)

But he does not offer any independent support for this claim.

[26] John Wallace proposed this example in conversation.

[27] This example is due to Carnap (*1963*, pages 899–900), where he observes that the commutativity of conjunction is commonly taken to preserve meaning in ordinary language "even in its most careful use of indirect discourse, for example, in the statement of a witness under oath." A reply along the lines of *King 1996* (*op. cit.*), that the equivalence depends on special properties of "said that", takes some of the sting out of these examples.

[28] In the same general class are examples where the commutation involved is not confined to sentences or singular terms, e.g., 'The yellow car ran down the street' ⇎ 'The car, which was yellow, ran down the street'. (Richmond Thomason proposed this example to me in conversation.) But here there is alteration in the rest of the sentence and hence some room to argue non-synonymy. There are however similar examples which do not involve such additions: 'Almighty God created the universe' ⇎ 'God Almighty created the universe' (cf. also *Richard 1990*, page 31).

We do not treat these things explicitly below, but it will be fairly clear how we would deal with them, once questions of logical form have been settled. Some such examples, anyway, might be taken as having the form: '$\ldots (\iota x)(\phi x \,\&\, \psi x) \ldots$' ⇎ '$\ldots (\iota x)(\psi x \,\&\, \phi x) \ldots$', so that the treatment should follow that of conjunction.

[29] See *Richard 1990* and *King 1996* for some considerations supporting this view. I remain suspicious of arguments that rely on the differences between such things as 'John

have the force of a synonymy based on a stipulation—our paradigm case.

The plot thickens a bit when we consider also technical extensions of English, mathematical English say, and examples of (interpreted) formalized languages. There one finds such possible counterexamples as these:

'$a < b$' $\approx$ '$b > a$',
'$P \supset Q$' $\approx$ '$Q \subset P$'[30] (commutation and conversion of non-symmetricals),

and '$a = b$' $\approx$ '$b = a$',
'$P \,\&\, Q$' $\approx$ '$Q \,\&\, P$' (commutation of symmetricals).

The technical extensions and formalized languages are actually more difficult to deal with than examples from the informal natural language because *stipulation* is entirely permissible in those contexts. Even if for English we can somehow establish that '*P and Q*' and '*Q and P*', say, are not really synonymous, does it *have* to be so, in every language, that '$P \,\&\, Q$' is non-synonymous with '$Q \,\&\, P$', *no matter what concept of conjunction we assign to* '&'? Can't we just stipulate that in our technical extension of English or in my constructed language, *order does not count* in this case?[31]

And if the principle should just happen to hold for English, that fact may be best seen as psychological or even accidental. What does this tell us about propositions and other complex concepts?

We are thus well advised then to consider some technical uses of notation. Some may plausibly be seen as involving the synonymy of expressions related by commutation with conversion. One commonly occurring candidate, mentioned already, involves the relations of *less* and *greater* among numbers:

(1a) But when we wish to indicate that a precedes and b follows in the natural scale, we employ one of the **inequalities**:

$a < b$, read "a is less than b";
$b > a$, read "b is greater than a." (*Fine 1906*, page 8)

(1b) Definition: *In an ordered domain, the two equivalent statements $a < b$ (read "a is less than b"), $b > a$ ("b is greater than a") both mean that $b - a$ is positive.* (*Birkhoff and MacLane 1944*, page 8)[32]

deduced that P and Q' and 'John deduced that Q and P', and the like, to conclude non-synonymy. These arguments come quite close to begging the question and depend on an assumption about the intensional operator in question which does not hold in all cases. Namely, that the proposition expressed does not depend on syntactical features of the operand—and this assumption fails in cases not necessarily amounting to hidden quotation.

An argument in *Linsky 1967*, page 34, similarly depends on such an assumption about the syntactical insensitivity of 'Jones denied explicitly that'. Church's "Translation Argument" does not seem to me to settle the matter (cf. *Linsky op. cit.*, note 1, page 34).

[30] This last is Church's notation for converse implication. See *Church 1956*, page 37.

[31] No, we cannot, as will emerge from our discussion below.

[32] Essentially similar to this last is:

Of course, we suppose that the definitions are not intended for just the cases of the particular constants or variables exhibited. And granted that the examples are somewhat different, we may ask, what theoretical account are we to give of these procedures? If we can answer this question, we will be in a position to better decide their force as proposed counterexamples to synonymous isomorphism.

A fairly clear candidate for commutation of symmetricals involves a notation common in set theory:

(2a) A set having just a few elements is usually denoted by putting braces around a list of symbols denoting its elements. *No significance is attached to the order in which the symbols of the list are written down*, and several symbols in the list may denote the same object. (*Warner 1965*, page 5, my emphasis)

Again, but less clearly referring to the meaning of notation:

(2b) For a set, the order of succession of its elements shall not matter, provided that nothing is said to the contrary. (*Kamke 1950*, page 1)

If similar things are not actually often explicitly said in connection with identity or conjunction, it is easy to imagine that they might have been.

5. CONTEXTUAL AND EXPLICIT DEFINITIONS; CONNECTIVES AND PROPER SYMBOLS

To see our way clearly through these matters, we must distinguish between *contextual definitions* and the more usual sort, let us call the latter, *explicit definitions*. For clarity on such topics, one naturally turns to the writings of Alonzo Church:

> An expression A introduced by *contextual definition*—i.e., by a definition which construes particular kinds of expressions containing A, as abbreviations or substitutes for certain expressions not containing A itself, but provides no such construction for A itself—is an incomplete symbol in this sense. (*Church 1942*)

In *Church 1956* he adds:

> In such a case, where a complex notation introduced by definition carries the false appearance or suggestion that some part of the notation

Now we can define the symbols $<$, $>$, $\leq$, $\geq$, called, respectively, *less than*, *greater than*, *less than or equal to*, and *greater than or equal to*, as follows:

$x < y$ means that $y - x$ is positive;
$y > x$ means that $x < y$;
$x \leq y$ means that either $x < y$ or $x = y$;
$y \geq x$ means that $x \leq y$. (*Apostel 1967*, page 20)

is to be taken as denoting (or otherwise as having meaning in isolation), it is usual to speak of contextual definition ....

On the other hand, D6 $[[A \equiv B] \rightarrow [A \supset B][B \supset A]]$ (for example) would not ordinarily be called a contextual definition of the sign $\equiv$.... The difference is that, in the notation ... introduced by D6 ... there is nothing to suggest that ... the sign ... $\equiv$ standing alone is significant in any way (e.g., as an abbreviation of a formula of a logistic system). (*Church 1956,* page 323, my insertion. The arrow '$\rightarrow$' is Church's sign for equality by definition.)

For lack of a better term, I shall include such things as D6 as contextual definitions. Such definitions are entirely permissible because they are in fact, if not in motivation, stipulations. And this is so even if we regard the definition as extending the object language so as to include the new notation. In Church's terminology (*Church 1956,* page 12, note 30) we may say that a contextual definition is a *stipulated sense-concurrence.* The author giving the definition has stipulated that this expression (and its alphabetic variants and instantiations by terms) is in every case to mean the same as the corresponding expression given as the definiens.

We shall also have to attend carefully to the related distinction between *proper symbols* and *connectives,* again as explained by Church. A proper symbol is a symbol

... having meaning in isolation, the primitive names as denoting (or at least purporting to denote) something, the variables having (or at least purporting to have) a non-empty range. (*Church 1946,* page 32)

Improper symbols do not have meaning in isolation; they are *syncategorematic,* but they can be combined with proper symbols to form longer expressions which do have meaning.

Connectives are combinations of improper symbols which may be used together with one or more constants to form or produce a new constant. ... [I]f we replace one or more of the constants each by a form which has the denotation of that constant among its values, the resulting expression becomes a form (instead of a constant); and the free variables of this resulting form are the free variables of all the forms (one or more) which were united by means of the connective (with each other and possibly also with some constants) to produce the resulting form. In order to give completely the meaning-producing character of a particular connective in a particular language, not only is it necessary to give the denotation of the new constant in every permissible case that the connective is used together with one or more constants to form such a new constant, but also, for every case that the connective may be used with forms or forms and constants to produce a resulting form, it is necessary to give the complete scheme of values of this resulting form for values of its free variables. (*Church 1956,* pages 32–33)

Observe that, on this account, contextual definitions may in all cases be seen as introducing (one or more) connectives—although this terminology is not actually used in the literature.

To illustrate and emphasize these distinctions, let us consider a simple form of a possible definition:

(CD) $F(x) =_{\text{df}} G(x) \& H(x)$.

This is a more or less standard form, among philosophers at any rate, of presenting definitions of predicates. But the difference between this, seen as a contextual definition, and as an explicit definition of a predicate is not usually noted—indeed, it is often of no real consequence.[33]

Taken as a contextual definition, this does not introduce a meaning, strictly so-called, for a predicate. Note well that if juxtaposition of a predicate with a variable (or other term) enclosed in parentheses has already been assigned a meaning (say application of function to argument or converse elementhood), then we are *not* at liberty to offer this as a stipulative definition of a predicate 'F'. We cannot, by fiat, make it the case that 'F' has a meaning which, when combined with our notation shall preserve the meaning of juxtaposition in '$F(a)$', say, and also make that meaning identical with that of '$G(a) \& H(a)$'.[34] But we can certainly introduce a new predicate by a purely stipulative, explicit definition in some such way as:

(D) $F =_{\text{df}} \lambda x(G(x) \& H(x))$.

In (CD) the expression 'F' is best seen as a *connective* (admitting the oddity of the terminology). In (D) we have a quite different case—a predicate is explicitly defined as a proper symbol and has been assigned a meaning in isolation. Both of these are of course entirely permissible methods of definition, but in the present connection, they must not be conflated.

Synonymous isomorphism, according to its intuitive motivation, should be understood to allow substitutions on the basis of both explicit and contextual definitions.

---

[33] The distinction becomes important when intensional meanings are under consideration. George Bealer, for example, does not seem to recognize the distinction in *Bealer 1982* and thus there seems to arise for his treatment a problem due to F. P. Ramsey (*1925*, in *Mellor 1990*, pages 14–15). (Cf. *Anderson 1987*, page 120, for details.) Once the difference between the two types of definition is clearly seen, the puzzle dissolves. Bealer partly sets matters right in *Bealer 1994*, but glides over the fact that his previous failure to distinguish the two types of definition actually leads to problems for the application of his suggested analyses of various notions—see further his note 3, page 165.

For the record, I do not, and did not (in *Anderson 1987*), accept the argument which he uncharitably attributes to me (*Bealer 1994*, page 142). I was using something analogous to the schema he calls, in the later piece, "highly intuitive" (*Bealer 1994*, page 154), viz. "If Fx *iff*$_{\text{df}}$ $\ldots x \ldots$, then the proposition that Fa is the proposition that $\ldots a \ldots$".

[34] Alonzo Church brought me to see the problem about the possible clash of meanings in such cases.

6. COMMUTATIVE MEANINGS

Armed with these distinctions, let us return to the examples. Consider first the commutation with conversion cases, (1a) and (1b).

There are two immediately plausible reconstructions for (1a), it being indifferent to the case whether we see them as explicit or contextual definitions. On the one, '$a < b$' is being set synonymous with 'a precedes and b follows' and '$b > a$' is being set synonymous with 'b follows and a precedes'—the not-implausible-assumption being that these last two English expressions are synonymous. On this showing, the "readings" are just that, suggestive but not necessarily intended as themselves synonymies. In this case, the matter devolves to the natural language, the stipulation having succeeding only in transferring the question of identity of meaning. Similarly if we take the displayed readings in English to literally give meaning to the formal expressions, the question is then whether the expression 'a is less that b' is synonymous in English with 'b is greater than a', as it arguably is not. In either case the stipulation has only succeeded in making the synonymy of formal expressions depend on that of natural language counterparts.

The example (1b) is most naturally seen as involving contextual definitions. Both '$a < b$' and '$b > a$' are being treated as connectives and are stipulated to be sense-concurrent with '$b - a$ is positive'.

We might regard '$a > b$' as asserting that a is greater than b and see '$b < a$' as asserting the same, except that the convention is that the latter is to be read from right to left. (Of course, we might also see both of them as asserting that b is less than a, with the first formula being read from right to left.) There is here really no difference from the corresponding contextual definitions.

Both of these are perfectly legitimate, but observe that the corresponding *explicit* definitions would require an additional assumption if we are to conclude the synonymy of '$a < b$' and '$b > a$'. In effect, the assumption is that there are relational attributes or meanings which can serve as simultaneous "solutions" for '$<$' and '$>$' in the two equations '$a < b =_{\mathrm{df}} b - a$ is positive' and '$b > a =_{\mathrm{df}} b - a$ is positive'.[35] And this does not go without saying.

Within the context of mathematics, the difference between obvious logical equivalence and synonymy is often little significance. [36] So we might view, in all these cases, the definitions as only requiring that relation between definiens and definiendum and thus as not really bearing directly on the question of synonymy.

Given these observations and considering the intuitive character of the

[35]Such a solution is possible if we regard lambda-conversion as preserving sense and (hence) take contextual and explicit definitions to amount to essentially the same thing in these cases. See the discussion of Alternative (1) below.

[36]It cannot be the same since synonymy is transitive and obvious logical equivalence is not.

alleged counterexamples, it does not seem to be required that the theory of meaning make special provisions for such cases.

The alleged counterexamples in cases of commutativity of relations and connectives which are (necessarily) symmetrical are less clearly defeasible. This is partly because of the intuitive force of the examples: '*a* equals *b*' and '*b* equals *a*' just do not seem to differ in meaning at all—even recognizing the fact that not all relations are symmetrical. And the matter is equally troubling in the case of connectives, for example '*P* and *Q*'.

Let us consider first the more difficult case of a formalized language with a connective for conjunction. One of our questions is whether one can produce by a definition together *with a stipulation that order is not significant*, say, a "commutative" concept of conjunction. One cannot.

Once we have assigned a meaning to, for example, "&", or to contexts in which it occurs, it is not clearly within our power to make that meaning commutative. Suppose, for example, that conjunction is defined (or explained in the metalanguage) using implication (or disjunction) and negation—along the lines indicated by the contextual definition:[37]

$$[A \mathbin{\&} B] =_{\mathrm{df}} {\sim}[B \supset {\sim}[B \supset A]].$$

Surely there are ways of giving appropriate meanings to '$\sim$' and '$\supset$' so as to make the corresponding law of the commutativity of conjunction quite significant.

So we must ask instead whether or not *there is* a permissible connective (or concept of conjunction) which we could introduce with (or express by way of) a symbol, say, '$\widehat{\&}$', and such that '$[A \mathbin{\widehat{\&}} B] \equiv [B \mathbin{\widehat{\&}} A]$' is not only a truth, but is a "meaning identity"—the two sides of the equivalence are synonymous.

Upon reflection, it appears that there are entirely permissible "symmetrical ways" of introducing such a connective. Together with an appropriate explanation for computing values, we can take it that the truth-table specification of a connective gives its "meaning-producing character", to which Church alludes in the quotation above—of course it depends on exactly how the truth-table is given. To introduce a meaning-symmetrical connective, we might give two truth-tables:

$\widehat{\&}$	t	f
t	t	f
f	f	f

[37]I am being a bit lax here about the difference between contextual definitions of connectives and explicit definitions of corresponding truth-functions. If such a commutative concept of the truth-function is possible, then both kinds of definition can be used in connection with it.

$\widehat{\&}$	f	t
f	**f**	**f**
t	**f**	**t**

The explanation is to be added that these are entirely on a par, even though one is here given before the other. Better still, we could give a symmetrical truth-table:[38]

$\widehat{\&}$	t	f	
t	**t**	**f**	t
f	**f**	**f**	f
	t	f	$\widehat{\&}$

Or we might more directly just construe such a table as defining a truth-function, the "associated function" of the connective. This may not be entirely persuasive.[39] But consider the following. First observe that it is possible to introduce a commutative *connective*. Consider a system of the propositional calculus which has disjunction as a connective, introduced in any way at all—not necessarily as having a commutative meaning. For simplicity suppose that all wffs are either sentential constants, $P_1, P_2, \ldots$, or $(\phi \vee \psi)$, where ϕ and ψ are wffs. Let there be given an interpretation, so that each sentential constant expresses a proposition and the complex wffs express propositions involving whatever concepts of disjunction are chosen for the purpose. Choose a particular effective enumeration of the wffs of S. Now consider an extension of S, S^*, whose wffs are determined by the rules of S together with the rule: If ϕ and ψ are wffs, then $A\phi\psi$ is a wff, where 'A' is a new binary connective. Assign to each wff ϕ of S^*, a wff ϕ^* of S determined by the following recursive rule: If ϕ is a wff of S, then ϕ^* is just ϕ. $(A\phi\psi)^*$ is the wff $(\phi^* \vee \psi^*)$ or the wff $(\psi^* \vee \phi^*)$, whichever comes first in the effective enumeration. The meanings of the wffs of S^* are determined by the rule that ϕ is to be synonymous with ϕ^*. By any plausible criterion, 'A' is a connective representing disjunction and $A\phi\psi$ is fully synonymous with $A\psi\phi$.

It might be maintained that a genuine counterexample to Alternative (0) must involve a proper symbol actually denoting the truth-function Disjunction, and not merely a connective which has such as its associated function.

[38] Even the repetition of the connective could be eliminated if we could use three dimensions.

[39] (Added in proof) Indeed, it doesn't persuade me since I presented and discussed this material at a Philosophy of Language workshop at UCLA. The counter example to follow was added in proof and owes much to the participants of that workshop, expecially David Kaplan, Terence Parsons, Louis deRosset, and Calvin Normore.

Well, consider again a simple system of propositional calculus, but instead of a connective for disjunction, let 'δ' denote the truth-function Disjunction (under some concept). Then $\delta(\phi, \psi)$ is a wff if ϕ and ψ are, and this means the application of the truth-function Disjunction to the truth-values of the wffs enclosed in parentheses. We can proceed as before by effectively enumerating the wffs and extending to a system containing wffs of the form $\delta[\phi, \psi]$—explained as being synonymous with that one of $\delta(\phi^*, \psi^*)$ and $\delta(\psi^*, \phi^*)$ which comes first in the enumeration. In this case we may regard 'δ' in the extended system as denoting Disjunction, and under the concept expressed by 'δ' in the system extended. What we have done here is to define a *connective* represented by square brackets in terms of Disjunction and ordinary application of function to argument, here represented by parentheses. And we may, if we choose, regard the wffs of the extended system as actually containing wffs of the form $\delta[\phi, \psi]$, rather than, as is more usual, construing the definition as a metalinguistic abbreviation.

If it is objected that the square brackets still represent a connective, then the reply is that this is correct but that it is in principle impossible to eliminate all connectives.[40] And in any case the observation does nothing to diminish the force of the proposed counterexample. If something more is required for a counterexample to synonymous isomorphism, then it is difficult to see what it might be.

Similarly, we could introduce or define other meaning-symmetrical connectives and functions, e.g. meaning-symmetric material equivalence, $\overline{\equiv}$, and accommodate the other counterexamples (taking them as genuine).

$$a \mathrel{\overline{=}} b =_{\text{df}} (F)[F(a) \mathrel{\overline{\equiv}} F(b)],$$

and

$$\{\widehat{a, b}\} =_{\text{df}} \{x : x = a \mathbin{\overline{\vee}} x = b\}$$

are definitions (one contextual, one explicit) of meaning-symmetrical notations for identity and pair-sets.

This last yields a plausible and satisfying account of the "definitions" introduced in the informal examples (2a) and (2b).

Of course there are other ways of reconstructing those examples. As already noticed, we might see them as employing a notion of definition which does not require strict meaning identity, obvious logical equivalence might be enough for many mathematical purposes—if certain other conditions are met. The matter is different if we try to embed the notation into a logic of belief, say.

Given the idea of symmetrical truth-tables, some general argument will be needed to show that there must be a difference in meaning conveyed by order and which is somehow lurking in the truth-table method we have used.[41]

---

[40] For a compelling argument for this, see *Church 1956*, page 35, footnote 87.

[41] Church's preference would presumably be to have the values of the "truth-tables" for the connectives given by way of semantical rules (see *Church 1956*). On some ways of

Once we admit symmetrical conjunction (or disjunction), it becomes possible to prove that there are symmetrical concepts of other definable binary functions—not confined to truth-values. To illustrate, suppose that a necessarily symmetrical binary function f is set-theoretically defined, say by:

$$f =_{\text{df}} \{\langle x, y, z\rangle : A[x, y, z])\},$$

where $A[x, y, z]$ is a well-formed formula with the indicated free variables. Then a meaning-symmetrical concept of the same function is expressed by the definition:

$$\widehat{f} =_{\text{df}} \{\langle x, y, z\rangle : A[x, y, z] \mathbin{\widehat{\vee}} A[y, x, z]\}.$$

The idea obviously extends to functions of other arities and with other permutations. The binary case, and possibly even the ternary, seems to have application to natural language. It is very unlikely that such things appear explicitly in any natural languages even for quaternary functions, for obvious reasons of computational complexity. Still, the general theory might have some intrinsic mathematical interest.

On the basis of the foregoing considerations, we propose to formulate a semantical criterion of identity which allows for symmetrical or commutative meanings.

7. QUINE'S NOTATION

There are some interesting observations by Quine about the possibility of various kinds of notation which might be thought to swiftly settle the questions we have been discussing.

> Let us tentatively suppose, contrary to Whitehead as of 1898, that '$x + y = y + x$' does hold as a genuine identity; i.e., that the order of summands is wholly immaterial. A notation of addition that is more suggestive than '$x + y$', then, would consist in simply superimposing 'x' and 'y' in the manner of a monogram. This notation quite properly refrains from suggesting any order of summands; and there ceases to be any analogue of '$x + y = y + x$', the nearest approach being of the vacuous type '$z = z$'. Now the one objection to this procedure is the expense of casting monograms; and thus it is that we revert to a linear notation which imposes an arbitrary notational order on summands. The printer foists upon us a redundant notation, issuing in synonyms. One and the same sum can now be expressed in two ways, '$x + y$' and '$y + x$'. The law '$x + y = y + x$' comes to be needed as a means of neutralizing this excess of notation over subject matter. The law thus preserves a consequential role, even with '$=$' construed in the strict sense of identity.

doing this, it is a bit difficult to "trivialize" commutativity, but it still will be possible.

> It is not easy to imagine a notation capable of absorbing various other laws, e.g., '$x(y+z) = xy+xz$', in the manner in which the monogram notation absorbed '$x+y=y+x$'; nor is it easy to imagine a notation capable even of absorbing '$x+y=y+x$' and yet leaving '$x+x=x$' unprejudiced. For most purposes, notations teeming with synonyms are forced upon us by circumstances yet more compelling than the cost of monograms. Hence the utility of the identity concept.
>
> (*Quine 1941*, pages 128–129)

If we distinguish the meaning of an equation from its truth, we need not be so generous with synonymy.[42] And we must object that *for certain concepts of addition*, the new notation will simply leave us with no way of expressing a significant law of commutativity. Suppose that we define addition by the recursion equations:

(a1) $x+0=x$,

(a2) $x+y' = (x+y)'$

(or better, by an explicit second-order definition). If addition is introduced in this way, the equation '$x+y=y+x$' is both informative and needing of proof.

For other (symmetrical) concepts of addition, it may be that a monogram notation is perfectly appropriate, considerations of the publisher's and typesetter's lots being set aside. We could prevent automatic reflexivity by making the monogram of 'x' with itself darker thus: $\boldsymbol{x}$. No doubt long sums will be of doubtful legibility and such a notation would not work for non-associative operations, but even these defects might be overcome with more notational ingenuity.

Nevertheless, the fact that such notations are possible does not settle, or even tend to settle, our questions. Not every aspect of the meaning assigned to a given notation need be made explicit. We might avoid the explicit expression of concepts in the object language, even if such are clearly present nevertheless—as indeed might be expressed explicitly in giving the metalinguistic semantical rules for the notation. So the exercise of constructing notations wherein various things cannot be expressed does not seem to offer any independent evidence for the alleged synonymies.

8. ASSOCIATIVE MEANINGS?

One may see a parallel in the case of *associative* functions and connectives. Given that a certain operation is associative, one often just drops parentheses or brackets. Here is a typical example:

---

[42]Note that the Whitehead volume first appeared in 1941—the sense-denotation distinction was not at that time commonly recognized.

> Here the order of combining the numbers has no influence on the result. This *law of the (sequence) of combination* also holds for all natural numbers.
>
> It means that brackets may be omitted: $5+3+4 = 12$. Similarly, additions of more than three summands may be written without brackets. (*Gellert, Küstner, Hellwich and Kästner 1977,* page 20)

One finds a similar sort of thing in connection with associative connectives in the propositional calculus. For example:

> On account of the commutative and associative laws, conjunctions and disjunctions with several components may be written without parentheses. (*Hilbert and Ackermann 1950,* page 7)

Here it is most natural to allow that the new notation, with no parentheses, really conveys a new concept. In spite of the fact that commutativity, associativity (and reflexivity) are often introduced and discussed together in elementary mathematics texts, we have here really a very different sort of case from commutativity. The examples are plausibly seen as requiring, instead of a special sort of meaning, only the recognition of concepts (of functions) which take a variable number of arguments. In such cases, one might see '$x + y + z$' and '$x + y + z + w$' as applications of a single summation function, perhaps construed as applicable to sequences. Better then to write such things as '$\Sigma(x, y, z)$' and '$\Sigma(x, y, z, w)$,' where 'Σ' is "anadic" (cf. *Grandy 1976*), or applicable to sequences of arguments of arbitrary length. Notice that if '+' is assigned a binary function concept as its meaning, it will not be up to us to stipulate that the true equation:

$$x + (y + z) = (x + y) + z$$

expresses a synonymy. This is slightly more evident if we use a notation whereby the function is written before the arguments, say:

$$\mathsf{Plus}(x, \mathsf{Plus}(y, z)) = \mathsf{Plus}(\mathsf{Plus}(x, y), z).$$

Indeed, it is to be observed that something is explicitly denoted[43] on the left side of the equation which need not be explicitly denoted on the right, namely, y plus z. It could still be maintained that the entity is "implicitly denoted" by way of being determined by a concept which must be grasped if the right hand side is understood, and hence that we do not really have distinctness of denotations. But the claim is strained, at best.[44]

On the account here being urged, the truth (and proof) of such an identity can motivate us to introduce a concept of the corresponding anadic function

[43]Really, the entity is the value of an expression on the left for an assignment of values of the variables, but the point holds if we use constants instead of variables in our example.

[44]The matter is of course different if "Plus" has been introduced by contextual definition and is not really a denoting expression.

Σ. And of course such a concept could certainly be introduced, using set theory or along the lines suggested by the "recursion" equations:

$$\Sigma(x,y) = x + y;\ \Sigma(x_1, x_2, \ldots, x_{n+1}) = (\Sigma(x_1, x_2, \ldots, x_n) + x_{n+1}).$$

Commonly, one suspects, the notation 'x + y + z' is just used ambiguously.[45] Within mathematical contexts it is an ambiguity which makes for very little difference.[46] But we are supposing generally that equivocations are to be eliminated in the base language before the theory is applied.

I conclude that the phenomenon of associativity does not require any modification of our criterion of synonymy, but points instead to the recognition of functions of variable numbers of arguments, and concepts thereof.

9. REDUNDANT MEANINGS?

Suppose that '$\ominus$' denotes a reflexive operation (in some domain), so that: $x \ominus x = x$ holds for all x (in that domain). Is it possible that there is, corresponding to this, a meaning for '$\ominus$' of such a sort that this equation expresses a meaning identity? Even though one can find in the informal mathematical discussions of such things examples which (vaguely) suggest that the author takes this to introduce a synonymy, it does not seem to be a real possibility. If '$\ominus$' expresses a certain function concept in the context '$a \ominus b$' and the same function concept in '$a \ominus a$', it is impossible to see this latter as synonymous with 'a' (except by the introduction of some ambiguity)—one must grasp the function concept to understand the former, not so the latter. Indeed, as in the case of associativity, something is denoted by an expression in the former which is not (explicitly) mentioned using the latter, namely the function $\ominus$.[47]

A related case arises in connection with the identity function, that function which yields as value the entity which is assigned as argument. In treatments of the lambda-calculus (such as *Church 1941*), one finds such things as: $Ix = x$, where I is the identity function, representable using the lambda operator as (λxx). Let 'a' be a constant and consider the identity $(\lambda xx)a = a$ It is very difficult to see this as expressing an identity of meaning. The concept of a function, even the relatively trivial identity function, requires a mathematical abstraction of some difficulty.

---

[45]One can of course introduce a convention of, say, association to the left and drop the parentheses—but this introduces no new concept.

[46]This is not so obvious in the context of a *proof.* Observe that Frege (*1884*, as translated in *Austin 1978*, page 7[e]) faults Leibniz' purported proof of $2+2=4$ exactly on the ground that it implicitly assumes the associativity of addition. The reply is not very plausible that there is being used here an "associative meaning" for the function of addition and hence that such a step is not really an inference.

[47]I am ignoring here the possibility that '$\ominus$' is a non-denoting expression, yet having a definite function concept as meaning. Here only the first point holds. Also it is obvious that the matter is quite different if '$\ominus$' has been contextually defined or is otherwise syncategorematic.

Another possible counterexample we must consider is produced by the iteration[48] of functions or connectives. Let us take as our example the iteration of negation. There is disagreement in the literature as to whether or not a double negation, say, '$\sim\sim P$' is to be regarded as synonymous with 'P'. F. P. Ramsey and Herbert Hochberg (*Hochberg 1977*)[49] say that these are mere symbolic variants. Arthur Prior (*1963*)[50] and Church regard them as distinct. If this former opinion is correct, then there is still another case which must be accommodated in the formulation of a correct criterion of synonymy.

From the present point of view,[51] the disputed question is not well-posed. There certainly are concepts of negation where it is very difficult to see their double iteration as producing the very same proposition. For example, in a formulation of the propositional calculus in which the primitives are '$\supset$' and constant falsehood 'f', one may define a concept of (or connective for) the truth-function of negation thus:

$$\sim\mathbf{A} =_{\mathrm{df}} (\mathbf{A} \supset f)$$

Then the corresponding law of double negation is

$$\mathbf{A} \equiv ((\mathbf{A} \supset f) \supset f),$$

and for natural ways of assigning meaning to '$\supset$', this need not express a synonymy. The real question is this: Is there *any concept of negation* whereby the corresponding law of double negation expresses an identity of meaning?

I know of only one argument on the matter. F. P. Ramsey argues thus:

> We might, for instance, express negation not by inserting a word 'not,' but by writing what we negate upside down. Such a symbolism is only inconvenient because we are not trained to perceive complicated symmetry about a horizontal axis, but if we adopted it we should be rid of the redundant 'not-not', for the result of negating the sentence 'p' twice would simply be the sentence 'p' itself.[52]
>
> It seems to me, therefore, that 'not' cannot be a name (for if it were, 'not-not-p' would have to be about the object not and so different in meaning from 'p'), but must function in a radically different fashion.
>
> (*Ramsey 1927*, in *Mellor 1990*, pages 42–43)

[48]The case of the identity function, already discussed, may be usefully seen as a degenerate case of the present sort of example.

[49]Hochberg there anticipates some of the property identities we endorse, but offers no justifications for his choices.

[50]Prior states as a general principle that "... no proposition *be* a logical *complication* of itself ..." (page 192), and thus also rejects the reflexitivity of conjunction as meaning preserving.

[51]That is, the approach according to which semantics and intensional logic should not be constructed in such a way to give preference to one concept of a function over any other. See further, *Anderson 1989*, section VII.

This argument may show something, but it does not show that there is a concept of negation such that the double negation of a proposition, formed by way of that concept, is identical with the proposition itself. One might concede that symbols for negation, 'not' or '$\sim$', say, may be, and perhaps typically are, treated as connectives—non-denoting, but having only an associated truth-function (see further, *Church 1956,* section 5). The argument certainly has no tendency to show that there is no truth-function which takes **t** into **f**, and **f** into **t**. If we accept the existence of such a function, the possibility arises of various concepts of it, expressed by definite descriptions or (perhaps) names. Indeed, formalized languages with such interpretations have been actually constructed. If we consider the notation proposed by Ramsey and assign to 'p' a proposition, say that snow is white, then a negation thereof will be expressed by 'd'. Operating on this latter symbol, we get back 'p'—which, we suppose, still expresses that snow is white. The only conclusion[53] to be drawn seems to be that one need not have a notation for expressing a double negation—the analogue of the situation with Quine's monogram notation.

I conclude that we need not countenance a concept of negation whose double application to a proposition produces that very proposition. If 'Neg', say, expresses a concept of negation in a given language, then it would seem that '$\mathsf{Neg}\text{-}\mathsf{Neg}\text{-}p \equiv p$' will express some information, perhaps obvious, but not on that account entirely without content. So too, it would seem, for the general iteration of function-concepts. Here no amendments are suggested for our formulation of a criterion of identity.

Can it happen that '$\ominus(a, a) = \ominus(a)$', where '$\ominus$' expresses a function concept, expresses a meaning identity so that there is a corresponding ontological possibility? Given that we allow functions with variable numbers of arguments, the thing is not automatically ruled out as ill-formed or as immediately conceptually incoherent. This case is actually the most difficult of the "redundant meaning" examples to decide. Many of the arguments used to help decide other cases do not apply here.

On balance, I believe that we should not grant the real possibility of an additional exception to synonymous isomorphism. First, it is not obvious how we could specify such redundant meanings by semantical rules, or other ways

---

[52] It is interesting to recall Augustus De Morgan's (*1860*) notation for producing the "contrary of a term"—an expression denoting the complement of the class denoted by the term. It consists of replacing a capital letter, 'A' for example, by its small letter counterpart 'a'. If we give no particular priority to either of these, then we have a notation to which a version of Ramsey's argument could be applied.

[53] The argument does, though, highlight a relation, **Neg**, say, which holds only between propositions about which we would normally say that the one is the negation of the other. At a certain level of abstraction, one could even maintain that $\forall p \exists ! q \mathsf{Neg}(p, q)$ and $\forall p \forall q(\mathsf{Neg}(p, q) \supset \mathsf{Neg}(q, p))$. We can view this as involving a notion of proposition which abstracts away from the common element of structure between what is expressed by a sentence and an explicit negation of it.

of specifying values, so as to have something analogous to our (truth) tables for symmetric meanings. Also, there are so far no intuitive examples supporting the proposed modification, as there are in the case of the symmetry exception.

More importantly, it is difficult to see that the case could have any mathematical or philosophical interest. Even if this is a real possibility, nothing significant seems to be lost if we restrict our theory to ignore them so as to deal only with those kinds of meanings which might have some independent interest. This might be due to a failure of imagination, but if some argument has been overlooked or some mathematical or philosophical interest develops for this case, then one can see how to modify (and complicate) the criterion which we shall propose.

10. OTHER CRITERIA AND ALTERNATIVE (1*)

In his work on intensional logic, Church used three different heuristic principles to guide the construction of the logic of sense and denotation. As we have said, Alternative (0) is the approach which uses synonymous isomorphism as (semantical) criterion. Church attempts to incorporate a(n ontological) criterion into a formalized language corresponding to this—in just the sense we have explained. The logic of sense and denotation itself does not speak of expressions and semantical notions,[54] but treats intensions directly in such a way that it is intended that the semantical criterion shall be entailed in any particular case where the theory is used to give the intensional semantics of a language.

Alternative (1) is a heuristic principle according to which expressions which are lambda-convertible are to be synonymous.[55] It is interesting to note that Alternative (1) entails, under natural assumptions, the synonymy of our commutation and conversion of non-symmetrical examples. Thus it may be regarded as being confirmed to the extent that the examples are forceful. On this alternative the distinction between the explicit definition of a predicate and the corresponding contextual definition essentially disappears—and with it any supposed advantage of the latter for this case. The argument used in Ramsey's Problem (cf. note 33) is a simple proof that this alternative is incompatible with the idea that complex concepts are built up in a natural way from a finite set of constituents.

Unfortunately it entails too much. The numerals and the rules for multiplication in arithmetic, for example, can be coded into the lambda-calculus (see *Church 1941*). Thus equations of the form '$k = m \times n$' becomes trivialities, even where m and n are fairly large numbers. (If desired, you may imagine

[54]Contrary to what the name suggests, as Rulon Wells (*1952*) points out.

[55]See ***Anderson 1998*** for more details on the various criteria. Church (*1993*) proposes a slight modification of Alternative (1) as being possibly worthy of investigation.

that we learn our numerals directly from the lambda calculus).[56] This is not the correct conclusion if propositions are to be the objects of belief.

Alternative (2) is based on the idea that logically equivalent expressions are to be synonymous. Church clearly recognized that this last, while adequate for treatments of modal logic, would not do if applied to such things as the logic of belief.

In the light of the present investigation, I propose the following (as applied to a given language):

$(\mathrm{CI}_2)^{\mathrm{P}}$ If x is a sentence and y is a sentence and if x and y express z_1 and z_2, respectively, then $z_1 = z_2$ if and only if x and y are synonymously isomorphic *modulo permissible permutations.*

In order for this to be generally applicable, we must suppose that the intensional semantical rules of the language may specify rules of synonymy in such a way as to allow to permutations in certain cases. Or, if the semantics is given by rules of sense, then it shall be permitted that commutative meanings of the sort we have discussed are assigned to certain expressions of the language.

The supporting theory here envisioned will employ a notion of the *composition* of concepts, complex concepts being capable of being constructed by way of this operation.[57] It will be supposed that every complex concept has a *principal* function-concept, that is, that there is a certain function concept involved in its composition, which may be taken to be that operation (concept) application of which immediately yields the complex concept. It is also supposed that a function-concept carries with it a *permutation set,* a set of operations which it is permissible to apply to the sequence of argument concepts with preservation of identity of the resulting complex concept. In order to accommodate our observations about associativity, it is assumed that functions (and their concepts) are permitted which are applicable to a variable number of arguments. The corresponding ontological criterion is then:

$(\mathrm{CI}_1)^{\mathrm{P}}$ If z_1 and z_2 are propositions, then $z_1 = z_2$ if and only if z_1 and z_2 have the same principal function concept and the same sequence of argument concepts—modulo permissible permutations.

Let us use capital Greek letters as variables ranging over function concepts and lower case Greek letters with an arrow above as ranging over sequences

[56] Alonzo Church pointed out this consequence of Alternative (1) to me (in correspondence) and thereby persuaded me that it won't do for belief and the like.

[57] Composition is the intensional analogue of the extensional relation of set membership. David Kaplan (*1975*) has used a version of this idea in connection with Alternative (2). On the assumption that all complex intensions are obtainable by composition, a notion of simplicity is easily defined.

of concepts. Then, with bold curly brackets to express composition, we can state our criterion a bit more precisely:

$(\mathrm{CI}_2)^{\mathrm{P}}$ $\{\Phi, \vec{\alpha}\} = \{\Psi, \vec{\delta}\}$ if and only if (i) $\Phi = \Psi$, and (ii) where S_1 and S_2 are the permutation sets for Φ and Ψ respectively, there are permutations $P \in S_1$ and $P' \in S_2$ such that $P(\vec{\alpha}) = P'(\vec{\delta})$.

(The appropriate identity permutation will be postulated to belong to every permutation set.) The adequacy of this as a general ontological criterion of identity for propositions depends on the plausible postulate that every proposition is a composition.

The analogous criterion of identity for complex function concepts Φ and Ψ may be supplied by analogy. For simple function concepts (and, as a special case, simple attributes), the things are identical if there is just one, otherwise not. Semantically, as applied to a particular language, it will be a question of whether two simple expressions are synonymous—a question which we regard to be (ideally) settled for anyone who has mastered the language. (Compare the discussion of the identity of truth-values at the end of section 1).

Because of its relationship to the other criteria Church proposed, we call this "Alternative (1*)". The corresponding criterion for Alternative (0), which may still to be tenable, results in the special case where the only permutation involved is the identity. Alternative (0) has simplicity[58] in its favor and given the alternative explanations available for its apparent failure, it does not seem to be definitely refuted.[59]

The important (and difficult) remaining task is to supply a detailed logistic treatment of the intensional entities in question to embody and clarify this last, and to investigate its consequences. Any further discussion really seems to call for a precise formulation of that underlying theory. Technical details of a formalized theory of intensions embodying these ideas and the corresponding formalized version of this criterion of identity are planned to be forthcoming in full. It is hoped that the present discussion will serve to expose the project to any remaining general and principled philosophical objections,[60] and to clarify the general debate about intensional logic.[61]

[58] King (*1996*) emphasizes this advantage.

[59] The objection of Linsky (*1949*) (to intensional isomorphism) fails against Alternative (0) once contextual definitions are clearly recognized. And the objection of Myhill (*1958*, page 82) is based on assumptions that we reject—briefly, that there is such a thing as the set of all propositions, so that one can construct the Cantor diagonal set involved.

[60] I have been careful to avoid engaging the question of interlinguistic criteria, sometimes with considerable difficulty. To fully deal with these, requires, I believe, a well developed theory of (possible) languages as well as considerable further discussion. And questions pertaining to various forms of physicalism or behaviorism as applied to theories of meaning have not been here discussed. I believe that such objections can be fully answered and hope to address them in another place.

[61] I am indebted to Tyler Burge for an objection to an argument of mine about commutative meanings. The objection forced me to re-think the conceptual foundations of Alternative (1*).

REFERENCES

Almog, J., J. Perry, and H. Wettstein

1989 (editors), *Themes from Kaplan,* Oxford University Press, Oxford.

Anderson, A. R., R. B. Marcus, and R. M. Martin

1975 (editors), *The logical enterprise,* Yale University Press, New Haven.

Anderson, C. A.

1980 Some new axioms for the logic of sense and denotation: Alternative (0), *Noûs,* vol. 14, pp. 217–234.

1987 Bealer's Quality and Concept, *Journal of Philosophical Logic,* vol. 16, pp. 115–164.

1989 Russellian intensional logic, in *Almog, Perry and Wettstein 1989,* pp. 67–103.

1991 Review of *Cresswell 1985, Philosophical Review,* vol. 100, pp. 476–479.

1998 Alonzo Church's contributions to philosophy and intensional logic, *Bulletin of Symbolic Logic,* vol. 4, no. 2.

Angelelli, I.

1984 Frege and abstraction, *Philosophy Naturalis,* vol. 21, pp. 453–471.

Apostel, T. M.

1967 *Calculus,* vol. 1, second edition, Xerox College Publishing, Lexington.

Austin, J. L.

1978 (translator), *Gottlob Frege, The foundations of arithmetic,* Basil Blackwell, Oxford; English translation of *Frege 1884.*

Bealer, G.

1982 *Quality and concept,* Oxford University Press, Oxford.

1994 Property theory: The type-free approach v. the Church approach, *Journal of Philosophical Logic,* vol. 23, pp. 139–171.

Birkhoff, G., and S. MacLane

1944 *A survey of modern algebra,* Macmillan, New York.

Black, M.

1945 The "paradox of analysis" again: a reply, *Mind,* n.s. 54, pp. 272–273.

1946 How can analysis be informative?, *Philosophy and Phenomenological Research,* vol. 6, pp. 628–631.

Carnap, R.

1947 *Meaning and necessity,* University of Chicago Press, Chicago.

1963 Replies and systematic expositions, in *Schilpp 1963,* pp. 859–1013.

Church, A.

1941 *The calculi of lambda-conversion,* Annals of mathematics studies, vol. 6, Princeton University Press, Princeton, New Jersey.

1942 Incomplete symbol, in *Runes 1942,* p. 143.

1946a A formulation of the logic of sense and denotation (abstract), *The Journal of Symbolic Logic,* vol. 11, p. 31.

1946b Review of White 1945a, Black 1945, White 1945b, and Black 1946, *The Journal of Symbolic Logic,* vol. 11, pp. 132–133.

1951 A formulation of the logic of sense and denotation, in *Henle, Kallen and Langer 1951,* pp. 3–24.

1954 Intensional isomorphism and identity of belief, *Philosophical Studies*, vol. 5, pp. 65–73.

1956 *Introduction to mathematical logic, vol. I,* Princeton University Press, Princeton, New Jersey.

1993 A revised formulation of the logic of sense and denotation, Alternative (1), *Noûs*, vol. 27, pp. 141–157.

Cresswell, M. J.

1985 *Structured meanings: The semantics of propositional attitudes,* MIT Press, Cambridge, Massachusett.

Davidson, D.

1963 The method of extension and intension, in *Schilpp 1963,* pp. 311–349.

De Morgan, A.

1860 Syllabus of a proposed system of logic, in *Heath 1966,* pp. 147–207.

Frege, G.

1884 *Die Grundlagen der Arithmetik,* Wilhelm Koebner, Breslau.

1919 Negation, *Beiträge zur Philosophie des deutschen Idealismus 1,* pp. 143–157; English translation in *Geach and Black 1966,* pp. 117–135.

Fine, H. B.

1906 *College algebra,* Ginn and Company; reprinted by Dover, New York, 1961.

Geach, P.

1957 *Mental acts, their content and their objects,* Routledge & Kegan Paul, London; Humanities Press, New York.

Geach, P., and M. Black

1966 (editors), *Translations from the philosophical writings of Gottlob Frege,* Basil Blackwell, Oxford.

Gellert, W., H. Küstner, M. Hellwich, and H. Kästner

1977 *The VNR concise encyclopedia of mathematics,* Van Nostrand Reinhold, New York.

Grandy, R.

1976 Anadic logic, *Synthese*, vol. 32, pp. 395–402.

Grim, P.

1991 *The incomplete universe. Totality, knowledge, and truth,* MIT Press, Cambridge, Massachusetts.

Hahn, L. E.

1986 (editor), *The philosophy of W. V. Quine*, Library of Living Philosophers, vol. 25, Open Court, La Salle.

Heath, P.

1966 (editor), *On the syllogism and other logical writings by Augustus De Morgan,* Yale University Press, New Haven.

Henle, P., H. M. Kallen, and S. K. Langer

1951 (editors), *Structure, method and meaning, Essays in honor of Henry M. Sheffer,* The Liberal Arts Press, New York.

Hilbert, D., and W. Ackermann

1938 *Principles of mathematical logic*, English translation of the second German edition, Chelsea, New York, 1950.

Hochberg, H.

1977 Properties, abstracts, and the axiom of infinity, *Journal of Philosophical Logic*, vol. 6, pp. 193–207.

Kamke, E.

1950 *Theory of sets*, English translation of the second German edition by F. Bagemihl, Dover, New York.

Kaplan, D.

1975 How to Russell a Frege-Church, *The Journal of Philosophy*, vol. 72, pp. 715–729.

1986 Opacity, in *Hahn 1986*, pp. 229–289.

King, J.

1996 Structured propositions, *Journal of Philosophical Logic*, vol. 25, pp. 495–521.

Linsky, L.

1949 Some notes on Carnap's concept of intensional isomorphism and the paradox of analysis, *Philosophy of Science*, vol. 16, pp. 343–347.

1967 *Referring*, Humanities Press, New York.

Mellor, D. H.

1990 (editor), *F. P. Ramsey, Philosophical papers*, Cambridge University Press, Cambridge, England.

Myhill, J.

1958 Problems arising in the formalization of intensional logic, *Logique et Analyse*, vol. 1, pp. 78–83.

1963 An alternative to the method of extension and intension, in *Schilpp 1963*, pp. 299–310.

Prior, A. N.

1963 Is the concept of referential opacity really necessary?, *Acta Philosophicia Fennica*, vol. 16, pp. 189–198.

Quine, W. V.

1941 Whitehead and the rise of modern logic, in *Schilpp 1941*, pp. 127–163.

1953 *From a logical point of view*, Harvard University Press, Cambridge, Massachusetts; Harper & Row, New York.

1953a Two dogmas of empiricism, in *Quine 1953*, pp. 20–46.

1970 *Mathematical logic*, revised edition, Harvard University Press, Cambridge, Massachusetts.

1975 On the individuation of attributes, in *Anderson, Marcus, and Martin 1975*, pp. 3–13.

1981 *Theories and things*, Harvard University Press, Cambridge, Massachusetts.

Ramsey, F. P.

1927 Universals, in *Mellor 1990*, pp. 8–30.

1927 Facts and propositions, in *Mellor 1990*, pp. 34–51.

Richard, M.

1990 *Propositional attitudes, An essay on thoughts and how we ascribe them*, Cambridge University Press, Cambridge, England.

Runes, D. D.
1942 (editor), *Dictionary of philosophy,* Philosophical Library, New York.

Russell, B.
1903 *The principles of mathematics,* Allen and Unwin, London.

Schilpp, P. A.
1941 (editor), *The philosophy of Alfred North Whitehead,* Library of Living Philosophers, vol. 3, Northwestern University, Evanston: second edition, Tudor, New York.
1963 (editor), *The philosophy of Rudolf Carnap,* Library of Living Philosophers, vol. 11, Open Court, La Salle; Cambridge University Press, London.

Warner, S.
1965 *Modern algebra,* Prentice-Hall, Englewood Cliffs; reprinted by Dover, New York, 1990.

Wells, R.
1952 Review of Church 1951, *The Journal of Symbolic Logic,* vol. 17, p. 133.

White, M. G.
1945a A note on the "paradox of analysis", *Mind,* n.s. 54, pp. 71–72.
1945b Analysis and identity: A rejoinder, *Mind,* n.s. 54, pp. 357–361.

Williamson, T.
1986 Criteria of identity and the axiom of choice, *The Journal of Philosophy,* vol. 83, pp. 380–394.
1990 *Identity and discrimination,* Basil Blackwell, Oxford & Cambridge.

JOHN P. BURGESS

NOMINALIST PARAPHRASE AND ONTOLOGICAL COMMITMENT

For Alonzo Church on the occasion of his ninetieth birthday.

Abstract. Several *nominalist paraphrases,* in the sense of methods by which assertions of conventional mathematics that give the appearance of implying that there are such things as numbers can be systematically replaced by other assertions not giving this appearance, have been proposed in the literature. Can the success of such methods help show that the original assertions are not really *ontologically committed* in the sense of not really being incompatible with the denial that there are such things as numbers? I argue against the claim that it can.

1. COMPATIBILIST NOMINALISM

The debate over intuitionism, which reached its climax in the early 1930's, just as Alonzo Church was beginning his career, is sometimes said to have been about existence theorems. This is not quite accurate, since there was no debate over such existence theorems as:

(1) There are numbers that are greater than $10^{10^{10}}$ and are prime.

It would be more accurate to say that the dispute was over existence theorems with nonconstructive proofs; most accurate to say it was over nonconstructive proofs. This debate reached a resolution of sorts: The mathematical community did not cease to accept nonconstructive proofs, as demanded by the intuitionists, but did learn, through Church's Thesis, how to distinguish within classical mathematics between constructive and nonconstructive, and so to partially address the concerns that motivated intuitionism.

A debate that *can* accurately be said to have been about existence theorems, and to which Church also contributed, began in the 1940's, namely, the debate over *nominalism* in its modern sense that was introduced in an article by Goodman and Quine (*1947*) in the journal founded and edited by Church. Nominalism in this sense denies:

(2) There are (such things as) numbers.

In denying (2), nominalists are not denying anything explicitly asserted in conventional mathematics, since, as pointed out by Carnap (*1950*), conventional mathematicians spend little time speaking (and presumably little time

C. Anthony Anderson and M. Zelëny (eds.), Logic, Meaning and Computation, 429-443.

thinking) about such characteristically philosophical theses as (2). Nonetheless, Goodman and Quine did not doubt that in denying (2) they were denying something implicit in assertions of conventional mathematics, such as (1).

This is not to say that their position was straightforwardly *revisionist* like that of the intuitionists: They did not take it to be a task for mathematicians to seek a nominalistic reformation of mathematical theory. On the contrary, Goodman (*1956*), in his position paper in a volume on nominalism edited by Church and others, devotes a section (§3) to answering objections and in answering one of them (viii) explicitly rejects revisionism:

> *Objection*: Nominalism would hamper the development of mathematics and the other sciences by depriving them of methods they have used and are using to achieve some of their most important results.
>
> *Answer*: Not at all. The nominalist does not presume to restrict the scientist. The scientist may use platonistic class constructions, complex numbers, divination by inspection of entrails, or any claptrappery that he thinks may help him get the results he wants.

Indeed, insofar as Goodman and especially Quine themselves worked in mathematics (mathematical logic), they spoke like conventional mathematicians while doing mathematics; though when doing philosophy they spoke, as above, of such "platonistic claptrappery" as, e.g., complex numbers being comparable to "divination by inspection of entrails". This is not to say their position was straightforwardly *instrumentalist* like that of the formalists. They did not take it to be *sufficient* merely to deny while doing philosophy what is asserted while doing mathematics, saying, "We don't really believe what we said," and nothing more.

Rather, they took it to be a task for philosophers to seek a nominalistic reformulation of mathematical language, so that they would be able to say while doing philosophy, "We don't really mean what we said while doing mathematics," and to say something more, "What we really mean is rather the following ...," where the reformulation or paraphrase of what they said would follow. Thus Goodman continues:

> But what he produces then becomes raw material for the philosopher, whose task is to make sense of all this: to clarify, simplify, explain, interpret in understandable terms. The practical scientist does the business but the philosopher keeps the books. Nominalism is a restraint that a philosopher imposes upon himself, just because he feels he cannot otherwise make real sense of what is put before him.

Goodman and Quine did not find more than a few of the paraphrases they were seeking, and Goodman gradually became a less vocal proponent of nominalism, while Quine soon became a quite vocal opponent. Already in the paper just quoted, written within a decade of their first nominalistic collaboration, Goodman writes of Quine's "recent as still somewhat tentative defection from nominalism". But the opposition voiced in Quine's later writings has tended to incite rather than deter later nominalists. Thus the Introduction to *Chihara 1973* begins by quoting an anti-nominalist passage from one of Quine's works, and continues:

> When I first read these words in Quine's *Word and Object* several years ago, I wrote in the margin: "This philosophical doctrine should be soundly refuted."

By the final chapter V and technical Appendix of the book, a strategy of paraphrase is been suggested, which has since been further developed in Part I of *Chihara 1990*. Part II, a survey of recent literature, mentions several competing strategies.

The competing strategies in the recent literature tend to share two common features distinguishing the later nominalists from Goodman and Quine. First, they tend to use apparatus the original nominalists would not have allowed themselves. Thus Chihara uses *modality*, a distinction between what there actually is and what it would be possible in principle, if not feasible in practice, to make. While for specialists, the most interesting aspect of Chihara's work is his development of a modal paraphrase for mixed contexts involving both physical and mathematical expressions and for contexts involving analytical as well as arithmetical expressions. For present purposes it will be enough to indicate how he treats these last. Assertions like (1) are systematically replaced by assertions like:

(3) There could be constructed numerals that were greater than $10^{10^{10}}$ and were prime.

Whereas (1) appears to imply (2), which Chihara denies, (3) merely appears to imply:

(4) There could be constructed (such things as) numerals.

which Chihara asserts.

Second, later nominalists tend not to be explicitly *incompatibilist*. They tend not to concede that the thesis that they as philosophers deny, *viz.* (2), is implicit in theorems that mathematicians assert, e.g., (1). This is not to say that they tend to be explicitly *compatibilist*, insisting that despite appearances (1) does not really imply (2). They tend rather simply to be inexplicit. It is unclear to me, for instance, whether Chihara wishes to deny (1) or to deny that (1) implies (2).[1] In his survey, he is critical of those rival nominalists who tend towards more explicit compatibilism, but also of those

[1]The present note derives (very indirectly) from (the first half of) the author's contribution to a symposium on "Modality and Existence" held at Princeton University, April, 1992. The author is grateful to the other symposiast, Professor Charles Chihara of the

who tend towards more explicit incompatibilism and of the incompatibilist arguments of the anti-nominalist Burgess (*1983*).

My aim in this note will be to subject compatibilism to explicit scrutiny. The issue is not whether nominalists like Goodman and Quine or Chihara can dissociate themselves from the implications of what they may say while doing mathematics by subsequent retractions while doing philosophy (or perhaps silent reservations even while doing mathematics). No one denies that (in some sense) they can. The issue is, rather, what those implications *are*. The issue is whether conventional mathematical theory, including (1), implies (2).

2. ONTOLOGICAL COMMITMENT

Questions of this kind are often discussed under the rubric "ontological commitment", which Quine introduced (*Quine 1948*) and explained (*Quine 1951*). Chihara (in the §III.3 of his first book) has been critical of Quine's explanations of the jargon "ontological commitment", and he is not the only one to have found them obscure. No one has done more to clarify the notion of "ontological commitment" than Church (*1958*). The notion of "ontological commitment" that is relevant to the issue of compatibilism, and roughly the one adopted by Church, whose formulation not even Chihara has been able to criticize in point of clarity,[2] is the following: T is *ontologically committed* to S's if the following is implied by T: "There are (such things as) S's, [or there is (such a thing as) an S]". In the instance relevant to the issue of compatibilism, T is a body of scientific theory (conventional mathematics), involving a distinctive general kind of expression, S expressions (number expressions), and it is debated among philosophers whether there are correspondingly things of a distinctive general kind, S's (numbers).

Now, there is no difficulty (apart from Church's Theorem) in determining the ontological commitments of a theory, when the theory is formalized. But in connection with the issue of compatibilism, we are concerned with

University of California at Berkeley, and to participants (too numerous to list individually) in the question and answer session that followed our talks for useful comments. Professor Chihara's contribution will appear elsewhere. The question whether he wished to deny (1) or to deny that (1) implies (2) was put to Professor Chihara during the question and answer session after his symposium talk. One school of thought doubts whether it is appropriate to quote in print what a philosopher says on such an occasion, given that it may represent no more than an initial reaction to an unexpected objection, rather than a considered opinion. But since Professor Chihara's practice in his critical survey plainly shows that he is not of this school of thought, it may be reported that he said he was undecided.

[2]Chihara does note the following: Suppose some mathematical or linguistic theory involving such scientific terminology as "number" or "proposition", but not the philosophical term of art "abstractum", is ontologically committed to numbers or propositions. Suppose some philosophical theorists defines "abstractum", as many do, to mean "number, proposition, or the like". Still, according to the above definition, the mathematical or linguistic theory cannot, strictly speaking, be said to be "ontologically committed" to abstracta.

an unformalized body of theory. The issue is one as to whether or not a certain theory in unformalized language implies a certain thesis also in unformalized language, with the further complication that the theory consists of assertions that are *ordinary* in the sense that they might well be made by a philosophically unselfconscious scientist, while the thesis is an assertion that is *extraordinary* in the sense that it would hardly be made except by a philosopher.

Now Church states his definition for formalized theories, but also wishes to apply to it to unformalized bodies of theory, namely, to the views of certain philosophers of the Oxonian ordinary language school about "propositions". (The status of propositions had also been what most concerned Carnap, who feared nominalist objections to such things would hamper the development of linguistics.) Church quotes assertions of the kind:

"there is a proposition that ...",

from several authors. Do such assertions imply:

"there is (such a thing as) as proposition",

which the authors whom Church quotes deny? Certainly the answer is yes *if* it is appropriate to formalize the "there is" in both cases as "$\exists$". For:

$$\exists x(Px \wedge \ldots)$$

implies:

$$\exists x(Px).$$

But should the "there is" here be formalized as "$\exists$"? In this connection, Church enunciates (at the end of his paper) a principle that might be called "Church's Thesis", were not that label already applied to another and more famous doctrine:

> [T]hose philosophers who speak of "existence," "reality," and the like are to be understood as meaning the existential quantifier, and are to be condemned as inconsistent if on this basis inconsistency appears in their writings. The justification is that no other reasonable meaning of "existence" has been provided (which would fit into the contexts of the kind that were quoted), and the burden of providing such a second meaning of "existence" rests on those whose writings or whose philosophical views require it.

I interpret this passage roughly as follows: Perhaps the authors Church quotes don't really mean what they say when they say, "There is a proposition that ...": In that case, the burden is on them to provide a paraphrase indicating what they *do* really mean. Or perhaps the authors in question only mean to deny that there "are" (such things as) propositions in some extraordinary sense of "are": In that case, the burden is on them to clarify what this sense may be.

The principle to be found in the early writings of Quine, and more clearly in the passage above, seems to be that there is a presumption to the effect that:

(5) There are S's that ... [or: there is an S that ...],

implies:

(6) There are (such things as) S's [or: there is (such a thing as) an S].

and that the burden is on a would-be compatibilist to show how denial of (6) can be consistent with assertion of (5).

But Quine's critics, including Chihara, have noted that this principle provides no fully reliable criterion. Consider, for instance:

(7) There is a strong chance that Moriarty is responsible for this outrage.

Surely this does not imply:

(8) There is (such a thing as) a strong chance.

This is surely an exceptional or idiomatic use of "there is", in contrast to a normal or paradigmatic use as in:

(9) There is a sturdy chair that Moriarty was reposing in this afternoon,

which surely this does imply:

(10) There is (such a thing as) a sturdy chair.

Can a fully reliable criterion be obtained by *amending* the principle above to exclude idiomatic exceptions? There is some difficulty in formulating the amendment in such a way as to provide a widely applicable and genuinely useful criterion. A formulation in terms of a distinction between occurrences of "there is" that are appropriately formalized as "$\exists$" and ones that are not would be useless, since it would leave one with the question which occurrences of "there is" these are. A formulation involving only notions from traditional grammar can distinguish (7) from (9), since in the latter, what comes after the "that" is a verb phrase; in the latter, a complete sentence; but one needs an applicable formulation, not just the example, of the *chance* that Moriarty is responsible, but also all the others considered in the literature (the *sake* of the illustrious client, the *whereabouts* of the missing treaty, the *identity* of the young lady's mysterious suitor, the *marital status* of Doctor Watson).

Whatever the correct interpretation of his early writings, Quine in the end made no claim to be able to provide a criterion. In a comparatively late retrospective, the opening and almost-title essay of an anthology (*Quine 1981*) of his writings from the 1970's, he devotes a section (§II) to the problem, and writes thus of the question where to draw the line between ontologically committed and uncommitted:

> [T]here is no line to draw. Bodies are assumed, yes; they are the things, first and foremost. Beyond them there is a succession of dwindling analogies. Various expressions come to be used in ways more or less parallel to the use of the terms for bodies, and it is felt that corresponding objects are more or less posited, *pari passu*; but there is no purpose in trying to mark an ontological limit to the dwindling parallelism.
>
> My point is not that ordinary language is slipshod, slipshod though it be. We must recognize this grading off for what it is, and recognize that a fenced ontology is just not implicit in ordinary language. The idea of a boundary between being and nonbeing is a philosophical idea ...

For Quine, the difference between the ontologically committed and uncommitted has become one of degree, not kind. It is a matter of the degree of resemblance between the pattern of usage of S expressions in T and the norm or paradigm, the pattern of usage of table and chair expressions in everyday discourse. For Quine, though there is no line to be *found*, yet one can be *made*, by formalizing (or "regimenting") the theory; but owing to the absence of a sharp distinction between committed and uncommitted in unformalized language, there is an absence of a sharp distinction between appropriate and inappropriate formalizations (or "regimentations").

While this view of Quine's is controversial, given a strong resemblance in pattern of usage to the norm or paradigm, even those tending towards compatibilism will tend to concede that there is at least a strong *appearance* of ontological commitment. Insofar as the burden of proof is generally on the side maintaining that reality is very different from appearance, the burden of proof will be on the would-be compatibilist.

This conclusion resembles the principle mentioned above, except that it involves consideration not just of one feature of usage, but rather of several. Those features emphasized in the writings of Quine, Church, and others include:

Nominal:	The usage of S expressions as nouns
Quantificational:	The making of assertions of the form: "There is an S that ..."
Identificational:	The making of assertions of the form: "This S is the same (one) as that S"

In the case of the use of natural number expressions in conventional mathematical and scientific and everyday discourse, the resemblance is very strong as regards all the features of usage cited. As for the nominal criterion, we have an elaborate system of numerals:

one, two, three, ...;

and besides these, compound phrases are formed using such expressions as:

the number of ..., the sum of ..., the product of ....

The numerals function sometimes as adjectives, as in:

1001 nights passed,

but often as nouns, as in:

1001 is composite,

while the compound phrases function only as nouns, and may occur as subject, direct object, or indirect object in a sentence.

As for the quantificational criterion, we have both the particular and the universal:

there is a number ..., for every number ...,

as well as complicated iterations, alternations, and nestings:

for every number m there is a number n ....

These appear not only in the indicative in theorems, but also the interrogative in problems and the imperative in algorithms:

is there any number n ...? take any number n ...!

As for the identificational criterion, we have both equations and inequalities:

... equals ..., ... is greater than ..., ... is less than ...,

as well as assertions of unicity and multiplicity:

there exists a unique number ...,

there exist several distinct numbers ....

(And to mention less philological considerations, there is something explicitly called "the theory of numbers" cultivated as an important branch of mathematical science, whereas there is no "theory of snags" or "theory of chances". Or rather, there *is* a theory of chances, the theory of probability, but in this theory a chance or probability is simply a *number*.) It is such features that constitute at least a strong *appearance* of ontological commitment, a strong *prima facie* case for the incompatibilist to answer.

3. NOMINALIST PARAPHRASE

Can the availability of a nominalist paraphrase help the compatibilist answer this case? Can the availability of a nominalist paraphrase help show that conventional mathematics, including (1), does not imply (2)? One early discussion of "ontological commitment", from about the same period as the early contributions of Quine and Carnap and Church, namely, Alston (*1958*), seems to suggest that paraphrase can *never* be relevant to questions of ontology. This is surely going too far, and it may be well to consider an instance constituting a counterexample to this suggestion.

Consider the method of paraphrase that replaces "there is a strong chance that ..." by "it won't be surprising if it turns out that ...". Applied to something that is ordinary in the sense indicated above, like (7), this method produces something like:

(11) It won't be surprising if it turns out that Moriarty is responsible for this outrage.

And the method seems to be *ordinarily adequate* in the sense that an ordinary original and its paraphrase seem to be usable more or less interchangeably in the context of philosophically unselfconscious discourse. Applied to the extraordinary (8), however, the method produces something like:

(12) It won't be surprising if it turns out

which is not a proper English sentence. This outcome may very well help lead us to view (8) itself as being, not an implication of (7), but rather a piece of nonsense.

The situation seems to be different with the modal paraphrase of conventional mathematics, which turns the ordinary (1) into (3). Again, the method of paraphrase seems ordinarily adequate in the sense indicated above, but in this instance, the method, applied to (2), produces something that seems to be proper English, to be implied by the paraphrase (3) of (1), and to be asserted by modal nominalists themselves, namely, (4). If the argument of compatibilist nominalists is supposed to be that they are warranted (despite their professed nominalism) in asserting (1) because its paraphrase (3) is assertable, parity of reasoning would suggest that one is warranted also in asserting (2), since its paraphrase (4) is assertable. In any case, the outcome that the paraphrase of (1) appears to imply the paraphrase of (2) can only reinforce, not undermine, the appearance that the original (1) implies the original (2).

This is not to say that the modal method of paraphrase can *never* be relevant. Indeed, applied to:

(13) There is no number greater than $10^{10^{10}}$ but there could have been,

the method produces something like:

(14) There could not have been a numeral greater than $10^{10^{10}}$, but there could have could have been.

And this outcome may very well help lead us to view (13) as a piece of nonsense, since (14) seems not to be proper English. This in turn might well help lead us to view numbers as appropriately classifiable as "abstract", since the inapplicability of the modal distinction between what is and what could have been seems to be part of what is meant by "abstract" (when this term is not just defined by a list of examples). (Note that Chihara's method of paraphrase applies the distinction *not* to "abstract" numbers, but rather to "concrete" numerals.) But abstractness is one thing, and nonexistence another.

The common feature of the examples considered so far is this: If an ordinarily adequate method of paraphrase is applied to a body of assertions T (which may consist of a single assertion, $T = \{\tau\}$) to produce $T*$ (which again may consist of a single assertion, $T* = \{\tau*\}$) and to an extraordinary assertion σ to produce $\sigma*$, then the paraphrase will tend to undermine or reinforce the appearance that T implies σ according as $T*$ does or does not appear to imply $\sigma*$. The issue of appropriate formalization may be reconsidered in the light of the principle just suggested.

As already indicated, were it agreed that (1) is to be formalized as something like:

(15) $(\exists x)(Nx \wedge \ldots)$,

compatibilism would not be an issue. For (15) implies:

(16) $(\exists x)(Nx)$,

and (16) may be "deformalized" as something like (2). But as already indicated, the appropriateness of this formalization is just what is *not* agreed. Whether *any* formalization is *uniquely* appropriate seems to be disputed by Quine, who, as indicated above, holds that there is no sharp line between appropriate and inappropriate formalization.

But even if one accepts this not uncontroversial view of Quine's, there is at least a *conventional* method of formalization for conventional mathematics, which would be used by any philosophically unselfconscious logician. This method is also ordinarily adequate, not quite in the sense indicated above, but in the sense of the following claim, sometimes called Hilbert's Thesis: For *ordinary* τ and θ, logicians' judgments about when the formalization $\theta*$ of θ is implied by or deducible from the formalization $\tau*$ of τ tend to agree with mathematicians' judgments about when θ is deducible from or implied by τ. (This agreement underlies the utility of the logicians' incompleteness or undeducibility results for mathematicians, much as Church's Thesis underlies the utility of incomputability or undecidability results: A result to the effect that $\theta*$ cannot be deduced from $\tau*$ is supposed both to explain the failure of past efforts to deduce θ from τ and warn of the futility of such future efforts.) Since this conventional method does replace (1) and (2) by something like (15) and (16), it tends, in accordance with the principle suggested above, to reinforce the appearance that (1) implies (2).

It is no accident that Chihara's method of paraphrase, like others in the literature, provides (2) with a paraphrase apparently implied by the paraphrase provided for (1). For indeed, the modal paraphrase is obtained *by way of the* conventional formalization, in something like the following sequence of steps:

(i) given an original like (1),

(ii) formalize it conventionally, obtaining something like (15);

(iii) replace the existential quantifier "$\exists x$" with variable x for numbers by a constructibility quantifier "$\odot\xi$" with variable ξ for numerals;

(iv) "deformalize" the result, obtaining a paraphrase like (3).

It is almost inevitable that a method of paraphrase that proceeds in this way will produce something like (4) as paraphrase for (2).

4. SPECIAL RELATIONSHIPS

Sometimes the compatibilist appeal to paraphrase is supplemented by a suggestion about the closeness of the relationship between paraphrase and original, best expressed in the terminology of modern linguistic theory. In the terminology that is perhaps best known to philosophers, the relationship suggested would be that (3) uncovers the "deep" structure underlying the "surface" structure of (1). This terminology derives from an early phase of the work of Noam Chomsky,[3] and has been superseded in later phases; but under one terminology or another, a distinction between the "superficial" analysis of traditional grammar and the "depth" analysis of modern linguistics has survived in many later theories. In another terminology, the claim would be that (3) uncovers the "logical form" underlying the "grammatical form" of (1). (This latter formulation of the claim suggests that Quine's views about formalization are being rejected, and the existence of a unique "logical form" and hence a unique appropriate formalization, is being assumed.)

[3]Formulated in urChomskyite terminology, the claim that "(3) reveals the 'depths' underlying the 'surface' of (1)" would be the claim that "(1) and 3 have different surface structures but the same depth structure, and this common depth structure is much more like the surface structure of (3) than like that of (1)".

There would be considerable historical irony in nominalists thus appealing to a hypothesis of covert "depth" forms underlying overt "superficial" forms of the kind made familiar by Chomsky. Historically, it has been characteristic of pro-intuitionist *anti-nominalists* from Dummett (*1956*) onwards to claim that sentences about what numerals could be constructed "underlie" sentences about sentences about what numbers exist (not so much in a Chomskyite sense as in the sense of providing their "verification conditions"). (Dummett derives his outlook from the same anti-nominalist source from which several of Church's views derive, namely, Frege.) Originally, it was the *nominalists*, Goodman and Quine, who were sympathetic with "structuralist" and "behaviorist" approaches to language; and when a Chomsky theorized about a "deep" level of language beneath the observed "surface", as when a Carnap or a Church wished to introduce abstract "propositions", they were skeptical and critical.

Irony aside, there is a considerable problem of *evidence* for any claim about "depth" analysis. And perhaps one needs to be *especially* careful about evidence when the claim is made, not by a linguist for whom an understanding of the hidden mechanisms of language is an end in itself, but rather by a philosopher with ulterior ontological motives. Evidence aside, there is also a considerable problem of *relevance*. Suppose there were the best possible evidence that (3) and (4) reveal the "depths" underlying the "superficial" (1) and (2). Would not this strong claim about the closeness of the relationship between paraphrase and original serve only to yet further reinforce the appearance that (1) implies (2), just as (3) implies (4)?

At any rate, it is hard to see how the claim that (4) reveals the "depths" underlying the "superficial" (2), and that the "depth" structure underlying (2) is quite different from its "superficial" structure, could serve to motivate a *denial* of (2). Chomsky, after all, held in his early work that:

(17) Chomsky has refuted the structuralist Bloomfield and behaviorist Skinner,

reveals what goes on in the "deeps" underlying the "surface" of:

(18) The structuralist Bloomfield and behaviorist Skinner have been refuted by Chomsky;

and hence that the "depth" structure of (18) is quite different from its "superficial" structure. But he emphatically did not take this to show that the structuralist Bloomfield and behaviorist Skinner had *not* been refuted by Chomsky!

Debates over ontology take place entirely in "superficial" language: Philosophers are never heard speaking nor seen writing "depth" language, which is an unobservable theoretical posit of technical linguistic science. The "superficial"/"depth" distinction therefore has no obvious relevance to the old issues of ontology, nominalism, and compatibilism. But acceptance of such a distinction may make it possible to introduce new issues, which might be called *bathontology*, *infranominalism*, and *infracompatibilism*.

Call (both paradigmatic and idiomatic occurrences of) the "there is" that occurs in audible, legible "superficial" language *top-quantification*, and the counterpart thereof in unobservable, theoretical "depth" language *bottom-quantification*. Let S be some common noun. Ontology was concerned with questions of the kind: Is one warranted in asserting (6)? In the present terminology this becomes the question: Is one warranted in asserting, "[top-quantification] S"? Nominalism was the negative answer to this question in the special case of (1) or the like, where S is "number" or the like. Compatibilism was the claim that endorsing nominalism is not incompatible with making certain assertions of the kind "[top-quantification] an S that ..." because (the occurrence of top-quantification here being idiomatic rather than paradigmatic) these despite appearances do not really imply "[top-quantification] S".

Bathontology would be concerned with questions of the kind: Is one warranted in asserting *something whose depth version is* "[bottom-quantification] S"? Infranominalism would be the negative answer to this question in the special case where S is "number" or the like. Infracompatibilism would be the claim that endorsing infranominalism is not incompatible with making the assertion (6) that "[top-quantification] S", because its depth version is something quite different from "[bottom-quantification] S". As the historical remarks above indicate, in the early debate over nominalism, some at least (on the *anti*-nominalist, not the nominalist, side) took a position not utterly unlike bathontological infranominalist infracompatibilism (namely, the position that the "verification conditions" for existence theorems in mathematics involve constructions on numerals denoting or pluralities instancing numbers and not directly on the numbers themselves).

Now all this involves a substantial presupposition, for which evidence would be required, namely, the presupposition that there *is* a "depth" counterpart of the "surface" expression "there is". The presupposition here is a double one: First, the "depth" versions underlying *most* "superficial" occurrences of "there is" must involve some *one* construction. (Those "superficial" occurrences of "there is" that are distinguishable as idiomatic even by the coarse criteria of traditional grammar, and presumably also some occurrences of "there is" that are distinguishable as not paradigmatic only by the refined criteria of modern linguistics need not do so, but *most* must.) Second, *most* occurrences of this "depth" construction must be normally expressible "superficially" using "there is". (Some may be normally expressed by some other expression, but *most* must be expressible by this one expression.) For if this is not the case, either because the "surface" expression "there is" *splits* into a great variety of different "depth" constructions, or because it is *subsumed with* a great variety of other "surface" expressions under some one "depth" construction, then there would seem to be no basis for distinguishing some one depth construction as the *analogue* or *counterpart* specifically of the expression "there is", and hence no basis for recognizing a topic of bathontology underneath the topic of ontology.

The question whether these presuppositions do indeed hold, and with it the status of more particular claims about bathontological commitment and the compatibility of infranominalism with conventional science, is one I am prepared to leave to the professional linguists and will not attempt to settle here. What I would like to emphasize here is that the kind of distinction of levels of language recognized in modern linguistic theory provides the most (and perhaps the only) obvious conceptual framework in which one might try to make sense of the distinction assumed, in one terminology or another, by those philosophers who try to distinguish "mere" from "ultimate" being or existence, asserting (1) (and for that matter (2)) in the "mere" sense, and denying (2) (and for that matter (1)) in the "ultimate" sense.

Such positions, which may be considered versions or variants of compatibilism, have been taken, for instance, in the closing paragraphs of *Bonevac 1980,* and in the opening paragraph of *Hodes 1990* (another author influenced, like Church and Dummett, by Frege), where one reads:

> The answers to questions like "Are there numbers?" and "Do sets exist?" are, trivially, "Yes." To not see these answers as trivialities bespeaks a misunderstanding of mathematical discourse. But to go on and say that there is a *realm* of mathematical objects is to engage in obscurantist hyperbole. Mathematical objects are second-rate; they are not among the "furniture of the universe."

These are authors of the kind described earlier, in connection with the passage from Church, as only meaning to deny that there "are" such things as numbers in some extraordinary sense of "are" (as contrasted with authors of the kind who when they say, "There is a number that ..." don't really mean what a philosophically unselfconscious mathematician who said this would mean).

The distinction here between "second-rate" (or "mere") and "first-rate" (or "ultimate") things or objects does not make any immediate sense. The first, "mere" sense of "existence" makes sense, since it is supposed to be this sense (only) that occurs in ordinary scientific discussion. In the scientific community there is considerable agreement about which assertions of existence in this first sense are warranted, suggesting that "existence" in the first sense is being used in that community according to fairly definite rules, giving it a fairly definite meaning. The problem is with the supposed second, "ultimate" sense that occurs (only) in extraordinary philosophical debate. The burden of providing such a second meaning of "existence" rests, as Church so long ago said, on those whose writings or whose philosophical views require it.

REFERENCES

Alston, W.

1958 Ontological commitments, *Philosophical Studies*, vol. 9, pp. 8–17.

Bonevac, D. A.

1980 *Ontological reduction in the abstract sciences*, Hackett.

Burgess, J. P.

1983 Why I am not a nominalist, *Notre Dame Journal of Formal Logic*, vol. 24, pp. 93–105.

Carnap, R.

1950 Empiricism, semantics, and ontology, *Revue Internationale de Philosophie*, vol. 4, pp. 20–40.

Chihara, C.

1973 *Ontology and the vicious circle principle*, Cornell University Press, Ithaca, New York.

1990 *Constructibility and mathematical existence*, Clarendon, London.

Church, A.

1958 Ontological commitment, *The Journal of Philosophy*, vol. 55, pp. 1008–1014.

Dummett, M.

1956 Nominalism, *Philosophical Review*, vol. 65, pp. 491–505.

Goodman, N.

1956 A world of individuals, *The problem of universals* (I. M. Bochenski, A. Church, and N. Goodman, editors), Notre Dame University Press, Notre Dame, pp. 155–172.

Goodman, N., and W. V. Quine

1947 Steps toward a constructive nominalism, *The Journal of Symbolic Logic*, vol. 12, pp. 105–122.

Hodes, H.

1990 Ontological commitment: Thick and thin, *Meaning and method: Essays in honor of Hilary Putnam* (G. Boolos, editor), Cambridge University Press, Cambridge, Massachusetts pp. 235–260.

Quine, W. V.

1948 On what there is, *Review of Metaphysics*, vol. 2, pp. 21–38.

1951 On Carnap's view on ontology, *Philosophical Studies*, vol. 2, pp. 65–72.

1981 *Theories and things*, Harvard University Press, Cambridge, Massachusetts.

MICHAEL DETLEFSEN

PEACE, JUSTICE AND COMPUTATION: LEIBNIZ' PROGRAM AND THE MORAL AND POLITICAL SIGNIFICANCE OF CHURCH'S THEOREM*

1. LEIBNIZ' PROGRAM

Throughout his life, Leibniz gave himself to projects he saw as having the potential to improve the conditions of human life. Among these, none was more important to him than the development of his *ars combinatoria*. This ambitious project, sketched in his earliest writings (cf. *De Arte Combinatoria*, 1666) and returned to again and again throughout the remainder of his life (cf. letter of January 10, 1714 to Nicolas Remond[1]), was originally divided into three sub-projects: (i) the *calculus ratiocinator* (calculus of reasoning), in which he hoped to codify, in mechanical form, all acceptable forms of logical reasoning, (ii) the *characteristica universalis* (or universal characteristic), intended to serve as a logically perspicuous language for the expression of all rational thought, and (iii) the *encyclopedia* of human knowledge, intended to catalog the whole of received human knowledge.[2] Leibniz believed that,

* I would like to thank audiences at the University of Notre Dame, the University of Queensland, the University of Sydney, LaTrobe University, Monash University and the University of Montreal for useful discussions of this paper. Among individuals, I am particularly indebted to Alasdair MacIntyre, Graham Priest and the late Ian Hinckfuss for thoughtful comments.

[1]English translation in *Leibniz 1969*, page 654. There Leibniz writes: "... if I had been less distracted, or if I were younger or had talented young men to help me, I should still hope to create a kind of *universal symbolistic* in which all truths of reason would be reduced to a kind of calculus. ... this could be a kind of universal language or writing ... for the characters and the words themselves would give directions to reason, and the errors—except those of fact—would be only mistakes in calculation."

[2]Later, Leibniz consolidated the three sub projects into two by combining the calculus ratiocinator and the characteristica universalis into a single project called 'the general science'. Thus, in the "New Proposals" he wrote: "I believe two things to be necessary for men to take advantage of their opportunities and to do everything they could to contribute to their own happiness, at least in the matter of knowledge .... These two things are, first an exact INVENTORY of all the knowledge dispersed and badly arranged (at least of that knowledge which appears to be most important at the beginning), and secondly, the GENERAL SCIENCE which should give us not only the means to use knowledge already acquired but also the method of judging and discovering, in order to go further and supply what we want. This inventory I speak of would be very different from systems and dictionaries, and would be composed only of Lists or enumerations, Tables, or Progressions which would serve to keep always before us, during some reflection or deliberation of any

C. Anthony Anderson and M. Zelěny (eds.), Logic, Meaning and Computation, 445-467.

taken together, these three devices would provide the basis for significant improvements in human knowledge and in the condition of human life generally.

He even believed that they could be used to serve the ends of peace and justice. Peace would be served by developing a mechanical means of resolving epistemic disputes. Justice would be served by using the mechanization of thought to effect a more equitable distribution of epistemic goods, one which would increase the shares of the epistemically least advantaged.

The mechanization of thought would serve peace by allowing opposing sides in an epistemic dispute to peacefully 'calculate' their way to a resolution. It would serve justice by enabling those not so richly endowed in native or 'intuitive' powers of reasoning to come closer to epistemic parity with those more richly endowed with such. In addition, it would make the social benefits of knowledge available to those to whom it could not make the knowledge itself available.

We will refer to this general program of promoting peace and justice through the formalization or mechanization of reason as *Leibniz' Program.* In doing so, we do not however insist that it agree in every particular with what the historical Leibniz himself may have had in mind. It is the ideas themselves, rather than perfect accuracy in attribution that is central to our concerns. That notwithstanding, we believe that what we here describe as Leibniz' Program is a plausible interpretation of Leibniz' own views.

Our aim is to assess the plausibility of Leibniz' Program. The assessment we offer is based on Church's celebrated proof of the undecidability of classical first-order validity. It is largely, though not entirely, pessimistic. We argue that Church's Theorem poses a serious theoretical limit on the extent to which programs like Leibniz' Program can be successful. If we are correct, Church's Theorem would appear to have non-trivial implications in the realm of social and political philosophy. These implications seem to have gone largely unnoticed.

Our assessment of Leibniz' Program rests upon the identification of what we take to be its key element; namely, a condition we shall refer to as the *Computability Requirement.* This requirement takes the form of a tenet of epistemic social policy concerning legitimate attempts by individuals to shape the beliefs and actions of others who are co-members with them in a given epistemic community. Specifically, it states that to be legitimate (i.e., to be the sort of thing that produces an obligation to believe that the community should regard as enforceable), an attempt to shape or conform the beliefs of others in an epistemic community must be backed by a supporting

sort, the catalogue of facts and circumstances and the most important assumptions and maxims which ought to serve as the basis of reasoning. ... [in a disorderly inventory] our very riches make us poor, somewhat like what would happen in a big store which lacked the order necessary to find anything one needed ... the general science helps to make up an orderly inventory ... it is with the general science that we shall have to begin." (English translation from *Leibniz 1951,* page 581, emphases Leibniz', square brackets mine).

'epistemic computation'. By an 'epistemic computation' of a given proposition, we mean, roughly, an effectively verifiable formalized demonstration[3] of the proposition from premises located in the community's encyclopedia or inventory of accepted knowledge.[4]

Leibniz advocated guidance by such a condition because he believed computational ability to be a kind of cognitive 'lowest common denominator' among human beings—a cognitive capacity that lies (or at least comes close to lying) within the range of competence of all human beings equally. This belief in the cognitive elementarity or commonality of computational thinking is what we shall refer to throughout this paper as *Leibniz' Thesis*. Its significance lies in the fact that the workings of such a common capacity are plausibly seen as enabling even the cognitively least advantaged to achieve a certain epistemic-political objective—namely, epistemic autonomy, the conduct of their epistemic affairs according to their own epistemic lights.

Leibniz' Thesis gives the reason why institution of the Computability Requirement is to be seen as recommended epistemic policy for a community. Given the truth of Leibniz' Thesis, enforcement of some condition like the Computability Requirement is necessary for the adequate protection of the epistemic autonomy of the cognitively least advantaged members of a community. This, at any rate, is true if the community in question represents a reasonably diverse cross-section of human cognitive capacities.

To put the point somewhat more accurately, enforcement of a condition like the Computability Requirement would appear to be necessary to protect what we will refer to as *weak epistemic autonomy*. Weak autonomy consists in the limited right of an individual not to believe and/or to act except out of her own conviction. The Computability Requirement protects the weak autonomy of the epistemically least advantaged individuals in a community by enforcing a high standard of elementariness for claims calling for their epistemic conformity. By enforcing this standard, even the conviction of the epistemically least gifted should be moved to autonomous conformity and their weak individual autonomy thus preserved.

Enforcement of the Computability Requirement is thus desirable as a means of securing the epistemic autonomy of the full range of citizens in a cognitively diverse community. What we will argue, however, is that it also threatens another type of epistemic autonomy—*strong epistemic autonomy*—for those who are cognitively more gifted.

Strong epistemic autonomy consists in the limited right of an individual

[3]In calling a formal demonstration 'effectively verifiable', I mean to say that its status as inferentially well-formed, according to the community's standards of well-formedness, is effectively decidable.

[4]The encyclopedia is somewhat abstractly and ideally conceived. Specifically, it is conceived as containing not only all actual knowledge confirmable by means commonly possessed and accepted within a given community, but, indeed, all knowledge, non-actual as well as actual, that is confirmable by such means.

to shape the beliefs and/or actions of others in ways she believes they ought rationally to be shaped. We will argue that there is a limit to which the interests of both weak and strong epistemic autonomy can be served, and that this limit is plausibly attributable to Church's Theorem. We thus claim that Church's Theorem poses limitations on the extent to which an enduring state of epistemic peace (i.e., a state during which epistemic conflicts are peacefully resolvable) can be expected. We will also argue that Church's Theorem (and a family of related results having to do with the 'Decision Problem') pose limits to the extent to which epistemic justice—in particular, an equitable distribution of basic epistemic goods—can be advanced by means broadly similar to those proposed by Leibniz. We believe, therefore, that Church's Theorem has not only social and political but also moral ramifications.

2. PEACE AND EPISTEMIC CONFLICTS

As we will treat of them, epistemic conflicts arise when (i) some member of a community makes a demand of epistemic conformity on the part of the rest or some part of the rest of the community, and (ii) some other member(s) of the community does not accede to this call for epistemic conformity. We assume that the community in question has a defined and mutually agreed-upon understanding of (1) what constitutes epistemic conformity/non-conformity, (2) what are the conditions under which conformity can rightly be called for, and (3) what are the measures that may rightly be taken in order to secure conformity. The specifics of these assumed elements of agreement need not concern us here.

On our conception, every epistemic conflict thus begins with what we will refer to as an 'epistemic claim'—that is, a call to a community to conform their belief to a certain proposition or propositions. Ordinarily, these will be propositions that the agent issuing the call herself believes. This, however, is not essential to the issuing of an epistemic claim. What is essential is that they are directed at communities and call for compliance as regards certain matters of epistemic commitment.[5]

[5]To avoid confusion, let me offer a couple of clarificatory remarks at this juncture. The first is that I take a rather broad and flexible understanding of the notion of community. In particular, I am assuming that a given individual may live in several different epistemic communities at once, and that these communities may be small. Indeed, I am prepared even to count the personae of a single divided mind as comprising a community of sorts provided that such a scheme of individuating epistemic agency should prove to be cogent (which it very well may not). Secondly, let me say that I am well aware that a great many epistemic conflicts do not have so much to do with the truth or falsity of a given belief as with the strength of evidence for or against it. I am assuming that all such disagreements can, however, be represented as disagreements regarding the truth-value of some proposition (e.g., such a propositions as "The evidence for p is conclusive" or "The evidence for p is stronger than the evidence for q."). Hence, I believe that the basic forms of dispute outlined here can serve as a basis for representing such seemingly more recondite types of conflict as those centering on differences in the strength with which the disputing

To put it more exactly, an epistemic claim is a quintuple of elements. The first is the claimant (the *conformant*) issuing the claim. The second is the target community to whom the claim is addressed. The third is a propositional element p giving the propositional content of the attitude-taking being proposed by the claimant. The fourth is an attitudinal element A specifying the particular epistemic attitude(s) and/or actional stance(s) towards p that the claimant is saying should be taken. The fifth, finally, is a modal element. In it the claimant specifies a normative modality (specifically, an obligation or an entitlement of some kind) which she takes as defining the responsibilities of the target community with regard to the epistemic attitude-taking featured in her claim.

Understood in this general way, epistemic claims can be seen as divided into two broad types, depending upon whether the normative or modal element of the claim is an obligation or an entitlement. We will call claims registering an obligation to believe (or to act as if one believes) *conformal* claims. Non-conformal claims, on the other hand, assert no such obligation, but merely register an entitlement to believe and or to act in a certain way. They thus serve primarily either to inform those in a given target community of some feature of the claimant's epistemic holdings or to facilitate sharing of non-conformal holdings amongst community members for the purposes of general epistemic enrichment and the like.

The distinction between conformal and non-conformal claims is essential to our notion of epistemic conflict. All conflicts, as we see them, require the lodging of a conformal claim (to hold epistemic attitude or actional stance A with respect to proposition p) to which some member (the non-conformant) of the conformant's target community does not then conform.

In order for a conformal claim to be legitimate, the conformant must produce a ground capable of moving the members of the target community autonomously to adopt the position (epistemic attitude or actional stance) she is calling for in her claim. The presentation of a conformal claim to a community thus incurs a debt on the part of the conformant—a debt to provide its members with a means of moving them autonomously to conformity. Indeed, the asserted obligation of conformity presented in her conformal claim does not go into effect until this debt has been discharged.

This is a key element in our conception of epistemic conflict. It institutes what might be called a *rule of conservatism* in relationships between conformal claimants and their target communities. Accordingly to this rule, no change in the epistemic or actional positions of the members of a target community is required until the conformant provides them with a compensating warrant for that change—a warrant whose purpose is to preserve their epistemic and actional autonomy whilst securing conformity. The rule of conservatism is thus rooted in a concern for the protection of the individual

parties hold beliefs.

from authoritarianism in her epistemic and actional life and is designed to insure that due consideration of the epistemic autonomy of the individual be included in any legitimate demand for epistemic conformity.

The issuing of a conformal claim thus requires compensation from the conformant. Such compensation must generally take the form of an argument or ground that is adequate to rationally move the members of the target community to autonomous adoption of the attitude or actional stance called for in the conformal claim. Proper conformant compensation should also make non-conformity actionable by making a properly compensated non-conformant guilty of a breach of rationality.

The particular features of a community's scheme of compensating grounds will be partly determined by how deeply into the epistemic and actional lives of individuals it believes the mechanisms for epistemic conflict-resolution ought to be permitted to go. Those who believe that relatively deeply and strongly held beliefs and actional stances should be open to conformant demands will require that compensatory grounds offered by conformants possess a high potency for autonomously moving rational agents to conformity. Those who believe that only less deeply held convictions should be open to demands of conformity will be free to adopt a scheme of compensating grounds of lesser potency.[6]

This way of thinking of conformant compensation puts substantial weight on the epistemic and actional autonomy of the individual. This concern for autonomy may be seen as based, in turn, on a concern for authenticity. The chief idea is that an individual who is called upon to believe a given proposition *p* must be given something that is capable of making her belief in *p* authentic, or a natural rational outflowing of convictions and inclinations that are, in the truest sense, her own. It is only through being given something capable of inclining her autonomous epistemic self towards acceptance of *p* that belief in *p* by someone not previously inclined by her convictions to believe in *p* can be authentic. Similarly for an individual's actional stances.

We emphasize this type of autonomy—freedom not to believe and/or to act other than out of one's own convictions and inclinations—because we believe it to be crucial to the issue of peaceful resolution of epistemic conflicts.

[6]It should not be assumed that there is automatically any correlation between what we are thinking of here as the "depth" of individual belief and the strength of the evidence required for its enforceable revision. What a community counts as evidence and, so, what must be accepted as evidence among its individual members, must be something that they can agree upon. At the same time, however, it might turn out that what can be agreed upon as evidence is not what any individual member of the community would herself count as belonging to the most powerful evidence at her disposal. In such cases, there will develop an important distinction between what might be thought of as public and what might be thought of as private evidence with at least some of what the various individual members of the community count as among their most powerful evidence being beyond the pale of that which is public. This notwithstanding, there may still be enough (and important enough) agreement between the members of a community for their formation into a community to make sense and to have a purpose.

Threats to it can serve as potent and legitimate motives for the use of force against a community by an individual. Demanding of a rational individual, and one that legitimately sees herself as rational, that she believe something that her own well-examined and responsibly developed convictions do not incline her to believe is calling upon her to do something that, in an important sense, *she*, as a rational individual, cannot do. Such a demand requires either that she become an individual other than the one she is or that she cease existing as an individual altogether. Such calls are not always illegitimate. However, when directed towards individuals who are under no obligation to reform or reconfigure their beliefs, they are legitimate only if the one issuing the call also provides a means (i.e., a grounds or reason) capable of making the change an authentic one.

Preservation of a similar type of autonomy—what we will refer to here as *integrity*—in the sphere of action (as contrasted with that of mere belief) is also important to the peaceful resolution of epistemic conflicts. Integrity concerns that integration of belief and action which makes the actions of a person an accurate and full rational expression of her beliefs. Only by preservation of this type of relationship between an individual's beliefs and her actions is she protected against the sort of violation of the actional self that occurs when she is not at liberty to act in accordance with her own convictions and is pressured to make her actional self a vehicle of expression for the will or beliefs of another. When done, this converts the active self of one person into an actional appendage or automaton of another and actional autonomy (integrity) is lost.

Together, authenticity and integrity comprise the two facets of that type of autonomy that we are referring to here as weak autonomy—that negative liberty not to believe and/or to act except out of one's own convictions.

Leibniz, we believe, was concerned with the preservation of this type of autonomy. In this, he seems to have followed the lead of his Protestant predecessor Luther.[7] There is a line of reasoning that supports the idea that the use of 'computations' of the sort envisioned by Leibniz in his *Ars Combinatoria* should provide a means of securing conformity to conformal claims while also preserving autonomy. This reasoning proceeds as follows.

By a 'computation' let us understand an argument whose premises can be seen by all community members to belong to their common 'encyclopedia' of accepted propositions and whose validity can be determined by a very rudimentary type of character-recognition and combinatorial judgment that is within the reach of all. Understood in this way, each community member would be able to verify the premises of a computation by consulting the community encyclopedia. Likewise, every community member would be able to

[7]This is not intended to suggest, however, that Leibniz was entirely Lutheran in his theology. He was heavily influenced by Catholic thinkers as well—in particular by the Jesuit Friedrich Spe (cf. F. W. C. Lieder, "Friedrich Spe and the Théodicée of Leibniz", *Journal of English and Germanic Philology,* 1912: 149–172; 329–354).

verify the validity of the computation because the evidence needed for such verification is so extremely elementary—a kind of rudimentary capacity to recognize and distinguish a finite variety of basic shapes and to tell whether they lie some type of elementary order.[8] Reliance on such a capacity, Leibniz said, should provide "a sensible and, as it were, mechanical direction of the mind which everyone, even the most stupid, could recognize."[9] Hence, it should prove capable of moving the members of a community to autonomous belief in the validity of the computation, at least on the assumption that their assent to the community standards is itself autonomously taken. This together with autonomously taken belief in the premises (i.e., the encyclopedia) should then provide for autonomous belief in the conclusion. Implementation of the Computability Requirement should thus preserve the weak autonomy, and so contribute to the peaceful epistemic existence, of that segment of the community made up of those who authentically assent to the propositions of the community's encyclopedia and who accept the inferences of the community's agreed-upon stock of 'valid' inferences.[10]

It should, however, also contribute to the epistemic peace of the entire community in another way—namely, by making the social benefits of knowledge or authentic belief available to all. Crucial to the peaceful resolution of conformal disputes is that all members of the community play by a common set of rules and that each be able to determine for him- or herself whether a given piece of reasoning is in conformity with those rules. This means that there should be a universally operable means of determining whether a given piece of reasoning is counted by a community as valid and, hence, as reasoning which the members of that community are obliged to honor. Implementation of the Computability Requirement should provide for this by offering a combinatorial or computational means of making such determinations. Hence, even for those not able authentically to assent to the tenets of the encyclo-

[8] Algebraic thought, based on what we are calling character recognition, was contrasted in Leibniz to intuitive knowledge (cf. *Leibniz 1961*, pages 422–425, the essay entitled 'Meditationes de Cognitione, Veritate et Ideis'). The latter was concerned with notions directly, while the former concerned itself with the signs used to signify notions. Thus, in intuitive thought, notions themselves are the direct objects of awareness and manipulation, while in algebraic thought (which Leibniz also referred to as 'blind thought'), it is symbolic expressions that play these roles. Cf. *Leibniz 1965*, pages 17–19, for a general discussion of the nature and utility of symbolic reasoning.

[9] Cf. Letter to Oldenburg (undated), in *Leibniz 1961*, page 14.

[10] Is idealization to communities all of whose (normal) members are taken to have the ability to accurately determine the validity of the formal rules used in their accepted calculus so restrictive as to be uninteresting? We don't think so. We may grant that the ability of a calculus of reasoning to preserve weak epistemic autonomy depends upon more than strictly "sensible" or "algebraic" knowledge without thereby restricting the extent to which it is a useful means of increasing the knowledge of those not gifted in intuitive logical power and of bringing within the purview of their logical judgment many cases of valid (or invalid) reasoning they would otherwise never discover because they would never think of them. This extension of the powers of the intuitively disadvantaged would, moreover, be at least in significant part due to the "mechanical" character of the calculus.

pedia and the inferences of the logical calculus, there will be the opportunity to take advantage of the benefits offered by conformal claim-making while at the same time protecting oneself from being taken advantage of by other such claim-makers.

Key to keeping the epistemic peace, then, is existence of a common and commonly operable means for determining whether a given piece of reasoning conforms to community standards of validity. Use of computations to back conformal claims—the 'policy' of the 'Computability Requirement'—should provide such a means. Hence, it should also provide a means of peacefully securing at least ostensible (as distinct from authentic) epistemic conformity. Such, at any rate, is the conclusion of the Leibnizian argument with which we are concerned. As we shall see later, Church's Theorem has something to say both about this and about the ability of the Computability Requirement to simultaneously preserve weak and strong autonomy.

3. EPISTEMIC JUSTICE

The elementary character of symbolic calculation is thus that feature of Leibniz' calculus of reasoning which is supposed to promote epistemic peace. This same feature suggests that implementation of the Computability Requirement might also aid in increasing *epistemic justice*—that is, the distribution of epistemic goods to those lying in a wide variety of different natural or non-self-engendered categories of cognitive capacity.

Epistemic goods can, of course, be of great private benefit to an individual. At the same time, they can also be of great social value. The ability to acquire and publicly to 'expend' epistemic goods seems clearly to be an important determinant of human welfare. Equally clearly, the distribution of such goods through genetic and (broadly speaking) societal forces is anything but equitable. Birth, upbringing and the use and abuse of personal, social and political power often result in unjustified epistemic advantage for some and unjustified epistemic disadvantage for others.

Leibniz was concerned with these inequities and saw their redress as a demand of justice. He seems to have believed that development of his combinatoric art would provide a means of making such redress. The key, once again, was its computational character. Being of elementary epistemic character, it could be driven by capacities that are more equitably distributed across the human community than are native capacities of intellect and favorable circumstances of upbringing.

Roughly, the idea was that the calculus of thought would run off will-power rather than intellectual ability. It would thus, in effect, reduce intellectual differences to differences in will-power, and such differences could then be removed by exercise (albeit, perhaps, diligent exercise) of the will.[11] In addi-

[11]Leibniz admitted, in the end, that there are even some minor inequities of will-power

tion, Leibniz believed, one's computational proficiency could be so improved through practice that the dull but diligent epistemic laborer could, by force of determination, match and even exceed the epistemic holdings of the intellectually more gifted. Such, at any rate, was his view of how development of his combinatoric art might promote greater justice in the distribution of epistemic goods.

To understand this better, let us briefly consider Leibniz' general conception of justice. He characterized it as "the charity of the wise man (caritatem sapientis)".[12] Charity, in turn, was defined as "universal benevolence" (ibid.), benevolence as the "habit of loving" (ibid.), loving as the taking of pleasure or delight in the "happiness of another" (ibid.) (or as the regarding of another's happiness "as one's own" (ibid.)), and pleasure or the taking of delight as a "sense" or recognition of perfection.[13]

Given the above, it follows that, since God is the most perfect or beautiful of all beings,[14] we find our greatest happiness or pleasure in loving Him.[15] One can not, however, love God without at the same time loving her fellow human beings. The reason is that in loving God what we love are His perfections. Hence, since our fellow human beings are capable of growing in the same perfections, we must desire to see those perfections reproduced to the highest possible extent in them.[16] It thus follows that, to the extent that we truly love God (i.e., to the extent that we truly "sense" His perfection), we must also cultivate the increase in human beings of those of God's perfections which can (though only to an imperfect extent) be realized in them.[17]

For Leibniz it was thus a demand of justice that we seek the perfection of our fellow human beings. Furthermore, among the ways of perfecting humankind, none was more important than the perfection of knowledge and reason. All other gifts, said Leibniz, may corrupt a human being; reason and reason alone is unconditionally wholesome for them.[18] With improvement of reason comes improvement in the ability to recognize perfection generally and to recognize rational perfection in particular. Improvement of reason should thus not only increase love of God, but also love of one's fellow human

amongst human beings. Hence, justice—epistemic justice—demands the encouraging of the discouraged too. Cf. *Theodicy*, pt. 120.

[12]Cf. the preface of *Codex Juris Gentium Diplomaticus* (1693), an English translation of an excerpt from which appears in *Leibniz 1951*, pages 559–563. Cf. also "Elements of Natural Law" (1670-1), pp. 133–4 in *Leibniz 1969.*

[13]Cf. *Letter to Nicaise 1698,* pages 567–568 in *Leibniz 1951.*

[14]Cf. *Leibniz 1969*, page 134, and *Leibniz 1951*, page 560.

[15]This is so, at any rate, provided that we are as capable of recognizing or sensing the perfections of God as we are of sensing the perfections of other beings.

[16]And ourselves too, of course.

[17]Cf. *Leibniz 1951,* pages 567–570.

[18]Cf. *Leibniz 1951,* pages 23–25, and *Leibniz 1969,* page 224. Wiener gives the date of this fragment describing the Universal Characteristic as 1677, Loemker as ca. 1679. Cf. *Leibniz 1961,* pages 184–189 for the original Latin.

beings.[19] As Leibniz himself put it:

> the religion I follow closely assures me that the love of God consists in an ardent desire to procure the general welfare, and reason teaches me that there is nothing which contributes more to the general welfare of mankind than the perfection of reason[20]

Leibniz' "computerization"[21] of knowledge was thus intended to distribute the perfections of reason and knowledge to agents of every (or nearly every) natural cognitive stripe.[22]

There are, however, limitations on how effective distribution of knowledge by such means can be. In particular, there are limits on how effectively it might be used to distribute authentic knowledge. The reason for this is that there is more to authentic knowledge or belief than merely the ability to operate a calculus. It requires as well the capacity authentically to judge the propositions of the encyclopedia to be true and the inferences of the calculus to be valid. Hence, implementation of a calculus of thought can be expected to distribute authentic epistemic goods only to those having such capacities.

We should perhaps also note that Leibniz, like Luther before him, was deeply concerned for authenticity, since he took it to be an indispensable ingredient of justifying religious belief. The basis for this concern was the Lutheran conception of justification before God, the critical moment of which

---

[19]Cf. *Leibniz 1951,* pages 24–25.

[20]Cf. *Leibniz 1951,* page 17 (from the preface to "The Universal Science"). Another statement to the same effect, though not one occurring in the context of a larger discussion of the *ars combinatoria* and its uses is given in the *Theodicy* where Leibniz writes:

> ... there is no greater individual interest than to espouse that of the community, and one gains satisfaction for oneself by taking pleasure in the acquisition of true benefits for men. ... Some Christians have imagined that they could be devout without loving their neighbor, and pious without loving God; or else people have thought that they could love their neighbor without serving him and could love God without knowing him. Many centuries have passed without recognition of this defect by the people at large; and there are still great traces of the reign of darkness. (*Leibniz 1966,* pages 2–3)

[21]By 'computerization', I mean the giving of a computational means of deciding the truth-values of propositions. One important issue bearing on the question of whether epistemic goods might be redistributed by means of such a device as Leibniz' combinatorial art is that concerning whether certain key epistemic goods are lost in the "translation", so to speak. This would happen if certain epistemically valuable features of non-combinatorial justifications were lost when they were exchanged for combinatorial ones, and no compensating gains were generally to be expected. One might, for example, be concerned about the loss of (certain kinds of) simplicity and also of such hard-to-get-at things as epistemic authenticity. More on this in a moment.

[22]Complexity is a legitimate concern here. There may be a tendency for the complexity of justifications to increase in dramatic and troublesome ways when non-combinatorial proofs are replaced by combinatorial ones. This is a well-known phenomenon in number theory, and, indeed, is the primary inspiration of the field known as analytic number theory. This is a phenomenon which Hilbert drew upon in motivating his Program.

was "belief from the heart" (Glauben von Herzen), a moment emphasized by both the prophet Habakkuk[23] and St. Paul: "He who *and it is one thing through faith* is righteous shall live ... for man believes *with his heart* and so is *justified.*"[24]

Luther's targets were those who tried to 'mechanize' or automate justification by reducing it to the performance of certain ritualistic acts or, even worse, certain acts of financial exchange. Such practices, he said, pose a threat to true justification before God. Specifically, they can deceive the practitioner into thinking that she has satisfied conditions sufficient for justification when, in fact, she has not. This was precisely the basis of Luther's opposition to such practices as the selling of indulgences.[25] In general, any attempt to ritualize or mechanize faith—any attempt to substitute a 'recipe' of practices for the essential element of belief from the heart—was to be rejected. Justification requires authenticity of belief; anything less constitutes a deficiency of involvement with God on the part of the believer.

Authenticity of belief (and action) was thus a concern of Leibniz'. It is perhaps curious, then, that he thought to distribute belief through the operation of a calculus of thought. Such calculation could perhaps allow one to take on the appearance of one who authentically believes, but truly authentic belief would require as well that one be genuinely convinced of the truth of the calculus' encyclopedia and the soundness of its logic. Neither of these can be obtained by purely computational means.[26]

A calculus of thought can thus be used to distribute authentic epistemic goods only to those who believe in the truth of its encyclopedia and the soundness of its logic. This suggests that Leibniz should also have regarded such belief as capable of widespread distribution among human beings.

Authenticity aside, a significant part of what makes (at least much) authentic knowledge valuable is its capacity for social expenditure—that is, its utility in securing epistemic and actional conformity and in fending off un-

[23] Habakkuk 2:4: "... he whose soul is not upright in him shall fail, but the righteous shall live by faith."

[24] Romans 1:17; 10:10, emphases mine. Cf. also Galatians 3:11, Philipians 3:9 and Hebrews 10:37–38.

[25] This was especially true of the so-called "Jubilee Indulgences" declared by Pope Leo X in 1507 and again in 1513 as part of a plan to raise money to rebuild St. Peter's in Rome. One particularly well-known fund-raiser, a Dominican friar named John Tetzel, is represented in documents printed in 1521 as having employed a jingle giving the essentials of the mechanism relating the purchase of an indulgence to the state of the soul of a dead loved one for whom the indulgence was purchased: "As soon as the coin in the coffer rings, immediately the soul to heaven springs." (So bald der Gulden in Becker klingt/ Im hui die Seel im Himmel springt/)

[26] It is not inevitable that this be so, however. An encyclopedia could consist wholly of truths apparent to even the least powerful intellect. Likewise, the inferences accepted as valid/invalid by a given community could be self-evidently so. We would not, however, expect that a calculus of thought capable of settling such contentious and difficult matters as the existence of God and the divinity of Christ (and, indeed, even much less contentious matters of ordinary science and everyday life) to consist wholly of such things.

wanted such calls from others. In other words, a significant part of what makes knowledge valuable is that which makes it socially valuable, and that consists in the capacity it gives to exercise strong autonomy (i.e., to effectively lodge conformal claims) and to protect weak autonomy (i.e., to effectively ward off unwanted conformal claims).

For social purposes, then, what is most important is not authenticity but the ability to identify those beliefs and inferences to which the community is committed. This follows from the fact that, in the first instance, one prevails in an epistemic dispute not by showing her claim to be logically implied by premises she authentically believes to be true or by showing her opponent's claim not to be implied by premises she (the opponent) authentically believes. Rather, she prevails either (if she is the conformant) by showing her claim to follow from premises and inferences that even her opponent must acknowledge as compelling, or (if her opponent is the conformant) by showing her opponent's position not to follow from such.

Development of a calculus that could bring effective manipulation of the accepted thought of a community within the reach of all would thus aid in effecting an equitable distribution of the social benefits of knowledge or authentic belief. The question, then, is whether there is such a calculus. One might think that there could not be, and for reasons similar to those given above for the non-mechanizability of genuine belief or knowledge. Genuine belief requires not only a calculus, but belief in its soundness. Likewise, it might be said, social use of a calculus requires not only that calculus but belief in its acceptance by the community. This latter, however, is no more securable by means of the calculus itself than is the soundness of the calculus. Hence, the reasoning continues, the social uses and benefits of knowledge are no more open to mechanization than is knowledge itself.

We believe this reasoning to be faulty. It overlooks the important fact that membership in an epistemic community presupposes a basic grasp (though not necessarily a perfect grasp) of whatever specification it is that the community gives of its accepted (resp. rejected) propositions and inferences. Anyone who does not or can not do so can not properly be said to belong to the community because they can not authentically affirm in the given basic way what the community standards are. Such a one not only can not function as an authentic member of the community (i.e., as one who authentically accepts the community's practices), she can not even act *as if* she were.

If this reasoning is correct, implementation of a calculus of thought might at least secure the social benefits of knowledge. To the extent that this is so it should also be a means of improving epistemic justice. Later, we will consider what if any effects Church's Theorem has on this.

4. CHURCH'S THEOREM

In their 1928 book, *Grundzüge der Theoretischen Logik*, Hilbert and Ackermann posed a fundamental problem of logic, to wit: "Is it possible to determine whether or not a given statement pertaining to a field of knowledge is a consequence of the axioms [given for that field—MD]?"[27] Hilbert and Ackermann dubbed this the *Decision Problem*. Hilbert's interest in the problem, and his reason for regarding it as fundamental, was that its positive solution, together with the provision of a *complete* system of logical reasoning and a *complete* axiomatization of each field of knowledge, would make it in principle possible to solve all problems in that field in a purely mechanical manner (where the solution to a problem P pertaining to a given field F is to be understood as consisting in the logical derivation of an answer to P from the class of acknowledged truths of F).

The requirement of axiomatic completeness for F[28] is essential since, without it, it is not legitimate to identify the question of the solvability of a problem of F with the question of the extraction by purely logical means of an answer to that problem from the axioms for F. For similar reasons, the requirement of completeness for the logical apparatus[29] (also a problem that Hilbert and Ackermann succeeded in giving definite shape to) is essential. If the logical apparatus of a system is not complete, then, even supposing its axiomatic basis to be complete, non-deducibility of a given statement will not imply unsolvability of the corresponding problem.

[27] Cf. page 108 of the English translation of the second edition of Hilbert and Ackermann (i.e., *Principles of mathematical logic*, edited by R. E. Luce).

[28] In calling an axiomatic basis for F complete, we mean that an answer to every problem of F (i.e., every problem formulable in a language in which the subject-matter of F can be expressed) is obtainable from the axioms of F by purely logical means. This is, of course, an informal definition of completeness, but it represents what the more formal definitions are intended to capture. The usual formal definition of completeness for a formal axiomatic system T is this: for every sentence S formulable in the language of T, either S is a theorem of T or $\neg S$ is a theorem of T. This, of course, presumes that the problems of F for which one seeks solutions are exactly coextensive with what might be called the *classical truth problems* of F—that is, the problems concerning the classical truth or falsity of a given sentence or proposition. We call such problems classical, because the classical conception of truth is presupposed in the definition—of every $(S, \neg S)$ pair, it is presupposed that exactly one will be true. There are, however, other kinds of problems—for instance, *classical validity problems*. These problems concern the classical logical validity of a given sentence or proposition. For such problems, the above formal definition of completeness is inappropriate since it is not classically the case that one member of every $(S, \neg S)$ pair is logically valid. Thus, formal definitions of completeness depend not only upon the choice of formalizations for F but also upon the *kind of problem* whose solvability is being queried. The only generic requirement of completeness, it seems, is that it solve all *problems* ... it being understood that differences in the kind of problem to be solved will result in differences in the formal definition of solvability.

[29] By the completeness of a logical apparatus we mean that every logically valid reasoning is represented by an acceptable derivation in it. Again, this is an informal definition, but it is that which the more formal definitions of the logical completeness are intended to capture.

Gödel, of course, demonstrated the existence of a complete formal apparatus for first-order classical logical consequence in 1930. He followed that up in 1931 with a proof of the incompleteness of consistent formal-axiomatic bases for first-order theories from elementary arithmetic on up. Neither of these settled the Decision Problem, however. Nor would denying Gödel's incompleteness theorem and adding as an assumption the completeness of the axiomatic bases with which one might be concerned.[30] Axiomatic completeness requires only that for every statement S pertaining to a given field F, either S or its denial is a consequence of the axioms for F—and it is one thing to know *that* either S or its denial is a logical consequence of the given axioms, and quite another thing to know *of* either S or its denial that it is. It is another thing still to be able to determine of any given S that either it or its denial is a logical consequence of the axioms for F by applying some general effective procedure or algorithm capable of deciding all such questions.

It is this latter type of problem—namely, problems concerning effective determination by general algorithms—with which Church dealt in his landmark 1936 paper "A Note on the Entscheidungsproblem". He gave a negative solution for the general problem of determining classical first-order validity. That is, he showed that there is no general algorithm for determining whether or not a given statement pertaining to a given field of knowledge is a classical first-order logical consequence of the axioms given for that field.[31] More precisely, and more generally, he showed that there is no effective procedure P such that for an arbitrarily selected set of sentences Σ and an arbitrarily selected sentence S of a non-monadic first-order language (i.e., a first-order language that contains at least one predicate of adicity greater than or equal to 2),[32] execution of P determines whether or not S is a logical consequence of Σ. Assuming the logical completeness of a given formal axiomatic theory F, this implies that there is no effective procedure P for generating, for an arbitrarily selected sentence S of the language of F, either a demonstration or a refutation of S in F.

We would now like to consider the significance of this discovery for projects of the sort discussed in the preceding sections of this paper; namely, those concerning the use of a computational device such as Leibniz' for (i) peacefully resolving epistemic disputes, and (ii) improving upon the injustices of the natural distribution of epistemic goods.

---

[30] The effective enumerability of the theorem-set is also required.

[31] Actually, there was a mistake in Church's original proof, but it was corrected and the correction published in the same volume (namely, the first) of *The Journal of Symbolic Logic* as carried the original proof.

[32] The restriction to non-monadic languages is necessary because the Decision Problem for the monadic predicate calculus has a positive solution. According to Ackermann, this was proved for the first time by Löwenheim in his 1915 paper "Über Möglichkeiten im Relativkalkül" (cf. Ackermann, *Solvable cases of the decision problem*, page 34).

5. PEACE: WEAK AUTONOMY, STRONG AUTONOMY AND CHURCH'S THEOREM

We have suggested that the capacity of a given mechanism for peaceful conflict-resolution is dependent upon its ability to preserve weak autonomy. We would now like to consider a related type of autonomy, which we shall refer to as *strong autonomy.* It concerns the freedom of an individual to attempt to conform a community to her beliefs or actions. This freedom is crucial to the ability of an individual to realize her goals for her life within a community while remaining a member of it.

Infringements on this type of freedom, like infringements on weak autonomy, pose a significant threat to peace, and for similar reasons. Persistent refusal to allow a person to pursue her goals through the making of conformal claims is tantamount to denying her 'conformal self' a place in the community. This is not to say that there might not be legitimate reason for restricting or denying attempts at conformal action. There very well might be. That this is so, however, does not change the fact that such restrictions might also provide a motive for violence against or defection from the restricting community.[33]

Concern for peace would thus seem to dictate the need for at least some freedom to exercise strong autonomy. At the same time, however, attempts to conform the beliefs and actions of others should be enforceable only to the extent that doing so provides adequate protection for the weak autonomy of those upon whom the call to conformity falls. The question, then, is how far the freedom to make conformal claims should be extended in a system in which protection of weak autonomy is to be provided by enforcement of the Computability Requirement.

It is at this point that Church's Theorem enters the picture. It poses a theoretical limit on the extent to which individuals living under the rule of the Computability Requirement are free to make conformal claims. It does this specifically when (a) the claim in question asserts the invalidity of some form of reasoning, and (b) the invalidity of this form of reasoning cannot, by Church's Theorem, be 'computed' in the given calculus. This limitation would not exist were there, contrary to Church's Theorem, a general effective procedure for determining of any form of inference whether or not it is valid.

By Church's Theorem there is thus no general protection against situations in which a would-be conformant is in a position to want to make a conformal claim regarding invalidity (through her knowing or justifiedly believing that an inference is invalid) but not in a position to be permitted to make it.[34]

[33]Throughout this paper when we speak of peaceful resolution of conflicts between an individual and a community, we mean resolutions which not only avoid violence but which also leave the individual's community membership intact (i.e., which find a way for the conflict to be resolved short of "defection" from the community by the individual).

[34]We do not, of course, suppose that authentic belief itself, or any action requiring something like direct activation from genuine convictions, will be elective or voluntary in

Such situations will, of course, frustrate the desire for rational epistemic community on the part of the would-be conformant. Should such frustration become persistent enough or, though perhaps not persistent, should it arise in serious enough scattered cases, it would have the same potential to disrupt the peace of a community as violations of weak autonomy. Both alienate an individual from her community and thus provide her with a motive for disrupting its peace.

Assuming, then, that the Computability Requirement (or some requirement sufficiently like it) is necessary for the adequate protection of weak autonomy in the face of possible demands for epistemic conformity, there develops an in-principle tension between weak and strong autonomy. The basis of this tension is Church's Theorem.[35]

6. JUSTICE: THE EFFECTS OF CHURCH'S THEOREM ON THE DISTRIBUTION OF THE SOCIAL BENEFITS OF KNOWLEDGE

We noted in section 3 that the Computability Requirement is not, in and of itself, an adequate means of producing an equitable distribution of authentic knowledge. At the same time, however, we noted that it fares somewhat better with respect to distribution of the social benefits of authentic knowledge. We will now consider what, if any, effect Church's Theorem has on these two points.

Let's consider first the distribution of authentic knowledge. In this connection, let us suppose, for the sake of argument, that authentic belief in a community's encyclopedia and its accepted logic can be widely distributed across it. Our question then is this: Granted such favorable assumptions for the distribution of authentic knowledge, could development of a calculus of

the appropriate sense.

[35]This does not, of course, mean that there are not sources of tension between individuals and communities other than Church's Theorem. There is a parallel type of tension sponsored by Gödel's incompleteness theorems. This would occur should one notice that, though they are not logically implied by the community's encyclopedia, commitment to certain sentences (viz. Gödel sentences for formal systems based on the encyclopedia) is nonetheless necessary given commitment to it. We believe, however, that the tension arising from Church's Theorem is generally more serious. This is due to the fact it applies to a broader range of sentences. It was only in 1977 (cf. "A mathematical incompleteness in Peano Arithmetic", in *Handbook of mathematical logic*, edited by J. Barwise) that Paris and Harrington found the first example of an undecidable yet mathematically interesting sentence. Paris and Harrington also showed, however, that the undecidability of their interesting sentence is implied by the undecidability of Gödel's sentence. In this sense, they refuted any general claim to the effect that Gödel's construction fails to capture anything of mathematical interest.

Another source of tension, and one that may be of greater practical importance, is that concerning the complexity/tractability of decidable cases of the Decision Problem. The length and/or other complexity of the computation might make it humanly impossible to produce or to comprehend it. Indeed, at present it appears that the Decision Problem for classical consequence is intractable for all but a relatively small class of argument-forms.

thought be expected to effect an equitable distribution of authentic belief or knowledge across the community?

Church's Theorem suggests that the answer is 'no', at least to the extent that certain logical abilities are not evenly distributed across the community. The specific logical ability we have in mind is the ability justifiedly to make judgements of the form 'S does not follow logically from Σ'. Let us suppose that capacity to make such judgements without the help of a backing computation is not distributed evenly across a community. By Church's Theorem, those who can justifiedly make them will have access to epistemic goods to which those requiring a confirming computation will not. Supposing (as seems reasonable) that possession of such goods constitutes some kind of private epistemic perfection in the one possessing it, there will be epistemic perfections within the reach of those of higher logical skill which will not be sharable with those of lower logical skill. Were it not for Church's Theorem, such asymmetries of logical knowledge would at least be in-principle removable through the use of a general algorithm for deciding questions of validity (invalidity). Church's Theorem, however, prohibits this happening. It thus points to a type of epistemic inequity that can not be removed through the development of any calculus of (classical) logical thought.

The inequity just described does not, however, confer any special significance on Church's Theorem as a limitation on epistemic justice. The reason is that similar inequities can be expected to arise from sources having nothing to do with Church's Theorem, for example inequities between community members concerning genuine knowledge of the community's encyclopedia and its logic. Limits on genuine factual knowledge of any type will pose the same general type of limitation on epistemic sharing that Church's Theorem poses with respect to logical knowledge. That being the case, the limitative effects of Church's Theorem on just distribution of genuine knowledge/belief can hardly be seen as distinctive.

Much the same must be said of the effects of Church's Theorem on the just distribution of the social benefits of genuine knowledge/belief. Indeed, Church's Theorem seems to lack even non-distinctive effects in this area. The general reason is this: inequities concerning the distribution of the social effects of genuine knowledge/belief arise not from inequities of authentic knowledge/belief, but from inequities in ability to operate or manipulate the *specified* calculus of a community to suit one's purposes (whether or not it is authentically possessed by the operating/manipulating agent). Church's Theorem implies no inequities of this latter type, however, since they have to do not with the linkage between a logical calculus and the logical reality it computes (i.e., the actual validity or invalidity of inferences), but only with the operation of the calculus itself.

All in all, then, the effects of Church's Theorem on that part of Leibniz' Program having to do with epistemic justice seem to be both less severe and less distinctive than those having to do with epistemic peace.

7. CONCLUSIONS

One might, of course, say that the culprit in all of the above is not Church's Theorem but the Computability Requirement. It is, after all, a necessary part of both the argument concerning the tension induced by Church's Theorem between weak and strong autonomy (and, so, of the threat that it poses to peaceful epistemic conflict-resolution) and the arguments concerning the inequitable distribution of authentic epistemic goods. This, together with the fact that it is a more 'negotiable' claim than Church's Theorem, might thus suggest that it rather than Church's Theorem should be seen as the true source of the moral and political significance that I have attributed to the latter.

I have two responses to this claim. The first is a clarification; namely, that I am not presenting Church's Theorem as *by itself* implying limits of any sort. My claims are rather that (1) Church's Theorem *in conjunction with* a condition like the Computability Requirement has such consequences, and that (2) this other condition by itself does not.

This would not, of course, be a very interesting claim were the Computability Requirement an implausible means by which a community might seek to promote epistemic justice and justice. This, however, does not appear to be the case, at least not in advance of Church's Theorem. We are thus led to a perhaps more surprising second point—namely, that institution of a condition like the Computability Requirement is not an implausible means of trying to promote epistemic peace and justice in a community. Allow me to elaborate.

To begin with, note that the Computability Requirement is right in precept. It, or some condition like it, would have to be enforced if epistemic peace and justice were to be achieved in any world in which (i) Leibniz' Thesis were true, (ii) restrictions on weak and strong autonomy tended to work against peace in the ways indicated above, (iii) it is allowed that non-computational knowledge of a case of non-consequence might count as an epistemic good, and (iv) there are differences between individuals as regards their general ability to grasp cases of non-consequence non-computationally.

I do not think that (i)–(iv) lack plausibility. With regard to (i), my belief in its plausibility is restricted to communities in which the individual members span a fairly wide range of human cognitive capacities. This, however, seems to me to be a pretty fair model of what the typical human epistemic community is like. Even if not all epistemic communities are of this type, it still seems right to say that one cannot altogether avoid epistemic communities containing persons of inconveniently low logical ability generally and still meet her obligations concerning epistemic justice. As with other sorts of goods, so, too, with epistemic goods, justice demands that we try to find ways of righting inequities of birth and upbringing and other accidental determinants of epistemic (mal)distribution. Entering into communities inclusive of logically challenged individuals would thus seem to be a likely consequence of seeking epistemic justice in a world such as ours.

More likely, though, misgivings regarding the Computability Requirement will come from other quarters—namely, (1) those who doubt the ability of a community to arrive at a satisfactory form of encyclopedia, (2) those who doubt the existence of an algorithm for arriving at plausible candidates for the logical forms of sentences framed in a natural language (assumed to be the, or at least a, type of language in terms of which the epistemic transactions of the typical community with which we are concerned are to be conducted), and (3) those who see the use of computation to back conformal claims as standing in need of augmentation by certain non-computational knowledge. It might be thought that these problems are deeper or more serious than those caused by Church's Theorem. For reasons I will now state, I do not share this view.

Let's begin with the question of the encyclopedia. To avoid misunderstanding, let me say from the start that I share the belief that it is highly unlikely that we, as members of the total community of human beings, will ever arrive at a shared encyclopedia that is at once sufficiently robust to yield perceptible resolutions of most matters (even most of the most important matters) of moral and religious significance and, at the same time, a matter of common agreement.

This does not mean, however, that smaller communities, communities capable of bringing us most of the advantages for which community membership is wanted in the first place, must suffer the same fate. At any rate this is so so long as the holding of such views does not carry with it a general absence of ability on the part of community members to provide goods and services important to their general welfare.

It seems doubtful that one could say the same about communities based on differentials in logical skill. Restricting oneself to communities where others do not exceed her logical ability nor her theirs would, to be sure, promise to minimize clashes of weak and strong autonomy. At the same time, however, there might be connections between logical ability and cognitive ability generally that would dictate that restricting oneself to membership in communities consisting of those of similar logical skill would mean a lower level of availability of goods and services generally. To the extent that this is so, there is reason (at least for the non-maximally gifted) to join communities with non-uniform levels of logical skill. Church's Theorem, however, implies that these are communities at risk for clashes between weak and strong autonomy. Hence, Church's Theorem would seem in some ways to be a deeper and more persistent threat to community than the difficulty of obtaining a common encyclopedia. Even plausible utopian communities (e.g., communities in which there is agreement on such generally controversial matters as religious and/or political belief and practices but significant differences in logical ability) would have to worry about its effects.

Let us now consider the second of the three above-mentioned objections to our treatment of the Computability Requirement—namely, that which sees the obstacle posed by Church's Theorem as minor in comparison to that posed by the difficulty of determining the correct logical forms of conformal claims. The thinking behind this objection may be presented as follows. (a) Conformal claims are to be presented in a natural language. (b) To be fit targets for computation, conformal claims must be expressed in a logically perspicuous language (i.e., in a language which clarifies their logical form). (c) The logical form(s) of natural language claims are often not perspicuous, are generally context dependent and, in any event, are not the type(s) of thing(s) the determination of which can be guided by an algorithm. Therefore, (d) the Computability Requirement will be violated before we ever get to Church's Theorem (i.e., before we ever get to the point of having to produce a backing computation for a conformal claim). Non-algorithmic means will have to be employed in order to express conformal claims in a logically perspicuous form (i.e., in the language of a first or higher order predicate calculus). Thus, it is not Church's Theorem but the lack of an algorithm for translating natural language claims into the language of a logical calculus which poses the more fundamental threat to Leibniz' Program.

I will accept, for the sake of argument, the point made in premise (b) and focus my attention upon premise (a)—or, better, upon the relationship between (a) and (c). (a) is not, I believe, rightly regarded as a commitment of Leibniz' Program. At least not in the presence of (c). Indeed, it seems altogether proper that the makers of conformal claims be required to *present* them in logically perspicuous form; e.g., in the language of a logical calculus (a *characteristica universalis*). Failing this, a conformant would not have presented a community with a call for conformity that is clear enough to give potential non-conformists a suitably clear target against which to make an effective counter-claim. It would therefore seem that presentation in logically perspicuous terms is rightly viewed as a condition for making a conformal claim.

To the extent that this is so, however, the present objection vanishes. Either (c) is correct, in which case natural language claims are not typically logically perspicuous and (a) ought not to be accepted, or natural language claims do possess an appropriate level of logical perspicuity, (c) is incorrect and there is no problem of translating from natural language into logically perspicuous form. Attainment of persipicuity in logical form may be a real difficulty. It is, however, something that should accompany a conformant's becoming convinced of a logical judgment. (How, rationally, could a conformant become convinced of a judgment of validity or invalidity without being similarly convinced of the (relevant) logical form of the item(s) she judges to be valid or invalid?) Hence, it should not interfere with any rational desire to make conformal claims.

Finally, let's consider the third of the objections stated above—namely, that concerning the need for a certain body of non-computational knowledge to confirm the correctness of any calculus of thought. Without such knowledge, the argument maintains, the computations produced by a calculus of thought would be incapable of shaping the beliefs of rational agents in the ways that Leibniz' Program (with its concern for weak autonomy) requires. The elementary character of the computations of the *calculus ratiocinator* is not enough to ensure the overall elementary character of one's knowledge of computed claims. In particular, it seems that knowledge of the soundness of the calculus would also have be elementary.

We agree that there is a need for a general belief in the soundness of a community's calculus within that community. The question is what is required in order for that need to be met. We believe that it could be met without a grasp of (or even an acquaintance with) a *proof* of soundness. Specifically, we believe it could be based on a rudimentary inductive acquaintance with the calculus and its products—an acquaintance that might be within the grasp of even those of relatively low cognitive ability. A capacity for basic inductive learning must, after all, extend as far (or very nearly as far) down the scale of cognition as does the capacity for basic symbolic manipulation. This being so, rudimentary inductive knowledge of the soundness of a calculus could be basically as widespread as the capacity to operate that calculus. Accordingly, the need for knowledge of the soundness of a calculus would pose no essential obstacle to the execution of Leibniz' Program.

We have presented Leibniz' idea (Leibniz' Program) for bringing epistemic peace and justice to human communities. We have also argued that Church's Theorem constitutes a fundamental theoretical limitation on the extent to which it and other programs like it can be carried out. It would be no surprise, however, if more detailed consideration of sub-cases of the Decision Problem for classical validity and of related issue of computational complexity proved capable of sharpening our understanding of these matters. Such work might even point the way to a revival of Leibniz' Program by identifying subdomains of logical reasoning within which large and important tracts of human reasoning might properly be seen as being confined, and by showing that the validity of reasoning within those domains is decidable, perhaps even feasibly so.

REFERENCES

Leibniz, G. W.

1951 *Leibniz selections* (P. Wiener, editor), Charles Scribner's Sons, New York.

1961 *Die Philosophische Schriften von Gottfried Wilhelm Leibniz*, vol. VII (C. I. Gerhardt, editor), Georg Olms, Hildesheim.

1965 *Monadology and other philosophical essays* (P. Schrecker, translator), Bobbs-Merrill, New York.

1966 *Theodicy* (abridged) (D. Allen, editor), Bobbs-Merrill, New York.
1969 *Leibniz: Philosophical papers and letters,* second edition (L. Loemker, translator and editor), Reidel, Dordrecht.

M. RANDALL HOLMES

TARSKI'S THEOREM AND *NFU*

Abstract. The Tarski paradox of the undefinability of truth is proved by a diagonalization argument similar to the argument of Russell's paradox. In *ZFC*, Russell's argument shows that the universal class (and large classes generally) do not exist. In other set theories, such as Jensen's variant *NFU* of Quine's "New Foundations", large classes such as the universe may exist; the diagonalization arguments lead to somewhat different restrictions on the existence of sets in the presence of different axioms. In this paper, we explore the possibility that semantics expressed in *NFU* may have somewhat different restrictions imposed on them by the diagonalization argument of Tarski. A language L is definable in *NFU*, in which the stratified sentences of the language of *NFU* can be encoded (but, it should be noted, as a proper subclass of L). Truth for sentences in L is definable in *NFU*, and the reason that a suitably adapted Tarski argument fails to lead to paradox is not that truth for L is undefinable in *NFU*, but that quotation becomes a type-raising operation, causing the predicate needed for the "Tarski sentence" to be unstratified.

1. INTRODUCTION

The well-known theorem of Tarski that truth of sentences in any reasonably expressive language L cannot be defined in the language L itself is proven by a diagonalization argument similar to the argument involved in Russell's paradox. The paradox of Russell shows us that some restriction on comprehension axioms in set theory is required, but it does not prescribe the restriction. The traditional approach involves "limitation of size", and is embodied in Zermelo set theory and its extensions. It is usual to think that Russell's paradox excludes "large" sets like the universe, but this is actually not the case. An alternate solution to Russell's paradox (and other paradoxes) was proposed by Quine (*1937*) in his system "New Foundations" (*NF*): comprehension restricted to stratified formulae. The consistency question for this theory remains open (as is generally known); what is less generally known is that the validity of the general approach of Quine to resolving the paradoxes has been demonstrated. Jensen (*1969*) showed that the theory *NFU* ("New Foundations" with the extensionality axiom weakened to allow urelements, but with the same comprehension axiom as "New Foundations" itself), is consistent relative to the usual set theory and remains consistent if the axioms of Infinity and Choice are adjoined. In these set theories, "large" sets like the universal set are provided by the comprehension axiom, and the resolution of the set-theoretical paradoxes proceeds along a different route. An all-purpose reference for set theories of this type, which can serve as a substi-

C. Anthony Anderson and M. Zelëny (eds.), Logic, Meaning and Computation, 469-478.

tute for most of the specific references cited in the paper, is Thomas Forster's excellent book *Set theory with a universal set* (*Forster 1992*), although the emphasis there is on *NF* rather than *NFU*.

Analogously, we have discovered that approaching the Tarski paradox of the definability of truth in a Quine-style set theory allows a different resolution of the problem; the paradox does not preclude the definability of truth any more than the Russell paradox precludes the existence of a universal set. We will present an infinitary language (with sentences of finite length but with infinitely many primitive predicates) which does define its own truth predicate, but avoids the paradox of Tarski because *quotation* turns out to be a "type-raising operation"! This appears to suggest an alternate approach to semantic paradoxes in general, analogous to the alternate approach to set-theoretical paradoxes embodied in *NFU*.

It should be noted that the infinitary nature of the language used is not accidental. Consideration of the paradox packaged succinctly in the phrase "the smallest natural number not describable in less than a billion words" reveals that any language which expresses its own semantic relations can be expected to have short names for each and every natural number, and so infinitely many atomic names (in a language without names, we can say, equivalently, "definite descriptions" in Russell's sense).

2. TARSKI'S THEOREM

We briefly (and informally) review the proof of the theorem of Tarski.[1] Suppose that we have encoded the formulae of a language L in such a way that they can be discussed in L (as numbers, for example). For each formula ϕ with one free variable x, let "ϕ" be the code for ϕ. We can think of formulas with one free variable as "definable predicates", holding of an object if substitution of a name for that object for the free occurrences of x in the formula yields a true sentence. Suppose, moreover, that truth of encoded sentences of L is a predicate definable in L.

We then consider the following, which should be expressible formally as a formula in one free variable x:

> The formula encoded by x, when each free occurrence of the variable "x" is replaced with the code x itself, yields the code of a sentence which is not true.

Less formally, for this to be true for an encoded formula "ϕ" in place of x means:

> The formula ϕ (as a predicate) does not hold of "ϕ".

[1]For details, see *Andrews 1986*.

Call the formula informally described above ψ; replacing the variable x in ψ with the code "ψ" yields a sentence which asserts its own falsehood! Roughly speaking, the resulting sentence says

> The predicate ψ does not hold of "ψ",
> but for ψ to hold of "ψ" means exactly:
>
> The predicate ψ does not hold of "ψ",
> so the sentence denies itself.

An obvious requirement for the argument to work is that the notion of substitution of a specific object for a variable in (the coded version of) a formula be definable in L; this holds (for the usual kind of coding using numbers) in any theory as strong as arithmetic. The resolution of the contradiction apparently must be (and indeed must be in the usual context) that the predicate "is true" of encoded sentences of L cannot actually be defined in L.

3. PRELIMINARIES IN *NFU*

NFU is a first-order, one-sorted theory with equality, membership, and a unary predicate of sethood.[2] The axioms of *NFU* are as follows:

***Extensionality*:** Sets with the same elements are the same.

***Urelements*:** Objects which are not sets have no elements.

***Stratified Comprehension*:** For each variable x and formula ϕ which is "stratified", $\{x|\phi\}$, the set of all x such that ϕ, exists.

A formula is said to be "stratified" if types can be assigned to each variable occurring in the formula in such a way as to obtain a formula of the simple theory of types.

Observe that the defining formula $x \notin x$ of the Russell class is not stratified, but the defining formula $x = x$ of the universe *is* stratified, so there is a universe V.

We usually introduce the Axiom of Infinity by introducing the projection relations for an ordered pair which has the same relative type as its projections; this is equiconsistent with the Axiom of Infinity in a more usual form, and implies it, although it is inessentially stronger. Functions and relations can then be defined in the usual way, and the Axiom of Choice can be stated in the usual equivalent forms: our favorite form asserts that disjoint partitions of sets have choice sets. Another form worthy of notice is the assertion that V can be well-ordered.

[2]This form of presentation of *NFU* was suggested by Quine himself in his remarks accompanying *Jensen 1969*.

Note that the usual Kuratowski definition of the pair ($\langle x, y \rangle = \{\{x\}, \{x, y\}\}$), which can be used in *NFU* without the assumption of Infinity, yields a pair with relative type two higher than the types of its projections. This is inconvenient but not impossible to work with; it has odd effects on the relative types of functions and their arguments, for example.

Cardinal numbers (including natural numbers) are defined as equivalence classes of sets under the obvious equivalence relation; ordinal numbers are defined as equivalence classes of well-orderings under similarity. The objects which occasion the paradoxes of Cantor and Burali-Forti in naive set theory (the cardinality of the universe and the order type of the ordinals) actually exist in *NFU* but do not have quite the expected properties.

The cardinality of the universe is clearly not less than the cardinality of the power set of the universe (the set of all *sets*), so Cantor's theorem in its usual form cannot hold. The situation can be clarified by considering the form of Cantor's theorem which can be proven in the theory of types: there, we cannot even ask whether the cardinalities of a set A and its power set $\mathcal{P}\{A\}$ are the same, because the types of these two sets are different. What can be proven, in *NFU* as in the theory of types, is that the cardinality of $\mathcal{P}_1\{A\}$, the set of one-element subsets of A, is strictly less than the cardinality of $\mathcal{P}\{A\}$. In *NFU*, we can draw the further conclusion that $|\mathcal{P}_1\{V\}| < |\mathcal{P}\{V\}| < |V|$ (there are "fewer" singletons of objects than there are objects). This should not be too surprising, since the function which takes each object to its singleton has an unstratified definition.

The role of the set of one-element subsets of A in the argument above inspires the definition of a parallel operation on cardinals: $\mathrm{T}\{|A|\}$ is defined as $|\mathcal{P}_1\{A\}|$. It is straightforward to show that this operation on cardinals does not depend on the choice of A; it is unstratified and does not define a (set) function. For each cardinal $|A|$, the cardinal $\exp(|A|) = 2^{|A|}$ is defined (following Marcel Crabbé rather than Specker to obtain a slightly stronger definition) as $\mathrm{T}^{-1}\{|\mathcal{P}\{A\}|\}$; the function exp has a stratified definition, but it is partial (because T^{-1} is partial). Observe that exp thus defined is also the natural exponentiation function for the theory of types; the operations T and T^{-1} can be thought of in the context of the theory of types as projecting cardinals to "the same" cardinals in higher or lower types (from the standpoint of the usual set theory; we have seen that $\mathrm{T}\{|V|\} \neq |V|$ in *NFU*, so this cannot be our position from the *NFU* standpoint).

The Burali-Forti paradox does not afflict *NFU*, because it depends on the theorem of naive set theory or Zermelo-style set theory that the order type of the ordinals below α is equal to α itself. Observe that the order type of the ordinals below α is an object two types higher than α in the theory of types; one will then not be surprised to find that the corresponding theorem of *NFU* asserts that the order type of the ordinals below α in the natural order is $\mathrm{T}^2\{\alpha\}$, where $\mathrm{T}\{\alpha\}$ is defined as the order type obtained by replacing the objects ordered by any order of type α with their singletons. Now the reasoning of the Burali-Forti paradox proves that $T^2\{\Omega\} < \Omega$, where Ω is the order type of the natural order on the ordinals; Ω proves to be greater than the order type of the segment below it and less than the largest ordinals. The sequence of iterated images of Ω under T can have no smallest element, but it is not a set. Such "sequences" (failing to be sets) of iterated images of objects under type-raising operations will play an important role below.

Note the role of type-raising operations such as the T operations on cardinal and ordinal numbers in avoiding paradox. The approach to avoiding semantic paradox here will allow the definition of notions such as truth and synonymy, at the price of treating *quotation* as a type-raising operation of this general kind and reference as an external relation with stratification restrictions similar to those on membership.

4. SEMANTICS IN *NFU*

The theory *NFU* + Infinity + Choice cannot describe its own semantics. This is fortunate, because the inconsistency of the theory would follow! However, it can describe the semantics of an infinitary language which captures the semantics of all *stratified* sentences of *NFU* in a certain guarded sense (if it did this unqualifiedly, it would still imply inconsistency of the theory!).

The obstruction to *NFU* introspecting on its own semantics is that the relation $\in$ is not a set; it should be clear that for any relation R which is a set, the set $\{x \mid \sim xRx\}$ is definable, and so the existence of the set $\in$ would imply the existence of the Russell class.

We use an idea of Grishin (*1972*) for a rather different purpose[3]. Observe that the relation of *inclusion* is a set; it has a stratified definition in which the two related objects are of the same type. Then observe that any formula $x \in y$ can be expressed as $\{x\} \subseteq y$, in which the non-relation $\in$ has been replaced by the relation $\subseteq$. One can then replace the expression $\{x\}$ by a variable X restricted to the set of singletons $\mathcal{P}_1\{V\}$.

In general, a formula in equality, membership, and the primitive projection relations can be translated into a formula in equality and the relations induced by inclusion and projection relations on n-fold singletons for each n in such a way that each variable of relative type i is replaced by a variable restricted to the set $\mathcal{P}_1^{N-i}\{V\}$ for a fixed natural number N large enough that $N-i$ is nonegative in each case needed. Note that any such formula can be expressed in the language with primitive unary predicates corresponding to each set in the universe and binary relations corresponding to each set relation in the universe. The specific unary and binary predicates we need are much more restricted, being simply the predicates of being an n-fold singleton for each n, equality, and the relations induced on n-fold singletons by inclusion and the projection relations; so it may seem to be overkill to use *all* sets and relations,

[3]See also *Forster 1992,* pages 64–66.

but there is no small set in *NFU* which contains all of the sets we need to encode these predicates (even though the class of sets we need is externally countable!).

If the sentence to be translated is not stratified, the translation process yields a formula in which some particular variable is replaced by variables representing different iterated singletons of objects represented by that variable; since the relation between the mth and nth iterated singletons of the same object for $m \neq n$ is not captured by any set relation in *NFU*, the translated sentence will not be usable in the construction below.

We now introduce an infinitary language L, having a primitive predicate "membership in x" for each object x in the universe and a primitive binary predicate R for each set of pairs such that xRy is to hold exactly when $(x, y) \in R$. The logical operations allowed in L are the usual propositional connectives and quantifiers (L is infinitary only in having infinitely many constants; sentences of L are finite in length). The terms of L are variables indexed by the natural numbers.

As we indicated above, each sentence of *NFU* can be translated into a sentence of the language L. There are *countable* sublanguages L_i of L capable of expressing all sentences of *NFU* using no relative types other than 0-$(i-1)$, but the "union" of the L_i's is not a set, and there is no set of exactly the sentences of L which encode sentences of *NFU*. If there were such a set, paradox would ensue.

We develop the semantics of L in some detail, and indicate how truth of coded sentences of L is definable in *NFU*. Use of the type-level pair of the previous section will be essential to allow induction on structures built up using pairing.

Moreover, we will make an assumption strengthening our set theory as well. Observe that $\mathrm{T}\{n\} = n$ holds for $n = 0, 1, 2, \ldots$, where T is the type-raising operation on cardinal numbers of the previous section. It is impossible to prove the assertion "$\mathrm{T}\{n\} = n$ for each natural number n" in *NFU* + Infinity + Choice; the attempt to prove it by induction fails because the condition on which induction is to be carried out is unstratified and fails to define a set. Nonetheless, this assertion, called Rosser's Axiom of Counting,[4] is consistent with *NFU* + Infinity + Choice (and strengthens it essentially).

Atomic terms of L are variables. There will be countably many variables v_i for i ranging through the non-negative integers. The fact that each singleton corresponds to a predicate means that we could just as well have a name for each constant in the universe, but we will refrain from this extravagance.

Atomic formulas of L are statements of membership in sets and participation in binary relations. The sentence "$v_i \in A$" is encoded as $((i, \{A\}), 0)$, while the sentence "$(v_i, v_j) \in R$" is encoded as $(((i, j), \{R\}), 1)$. The use of the singleton operation is to preserve stratification; the relative type of a

[4]Proposed for *NF* in *Rosser 1953*.

coded sentence is to be the same as the intended relative type of its variables, while the sets represented by predicates, unary or binary, are one type higher.

Similarly, we define codes for formulas inductively:

$$\text{“}\sim\phi\text{”} = (\text{“}\phi\text{”}, 2); \quad \text{“}\phi \wedge \psi\text{”} = ((\text{“}\phi\text{”}, \text{“}\psi\text{”}), 3); \quad \text{“}(\forall v_i)(\phi)\text{”} = ((i, \text{“}\phi\text{”}), 4).$$

It is straightforward to define the predicate "x is a code for a formula" and the operation of substitution of a term for a variable by induction; *NFU* provides more than enough set theoretical machinery for this, but it is crucial that the type-level pair is used to allow structural induction on pairs.

Notice that a type raising operation T on coded sentences can be defined: replace each set x which appears representing a unary relation with $\mathcal{P}_1\{x\}$ and each relation R with the corresponding relation on singletons. Our assumption of the Axiom of Counting ensures good behavior of this type-raising operation; in particular, it implies that coded sentences which do not involve any constants will actually be sent to themselves, as one would expect! T{"ϕ"} can be obtained from the singleton of "ϕ" by an inductively defined set function. If this operation is applied to a (doubly) encoded sentence of *NFU*, it actually has the effect of systematically *lowering* types by one.

One proceeds to define "satisfaction" as a predicate of pairs ({"ϕ"},f), where f is a function taking each natural number i to an intended value for v_i, in the usual way; the singleton operation is used to preserve stratification. The satisfaction predicate, which is stratified and does define a set, can then be used to define the set of true closed sentences of L in the obvious way. Notice that the closed sentences of each L_i can actually be encoded as numbers; there are only countably many sentences in each of these languages.

5. THE TARSKI "PARADOX" IN THE LANGUAGE L

The "paradoxical" situation which now arises is this. We have defined the notions of "sentence of L" and truth of sentences of L via stratified constructions in *NFU*. But we know that L itself can express any stratified sentence of *NFU*. Thus, it appears that L may be able to capture its own truth predicate.

To see what actually happens, it is useful to look at the world of *NFU* as it is seen from the standpoint of L. Sentences of *NFU* are effectively translated into sentences of a type theory (not internally describable in L). The types are the iterated images of V under $\mathcal{P}_1$; the membership relations are inclusion (of singletons in general sets, between the top type and the type below it) and the relations induced by inclusion on iterated singletons (between successive lower types). Notice that this is a "downward" type theory in which there is a top type and no bottom type; in all other respects it is precisely analogous to the usual kind of type theory. A consequence of our assumption of the Axiom of Counting is that the type-raising operation T on "pure" translated sentences of *NFU* (involving no constants other than those used to represent types) preserves truth value.[5]

Our construction of the satisfaction predicate for L involved the relative type of coded sentences of L and two types above that. In the translations of sentences involving satisfaction of the sentence ϕ into L, the double singleton of the code "ϕ" will appear in place of the code itself. Moreover, there is a further reflection into lower types involved if we consider the translation of a sentence *about* satisfaction into L; this will mention the relation of satisfaction using its double singleton and any formulas to be satisfied via their *quadruple* singletons.

Suppose we try to replicate the argument of Tarski. We would use the predicate:

The predicate represented by formula ϕ is not true of "ϕ".

But this predicate is either unstratified or senseless (depending on how it is read). Certainly the reference to the formula ϕ outside of quotes in the predicate is, for L, a reference to the double singleton of ϕ. Then it is necessary to realize that the truth predicate appearing in a sentence interpreted inside of L now refers to a property not of double singletons of formulas but of double singletons of double singletons of formulas; the question is then whether the term "ϕ" is to be understood as referring to the double singleton of ϕ, in which case it is not understood as a code for a formula (so we have the "senseless" interpretation) or as a quadruple singleton, in which case the formula is unstratified and so does not define a set in *NFU* or predicate in L.

The Tarski paradox is blocked by the fact that quotation of a formula is a type-raising operation, preventing diagonalization. If we had terms representing constants, quoting them would raise types in the same way. Since the diagonalization is blocked by stratification, it is not a problem that the set of true sentences of L (as represented by their double singletons) proves to be definable in L by following our development above in *NFU*.

6. MODELS USING EXTERMAL AUTOMORPHISMS

The same situation can be modelled in the usual set theory using models of initial segments of the cumulative hierarchy with external automorphisms. In fact, this is how *NFU* itself is best modelled.

We work in a nonstandard model of the usual set theory *ZFC* (or of "enough" axioms of *ZFC*) with an external automorphism j and nonstandard infinite ordinal α such that $j(\alpha) > \alpha$. We consider V_α, stage α in the cumulative hierarchy.

---

[5]See *S. Orey 1964* for a discussion of the need for the Axiom of Counting here.

It is shown elsewhere[6] that V_α is readily interpreted as a model of *NFU*. The trick is to observe that V_α contains the much earlier stage $V_{j(\alpha)}$ of the cumulative hierarchy which looks exactly like it. Construe each element of $V_{j(\alpha)+1}$ as a set with its elements replaced by the inverse images under j of its actual elements and each element of $V_\alpha - V_{j(\alpha)+1}$ as an urelement (notice that there are a lot of urelements!). This is achieved by defining a new membership relation "$x \in_{new} y$" as "$x \in j(y)$ and $j(y) \in V_{\alpha+1}$". It is straightforward to prove that V_α with membership relation $\in_{new}$ is a model of *NFU*.

The semantics we have been doing in this paper can be clarified in the same context. It is possible to define the semantics for the "full language" (in which each set corresponds to a unary predicate and each relation to a binary predicate) on V_α in $V_{\alpha+2}$. If we replace elements of sets with their images under j as in the model construction for *NFU* and inflate power sets with urelements in the same way (setting up the construction to be carried out in *NFU* as above), we observe that the semantics for the "full language" on $V_{j^2\alpha}$ can be expressed in V_α in a way which translates successfully into terms of $\in_{new}$. The construction is in fact exactly that of the previous sections.

If L is the "real" full language on V_α, the language, which we see internally to V_α in the construction above, is $j^2(L)$. We cannot derive the Tarski paradox here because the needed predicate would be:

Predicate ϕ does not hold (in $j^2(L)$) of $j^2(\text{“}\phi\text{”})$,

which cannot be expressed in our working model of set theory because j is external. This is precisely analogous to the way in which the argument for Tarski's paradox fails above.

A system resembling the system of this paper in its semantic features, although not apparently motivated by work in Quine-style set theory, is discussed by Hiller and Zimbarg (*1984*). Their system uses additional assumptions which make it far stronger than the system *NFU* + Infinity + Choice used here, even when this is further extended with the Axiom of Counting.

REFERENCES

Andrews, P.

1986 *An introduction to mathematical logic and type theory*: *To truth through proof*, Academic Press, New York, pp. 264–265.

Forster, T.

1992 *Set theory with a universal set*, Oxford Logic Guides, no. 20, Clarendon Press, London.

Grishin, V. N.

1972 The equivalence of Quine's *NF* system to one of its fragments (in Russian), *Nauchnotekhnicheskaya Informatsiya*, series 2, vol. 1, pp. 22–24.

[6]See *Forster 1992*, pages 67–68.

Hiller, A. P., and J. P. Zimbarg
1984 Self-reference with negative types, *The Journal of Symbolic Logic*, vol. 49, pp. 754–773.

Jensen, R. B.
1969 On the consistency of a slight(?) modification of Quine's *NF*, *Synthese*, vol. 19, pp. 250–263.

Orey, S.
1964 New Foundations and the axiom of counting, *Duke Mathematical Journal*, vol. 31, pp. 655–660.

Quine, W. V.
1937 New foundations for mathematical logic, *American Mathematical Monthly*, vol. 44, pp. 70–80.

Rosser, J. B.
1953 *Logic for mathematicians*, McGraw-Hill, New York; and Chelsea, New York, 1978.

GARY MAR

CHURCH'S THEOREM AND RANDOMNESS*

A good work of art can in its entirety be expressed only by itself.
—Tolstoy

Abstract. Alonzo Church demonstrated in a logically rigorous way that Leibniz's dream of a logical calculus that could decide all truths was not only unfulfilled but unfulfillable. According to Church's Theorem (*1936a*), even first-order logic is undecidable. Kleene's (*1936*) proof of the equivalence of the Church-Kleene notion of λ-definability and the Herbrand-Gödel (*1931*) notion of general recursiveness and Turing's (*1937*) proof of the equivalence of λ-definability and computability were regarded by Church as evidence for Church's Thesis (*1936*). Church's Thesis is the philosophical claim that a function of positive integers is effectively calculable if and only if it is λ-definable.[1] Using the notion of effective calculability, Church (*1940*) proposed a definition of randomness. Church's definition can be seen as a precursor to algorithmic definitions of randomness formulated independently by Kolmogorov (*1963, 1965*) and Chaitin (*1966*). In this paper we sketch one way of relating Church's Theorem to randomness by way of a recent generalization of the paradox of the Liar (*Mar and Grim 1991*) that uses the mathematics of chaos.

1.

The *Classical Liar* is a sentence that asserts its own falsity:

The boxed sentence is false.

According to the Tarskian (T) schema, a sentence stating that a sentence p is true has the same truth-value as p itself. Hence,

(1) 'The boxed sentence is false' is true if and only if the boxed sentence is false.

---

* This paper is a concise presentation of the foundation of collaborative research with Patrick Grim and Paul St. Denis. See *Mar and Grim 1991*, and *Mar and St. Denis 1994* and *1996*.

After the paper was submitted for this volume, an additional article (*Mar and St. Denis 1999*) has been published that gives an elegant characterization of the formal system used here.

[1] *Church 1936*. The precise epistemological status of Church's Thesis is a subject of considerable philosophical discussion. See the *Notre Dame Journal of Formal Logic*, vol. 28, no. 4, 1987, devoted to this topic.

C. Anthony Anderson and M. Zelëny (eds.), Logic, Meaning and Computation, 479-490.

But since it is empirically true that

(2) 'The boxed sentence is false' is identical to the boxed sentence,

we may infer by Leibniz's Law that

(3) The boxed sentence is true if and only if the boxed sentence is false.

The assumption that the boxed sentence is true leads to the conclusion that it is false, and the assumption that it is false leads to the conclusion that it is true. The semantic behavior of the Classical Liar can therefore be represented as an infinite oscillation between the classical truth-values true and false. Suppose we do not restrict ourselves to the assumption of bivalence. What form could the Classical Liar take when generalized to the continuous real-valued case?

First, we generalize the logical connectives by adopting a Lukasiewiczian infinite-valued logic.[2] Following Rescher (*1969*), we use the notation '$/p/$' to represent the value of the sentence p, which may now have as a value any real number in the $[0, 1]$ interval. A natural infinite-valued negation rule is:

$$/{\sim}p/ = 1 - /p/.$$

Intuitively, the negation of p is true to the extent that p is untrue, i.e., to the extent the value of p *differs* from the value of 1 or complete truth. The infinite-valued Lukasiewiczian biconditional rule is:

$$/(p \leftrightarrow q)/ = 1 - |/p/ - /q/|.$$

(Here the vertical slashes represent the *absolute value* of the difference.) Intuitively, the biconditional is true to the extent that its constituents do *not differ* in truth-value. Both these infinite-valued rules are *faithful* to classical logic: when the values of the sentences are restricted to the classical truth-values, we obtain the classical truth tables.

Secondly, the Tarskian (T) schema can also be generalized to an infinite-valued context. The (T) schema implies the truth of each biconditional whose constituents are the sentence asserting that p is true and the sentence p itself. Using the biconditional rule, we can express this truth by $1 = 1 - |/\mathrm{T}\ulcorner p \urcorner/ - /p/|$. Alternatively stated, we have $/\mathrm{T}\ulcorner p \urcorner/ = 1 - |\mathbf{t} - /p/|$, where $\mathbf{t}$ is the value of 1 or complete truth. Replacing the constant $\mathbf{t}$ with a parameter $\mathbf{v}$ (which ranges over the [0,1] interval) and replacing the bivalent truth predicate 'T' with a multi-valued relation '$\mathrm{V}\mathbf{v}p$' (which is to be read '$\mathbf{v}$ is the truth-value of the sentence p'), we obtain Rescher's (*1969*) schema for his parametric-operator development of many-valued logics:[3]

$$/\mathrm{V}\mathbf{v}p/ = 1 - |\mathbf{v} - /p/|.$$

---

[2] A Lukasiewiczian infinite-valued logic is characterized by its rule for the conditional: $/(p \rightarrow q)/ = \textsc{Min}\{1, 1 - /p/ + /q/\}$. See *Rescher 1969*, page 36.

[3] *Rescher 1969*, page 81.

Intuitively, Rescher's schema states that the sentence V**v**p, namely, the sentence that **v** is the truth-value of the sentence p, is true to the extent to which the value of p does not differ from **v**.[4]

Consider again, for example, the Classical Liar λ, which asserts its own falsity. To determine the value of the sentence asserting that the Liar is true, a calculation using Rescher's schema yields

$$/\mathrm{V}\mathbf{t}\lambda/ = 1 - |f - /\lambda/|,$$

where **t** and **f** are the values completely true and completely false, respectively. How can we represent the self-referential element in the Liar? One way is to use *functional iteration*: a value for /V**t**λ/ can be calculated on the basis of an estimate for /λ/, and then the calculated value for /V**t**λ/ can then be recycled as a revised estimate for /λ/, and so forth. Formally, we represent the process of iteration by replacing the sentence letter 'p' in the V**v**p schema by 'x_n' and by replacing '/V**t**λ/' by 'x_{n+1}'. Making these replacements and using the value 0 for **f**, we can now represent the semantics for the Classical Liar by the algorithm:

$$x_{n+1} = 1 - |0 - x_n|.$$

Given an initial estimated value of x_0, the successive estimated values for the continuous-valued Classical Liar will be an alternating cycle of period 2 between the values x_0 and $|1 - x_0|$. The single exception is the value of 1/2 which is a fixed point.

Suppose we want to evaluate sentences that do not attribute to themselves a *particular* truth-value **v** but rather attribute to themselves a truth-value expressed as a *function* of its previous estimated value.[5] To evaluate such sentences, we replace the '**v**' in the V**v**p schema with a functional expression $\mathrm{S}(x_n)$ that expresses the value the sentence attributes to itself in terms of its previously estimated value x_n:

$$x_{n+1} = 1 - |\mathrm{S}(x_n) - x_n|.$$

[4]The Tarskian (T) schema was stated above in terms of sentences; Rescher states his V**v**p schema in terms of propositions. For the sake of consistency we have reformulated Rescher's schema in terms of sentences. For present purposes, we set aside the philosophical controversy as to what should properly be regarded as the bearers of truth. See *Church 1956*, page 27, footnote 72.

[5]Less restrictive options are possible. For example, given a sequence $x_0, x_1, \ldots, x_n$ of length $n+1$, let us define the *middle* of the sequence $m(n)$ to be $n/2$ if n is even or $(n-1)/2$ if n is odd. Then define x_n to be $x_{m(n)}$ if n is even and to be $1 - x_{m(n)}$ if n is odd. Then given $x_0 = 0$ and $x_1 = 1$, the sequence $\{x_n\}$ for $n \in \omega$ generates the famous Morse-Thue (MT) sequence from symbolic dynamics. The MT sequence is of interest because it is an effectively calculable binary sequence which is aperiodic, infinitely self-similar and intimately related to the period-doubling route to chaos (see *Schroeder 1991*, pages 264ff.). Consider the infinitely nested sequence of *Biconditional Samesayers*: $\beta_0 := {\sim}p_0$; $\beta_1 := p_1 \leftrightarrow \beta_0$, and in general, $\beta_{n+1} := (p_{n+1} \leftrightarrow \beta_0)$. Then given a standard truth-table assignment to the sentence letters, the final column for each of the nested biconditionals β_n is precisely the initial binary block of length 2^n of the MT sequence.

We can illustrate the use of this modified V*vp* schema by comparing the semantic behavior of two Liar-like sentences.

Consider first the sentence: *This sentence is half as true as it is estimated to be true*, which we call the *Cautious Truth-Teller.* Using the modified V*vp* schema, the successive values for the Cautious Truth-Teller are given by $x_{n+1} = 1 - |x_n/2 - x_n|$. Compare this with a second sentence, *This sentence is as true as the estimated value of the conjunction of itself and its negation*, which we call the *Contradictory Liar.* The value of a conjunction, in a Lukasiewiczian logic, is given by the minimum of the values of its conjuncts: $/(p \wedge q)/ = \text{Min}\{/p/, /q/\}$. Hence, the sequence of estimated values for the Contradictory Liar is given by $x_{n+1} = 1 - |\text{Min}\{x_n, 1 - x_n\} - x_n|$.

The difference between the semantic behaviors of these two sentences can be made visually perspicuous using a *web diagram.* Beginning by plotting a line vertically from $(x_0, 0)$ to (x_n, x_{n+1}), the web diagram continues the line horizontally from (x_n, x_{n+1}) to (x_{n+1}, x_{n+1}) and then iterate the process by using x_{n+1} for x_n. The semantics for the continuous-valued Classical Liar, for example, appears in a web diagram as nested series of simple squares.

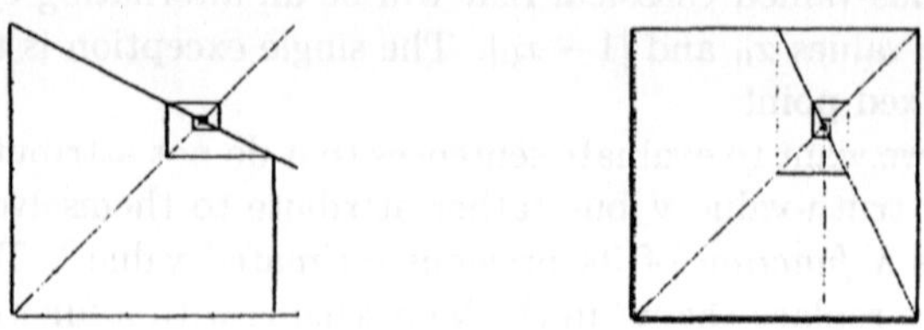

Figure 1: Web diagrams for the Cautious Truth-Teller
and the Contradictory Liar.

The Cautious Truth-Teller, on the other hand, yields a *fixed-point attractor*: no matter what initial value with which we begin, the successively revised estimated values are inevitably drawn toward the fixed point of 2/3. The Contradictory Liar, in contrast, yields a *fixed-point repellor*: for any values other than the fixed point 2/3, the successively revised values for the Contradictory Liar are repelled away from 2/3 until the values settle on the oscillation between 1 and 0, characteristic of the Classical Liar.

To sum up, the semantic behavior of the Classical Liar, though identical to the Contradictory Liar on classical values, diverges from the Classical Liar on the range of values between 0 and 1. Sentences like the Cautious Truth-

Teller and the Contradictory Liar, on the other hand, have precisely the opposite semantic behavior in terms of being attractors and repellors around the same fixed point. Infinite-valued logic can, therefore, reveal intriguing new patterns of paradox that have remained hidden in a classical bivalent setting.[6]

2.

Consider next a more intriguing, infinite-valued generalization of the Classical Liar, which asserts of itself not that it is simply false, but that it is true to the extent that it is false:

The boxed sentence is as true as it is estimated to be false.

We call this sentence the *Chaotic Liar* for reasons which will soon become apparent. What the Chaotic Liar asserts of itself is that it is false; hence, $S(x_n) = 1 - |0 - x_n|$. Since $x_n \geq 0$, we have $S(x_n) = 1 - x_n$. The algorithm for the value the Chaotic Liar attributes to itself, therefore, is identical to the algorithm for the Classical Liar. The successive values for the Chaotic Liar will be given by $x_{n+1} = 1 - |(1 - x_n) - x_n|$. The web diagram for the Chaotic Liar makes it clear that the complexity of the semantic behavior

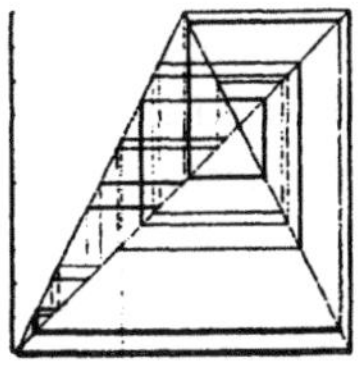

Figure 2: The Chaotic Liar has chaotic semantic behavior.

[6]Consider the *Contingent Liar* based an infinite-valued generalization of the paradox due to Kleene and Rosser (*1935*), discussed by Curry (*1941,*) and also known as Löb's paradox (see *Barwise and Etchemendy 1987,* footnote 14). The Contingent Liar asserts *This sentence is as true as the conditional: if this sentence is true then q*, where *q* is some contingent sentence. Using the infinite-valued rule for the Lukasiewiczian conditional, we have that the iterated values for the Contingent Liar is given by the sequence: $x_{n+1} = 1 - |\text{MIN}\{1 - x_n + /q/, 1\} - x_n|$. The Contingent Liar exhibits the fixed-point semantic behavior of the *Truth-Teller* (which asserts *This sentence is true*) on the interval $[0, /q/]$ and exhibits chaotic semantic behavior on the interval $[/q/, 1]$.

of the Chaotic Liar far surpasses the monotonous regularity of the Classical Liar (see Figure 1).

The Chaotic Liar is so-called because its algorithm is *chaotic* in a precise mathematical sense.[7] Known more familiarly as the chaotic tent function, the algorithm for the Chaotic Liar is paradigmatically chaotic.[8] The semantic patterns inherent in the Chaotic Liar, therefore, exhibit all the intriguing properties of chaotic functions. Among these properties will be:

- *unpredictability*: the sensitive dependence on initial conditions of chaotic functions entails that prediction will fail if the initial conditions are not known with infinite accuracy;
- *infinitely many periodic cycles*: since the Chaotic Liar has a cycle of period three it will have cycles of all other periods according to Sarkovskii's ordering (see *Devaney 1989,* pages 60–62);
- *fractal images of the semantics of paradox*: *fractals,* sets with a fractional Hausdorff dimension (see *Peitgen and Saupe 1988,* pages 28–29), are characterized by self-affinity at increasing powers of magnification. Within the patterns of paradox for Dualist forms of the Liar, for example, there are infinitely complex fractal patterns (see *Mar and Grim 1991*).

The semantic paradoxes have been a trap for logicians who, in their attempts to solve the paradoxes, have tended to view the patterns of paradox as simpler and more predictable than they actually are. Even in the sophisticated work of Barwise and Etchemendy on the Liar (*1990*), the cyclical regularity of the semantical paradoxes has been obvious but their incalculable complexity has remained hidden. Here, instead of searching for simple patterns of semantic *stability* (as in *Gupta 1982* and *Herzberger 1982*), we have tried to exhibit infinitely complex and chaotic patterns of semantic *instability* which have gone virtually unexplored.

---

[7] *Devaney 1989,* page 50, notes that there are stronger and weaker definitions of chaos. Devaney's definition of chaos is as follows. A function $f : \mathrm{I} \rightarrow \mathrm{I}$ is *chaotic* on a set I if all three of the following conditions hold:

(i) f has *sensitive dependence on initial conditions*: there exists points arbitrarily close to x which eventually separate from x by at least δ under iteration of f, i.e., $\exists\delta > 0$ $\forall x \in \mathrm{I}$ $\forall$ neighborhood N of x $\exists y \in \mathrm{N}$ $\exists n \geq 0$ $|f^n(x) - f^n(y)| > \delta$ (here '$f^n(x)$' represents the nth iteration of the function f);

(ii) f is *topologically transitive*: f has points which eventually move under iteration from one arbitrarily small neighborhood to any other, i.e., $\forall$ opens sets U, V $\subset$ I $\exists k > 0$ $f^k(\mathrm{U}) \cap \mathrm{V} \neq \phi$;

(iii) the *periodic points are dense* on I: there is a periodic point between any two periodic points in the interval I, where a point x is *periodic* if $\exists n f^n(x) = x$.

[8] The tent function is mentioned in Robert May's (*1976*) ground-breaking paper as a "mathematical curiosity". Here, however, the tent function appears as perhaps the simplest and most natural generalization of the Classical Liar that yields chaotic semantic behavior.

Figure 3: Infinite-valued dualist forms of the Chaotic Liar generate fractal images.

3.

Variations of the *Strengthened Liar*, namely the sentence: *This sentence is false or neither true nor false*, can be used to demonstrate that the move to truth-value gaps is at best a temporary expedient in blocking Liar-like paradoxes.[9] Here we use a variation of the Strengthened Liar to motivate a limitative theorem regarding the calculability of chaos. Combining the Strengthened Liar with the notion of chaotic semantic behavior, we arrive at the *Strengthened Chaotic Liar*:

Either the boxed sentence has a chaotic semantic behavior or it is as true as it is estimated to be false.

Does the Strengthened Chaotic Liar have chaotic semantic behavior or not? Assuming it *does* have a chaotic semantic behavior, it will be completely true in virtue of its first disjunct. However, the semantic behavior of a sentence that is completely *true* will not qualify as chaotic since it will have the constant value true regardless of previous estimates. Assuming the Strengthened Chaotic Liar does *not* have a chaotic semantic behavior, its truth-value will

[9] *Mar 1985* (completed under the direction of Alonzo Church) demonstrates this claim for three truth-value gap solutions, namely, for Martin's (*1970*) Category Solution, van Fraassen's (*1968*) Presupposition Solution, and Kripke's (*1975*) Semantic Grounding solution to the Liar.

depend entirely on its second disjunct. However, the semantic behavior of the second disjunct will mimic the behavior of the Chaotic Liar, which is known to be semantically chaotic. We have derived a contradiction in either case. This paradox can be used (in much the same way as Gödel (*1931*) used the Richard paradox, Tarski (*1936*) used the Liar paradox, and Chaitin (*1966*) used the Berry paradox) to motivate a limitative theorem.

Non-Calculability of Chaos. Let C be the set of chaotic functions defined on the set of partial recursive functions f on the real interval $[0, 1]$. Assume that $\lambda x[1-|(1-x)-x|]$ is in C but that $\lambda x[1]$, the constant function identical to 1, is not in C. Then the index set $\mathrm{I(C)} = \{i : f_i \in \mathrm{C}\}$ is not effectively calculable.

Assume, for a proof by contradiction, that the index set of chaotic functions is λ-definable. We have that $\lambda x[1-|(1-x)-x|]$ is some f_i in C and that $\lambda x[1]$ is some f_j not in C. Then we may define the diagonal function $d(x)$ to be j, if $f_x \in$C, and $d(x)$ to be i otherwise. By the fixed point lemma, there will be a k such that $f_k = f_{d(k)}$. Hence, by the definition of $d(x)$, we have $f_k \in$C if and only if $d(k) = j$; but since $f_j(x)$ is the non-chaotic constant function identical to 1, we have $f_k \in$C if and only if $f_k \notin$C, which is a contradiction. Therefore, contrary to our assumption, I(C) is not λ-definable. It follows by Church's thesis that the index set of chaotic functions is not effectively calculable.[10]

Church's definition of randomness (*1940*) for an infinite series of 0s and 1s uses the idea that any effectively calculable way of choosing elements of the sequence will not alter the limit of the ratio of 1s in arbitrarily long finite sequences of the series. Intuitively, there is no effectively calculable way of capturing the information contained in the binary series. Church's idea can be seen as a precursor to the algorithmic definition of randomness proposed by Solomonoff (*1960*) and developed independently by Kolmogorov (*1965*) and Chaitin (*1966*).

The informational content of a computer program that computes a real number in the continuum between 0 and 1 can itself be represented by a binary string, which itself represents a real number in binary decimal notation. The length of a string can therefore be compared to the length of the string which it is supposed to compute. The *complexity* of a number x, $\mathrm{K}(x)$ can then be defined to be the length of a program φ_e of minimal length that computes x on the basis of a fixed input, say 0. Then a number x is defined to be *random* if $\mathrm{K}(x) \geq x$. This definition of randomness is not completely precise insofar as we have not given a specific definition of a computer. However, the problem of defining an idealized computer was solved by Turing (*1936*) and Post (*1936*) in a very intuitive way. The complexity of a string

[10]This form of the proof makes the result an application of Rice's Theorem (*1953*). For a proof using diagonalization and Gödel numbering, see *Mar and Grim 1991*.

is a mathematical concept that is not dependent on any particular definition of a computer.

Chaos and randomness are two sides of the same conceptual coin. Chaotic dynamical systems are intuitively random. Due to sensitive dependence on initial conditions, a chaotic dynamical system can require infinite-precision of the initial input; and hence programs for chaotic systems will have irreducible information.[11]

Chaitin's Theorem (*1965*) states that for any consistent sufficiently expressive formal axiomatic system T including number theory there will be numbers whose randomness cannot be proved. In particular there will be some κ such that T cannot prove that any specific binary string has complexity greater than κ. Church's theorem can be seen as a consequence of this theorem: for any sufficiently expressive formal axiomatic system T there will be binary strings whose complexity is greater than κ for which T cannot prove that they have complexity greater than κ. Formal systems are therefore informationally limited: even though there must be strings of unbounded complexity, there is an upper bound to the complexity it is possible to prove any particular string to have.

Church theorem established that Leibniz's dream of reason was logically unfulfillable, and Church's definition of randomness ultimately held the key to understanding why formal axiomatic theories will inevitably fall short. In retrospect, the information-theoretic interpretation of the limitative theorems suggests that these results do not depend on anomalous or pathological singularities. From this perspective Church's theorem can be seen as the inevitable outcome of incorporating the notions of chaos and randomness within formal systems.

---

[11] Chaos emerges naturally from an infinite-valued variation of Russell's paradox that uses a notion of degrees of randomness. Consider programs for computing binary strings together with binary representations of these programs. Rather than determining whether the complexity of a string is simply greater than or equal to the string itself, we can introduce a measure $\mathrm{m}(x)$ of how much greater the complexity of the string is. (For example, $\mathrm{m}(x)$ could be a partially defined function which is the difference between the length of the minimal program $\mathrm{K}(x)$ for x and x itself.) Then define $R^n = \{x : \mathrm{m}(x) \geq n \,\&\, \mathrm{K}(x) \text{ is defined}\}$, and define $\mathrm{R}^\omega = \bigcup \mathrm{R}^n$ for every $n \in \omega$. Note that for $\forall n \in \omega$, $\exists \mathrm{N} > n$, $\mathrm{R}^{\mathrm{N}} \neq \phi$ and R^ω is a proper subset of the set of strings since R^ω is countable. If x is a string such that n is the least n such that $x \in \mathrm{R}^n$, then we can assign a degree of randomness to x in $[0,1]$ equal to $1 - 1/n$. Now assume, for a proof by contradiction, that there is a minimal program P, represented by a string $\#\mathrm{R}^\omega$, such that P outputs a string encoding the members of R^ω. Is $\#\mathrm{R}^\omega \in \mathrm{R}^\omega$ or not? If $\#\mathrm{R}^\omega \in \mathrm{R}^\omega$, then $\#\mathrm{R}^\omega$ is a member of some R^n. Then P is a program of finite complexity which is generating the numbers of programs of arbitrarily high complexity. Hence, $\#\mathrm{R}^\omega \notin \mathrm{R}^\omega$. If $\#\mathrm{R}^\omega \notin \mathrm{R}^\omega$, then given that the complexity of program P is n_0, we have $\mathrm{K}(\#\mathrm{R}^\omega) < n_0$, contradicting that P is a minimal program for R^ω; hence $\#\mathrm{R}^\omega \in \mathrm{R}^\omega$. The Russellian contradiction $\#\mathrm{R}^\omega \in \mathrm{R}^\omega$ iff $\#\mathrm{R}^\omega \notin \mathrm{R}^\omega$ can be modeled in the infinite-valued logic above again yielding the algorithm for the Chaotic Liar. This variation of Russell's paradox in terms of degrees of randomness suggests that infinite-valued generalizations of Rice's Theorem are worthy of further investigation.

REFERENCES

Barwise, J., and J. Etchemendy
1987 *The Liar,* Oxford University Press, New York.

Chaitin, G. J.
1966 On the length of programs for computing finite binary sequences, *Journal of the Association of Computing Machinery,* vol. 13, pp. 547–569.
1974 Information-theoretic limitations on formal systems, *Journal of the Association of Computing Machinery,* vol. 21, pp. 403–424.

Church, A.
1935 An unsolvable problem of elementary number theory (abstract), *Bulletin of the American Mathematical Society,* vol. 41, pp. 332–333.
1936 An unsolvable problem of elementary number theory, *American Journal of Mathematics,* vol. 58, pp. 345–363.
1936a A note on the Entscheidungsproblem, *The Journal of Symbolic Logic,* vol. 1, pp. 40–41; correction, ibid, pp. 101–102.
1940 On the concept of random sequence, *Bulletin of the American Mathematical Society,* vol. 46, pp. 130–135.
1956 *Introduction to mathematical logic, vol. 1,* Princeton University Press, Princeton, New Jersey.

Curry, H. B.
1941 The paradox of Kleene and Rosser, *Transactions of the American Mathematical Society,* vol. 50, pp. 454–516.

Devaney, R. L.
1989 *An introduction to chaotic dynamical systems,* second edition, Addison-Wesley, Menlo Park.

Gödel, K.
1931 Über formal unentscheidbare Sätze der Principia Mathematica und verwandter Systeme I, *Monatshefte für Mathematik und Physik,* vol. 38, pp. 173–198.

Gupta, A.
1982 Truth and paradox, *Journal of Philosophical Logic,* vol. 11, pp. 1–60.

Herzberger, H.
1982 Notes on naive semantics, *Journal of Philosophical Logic,* vol. 11, pp. 61–102.

Kleene, S. C.
1936 λ-definability and recursiveness, *Duke Mathematical Journal,* vol. 2, pp. 340–353.

Kleene, S. C., and J. B. Rosser
1935 The inconsistency of certain formal logics, *Annals of Mathematics,* vol. 36, pp. 630–636.

Kolmogorov, A. N.
1964 Three approaches to the definition of the concept of 'amount of information', *Problemy Peredachi Informatsii,* translated as "Problems in Information Transmission", vol. 1, 1965, pp. 3–11; translated in vol. 1 1965, pp. 1–7.

Kripke, S.

1975 Outline of a theory of truth, *The Journal of Philosophy.*, vol. 72, pp. 690–716.

Mar, G.

1985 Liars, truth-gaps, and truth: A comparison of formal and philosophical solutions to the semantical paradoxes, *Dissertation*, University of California at Los Angeles, University Microfilms, Ann Arbor.

Mar, G., and P. Grim

1991 Pattern and chaos: new images in the semantics of paradox, *Noûs*, vol. 25, pp. 659–693.

Mar, G., and P. St. Denis

1994 Chaos in cooperation: continuous-valued prisoner's dilemmas in infinite-valued logic, *International Journal of Bifurcation and Chaos*, vol. 4, pp. 943–958.

1996 Real life, *International Journal of Bifurcation and Chaos*, vol. 6, pp. 2077–2086.

1999 What the Liar Taught Achilles, *Journal of Philosophical Logic*, vol. 28, pp. 29-46.

Martin, R. L.

1970 A category solution to the Liar, *The Liar Paradox* (R. L. Martin, editor), Yale University Press, New Haven.

May, R.

1976 Simple mathematical models with very complicated dynamics, *Nature*, vol. 261, pp. 459–467.

Peitgen, H.-O., and D. Saupe

1988 (editors), *The science of fractal images*, Springer-Verlag, New York.

Post, E. L.

1936 Finite combinatory processes–formulation 1, *The Journal of Symbolic Logic*, vol. 1, pp. 103–105.

Rescher, N.

1969 *Multi-valued logic*, McGraw-Hill, New York.

Rice, H. G.

1953 Classes of recursively enumerable set and their decision problems, *Transactions of the American Mathematical Society*, vol. 74, pp. 358–366.

Schroeder, M.

1991 *Fractals, chaos, power laws*, W. H. Freeman, New York.

Solomonov, R.

1964 A formal theory of inductive inference, *Information and Control*, subsequently renamed to *Information and Computation*, vol. 7, pp. 1–2.

Tarski, A.

1936 Der wahrheitsbegriff in den formalisierten sprachen, *Studien Philosophica*, vol. I, pp. 261–405.

Turing, A.

1936 On computable numbers with an application to the Entscheidungsproblem, *Proceedings of the London Mathematical Society,* vol. 42, pp. 230–265; corrections ibid., vol. 43 (1937), pp. 544–546.

1937 Computability and λ-definability, *The Journal of Symbolic Logic,* vol. 2, 153–163.

Van Fraassen, B. C.

1968 Presupposition, implication, and self-reference, *The Journal of Philosophy,* vol. 65, pp. 136–152.

ENRICO MARTINO

RUSSELLIAN TYPE THEORY AND SEMANTICAL PARADOXES

Abstract. We compare Grelling's paradox with Russell's paradox about propositions, in order to illuminate their different natures. We criticize Church's reconstruction of Grelling's paradox, according to which it would be an *intensional* antinomy arising within the simple type theory and suitably solved within the *ramified* type theory. We claim, on the contrary, that this paradox is not genuinely intensional and that ramification is of no help for its resolution. We argue that the paradox rests on a highly problematic assumption, which can be rejected within the simple theory itself, since its alleged evidence has nothing to do with the lack of ramification. In contrast, we show that a reconstruction, similar to the one proposed by Church for Grelling's paradox, is appropriate for Russell's paradox about propositions, which turns out to be really intensional and capable of an adequate solultion by means of ramification.

1. INTRODUCTION

According to Church (*1976*), Myhill (*1979*), and Hazen (*1983*), the Russellian ramified type theory supplies a solution to Grelling's paradox, which arises in the simple type theory.

In particular, Church (*1976*) provides an interesting detailed reconstruction of the paradox. By using certain semantical constants, Church expresses the term "heterological" within the simple type theory and shows how, within the ramified theory, it splits into infinitely many terms "n-heterological" for every level n. He then shows that, while in the simple theory the question of whether "heterological" is heterological leads to contradiction, in the ramified theory the infinitely many questions of whether "n-heterological" is m-heterological are all capable of precise and coherent answers. In this sense, Church concludes, the ramification solves Grelling's paradox of the simple type theory.

Though the Russellian notion of propositional function described by Church is distinctively intensional, the axioms he uses are all extensional, so that the paradox in question seems to have little to do with the intensionality of propositional functions. On the other hand, the superiority of the ramified logic, compared with the simple one, seems to emerge just in dealing with intensions. In this situation, it is hardly plausible that a paradox, not essentially intensional, be solved by means of ramification. According to this observation, we shall hold that, in the simple type theory, Grelling's paradox rests on an erroneous assumption, quite extraneous to matters of intensionality,

C. Anthony Anderson and M. Zelëny (eds.), Logic, Meaning and Computation, 491-505.

and gets therefore its appropriate solution by means of the mere rejection of that assumption within the simple theory itself. Furthermore, we shall show how, even within the ramified theory, a similar assumption leads, with the help of the axiom of reducibility, to a similar paradox.

In contrast, a genuinely intensional paradox of the simple type theory, suitably solved by means of ramification, is *Russell's paradox about propositions*, stated informally by Russell in appendix B of the *Principles* (*1903*). It exploits essentially Russell's conception of the identity of propositions and is perhaps the source of the very idea of ramification. An interesting formalization of this paradox is expounded by Church (*1984*) himself in which he no longer employs the method of semantical constants (cf. *Church 1976*), but introduces a special primitive connective in order to express an intensional identity of propositions.

We shall also show that the method of semantical constants, suitably refined, is adequate for treating Russell's paradox about propositions.

We believe that setting up the two paradoxes within one and the same formal context may help to better understand their different natures.

2. RAMIFIED TYPE THEORY

We assume acquaintance with Church's (*1976*) system of ramified type theory, which will be here summarized.

Ramified types, shortly *r-types*, are recursively defined:

(a) i is an r-type,

(b) if $\beta_1, \ldots, \beta_m$ $(m \geq 0)$ are r-types, so is $(\beta_1, \ldots, \beta_m)|n$ $(n > 0)$.

If $k < n$, $(\beta_1, \ldots, \beta_m)|k$ is said to be *less* than $(\beta_1, \ldots, \beta_m)|n$.

i is the r-type of individuals; $(\beta_1, \ldots, \beta_m)|n$ is the r-type of m-place propositional functions of *level* n with arguments of r-types $\leq \beta_1, \ldots, \beta_m$.

Types of the form $(i, i, \ldots, i)|n$, with m i's, are indicated briefly by $m|n$ ($0|n$ is the r-type of propositions of level n).

To every r-type is assigned an *order*:

(a) $\mathsf{ord}\, i = 0$,

(b) $\mathsf{ord}(\beta_1, \ldots, \beta_m)|n = N + n$, with $N = \max[\mathsf{ord}\beta_1, \ldots, \mathsf{ord}\beta_m]$.

For every r-type, there are infinitely many variables and possibly some constants.

Atomic formulas are of the form $F(x_1, \ldots, x_m)$, with F being a variable (or constant) of type $(\beta_1, \ldots, \beta_m)|n$ and where the x_i variables (or constants) are of the r-types $\beta_1, \ldots, \beta_m$, respectively.

Molecular formulas are built up in the usual way.

We will also use the following definition of *order of a formula* (not used by Church, *1976*): if h is the greatest order of free variables and constants, and k the greatest order of bounded variables in the formula P, then $\mathsf{ord}P = \mathsf{max}[h, k+1]$.

We will use metavariables p, q for propositions, F, G for propositional functions, P, Q for formulas (the r-types will often be understood).

We presuppose some standard system of rules and axioms for logical constants. These will never be explicitly used, our proofs being presented in a *semiformal* style.

A central role is played by the comprehension axioms:[1]

AXIOM 2.1. $\exists p \,.\, p \equiv P$,

where p is a propositional variable of r-type $0|n$ not free in P, with $\mathsf{ord}P \leq n$;

AXIOM 2.2. $\exists F \,.\, F(x_1, \ldots, x_m) \equiv_{x_1,\ldots,x_m} P$,

where F is a variable of type $(\beta_1, \ldots, \beta_m)|n$ not free in P; $x_1, \ldots, x_m$ are distinct variables of types $\beta_1, \ldots, \beta_m$, and $\mathsf{ord}F \geq \mathsf{ord}P$.

Finally, we recall the *axioms of reducibility,* assuring that every propositional function is coextensive with some *predicative* (i.e., of level 1) propositional function:

AXIOM 2.3. $\forall F \exists G \,.\, F(x_1, \ldots, x_m) \equiv_{x_1,\ldots,x_m} G(x_1, \ldots, x_m)$,

with G predicative.

In virtue of 2.3, *identity* is definable as *indiscernibility* with respect to predicative propositional functions:

DEFINITION 2.4. $x = y \ =_{\text{def}}\ F(x) \supset_F F(y)$

where x, y are of the same r-type β, and F is of type $(\beta)|1$.

3. CHURCH'S FORMULATION OF GRELLING'S PARADOX

Here we sketch Church's reconstruction of Grelling's antinomy. Given a natural number $n \geq 1$, every formula of order n with a single individual free variable (and without any other free variables) represents (according to 2.2) a propositional function of order $\leq n$. Regarding linguistic expressions as individuals, such relations between formulas and propositional functions are 3-place propositional functions, which we can express in the language by means of special constants.

[1] We follow Church's convention of using a bold dot as a left bracket with a corresponding right bracket as far to the right as possible (while the normal dot stands for logical conjunction).

In fact, for all $n > 0$, we introduce the constant val^{n+1} of r-type $(i, i, 1|n)|1$. $\mathsf{val}^{n+1}(a, v, F)$ is intended to mean that a is an individual variable, v a formula of order $\leq n$ whose unique free variable is a, and F the propositional function expressed by v.

In accordance with this interpretation, we assume the following axioms:

AXIOM 3.1. $\mathsf{val}^{m+1}(a, v, F) \supset_{a,v,F} . \mathsf{val}^{n+1}(b, v, G) \supset_{b,G} F = G.$

AXIOM 3.2. $\exists a \exists v \exists F . \mathsf{val}^{n+1}(a, v, F) \cdot F(x) \equiv_x P$, *with* P *a formula of order* $\leq n$ *without free variables except* x.

AXIOM 3.3. $\mathsf{val}^{n+1}(a, v, F) \supset \mathsf{val}^{m+1}(a, v, F)$, *if* $m \geq n$.

From 3.1 and 2.4, it follows that

COROLLARY 3.4. $\mathsf{val}^{m+1}(a, v, F) \supset_{a,v,F} . \mathsf{val}^{n+1}(b, v, G) \supset_{b,G} . F(x) \equiv_x G(x).$

By means of such constants, the predicate "heterological" of Grelling's paradox is interpreted into an infinite hierarchy of predicates, according to the definitions

DEFINITION 3.5. $\mathsf{het}^{n+1}(v) =_{\mathrm{def}} \exists a \exists F . \mathsf{val}^{n+1}(a, v, F) \cdot \neg F(v).$

From 3.3, it follows immediately that

COROLLARY 3.6. $\mathsf{het}^{n+1}(v) \supset \mathsf{het}^{m+1}(v)$, *if* $m \geq n$.

The following theorems hold:

THEOREM 3.7. $(\mathsf{val}^{m+2}(a, v, G) \cdot G(x) \equiv_x \mathsf{het}^{m+1}(x)) \supset$
$\neg \mathsf{het}^{n+1}(v)$, *if* $m \geq n$.

Proof. Let

(1) $\mathsf{val}^{m+2}(a, v, G)$, and

(2) $G(x) \equiv_x \mathsf{het}^{m+1}(x)$.

Assuming, for a reductio, $\mathsf{het}^{n+1}(v)$, with $m \geq n$, we get, by 3.6, $\mathsf{het}^{m+1}(v)$ and therefore, by 3.5, $\exists a \exists F . \mathsf{val}^{m+1}(a, v, F) \cdot \neg F(v)$. Using (1) and 3.1, one obtains that $\neg G(v)$ and hence, by (2), $\neg \mathsf{het}^{m+1}(v)$, which is absurd. ⊣

THEOREM 3.8. $(\mathsf{val}^{m+2}(a, v, G) \cdot G(x) \equiv_x \mathsf{het}^{m+1}(x)) \supset \mathsf{het}^{n+1}(v)$ *if* $m < n$.

Proof. Let

(1) $\mathsf{val}^{m+2}(a, v, G)$, and

(2) $G(x) \equiv_x \mathsf{het}^{m+1}(x)$.

By 3.7 we have $\neg\,\mathsf{het}^{m+1}(v)$ and hence, by (2),

(3) $\neg\, G(v)$.

From the conjunction of (1) and (3) and from 3.5 if follows that $\mathsf{het}^{m+2}(v)$, whence, by 3.6, $\mathsf{het}^{n+1}(v)$, for all $n < m$. ⊣

Finally, as an instance of Axiom 3.2, we have

COROLLARY 3.9. $\exists a \exists v \exists G\,.\,\mathsf{val}^{m+2}(a, v, G) \cdot G(x) \equiv_x \mathsf{het}^{m+1}(x)$.

3.7, 3.8, and 3.9 say, intuitively, that the formula $\mathsf{het}^{m+1}(v)$ is $n+1$-heterological iff $n > m$.

We believe that all of this is perfectly correct. What we would like to discuss is Church's claim that such results constitute the solution, by means of ramification, of Grelling's antinomy of simple type theory. Let's quote Church's (*1976,* at the end of §2) argument:

> If we reduce to simple type theory by dropping all level indicators, the infinitely many constants val^{n+1} coalesce into a single constant, val, whose type (in the sense of simple type theory) is $(i, i, (i))$. The informal explanation of the meaning of val^{n+1} then becomes an explanation of the meaning of val, and the postulates (1), (3) [our 3.1, 3.2] still seem to be evident from this intended meaning; moreover, (2) [3.4] is still a consequence of (1) [3.1], and (4) [3.3] becomes tautologous. The proofs of theorems (5)–(8) [3.7–3.9], still hold, after dropping the level indicators, and the last three theorems then constitute a contradiction. This is Grelling's antinomy, as it arises in simple type theory.
>
> The resolution of the antinomy by ramified type theory consists not merely in the fact that, after restoration of the level indicators, theorems (6)–(8) [3.7–3.9] are no longer a contradiction, but also in that the question "Is the proposition from $\mathsf{het}^{m+1}(x)$ autological or heterological?" can now be answered: namely it is (by (6)) [3.8] autological at all levels $\leq m+1$, and it is (by (7)) [3.7] heterological at all levels $> m+1$.

4. CRITICISM OF CHURCH'S ARGUMENT

Though Church presents Grelling's antinomy as an intensional paradox, none of the axioms he uses is of an intensional character. In fact, the alleged evidence for these axioms does not seem to be undermined, even if propositional functions are interpreted as sets and propositions as truth values. This makes the argument in question somewhat suspicious. The resolution of a paradox by means of ramification suggests that the source of the contradiction consists in a confusion of levels. So, if the paradox arises also within an extensional context, such a resolution could be advanced as an argument against the adequacy of simple type theory also to handle set-theoretical notions. There is certainly a well-known notion of class which could support such a claim of inadequacy: it is the *derivative* (sometimes called *logical*) notion of class, according to which classes are conceived of as extensions of propositional functions, so that a class cannot exist independently of some generating propositional function. This conception finds classes to follow the hierarchy of propositional functions, with the effect that, if simple type theory is inadequate for the latter, so it is inadequate for the former. This is not so, however, for the usual and more successful *primitive* notion of set, according to which a set is built up only by its members, no other entities being involved in its existence. In particular, according to this conception, a set exists quite independently of any propositional function we may use to describe it. Simple type theory is certainly adequate for this notion, since the simple hierarchy of sets is just devised to guarantee that the members of a set are pre-existing to the set itself (Zermelo-Fraenkel set theory may be regarded as a simple type theory extended up to transfinite levels). Though the derivative notion is certainly the appropriate one for Russell's logistic attempt to reduce classes to propositional functions, nevertheless the primitive notion seems to be quite compatible with Russell's ramified theory of propositional functions. Moreover it provides a good ontological tool for a satisfactory justification of the axiom of reducibility. Russell observes that, if the existence of classes is assumed from the outset, the axiom of reducibility immediately follows: in fact, if F is any propositional function, a coextensive predicative G is given by the property of belonging to the extension of F. And, although Russell seems to believe that axiom to be an assumption weaker than the existence of classes, he does not offer any other intuitive support for his axiom but the intuition of sets as primitive entities.[2] For these reasons, even from the point of view of ramified hierarchy for propositional functions, the possibility of arguing against the simple hierarchy of sets seems to be highly implausible. Thus, since Church's derivation of Grelling's paradox is logically unexceptionable, it must rest on some erroneous axiom, whose fallacy should be recognizable even within the perspective of simple type theory. All that Church says about the plausibility of his assumptions is that the axioms on constants val^{n+1}, when passing from ramified to simple theory, seem to be still justified by the intended meaning of the single constant val. On the contrary, we will contend that, within the perspective of simple types, schema 3.2, unless it is suitably restricted, loses any plausibility.

What makes val a well-determined relation is the presupposition that every *propositional form* $P(x)$ (i.e., formula with the single free variable x) represents a well-determined propositional function F. Besides the intended meaning of logical constants and the intended domains of quantification, F

[2] An interesting model for ramified type theory, where predicative functions are explicitly grounded on set-theoretical membership, has been recently constructed by C. A. Anderson (*1989*, §V).

is determined by the meaning of non-logical constants (if any) occurring in $P(x)$. It follows that val must be relative to some understood assignment of values (of appropriate types) to all non-logical constants of the given language $\mathcal{L}$. When introducing a specific constant "$\mathsf{val}_{\mathcal{L}}$" for the valuation function relative to $\mathcal{L}$, we get an extended language $\mathcal{L}' = \mathcal{L} \cup \{\mathsf{val}_{\mathcal{L}}\}$. Grelling's antimony exploits essentially the fact that Axiom 3.2, once the restriction on the order of P is dropped, allows val to occur in P. This amounts to assuming $\mathsf{val}_{\mathcal{L}} = \mathsf{val}_{\mathcal{L}'}$, but no reason or intuition seems to support such identification.

Observe that, after extending $\mathcal{L}$ to $\mathcal{L}'$ by means of a 3-place relational symbol R, it would be patently circular any attempt to assign *stipulatively* to R the valuation function relative to $\mathcal{L}'$, the latter being determined only after having fixed a value for R. Thus, within simple type theory, Grelling's paradox is not an authentic antinomy. It is rather a proof by contradiction that, however you choose a value for R (so to determine $\mathsf{val}_{\mathcal{L}'}$), that value cannot be coextensive with $\mathsf{val}_{\mathcal{L}'}$.[3]

Church remarks, in the above quoted passage, that the resolution of Grelling's paradox by means of ramification does not simply consist in the fact that, after restoration of the levels, the paradox is no longer derivable, but also in the possibility to answer precisely the questions whether $\mathsf{het}^{m+1}(v)$ is or not $n+1$-heterological. But, if our analysis is correct, also in simple type theory, once the incriminating assumption has been rejected, the intended meaning of val supplies a precise answer to the question whether $\mathsf{het}(v)$ is or is not heterological.

Since the relation val is relative to the given language $\mathcal{L}$ (without constant "val"), the proper meaning of $\mathsf{val}(a, v, F)$ is: a is an individual variable, v is a formula *of* $\mathcal{L}$ whose unique free variable is a, F is the propositional function expressed by v. We must therefore restrict axiom schema 3.2 to formulas P *of* $\mathcal{L}$. So 3.9 (in its simple type reformulation) is not an instance of 3.2, since $\mathsf{het}(v)$ is not in $\mathcal{L}$, and the proof that $\mathsf{het}(v)$ is heterological is blocked. Furthermore, the formula $\mathsf{val}(a, v, F)$, by its very meaning, can be satisfied by no other value of v but a formula of $\mathcal{L}$, so that $\exists a \exists F.\mathsf{val}(a, v, F)$ is false when v is assigned "$\mathsf{het}(v)$". Thus, by Definition 3.5, $\mathsf{het}(v)$ is not heterological.

This conclusion cannot be formally derived from the axioms on val corresponding to 3.1–3.3. After dropping levels, 3.3 becomes superfluous and 3.1, 3.2 (the latter restricted as explained) do not say anything about $\mathsf{val}(a, v, F)$, when v is not a formula of $\mathcal{L}$. It is easy, however, to reinforce the axioms, so as to derive results quite analogous to those in ramified theory. To that

[3]These considerations presuppose that, in order for a language to be well-determined, all its non-logical constants should be given. This seems to be in contrast with Church's (*1976*, note 5) observation that Russell's formal language would be an *open* language (i.e., without any prefixed vocabulary) to which new constants could be added step by step. Observe, however, that adding new constants affects the valuation function. We believe therefore that such an *opening* (present or not in Russell) should be banished from the reconstruction in question, on pain of the indeterminacy of val.

end, restate (the simple type versions of) 3.1, 3.2 both for $\mathsf{val}_{\mathcal{L}}$ and for $\mathsf{val}_{\mathcal{L}'}$, where $\mathcal{L}' = \mathcal{L} \cup \{\mathsf{val}_{\mathcal{L}}\}$, and replace 3.3 by

$$\mathsf{val}_{\mathcal{L}}(a, v, F) \supset \mathsf{val}_{\mathcal{L}'}(a, v, F).$$

The analogue of 3.5, for $\mathsf{val}_{\mathcal{L}}$ and $\mathsf{val}_{\mathcal{L}'}$ respectively, supply definitions of $\mathsf{het}_{\mathcal{L}}(v)$, $\mathsf{het}_{\mathcal{L}'}(v)$; and, just as in ramified theory, one proves that $\mathsf{het}_{\mathcal{L}}(v)$ is not $\mathcal{L}$-heterological but is $\mathcal{L}'$-heterological.

This solution is essentially that of Tarski and leads to a hierarchy of semantical languages. Church observes that also Russell's hierarchy of orders may be viewed as a particular hierarchy of languages, so that Russell's resolution of semantical antinomies is not really different from Tarski's but rather a special case of it. Indeed, our restriction on Axiom 3.2, forbidding occurrences of val in P, is automatically assured by the original restriction on the order of P. But this is exactly the reason why 3.2 is evident within ramified theory, while its unrestricted simple type version is not, even within the simple type perspective. And when trying to correct it, we are naturally led to the Tarskian hierarchy of languages. In this sense Tarski's solution turns out to be the appropriate one, since it acts only on the unique wrong assumption, leaving the general framework unchanged. On the contrary, Russell's solution imposes a drastic revision on the whole framework, which seems to be unjustified, since the credibility of that framework was not affected at all by the paradox.

Since, however, quite independently of semantical paradoxes like Grelling's, there are good reasons for rejecting simple type logic in favor of ramified logic, as a framework for Russellian propositional functions (see next section), one might think that, after all, ramified logic provides, as a by-product, a good alternative to Tarski's resolution of semantical antinomies. This is not the case, however. We shall show that, even within ramified theory, an assumption quite similar to unrestricted 3.2, but in agreement with the order hierarchy, leads, with the help of the axiom of reducibility, to a reconstruction of the paradox.

To this end, define, within ramified theory,

DEFINITION 4.1. $Q =_{\text{def}} \exists F.\, \mathsf{val}^3(a, v, F) \cdot G(x) =_x F(x)$, with G *predicative.*

By the comprehension and reducibility axioms, there is a relation R of type $(i, i, 1|1)|1$ such that $R(a, v, G) \equiv_{a,v,G} Q$. Introduce into the object language a new constant val^* of type $(i, i, 1|1)|1$ for R and define

$$\mathsf{het}^*(v) =_{\text{def}} \exists a \exists G\,.\, \mathsf{val}^*(a, v, G) \cdot \neg G(v).$$

Now, assume that Axiom 3.2 allows val^* occurring in P (keeping fixed the restriction on the order of P). We have then, as an instance of 3.2,

$$\exists a \exists v \exists F\,.\, \mathsf{val}^3(a, v, F) \cdot F(x) \equiv_x \mathsf{het}^*(x).$$

It turns out $\text{het}^*(v) \equiv_v \text{het}^3(v)$ and the paradox follows from the question of whether $\text{het}^*(v)$ is or not 3-heterological.

Here the appropriate solution is still the same as in simple type theory: if $\mathcal{L}_0$ is the initially given language, val^{n+1} is the valuation function relative to propositional forms of order $\leq n$ of language $\mathcal{L}_n = \mathcal{L}_0 \cup \{\text{val}^1, \ldots, \text{val}^n\}$, so that only formulas P of $\mathcal{L}_n$ are allowed in 3.2. But val^* is not in $\mathcal{L}_n$ and the paradox becomes a proof by contradiction that no constant of $\mathcal{L}_n$ of type $(i, i, 1|1)|1$ can be coextensive with val^*. In this version of the paradox, the crucial restriction on 3.2, imposed by the intended meaning of val^{n+1}, is not automatically respected in virtue of the order theory, which turns out to be unable to offer any alternative to the Tarskian solution. So the latter is to be adopted even within ramified type theory.

Our considerations seem to be in agreement, to a certain extent, with Myhill's (*1979*) argument in defense of the axiom of reducibility. The argument is addressed to those who brought against the axiom of reducibility the charge of allowing a reconstruction, within the ramified theory, of Grelling's paradox. Myhill's reply is that such alleged reconstruction of the paradox is nothing but a *reductio ad absurdum* of the unsupported assumption that a certain propositional function (similar to our val^*) is expressible in the object language. We certainly agree with this conclusion. Nevertheless, Myhill too, like Church, regards the simple type version of Grelling's paradox as an authentic antinomy, of which the ramified theory would be free, thanks to the order hierarchy. He seems not to be aware that his own argument for the axiom of reducibility immediately solves the paradox even in the case of simple types. What really makes that axiom harmless is not the fact that it does not allow the reconstruction of the paradox: indeed, as we have seen, it does. What makes it harmless is rather the fact that the paradox, once reproduced in the ramified theory, is quite satisfactorily resolved in the same way as in the simple theory.

5. RUSSELL'S ANTINOMY ABOUT PROPOSITIONS

Unlike Grelling's paradox, Russell's antinomy about propositions is really an intensional antinomy of simple type theory: it shows the inadequacy of simple logic in dealing with the Russellian notion of propositional function and finds its proper solution in ramified logic. From now on we shall, for brevity, refer to this antinomy as *Russell's antinomy* (not to be confused with the more popular set-theoretical one). The antinomy is presented informally by Russell in Appendix B of *Principles* (*1903*) and, as Church (*1984*, note 5) observes, is of historical importance as Russell's first indication of the need for something beyond simple type theory to resolve antinomies. Church's (*1984*) reconstruction of Russell's antinomy is radically different from that of Grelling's antinomy (*Church 1976*): instead of the semantical constant val, a new propositional connective '$\equiv$' expressing intensional identity between propositions is used. For a better comparison of the two antinomies, we shall here propose an alternative reconstruction of Russell's antinomy, by suitably generalizing the method of semantical constants. Church's treatment refers mostly to Russell's version of the paradox in §500. But a simpler version is already present in the following short passage at the end of §488:

> That propositions are a type results from the fact—if it be a fact—that only propositions can significantly be said to be true or false .... But if so, the number of propositions is as great as that of all objects absolutely, since every object is identical with itself, and "x is identical with x" has a one-one relation to x .... If it is possible, as it seems to be, to form ranges of propositions, there must be more such ranges than there are propositions, although such ranges are only some among objects.

Really, that the number of propositions is as great as that of all objects absolutely is not a meaningful assertion of simple type theory, since there is no type of all objects. In any case, the paradox does not exploit this assertion in its full generality, but only in the restricted form that the number of propositions is as great as that of all ranges of propositions. This is all that is needed to contradict Cantor's theorem (provable within simple type theory). So the paradox rests on Russell's claim that, by associating with every range x of propositions, the proposition "x is identical with x", we get a one-one relation between ranges of propositions and propositions. Thus, in order to get a formal reconstruction of the paradox, we have to derive that claim from some general principle, fitting with Russell's conception of propositional function. A suitable principle, which seems to underlie several of Russell's arguments, is stated by Church (*1984,* page 518) himself. Let's call it Int (for *principle of intensionality*):

Int. If n and m are names of different things, but are of the same type, and if a sentence is altered by replacing one occurrence of n by m, the original sentence and the altered sentence represent different propositions.

Church does not formalize, however, this principle in his system: he prefers to derive the paradox from other intensional axioms involving the logical constants. Yet, his axiomatization, as observed by Anderson (*1986*), turns out to be incompatible with Int (see also our §5).

We want to formalize Int by means of a natural generalization of the method of semantical constants.

Church (*1976,* §3) introduces semantical constants, analogous to the val^{n+1}, for formulas with any number of free variables of any r-types. In fact, for every r-type $\mu = (\beta_1, \ldots, \beta_m)|n$, he introduces a constant val_μ of type $(i, \ldots, i, \mu)|1$, with $m+1$ i's. According to the intended meaning, $\mathsf{val}_\mu(a_1, \ldots, a_m, v, F)$ says that $a_1, \ldots, a_m$ are variables of types $\beta_1, \ldots, \beta_m$,

v is a formula whose variables are among $a_1, \ldots, a_m$ and F is the propositional function obtained from v by abstraction with respect to the variables $a_1, \ldots, a_m$.

So Church's semantical relations are only those associating with a formula P the propositional function obtained by abstraction relative to a sequence of variables including *all* free variables of P. Starting from P, it is possible, however, to obtain a propositional function also by abstraction relative to *some* (not necessarily all and possibly none) of the free variables of P and assigning to the others arbitrary values (of appropriate types). It is natural, therefore, to generalize Church's relations as follows.

Let L be the language of ramified type theory *without non-logical constants.* For every sequence of r-types $\pi_1, \ldots, \pi_k$, $\mu = (\beta_1, \ldots, \beta_m)|n$, introduce the constant $\mathsf{val}_{\pi_1,\ldots,\pi_k,\mu}$ of type $(i, \ldots, i, \pi_1, \ldots, \pi_k, \mu)|1$ (with $m+1$ i's). The formula

$$\mathsf{val}_{\pi_1,\ldots,\pi_k,\mu}(a_1, \ldots, a_m, v, u_1, \ldots, u_k, w)$$

says that $a_1, \ldots, a_m$ are distinct variables of types $\beta_1, \ldots, \beta_m$, v is a formula *of* $\mathcal{L}$ whose free variables, different from $a_1, \ldots, a_m$, are, in order of first occurrence, $b_1, \ldots, b_k$; and w is the propositional function obtained from v by abstraction with respect to $a_1, \ldots, a_m$ and assignment of $u_1, \ldots, u_k$ respectively to $b_1, \ldots, b_k$.

To begin with, we shall assume for these constants the analogues of Axioms 3.1–3.3:

AXIOM 5.1. $\mathsf{val}_{\pi_1,\ldots,\pi_k,\mu}(a_1, \ldots, a_m, v, u_1, \ldots, u_k, w) \supset$
$. \mathsf{val}_{\pi_1,\ldots,\pi_k,\mu'}(a_1, \ldots, a_m, v, u_1, \ldots, u_k, w') \supset$
$w = w'$ *(with μ and μ' comparable).*

AXIOM 5.2. $\exists a_1, \ldots, \exists a_m, \exists v \forall u_1, \ldots, \forall u_k, \exists F$. $\mathsf{val}_{\pi_1,\ldots,\pi_k,\mu}(a_1, \ldots, a_m, v, u_1, \ldots, u_k, F) \cdot F(x_1, \ldots, x_m) \equiv_{x_1,\ldots,x_m} P$, *where $x_1, \ldots, x_m$ is a sequence of variables of types $\beta_1, \ldots, \beta_m$, $\mathsf{ord}P \leq \mathsf{ord}\mu$ and $u_1, \ldots, u_k$ are all free variables of P different from $x_1, \ldots, x_m$.*

AXIOM 5.3. $\mathsf{val}_{\pi_1,\ldots,\pi_k,\mu}(a_1, \ldots, a_m, v, u_1, \ldots, u_k, w) \supset$
$\mathsf{val}_{\pi_1,\ldots,\pi_k,\mu'}(a_1, \ldots, a_m, v, u_1, \ldots, u_k, w)$, *if $\mu' \geq \mu$.*

We can now formalize Int by the axiom

AXIOM 5.4. $\mathsf{val}_{\pi_1,\ldots,\pi_k,\mu}(v, u_1, \ldots, u_k, p) \cdot \mathsf{val}_{\pi_1,\ldots,\pi_k,\mu}(v, u'_1, \ldots, u'_k, p) \supset$
$u_i = u'_i$, $(i = 1, \ldots, k)$, *where μ is a* propositional *r-type.*

If propositional functions, conceived of as intensional entities in the sense of Int, are supposed to satisfy the simple type hierarchy, the analogues for simple types of Axioms 5.1–5.4 (obtained by dropping the level indicators) seem to be quite plausible. They lead, however, to Russell's antinomy.

Within simple type theory, the following intensional version of Cantor's theorem is provable, which asserts that the number of propositional functions of any type is greater than that of their arguments:

THEOREM 5.5. *For every type μ, there is no relation R of type $((\mu),\mu)$ such that*

(a) $\forall F \exists u \,.\, R(F,u)$,

(b) $(R(F,u)\cdot R(G,u)) \supset_{F,G,u} F = G$.

Proof. Let, for a reductio, R be a relation satisfying (a) and (b). By applying the comprehension principle 2.2 to the formula

$$P =_{\text{def}} \exists F \,.\, R(F,u) \cdot \neg F(u),$$

we get a G of type (μ) such that

(1) $G(u) \equiv_u \exists F \,.\, R(F,u) \cdot \neg F(u)$.

By (a) and (b), there is a v such that G is the unique propositional function satisfying the condition $R(G,v)$. From (1), it follows then $G(v) \equiv \neg G(v)$, which is absurd. ⊣

By the way, the following corollary is worth noting:

COROLLARY 5.6. *There is no formal language in which all propositional functions of type (i) can be represented.*

Proof. Suppose, for a reductio, that the propositional functions in question are all representable in one and the same formal language. The relation of type $((i),i)$, which with every propositional function associates its own representing expression, satisfies clauses (a) and (b) of the preceding theorem, which is absurd. ⊣

On the other hand, Russell's counter-example to Theorem 5.5 is derivable from Axiom 5.4. By applying 5.2 to the formula $P =_{\text{def}} F = F$, with F of type (0), we get

COROLLARY 5.7. $\exists v \forall F \exists p \,.\, \mathsf{val}_{(0),0}(v,F,p)$,

and from the comprehension axiom, applied to formula $\mathsf{val}_{(0),0}(v,F,p)$, it follows that

COROLLARY 5.8. $\forall v \exists R \,.\, R(F,p) \equiv_{F,p} \mathsf{val}_{(0),0}(v,F,p)$, *with R of type* $((0),0)$.

Thus from 5.7 and 5.4 we get a relation satisfying clauses (a) and (b) of Theorem 5.5. This is our reconstruction of Russell's antinomy about propositions.

We want to stress the main difference between Grelling's and Russell's antinomies emerging from our treatment. As we have seen, the crucial step in the deduction of Grelling's paradox is an application of Axiom 5.2, which is in conflict with the intended meaning of val: while val is to be understood as relative to the initial language $\mathcal{L}$ (without val), it is used as if it were relative to the extended language $\mathcal{L}' = \mathcal{L} \cup \{\mathsf{val}\}$. On the contrary, the use of Axiom 5.2 in the deduction of Russell's antinomy is in perfect agreement with the intended meaning of relations val's: these are conceived of as relative to the language $\mathcal{L}$ of simple type theory *without non-logical constants* and 5.2 is, accordingly, restricted to formulas P of $\mathcal{L}$. Here, the crucial step towards the contradiction is the use of 5.4 for a counter-example to Cantor's theorem. And since the latter is certainly appropriate for simple type theory, while 5.4 expresses the intensionality principle Int, the paradox shows that Int is incompatible with simple type theory. In this sense we can conclude that Russell's antinomy, unlike Grelling's, is an authentic *intensional* paradox.

Within the ramified theory, we can still derive, by means of the axiom of reducibility, a theorem analogous to 5.5, namely

THEOREM 5.9. *For every r-type μ, there is no relation R of r-type $((\mu)|1, \mu)|1$*

(a) $\forall F \exists u \,.\, R(F, u)$,

(b) $\big(R(F, u) \cdot R(G, u)\big) \supset_{F,G,u} F = G$.

The counter-example is blocked, however. In fact, if F is of r-type $(0|1)|1$, the order of the formula $F = F$ is, by definition of identity, $n + 3$. So, by applying 5.2, we get

$$\exists v \forall F \exists p \,.\, \mathsf{val}_{(\dot{0}|n)|1,0|n+3}(v, F, p).$$

By comprehension, we can obtain a relation R satisfying clauses (a) and (b), but its type turns out to be $\big((0|n)|1, 0|n + 3\big)|1$, so that it is not of the form specified in 5.9.

Thus, Russell's antinomy is happily resolved by means of ramification.

6. SOME FINAL REMARKS

We want to make here some comments on the notion of intensionality expressed by Axiom 5.4.

Taking $P =_{\mathrm{def}} F(x)$, with F and x variables of types $(\pi)|n$ and π respectively, we have, as an instance of 5.2,

COROLLARY 6.1. $\exists v \forall F \forall x \exists p \,.\, \mathsf{val}_{(\pi)|n,\pi,0|m}(v, F, x, p) \cdot p \equiv F(x)$, *where* $m = \mathsf{ord} P$.

This says, intuitively, that p is the proposition expressed by $F(x)$.

As an instance of 5.4, we have

COROLLARY 6.2. $\mathsf{val}(v, F, x, p) \cdot \mathsf{val}(v, G, y, p) \supset F = G \cdot x = y$.

There follows a principle of *strong extensionality* for propositional functions:

PRINCIPLE 6.3. *Two propositional functions are identical, provided they have one value in common.*

Identifying F and G in 6.2, we have, as a corollary of strong extensionality,

COROLLARY 6.4. *Every propositional function is* injective*: it maps different arguments into different values.*

The fact that propositional functions, precisely in virtue of their intensional character, do not satisfy a principle of extensionality, even stronger than the usual one for extensional set-theoretical functions, might seem an oddity. This is not so, however. In fact, according to the notion of propositional function underlying Int, propositional functions are *constituents* of (those propositions which are) their own values, in a similar sense as symbols for propositional functions are constituents of those formulas in which they occur. So, for two propositional functions, to share a value means to be the same constituent (in the same role) of a certain proposition. This is what justifies strong extensionality.

Anderson (*1986*) has shown that 6.4 cannot be consistently added to Church's (*1984*) system. In fact, 6.4 is incompatible with Church's axiom

AXIOM 6.5. $\exists F . F(x_1, \dots, x_m) \equiv_{x_1,\dots,x_m} P$

with $x_1, \dots, x_m$ all distinct, $\mathsf{ord}F \geq \mathsf{ord}P$, F not free in P, and where '$\equiv$' is Church's operator of intensional identity between propositions. If $x_1, \dots, x_m$ do not occur free in P, the F satisfying 6.5 is a constant function, so that it fails to be injective.

The principle 6.5, however, does not fit with the notion of propositional function just sketched above. In fact, the natural candidate for F in 6.5 should be the propositional function obtained from the proposition (expressed by) P by abstraction on its arguments. But such an F, because of its very genesis, cannot be a constituent of P, while, according to that conception, it is a constituent of the proposition $F(x_1, \dots, x_m)$, so that the two propositions fail to be identical. Accordingly, in our Axiom 5.2, $F(x_1, \dots, x_m)$ and P are asserted only to be logically equivalent and they are not expected to be strictly identical.

The adoption of 6.5 entails the rejection of the feature of propositional functions being constituents of their values. This is a possible approach and, in fact, it is the one chosen by Anderson (*1989*). He brings out some arguments in favor of 6.5 and proposes a revision of Church's theory, in which that principle is essentially saved, while extensionality and injectivity in their general form are banished.

Indeed, we believe that more than one conception of proposition and propositional function may be suggested by Russell's writings. The one worked out by Anderson is certainly praiseworthy in various respects. We think, however, that also the alternative choice of espousing 5.4 and rejecting 6.5 is coherent and worth investigating.

ACKNOWLEDGEMENTS

I am grateful to Anthony Anderson and Nino Cocchiarella for their useful comments to the first draft of this paper and to Neil Nelson for a number of stylistic improvements.

REFERENCES

Anderson, C. A.

1986 Some difficulties concerning Russellian intensional logic, *Nous,* vol. 20, pp. 35–43.

1989 Russellian intensional logic, *Themes from Kaplan,* Oxford University Press, Oxford.

Church, A.

1984 Russell's theory of identity of propositions, *Philosophia Naturalis,* vol. 21, pp. 513–522.

1976 Comparison of Russell's resolution of the semantical antinomies with that of Tarski, *The Journal of Symbolic Logic,* vol. 41, pp. 747–760; reprinted in *Recent essays on truth and the Liar Paradox,* (R. L. Martin, editor), Clarendon Press, Oxford, pp. 289–306.

Hazen, A.

1983 Predicative logics, *Handbook of philosophical logic 1* (D. Gabbay and F. Guenthner, editors), Reidel, Dordrecht, ch. I.5, pp. 331–347.

Myhill, J.

1979 A refutation of an unjustified attack on the axiom of reducibility, *Bertrand Russell memorial volume* (G. W. Roberts, editor), Allen and Unwin, London and New York, pp. 81–90.

Russell, B.

1903 *The principles of mathematics,* Cambridge University Press, Cambridge, England.

TERENCE PARSONS

THE LOGIC OF SENSE AND DENOTATION: EXTENSIONS AND APPLICATIONS

Abstract. The purpose of this paper is to extend Alonzo Church's (*1951*) investigation of the logic of sense and denotation in order to address some issues that have arisen since his classic paper on this topic.

After some background material in sections 1 and 2, a formulation of the system is given in section 3 that treats empty terms and functional senses, following out some ideas developed by Church (*1973, 1974*).

De re contexts are discussed in section 4, where I suggest that these are essentially already present in the system. In section 5 the issue of language learnability is addressed in the context of Frege's infamous linguistic hierarchy of senses. In section 6, I suggest that the problematic linguistic hierarchy is superfluous, but that this leaves untouched the ontological hierarchy of senses that is the subject matter of Church's work.

1. INTRODUCTION

According to Frege (*1892* and elsewhere), words have both sense and denotation. The word 'Socrates' denotes the man so named, and it expresses a sense, the meaning of that word.[1] Whole sentences denote truth values, and express propositions; the sentence 'Socrates was a philosopher' denotes the truth value True and expresses the proposition that Socrates was a philosopher. In each case the sense of a word *determines* its denotation. This is because every sense "presents," or "is a concept of" something. The sense expressed by 'Socrates' is a concept of the man, and the sense expressed by 'Socrates is a philosopher' is a concept of the True.[2] Being a concept of is a contingent, non-linguistic relation.[3] What a word or phrase denotes is whatever the sense it expresses is a concept of (see Figure 1). Frege thought that certain contexts cause a shift in these semantic relations. In indirect quotation, such as 'Agatha said that Socrates is a philosopher', the phrase 'Socrates is a philosopher' comes to denote the proposition that it normally

[1] I use 'denote' and 'refer to' interchangeably for Frege's term 'bedeuten'.

[2] Here I follow Church's usage of 'is a concept of'. Frege, uses 'concept' differently, for the denotation of a predicate, which is a function from things to truth-values.

[3] If the world had been different, the sense that 'Socrates' actually has might have been a concept of someone else, or of nobody at all, and the proposition that Socrates is a philosopher might have been a concept of the value False. If the language had been different, the words 'Socrates is a philosopher' might have expressed a different proposition, but the proposition that we now express by 'Socrates is a philosopher' would not for that reason change which truth value it is a concept of.

C. Anthony Anderson and M. Zelëny (eds.), *Logic, Meaning and Computation*, 507-543.

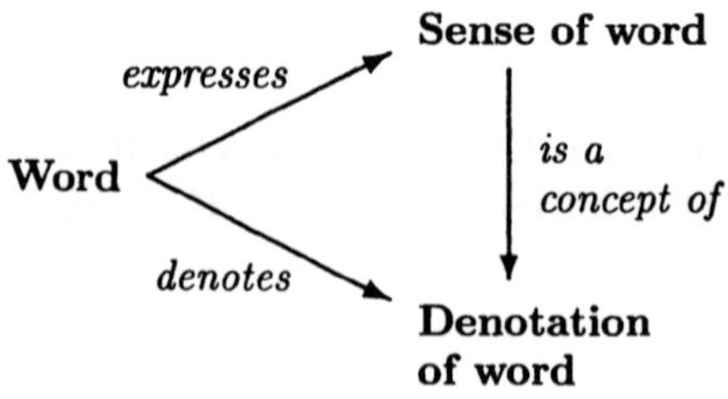

Figure 1.

expresses. Since the phrase has acquired a new denotation, and since its denotation is always whatever its sense is a concept of, it must also have acquired a new sense, one that is a concept of the proposition that Socrates is a philosopher. Thus the semantics requires concepts of propositions in addition to the propositions themselves. Frege (*1902*) also assumed that when such a phrase is *re*embedded, as in 'John thinks that Agatha said that Socrates is a philosopher', there is another shift of denotation, which requires another shift in what sense is expressed. Thus we also have concepts of concepts of propositions (which are concepts of truth values). Any phrase of natural language is thus automatically associated with an infinite number of distinct senses.

Frege assumed compositionality at both levels: The sense of a complex phrase is a function of the senses of its parts, and the denotation of a complex phrase is a function of the denotations of its parts. Thus the semantic shifts mentioned above also apply to the parts of sentences. As a consequence of this shifting, every meaningful word or phrase of the language has associated with it an infinite number of senses and denotations. Frege thought of this infinite semantical shifting as a defect of natural language, and he urged the use of notation in which words never shift either sense or denotation.[4] Instead, when we wish to say 'Agatha said that Socrates is a philosopher' we would use new words with fixed sense and denotation instead of the same words with shifted sense and denotation; for the ordinary use of 'Socrates is a philosopher' we would write:

Socrates$_0$ is a philosopher$_0$

and for the longer sentence we would write:

Agatha$_0$ said$_0$ that Socrates$_1$ is a philosopher$_1$

where 'Socrates$_0$' and 'Socrates$_1$' are distinct words, with the latter denoting

[4] *Frege 1902*: "To avoid ambiguity, we ought really to have special signs in indirect speech, though their connections with the corresponding signs in direct speech should be easy to recognize."

the customary sense of the former and having a new sense of its own. The similarity in word root would let us keep track of the similar sources of these words in natural language.

Frege never carried out the task of designing such a language, but Alonzo Church (*1951*) took up the task in his classic paper "A Formulation of the Logic of Sense and Denotation." Church did not develop a semantic theory; the terms 'denote' and 'express' do not occur in the symbolism.[5] Instead, his task was to construct and axiomatize the scientific notation itself. This notation supplies us with ordinary logical notation for the simple theory of types together with ample vocabulary to denote customary senses, concepts of customary senses, and so on, and it also contains terminology for the non-linguistic, non-semantic notion *being a concept of.* It is a useful scientific notation of the sort Frege envisaged, and it is simultaneously an axiomatic description of the ontology of senses and denotations appealed to by any language (including an imperfect natural language). As such, it has wide interest. The goal of this paper is to explore and extend the usefulness of Church's system construed as a general theory of senses that is applicable to the semantics of natural language, and as a scientific notation for the study of Frege's ontology.

One of Church's goals was to say more than Frege about the identity conditions for senses. He explored three alternatives. Alternative 0 requires (roughly) that two phrases have different senses whenever the phrases themselves differ (except for the choice of bound variables), alternative 1 is an intermediate system (which permits lambda-conversion without alteration of sense), and alternative 2 identifies the senses of phrases whenever they are necessarily equivalent. Difficulties have been found with all of these options. Alternatives 0 and 1 identify senses so closely with the phrases that express them that they tend to run afoul of analogues of the semantic paradoxes, and complications need to be introduced to guard against this. Alternative 2 is inconsistent with a principle about functional senses (discussed below).

My concerns in this paper are not with a choice among these alternatives, but rather with a number of issues that bear on the general usefulness of any of these options in doing semantics. After describing the ontology and symbolism (section 2) my first major goal (section 3) is to provide a criterion for when functions of the appropriate types can themselves be senses. This requires a discussion of empty concepts. Following that I show (section 4) how to extend the system to include *de re* contexts. Next (section 5) I show how to extend the system further to avoid criticisms based on the purported unlearnability of the vocabulary. Finally (section 6) I discuss reasons for rejecting the view that attributes to each word and phrase of natural language an infinite hierarchy of senses. I argue for a simpler view, which is one endorsed by Church. My overall goal is to show that Church's approach can

[5]Church took up these issues elsewhere.

be less daunting than it sometimes appears, and that it has applications that give it a much wider appeal than is often assumed.

2. THE ONTOLOGY: TYPES AND FUNCTIONS

Church's ontology is meant to be Frege's, with the one exception that the types of sentence-related entities (truth values, propositions, concepts of propositions, ...) are distinguished from the types of individual-related entities (individuals, concepts of individuals, ...).[6] I use mostly the same types as Church, expressed in a less compact but easier to read notation. First, there are non-functional types:

> *Sentence-related types*: the type of truth values is o ("omicron"), the type of concepts of truth values (the type of propositions) is o_1, the type of concepts of propositions is o_2, and so on. The basic type o is also written as o_0.
>
> *Name-related types*: the type of individuals is ι (also written ι_0), the type of concepts of individuals is ι_1, the type of concepts of concepts of individuals is ι_2, and so on.

Second, there are functional types, the types of (total) functions from entities of one type to entities of another type:

> The type of total one-place functions from entities of type α to entities of type β is $(\alpha \to \beta)$. The types of two-place functions are $(\alpha, \beta \to \gamma)$, and so on for functions of additional places.[7]

To illustrate the use of these types:

> The name 'Socrates' denotes an individual (of type ι) and expresses a sense of type ι_1. (The sense expressed is thus the right type to be a concept of the individual denoted.)
>
> The predicate 'is wise' denotes a function from individuals to truth-values, which is an entity of type $\iota \to o$, and it expresses a sense of type $\iota_1 \to o_1$, the type of functions from concepts of individuals to propositions. (Again, the function expressed by the predicate is a concept of the function denoted by it.)

---

[6]Frege explicitly included truth values together with individuals in the *same* type, which he called objects: "A statement contains no empty place, and therefore we must regard what it stands for as an object." (*Frege 1981*; translated in *Frege 1960*.) This is unnatural and inessential to his/our purpose.

[7]This departs slightly from Church, who disallowed multiplace functions. Instead of a two-place function from pairs of individuals to truth values, he uses a one-place function from individuals to one-place functions from individuals to truth values. Nothing of importance hangs on this choice.

> The sentence 'Socrates is wise' denotes a truth value (of type o) and expresses a proposition (of type o_1). (The proposition expressed is a concept of the truth value denoted.)

The truth value of 'Socrates is wise' is doubly-determined. It is the unique truth value that the proposition expressed by the sentence is a concept of. This proposition in turn is the value of the function expressed by 'is wise' for the argument that is the individual concept expressed by 'Socrates'. The former function is a concept of the function denoted by 'is wise', and the latter concept is a concept of the object denoted by 'Socrates'. Finally, the truth value of the sentence is determined at the level of denotation by being the value of the function denoted by 'is wise' for the argument that is the individual denoted by 'Socrates'. These multiple relations are part of what gives the theory its logical complexity and interest.

It is sometimes convenient to speak of a type that is a certain number n of levels "above" another type in a hierarchy. For the non-functional types this is achieved by raising the subscript on the type symbol by n. If α is the type symbol β_n, Church uses α_m to abbreviate β_{n+m}. He introduces this notation for functional types too: if α is the type $\beta_3 \rightarrow \gamma_1$ then α_2 abbreviates the type $\beta_5 \rightarrow \gamma_3$. This is a mere convention, and it is important to keep that in mind. Otherwise one might assume that if α is a functional type, then α_1 is a type of entity all members of which are senses. In section 3 we will see that the entities of functional type α_1 always outrun the senses of that type.

We can use these types in discussing ordinary language. There, when a sentence is embedded in an indirect context (such as 'Mary believes that ____') there is a semantic shift, in which the words come to refer to the senses they customarily express, and they also acquire new senses. Frege thought that in such a context the name 'Socrates' refers to its customary sense and acquires a new higher-level sense (of type ι_2). Likewise, 'is wise' acquires a new sense of type $(\iota_1 \rightarrow o_1)_1$, that is, of type $\iota_2 \rightarrow o_2$. Since the complement sentence denotes the proposition that it customarily expresses (the proposition that Socrates is wise, which is of type o_1) it acquires a new sense of type o_2. As previously noted, Frege saw this shift of semantic relations in indirect contexts as a defect of language, and he proposed that in a scientific notation words would never change their semantics (not even systematically). Instead, we should use different words in these contexts, but the different words in a scientific notation that represent different uses of the same ordinary word should have some identifiable link with one another. Church's notation may be used to accomplish this neatly, since each of his proper symbols has two parts; a root that resembles the word of the pre-scientific notation, and a type symbol that indicates the type of entity denoted by that symbol in a certain context. Thus, if Frege is right about natural language,[8] the pre-scientific 'Socrates' can be represented by the chain of

[8] Church did *not* intend his scientific notation to be used for this purpose. See discussion

scientific symbols 'Socrates$_{\iota_0}$', 'Socrates$_{\iota_1}$', 'Socrates$_{\iota_2}$', ..., where the first scientific symbol mimics the semantic behavior of 'Socrates' when it is not embedded in an indirect context, the second mimics the semantic behavior of 'Socrates' when it is embedded in a single, indirect context, the third mimics the semantic behavior of 'Socrates' when it is embedded in an indirect context which is itself in an indirect context (as in 'Agatha believes that Harry thinks that Socrates is wise'), and so on.

I use the symbol '►' for 'is a concept of'.[9] So the following are true:

Socrates$_{\iota_1}$ ► Socrates$_{\iota_0}$
Socrates$_{\iota_2}$ ► Socrates$_{\iota_1}$
....

(Strictly, '►' represents an infinite number of two-place predicate symbols, each with a type of its own. The first occurrence of '►' above should be '$►_{\iota_1,\iota_0 \to o}$' and the second should be '$►_{\iota_2,\iota_1 \to o}$'. I omit the type symbols on '►' because in such a context they can be uniquely recovered from those of its arguments when it is used to make a sentence (of type o).)

I also follow Church in assuming that every primitive symbol of the language occurs with an infinite number of different type subscripts; its basic subscript α, and also α_1, α_2, and so on, with each successive occurrence identified by the axioms as a concept of the previous one. Thus, if we have a primitive negation sign $\neg$ of type $o_0 \to o_0$, we also have the following signs and axioms:

$\neg_{o_1 \to o_1}$ ► $\neg_{o_0 \to o_0}$

$\neg_{o_2 \to o_2}$ ► $\neg_{o_1 \to o_1}$

$\neg_{o_3 \to o_3}$ ► $\neg_{o_2 \to o_2}$

....

Church's axioms come in three sets:

I. There are axioms for the basic logic of the functional calculus with identity, including unrestricted λ-abstraction and -concretion. I will take for granted that there is some adequate set of such axioms in the background.

II. There are three alternative sets of axioms for the three major alternative assumptions (Alternatives 0, 1, and 2) about the identity conditions for senses. Most of my discussion is intended to be neutral among these.

in section 6.

[9] I use this notation because it is easy to read; 'X ► Y' is read 'X is a concept of Y'. Church's notation for this is 'ΔYX', which can be read (in order) as 'being a concept of Y holds of X'.

III. Finally, there are some sets of axioms governing the predicate '►' for *being a concept of*. These are my primary focuses throughout this paper. One is axiom set 14, which says that any given concept is a concept of at most one thing; I take this for granted throughout my discussion. The controversial ones are axiom sets 15 and 16, which are explained in the next section.

3. WHICH FUNCTIONS ARE SENSES?

Church assumes that senses of functional signs are themselves functions; e.g., that the sense of the sign 'F' is a function that maps the sense of 's' to the sense of the sentence '$F(s)$'. Here is a reason why. Suppose that senses of functional signs are not functions. We still want the sense of a complex expression to be determined by the senses of the component expressions and the way in which they are put together. Thus there will *be* a function that maps the sense of 'F' and that of 's' to the sense of '$F(s)$', and this will be systematic for all sentences of the form '$F(x)$'. So in terms of this we can define a function that maps any individual concept to the proposition expressed by '$F(a)$' when 'a' expresses that concept. This defined function already exists in the ontology we are presuming, and so the ontology is simpler if we identify the sense of 'F' with that function than if we presume a completely new kind of entity to serve as the senses of functional expressions. In addition, we can exploit the logical function-argument relation in explaining how senses combine with one another, and this simplifies the logic.

There are several objections to identifying functional senses with functions of senses (though I think that all of them can be overcome); in the following few sections I focus on the one that historically led to problems with Church's system.

3.1. Characterization

Not all functions from senses to senses can be senses. Suppose that E and M are two individual concepts that are concepts of the same individual i. For example, E might be the sense of 'the evening star', M the sense of 'the morning star', and i the planet Venus. Let F be some function that maps E to a true proposition and M to a false proposition. Then F cannot be the sense of any predicate. This is because F is not "predicational"; it cannot be the concept of a predicate reference f, which is a function from individuals to truth values. Any such predicate reference maps the individual i to a truth value, and that truth value is the reference of *any* sentence whose subject refers to i and whose predicate refers to f. If F were the sense of a predicate P, then there could be two sentences containing the predicate P, one whose subject expresses E and one whose subject expresses M. These sentences would then have the same truth value while expressing propositions with different truth values, thus contravening the principle that the truth value of a sentence is the truth value which the proposition it expresses is a concept of.

Church's solution is to require that senses of predicates be predicational, and to require that senses of other types of functional expressions have the analogous property; he introduces the term 'characterization' for the general case. A functional sense F *characterizes* a function f iff whenever F's argument A is a concept of some thing a, F maps A to a concept of what f maps a to:

> Suppose that the type of F is $\alpha_1 \to \beta_1$ and the type of f is $\alpha \to \beta$. Let A range over entities of type α_1 and a over entities of type α. Then:
>
> $$F \textit{ characterizes } f =_{\text{df}} \forall A \forall a[A \blacktriangleright a \to F(A) \blacktriangleright f(a)].$$

This is a necessary condition for a function to be a concept of a function. What more is needed? Church (*1951*, page 16) supposes that nothing more is needed. This assumption leads to the adoption of his axiom-sets 15 and 16, which hold on the assignment of any types α and β as described above:

15: If $F \blacktriangleright f$, then F characterizes f.

16: If F characterizes f, then $F \blacktriangleright f$.

With the definition eliminated, these say:

15: If $F \blacktriangleright f$, then $\forall A \forall a[A \blacktriangleright a \to F(A) \blacktriangleright f(a)]$.

16: If $\forall A \forall a[A \blacktriangleright a \to F(A) \blacktriangleright f(a)]$, then $F \blacktriangleright f$.

3.2. The Reduction to Extensionality

There is a problem with axiom set 16. For some types $\alpha_1 \to \beta_1$ of functions there are no functions of that type that are defined *only* for senses. (At least, this is true in the intended applications.) Some functions have a domain α_1 that contains some things that are, and some that are not, senses. Since all functions are total on their domains, these functions cannot be defined only for senses. As a result, there will often be several different functions that are equally good at characterizing a given function f, different because they differ (only) in how they treat arguments that are not senses. This seemingly irrelevant behavior turns out to conflict with axiom set 16 in the context of alternative 2. In the remainder of this section I show how this happens, and in the next section I propose an improvement, similar to that of Church (*1973, 1974*).

The difficulty takes this form. Church's axioms for Alternative 2 in his original paper permit the derivation of the unpalatable result that there is exactly one true proposition. This result is due to the interaction of axiom set

16 for functional senses with the axioms that identify necessarily equivalent propositions. Here is how that happens.[10]

In the following, I use the term 'concept' in connection with a function to mean 'concept of something'. The proof exploits the fact that in alternative 2 necessarily equivalent concepts are identical. That is, two concepts that are necessarily concepts of the same thing are identical. The proof proceeds by defining two concepts that can be proved to be concepts of the same function, because they can be proved to characterize the same function f; axiom set 16 thus makes them (provably) concepts of f. Thus, in alternative 2 they are identical. However, they are defined in such a way that they are *not* identical, because they differ for certain arguments that are not concepts of anything (recall that axiom set 16 ignores what functions do to non-concepts). At least they differ if such arguments exist. And such arguments can be shown to exist whenever there is more than one true proposition. The only escape from the inconsistency then is to conclude that there is not more than one true proposition.

In the following version of the proof, I use 'T' to name the truth-value true, 'N' to name some necessary proposition, and 'I' to name an impossible proposition.[11] We begin by defining a function f and the two functions V and W that will turn out to be concepts of f. For convenience here I call things of type $o \rightarrow o$ *unary truth-functions*, and things of type $o_1 \rightarrow o_1$ *unary propositional functions*.

> $f =_{\mathrm{df}}$ the function of type $(o_0 \rightarrow o_0) \rightarrow o_0$ that maps every unary truth-function to T.
>
> $V =_{\mathrm{df}}$ the function of type $(o_1 \rightarrow o_1) \rightarrow o_1$ that maps every unary propositional function to N.
>
> $W =_{\mathrm{df}}$ the function of type $(o_1 \rightarrow o_1) \rightarrow o_1$ that behaves as follows:
>
> If there is only one true proposition, then W maps every unary propositional function to N.
>
> Otherwise, W maps M to I and maps all other unary propositional functions to N, where M is the unary propositional function that maps N to itself and all other propositions to I.

We want to prove first that both V and W characterize f. Recall that this means:

$$F \text{ characterizes } f \equiv \forall A \forall a[A \blacktriangleright a \rightarrow F(A) \blacktriangleright f(a)].$$

[10]This proof was apparently discovered independently by A. F. Bausch (reported in *Church 1973*) and by David Kaplan (personal communication), and perhaps by others. Apparently, the proof has never been published. The present version is based on a version by Alonzo Church (personal communication).

[11]In alternative (2) there is exactly one necessary proposition, and exactly one impossible one, but those facts are not relevant to the proof.

The proof is simple for V, since both V and f are constant functions, and the former maps everything in its range to N, which is a concept of what the latter maps everything in its range to (namely, T). So V characterizes f.

W also characterizes f, because it either maps everything to N, and so it is identical with V (which characterizes f), or else it differs from V only for the argument M in a situation in which there is more than one true proposition. But in the latter situation, M is not a concept of anything, and so W still characterizes f.

Since both V and W characterize f, axiom 16 says that they are both concepts of f. Now suppose there are signs '$P_{(o\rightarrow o)\rightarrow o}$' and '$Q_{(o\rightarrow o)\rightarrow o}$' whose "senses" are respectively V and W; that is, '$P_{(o_1\rightarrow o_1)\rightarrow o_1}$' names V and '$Q_{(o_1\rightarrow o_1)\rightarrow o_1}$' names W, and '$P_{(o\rightarrow o)\rightarrow o}$' and '$Q_{(o\rightarrow o)\rightarrow o}$' name what V and W are concepts of, namely, f. From the definitions above we have proved

$$P_{(o\rightarrow o)\rightarrow o} = Q_{(o\rightarrow o)\rightarrow o}.$$

Since this sentence is provable, its sense must be a necessary proposition. A name of its sense is got by increasing the types of all symbols in the sentence by 1. Using subscripts to indicate this raising, the following may thus be proved:

$$\mathrm{Nec}[P_{o_1} =_1 Q_{o_1}],$$

where 'Nec' stands for 'is necessary' (of type $o_1 \rightarrow o$). The distinctive principles for alternative 2 (embodied in the second displayed formula on page 21 of *Church 1951*) then allow us to say that the terms flanking the sense-of-identity sign in the just-given formula are themselves identical:

$$P_{(o_1\rightarrow o_1)\rightarrow o_1} = Q_{(o_1\rightarrow o_1)\rightarrow o_1}.$$

That is, $V = W$. But $V = W$ only if there is not more than one true proposition. ⊣

This is only a sketch of a proof; in order to turn it into a proof one would need to actually define '$P_{(o\rightarrow o)\rightarrow o}$' and '$Q_{(o\rightarrow o)\rightarrow o}$' within the formal system, and I have not been able to do this. Indeed, I am not certain that it can be done. But I agree with Church (*1973*) that it would be undesirable to have senses in the theory that cannot be assigned to any signs on pain of a collapse to extensionality.

3.3: The Problem is Independent of Alternative 2

Not every function of the appropriate type can be a sense; a necessary condition is that it must *characterize* some function. But even that is not enough. In the last section we saw that how functions behave with regard to mapping non-concepts to things can disqualify them as senses, in the context of alternative 2. One could conclude from this that the fault lies with the modal identity conditions of alternative 2, which are independently implausible. But I think the problem is more general. In order to get at the heart of the issue in its full generality we need to consider a related phenomenon: how to treat empty terms and their senses. In a scientific notation we can avoid empty terms, but they occur in nonscientific discourse, and so their senses must be in the ontology. This was Church's intent (*Church 1951*, page 15, note 16):

> "... tentatively, we shall allow that in some language, ... there may be names which have a sense but no denotation. Hence we also admit concepts that are not concepts of anything ..."

We need to be careful to have our axioms treat such concepts appropriately. For it is easy to show, without any of the special assumptions about Alternatives 0–2, that axiom set 16 (which says nothing about empty terms) is incompatible with the existence of empty terms in any language, at least if their senses are included in the ontology.[12] Here is an illustration.

Let s be the sense expressed by 'Superman', and assume that s is not a concept of anything.

Let C be the sense of 'is clever', and let c be the function that maps clever things to truth and all other objects to falsehood. Then C is a concept of c, and $C(s)$ is the proposition that Superman is clever, which is not a concept of anything.

Let F be the same function as C except that F maps s to the proposition that snow is white. So $F(s)$ is a concept of truth.

Then, by axiom set 16, F is a concept of c. This is because F coincides with C on all individual senses that are actually concepts of something. So F characterizes c:

$$\forall A \forall a[A \blacktriangleright a \rightarrow F(A \blacktriangleright c(a)].$$

By axiom set 16 this entails that F is a concept of c. But this leads to absurdity if there exists any language with empty terms and a predicate that expresses F. Assume that F is expressed by the predicate 'is such-and-such'. Then 'Superman is such and such' expresses $F(s)$, which is the proposition that snow is white, and so the sentence is true. But 'is such and such' expresses F, which is a concept of c, so the reference of 'is such and such' is c. But 'is clever' also refers to c. So you can substitute 'is clever' for 'is such and such' without changing the reference of the whole. This yields 'Superman is clever', which, since the reference couldn't have changed, must be true. This contradicts the fact that 'Superman is clever' expresses a proposition that is not a concept of anything.

My diagnosis is that this problem arises not just because senses of predicates are functions, but because axiom set 16 makes functions be senses without looking at what they do to arguments that are not senses. The view

[12] Church predicts this in *Church 1951*, footnote 4, and comments on its importance in *Church 1973*, page 25. I think I am following out Church's line of thought here.

I wish to defend is that a functional sense should be something like a partial function, defined on senses and not defined at all for non-senses. This idea is articulated in *Church 1973, 1974,* and is repeated in *Church 1993,* in a somewhat different form. It is also adopted in *Anderson 1984.* I will implement it differently from the way Church and Anderson do, within a slight modification of Church's original theory, a modification that permits intuitively empty terms. This will also permit a simple reformulation of axiom sets 15 and 16.

3.4. Empty Terms and Empty Senses

Frege thought that in natural language some terms lack reference, but that in a good scientific notation this would not happen. Instead, we would take each term that lacks reference, and *give* it one. He presents this as an improved way to treat a single notion of reference, but it is better to view it as a correct way to compare two different reference notions. The first is our ordinary intuitive notion;[13] in this ordinary sense, some words and phrases lack reference, and no theorist can alter this by decree. Instead, the theorist can define a second, technical notion of reference for which reference failure does not occur. The technical relation holds between a term and a thing whenever the intuitive one holds; in case the intuitive notion does not relate a word to anything, the technical notion relates the word to a special "supplied" object. This is the idea I wish to explore.

The "supplied object" idea is similar to Frege's "chosen object" option, except that if intuitively empty terms are made to refer technically to a specially "chosen" *ordinary* object, the technical notion of reference does not encode the difference between (intuitive) reference failure and successful (intuitive) reference to the chosen object. It is better to select for this purpose a "supplied object," that is, some entity that is not already in the domain of reference for entities of the given type. For example, for reference to individuals, one can extend this to a technical relation between terms and things in an expanded domain consisting of individuals plus something else (e.g., plus the *set* of all old individuals, assuming that it is not already among the individuals). So assume that each domain of things of a given type is expanded by the addition of a new thing, called a "zip." We *stipulate* that technical-reference relates terms of the language to zips when the terms really refer to nothing at all in the intuitive sense of 'refer'. We use the technical extension of 'refer' to emulate real reference failure by the account:

w refers to $x =_{\text{df}} w$ refers$_{\text{technical}}$ to $x \,\&\, x$ is not a zip.

w fails to refer $=_{\text{df}} w$ refers$_{\text{technical}}$ to a zip.

---

[13]I speak here as if our ordinary intuitive notion of denotation included the assumption that sentences denote truth values. This is not what people normally assume. The point can be made without this assumption; it is just a bit more cumbersome.

This gives us all of the neatness of no-failure-of-reference at the technical level, while allowing us to be true to the fact that words sometimes fail to refer in the intuitive sense. It resolves Frege's two approaches to reference by making them mutually complementary aspects of an analysis employing two relations.

In more detail: we add to each domain of non-functional types a special object, a zip, symbolized ✿$_\alpha$ for each non-functional type α. We decree that no zip is a concept of anything. Nonfunctional senses of type α_1 that are not intended to be concepts of anything (in the intuitive sense of 'concept of') are decreed to be concepts of ✿$_\alpha$ (in the technical sense of 'concept of' symbolized by '$\blacktriangleright$'). For example, in the type of truth values there are now three truth values, truth, falsity, and zip. Any proposition (of type o_1) that in the intuitive sense lacks truth value is now a concept of ✿$_{o_0}$.

In the functional types there are now automatically the appropriate functions to play the role of zips. We say that x is *the* zip of type $\alpha \to \beta$ iff x maps every entity of type α to the zip of type β. The notation is neatest if we use the same symbol for this, so that the zip of type $\alpha \to \beta$ is denoted by '✿$_{\alpha\to\beta}$'. (We do not actually need to introduce new symbols for the zips; we could use '$\iota x_\alpha(\neg x_\alpha = x_\alpha)$' for the zip of type α.)[14]

Lastly, we need to take care in discussing empty *concepts*, for an entity of a conceptual type can be empty in two different senses. In one sense it is empty if it is a concept of the appropriate zip of lower type. In another sense it is empty if it is not a concept of anything at all. The former sorts of entities are the things that are suited to be senses of singular terms that, in the intuitive sense, lack reference. The latter are ... what? In the next section we address this question, and give the final form to the axioms.[15]

3.5. Supercharacterization

Earlier I suggested that the idea to be pursued is to get the effect of making functional senses be partial functions defined only on senses. But first we need to know what a sense is. In particular, we need to be able to tell the difference between a sense that is a concept of nothing, and something that is a concept of nothing because it is not a sense at all. This problem can now be solved. In the revised system we can define a sense as anything that

[14]This approach bears strong parallels to the constructions in *Church 1973, 1974*. The principal difference is that Church does not allow the zips to be members of the domains of things of a given type. Thus he need not distinguish the intuitive from the technical senses of referring; when a term "refers" to a zip it is referring to something outside the domain of quantification, so we needn't say that it refers to something (implicitly: to something in the domain). The price he pays is a fairly complicated set of replacements for axiom sets 15 and 16.

[15]Distinguishing these two kinds of emptiness is part of the process that Kaplan (*1973*) called "garbage control." I am indebted to this talk for introducing me to the complexity of these issues.

is a concept of something in the technical sense of *concept of*. Previously this was inappropriate because there could be empty senses. But now an intuitively empty sense is a concept of the appropriate zip, and so we can construe nonconcepts (= things that are not *technically* concepts of anything) as things that are not senses at all.

We get the right treatment of functional senses by saying that F is a functional sense iff it characterizes some function *and* it maps every non-sense (every thing in the domain of the function that is not a concept of something) to the zip of appropriate type. I call this *supercharacterization*. This notion permits us to replace both axiom sets 15 and 16 with what is in effect a recursive definition of the technical notion of "concept of":

> SUPER-15-16: Suppose that the type of F is $\alpha_1 \rightarrow \beta_1$ and the type of f is $\alpha \rightarrow \beta$. Let A range over entities of type α_1 and a over entities of type α. Then:
>
> $$F \blacktriangleright f \equiv \forall A \forall a[A \blacktriangleright a \rightarrow F(A) \blacktriangleright f(a)] \,\&\, \forall A[\neg \exists a A \blacktriangleright a \rightarrow F(A) = ✿_{\beta_1}].$$

This proposal has the virtue of simplicity and of being based on an intuitive conception of senses. Whether it escapes the difficulties of the earlier account is to be seen.

With the introduction of zips, we now have terms in our scientific notation that, intuitively, lack denotation—because they (technically) denote zips. With this in mind, let us rehearse the Superman example of the last section:

> Let s be the sense expressed by 'Superman', and assume that s is not intuitively a concept of anything. So s is technically a concept of $✿_\iota$.
>
> Let C be the sense of 'is clever', and let c be the function that maps clever things to truth and all other objects to falsehood. Then C is a concept of c, and $C(s)$ is the proposition that Superman is clever, which is not intuitively a concept of anything, but is now technically a concept of $✿_o$.
>
> Let F be the same function as C except that F maps s to the proposition that snow is white. So $F(s)$ is a concept of truth.
>
> Previously, we could show that F is a concept of c. This is because F coincides with C on all individual senses that were (formerly) concepts of something (ignoring s). So previously F characterized c, and then by axiom set 16, F was a concept of f. But now s is technically a concept of something, the individual zip, $✿_\iota$. In order for F to characterize c, F must map s to a concept of what c maps $✿_\iota$ to, which is $✿_o$. But F maps s to a concept of truth. So F does not characterize c, and the proof breaks down.

We should also review how the revised system escapes the reduction to extensionality that plagued the original system. Here is an attempted reconstruction of that proof as given earlier. We begin with the same definitions as before:

> $f =_{df}$ the function of type $(o_0 \rightarrow o_0) \rightarrow o_0$ that maps every unary truth-function to T.
>
> $V =_{df}$ the function of type $(o_1 \rightarrow o_1) \rightarrow o_1$ that maps every unary propositional function to N.

We can see right away that V cannot be shown to be a concept of f, because V maps all non-concepts to N, not to a zip. And there *are* nonconcepts in the range of V if there is more than one true proposition; any function that maps some true propositions to true ones and others to false ones is not a concept of anything. (We can't assume at this point that there is only one true proposition.) So with SUPER-15-16 in place of axiom set 16 the proof that V is a concept of f breaks down.

But we can fix up this part of the proof. Suppose we replace the definition of V with this:

> $V^o =_{df}$ the function of type $(o_1 \rightarrow o_1) \rightarrow o_1$ that maps every unary propositional function F to N if F is a concept of something, and to ✿_{o_1} otherwise.

Then SUPER-15-16 lets us show that V^o is a concept of f. And so the proof goes on as before. So we have to examine the other functional sense, which is:

> $W =_{df}$ the function of type $(o_1 \rightarrow o_1) \rightarrow o_1$ that behaves as follows:
>
> If there is only one true proposition, then W maps every unary propositional function to N.
>
> Otherwise, W maps M to I and maps all other unary propositional functions to N, where M is the unary propositional function that maps N to itself and all other propositions to I.

In the original proof, W characterizes f, and is a concept of f. With SUPER-15-16 replacing axiom set 16 the proof again breaks down, because W maps some nonconcepts to N, not to zip. (At least it maps them to N if there are any nonconcepts of the appropriate type, and again there are some if there is more than one true proposition.)

So let us fix up W as we did V. Make W^o map non-concepts to zip, and the rest as above. Then it is easy to show that W^o is a concept of f, because W^o is identical to V^o, no matter how many true propositions there are.

So we can continue the proof. It is easier now, since we can skip the part that uses the special axioms of alternative 2—since we have already shown without them that $V^o = W^o$. (This is a reason why I do not see the specialness of alternative 2 as being relevant to the key issue.)

But now the final part of the proof does *not* go through. When W^o and V^o were the same on their original definitions, we could prove that there is only one true proposition, for that was the only way to defeat the consequence that they otherwise map M to different things. But on the revised definitions, they do not differ on M; they both map it to zip if it is a non-concept and to N if it isn't. So the inconsistency in the original proof has not been duplicated.

3.6. Models for the Basic Axioms

It would be nice at this point to have a proof of consistency for this approach. It is not difficult to sketch a model theory for the system without the special alternative 2 axioms; it is then possible to piggyback a possible worlds account on that model (as in *Church 1973, 1974*) to secure the consistency of alternative 2. (This section may be skipped without loss of continuity.)

3.6.1 Notation

Assume that the notation is the same as *Church 1951* with the addition of names of zips of the form ✪_α for every type α.

3.6.2 Model Theory

An arbitrary model is subject to the following conditions. First, we assume that there are disjoint domains for every nonfunctional type. Each domain of non-functional type is to contain a single primitive zip of that type. The domain of things of type o contains exactly three things: truth, falsity, and the zip of type o. The domain of things of type ι contains the zip of type ι along with any number (possibly zero) of other things. The domain of nonfunctional things of type α_1 is stipulated to contain more things than the domain of things of type α. The domain of type $\alpha \rightarrow \beta$ consists of all total functions from the things of type α to the things of type β. The zip of type $\alpha \rightarrow \beta$ is that function that maps every member of type α to the zip of type β.[16]

Lambda abstracts are to be interpreted as standing for (total) functions in the usual way.

The universal quantifier of type $(\alpha \rightarrow o) \rightarrow o$ stands for that function that maps a function of type $\alpha \rightarrow o$ to T if that function maps everything in its domain to T, to F if it maps something in its domain to F, otherwise to ✪_o.

[16]In the terminology of *Church 1951*, every type is now a preferred type, and the zips are the designated members.

The definite description operator iota of type $(\alpha \to o) \to \alpha$ stands for the function that maps a function f of type $\alpha \to o$ to x if x is the unique thing of type α that f maps to T; otherwise it maps f to $\circledast_\alpha$.

The conditional is interpreted by this truth table:

$A \supset B$	T	$\circledast_o$	F
T	T	$\circledast_o$	F
$\circledast_o$	T	T	T
F	T	T	T

Negation is defined by Church as

$$\neg S =_{\mathrm{df}} S \supset \mathrm{F}.$$

As a consequence, negation has the following truth table:

A	$\neg A$
T	F
$\circledast_o$	T
F	T

There is a binary relation, *being a concept of*, defined on domains. For every nonfunctional type α, *being a concept of* is any many-one relation from things of type α_1 onto things of type α (so everything of type α is "concepted"), with the zip of nonfunctional type α_1 required to be a concept of the zip of type α. For functional types of the form $\alpha_1 \to \beta_1$ the *concept of* relation is extended by recursion in accord with the formula:

$F_{\alpha_1 \to \beta_1}$ *is a concept of* $f_{\alpha \to \beta}$ iff

$\forall A \forall a[A \text{ is a concept of } a \to F(A) \text{ is a concept of } f(a)]$ &

$\forall A[\neg \exists a[A \text{ is a concept of } a] \to F(A) = \text{the zip of type } \beta_1]$.

The predicate '$\blacktriangleright$' is interpreted as denoting the relevant *concept of* relation, depending on its subscript.

3.6.3 Axioms and rules

The goal of the axioms and rules is to state and preserve truth. The logical ones (Rules I–V and Axiom sets 1–10 of *Church 1951*) are all valid in the models described above except for axiom 8, which needs revision. Axiom 8 says:

$$(p_o \supset q_o) \supset ((q_o \supset p_o) \supset p_o = q_o),$$

which presumes that there are two truth values. Now there are three. Axiom 8 can be replaced with the following four axioms:

$$\mathrm{T}_o \neq \mathrm{F}_o$$

$$\mathrm{T}_o \neq \text{✿}_o$$

$$\text{✿}_o \neq \mathrm{F}_o$$

$$\forall p_o[p_o \neq \mathrm{T}_o \supset [p_o \neq \mathrm{F}_o \supset p_o = \text{✿}_o]].$$

So the logical axioms and rules are all satisfied by any model. In addition, we should add more axioms and/or rules to take account of the presence of new truth values. This is for completeness, and does not impact the fact that the existing logical axioms and rules are all validated by the model theory. I do not attempt here to state the additional rules.[17]

It is also apparent that the model theory validates the axioms governing '►'. Axioms 11–13 merely guarantee that when the type subscript α on a primitive symbol is replaced by α_1 then the new symbol names a concept of what the old one names. Since everything in any model is concepted, and since there are no special axioms for these elevations of the primitive symbols, this may be stipulated. Axiom set 14 says that "concept of" is many-one, and this is guaranteed by the characterization of our models. And finally, our new axiom Super-15-16 just puts into words the model theoretic characterization of the *concept of* relation for functional types.

I conclude that the model theory validates the logical rules and axioms as well as those governing '►'.

3.6.4 Possible Worlds Model Theory

We can extend the model theory given above to conform to alternative 2 by the possible worlds technique employed in *Church 1973, 1974*, which is a bit easier for us because of our logical treatment of zips as members of their domains.[18] This lets us construct models for alternative 2 that do not entail that there is only one true proposition. Details are not pursued here.

It is also open to us not to do this, and to leave the theory as stated, without commitment to any formal or mechanical analysis of the identity conditions of senses. (This is my preferred approach.)[19]

---

[17]This may not be possible with the given notation. The connectives as newly interpreted appear not to be truth-functionally complete (e.g., there appears to be no way to define a conjunction that is true iff both conjuncts are true, false iff either conjunct is false, and otherwise zip). Thus, normal ways of filling out the rules to "three-valued" completeness do not necessarily work in the usual way. If this is the case, the natural development would be to add a primitive conjunction or disjunction with its own semantical clause in the model theory; there then seems to be no difficulty in principle of completing the rules and axioms.

[18]Church (*1973, 1974*) presumed that the zip of a given type is not a member of any domain; this required special treatment of them in the model theory, and required complicated replacements for axiom sets 15 and 16.

[19]If modalities are present in the language, then there is a difficulty with axiom 16 in

4. DE RE CONTEXTS

Frege seems to have disavowed *de re* constructions. When Russell suggested that the proposition that Mont Blanc is over 1000 meters high has Mont Blanc as one of its constituents, Frege balked at the idea that a concrete object could be *part* of a sense. He apparently concluded from this that no *de re* construction makes sense. I say "apparently" because he probably did not think in these terms, and to say this attributes to him a general view that goes beyond any particular one that he had.[20]

Church follows Frege in this; his discussion of concepts is framed entirely in terms of a scientific notation that captures *de dicto* readings, but not *de re* ones. But this is a matter of focus, and is not required by the ontology at hand. Indeed, the ontology at Church's disposal contains all the ingredients one would need to give a semantics for *de re* constructions. This is what I argue here.

Consider the ambiguity in 'Agatha thinks someone is a spy'. The *de dicto* reading links Agatha with the proposition that someone is a spy. The *de re* reading says that there is someone whom Agatha believes to be a spy. Let us consider the types of the symbols in the *de re* reading. The whole sentence denotes a truth value, and is of the form:

$$\exists x[\text{Person}(x)\,\&\,\text{Believes}(\text{Agatha}, \text{Spy}(x))].$$

This is true iff the following is true on some assignment to 'a':

$$\text{Person}(a)\,\&\,\text{Believes}(\text{Agatha}, \text{Spy}(a)).$$

But what is assigned to a in the first conjunct must be an individual, and the object of 'believes' denotes a proposition. So we have a constituent of the form:

$$\text{Spy}(a)$$

which, in this context refers to a proposition (of type o_1) when 'a' refers to an individual. This can happen only if 'Spy' refers, in this context, to something of type $\iota \rightarrow o_1$, that is, to a function from individuals to propositions. In the *de dicto* reading, on the other hand, 'Spy' refers to a function of type $\iota_1 \rightarrow o_1$, from concepts of individuals to propositions. So to mirror a *de re* indirect

addition to the one discussed in this paper. Kaplan (*1964*) points out that in order for a function to be a concept of another it should necessarily characterize it. In terms of the present proposal, one needs to add to the right-hand side of SUPER-15-16 the condition:

$$\forall A \forall B\, \text{Nec}[A =_1 B \rightarrow_1 F(A) =_1 F(B)]$$

where 'Nec' maps necessary propositions to truth. I am indebted to David Kaplan for pointing this out to me.

[20]The Mont Blanc discussion is in *Frege 1904*; there is a similar example using Mt. Etna in *Frege 1914?*. The general rejection of *de re* constructions is consistent with all of Frege's stated views; see especially his response to Russell on the paradox about sets of propositions in *Frege 1902*.

context in a scientific notation we need to use a different symbol from the one that we need to use to mirror a *de dicto* indirect context. This should be no surprise; the two contexts can have different truth values, so something must have changed. Maybe this is it.

This hypothesis meshes nicely with Church's ontological development, which allows us to bifurcate senses of words. Predicates of natural language must express *two* senses, a *de re* one and a *de dicto* one.[21] When a predicate of natural language is embedded in what we normally call a *de dicto* indirect context, it comes to refer to its *de dicto* sense, and when it is embedded in what we normally call a *de re* indirect context it comes to refer to its *de re* sense. In a scientific notation there is no reference shifting, and so we need additional symbols to denote these senses. In Church's original approach we needed an infinite chain of symbols for each predicate of natural language: one symbol for the predicate in direct contexts, one for the term in a singly embedded *de dicto* context, one for the term in a doubly embedded *de dicto* context, and so on. When *de re* indirect contexts are taken into account, each predicate of natural language needs at least a double chain: one symbol for the predicate in direct contexts, two for the term in a singly embedded context (one for the *de dicto* use and one for the *de re*), two for the term in a doubly embedded context (one for the *de dicto*, and one for the *de re*), and so on. (Actually, as we see below, it will be a little more complicated than this, to accommodate, e.g., *de dicto* beliefs about *de re* beliefs.) The multiplication of symbols becomes a bit more rococo, but there is no difference in principle than with the straight *de dicto* approach: with each different context in which the natural language symbol appears we get a different symbol in the scientific notation. This is the standard tradeoff; one avoids the referential ambiguity of natural language in favor of multiple symbols in the scientific notation.

There is no worry about finding the senses themselves, for they are already present in the ontology. That is, there are already functions of the appropriate types, functions from individuals to propositions. What is new is to appeal to some of them as senses of symbols.

We can easily extend the "concept of" relation to these new entities. There is no danger of confusing the use of this relation in *de dicto* cases with uses in *de re* cases, for the types are different. The original "concept of" relation introduced by Church relates a function of type $\alpha_1 \to \beta_1$ to a function of type $\alpha \to \beta$. Call these "Fregean" instances of the *concept of* relation. The other instances of the *concept of* relation, which I call "Russellian," are those that relate a function of type $\alpha \to \beta_1$ to a function of type $\alpha \to \beta$. For easy readability I use '$\triangleright$' for the Russellian instances of the *concept of* relation,

[21] This means that simple sentences of natural language, such as 'Holmes is clever', are ambiguous. I think on independent grounds that this is correct; the ambiguity is essentially that of the "referential/attributive" distinction.

to call attention to the difference between these instances and the Fregean ones, though it is just as correct to use '$\blacktriangleright$' as the root symbol and leave it to the type subscripts to distinguish them.

The Fregean and Russellian instances are identical for nonfunctional types. For functional types the Russellian ones can be defined by a simple version of super-characterization:

RUSSELL-15-16: $F_{\alpha\to\beta_1} \rhd f_{\alpha\to\beta}$ iff $F_{\alpha\to\beta_1}$ Russell-characterizes $f_{\alpha\to\beta}$,

where "Russell-characterize" means:

$\forall a[F(a) \blacktriangleright f(a)]$.[22]

The two versions of Super-15-16 now apply automatically to multiply embedded contexts, included mixed *de re/de dicto* contexts.

As an illustration of how this should work, here are some sample sentences and their types when represented in scientific notation, (in which for simplicity I omit notation for the subjects of 'believe').

Agatha is a spy.

$$\mathrm{Spy}_{\iota\to o}(\mathrm{Agatha}_{\iota}).$$

Mary believes (*de dicto*) that Agatha is a spy.

$$B_{o_1\to o}(Spy_{\iota_1\to o_1}(\mathrm{Agatha}_{\iota_1})).$$

Sue believes (*de dicto*) that Mary believes (*de dicto*) that Agatha is a spy.

$$B_{o_1\to o}(B_{o_2\to o_1}(\mathrm{Spy}_{\iota_2\to o_2}(\mathrm{Agatha}_{\iota_2}))).$$

Mary believes (*de re*) that Agatha is a spy.

$$B_{o_1\to o}(\mathrm{Spy}_{\iota\to o_1}(\mathrm{Agatha}_{\iota})).$$

Sue believes (*de re*) that Mary believes (*de re*) that Agatha is a spy.

$$B_{o_1\to o}(B_{o_2\to o_1}(\mathrm{Spy}_{\iota\to o_2}(\mathrm{Agatha}_{\iota}))).$$

Sue believes (*de dicto*) that Mary believes (*de re*) that Agatha is a spy.

$$B_{o_1\to o}(B_{o_2\to o_1}(\mathrm{Spy}_{\iota_1\to o_2}(\mathrm{Agatha}_{\iota_1}))).$$

The relations among the various representations of 'Spy' are then:

$$\mathrm{Spy}_{\iota_2\to o_2} \blacktriangleright \mathrm{Spy}_{\iota_1\to o_1} \blacktriangleright \mathrm{Spy}_{\iota\to o},$$

$$\mathrm{Spy}_{\iota\to o_2} \rhd \mathrm{Spy}_{\iota\to o_1} \rhd \mathrm{Spy}_{\iota\to o},$$

$$\mathrm{Spy}_{\iota_1\to o_2} \blacktriangleright \mathrm{Spy}_{\iota\to o_1} \rhd \mathrm{Spy}_{\iota\to o}.$$

[22] One can ask about using '$\rhd$' in defining Russell-characterization instead of '$\blacktriangleright$'. If the type β is nonfunctional, this is merely a change in notation. Otherwise the question arises of there being a multiplicity of types of instances of *concept of* relations, not just the Fregean and Russellian ones I am discussing. I haven't explored this.

Notice that on this account a concept can be a concept of more than one thing. But this does not really violate the idea that *concept of* is functional, because the types are different; a concept can never be a concept of two different things in the same sense of 'concept of'. A simple illustration of this is that the *de dicto* sense of an ordinary predicate (say 'is a giraffe') is a Frege-concept of the denotation of that predicate, which is a function from giraffes to truth and from nongiraffes to falsehood (and perhaps from zip to zip). But it is also automatically the *de re* sense of another (possible) predicate, one whose denotation would map *concepts* of giraffes to truth and concepts of nongiraffes to falsehood (and perhaps the concept of a zip to zip). Thus it is a Russell-concept of that denotation. So it is a Frege-concept of one thing, and a Russell-concept of another.

5. LEARNABILITY AND THE HIERARCHY

5.1. Learnability

Frege's semantics with its infinite hierarchies of senses has often been used to illustrate a problem about learnability of language (cf. *Davidson 1965*). If people learn language by learning the meanings of words, and if senses are meanings, and if each word has an infinite number of different senses, then language is apparently unlearnable. Having learned the meaning of 'giraffe' for 'Sue is a giraffe', it would appear that we would not yet know the meaning of 'giraffe' in 'George thinks that Sue is a giraffe', let alone 'She didn't realize that he thought that Sue is a giraffe'. But this seems to fly in the face of the data; multiple embeddings do not appear to cause difficulty for the understanding of the simple words therein.

This difficulty is reflected in Church's development by the fact that to every word of natural language there corresponds an infinite number of distinct primitive words. Church's language is (apparently) unlearnable because it has an infinite number of primitives, and this captures the idea that its Fregean source is a theory of natural language that, were it true, would make natural language unlearnable too. This is the issue to be discussed in this section.

The issue arises from the most usual interpretation of Fregean semantics according to which when a word or phrase that is already in an indirect context is embedded in an additional indirect context, it comes to refer to the sense that it expressed before reembedding, and it thus comes to express a new sense; this is a consequence of the ubiquitous principle that a word or phrase always refers (in any given context) to whatever its sense (in *that* context) is a concept of. Church's development of Frege's *scientific* notation is independent of this hypothesis (as I will argue in the next section), and the hypothesis itself is subject to doubt (as I will also argue in the next section). But the issue is worth discussing in any event. So for purposes of discussion here I adopt two hypotheses. The first is that natural language obeys Frege's principle that reembeddings in indirect contexts cause additional shifts in sense and reference. The second is that it is desirable to study a scientific notation that reflects this process by assuming that whenever we have the same root symbol occurring in the scientific notation with varying type subscripts, these combinations represent the uses of a single word of natural language in various degrees of embeddings. Then an example would be the string of symbols:

$$\mathrm{Spy}_{\iota \to o}, \mathrm{Spy}_{\iota_1 \to o_1}, \mathrm{Spy}_{\iota_2 \to o_2}, \ldots.$$

(For simplicity here I ignore *de re* senses.) The first denotes what an unembedded use of 'is a spy' denotes, the second denotes the sense that an unembedded use of 'is a spy' expresses (which is what a singly embedded use of 'is a spy' denotes), the third denotes the sense that a singly embedded use of 'is a spy' expresses (which is what a doubly embedded use of 'is a spy' denotes), and so on. This linkage between the scientific notation and natural language is not inherent in the scientific notation itself, but it is worth exploring as a possible application.

The claim that natural language is unlearnable according to Frege's theory of natural language is easily undercut by presuming that humans have an innate ability to predict the reembedded sense of a word of natural language from its unreembedded use. The fact that a sense of a newly embedded sign is itself new to the hearer is no more a problem than is the fact that the sense of a novel combination of words is new to the hearer; in both cases we have a (partially) innate ability to get the new sense from the already familiar inputs. Language understanding could work this way, and if it does, there is no special problem of learnability; there is just the usual situation that speakers need to be able to project the meanings of novel constructions from familiar ones.

Interest then shifts to this innateness hypothesis, which has consequences that are worth studying. One consequence seems to be that if this innate ability exists, it must be independent of the particular human language in question. So let us consider a "context free" form of that hypothesis:

> Given any sense whatsoever, there is a unique ascendent of that sense, such that if any word in any language expresses that sense in any context, then if that context is reembedded one more time, that word in that language in that reembedded context expresses that ascendent.

If this is so, there exists a function that maps senses to their ascendents, which obey this hypothesized pattern. We can represent this function in the scientific notation, and ask what are the logical constraints on it. Suppose we introduce a symbol '$\Uparrow$' for this function. Suppose the (*de dicto*) sense expressed by 'is a giraffe' in ordinary contexts is something, g, of type $\iota_1 \to o_1$. Then when 'is a giraffe' is embedded, its sense must be $\Uparrow(g)$, which will be of type $\iota_2 \rightarrow o_2$. And when it is reembedded, its sense becomes $\Uparrow(\Uparrow(g))$, of type $\iota_3 \rightarrow o_3$. And so on.

With $\Uparrow$ in the scientific notation we no longer need multiple occurrences of any *other* symbol with the same root and different type subscripts to represent primitive non-logical predicates. Except for the lowest subscript of the form α_1, all symbols with higher subscripts can be replaced by occurrences of the sign with the lowest-sense subscript preceded by applications of $\Uparrow$. For example, we can replace the infinite sequence of scientific symbols:

$$\mathrm{Spy}_{\iota\rightarrow o}, \mathrm{Spy}_{\iota_1\rightarrow o_1}, \mathrm{Spy}_{\iota_2\rightarrow o_2}, \mathrm{Spy}_{\iota_3\rightarrow o_3}, \ldots$$

by: $\quad \mathrm{Spy}_{\iota\rightarrow o}, \mathrm{Spy}_{\iota_1\rightarrow o_1}, \Uparrow(\mathrm{Spy}_{\iota_1\rightarrow o_1}), \Uparrow(\Uparrow(Spy_{\iota_1\rightarrow o_1})), \ldots$ [23]

So the only remaining source of infinitude in the primitive vocabulary would be the infinite number of distinct occurrences of '$\Uparrow$' itself, with *their* distinct type subscripts. And these can all be reduced to one if '$\Uparrow$' is construed as what Church calls an "improper" symbol, that is, as untyped. This would be justified on the grounds that '$\Uparrow$' represents an innate combinatory ability, and is not the normal sort of item that mirrors the uses of items of natural language vocabulary.

What about complex signs; can these too be reduced? Yes, if we assume that the $\Uparrow$ function satisfies two principles of compositionality, one for functional application and one for abstracts. Suppose we define:

$$x \text{ is a sense } =_{\mathrm{df}} \exists y[x \blacktriangleright y].$$

Then the conditions we want are:

1. If $f_{\alpha_1\rightarrow\beta_1}$ is a sense and a_{α_1} is a sense then

$$\Uparrow(f_{\alpha_1\rightarrow\beta_1}(a_{\alpha_1})) = \Uparrow(f_{\alpha_1\rightarrow\beta_1})(\Uparrow(a_{\alpha_1})).$$

2. If $\lambda x_{\alpha_1}[M_{\beta_1}]$ is a sense then

$$\Uparrow(\lambda x_{\alpha_1}[M_{\beta_1}]) = \lambda x_{\alpha_2}[\Uparrow(M_{\beta_1})].$$ [24]

[23] The intent of the hypothesis is clearly that $\Uparrow(s)$ is always a concept of s.

[24] There is a complication here: the variable following 'λ' on the left hand side is 'x_{α_1}' and that on the right is 'x_{α_2}', which apparently does not bind the 'x_{α_1}' that remains inside 'M_{β_1}'. For example, an instance of (2) is:

$$\Uparrow(\lambda x_{\iota_1}[\mathrm{Spy}_{\iota_1\rightarrow o_1}(x_{\iota_1})]) = \lambda x_{\iota_2}[\Uparrow(\mathrm{Spy}_{\iota_1\rightarrow o_1}(x_{\iota_1}))],$$

which, after applying condition (1) to the inside yields:

$$\Uparrow(\lambda x_{\iota_1}[\mathrm{Spy}_{\iota_1\rightarrow o_1}(x_{\iota_1})]) = \lambda x_{\iota_2}[\Uparrow(\mathrm{Spy}_{\iota_1\rightarrow o_1})\Uparrow(x_{\iota_1})].$$

Here the rightmost 'x_{ι_1}' is apparently unbound. This oddity cannot be removed *merely* by changing the 'x_{ι_1}' to 'x_{ι_2}', for then the resulting '$\Uparrow(x_{\iota_2})$' is the wrong type to be an argument of '$\Uparrow(\mathrm{Spy}_{\iota_1\rightarrow o_1})$'. There are a number of ways of handling this issue. The simplest is the most artificial; this is to change 'x_{α_1}' to 'x_{α_2}' on the right, and decree that when the argument of '$\Uparrow$' is a variable the result is the same as if '$\Uparrow$' was not there; '$\Uparrow(x_{\iota_2})$' is automatically just the same as 'x_{ι_2}'. This *ad hoc* move seems to work adequately. (This approach requires a slight modification of condition (1).)

A good application of (1) tells us, e.g., that the ascendent of the customary sense of 'Agatha is a spy' is a function of the ascendents of the customary senses of its parts:

$$\Uparrow(\mathrm{Spy}_{\iota_1\to o_1}(\mathrm{Agatha}_{\iota_1})) = \Uparrow(\mathrm{Spy}_{\iota_1\to o_1})\big(\Uparrow(\mathrm{Agatha}_{\iota_1})\big).$$

Recall Church's policy that uniformly raising the type subscripts on any primitive symbol by 1 produces a name of the (*de dicto*) sense of the original symbol. If this is combined with the assumptions above, we can derive that if F is a closed formula then we have:

$$\Uparrow(F) = F_1,$$

where F_1 is what you get by raising every type subscript on every symbol in F by 1, *provided that* every primitive symbol in F denotes a sense; otherwise this is not true.

The indirect sense of 'what John said' is thus the ascendent of its customary sense. But this does *not* mean that the ascendent of the customary sense of 'what John said' is represented by placing '$\Uparrow$' in front of 'what John said':

$$\Uparrow((\iota_{(o_1\to o)\to o}x_{o_1})[\mathrm{Said}(j, x_{o_1})]).$$

(This does not satisfy the restrictions stated for condition (1) above.) Suppose that what John said is that Agatha is a spy, which is also what Mary said. Then the ascendent of what John said is the ascendent of the proposition that Agatha is a spy, which is $\Uparrow(\mathrm{Spy}_{\iota_1\to o_1}(\mathrm{Agatha}))$:

the ascendent of what John said = the ascendent of what Mary said
= $\Uparrow$(the proposition that Agatha is a spy),

But the ascendent of the *sense* of 'what John said':

$$\Uparrow(\iota_{(o_2\to o_1)\to o_1}x_{o_1}[\mathrm{Said}(j, x_{o_1})]),$$

Another approach is both more natural and more complicated. We suppose that variables come in sequences, distinguished within the sequence by their type subscripts. So for any type α we consider not just a variable 'x_α' of that type, but the sequence $x_{\alpha_0}, x_{\alpha_1}, x_{\alpha_2}, x_{\alpha_3}, \ldots$ of variables with the common variable-root 'x'. Then we do not make assignments to variables, we make assignments to sequences of variables. An assignment to the sequence $x_{\alpha_0}, x_{\alpha_1}, x_{\alpha_2}, x_{\alpha_3}, \ldots$ assigns a chain of things to the sequence, assigning to each '$x_{\alpha_{i+1}}$' a concept of whatever is assigned to x_{α_i}. Further, it must assign to each $x_{\alpha_{i+1}}$ the unique concept which is $\Uparrow$(the thing assigned to x_{α_i}) for all $i \geq 0$. Then when we give the conditions for quantification or abstraction we do not speak of assignments that differ from a given assignment at most in what they assign to 'x_{α_1}', we speak of assignments that differ from a given assignment at most in what they assign to the whole 'x' sequence of variables. (We must take care in writing formulas then to use distinct variable-roots in distinct abstracts so as to avoid an extended kind of clash of bound variables.) Then we leave condition (2) exactly as is, since the variable 'x_{α_2}' following the 'λ' in effect binds the variable 'x_{α_1}' inside the abstract, and the condition yields the results we want. (If this approach is adopted in detail, it will probably require the refined types discussed in section 5.2 below.)

is something else entirely. The ascendent of the sense of 'what John said' $\neq$ the ascendent of the sense of 'what Mary said'.

The above considerations can naturally be duplicated for *de re* ascendents of senses alongside of their *de dicto* versions. This leaves each functional word having two basic senses, *de dicto* and *de re*, with the rest projected from these, and one wonders about reducing this pair to one. This is easily accomplished if one does not expect an informative analysis. One could simply hypothesize that the *de dicto* meaning of a word determines its *de re* meaning, and introduce a "sideways projection" function symbol for this. Then all of the various uses could be reduced to one.

5.2. *Refining the Type Hierarchy*

The above considerations suggest a more refined structure for the type hierarchy. Compare the senses expressed by '*snow is white*' and by '*what John said*' in these two examples:

Mary believes that snow is white,

Mary believes what John said.

Both are of type o_2, but the former is an indirect sense and the latter a customary sense. That is, the former is the sense of an embedded phrase, and the latter is the sense of a non-embedded phrase. In addition, the latter is *not* the indirect sense of any phrase in any language. This can be shown as follows.

Assume that what John said was that snow is white. It is apparent that the *de dicto* senses of

Mary believes what John said,

and

Mary believes that snow is white

are different. The matrices of these sentences are the same, and so the customary sense of 'what John said' must be different from the indirect sense of 'snow is white'. Represent the direct use of the sentence 'snow is white' by S_o, which refers to a truth value. Its singly indirect reference (its customary sense) is represented by S_{o_1}, which refers to the proposition that snow is white, and its doubly indirect sense is represented by S_{o_2}, which refers to a concept of the proposition that snow is white. Represent the direct use of 'what John said' by W_{o_1}, because the phrase refers to a proposition. (In fact, it refers to the proposition that snow is white, so $W_{o_1} = S_{o_1}$.) Its singly indirect reference is denoted by W_{o_2}, and we argued above that $W_{o_2} \neq S_{o_2}$.

Now suppose for *reductio* that the customary sense of 'what John said' is the indirect sense of some phrase X, so that $W_{o_2} = X_{o_2}$. Since the indirect sense of a phrase is the result of operating on its direct sense by $\Uparrow$, this entails that $X_{o_2} = \Uparrow(X_{o_1})$. However, $X_{o_1} = S_{o_1}$. (This is because X_{o_1} = whatever X_{o_2} is a concept of, which is whatever W_{o_2} is a concept of, which is W_{o_1}, i.e., S_{o_1}.) So, substituting S_{o_1} for X_{o_1}, we conclude that $X_{o_2} = \Uparrow(S_{o_1})$. But $\Uparrow(S_{o_1}) = S_{o_2}$. So we proved that $W_{o_2} = S_{o_2}$, contrary to the data from which we started. ⊣

If this argument is generalizable (or even if its conclusion is consistently generalizable by hypothesis) this leads naturally to the idea that senses are not just categorized into types by Church's type symbols, they also come stratified *within* these types by the linguistic level of symbols that could express them. E.g., a given sense might be inherently a kind of thing that is expressible by a word or phrase of natural language embedded n times, and not expressible by any word or phrase embedded m times, with $m \neq n$. This then suggests a different treatment of the "concept of" relation for functional types. A functional sense of type $\alpha_1 \rightarrow \beta_1$ should not be considered to be a partial function defined for concepts of type α_1, it should be defined for senses that are concepts of type α_1 *and* of linguistic level n. This suggests an interesting complication of the theory which could very well lead to fruitful results. I don't pursue it here partly for reasons of space, and partly because I suspect that there is a better and simpler way to proceed. This is articulated in the last section.

6. A SIMPLER ACCOUNT

The idea that there is a hierarchy of linguistic levels of sense corresponding to the levels of embedding in indirect contexts predicts distinctions that are not readily apparent in the data. Their absence needs to be accounted for. I suggest that the best account of natural language is one that is far simpler than the one contemplated in the last section.

Here is one small but important consideration that leads in this direction, discussed in *Bealer 1982*, pages 38–41.[25] The following argument seems to be well-formed and valid:

Mary believes everything John thinks she believes,

John thinks she believes that snow is white,

∴ Mary believes that snow is white.

The types of the symbols in the first premise are constrained by its form, which seems to be:

$$\forall x[j\text{Thinks}(m\text{Believes}(x)) \rightarrow m\text{Believes}(x)].$$

[25] Bealer argues that the argument is not representable in Church's system, but this is the version of the system as used by Church; in the previous section I argue that Church's system permits "Russellian" *de re* constructions, which is what Bealer urges should be used.

The first occurrence of x is in a doubly embedded context, and the second is in a singly embedded context. So how can the sentence be well-formed? The answer must be that the embedded 'believe' refers to the *de re* sense of its customary use. The type of x is o_1, and the types of the predicates are

$$x[j\text{Thinks}(m\text{Believes}(x)) \rightarrow m\text{Believes}(x)]$$

$$o_1 \rightarrow o \quad o_1 \rightarrow o_1 \qquad\qquad o_1 \rightarrow o$$

The second premise has the form:

$$j\text{Thinks}(m\text{Believes}(\text{Snow is white}))$$

$$o_1 \rightarrow o \quad o_1 \rightarrow o_1 \qquad o_1$$

and the argument is valid.

However, this example leads to a question about that-clauses. The logical symbolism seems to permit a construal of the second premise whereby it is well-formed and the argument is *not* valid. Yet this does not seem correct for the original argument in English. The invalid reading is easily illustrated by substituting for 'snow is white' the phrase 'what Sam said'. The argument is then

Mary believes everything John thinks she believes,

John thinks she believes what Sam said,

∴ Mary believes what Sam said.

It is now possible to read the second premise in such a way that the argument is not valid. The theory yields this reading by construing 'believes' in the second premise *de dicto*:

$$j\text{Thinks}(m\text{Believes}(\text{what Sam said}))$$

$$o_1 \rightarrow o \quad o_2 \rightarrow o_1 \qquad o_2$$

The fascinating question posed by Bealer is why such a reading is not available when we use a that-clause. That is, why is there *not* a construal of the second premise with these types:

$$j\text{Thinks}(m\text{Believes}(\text{Snow is white})),$$

$$o_1 \rightarrow o \quad o_2 \rightarrow o_1 \qquad o_2$$

which yields a form according to which the argument is not valid?

There are two natural answers to this question. One is that the invalid construal is somehow equivalent to the earlier one that makes the argument valid. The other is that the invalid construal does not exist in the language. I consider these in turn.

6.1. The Equivalence Hypothesis

Suppose that the English sentence does permit a doubly shifted reading, but one that is *equivalent* to the singly shifted reading. The assumption is that these are equivalent:

$$j\text{Thinks}\big(m\text{Believes(Snow is white)}\big)$$

$$o_1 \to o \quad o_2 \to o_1 \qquad o_2$$

$$o_1 \to o \quad o_1 \to o_1 \qquad o_1$$

If the pattern is general, it must be because the *de dicto* sense of 'believes' maps the doubly embedded sense of 'Snow is white' to the same thing that the *de re* sense of 'believes' maps the singly embedded sense of 'Snow is white' to. In symbols:

$$B_{o_1 \to o_1}(S_{o_1}) = B_{o_2 \to o_1}(S_{o_2})$$

If the example is further complicated by an additional embedding, the result is that these are equivalent:

$$a\text{Regrets}j\text{Thinks}\big(m\text{Believes(Snow is white)}\big)$$

$$o_1 \to o \quad o_2 \to o_1 \quad o_3 \to o_2 \qquad o_3$$

$$o_1 \to o \quad o_1 \to o_1 \quad o_1 \to o_1 \qquad o_1$$

Again, if the pattern is general, it must be because the *de dicto* sense of 'thinks' maps what the *de dicto* sense of 'believes' maps the doubly embedded sense of 'Snow is white' to to the same thing that the *de re* sense of 'thinks' maps what the *de re* sense of 'believes' maps the singly embedded sense of 'Snow is white' to to. In symbols:

$$\mathrm{T}_{o_1 \to o_1}\big(B_{o_1 \to o_1}(S_{o_1})\big) = \mathrm{T}_{o_2 \to o_1}\big(B_{o_3 \to o_2}(S_{o_3})\big)$$

It turns out that although 'Snow is white' has an infinite number of distinct *de dicto* senses, and 'believes' and 'thinks' do too, their combination yields exactly the same result as application of the singly embedded *de re* senses of 'thinks' and 'believes' to the singly embedded *de dicto* sense of 'Snow is white'. The infinite complications, after taking us once each around the barn, end up cancelling one another out. This strongly suggests that the simpler hypothesis is on the right track.

6.2. The Simpler Hypothesis

The simpler hypothesis denies that when words or phrases of ordinary language are reembedded in an indirect context they undergo an additional semantic shift. One shift makes words come to denote their senses, and additional shifts have no further effect. This hypothesis requires one minor alteration of Frege's principles for words in natural language. Instead of:

> A word or phrase in any context denotes the unique thing that its sense (in that context) is a concept of,

we need:

> A word or phrase in a direct context denotes the unique thing that its sense is a concept of; a word or phrase in an indirect context denotes its sense.[26]

In terms of the examples surveyed above, the problematic readings simply do not exist, and so there is no problem of accounting for why they are equivalent to the other readings. I think that this is the right answer; the additional semantic shift required of reembedded terms is entirely unmotivated by any data of natural language.[27] Further, the theory is simpler without it, because it requires that words have only customary sense and reference. This also bypasses the learnability issue entirely, since there are no new senses to understand.

This may be Church's view as well; he has expressed the opinion (personal communication):

> "... a name has just *two* kinds of meaning, a denotation and a sense. Indeed the sense of a given name may be denoted by another name, and this latter name may be expected to have a sense of its own. And so on *ad infinitum.* But this is not the same as saying that *one* name may have *three* kinds of meaning (such as, e.g., a denotation, and a sense, and after that also an *indirect sense*, ...)."

I agree entirely. Church's motivation for his axiomatic study does not come from the purported "reembedding causes reshifting" view of what happens in natural language. This view is not articulated in Church's writings. Instead, he simply says things like this (*Church 1951,* page 12, note 13):

> "The hierarchy of concepts of successively higher orders arises as soon as we suppose that a concept, like anything else which can be discussed at all, is capable of having a name given to it. For a sense of a name of a concept is a concept of the next higher order, and so on."

So Church's axiomatic study, with its ontological hierarchy of types is needed even if natural language lacks the linguistic hierarchy popularly attributed to Frege.[28]

---

[26]This change is required for the theory of natural language. No change is required for the scientific notation, which contains no indirect contexts.

[27]I argue this from different perspectives in *Parsons 1981* and *1996*.

[28]I thus take exception to Geach's (*1980,* page 95) claim that "With this view, Church's hierarchy of senses collapses like a house of cards." ("This view" = the view that an unambiguous term has a single sense, and its reference depends on this sense together with the kind of propositional context in which it is embedded.)

7. AFTERWORDS

The preceding paper was completed sometime before 1995. After it was typeset a number of issues arose from conversations with David Kaplan that require some developments in section 3 regarding the question of when a function on concepts is itself a concept. Rather than patching the paper,[29] it seems more useful to describe the problems that need addressing, since it is not clear whether certain changes are necessary, and since they turn on issues that will have to be faced by anyone working in the vicinity of Church's framework. Three issues are discussed here: the reduction to extensionality, the question of modal and epistemic uniformity, and the treatment of complex names having empty names as parts.

7.1. The Reduction to Extensionality

When I wrote the paper I did not know how to carry out the proof of a reduction to extensionality within the object language of *Church 1951*, alternative 2, without making an assumption that every sense can be assigned to a symbol of some language. In fact, a proof can be given, though along quite different lines than the one discussed there.

The proof is short. If there are only two truth-values, then there are only four one-place truth-functions, and they all have this property: that if you apply any one of them three times, it's the same as applying it only once. So the function that maps each monadic truth-function to the result of applying it to a thing once is the very same function that maps each monadic truth-function to the result of applying it to that same thing three times. So:

Let the variable f be of the type $o \rightarrow o$, the type of functions which map truth-values to truth-values. Let T_o be a constant name of type o; for convenience we use the existing name of truth (that is, T_o is a particular logically true sentence). Then we can prove:

$$\lambda f\, f(\mathrm{T}_o) = \lambda f\, f(f(f(\mathrm{T}_o))).$$

In alternative 2, the senses of a provable identity are themselves identical, and in *Church 1951* you produce the name of a sense of a formula by raising all of its type subscripts by 1. So the following must be true where F is of the type $o_1 \rightarrow o_1$, the type of functions that map propositions to propositions:

$$\lambda \mathrm{F}\, \mathrm{F}(\mathrm{T}_{o_1}) = \lambda \mathrm{F}\, \mathrm{F}(\mathrm{F}(\mathrm{F}(\mathrm{T}_{o_1}))).$$

But this is plainly false if there are more than two propositions.

This proof differs from the one discussed in the paper because it does not

[29] I did in fact make some changes in the paper, including correcting a crucial error about zips. The reader is indebted to David Kaplan for sparing him or her some confusion while reading the paper. I am responsible for whatever confusion remains.

even appeal to the troublesome axiom 16. Although Church (*1973, 1974*) does not indicate that any problems arise independent of axiom 16, they are all elegantly solved in the way proposed in his paper, which is to add a type subscript to 'λ' itself. The basic 'λ' of type zero obeys lambda-conversion; the type-raised versions of 'λ' do not—they thus require special treatment. With type subscripts in place, the proof given above has a different result; it yields:

$$\lambda_1 \mathrm{F}\, \mathrm{F}(\mathrm{T}_{o_1}) = \lambda_1 \mathrm{F}\, \mathrm{F}(\mathrm{F}(\mathrm{F}(\mathrm{T}_{o_1}))).$$

The effect of the subscript on 'λ' is as if it restricts the resulting function to one that is not defined on its whole natural domain; it is defined only for those things in its domain that are concepts of something. This restriction was accomplished in Church (*1973, 1974*) by designing the axioms in such a way that lambda-conversion is obeyed by the subscripted lambda for arguments that are concepts, and not for others. In the spirit of the zip formulation in my paper, the subscripted lambda can be introduced by definition as the function which behaves as expected if its argument is a concept of something, and otherwise yields the appropriate zip. That is, it can be defined in context in this way:

$$\lambda_1 f_{\alpha_1}\, \phi_{\beta_1} =$$
$$\iota \mathrm{F}_{\alpha_1 \to \beta_1} \forall f_{\alpha_1} [[\neg \exists g_\alpha f_{\alpha_1} \blacktriangleright g_\beta \,\&\, \mathrm{F}_{\alpha_1 \to \beta_1}(f_{\alpha_1}) = \text{✿}_{\beta_1}] \vee \mathrm{F}_{\alpha_1 \to \beta_1}(f_{\alpha_1}) = \phi_{\beta_1}].$$

It is then most efficient to follow Church's lead, and produce this modified axiom:

$$\text{SUPER-15-16: } \mathrm{F}^1 \blacktriangleright f \equiv \forall A \forall a [A \blacktriangleright a \to \mathrm{F}(A) \blacktriangleright f(a)]$$

where F^1 is defined to be $\lambda_1 x\, \mathrm{F}(x)$.

7.2. Kaplan's Argument for Necessary Uniformity

Kaplan (*1964*) indicates another problem with axiom 16. That axiom requires that for a function to be a concept of something it must actually characterize that thing - but the axiom does not require that this be necessarily true. Because of this, an argument can be given[30] which shows that this lack of what Kaplan calls "necessary uniformity" conflicts with certain natural principles about modality (provided that such a nonnecessarily uniform sense is actually assigned to some sign of the language). This problem differs from the one discussed above in that (1) it is not solved by the device discussed above, and (2) it arises independently of the special principles of alternative 2, and so it extends to a wide range of phenomena.

In support of point (2) is the consideration that all that is required by the argument is that provable sentences be subject to necessitation, and that necessitation be achieved by producing a name of the sense of the proved formula, which can be formed in some regular way (e.g., by type-raising). Then

[30] The argument was personally communicated to me by David Kaplan.

you need a couple of very plausible modal principles. Nowhere does the argument appeal to a principle that identifies necessarily equivalent senses. For this reason, the argument can be applied even when we distinguish necessarily equivalent senses. And we want to distinguish such senses in order to accommodate the well-known phenomenon that certain necessary truths, e.g., certain complex truths of arithmetic, are not automatically known by virtue of knowing simple ones.

Here is how a problem about knowledge arises, similar to the modal example discussed by Kaplan. Suppose that we have a system of axioms for sense and denotation and that we have already patched axiom 16 in accordance with the modal constraint. Then define a function gamma that maps each necessary proposition to the sense of a trivial tautology, maps each impossible proposition to the sense of a trivial impossibility, and maps contingent propositions to themselves. For example, let 'T_o' be some simple tautology, and assume that 'T_{o_1}' denotes its sense. Let gamma map necessary truths to T_{o_1}, necessary falsehoods to $\neg_1 T_{o_1}$, and all other proposiitons to themselves. (Notice that gamma is already necessarily uniform, as proposed by Kaplan.) Let G express this function gamma as its sense, so that 'G_1' denotes gamma, and 'G' denotes the unique function that gamma is a concept of (which is in fact the function that maps each truth-value to itself, though that detail is not relevant to the argument). Let 'S_o' be any sentence, and let 'K' stand for 'you know' (so 'K' denotes a function that maps the propositions that you know to truth, and the rest to falsehood).

Now notice that the following is an instance of Leibniz's Law:

$$S_o = T_o \to G(S_o) = G(T_o).$$

Reflecting on this fact, and understanding the senses of the ingredients, you should know the proposition expressed. So this should be true:

$$K[S_{o_1} =_1 T_{o_1} \to_1 G_1(S_{o_1}) =_1 G_1(T_{o_1})].$$

But the following schema seems to be a plausible principle of epistemic logic (at least it holds if you reflect on your knowledge for this case):

$$K[A_1 \to_1 B_1] \to \neg K[\neg_1 A_1] \to_1 \neg K[\neg_1 B_1].$$

This schema together with *modus ponens* applied to our result above yields:

$$\neg K[\neg_1 S_{o_1}] \to \neg K[\neg_1 [G_1(S_{o_1}) =_1 G_1(T_{o_1})]].$$

This argument holds good no matter what sentence 'S_o' is. So suppose that 'S_o' expresses some necessarily false proposition which you do not know to be false—that is, you don't know its negation; it is one of those recondite necessary falsehoods of arithmetic that only afficionados are aware of. Then, the antecedent of what we have proved is true. So by *modus ponens* we infer:

$$\neg K[\neg_1 G_1(S_{o_1}) =_1 G_1(T_{o_1})].$$

However, by hypothesis, 'G_1' denotes gamma, and so $G_1(S_{o_1}) = \neg_1 T_{o_1}$ and $G_1(T_{o_1}) = T_{o_1}$. Substituting, we get:

$$\neg K[\neg_1[\neg_1 T_{0_1} =_1 T_{o_1}]].$$

But this is factually wrong; you are not ignorant of the truth: $\neg_1[\neg_1 T_{o_1} =_1 T_{o_1}]$. So something has gone wrong somewhere.

One might think that we need to repair this by adding an epistemic condition to axiom 16 in a way parallel to Kaplan's suggestion about the modal case. We need not just necesssary uniformity, but also epistimic uniformity. The condition we add would have to be of the form:

$$\forall a_{\alpha_1} \forall b_{\alpha_1} K[a =_1 b \rightarrow_1 F_{\alpha_1 \rightarrow \beta_1}(a) =_1 F_{\alpha_1 \rightarrow \beta_1}(b)].$$

The problem with this strategy is that there appear to be hosts of other examples waiting in the wings, each of them requiring additional patches to axiom 16. We will need additional constr aints for deontic logic, for the logic of locutions involving desirability, perhaps also for probability. This worry motivates one to consider alternative solutions.

As an alternative, suppose that we do not expect our intensional logic to rule out the existence of concepts like gamma. There is another place to cut off the reasoning. In the modal case, perhaps we were too hasty in adopting unrestricted necessitation. The fact that a sentence is provable means that we are sure that it expresses a truth, but this need not be because it expresses a *necessary* truth. In fact, in the modal argument given by Kaplan, the provable sentence expresses a proposition that is *not* necessary, as the follow-up argument demonstrates. In the epistemic case something similar happens; there, even though you know *that* the sentence in question expresses a true proposition, you do not thereby know the proposition that it expresses. This is possible because you do not know *which* proposition the sentence expresses. Knowing that the sentence is true does not foster knowing the proposition it expresses if which proposition it expresses is not apparent from the wording of the sentence itself.

This argument, however, appears to cut both ways. The reason that you cannot infer knowledge of the expressed proposition from knowledge that the sentence expresses a truth is that a natural *apparency* condition is not satisfied: you do not know *which* proposition is expressed. And this is so even though you know the meanings of the parts of the sentence. The problem with the predicate G is not that you do not know its meaning; the problem is that knowing its meaning does not put you in a position to know what gets expressed when you combine it with other phrases whose meaning you also know. So it can be used to construct sentences whose meanings are not apparent from the meanings of their parts. But doesn't that explain why we want to rule out the meaning of G as a posssible sense? Perhaps. But we don't know how to develop a theory of the ontology of senses that will rule out all of the senses that generate problems like this. Instead, we might permit them in the ontology and then be careful to take their pathological behavior into account when formulating our rules of inference. For example, modal necessitation should apply to sentences that consist only of signs with modally uniform senses, epistemic ascent should apply to sentences that consist only of signs that satisfy the appropriate epistemic condition, and so on. This is as easy to implement as it is to introduce such constraints on the axiom for the existence of concepts.

The advantage in approaching the problem in this way is that we face the particular problems exactly when they naturally arise. The epistemic problem arises when we add to out axioms a special inference rule of epistemic ascent. When we add such a rule, we need to ask whether it is valid, and under what conditions. And here may be the best place to face and deal with special constraints.

7.3. Zip-ity Doo-Dah

The problem of empty names is raised by Frege's discussion of "Odysseus was set ashore at Ithaca while sound asleep". Frege argues there that if a name lacks reference, so must a complex sign containing that name; so the failure of reference of 'Odysseus' explains why the sentence containing it lacks an actual truth-value. This has consequences for the senses assigned to complex names. Church (*1973, 1974*) puts it succinctly in his footnote 5: "if the sense-value assigned to one of the variables is not a concept of anything ... then the sense-value of the entire formula must correspondingly not be a concept of anything ..."

I suggested addressing this problem by having intuitive failure of reference be emulated by technical-reference to a zip. However, I did *not* end up with a symbolism in which an intuitively empty name (one denoting a zip) forces a complex name containing it also to be empty (to denote a zip). A simple counterexample is negation, which produces a true sentence from one denoting a truth-value zip. A choice must be made here. If you insist that the language obey the principle that concepts of zips beget concepts of zips, then you end up with a working logic that is so impoverished it is hard to formulate strong general principles within it. I didn't see how to do this without departing substantially from Church's system of axioms, and so I introduced a three-valued logic that departs minimally from his, and is strong and useful in the way that it handles zips. But then it does not obey the principle that zips beget zips. (It does not obey the principle that a name must denote a concept of a zip if a part of that name denotes a concept of a zip.) A language that emulates English should then be constrained by being limited to vocabulary that obeys the "zips beget zips" principle. At least, it should be limited in this way if Frege is right. (And that is what I had in mind in the Superman example.)

Frege's insight should be divided into two precepts. One is that all failure of reference (as opposed to oblique contexts) should be treated alike, so that e.g., if one sentence that lacks truth-value has a true negation, they should all behave this way (if negation does not create obliquity). The other is that reference failure should propogate from simple names to complex names containing them. I implemented the first of these without the second. This is not clear from my paper because I was not clear when I wrote it.

ACKNOWLEDGEMENTS

I am indebted to Alonzo Church for his gracious replies to some naive questions about his systems, and to David Kaplan for various discussions about sense and reference over the years. I also thank the UCI L&M group: Peter Woodruff, Penny Maddy, and Jeff Barrett for criticisms and suggestions.

REFERENCES

Anderson, C. A.

1984 General intensional logic, *Handbook of philosophical logic II* (D. Gabbay and F. Guenthner, editors), Reidel, Dordrecht, pp. 355–385.

Bealer, G.

1982 *Quality and concept,* Oxford University Press, Oxford.

Church, A.

1951 A formulation of the logic of sense and denotation, *Structure, method, and meaning: Essays in honor of Henry M. Sheffer* (Paul Henle, H. M. Kallen, and S. K. Langer, editors), Liberal Arts Press, New York, pp. 3–24.

1973 Outline of a revised formulation of the logic of sense and denotation (Part I), *Nous,* vol. 7, pp. 24–33.

1974 Outline of a revised formulation of the logic of sense and denotation (Part II), *Nous,* vol. 8, pp. 135–156.

1993 A revised formulation of the logic of sense and denotation, Alternative (1), *Nous,* vol. 27, 141–157.

Davidson, D.

1965 Theories of meaning and learnable languages, *Proceedings of the 1964 international congress for logic, methodology and philosophy of science* (Y. Bar-Hillel, editor), North-Holland, Amsterdam.

Frege, G.

1891 Funktion and Begriff (Function and Concept); translated in *Frege 1960.*

1892 Über Sinn und Bedeutung (On Sense and Reference); translated in *Frege 1960.*

1902 Letter to Russell, 28 December 1902, in *Frege 1980,* pp. 152–154.

1904 Letter to Russell, 13 November 1904, in *Frege 1980,* pp. 160–166.

1914? Letter to Jourdain, undated, in *Frege 1980,* pp. 78–80.

1960 *Translations from the philosophical writings of Gottlob Frege* (P. Geach and M. Black, editors), Blackwell, Oxford.

1980 *Gottlob Frege, Philosophical and mathematical correspondence,* (G. Gabriel, H. Hermes, F. Kambartel, C. Thiel, and A. Veraart, editors; H. Kaal, translator), University of Chicago Press, Chicago.

Geach, P.

1980 Some problems about the sense and reference of proper names, *New essays in philosophy of language,* (F. J. Pelletier and C. G. Normore, editors), Canadian Association for Publishing in Philosophy, Guelph, Ontario, pp. 83–96.

Kaplan, D.

1964 Foundations of intensional logic, *Dissertation,* University Microfilms, Ann Arbor.

1973 The Church reformation, a talk given to the Western Division APA.

Parsons, T.

1981 Frege's hierarchies of indirect sense and the paradox of analysis, *Midwest Studies in Philosophy,* vol. VI, pp. 37–57.

1996 Fregean theories of truth and meaning, *Frege*: *Importance and legacy* (M. Schirn, editor), Walter de Gruyter, New York and Berlin.

TP Theory of meaning for natural language, forthcoming in a volume on Reference (A. Grayling, editor), Oxford University Press, Oxford.

MARK RICHARD

ANALYSIS, SYNONYMY AND SENSE

Abstract. The paradox of analysis is this: some analyses ought be informative; but since analysands and analysandum must be synonymous, a correct analysis will be trivial and thus non-informative. This paper begins by distinguishing two sorts of synonymy, phrasal and structural. The distinction resolves the paradox: only substitution of structural synonyms is guaranteed to preserve what is said; only phrasal synonymy is required in analysis. An account of analysis is sketched, and the solution defended against objections. Church's well-known Fregean solution is compared and two objections are brought against it. A problem, which the paradox presents for Fregeanism, is discussed. The Fregean can avoid the problem if he is willing to abandon the claim that sense is transparent (viz., the competent speaker knows on reflection, of terms with the same sense, that they mean the same). An argument against this claim is developed. An appendix compares the notion of structural synonymy with Church's notion of synonymous isomorphism, and discusses the exchange of Church, White, and Black on the paradox.

0.

Moore's concise statement of the paradox of analysis asks us to consider the statement one would make with

(1) To be a brother is to be a male sibling,

a statement he supposes to be a fair example of an analysis. Moore (*1942*, page 665) observes

> The paradox arises from the fact that, *if* this statement is true, then it seems as if it must be the case that you would be making exactly the same statement if you said:
>
> (2) To be a brother is to be a brother.
>
> But it is obvious that these two statements are *not* the same; and obvious also that nobody would say that by asserting "To be a brother is to be a brother" you were giving an analysis of the concept "brother".[1]

The problem seems to rest on four assumptions:

> Identity: An analysis is what is said by a sentence of the form *To be an A is to be a B* (or by one of the forms: *Being an A is being a B*, *The*

[1]For ease of future reference, I have taken the liberty of displaying and numbering the sentence 'to be a brother is to be a brother' in the citation; Moore quoted it.

C. Anthony Anderson and M. Zelëny (eds.), *Logic, Meaning and Computation*, 545-571.

concept A is the concept B, To V is to V′, or *To have an A is to have a B*). In such sentences 'is' names the relation of identity.

Synonymy: When such a sentence expresses a correct analysis, A and B (and thus 'being an A' and 'being a B', etc.) are synonymous.

Compositionality: Substitution of synonyms for synonyms, at least in a position like that of 'to be a B' in the above, does not affect what a sentence says.

Triviality: Genuine and correct analysis is not trivial, but what is said by a sentence of the form

To be an A is to be an A,

if 'is' therein names a relation of identity, *is* always trivial.[2]

These are inconsistent with the claim that there are some genuine, correct analyses.

I will begin by suggesting that we have two notions of synonymy, *phrasal* and *structural* synonymy. If I am correct about this, the paradox of analysis has a simple solution, since it turns out that if by 'synonymous' we mean 'phrasally synonymous', Synonymy is true but Compositionality is false; if we mean 'structurally synonymous', Compositionality is true, Synonymy false. Section 1 develops this response. Section 2 responds to some objections and discusses relations between analysis and synonymy. Section 3 criticizes a Fregean account of the paradox due to Church. Section 4 argues that a Fregean can coherently adopt the solution to the paradox of analysis sketched in section 1. It also poses a problem for Fregeans involving the paradox. I argue that a Fregean best solves the problem by abandoning such principles as

Transparency: It is impossible for a (normal, rational) person to understand expressions which have identical senses but not be aware that they have identical senses (or be such that he would immediately come to know this if he were to reflect on it).

I argue that Transparency is false, and that it in fact adds to the plausibility of Fregeanism if it is explicitly rejected. A corollary to this is that constraints such as:

Frege's Constraint: It is impossible for a rational person to believe an object to be F under a sense or mode of presentation m while simultaneously disbelieving it to be so under m,[3]

---

[2]This formulation of the paradox assumes that it is propositions or thoughts—the claims made by sentence utterances—which are trivial or not trivial.

[3]This is a simplification of a principle Stephen Schiffer calls 'Frege's Constraint'. For discussion, see *Schiffer 1990*.

can coherently, and should, be rejected by the Fregean.

1.

Synonymy, at least in the Frege-Russell tradition, is closely tied to what is said, with expressions synonymous if they inevitably make the same contribution to what is said by sentences in which they appear. Here, 'inevitably' indicates two degrees of variation, as an expression may appear in different sentences and may be used in different speech situations (*contexts*, as they are sometimes called). Such accounts of synonymy presuppose that there is a relation—call it the semantic value relation—which relates a contribution to what is said, an expression, and a context of use. Semantic value echoes syntactic kind to the extent that the class of semantic values of each syntactic kind forms a (more or less) natural "semantic kind". For example, the semantic values of nouns are all properties, where a property is (minimally) something which (relative to a way things might be) determines a set of individuals. Syntactic form and semantic values are supposed to determine what is said in a broadly compositional fashion. At the least, sentences that differ only in what lexical items they contain will say the same thing, if the lexical items on which they differ have the same semantic values. Given all this, a standard account of synonymy has it that expressions are synonymous if and only if they have in all contexts the same semantic value.[4]

This account of synonymy seems to me at best incomplete. In order to explain what is missing and to introduce the contrast between phrasal and structural synonymy, let us for the moment accept a broadly Russellian view of proper names, on which what a name contributes to what is said—the semantic value of a name—is identified with its bearer. Consider, now, the complex propositional name:

(3) that all men are created equal.

This contributes a certain proposition—"Jefferson's Doctrine", we might call it[5]—to what is said. For example,

(4) Scalia doubts that all men are created equal,

seems to say that Scalia doubts Jefferson's Doctrine; it says this, in part, because (3), as it occurs in (4), contributes Jefferson's Doctrine to what (4) says.

[4] I will henceforth ignore context sensitivity so far as possible. For example, I will suppress reference to context of use, when incorporating it would be a routine matter. I am also ignoring for the most part issues which have to do with the possibility of a shift in semantic values across different positions within sentences.

I am, of course, using 'semantic value' in a manner quite different from that in which it is used by many Fregeans (Dummett, for example).

[5] I am using this as a name, not a description, and assume it to have no semantically significant structure.

Now consider the proper name

(5) Jefferson's Doctrine.

What does this contribute to what is said? On a broadly Russellian (or "direct reference" or "naive") view, it contributes its bearer, Jefferson's Doctrine. So (5) and (3) make the same contribution to what is said. So they are synonymous.[6]

One might find this puzzling, for it seems that sentence 4 and

(6) Scalia doubts Jefferson's Doctrine

are *not* synonymous. As Scott Soames points out, a person might be acquainted with the proposition "by name" (she has been told that Jefferson's Doctrine was originally propounded by Locke, that it is one of the cornerstones of the Declaration of Independence, etc.), but she might not recognize the proposition when it is expressed canonically. Such a person might understand (4) and (6) but take different views on their truth.[7] This at least suggests that the sentences are not synonymous. But if (4) and (6) are not synonymous while (5) and (3) are, substituting synonyms for synonyms does not preserve synonymy.[8]

One reaction to the argument is that it simply draws yet another intolerable consequence from broadly Russellian assumptions; but we already knew that the idea that names contribute (only) their bearers to what is said leads to intolerable consequences. Such a response misses an interesting fact about (3) and (5). For (3) and (5) in some important sense have the same semantic function: they are devices for introducing the proposition that all men are created equal into what is said. And so, given the perspective on synonymy outlined above, there ought to be a sense in which they are synonymous. However, (3) seems to "contribute" various things to what (4) says (equality, for example) which (5) does not contribute to what (6) says. So while there is a sense in which (3) and (5) are responsible for the same contribution to what is said, there is a sense in which they are not; thus, there is a sense in which they are synonymous and a sense in which they are not.

It will be helpful to introduce some terminology. I will say that (3), as it occurs in (4), is *responsible* for certain contributions to what is said which it does not *properly make*. For example, (3) is responsible for contributing equality to what (4) says, although it does not properly make the contribution

[6]Speaking strictly, synonymy follows only if the contributions are the same in all contexts.

[7]See *Soames 1989.*

[8]Note, we are not substituting *within* an intensional context here. In any case, the same point can apparently be made with

Jefferson's Doctrine is true
That all men are created equal is true.

in question. (Rather, 'are equal' properly makes that contribution.) An expression will be responsible for the contributions properly made by all of its subexpressions, though it itself properly makes but a single contribution to what is said. In general, the proper contribution of an expression is what was called above its semantic value. So a noun's proper contribution is a property, a name's contribution (in broadly Russellian framework) is its bearer, and so on. In the example above, I suggest, both (3) and (5) make the same proper contribution to what is said (namely, Jefferson's Doctrine), but (3) is responsible, as it occurs in (4), for contributions to what is said for which (5), in (6), is not. It is this that explains how (4) and (6) come to say different things. Introducing a last piece of terminology, let us say that expressions are *phrasally synonymous* if they always properly make the same contribution to what is said: *structurally synonymous* if they are always responsible for the same contributions.[9]

It may sound as if the claim,

(a) e (as it occurs in S) is responsible for contributing x to what S says,

simply comes to this:

(b) e (as it occurs in S) has a part whose semantic value is x.

It's not that simple. Before explaining why, let me observe that both responsibility and proper contribution are relations between an occurrence of an expression in a sentence and an occurrence of a constituent in a proposition. Thus, the distinction between them can be drawn only if one accepts a view like Russell's, on which what a sentence says is a structured entity, with occurrences of meaningful expressions in a sentence contributing constituents to the proposition expressed.

Suppose we accept such a view of propositions. As we have analyzed the 'Jefferson's Doctrine' example, the name

(5) Jefferson's Doctrine

has no (semantically significant) syntactic structure. It makes the same proper contribution to what is said as the syntactically complex

(3) that all men are created equal,

but the two expressions, (3) and (5), are responsible for distinct contributions. I suspect that it is rare for a syntactically simple expression to be responsible for precisely the same contributions as a syntactically complex one. But one might hold that it is at least possible to introduce a simple expression—a predicate constant, for instance—as meaning the same as a

[9]Some of the themes of the balance of the section, as well as of the appendix, are developed in *Richard 1993*. That article does not, however, use the terms 'phrasally synonymous' and 'structurally synonymous'.

complex expression (such as a predicate abstract), so that interchange of the simple expression for the other would preserve identity of what is said. Church offers just this possibility in criticism of the idea that sentences say the same thing only if one can be turned into the other by alphabetic variation and exchange of constants for synonymous ones.[10] (That is, Church offers this possibility as an objection to the claim that intensionally isomorphic sentences say the same thing.) I see no argument that what Church says is possible is not.

Since structural synonymy is in some sense "complete synonymy", substitution of structural synonyms should preserve what is said, even if those synonyms are not syntactically isomorphic. It is not hard to design accounts, of the relations of sentences to what is said, which allow for this. I will sketch one, with the hope that by showing one way of fleshing them out, the notion of responsibility and the distinction between phrasal and structural synonymy, will be clearer.

Suppose, then, that propositions are structured like the sentences which express them.[11] In fact suppose, for simplicity, that a proposition is an "interpreted sentence structure"—what results from replacing lexical items in a sentence with referents, properties, and so forth, and annotating phrases with their semantic values. For example, suppose the proposition expressed by the sentence 'Tom loves Jill'—that is, let us suppose, the sentence

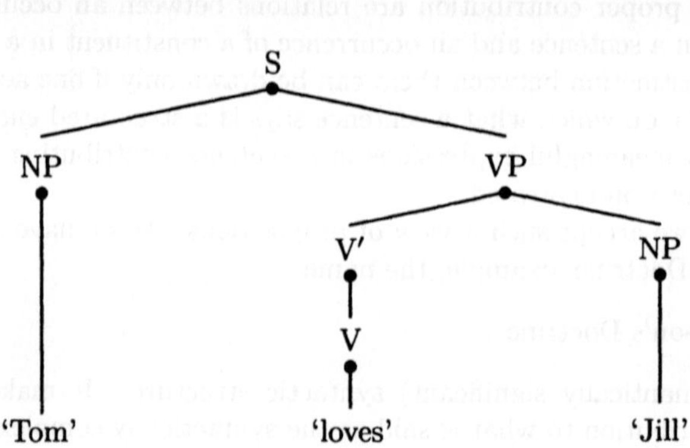

Figure A

is the structured entity

[10] See *Church 1954*. Page references are to the reprinted version. The criticism ascribed to Church appears at page 161.

[11] So far as I can tell, it is not possible to characterize the distinction between structural and phrasal synonymy without making some mildly contentious assumptions about propositional structure.

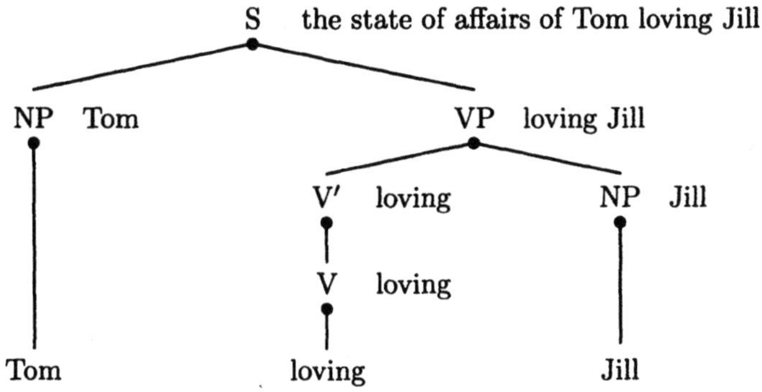

Figure B

On this picture of propositional structure, a proposition is something like a tree each of whose nodes is annotated with a constituent.

I assume we can give rules which generate the proposition a sentence expresses, working "from the bottom up", given information about the "propositional contributions" of lexical items and the sentence's structure. Let me mention a few such rules, enough so that we can discuss the sorts of contributions that simple and complex nouns, like 'brother' and 'male sibling', may make to what a sentence says. I will give the rules as rules which map labeled bracketings to labeled bracketings (or, equivalently, annotated tree structures to annotated tree structures). Some such rules—for example,

(R1) PC([$_{NP}$ 'Tom']) = [$_{NP}$ Tom Dector]

(R2) PC([$_{N}$ 'sibling']) = [$_{N}$ being a sibling]

(R3) PC([$_{A}$ 'male']) = [$_{A}$ being a male]

—associate propositional contributions directly with a lexical item. For example, the third rule maps the structure

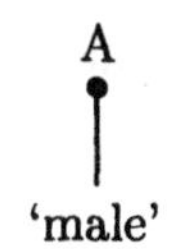

Figure C

to the structure

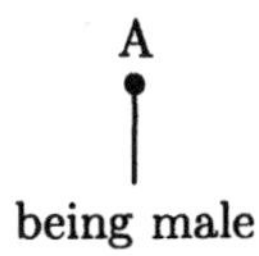

Figure D

Other rules take partially interpreted structures, such as

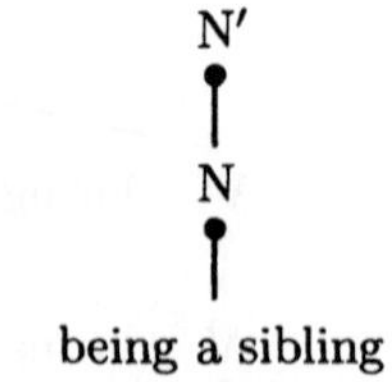

Figure E

and interpret their roots. For example, we will have rules which "pass values up the tree". One such rule will map the structure in Figure E to

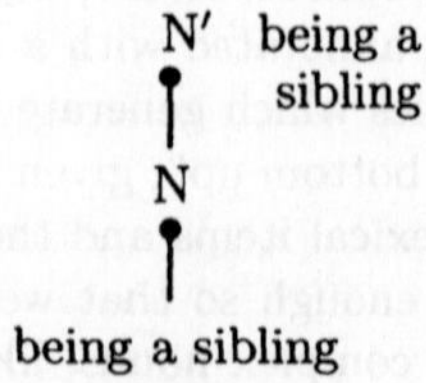

Figure F

Another rule passing values up the tree will conspire with (R3) to map

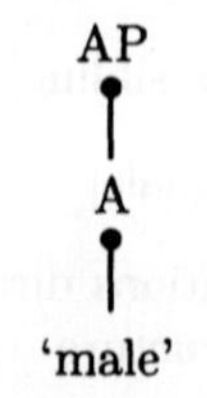

Figure G

to

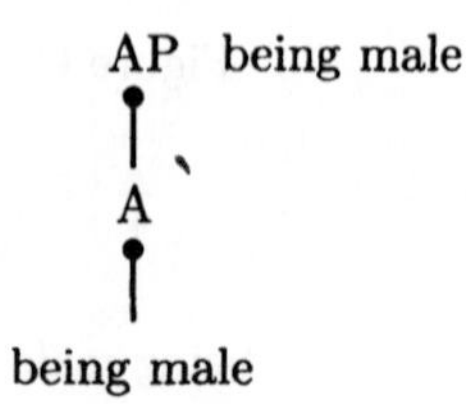

Figure H

And there will be rules which interpret adjectival modification. One such a rule will anoint the root of

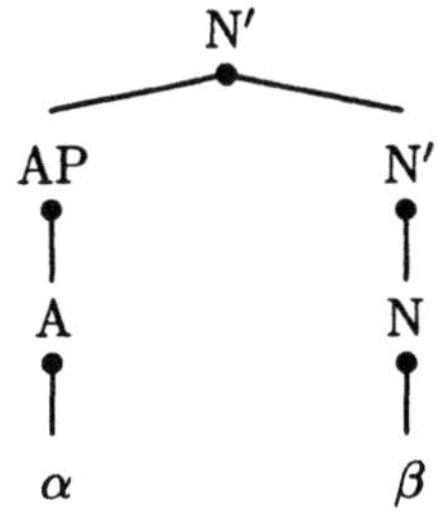

Figure I

with the result of applying an operation—let us call it conjunction—to what annotates the roots of the AP and the lower N′. Such a rule maps the phrase 'male sibling' to

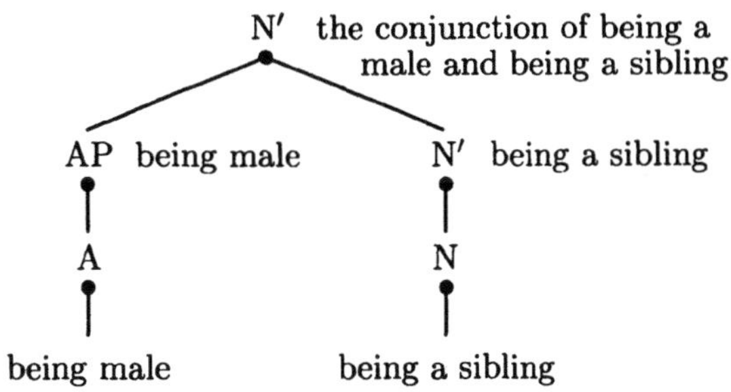

Figure J

My preferred way of using 'semantic value' identifies the semantic value of a phrase with what ends up annotating the phrase's root when we apply such rules. So the semantic value of the NP 'Tom' (given the above rules) is Tom Decter; the semantic value of the noun 'sibling' is the property of being a sibling; the semantic value of the noun 'male sibling' is the property of being a male sibling. If we use the term 'semantic value' in this way, we identify what a phrase properly contributes to what is said with its semantic value. And we can define *expression e is responsible for constituent c* inductively, by describing directly the responsibilities of lexical items ('male', for example, is responsible for the property of being a male), and by saying that a phrase of the form

(F) $[_X [_{Y_1} \ldots] [_{Y_2} \ldots] \ldots [_{Y_n} \ldots]]$

is responsible for its semantic value and for whatever the constituent phrases Y_1 through Y_n are responsible.[12]

I trust it is clear that within this framework there are at least two ways in which a property like that of being a male sibling can come to be a constituent of a proposition. First of all, it can be introduced *via* the introduction of some properties—say, being male and being a sibling—and an operation, such as conjunction, which maps the properties to the property of being a male sibling. Such is the way that being a male sibling enters into the proposition that Tom loves a male sibling. It is also perfectly possible for the property of being a male sibling to be a constituent of a proposition without the properties of being a male or being a sibling to be constituents. If we introduce the noun 'goy' with the stipulation:

> consider the property, being a male sibling ... to be a goy is to be *that*,

this plausibly has the effect of giving the noun goy the property of being a male sibling as its semantic value; the stipulation does not, however, result in the property of being a male being a constituent of the proposition, that Biff is a goy.

The framework just sketched allows for a third possibility. Since the semantic values and proper contributions of lexical items are assigned to them directly, there is no reason we might not have lexical items which were responsible for a number of constituents. We could in principle have an expression 'yog' which had the semantic effect of the phrase 'male sibling'. That is, what 'yog' contributes to a proposition is exactly what 'male sibling' does:

$$\mathrm{PC}([_{\mathrm{N}}\ \text{'yog'}]) = \mathrm{PC}([_{\mathrm{N}'}\ [_{\mathrm{A}}\ \text{male}][_{\mathrm{N}'}\ \text{sibling}]]),$$

which is to say that its propositional contribution is pictured in Figure J. If such is the semantics of 'yog' and 'goy', then the proposition that Tom loves a male sibling is the proposition that Tom loves a yog. But the proposition that Tom loves a goy is not the proposition that Tom loves a yog, even given that the property of being a goy *is* the property of being a yog. Thus we accommodate the possibility envisioned by Church.[13]

Let us return to the paradox of analysis. A simple line of response rests upon the distinction between phrasal and structural synonymy. Continuing for the moment to work within a broadly Russellian account, the response may be developed as follows. We are assuming that all nouns, whether syntactically simple like

(7) brother,

or complex like

---

[12]For further discussion, see the appendix.

[13]Again, for further discussion see appendix.

(8) male sibling,

have semantic values of the same sort, *properties*. It is reasonable to assume that associated with each syntactic operation which can be used to form a noun is an operation which takes the semantic values of its operands to a property. For example, associated with the syntactic operation which forms (8) from 'male' and 'sibling' is an operation ("conjunction") which maps the properties, being male and being a sibling, to the property of being a male sibling.

When a noun is complex like (8), it will be the result of applying some syntactic operation O to some of its constituents. The contribution the noun (properly) makes to what is said will be the property resulting from applying the semantic operation associated with O to the contributions made by O's operands. But a complex noun will be *responsible* for contributions beyond the one it makes. The contribution (8) properly makes to what is said by

(9) I am a male sibling

is the property of being a male sibling; (8) is responsible for this, as well as for contributions made by 'male' and by 'sibling'.

Suppose that it is true that being a brother *is* being a male sibling. (7) contributes being a brother to what is said; (8) contributes being a male sibling to what is said. So, given the identity of the properties, there is a sense in which the two nouns are synonymous. Precisely: (7) and (8) are phrasally synonymous. But they are not structurally synonymous. At least I can see no reason to say that are, and a compelling reason to say they are not. For the claim that all brothers are brothers is different from the claim that all brothers are male siblings. But the claims would be the same if (7) and (8) were structural synonyms, for then there would be no difference in the constituency or structure of the claim made by a sentence containing (7) and that made by replacing occurrences of (7) in the sentence with (8).

The application of all this to the paradox of analysis is as follows. The Paradox is that the four claims, Identity, Triviality, Compositionality, and Synonymy, are inconsistent with the claim that there are genuine analyses. The solution is that there is an equivocation in the notion of synonymy. We may accept Identity. And we may accept Triviality in the form: Analysis is not trivial, but what is said by a sentence *X is Y* is true if X and Y are phrasally synonymous and trivial if they are structurally synonymous.[14] We may accept

> Synonymy When *being an A is being a B* expresses a correct analysis, A and B (and thus *being an A* and *being a B*, etc.) are synonymous,

[14] *X* and *Y* here cannot be quantifier phrases like 'the man in the corner', but must be singular terms, with the notion of term construed broadly enough so that (for example) infinitives in subject position (as in 'to be happy is to be a Libertarian') count as terms.

if by 'synonymous' we mean 'phrasally synonymous', since in a correct analysis, the property assigned to the term which gives the analysandum will be the same as that assigned to the term giving the analysans, and thus the two terms will make the same proper contribution. But if we so understand 'synonymous', we must reject

> Compositionality: Substitution of synonyms for synonyms, at least in a position like that of 'to be a B' in the above, does not affect what a sentence says,

since only exchange of structural synonyms will preserve what is said. We may accept Compositionality but not Synonymy, if by 'synonymous' we mean structurally synonymous, since, as just argued, the terms in a sentence giving a correct analysis will not be structural synonyms.[15]

If we respond in this way to the paradox of analysis, we can maintain, with Moore, that when one gives an analysis, one is not talking about words, but about what words mean, or present, or designate. On this account, an analysis is an account of the nature of a property; in giving an analysis, one is in no way talking about words or the sentences in which they occur. And the account is in accord with the fact that if there are any analyses, they are *not* trivial (and is consistent with the claim that there are some analyses).

I am not suggesting that the distinction between phrasal and structural synonymy resolves every puzzle or problem about analysis. For example, nothing I have said provides an account of properties, or of operations, associated with syntax, which map properties to properties. Thus, I have said nothing which might answer someone who was skeptical about the truth of Identity. My concern in this section has been only with the broadly semantic question posed by the paradox of analysis, namely: How can sentences differing only by synonyms say different things?

[15] The account in the text needs to be generalized, for we would want to be able to apply it to analyses given by sentences like 'to be a brother is to be a male sibling' or 'being a brother is being a male sibling'. I shall not do that here, since I do not think that so generalizing, in and of itself, is much more enlightening than what I have thus far said. (What is said in the text also needs to be generalized to be applied to the analysis of propositions; for simplicity I pretend throughout the discussion that only properties and relations are the objects of analyses.)

Why not just say that 'brother' and 'male sibling' are not synonymous, or at least that they do not make the same contribution to what is said, and solve the paradox of analysis that way?

This "solution" ignores the fact that analysis is closely related to definition. Furthermore, it seems that the semantic value of a noun—what it contributes to what is said—is a property. If so, then the semantic value of 'brother' is the property of being a brother; the semantic value of 'male sibling' is the property of being a male sibling. These are presumably the same property. But the proposed "solution" is one on which we simply say that the semantic values of the two nouns are different.

The following section discusses further some of the relations between analysis and synonymy.

2.

The skeptical critic says: You accept Synonymy. You then distinguish two sorts of synonymy, phrasal and structural. But obviously enough, structural synonymy is *real* synonymy. At the least, the synonymy of structural synonyms is *greater* than that of phrasal synonyms, since only structural synonyms can be substituted for one another with a guarantee that what is said is preserved. So, since 'brother' is "more synonymous" with itself than with 'male sibling', shouldn't one say, on your view, that the claim

(B1) To be a brother is to be a brother,

provides a *better* analysis of the concept brother than does the claim

(B2) To be a brother is to be a male sibling?

Such criticism can only be answered by saying something about the nature of analysis. As I see it, there are two somewhat different reasons for wanting an analysis, and thus two somewhat different accounts one might offer of the purpose or point of analysis, and thus of its nature.

In some situations, an analysis is sought for explanatory purposes. We are "in touch" with a property or relation—that is, we are able to refer to and think about it, and able, in favorable situations, to ascertain or detect its presence. But though it is clear that we are in touch with the property, it is not clear how we come to be in touch with it. In *some* such cases, a proper analysis can provide the materials for explaining this. A correct analysis provides a collection of properties and relations (and "a mode of composition thereof"), which are (under the mode of composition) identical with the *analysandum,* and with all of which we are in touch. Given such, we may be in a position to answer the question

(Q) How is it that we are in touch with the *analysandum*?

A second reason to seek an analysis is classificatory: One wants an analysis because one has a property or relation which one believes is (in some sense) reducible to others and some mode of combination thereof. Giving an analysis, of course, is a way of specifying such a reduction.

This proto-analysis of analysis answers the critic's question about synonymy. Suppose an analysis is sought for explanatory purposes. Then a necessary condition on an analysis is that it provide the materials for an acceptable answer to (Q). (B1) and its ilk cannot provide such; (B2) and its ilk, *ceteris paribus*, can. Suppose an analysis is sought in order to provide a reduction. (B1) cannot provide a reduction, (B2), *ceteris paribus*, can.

I do not claim—and practitioners of analysis like Moore did not claim—that whenever a question like (Q) arises, analysis is possible. Likewise, I do not claim that whenever one property is reducible to another, there is an analysis which states the reduction. Besides the requirements just suggested, an analysis must exhibit a "decomposition", of a property or a relation, which was in some important sense "conceptually accessible to us" before the analysis was given. We might have an ability to refer to a property without having conceptualized it—that is, without anything which could count as decomposition of the property being in any plausible sense conceptually accessible to us. Natural kinds generate examples of this. It seems possible that in some such cases a property has a "nature"—it is a "complex property"—even though we do not conceptualize constituents into which the property can be "resolved". If so, the property's nature will not be conceptually accessible to us.[16] Thus we have the possibility of examples in which a question like (Q) may arise, or in which reduction is possible, but in which (Q)'s answer or the reduction is not to be obtained analytically.

Summarizing: Analyses are certain true claims of property identity. They differ from "trivial" property identities in terms of structure and constituency. An analysis has constituents which can "be combined" to yield the *analysandum* (using the "mode of composition" encoded by the syntax of a sentence expressing an analysis) and a structure accommodating such constituency; a "trivial identity" does not. Whether a claim which satisfies the above conditions gives an analysis is a relative affair: A claim is an analysis *for a person* x, or *for a population* P, only if it satisfies the above conditions and involves a decomposition of the *analysandum* which is conceptually accessible to x or to P.

If all this is correct, it explains why "complete"—that is, structural—synonymy is not required of the forms occurring in a sentence which gives an analysis. And so, if all this is correct, we have a response to the original objection. But one might at this point wonder about the status of Synonymy: Why should we suppose that synonymy of forms in *any* sense is required in a sentence giving an analysis?

Historically, I suspect, Synonymy was motivated by the idea that an analysis must reflect *conceptually accessible* structure, along with the presupposition that such structure would always (and only) be reflected by synonymous forms. The idea, put pretty roughly, is that analysis is "conceptually accessible" only if obtainable *a priori*; but if an identity is knowable *a priori*, the terms in any sentence expressing it must be synonymous, or at least analytically equivalent. (And if it be asked, why does analysis have to reveal conceptually accessible structure, the answer will be in terms of one of the purposes—the explanatory—which might inspire one to seek an analysis.)

The critic worried about Synonymy will not be satisfied with such historical defenses of it. I imagine her posing the question: If all this is correct about the "point" of analysis—if this is really what those who wanted philosophical analyses of concepts were after—why not simply *reject* Synonymy?

[16] Which is not to say that it could not come to be so accessible.

Suppose we want an explanation. All one really wants, or needs at any rate, to answer (Q), is a collection of properties and relations (and a mode of composition) which are such that (a) the properties (under the mode of composition) are identical with the *analysandum*, and (b) we are "in touch" with these properties (in the appropriate way). A sentence

(K) To K is to K′,

might make a claim which provided such a collection without K and K′ being synonymous. Indeed, consider the paradigm example of a concept which is a candidate for analysis, knowledge. Presumably there are non-trivial necessary and sufficient conditions for knowing something to which our practices of ascribing knowledge are sensitive. Should we solve the Gettier problem, we will do so by writing down a sentence of the form of (K) stating such conditions which gives an analysis of knowledge. It is quite unlikely that when we do this, K and K′ will be synonymous in *any* sense whatsoever. Thus, analysis does not require that *analysans* and *analysandum* be expressible by synonymous expressions.

Much the same, the critic continues, could be said about the reductive account of analysis: Given that a property is something which is to be specified independently of how it might be conceptualized (so that a property is, or corresponds to, something like a function-in-extension from worlds to individuals), there seems no reason why illumination on the nature of a property might not be cast by the claim made by (K), though its component infinitives were not synonymous.

In so arguing, our critic seems to assume that meaning is in some interesting sense *transparent*, so that expressions which have the same meaning must, if we understand them, be such that we know them to have the same meaning. For the reason she gives for rejecting Synonymy is this: something could do what an analysis is supposed to do (and hence be an analysis): but while we antecedently understood the terms in which it was given, we found it surprising. This is an objection to Synonymy only if antecedent understanding of synonyms must suffice for recognizing their synonymy, or at least their co-extensiveness.

That meaning is thus transparent is a familiar view. It is associated with Fregean views, on which meaning is (intimately related to) sense; and sense, since it is a way of thinking, *must* be transparent. I will argue directly against such views in section 4. At the moment, I observe only that there are coherent views of meaning on which: (a) expressions A and B may be synonymous in language L; (b) person X may be a competent speaker of L, and therefore understand sentences in which A and B occur; but (c) X does not, even on reflection, know that A and B are synonymous, or (supposing them to be predicates) that they apply to the same things. Almost any view of meaning which sees the meaning of a predicate as being determined, in part, by broadly social factors, will allow for this. Consider, for example,

Tyler Burge's (*1986*) view, on which the meaning of some terms is determined by what a community of speakers would find on reflection to best explain the application of the term. Suppose this is what makes 'sofa' mean 'upholstered piece of furniture meant for seating several individuals'. Then a competent speaker who understands 'sofa' need not know, even on reflection, that 'sofa' and the longer term are synonymous, since the reflection need not reveal what the reflection of the community would decide about the term.[17]

It is not true that to respond to the paradox of analysis as I have is, in effect, to give up Synonymy (and, therefore, a better response than mine is simply to admit that synonymy and analysis have nothing to do with each other.) So long as it is allowed that meaning may be opaque, the solution of section 1 is in principle open to us.

3.

For a Fregean, meaning, tied as it would seem to be to sense, is not supposed to be opaque. How does the Fregean respond to the paradox? In a well-known review of articles by Max Black and Morton White, Church (*1946*, page 133) made the following suggestion.[18]

> The paradox of analysis has an obvious analogy with Frege's puzzle ... as to how an equation, say '$a = b$', can ever be informative—because, it seems, if the equation is true then 'b' is replaceable by 'a', and hence '$a = b$' is the same in meaning as '$a = a$', ... the paradox of analysis is a special case of Frege's puzzle and is to be solved in the same way ... by the distinction of sense and denotation ....
>
> ... the analysis that a brother is a male sibling is not about the classes *brother, male,* and *sibling* but about the corresponding concepts (which are the senses of the class names 'brother', etc.). Hence a correct expression of the analysis must employ names of these concepts. In the expression '$b = ms$', the names 'b', 'm', and 's' must denote the concepts 'brother', 'male', and 'sibling' respectively—not the classes. Now like any other name, a name which denotes a concept must have, besides its denotation, a sense. From the truth of the equation '$b = ms$', it follows that the names 'b' and 'ms' have he same denotation, hence one may replace the other in a sentence without changing the denotation (truth value). But if the analysis expressed by '$b = ms$' is non-trivial, the names 'b' and 'ms' have different senses, hence the replacement of one by the other in a sentence may very well change the sense (proposition) expressed.

[17] Burge has, in fact, argued that one can actively doubt the synonymy without being incompetent.

[18] Limitations of space prevent me from discussing other Fregean accounts, such as that in *Anderson 1993*.

I assume that Church would accept the forms mentioned in Identity as expressing an analysis. If so, Church apparently accepts Identity, Compositionality, and Triviality,[19] as well as a modified version of Synonymy:

> Synonymy* If *being an A is being a B* expresses a correct analysis, A and B are synonymous.

What Church rejects is inferences such as

> The nouns *A* and *B* are synonymous,
>
> So *being an A* and *being a B* are synonymous.

Apparently, Church sees the forms last mentioned as creating an oblique context, in which the embedded nouns name their ordinary sense and express concepts of these senses; when an analysis is non-trivial, the senses expressed are distinct.[20]

It seems to me that there are two problems with this response. Given the claim made by a sentence of the form

(10) To V is to V*,

let us speak of its *corresponding biconditional,* meaning thereby the claim made by a sentence of (roughly) the form

(11′) Whatever Vs V*s, and vice versa.

And let us also speak of the *triviality* corresponding to (10), which is simply the claim made by a sentence of (roughly) the form

(11″) Whatever Vs Vs, and vice versa.

For example, the biconditional corresponding to the claim expressed by

(12) To know a proposition is to have justified true belief therein,

is the claim made by

(13′) Whoever knows a proposition has justified true belief therein, and vice versa,

while the corresponding triviality is that expressed by

(13″) Whoever knows a proposition knows it, and vice versa.

---

[19]Actually, Church's wording suggests that he thinks that in some cases an analysis might be trivial; note the conditional 'if an analysis is non-trivial'. I will ignore this complication, since it is orthogonal to what I have to say about Church's response, given the admission that not all analysis is such.

[20]Which is not to say that they express their indirect senses. As Nathan Salmon pointed out to me, Church rejects Frege's notion of indirect sense. For discussion, see *Salmon 1993,* especially note 10.

One objection to Church is that on his account, whoever knows the triviality corresponding to an analysis knows the biconditional corresponding to it—but this is simply not true. I can know, that all who know a proposition know it, without knowing that all who know a proposition have justified true belief plus X therein, where X is whatever you add to justified true belief to get an analysis of knowledge.[21] Analogously, Church must allow that either I cannot be surprised by B—where B is the biconditional corresponding to a correct analysis—or I can be surprised by T, where T is the triviality corresponding to the analysis. For a surprising analysis, neither horn of this dilemma looks promising.

A second problem with Church's account is this. Let us for the nonce pretend that knowledge is analyzed as justified true belief. Church wishes to respect the intuition that the claim

(A) the concept *knows proposition p* is the concept *knows proposition p*,

is trivial, though the claim

(B) the concept *knows proposition p* is the concept *has justified true belief in p*,

which gives an analysis of knowing that p, is not trivial.[22] But if (B) is not trivial, neither is the claim

(C) take whatever you like x: x falls under the concept *knows p* iff x falls under the concept *has justified true belief in p*.

If knowing (A) and other trivialities does not suffice (given normal logical prowess) for knowing (B), it should not suffice for knowing (C). However, it would seem the claims

(P) Whoever knows proposition p falls under the concept *knows proposition p*, and vice versa,

(Q) Whoever has justified true belief in p falls under the concept *has justified true belief in p*, and vice versa,

are just as trivial as (A). On Church's account, (Q) is identical with the claim

(R) Whoever knows p falls under the concept *has justified true belief in p*, and vice versa.

But surely a Fregean such as Church holds that anyone who knows (P) and (R) will, if they are at all rational and reflect for just a moment, know (C). So Church is committed to saying that knowledge of trivialities suffices for

[21] This point is made by Terence Parsons (*1981*).

[22] Here, p is supposed to be some fixed proposition.

knowledge of (C). And this seems to imply that knowledge of trivialities suffices for knowledge of (B). And this seems to imply that (B) itself is trivial—a claim which Church seems committed to denying.

4.

I would urge Fregeans such as Church to adopt the account of the paradox of analysis sketched in section 1. Of course, the account comes at a price. A relatively small cost of the account is the picture of thoughts as structured collections of senses. Once this is accepted, phrasal synonymy may be identified with (invariable) identity of senses of phrases as wholes; structural synonymy with (invariable) identity of senses contributed to thought expressed. Acceding to this, a Fregean can say that 'brother' is phrasally but not structurally synonymous to 'male sibling'; thus, the claim that being a brother is being a male sibling is distinct from the claim that being a brother is being a brother. The former, but not the latter claim, has the senses of 'male' and 'sibling' as constituents. Entertaining the former, one might say, requires *inter alia* thinking of brothers as males and siblings; entertaining the latter does not *require* that.

A somewhat greater cost may be paid, if one acknowledges the possibility of surprising analysis. For on a Fregean version of the present account, in a correct analysis the expressions giving *analysans* and *analysandum* have the same sense. That is, the sense of those phrases as wholes—their Fregean semantic values—are the same. Fregeans have generally held that when expressions have the same sense, a speaker who understands both will know that they are synonymous, at least if she minimally reflects on it. But a surprising analysis would be one given in terms antecedently understood, but in terms which we need not recognize, even after minimal reflection, to apply to the same things.

I do not think that what the Fregean ought to do here is reject section 1's account of the paradox of analysis; quite independently of whether or not it is accepted, the Fregean is going to have to abandon the idea that when expressions have the same sense, the understanding speaker can recognize this.

To make a case for this, I being by pointing out that analysis presents a quite general problem for a Fregean, whether or not she accepts my account of its nature. To see why, let us return to the pretense

(A) The claim, that knowledge is justified true belief, is an analysis of knowledge, and this claim is the claim expressed by the English sentence 'knowledge is justified true belief'.

It seems possible that there be someone—Freddie, call him—such that

(B) Freddie does not know that whatever has knowledge has justified true belief, and vice versa, but

(C) Freddie is a normal, competent speaker of English, with reasonable logical prowess, who understands 'knowledge' and 'justified true belief', and who has reflected upon the sentence 'whatever has knowledge has justified true belief, and vice versa'.

If (A) through (C) are jointly possible, then a Fregean view which accepts that exchange of expressions with the same sense (in the context of the exchange) preserves sense, must reject either

Synonymy$_P$ If *being an A is being a B* expresses a correct analysis, A and B (and thus 'being an A' and 'being a B', etc.) make the same contribution to what is said,

or Transparency$_M$ It is impossible for a (normal, rational) person to understand context insensitive predicates *A* and *B* which make the same contribution to what's said but not know what is said by *Whatever As Bs and vice versa* (or be such that he would come to know this if he were to reflect on it).[23]

I take it that rejecting Transparency$_M$ requires rejecting Transparency, the constraint mentioned in the introduction to this discussion.

Surely it *is* possible that someone should be surprised by an analysis, even though she antecedently understood the terms in which it was given. If so, then something like (A) through (C) are jointly possible.[24] If we think that a sufficient condition for understanding an expression is satisfying conventional standards for its use,[25] and we accept that philosophically interesting analysis is possible, the joint possibility of something like (A) through (C) borders on the obvious. It follows that a Fregean must reject either Synonymy$_P$ or Transparency. This is *quite independent* of whether she accepts the account of analysis proffered in section 1.

What should the Fregean do? Church's way was to reject the first claim. I have argued that Church's solution was unfortunate. I offer a direct argument against Transparency. Suppose that I introduce a predicate A by defining it thus

x is an A $=_{df}$ x is an BCD,

and that I introduce a predicate Z by defining it thus

x is a Z $=_{df}$ x is a YXW.

---

[23]I pretend that 'knowledge' and 'justified true belief' are context insensitive for present purposes.

[24]Speaking strictly: The sort of state of affairs for which (A) through (C) are representatives (in which there is an analysis of a property in terms which are antecedently understood, but which is surprising) is possible. (Given that knowledge *is not* justified true belief, it is presumably impossible that (A) through (C) themselves be true.)

[25]I borrow the phrase from Soames, *op. cit.*

The terms in the definiens may themselves be defined terms, all of which I understand, and all of which I have defined in terms of terms which are themselves defined terms. Surely it could be the case that by replacing defined terms with defined terms I might eventually transform the definition of A and the definition of Z into the same expression, though even after reflecting on the claim, that being an A is being a Z, I do not know that it is true. (Someone who says otherwise should take a course in analysis or topology.) Shouldn't we say that in defining A and Z as I did, I set the sense of 'x is an A' to be that of 'x is a BCD', and analogously for 'x is a Z'? After all, I *was* giving a definition. If this is so, the same may be true of the other definitions involved. So long as substitution of expressions with the same sense preserves sense, it follows that Transparency is false.[26] It should be stressed that the argument does not depend upon the idea that the meaning (and thus the sense) of one's terms might be set, in part, by social factors. The definitions may have been given by a solitary mathematician.

If sense is not transparent, one can understand sentences with the same sense but not realize that they have the same sense. And so, if understanding a sentence, assenting to it, and being willing to use it as a premiss in (relevant) reasoning suffices for one's believing what it says true—a claim which seems quite tempting when made about the sentences of deductive sciences like mathematics—it seems to follow that a rational person may believe the sense of a sentence S *and* believe the sense of its negation. So Frege's Constraint, mentioned in the introduction, is false.

Some will say that the denial of Frege's Constraint is absurd; only someone who does not understand what a sense is supposed to be could be tempted to deny it. But it is *not* absurd to hold that a rational person might believe and disbelieve a sense, *given* that senses are the sorts of things which (*inter alia*) can be and sometimes are introduced or articulated by definitions. Such definitions, to steal a phrase from Tyler Burge (*1979*), articulate conceptual structure, by spelling out interrelations among concepts. Conceptual structure, simply because it may be complex, need not be transparent to a thinker: The *identity* of conceptual structure is transitive; the *obvious identity* of conceptual structure is not.

Sense, *qua* cognitively accessible structure which may be captured or im-

---

[26]The substitutions are, we may suppose, substitutions of *structural synonyms* for structural synonyms.

One might respond that all the Fregean is committed to is that a speaker who understands expressions *can* (perhaps under idealization which rarely obtains) recognize their synonymy. If so, there is no bar to the Fregean accepting the solution of section 1. Of course, if this is the Fregean's position, it is not altogether clear that that position will support anything like Frege's Constraint, since in non-ideal circumstances a speaker may understandingly accept a sentence S and the negation of a sentence T (T and S with the same sense), thereby acquiring contradictory beliefs. And if the Fregean position will not support anything like Frege's Constraint, it is not altogether clear that Frege's puzzle can be solved (simply) by distinguishing between sense and denotation.

posed via definition, need not be transparent. One may apprehend the same sense twice over—may think about an object in the same way twice over—without realizing this. (Again, I stress that this point has nothing to do with the familiar point that meaning, since it is socially constructed, is not transparent.) Someone may believe and disbelieve one and the same sense because, for example, the conceptual structure which is that sense is accessible to one in different ways which do not allow one to see that it is the same thing which is accessible to each case.

The question naturally arises, Shouldn't we then identify sense with the "modes of access" we have to conceptual structure, thereby preserving the transparency of sense? It should be pointed out how utterly *unFregean* such a response is. The modes of access in question are (at least in many cases) something like the representational vehicles which we manipulate in thinking. That is to say, these modes of access to sense are really what *have* senses, not the senses themselves. They are the sort of things which Frege has in mind when he cautions his readers not to mistake senses for ideas.

What gives a word, in a natural language or a "language of thought", its sense, are the connections it has to our conceptual apparatus. But not *every* connection to our conceptual apparatus is relevant to sense. For example, differences in spelling, pronunciation, and time of acquisition are irrelevant to identity of sense. But these sorts of differences may be extremely relevant to our ability to "process" a sentence. In particular, such "orthographic" differences may block our ability to see that two sentences stand in one or another relation of equivalence, *even though* we understand both sentences. This much should be non-controversial. What I am urging (and what is somewhat controversial) is that we recognize that "differences in orthography" (in some suitably broad sense) may inhibit our ability to see that expressions are synonymous, even though such differences do not inhibit our understanding. At least sometimes, it is our finitude and lack of computational incisiveness—our inability to perfectly manipulate the symbols which realize our thoughts—which accounts for our failure to see identities of sense. At least some of the time, that we fail to see identity of meaning is not the result of our failing to understand the expressions whose meanings are identical.

Let me sum up what I have tried to do in this discussion. I began by introducing a distinction between two sorts of synonymy, phrasal and structural. Phrases are phrasally synonymous if "as phrases" they introduce the same constituent into what a sentence says; they are structurally synonymous if they make the same "total contributions" to what a sentence says. I argued that the distinction allows us to solve the paradox of analysis in the version I cited from Moore, since the ambiguity in the notion of synonymy induces an ambiguity in two of the claims, Synonymy and Compositionality, on which the paradox rests. I defended this solution against the objection that it amounted to abandoning the claim, that a sentence which gives an analysis is an identity whose flanking terms are "really" synonymous. I argued that the standard Fregean solution to the paradox, Church's solution, is defective, and that it is open to a Fregean to adopt the solution presented here.

Finally, I argued that since it seems we can be surprised by analysis, though we understand the terms in which it is given, the Fregean who accepts that analysis is possible has to give up either the idea that the sentence which expresses an analysis is an identity flanked with synonymous forms, or must abandon the view that one who understands synonymous forms will recognize their synonymy (and thus must abandon claims such as Frege's Constraint). I argued directly against the latter. Whether this argument, if successful, undermines the motivation for preferring a broadly Fregean account of meaning and thought to a broadly Russellian one is, perhaps, an issue better addressed on another occasion.[27]

APPENDIX

1. For those who like hair, I give statements of the rules for constructing propositions alluded to in section 1. In stating such rules, it is most convenient to understand labeled bracketings so that when Y1 in

(i) $[_X$ Y1 Y2 ... Yn$]$

is *not* itself of the form

$[_Z$ W1 W2 ... Wn$]$,

then Y1 in (i) is itself a "label", as is 'X' (and both are labels of the same node). This means, for example, that

$[_{N'}$ being a male sibling $[_{AP}$ being male $[_A$ being male$]]$ $[_{N'}$ being a sibling $[_N$ being a sibling$]]]$,

names the structure illustrated in Figure J above. Further, when a propositional contribution of the form $C = [_X\ Y^*_1\ Y^*_2 \ldots Y^*_n]$ is assigned to the phrase $P = [_X\ Y_1\ Y_2 \ldots Y_n]$, we say that Y^*_1 is the semantic value of P; we also call Y^*_1 the semantic value of C. Then some rules simply pass semantic values up the proposition:

$$PC([_{N'}\ [_N\ \alpha]]) = [_{N'}\ SV([_N\ \alpha]\ PC([_N\ \alpha])],$$
$$PC([_{AP}\ [_A\ \alpha]]) = [_{AP}\ SV([_A\ \alpha]\ PC[_A\ \alpha]].$$

Others interpret a structure $[_X\ \alpha_1, \alpha_2, \ldots, \alpha_n]$, given interpretations for the α's:

[27] Jody Azzouni, David Braun, Graeme Forbes, and Nathan Salmon made helpful comments on various drafts of this paper, for which I thank them. I also thank the Philosophy Department of the University of Rochester for helpful comments on an oral presentation.

$$\mathrm{PC}([_{N'}\,[_{AP}\,\alpha][_{N'}\,\beta]]) = [_{N'}\,\mathrm{CONJ}(\,\mathrm{SV}([_{AP}\,\alpha]), \mathrm{SV}([_{N'}\,\beta])), \mathrm{PC}([_{AP}\,\alpha]), \mathrm{PC}([_{N'}\,\beta])].$$

Here, CONJ is an operation which maps a pair of properties to a property; the value of CONJ(P,Q), of course, is to be "the conjunction" of P and Q (so that CONJ(being a male, being a sibling) is being a male sibling). It is clear, I hope, that these rules suffice to assign the structure pictured in Figure J to the phrase 'male sibling'.

2. Those familiar with Church's notion of synonymous isomorphism will recognize the notion of structural synonymy, as I have fleshed it out, as being kindred to Church's notion. Church (*1954*, page 161) proposed the latter as the relation in which sentences stand when they express the same belief. Speaking somewhat roughly, expressions are synonymously isomorphic provided one can be obtained from the other by a sequence of operations involving: alphabetic variation; exchange of either individual constants of descriptions for *either* synonymous individual constants or descriptions; exchange of either predicate constants or predicate abstracts for *either* synonymous predicate constants or abstracts. (Church's criterion for synonymy, in this definition, is identity of sense.) Just as structural synonymy does not require syntactic isomorphism, synonymously isomorphic expressions need not be syntactically isomorphic, since a predicate constant may be synonymously isomorphic with an expression with semantically significant structure.

Church, however, cannot draw a distinction, analogous to that between phrasal and structural synonymy, between synonymous expressions (that is, ones with the same sense) and those which are synonymously isomorphic. He is, for example, committed to saying that all predicates which have the same sense are synonymously isomorphic. For suppose that predicates F and G have the same sense, but are not synonymously isomorphic. If we suppose (as, in this context, is fair enough) that every predicate is either a predicate constant or a predicate abstract, it follows that F and G are both predicate abstracts, since a predicate constant is synonymously isomorphic with any predicate (constant or abstract) with which it is synonymous. "But ...", Church (*1954*, page 161) counsels "nothing prevents us from introducing (say) a predicator constant R as synonymous with the abstraction expression $(\lambda x)[\ldots x \ldots]$, and taking $R \equiv (\lambda x)[\ldots x \ldots]$ as an axiom. And if this is done, then R must be interchangeable with $(\lambda x)[\ldots x \ldots]$ in all contexts, including belief contexts, being synonymous with $(\lambda x)[\ldots x \ldots]$...." Thus, we could introduce constants F' and G' which were synonymously isomorphic with F and G respectively. Being constants with the same sense, they would be synonymously isomorphic with each other. But synonymous isomorphism is supposed to capture the equivalence relation *expresses the same belief as*, and is presumably transitive. Thus in a situation in which we were to make such introductions, F and G would be synonymously isomorphic. Since making the introductions would not affect the meanings of F and G, F and G

are in fact synonymously isomorphic.

To avoid this result, it is sufficient (and, I believe, necessary) to allow that propositions have a constituent structure, and that their constituents may have a constituent structure as well. If we allow this, then the following possibility presents itself: F is the predicate $(\lambda x)[Ax \,\&\, Bx]$, G is $(\lambda x)[Cx \,\&\, Dx]$. The senses of the predicates *as wholes* are the same, since the sense of & maps the senses of Ax and Bx to the sense to which it maps the senses of Cx and Dx. This makes F and G phrasally synonymous. So long as the sense of & can map different pairs of arguments to the same value, and the senses of F's syntactic constituents are constituents of what is said by a sentence in which F occurs, sentences which vary simply by occurrences of F and G need not express the same thought. For in this case, the predicates F and G will be responsible for contributing distinct senses to what is said, and will not be structurally synonymous.

3. Finally, I note in passing that, given that the notions of structural and phrasal synonymy are fleshed out in the way indicated, the solution offered here to the paradox of analysis does not run afoul of an objection to solutions like that of Max Black. This objection is due to Morton White and extended by Church in a well known review of papers by White and Max Black (*1946*). According to White, Black assumed that "no two sentences can express the same proposition if one expressly mentions a relation which is distinct from the relation expressly mentioned by the other". Call this *Black's Principle.* White (*1945*) offers

21 is 3 times 7,

21 is thrice 7,

as a counterexample; Church suggests that if B has the sense of

'$\lambda u \lambda x \lambda y(u = xy)$',

sentences related as are

$b = ms$,

$B(b, m, s)$,

constitute a counterexample.

The account of the Paradox of Analysis I am proposing is *not* committed to Black's Principle. An analysis is a claim which is an identity and so is *expressible* by a sentence of the form

To be an A is to be a B,

where 'is' functions to pick out an identity relation and B possesses syntactic complexity beyond that of A. The constituents of the analysis are not merely the property being analyzed but various properties which "go to make it up", properties which are mapped to the analysans by an operation associated with the syntactic structure (of B) in some sentence expressing the analysis. This does *not* in and of itself rule out the existence of nouns A* and B* which are structurally (and thus phrasally) synonymous, but are such that B* has, and A* does not have, semantically significant syntactic structure. One can allow that someone who introduced the expression 'thrice' via the definition

$$x \text{ is thrice } y =_{\mathrm{df}} x \text{ is three times } y,$$

insured that White's pair of sentences made the same claim. On the view I am proposing, an "atomic" expression *could* be responsible for a number of contributions to what is said by sentences in which it occurs. Of course, if 'thrice' were so introduced, the sentence 'to be thrice y is to be three times y' would not express an analysis, since it would make the same trivial claim as 'for x to be three times y is for x to be three times y'.[28]

REFERENCES

Anderson, C. A.

1993 Analyzing analysis, *Philosophical Studies,* vol. 72, pp. 199–222.

Burge, T.

1979 Sinning against Frege, *The Philosophical Review.*

1986 Intellectual norms and foundations of mind, *The Journal of Philosophy,* vol. 83, pp. 697–720.

Black, M.

1945 The "Paradox of analysis" again: A reply, *Mind,* vol. 54, pp. 272-273.

[28]One might reject White's original "thrice"-objection to Black. For, the response goes, the claim made by

For x to be thrice y is for x to be three times y

provides an analysis of the relation, x is thrice y. But if White's pair of sentences say the same thing, then the claim made by this last sentence is in some sense trivial, since it is the very same claim as that made by

For x to be three times y is for x to be three times y.

I have some sympathy with this objection, since I have some sympathy with the idea that semantic structure must be articulated (so that if x is a constituent of claim p, and S expresses p, there is a constituent of S whose proper contribution to what S says is x). But I think one may cogently defend White thus: If White's pair of sentences are synonymous, then 'to be thrice y is to be 3 times y' does not *quite* express an analysis. But it does express something (that for x to be three times y is for x to be three times y) from which an analysis can be quite easily "derived". For a person who knows this last claim can reason thus: "There is a relation, x is three times y. Let us introduce 'to drice' as a verb that indicates this relation (in the way that 'knows' indicates knowledge). So, for x to drice y is for x to be three times y". Someone who reasons in this way has reasoned her way to an analysis of the drice—i.e., the thrice—relation.

Church, A.

1946 Review of Black and White (*1945*), *The Journal of Symbolic Logic,* vol. 11, pp. 132–133.

1954 Intensional isomorphism and identity of belief, *Philosophical Studies,* vol. 5, pp. 65–73; reprinted in *Propositions and attitudes* (N. Salmon and S. Soames, editors), Oxford University Press, Oxford, 1988, pp. 159–168.

Moore, G. E.

1942 A reply to my critics, *The philosophy of G. E. Moore* (P. Schilpp, editor), Open Court, La Salle, p. 665.

Parsons, T.

1981 Frege's hierarchies of indirect sense and the paradox of analysis, *Midwest studies in philosophy,* vol. VI (French *et al.*, editors), University of Minnesota Press, Minneapolis.

Richard, M.

1993 Articulated Terms, *Philosophical Perspectives,* vol. 7 (J. Tomberlin, editor), Northridge Publishing, Atascadero.

Salmon, N.

1993 A problem in the Frege-Church theory of sense and denotation, *Nous,* vol. 27, pp. 158–166,

Schiffer, S.

1990 The mode-of-presentation problem, *Propositional attitudes* (C. Anderson and J. Owens, editors), CSLI, Stanford.

Soames, S.

1989 Semantics and semantic competence, *Philosophical Perspectives,* vol. 3 (J. Tomberlin, editor), Northridge Publishing, Atascadero, California.

White, M.

1945 A note on the "Paradox of analysis," *Mind,* vol. 54, pp. 71–72.

NATHAN SALMON

THE VERY POSSIBILITY OF LANGUAGE

A Sermon on the Consequences of Missing Church*

1.

An English speaker in uttering the words, (0) 'The earth is round', says, or asserts, the same thing as a French speaker uttering the words, (0′) '*La terre est ronde*'.[1] The thing asserted is a proposition, the proposition that the earth is round. That there are propositions, as distinct from the sentences that express them, is a commitment of psychology and other human sciences, which ascribe beliefs and other propositional attitudes. The existence of propositions is an integral part of our ordinary conceptions of consciousness and cognition, and therewith of our ordinary conception of what it is to be a person.

The evidence for this commitment to propositions, qua extra-linguistic entities expressed by intra-linguistic entities, is compelling. No one has done more to bring that evidence to the attention of philosophers than Alonzo Church, and nowhere does he do so with more force than in his elegant and farsighted paper "On Carnap's Analysis of Statements of Assertion and Belief."[2] That deceptively brief note presents a sharp and telling criticism of one possible attempt to do away with propositions (in the sense that Church intends) in favor of the cognitive dispositions of speakers *vis a vis* particular sentences of a language ("semantical system"). The argument has

*As a student, I had the privilege and distinct good fortune to take numerous courses from Alonzo Church, one of the truly great analytical thinkers of this century. My own intellectual development has benefited enormously from the masterful tutelge of this extraordinary logician and philosopher, in whose honor this essay was written, shortly before his death. I am grateful to my audiences at UCLA in 1994 and at UC Berkeley in 1996 for their comments, and to C. Anthony Anderson and Saul Kripke for discussion.

[1]These sentences are to be taken in the usual sense throughout, as expressing that the earth is spherical.

[2]*Analysis*, 10, 5 (1950), pp. 97–99; reprinted in L. Linsky, ed., *Reference and Modality* (Oxford University Press, 1971), pp. 168–170. See also Church's "A Formulation of the Logic of Sense and Denotation," in Henle, Kallen, and Langer, eds., *Structure, Method, and Meaning: Essays in Honor of Henry M. Sheffer* (New York: Liberal Arts Press, 1988), pp. 3–24 at 5–6*n*; "Intensional Isomorphism and Identity of Belief," *Philosophical Studies*, vol. 5 (1954), pp. 65–73, reprinted in N. Salmon and S. Soames, eds., *Propositions and Attitudes* (Oxford University Press, 1988), pp. 159–168; and *Introduction to Mathematical Logic I* (Princeton University Press, 1956), p. 62*n*.

C. Anthony Anderson and M. Zelëny (eds.), Logic, Meaning and Computation, 573-595.

a broad, generic sweep; it is equally applicable to any of a wide range of theories that posit a reference to sentences or other intra-linguistic items in the semantic content of a variety of constructions, especially attributions of propositional attitude, even including such theories as may be proposed that do this while simultaneously embracing propositions. (The particular theory Church criticizes, as Carnap intended it, arguably does just this.) The argument, which invokes the translation of various sentences between languages, has come to be called *the Church Translation Argument.*[3]

Some philosophers of language, many of them in pursuit of a propositionless world, have attempted to rebut the Translation Argument.[4] Numerous writers today dismiss the argument as weak, fallacious, or otherwise underwhelming, or they simply ignore it. Where Church wasted few words, some of his critics have minced even fewer. Michael Dummett, who does not himself reject propositions, nevertheless said that "it is difficult to treat this objection very seriously."[5] Peter Geach labelled the argument "frivolously bad."[6] In the judgment of the present author, the current dismissive attitude toward the Church Translation Argument constitutes a quantum leap backward, and is, from the point of view of the formal study of semantics,

---

[3]In all of his writings on the subject, Church reminds the reader that the basic insight of employing translation as a test for determining whether an expression is being used or mentioned is due to C. H. Langford. See also footnote 12 below.

[4] The following is a partial list: Tyler Burge, "Self-Reference and Translation," in F. Guenthner and M. Guenthner-Reutter, eds., *Translation and Meaning* (London: Duckworth, 1977), pp. 137–153, and "Belief and Synonymy," *Journal of Philosophy*, vol. 75, no. 3 (March 1978), pp. 119–138; Rudolf Carnap, "On Belief-Sentences: Reply to Alonzo Church," in Carnap's *Meaning and Necessity* (University of Chicago Press, 1947, 1956), pp. 230–232; Donald Davidson, "The Method of Extension and Intension," in P. A. Schilpp, ed., *The Philosophy of Rudolf Carnap* (La Salle, Ill., 1963), pp. 331–349, at 344–346; Michael Dummett, *Frege: Philosophy of Language* (Cambridge, Mass.: Harvard University Press, 1973, 1981), at pp. 372–373, and *The Interpretation of Frege's Philosophy* (Cambridge, Mass.: Harvard University Press, 1981), at pp. 90–94; Peter Geach, *Mental Acts* (London: Routledge, 1957), at pp. 89–92, and "The Identity of Propositions," in Geach's *Logic Matters* (Oxford: Blackwell, 1972), pp. 166–174; Steven Leeds, "Church's Translation Argument," *Canadian Journal of Philosophy*, vol. 9, no. 1 (March 1979), pp. 43–51 (I thank Mark Richard for providing this reference); Brian Loar, *Mind and Meaning* (Cambridge University Press, 1981), pp. 29–30, 152; Hilary Putnam, "Synonymy, and the Analysis of Belief Sentences," *Analysis*, vol. 14, no. 5 (April 1954), pp. 114–122, reprinted in N. Salmon and S. Soames, eds., *Propositions and Attitudes* (Oxford University Press, 1988), pp. 149–158; W. V. Quine, "Quantifiers and Propositional Attitudes," reprinted in L. Linsky, ed., *Reference and Modality* (Oxford University Press, 1971), pp. 101–111, at 110; Mark Richard, *Propositional Attitudes* (Cambridge University Press, 1990), at p. 156ff; Israel Scheffler, "An Inscriptional Approach to Indirect Quotation," *Analysis*, vol. 14, no. 4 (March 1954), pp. 83–90, and "On Synonymy and Indirect Discourse," *Philosophy of Science*, vol. 22, no. 1 (January 1955), pp. 39–44, at 43–44*n*. While no attempt is made here to respond adequately to each of these critiques, much of what will be said here is applicable to a number of them.

[5]Frege, *Philosophy of Language*, p. 372.

[6]"The Identity of Propositions," p. 167.

regrettable in the extreme.[7] The point at issue is by no means narrowly confined to the question of whether attributions of propositional attitude involve a commitment to propositions. Those who reject the Translation Argument have typically failed to comprehend the more general point on which it is based, and have thereby failed to appreciate one of the most fundamental facts concerning the phenomenon of understanding. This failure is especially dramatic in the case of Dummett. As I shall argue, Dummett's failure to grasp the larger import of Church's argument has led him and his followers to defend a seriously distorted version of Frege's theory, one that has the consequence (clearly unintended) that language, as a mode of communication and a vehicle of cognition, is altogether impossible. I shall endeavor here to explain the main thrust of the Church Translation Argument, especially in regard to those aspects of the argument that are misunderstood by its detractors. I shall do so by reference to a particular lacuna in Frege's philosophy of semantics, showing how Church's valuable insights help to resolve a longstanding controversy concerning Frege's notion of indirect sense.

Generalizing slightly on Church's presentation, consider the theory that the first of the following two English sentences is analyzable (definable with full preservation of meaning) by means of the second:[8]

(1) Chris believes that the earth is round.

(2) Chris accepts 'The earth is round'.

The analysans (2) may be expanded if necessary to focus on a particular meaning for the sentence it mentions, perhaps by adding a phrase like 'as a sentence of English', 'as a sentence of the same language as this very sentence', or 'as I, the present speaker of this very sentence, would mean it'.[9] It

---

[7] More generally, it is my considered view that the attempt to avoid an ontology of extra-linguistic abstract entities by an appeal to intra-linguistic substitutes is philosophically misguided. The reasons for this judgment are complex. As an excessively brief summary, I mention that the ontology of everyday discourse is replete with abstract entities other than propositions. The philosophical security that is supposed to be afforded by replacing propositions with sentences is largely an illusion, since sentences, no less than propositions, are abstract entities. Many sentences—infinitely many, in fact—are too long to be written or uttered by any conceivable creature. Moreover, principles concerning the identification of sentences are not nearly as "extensional" as is sometimes supposed. This last difficulty often manifests itself in the need to resort not merely to sentences as such, but to *sentences as sentences of a particular language*, or to *sentences as meaning that such-and-such*, etc.

[8] For many applications, the requirement of full preservation of meaning (whereby, as a consequence, one believes what is expressed by the analysandum if and only if one also believes what is expressed by the analysans) may be weakened to mere logical equivalence, and in some applications, to such wider equivalence relations as modal equivalence or *a priori* equivalence, with little effect on the overall force of the argument.

[9] Church's conception of linguistic expressions and their semantics is such as to require that such semantic attributes as sense, denotation, and truth value always be relativized to a particular language. Others favor a conception according to which relativization to a language is unnecessary or even inappropriate. Cf. Peter Geach's protests in *Mental*

should be noted that any possible expansion of (2) that apparently mentions the proposition that the earth is round, or otherwise apparently logically entails the existence of a proposition, would not be suitable as part of an attempt to relieve the author of (1) of his or her apparent commitment to propositions, although it may be appropriate for other purposes.[10] The particular word 'accepts' is used schematically; any of a variety of transitive verb phrases may be substituted. As an example, one may replace 'accepts' with a carefully formulated phrase like 'is disposed, on reflection, when sincere and nonreticent, to assent to some sufficiently understood translation or other of'. Alternatively, one may substitute Quine's word 'believes-true'[11] or even

---

Acts, pp. 88–89; and David Kaplan, "Words," ***Proceedings of the Aristotelian Society***, vol. 64 (1990), pp. 93–116. Although Church's presentation of the Translation Argument assumes his view on this issue, the argument can also be presented with the opposite view presupposed, or with neutrality (as it is here).

[10]One addendum that would obviously be inadmissible for the purpose of eliminating commitment to propositions is the phrase 'as expressing the proposition that the earth is round', or more simply 'as meaning that the earth is round'. (Cf. footnote 7 above.) Church explicitly considers a possible expansion of (2) that logically entails this inadmissible one, admitting that such an expanded version may yield (1) as a logical consequence. He does not note that the expansion in question would not be suitable for the elimination of propositions. (It may even lead to circularity, if the proposed analysis is extended to such constructions as 'Sentence S means in language L that the earth is round' in addition to (1).) Instead he notes, correctly, that the particular expansion of (2) he considers does not preserve the meaning of (1). Again he utilizes translation to crystalize the point. In addition, he considers embedded constructions like 'Jones believes that Chris believes that the earth is round', arguing that alleged analyses of this and of its translation may even differ in truth value in their respective languages. (Cf. footnote 8 above.) Carnap says in response (*op. cit.*, p. 230, in a highly compressed paragraph) that he intended precisely such an expansion of (2), while conceding that Church's objection is correct.

Carnap's overall response to Church is unclear and puzzling. In ***Meaning and Necessity***, pp. 63–64, scarcely a page after presenting his analysis of statements of belief, Carnap says of analysis in general that although analysandum and analysans must be logically equivalent, they need not be intensionally isomorphic—or as Carnap also puts it, the analysis must preserve intension but need not preserve intensional structure. Davidson (*op. cit.*) and Putnam (*op. cit.*) argue, in effect, that Carnap's analysis of statements of assertion and belief in particular, therefore, is not intended to capture their meaning but only something logically equivalent—thus making Church's objections inapplicable. In sharp contrast, Carnap concedes not only Church's objection, but futhermore that (1) is strictly not even a logical consequence of the intended version of (2) (pp. 230–231). Curiously, Carnap also endorses Putnam's response (p. 230), while proffering an alternative version of (2) as a scientific replacement (presumably now logically equivalent) for (1) (pp. 231–232). The textual evidence suggests, however, that Carnap confused Putnam's response on this point with Putnam's response to a separate objection by Benson Mates concerning embedded constructions. In "Intensional Isomorphism and Identity of Belief," Church extends the Translation Argument against Putnam's response to Mates. See also the introduction to N. Salmon and S. Soames, eds., ***Propositions and Attitudes***, (Oxford University Press, 1988), pp. 13–14*n*10.

The issues here are numerous and quite complex. Cf. footnote 28 below. Insofar as embedded constructions are involved, the dispute is intimately related to issues concerning Frege's notion of indirect sense. See footnote 25 below.

[11]Quine, *op. cit.*, at pp. 109–110. The expression 'believes-true' is, however, significantly misleading. For discussion, see my "Relational Belief," in P. Leonardi and M. Santambro-

the phrase 'believes the proposition expressed by'. The latter cannot be used if one's objective is to avoid commitment to propositions, but might be used, for example, by one (such as Dummett) who supposes that a 'that'-clause is to be understood by reference to its contained sentence. Moreover, the pair of words, 'believes' and 'accepts', may be replaced by other pairs of suitably related terms, e.g., 'asserted'-'uttered', 'disbelieves'-'rejects', etc.

We demonstrate that the proposed analysis fails as follows: Translating both analysandum and analysans into French, we obtain,

(1′) *Chris croit que la terre est ronde.*

(2′) *Chris accepte* 'The earth is round'.

(The word '*accepte*' is used here schematically for the literal French translation of whatever English phrase replaces 'accepts', subject to the same possible variations mentioned earlier.) We pause to note that the proper translation for (2) is not

(3′) *Chris accepte 'La terre est ronde'.*

This sentence mentions a particular French sentence not mentioned in (2), while lacking any mention of the English sentence mentioned in (2). It is thus (2′) rather than (3′) that captures the literal meaning of (2).[12] Likewise, (1) and (1′) are literal translations of one another. But it is evident that the French sentences (1′) and (2′) are not synonymous. For they "would obviously convey different meanings"—indeed, logically independent propositions—to a French speaker having no knowledge of English. Of the two only (1′) conveys the content of what Chris allegedly believes. Sentences (1) and (2) must therefore differ in meaning in English, contrary to the proposed analysis.

2.

As noted above, the scope of the argument is wide. It is equally applicable, for example, against the *redundancy* or *disquotational theory of truth*, according to which the English sentence ' 'The earth is round' is true in English' simply reduces to 'The earth is round'. The argument demonstrates that the two sentences are in fact logically independent. It is also directly applicable to a longstanding controversy in orthodox Fregean theory. Frege held that the English sentence (0) ordinarily denotes its truth value—in this case, the value truth ("the True")—but that, when occurring in an indirect or oblique (*ungerade*) context in English, as in (1), it instead has its indirect denotation, denoting what is ordinarily its English sense, the proposition that the

gio, eds., *On Quine* (Cambridge University Press, 1995).

[12]Cf. G. E. Moore, "Russell's 'Theory of Descriptions'," in his *Philosophical Papers* (New York: Collier Books, 1959), pp. 149–192, at 156–157, where he anticipates Langford's translation test.

earth is round.[13] Frege held further that when (0) occurs in any English indirect context, it takes on its indirect sense, which is a concept of its indirect denotation, or ordinary sense. But which of the myriad concepts of the proposition that the earth is round is thereby expressed? Carnap complained that "Frege nowhere explains in more ordinary terms what this third entity is."[14]

The matter remains controversial. Dummett argues that, for Frege, the indirect sense in English of (0) is the ordinary sense in English of the phrase

(4) the ordinary sense in English of 'The earth is round'.[15]

Dummett does not support this exegetical thesis by citing any passage from Frege's writings. Instead he provides quite general considerations in favor of the interpretation:

> There is nothing in what Frege says about direct [i.e., ordinary] and indirect sense and reference [denotation] to rule out the possibility that, although distinct expressions, in the same or different languages, may have the same sense, no sense can be given to us save as the sense of some particular expression; such a thesis would fit very well what Frege sometimes says to the effect that *we* can grasp thoughts only via words or other symbols. But, if so, then, if an expression stands for a sense, and does so in virtue of *its* sense, that sense must involve a reference, overt or covert, to some expression—the same or different—whose sense its referent is (*op. cit.*, p. 91).
>
> It is perfectly consistent to combine the thesis that the object of belief is a thought [proposition] with the thesis that we can apprehend a thought only as the sense of a sentence (in a verbal or symbolic language). It is therefore equally consistent to combine the thesis that a clause in *oratio obliqua* [indirect discourse], or an expression within that clause, stands for its ordinary sense with the thesis that, in understanding it as referring to that sense, we apprehend that sense as being the sense of that clause or expression. ... Frege [adheres] to both these theses (p. 94).

[13] "Über Sinn and Bedeutung", translated as "On Sense and Reference," in R. M. Harnish, ed., ***Basic Topics in the Philosophy of Language*** (Englewood Cliffs, N.J.: Prentice Hall, 1994), pp. 142–160, at 144. I follow Church in translating Frege's use of '*Bedeutung*' as 'denotation' rather than 'meaning' or 'reference', and Frege's use of '*Gedanke*' as 'proposition' rather than 'thought'. I also use the word 'concept' in Church's sense, which is very different from Frege's use of '*Begriff*'.

[14] *Meaning and Necessity*, §30, pp. 129–133.

[15] *The Interpretation of Frege's Philosophy*, at pp. 89–100. This represents a turnabout for Dummett. In his earlier work, *Frege*: *Philosophy of Language*, he considered the thesis that the indirect sense of (0) in English is the customary sense of (4), only to dismiss it as "rather implausible" (p. 267). Both this thesis concerning indirect sense, and the exegetical thesis that Frege's theory implies the former thesis, are defended in Gary Kemp, "Salmon on Fregean Approaches to the Paradox of Analysis," *Philosophical Studies*, vol. 78, no. 2 (May 1995), pp. 153–162, wherein specific references to Frege's writings are provided (at p. 160) in support of Dummett's interpretation.

Dummett adds further on that Frege thought that "we can grasp a thought only as expressed in some way" (p. 98), and that the thesis that the indirect sense in English of (0) is the ordinary sense of (4) "does appear to follow from the two theses, taken together: the thesis that a sense can be given only as the sense of some expression, and the thesis that an expression in *oratio obliqua* stands for its ordinary sense" (p. 95).[16] He continues, saying that "our grasp of the sense of the expression ..., on Frege's account, ... leaves us with no access to that sense save as the sense of some expression" (p. 97).[17]

Echoing Carnap, Dummett argues against Frege that it is preferable not to confer on an expression a concept of its ordinary sense to serve as the sense expressed in indirect contexts.[18] Despite his misgivings about the notion of indirect sense, Dummett explicitly endorses the idea that, as he puts it, "a sense can be given to us only as the sense carried by some particular expression." He explains that to say this "implies that the most direct means by which we can refer in English to the sense expressed by, say, 'the moon' is by using the phrase 'the sense of 'the moon''" (p. 95). Dummett says moreover that "we cannot refer to the sense of most expressions save by explicit allusion to the expressions" (p. 90). And he joins with Frege, as Dummett has interpreted him, in holding that

> we apprehend the indirect referent as being the sense of the expression which we perceive as occurring in the *oratio obliqua* clause. ... our way of grasping what the sense of an expression is, renders us incapable of detaching the sense from every actual or hypothetical means of expressing it. ... when we take the expression in *oratio obliqua* as standing for its own sense, we are conceiving of that sense as the sense of that very expression (pp. 97–98).

Although Dummett does not share Frege's belief in the need for indirect senses, it is clear from the foregoing passages that Dummett maintains the following thesis (which he cites in support of his interpretation of Frege on indirect sense): In order for us to conceive of any sense at all (in order to form a belief about it, or to speculate about it, or even merely to entertain a

[16]Dummett points out that the inference in question is actually invalid, but he allows (pp. 98–99) that it is validated by the addition of a third thesis (which Frege held and which Dummett rejects) as premiss, to wit, that "the sense of an expression is the way in which its [denotation] is given to us."

[17]Bishop Berkeley argued, against Locke's doctrine of "abstraction," that one cannot conceive of, for example, a color without also conceiving of some shape or other of that color. According to Dummett, Frege did not hold, in the same spirit as Berkeley, merely that one cannot conceive of a sense in any way without also conceiving of it as the sense of some expression or other. It is not that any conception of a sense we have, must be ***accompanied*** in cognition by a conception of it as the sense of some expression. The point, rather, is that in conceiving of a sense, we conceive of it only as the sense of e, for some expression e or other, since we have no conception of the sense other than as the sense of e, for some expression e, for such a conception to accompany. (This will be made more precise below. I am grateful to Charles Chihara for bringing this distinction to my attention.)

[18]Carnap, *op. cit.*, at p. 129; Dummett, *Frege: Philosophy and Language*, pp. 266–269.

thought about it, etc.), there must be some expression *e* (of some language *l*) such that we conceive of the sense in question (or as Dummett also puts it, the sense is "given to us") *as* the ordinary sense of *e* (in *l*); we are unable to conceive of a sense in any way other than as the ordinary sense of *e* (in *l*), for some expression *e* (of some language *l*).[19] Let us call this *Dummett's Thesis.*

The considerations cited by Dummett in favor of his interpretation of Frege are unconvincing, especially in light of Frege's explicit pronouncement (which both Carnap and Dummett may have overlooked) that the indirect sense in English of (0) is rather the ordinary sense in English of the alternative phrase

(5) the proposition that the earth is round.[20]

One may wonder whether the phrases (4) and (5) are not themselves synonymous in English. In fact, Geach explicitly argued that on Fregean theory, such phrases are completely interchangeable (*op. cit.*, p. 168–169). And Dummett evidently treats such phrases as synonyms.[21] Furthermore, Dummett's interpretation depicts Frege as having held that the ordinary sense of (5) in English presents the proposition in question as the ordinary sense of some particular linguistic expression. The only plausible candidate for that expression is the English sentence (0).[22] This would make (5) a mere English paraphrase of (4)—even if it is a "less direct means" of denoting the proposition denoted by each. If (4) and (5) are indeed synonymous in English, then Dummett's formulation of Frege's thesis concerning indirect sense is simply another way of saying the same as Frege.

Any suspicion that (4) and (5) are synonymous may be dispelled, however, by the Translation Argument. Of course, (4) and (5) denote the same proposition in English, but their literal French translations—'*le sens ordinaire en anglais de* 'The earth is round'' and '*la proposition que la terre est ronde*', respectively—clearly carry different meanings for the French speaker who knows no English. Hence, (4) and (5) are not ordinarily synonymous in English, i.e., they differ in ordinary English sense.

Furthermore, the Translation Argument demonstrates that the thesis Dummett erroneously attributes to Frege is in any case incorrect. For just

[19] Dummett does not explicitly speak in a relativized manner of the sense of an expression *in a language.* The parenthetical references to the language *l* are included here for the benefit of those whose view of linguistic expressions and their semantics requires for propriety the relativization to a language. See footnote 9 above.

[20] "*Uber Sinn und Bedeutung*," at p. 149 of Harnish.

[21] In *The Interpretation of Frege's Philosophy*, at p. 89, there occurs an otherwise inexplicable switch, wherein the discussion, suddenly and without notice, changes its focus from an identification of the indirect sense of a sentence like (0) with the ordinary sense of the corresponding analogue of (5), to an identification of the indirect sense of 'Aristotle' instead with the customary sense of 'the sense of 'Aristotle''.

[22] Cf. the argument given by Dummett *op. cit.*, p. 95, top paragraph. See also Kemp, *op. cit.*

as (0) expresses its English indirect sense in (1), (0′) expresses its French indirect sense in (1′). Yet unlike (2′), (1′) evidently makes no mention of (0). It is probable, furthermore, that the particular linguistic item creating the indirect context in (1) is not the English word 'believes' but 'that'.[23] For terms denoting the same proposition are typically interchangeable, with no affect on truth value, following occurrences of 'believes', 'asserted', 'doubts', etc. Consider for example the following inferences:

Carnap believed Frege's central doctrine in the philosophy of mathematics.
Frege's central doctrine in the philosophy of mathematics was logicism.
Therefore, Carnap believed logicism.

Church doubts logicism.
Logicism is the doctrine that mathematics is reducible to logic.
Therefore, Church doubts that mathematics is reducible to logic.

By contrast, the word 'that' is equally involved in substitution failures that do not involve belief ('Bert asserted *that* ...') and even in substitution failures that do not concern attitudes of any kind, for example in contexts concerning modality ('It is a necessary truth *that* ...'). The word 'that' is plausibly regarded as a device of sense-quotation, which when attached to an English declarative sentence forms the standard English name for the proposition ordinarily expressed by that sentence (in a manner analogous to that in which syntactic quotation marks form the standard name of the expression enclosed within). As such, 'that' would be a paradigmatic *ungerade* device ("oblique operator") of English, with the phrase 'the proposition' occurring in (5) functioning as a grammatical appositive to the 'that'-clause.[24] This hypothesis on Dummett's proposal implies that, since (0) is induced by the 'that' prefix to express its English indirect sense in (1), (1) itself may be rewritten without alteration of meaning in a form dispensing with any *ungerade* device (beyond ordinary quotation marks), simply by substituting (4) for the shortened version of its alleged synonym (5) within it:

(6) Chris believes the ordinary sense in English of 'The earth is round'.

Since this is a sentence of precisely the form of (2), the Translation Argument, as originally given, disproves the theory that it is an English paraphrase of

[23]It must be admitted that the construction 'Chris believes the earth is round', without the word 'that', is perfectly grammatical English, and (0) occurs nonextensionally therein. This phenomenon, however, does not immediately yield the result that 'believes' is a nonextensional (*ungerade*) operator. It is arguable that (0) is induced to shift to the indirect mode here by other (perhaps pragmatic) factors. (Indeed, linguists commonly refer to the phenomenon as "'that' deletion.")

[24]Cf. my *Frege's Puzzle* (Atascadero, Ca.: Ridgeview, 1986, 1991), at pp. 5–6; and "Reference and Information Content: Names and Descriptions," in D. Gabbay and F. Guenthner, eds., *Handbook of Philosophical Logic IV: Topics in the Philosophy of Language* (Boston: D. Reidel, 1989), pp. 409–462, at 440–441.

(1). Again by contrast, the hypothesis that the English word 'that' is a device for sense-quotation strongly supports Frege's actual claim concerning the indirect sense of a sentence like (0).[25]

3.

A few years after Church's paper appeared, Geach protested,

> Very often, what we count as correct translation will include translation of quoted expressions; a translator of *Quo Vadis* would not feel obliged to leave all the conversations in the original Polish, and we should count it as perversely wrong, not pedantically correct, if he did so.[26]

Others have since echoed Geach's complaint, notably Tyler Burge and Dummett.[27] It must be observed that each of Burge, Dummett, and Geach offers a reply different in various respects from the other two—and it is not obvious that any of the three agrees entirely with any other—but they share an emphasis on divergences between translation in actual practice and the special, sense-preserving translation proposed by Church. Translators would very likely proffer (3′) rather than (2′), they argue, as a correct rendering of the English (2) into French. No one denies that (3′) fails to preserve (2)'s mention of (0); (3′) mentions different words of a different language altogether. Insofar as preservation of meaning, or sense, requires the preservation of denotation, rendering the English (2) into French by means of (3′) does not preserve meaning.[28] But Burge, Dummett, and Geach each deny

---

[25] The Translation Argument is not supportive of all aspects of Fregean theory. As is shown in my article "A Problem in the Frege-Church Theory of Sense and Denotation," *Noûs*, vol. 27, no. 2 (1993), pp. 158–166, while the argument supports Frege's thesis concerning indirect sense, it thereby leads to an inconsistency when combined with a Fregean solution to the Paradox of Analysis, of a sort advocated by Church in his review of the Black/White exchange concerning the Paradox of Analysis, *The Journal of Symbolic Logic*, vol. 11 (1946), pp. 132–133. I argue in the article that relinquishing the Fregean solution to the Paradox of Analysis threatens the Frege-Church theory, by collapsing Frege's and Church's original argument for the pivotal distinction between the sense and denotation of an expression. (Whereas this difficulty exposes a potentially serious weakeness in the Frege-Church theory of sense and denotation, it does nothing to weaken the force of the Translation Argument.)

[26] *Mental Acts*, pp. 91–92. (I thank Saul Kripke for providing this reference.)

[27] Burge, "Self-reference and Translation," especially pp. 141–144; Dummett, *loc. cit.* Burge's response to the Translation Argument is also endorsed by James Higgenbotham, in "Linguistic Theory and Davidson's Program in Semantics," in E. Lepore, ed., *Truth and Interpretation: Perspectives on the Philosophy of Donald Davidson* (Oxford: Basil Blackwell, 1986), pp. 29–48, at 39*n*. (I thank Higginbotham for providing the last reference.)

[28] This point is explicitly acknowledged by Dummett (*Frege: Philosophy of Language*, p. 372; *The Interpretation of Frege's Philosophy*, p. 90). It is more or less acknowledged by Burge (*op. cit.*, p. 141), although his overall response to the Translation Argument is explicitly part of a program to avoid commitment to propositions (pp. 152–153). Geach espouses an extreme skepticism regarding the notion of synonymy. He argues that (3′) is "of the same force as" (his version of) (2), and that (3′) and (2) are "reasonably equivalent,"

that correct translation must preserve meaning—insofar as the preservation of meaning requires the preservation of denotation. Thus Dummett writes, "There is no ground for the presumption that the practical canons of apt translation always require strict synonymy. On the contrary, translations of fiction and, equally, of historical narrative (including the Gospels) always translate even directly quoted dialogue."[29] Amplifying the argument, Burge says, "translation of foreign quoted material aims at conveying the 'point' of the passage that contains it" (*op. cit.*, p. 145). And according to Burge, what is crucial to the point of sentences like (2)

> is not part of the grammar or semantics of the sentences themselves. It is better seen as involved in a convention presupposed in the use and understanding of such sentences. ... What is involved in rightly construing the expressions that are mentioned in problem sentences like [(2)] is ... the ability to understand those expressions as they would be intended if they were used by the person who uses the relevant token of the sentence in which they are mentioned (p. 146).

Burge argues that the operative nonsemantic, pragmatic convention in connection with (his version of) the English (2) directs one to interpret the sentence it mentions in the specified manner, and that this yields the result, contrary to Church, that (3′) is a better translation of (2) into French than is (2′).

This general line of criticism was cogently refuted independently by Herbert Heidelberger, Casimir Lewy, and Leonard Linsky.[30] It is completely inessential to the Translation Argument whether (2′) is deemed a correct

but specifically stops short of declaring that they are either the same, or different, in meaning. By the same token, Geach does not propose (2) as a meaning-preserving analysis of (1). It is unclear therefore why he does not concede that any author of (1) is committed to the existence of the proposition that the earth is round (this being what the Translation Argument is aimed at demonstrating), and let this be his reason for recommending that the author substitute (2), which lacks any such commitment. (Cf. Quine, *op. cit.*, and my "Relational Belief" for discussion.) In any event, Dummett's reply has the advantage over Geach's (and Quine's) that it does not depend (at least not to the same extent) on any implausible or otherwise controversial skeptical theses concerning synonymy.

Burge claims that the sentence he proposes for (2) involves self-denotation, in such a way that the best translation will preserve this feature at the expense of denotation. The details of Burge's argument will not be pursued here. One minor correction should be noted, however. Contrary to Burge's initial claim (which he credits to W. D. Hart) that translation of a self-denoting sentence either preserves self-denotation or denotation but not both, translations may be given that preserve neither one. (For whatever it is worth, it is even probable that such translations have been given in actual practice.) Burge's argument requires only the weaker claim that no translation of a self-denoting sentence preserves both denotation and self-denotation.

[29] *Frege: Philosophy of Language*, p. 372. But see also *The Interpretation of Frege's Philosophy*, p. 90.

[30] Heidelberger, Review of Dummett's *Frege: Philosophy of Language*, in *Metaphilosophy*, vol. 6, no. 1 (January 1975), pp. 35–43, at 42–43; Lewy, *Meaning and Modality* (Cambridge University Press, 1976), at pp. 64–66; Linsky, *Oblique Contexts* (University of Chicago Press, 1983), at p. 8. (I thank C. Anthony Anderson for providing the last two

translation into French of the English (2). The claim that an actual translator or interpreter would offer (3′) instead is similarly irrelevant. The Translation Argument is concerned exclusively with the semantics of (1) and (2), and not with any pragmatic "conventions presupposed in their use" or with any resulting "practical canons of apt translation" of texts containing either. As Heidelberger correctly notes, the construction "S' is a translation into L' of the L expression S" may even be replaced uniformly throughout the argument by "S' has the same meaning in L' that S has in L", thereby eliminating any allusion to translation. Any controversy concerning translation (in particular, whether apt or correct translation must preserve meaning) is thus seen to be entirely irrelevant.

Dummett appears to have conceded the irrelevance of the issue of whether the English (2) would be translated into French in actual practice by (2′) rather than (3′). Yet (1′) is clearly not synonymous in French with (2′), which explicitly mentions the English sentence (0). Once it is conceded that (2′) has the same meaning in French that (2) has in English, the critic of the Translation Argument has no alternative but to challenge the remaining premiss, that (1′) is a sense-preserving French translation of the English (1). In response to unspecified critics (presumably including at least Heidelberger, Lewy, and/or Linsky), Dummett does just that.[31] And indeed, Dummett's interpretation of Frege implies that Frege should have rejected this premiss. For according to that interpretation, Frege is supposed to have held that

references.) See also Saul Kripke, "A Puzzle about Belief," reprinted in N. Salmon and S. Soames, eds., *Propositions and Attitudes* (Oxford University Press, 1988), pp. 102–148, at 142*n*25.

In his discussion in *The Interpretation of Frege's Philosophy*, Dummett attributes both the thesis that the indirect sense of (0) in English is the ordinary sense of (4), and the exegetical thesis that Frege held the former thesis, to Heidelberger. Dummett also there accuses Heidelberger of inconsistently conjoining these theses with an endorsement of the Translation Argument (pp. 91, 94). These attributions are dubious. Although Heidelberger defends the Translation Argument against Dummett's criticism, he explicitly declines to endorse the argument and instead expresses sympathy for the alternative criticisms of it by Davidson and Putnam (p. 43*n*). Moreover, he does not straightforwardly propose either of the theses in question. Instead he correctly attributes to Frege the thesis that the indirect sense of (0) in English is the ordinary sense of (5), while mis-identifying this thesis with the alternative thesis which Dummett had branded "rather implausible" (p. 37; see footnote 15 above). As noted earlier, Dummett also fails to distinguish between these two theses concerning indirect sense, and therefore fails to distinguish properly between the corresponding exegetical theses concerning Frege on indrect sense. (It should also be noted that whereas Heidelberger explicitly attributes to Frege the thesis that the indirect sense of (0) in English is the ordinary sense of (5), he also attributes to Frege a fallacious argument for that thesis which Frege does not give, and which is in fact inconsistent with his views.)

[31] *The Interpretation of Frege's Philosophy*, pp. 90–91. The premiss is explicitly rejected by Kemp, *op. cit.*, and Stephen Leeds, *op. cit.* Curiously, Dummett ulitmately endorses the premiss (p. 94), on the ground that (1) has the same "conventional significance" in English that (1) has in French (p. 99). It is unclear why Dummett insists nevertheless that it is illegitimate for Church to assert this premiss, on the same or very similar ground, in the course of his argument against the proposed analysis.

(1) is synonymous in English with (6), which mentions the very expression (0), whereas the French (1′) instead mentions (0′). In the general case, the proponent of (2) as an English analysis of (1) will contend that (1′) is synonymous in French with (3′), which explicitly mentions (0′), and therefore does not have the same meaning in French that (1) has in English. Such a proponent might even enlist the support of translation at this point, since the sense-preserving translation of (3′) into English bears little resemblance to (1).[32]

Though the argument's appeal to sense-preserving translation may be thus thwarted by controversy, Church's general point might still be made without appealing to any contested translation. Church supports the premiss that (1′) and (2′) are not synonymous in French by observing that they "obviously convey different meanings" to a French speaker who knows no English, with only (1′) conveying the content of what Chris is supposed to believe. In a similar vein, one might object to the proposal that (1) is synonymous in English with (2) on the ground that different information is conveyed to an English speaker by (1) than is conveyed to a French speaker by (2′)—which, it has now been conceded, is a sense-preserving French translation of (2).

In response to this possible objection, Stephen Leeds objects that

> what information a sentence conveys to a hearer depends not only on what the sentence means but on what background information the hearer has. The mere ability to understand a language can constitute such background information; for example, '*Luther sprach*: '*Hier steh ich*'' will convey more to someone who speaks German than its strict translation into English: 'Luther said: '*Hier steh ich*'' will to a monolingual speaker of English (*op. cit.*, p. 46).

This last point might also be illustrated by (2) and (2′). The two have the same meaning in their respective languages, yet an English speaker obtains more information on the basis of the former than a French speaker obtains on the basis of the latter. Church himself noted, as part of his argument, that the knowledge of what (0) means in English would enable one to infer from (2′) the content of the belief attributed to Chris. Leeds contends that it is arguably this ancillary knowledge, rather than the content semantically expressed by (1), that likewise accounts for the information conveyed to an English speaker by (1) that is not conveyed to a French speaker by (2′). Indeed, as we have seen, Burge argues that the pragmatics involved in the use of (2) (and therefore also in the use of its alleged paraphrase (1)) directs one to interpret the mentioned sentence (0).

It is here that this general response to the Church Translation Argument betrays a failure to grasp the central point of the argument—or perhaps I

[32]For more on this sort of issue, especially as it relates to Burge's response to the Translation Argument, see my "A Problem in the Frege-Church Theory of Sense and Denotation," at p. 166n15.

should say, at least one of the central points of the argument—and in a sense, one of the main points also of Frege's philosophy of language.

4.

Just as the "practical canons of apt translation" are entirely irrelevant to the Church Translation Argument, so also is the information obtained on the basis of an utterance. To use a terminology I introduced in previous work, the information obtained from an utterance can involve not only *semantically encoded information*, but also *pragmatically imparted information*.[33] (What counts as apt translation, for that matter, may also be concerned to some extent with information of both kinds.) The Translation Argument, by contrast, is concerned exclusively with the former. One who knows the English language is able to infer additional information from an utterance of (2) not semantically contained within (2) itself. While this additional information is thereby pragmatically imparted (at least indirectly) to an English speaker by the utterance, it is not directly "conveyed" by (2) in the sense relevant to the Translation Argument. Rather, it is inferred from the semantically contained information taken together with the ancillary knowledge of what (0) means in English. When Church argues that (1′) and (2′) convey different information to a French speaker who knows no English, he is speaking of the information semantically encoded in each, the propositions that the French sentences convey to the French speaker solely in virtue of their literal French meanings. He explicitly contrasts this with information that may be inferred from this together with knowledge of English. And indeed, the very conclusion of the argument (the denial of which Leeds seeks to defend) is that (1) and (2) semantically contain different information, despite the fact that the content of (1) is easily inferred from an utterance of (2) given knowledge of the English meaning of (0).

Translation between languages is invoked merely as a device to facilitate our seeing the difference in semantic content that exists between (1) and (2)—or as a special case, between (1) and (6)—despite the ease with which one is inferred from the other, relying on our knowledge of English. The merely auxiliary role of translation in the Translation Argument was first noted explicitly by Church himself:

> The existence of more than one language is not usually to be thought of as a fundamental ground for the conclusions reached by this method. Its role is rather as a useful device to separate those features of a statement which are essential to its meaning from those which are merely accidental to its expression in a particular language, the former but not the latter being invariant under translation. And distinctions

[33] *Frege's Puzzle, loc. cit.* (footnote 24 above), pp. 58–60 and elsewhere (especially pp. 78–79, 84–85, 100, 114–115).

> (e.g., of use and mention) which are established by this method it should be possible also to see more directly.[34]

In the application of the Translation Argument to the question of indirect sense, the distinction that one should be able to see more directly is a special distinction between the meanings of (4) and (5). This same distinction accounts for the difference in meaning between (1) and (2).

What exactly is this special distinction, and how are we supposed to see it without resorting to translation? The answer to this is what I take to be the crucial point, and the very point that Church's detractors have failed to appreciate. Church puts this point by noting that the proposed analysis of (1)

> must be rejected on the ground that [(2)] does not convey the same information as (1). Thus (1) conveys the content of what [Chris believes] ... it is not even possible to infer (1) as a consequence of [(2)], on logical grounds alone—but only by making use of [an] item of factual information, not contained in [(2), that (0) means in English that the earth is round] ("On Carnap's Analysis of Statements of Assertion and Belief," pp. 168–169 of Linsky).

Consider (6), as a special case of (2), and the purely semantic differences that Church is limning between it and (1). Church's main point is that (1) gives the content of the belief attributed to Chris in a special manner, a manner in which (6) does not. In short, (1) *identifies* that content. To be sure, both (1) and (6) specify the content, but (6) does so only by describing it in the manner of (4), as *whatever sense* (0) *ordinarily expresses in English.* This is indeed a way of conceiving of the English ordinary sense of (0)—it is a concept of the proposition, in Church's terminology—but it is a concept of that sense that even one who has no understanding of (0), as an English expression, may possess with perfect mastery, provided he or she knows only that (0) is a meaningful English expression. For such a person is in a position to infer that (0) ordinarily expresses in English its own English ordinary sense. This inferred knowledge is trivial; it does nothing to further the quest to learn what, by contrast, one who understands (0), as a sentence of English, thereby knows, viz. that (0) specifically means *this* proposition in English: that the earth is round.

This is essentially a special case of Russell's more general distinction between *knowledge by acquaintance* and *knowledge by description.*[35] The lat-

[34] "Intensional Isomorphism and Identify of Belief," in Salmon and Soames, p. 168*n*22. In a similar vein, Church had written in "A Formulation of the Logic of Sense and Denotation," at p. 5*n*: "This device is not essential to the explanation, but is helpful in order to dispel any remnants of an illusion that there is something in some way necessary or transparent about the connection between a word or a sentence and its meaning, whereas, of course, this connection is entirely artificial and arbitrary."

[35] Russell, "Knowledge by Acquantance and Knowledge by Description," reprinted in N. Salmon and S. Soames, eds., *Propositions and Attitudes* (Oxford University Press, 1988), pp. 16–32.

ter is a distinction between two kinds of knowledge of things, as opposed to knowledge of facts (in French, two kinds of *connaissance* as opposed to *savoir*). Knowledge of a thing by acquaintance might be explained as a conception of the thing *qua* "that F," perhaps perceptually ostended or otherwise demonstratively selected, for a particular sortal 'F' (which might even be the universal sortal, 'thing'). Knowledge of a thing by description, by contrast, is a correct conception of the thing *qua* "whatever uniquely satisfies C," invoking some purely descriptive condition C. In Church's terminology, knowledge of a thing by description is exactly the conceiving of that thing through the apprehension of a concept of it.[36]

[36] In calling a condition "purely descriptive," I mean to preclude its being such a condition as might be expressed in the form 'the condition of being [identical with] that very F'. It is assumed here that Church's notion of a *concept* likewise excludes such conditions. Cf. my *Reference and Essence* (Princeton University Press and Basil Blackwell, 1981), pp. 14–23 (where the term 'descriptional' is employed instead of 'descriptive').

Russell construed acquaintance in a very strict sense which excluded the possiblity of acquaintance with particulars other than oneself or mental items directly contained in one's consciousness. The distinction itself can be drawn independently of this severe restriction, however, and is clearly legitimate with regard to more familiar notions of acquaintance. One such notion is that of having perceptual, or other natural or "real," cognitive contact with a particular person or object—the sort of connection that is sufficient to enable one to form beliefs or other attitudes *about* the object (in an ordinary sense). A somewhat stricter notion imposes the further condition of *knowing who or what* the object is, in an ordinary sense. (Some philosophers, having evidently overlooked the possibility of perceiving an object without knowing who or what the object is, have confused these two broader notions of acquantance, See my "How to Measure the Standard Metre," *Proceedings of the Aristotelian Society*, New Series, vol. 38 (1987/88), pp. 193–217, especially at 200–201*n*, 213ff.)

Throughout I use the term 'identification' for a notion of acquaintance implying knowledge of who or what the object is. Some contemporary neo-Russellian theories of meaning deny that one who knows the content of (1) automatically thereby knows what proposition it is that Chris believes (i.e., automatically thereby knows at least one proposition that Chris believes). Such theories may hold instead that one who knows the content of (1) is thereby acquainted with the believed proposition in some less familiar way, and may not know exactly what proposition is in question. Knowing what F so-and-so is (as special cases, knowing what proposition Chris is hereby held to believe, or knowing what person—or who—so-and-so is) may be a matter knowing of so-and-so, *de re*, that so-and-so is it (him/her), believing this fact about so-and-so while conceiving of it in a special, identifying way. (See my *Frege's Puzzle*, pp. 103–118, on the notion of believing a proposition while taking it a particular way; see also footnote 43 below.) Even such theories, however, will generally recognize an important epistemological difference between the contents of (1) and (6), such that knowing the fact described by (6) falls well short of knowing that fact described by (1). It may be acknowledged, for example, that one who knows the content of (1)—by contrast with one who merely knows the content of (6)—*ipso facto* kows something *de re* about the proposition that the earth is round, namely, that it is something Chris believes (even, perhaps, without knowing exactly what proposition is in question). This is sufficient for my primary purpose in the discussion to follow. For related discussion, see Mark Richard, "Articulated Terms," in J. Tomberlin, ed., *Philosophical Perspectives, 7: Language and Logic* (Atascadero, Ca.: Ridgeview, 1993), pp. 207–230.

It is possible that there is a gradation of notions of acquaintance. This would not make Russell's distinction between knowledge by acquaintance and knowledge by description untenable; on the contrary, it would make for a multiplicity of legitimate Russellian dis-

Russell thought of the two kinds of knowledge as mutually exclusive. To be sure, the very same thing may be simultaneously known in each of the two ways, but on Russell's conception of the distinction, no knowledge of a thing by description is also acquaintance with that same thing. Other philosophers embrace a strict representational epistemology according to which all knowledge of things is achieved only through the apprehension of concepts (in Church's sense), so that all knowledge of things is ultimately knowledge by description (in Russell's sense). Any knowledge that can be called 'acquaintance', on such a view, is merely a peculiar kind of knowledge by description, one in which the relevant descriptive condition or concept is of a special sort. It is arguable that Frege in particular held a strict representational epistemology. And indeed, this epistemological stance may lie behind the principal divergence between broadly Fregean and Russellian semantic theories.[37]

There is an infinite-regress argument against the tenability of this strict representational sort of epistemology: If conceiving of a thing invariably invokes a concept of that thing, then in order to conceive of any particular thing x, one must, in that very act of conceiving, apprehend some antecedently understood concept c_1 which is a concept of x. But by the same token, in order to know c_1, one would have to apprehend some further antecedently understood concept c_2 which is a concept of c_1, and so on. It seems to follow that in order to conceive of anything at all, one would have to apprehend each of an infinite sequence of antecedently understood concepts. One is invited to conclude from the threat of this infinite regress that all knowledge of things ultimately rests on "direct" knowledge of things, knowledge that is not mediated through concepts of those things.[38]

Textual evidence suggests that, in a sense, Frege instead embraced the infinite regress, via his infinite hierarchies of indirect senses (ordinary sense, indirect sense, doubly indirect sense, and so on).[39] It is a possibility also

---

tinctions. The distinction on which I rely in the discussion to follow need not be made completely precise, and will be sufficiently obvious to support the conclusions reached by its means.

[37]As explained in the previous note, Russell himself held a respresentational epistemology with regard to all particulars other than oneself and the mental contents of one's consciousness. Contemporary Russellians have typically favored a less restrictive epistemology on which one knows various concrete particulars "directly," i.e., without appealing to individuating qualitative concepts of those particulars.

[38]Cf. the infinite-regress argument in Russell, *op. cit.*, at pp. 28–29 of Salmon and Soames.

[39]One may not straightforwardly conclude from the infinite-regress argument that according to the strict representational sort of epistemology in question, in order to conceive of a thing x, one must *in that very act of conceiving* apprehend each of infinitely many concepts. For although the epistemology requires that the act of conceiving of x necessarily involves an act of apprehanding a concept c_1 of x, it does not require that the act of apprehending c_1 necessarily involves conceiving of c_1. One might even label the act of apprehending a concept without conceiving of it a kind of "direct acquaintance" with the concept.

One may not even conclude from the infinite-regress argument that according to strict

that Church holds a similar epistemological view.[40] Even so, both Frege and Church doubtlessly recognized that knowledge of a thing by acquaintance has a very different cognitive flavor from mere knowledge by description (i.e., from knowledge of the thing not by acquaintance).

The distinction is made evident in the familiar contrast found in pairs of color terms like 'white' and 'the color of snow'.[41] The former identifies the color in question in a way that the latter does not. One who, because of limited experience, is ignorant of both the color of snow and the color of Church's hair, may still know that Church's hair is the color of snow, i.e., that Church's hair has the same color that snow has, whatever color that may be. If such a person may be said to know what color Church's hair is, then he or she knows it merely by description. Just as he or she knows that Church's hair is the color of snow, he or she also knows that snow is the color of snow. What such a person lacks is the knowledge of *which color* that is, the identifying information normally taken to be contained in the words: 'Snow is *white*'. A person who has been deprived of sight since birth may be incapable of knowing any color except merely by description (e.g., as the property, visually manifested in some manner or other, of reflecting light of

---

representational epistemology, in order to conceive of anything one must already apprehend each of infinitely many concepts. What is required on Frege's doctrine of the hierarchy is that one be able to acquire an apprehension of any one of the infinitely many concepts on demand, so to speak, in order to comprehend a sentence that embeds an expression within a sufficiently large number of *ungerade* operators. Strict representational epistemology nevertheless yields a possible answer to Russell's famous "no backward road" observation and the difficulty it is supposed to raise for Frege's doctrine. It is true that on Frege's theory, for anything x there are countless concepts of x, no one of which can be singled out as privileged or as the "designated" or "standard" concept of x. But according to strict representational epistemology, in conceiving of a concept c_1 through one's acquaintance with it, one is thereby apprehending a special identifying concept c_2 of c_1. By attending to what one is apprehending, one can conceive of c_2 through one's acquaintance with it, and hence through one's apprehension of a special identifying concept c_3 of c_2, and so on. Thus it seems that one need only attend to what one is apprehending to generate a proper Fregean hierarchy starting from a single concept. Even if there is no "backward road" (privileged branch) from a thing that is not a concept to a concept of the thing, it seems there may be a backward road from a concept to a concept of the concept. (I am indebted here to remarks made in a seminar by Saul Kripke and to later discussion with C. Anthony Anderson.) Cf. my "Reference and Information Content: Names and Descriptions," *loc. cit.*, and "A Problem in the Frege-Church Theory of Sense and Denotation," at p. 163.

[40] It must be noted, however, that Church rejects Frege's notion of indirect sense. See "A Problem in the Frege-Church Theory of Sense and Denotation," pp. 164–165*n*10.

[41] A number of philosophers have noted differences of meaning beween such terms. For a sample of the relevant literature, see D. M. Armstrong, "Materialism, Properties and Predicates," *Monist*, vol. 56, no. 2 (April 1972), pp. 163–176, at 174; Jaegwon Kim, "On the Psycho-Physical Identity Theory," *American Philosophical Quarterly*, vol. 3, no. 3 (July 1966), pp. 227–235, and "Events and Their Descriptions," in N. Rescher, ed., *Essays in Honor of Carl G. Hempel* (Dordrecht: D. Reidel, 1969), pp. 198–215, at 205–206; Bernard Linsky, "General Terms as Designators," *Pacific Philosophical Quarterly*, vol. 65 (1984), pp. 259–276; N. L. Wilson, "The Trouble with Meanings," *Dialogue*, vol. 3, no. 1 (June 1964), pp. 52–64. (The last is evidently the ancestor of the other discussions.)

such-and-such wavelengths.) The legitimacy of this distinction between two different ways of knowing the color white (or of knowing the color of Church's hair, etc.) seems sufficiently obvious that it will be readily recognized even by those who hold that the knowledge that snow is white is ultimately knowledge to the effect that snow has whatever color it is that uniquely satisfies a certain, special visual condition (perhaps a phenomenological condition).

The analogy between the pair of contrasting color terms and (4)–(5) is striking. In particular, knowing the English meaning of (0) by the description (4) is not a way of *understanding* (0), in any ordinary sense. As noted above, someone who speaks no English, upon learning only that (0) is a meaningful expression of English, is thereby given its meaning by this description. The person still lacks information specifying what that meaning is—to wit, the identifying information that (0) expresses in English the proposition that the earth is round. (Compare again knowing what color snow is only by the description 'the color of snow' vs. knowing that snow is this color: white.)

Earlier we applied the Translation Argument against Dummett's apparent hypothesis that (4) and (5) are ordinarily synonymous in English. A more direct application of the argument is possible. The translations of the following metalinguistic sentences reveal a fundamental difference in meaning:

(7) 'The earth is round' ordinarily expresses in English the proposition that the earth is round.

(8) 'The earth is round' ordinarily expresses in English the ordinary sense in English of 'The earth is round'.

The meta-English sentence (8) is tautologous, or virtually so. By contrast, (7)—or more simply '(0) means in English that the earth is round'—"conveys the content," in Church's terminology, of (0) in English; it identifies the English meaning of (0), semantically specifying it in a way that (8) does not even come close to doing.

Translation into another language of both (7) and (8), and likewise the translation of both (1) and (2), is merely a pedagogical aid which more clearly reveals their divergent semantic properties. We have already seen that the question of whether the strict sort of translation that is invoked in The Translation Argument conforms with the practice of actual translators and interpreters is quite irrelevant to the purpose for which translation is pressed into service. What is essential for that purpose is that the translation be what Church calls 'literal', i.e., sense-preserving.[42] The relevant differences in meaning between the English (4) and (5)—which is the principal point of the argument—can be seen independently of translation, for example by appealing to the striking analogy with 'white' and 'the color of snow'. And as Church says, it should be possible also to see the point directly.

[42] "Intensional Isomorphism and Identity of Belief," footnote 25 (p. 168 of Salmon and Soames).

5.

Recall Leed's contention that it is arguably not the semantic content of (1) in English but the understanding of what (0) means (which, according to Burge, is pragmatically required by (1)), that accounts for the additional information that an English speaker obtains from (1) but a French speaker does not obtain from the sense-preserving translation of (6). This observation is meant to isolate the claim that, semantically, (1) specifies the content of the belief attributed to Chris (as does (6)) in precisely the manner of (4). But this claim flies in the face of Church's main point, that (1) semantically "conveys the content" of Chris's alleged belief, identifying that content in a special manner exhibited by neither (6) nor its French translation. The content of Chris's belief is not merely pragmatically imparted to an English speaker by an utterance of (1) *via* an inference. Contrary to Leeds, the content of Chris's belief is semantically encoded by (1). Indeed, Leed's rejoinder is nearly enough self-defeating. For it implicitly recognizes that there is a special way of thinking of the proposition that the earth is round—a way of being acquainted with it—such that one who knows that (0) in English expresses that proposition conceived in that way, knows what (0) means in English, whereas one who knows only that (0) in English expresses the English ordinary sense of (0) (conceived in that way, by that description) does not know what (0) means. This special identifying "way of thinking" of the proposition seems to be carried in English by (5), and it would seem to be precisely this that so markedly distinguishes (5) from (4). It is natural to assume, in fact, that the special way of thinking of the proposition is nothing less than the concept semantically expressed in English by (5). It is difficult to imagine how else one might express it.[43]

The foregoing considerations reveal the reasons for the failure of the theory that Dummett mistakenly imputes to Frege. For the hypothesis that the indirect sense of (0) in English is the ordinary sense of (4) yields the erroneous conclusion that one who understands no English beyond the words 'Chris believes that', and who is informed that (0) is a meaningful English sentence without being given its actual meaning, may thereby know what (1) means in English. In fact, it would seem that what he or she still needs in order to know the meaning of (1) is to be given the meaning in English of (0) *qua* the proposition that the earth is round, i.e., via the concept that is ordinarily expressed in English by (5) rather than that ordinarily expressed by (4). It is thus possible to see directly, without resorting to translation, the supe-

[43] Presumably understanding (0), as a sentence of English, entails knowing what (0) means in English, but it is arguable that understanding, in a strict sense, requires more than this (perhaps acquiring the knowledge of the meaning of (0) in a special computational manner). There are delicate issues, on which the present discussion is neutral, concerning the extent to which the English meaning of (5) captures the way of conceiving the proposition in question that is possessed by one who correctly understands (0) as a sentence of English. See footnote 36 above.

riority of Frege's thesis concerning indirect sense over the thesis Dummett erroneously attributes to him.

The foregoing also dramatically exposes a fatal error in Dummett's Thesis that the only way in which we can conceive of any sense is as the sense of some particular linguistic expression. Using Church's terminology (in combination with some of Dummett's), the thesis may be stated more precisely thus:

> For every concept c_1, if someone conceives of c_1, then there is some concept c_2 of c_1 such that that very act of conceiving of c_1 consists in apprehending c_2; and furthermore, for every concept c_2 of c_1 that we can apprehend, there is some expression e (of some possible language l) such that: (*i*) c_1 is the ordinary sense of e (in l); and (*ii*) c_2 presents c_1 *as* the ordinary sense of e (in l), i.e., to apprehend c_2 is exactly to conceive of c_1 as the ordinary sense of e (in l).[44]

Although Dummett attributes this thesis to Frege, it is in fact deeply out of sync with the fundamental character and structure of Frege's philosophy of language. Frege does say that we apprehend propositions through words.[45] And it seems clearly true that we do this with at least a great many propositions. It by no means follows, however, that we are unable to conceive of the proposition except as whatever sense it may be that those words ordinarily express. Frege would certainly have insisted, against Dummett, that there are infinitely many ways to denote the proposition that the earth is round without mentioning particular linguistic expressions, either explicitly or implicitly: 'the proposition that Columbus set out to prove by sailing westward to India', 'the proposition that Aristotle proved on the basis of the shape of the earth's shadow on the moon', 'Chris's favorite proposition', 'the first thing Columbus asserted upon sighting the mainland of America', etc. These unremarkable phrases, and many others, express graspable concepts that "give us" the proposition in ways other than as the sense of an expression.[46] More important, the very apprehension of a proposition, even by means of the words through which we apprehend it, crucially involves an identifying conception of the proposition, a knowledge of it by acquaintance and not merely as whatever it is that these words ordinarily express. This is

[44]See footnotes 19 and 36 above. The first conjunct expresses strict representationalism with regard to concepts. On a Fregean theory the word 'as' occurring in Dummett's constructions '... is given to us as —' and 'conceives of ... as —' must be itself regarded as an ***ungerade*** operator. Dummett's Thesis, as formulated here and as given by Dummett himself, thus involves quantification into a nonextensional context. (I am not claiming that Dummett would accept my specific formulation of his thesis, only that he explicitly endorses the thesis itself, which I formulate thus, and attributes the same thesis to Frege. The textual evidence for these claims is clear.)

[45]For example, in "On the Scientific Justification of a Conceptual Notation" (reproduced in his ***Conceptual Notation and Related Articles***, T. W. Bynum, ed., (Oxford University Press, 1972), pp. 83–89), Frege says that "we think in words ..., and if not in words, then in mathematical or other symbols" (p. 84). See footnote 15 above.

[46]In light of this, Frege undoubtedly rejected even the weaker, Berkeleyan thesis mentioned in note 17 above.

true in fact of understanding in general; there is no comprehension without identification.

Dummett's Thesis precludes the possibility of a language that works the way Frege thought all language worked, and even the very possibility of language itself. On Dummett's view, in taking (0) to express such-and-such, one must, in that very act of cognition, conceive of the concept in question as *whatever sense is ordinarily expressed by e*, for some particular linguistic expression *e*—there being no other way for us to conceive of any concept. This theory makes all of language unintelligible for us in principle. Dummett's Thesis leaves our grasp of language in a state of ignorance exactly analogous to someone who, from birth, sees the world only in black and white and shades of grey. A person who has always been completely color-blind may know that physical objects are colored, and that certain of them have the same color as certain others (snow and Church's hair, grass and emeralds, etc.), but in some sense has no way of learning what color anything is. Dummett's Thesis reduces us all to the state of the international traveler who knows that the words he or she sees and hears have meaning, but is completely ignorant of what those meanings are. Indeed, the theory renders us considerably worse off than the traveler. A tourist can identify the foreign meanings by consulting a phrase book. Dummett's theory has the consequence that we are unable even to understand expressions by translating them into our native language. For the only knowledge we have of the meanings of *our own* words is by the description: whatever it is that those words mean. On Dummett's theory, there is no identification of meanings, no knowledge of meaning by acquaintance, only knowledge by description. This makes the veil of syntax utterly opaque, and the wall of unintelligibility impenetrable even in principle. Since there can be no comprehension without identification, Dummett's theory has the unintended consequence that we comprehend nothing.[47]

The fundamental flaw in Dummett's theory can also be seen through the theory of definition, by means of a variation of the infinite-regress argument considered above. According to Dummett's Thesis, in knowing that the ordinary sense of an expression e_1 (in a language l_1) is such-and-such, one must conceive of that sense as the ordinary sense of some expression e_2 (of a language l_2). This is tantamount to the claim that every meaningful linguistic expression that the speaker understands is understood by means of a kind of verbal definition: The ordinary sense of e_1 is the same as the ordinary sense

[47] It is arguable that a person completely color-blind from birth may nevertheless learn of the color green (*de re*) that grass and emeralds are *that color*, by learning that they reflect light of such-and-such wavelengths under visually normal conditions. Even so, there still seems to be some knowledge that those of us who see green (as green) have, and that the completely color-blind person lacks, concerning the color of grass. In some sense, the completely color-blind person still does not know how grass looks, phenomenologically, with regard to color. By contrast, if Dummett's theory were correct, none of us could know *of* the proposition that the earth is round (*de re*), that (0) expresses it in English (let alone could we understand what (0) means in English—see footnotes 36 and 43 above).

of e_2, or as it is sometimes written, $e_1 =_{\text{def}} e_2$.[48] The definiens e_2 would likewise be understood by means of an analogous definition, $e_2 =_{\text{def}} e_3$, and so on. It is well known, however, that not all expressions of a language may be understood by means of verbal definition. There has to be a stock of primitives, whose sense is learned in some way other than by verbal definition (by ostensive definition, for example). Dummett's Thesis leads ultimately to a vicious circularity among the definitions. And this makes all understanding impossible even in principle.

The root problem with Dummett's Thesis is that it restricts all knowledge of meaning to knowledge by description, precluding even the possibility of our identifying the ordinary sense of any expression. Since there is no comprehension without identification, Dummett's Thesis makes comprehension utterly unattainable. And this, I submit, excludes the very possibility of the phenomenon we call *language*.

If I am correct, the main point of the Church Translation Argument is exactly the antithesis of Dummett's Thesis. The 'that'-clause in English expressions like (1) and (7) carries with it a special way of conceptualizing the content of the sentence following 'that', an identifying way of thinking of the proposition which constitutes acquaintance rather than mere knowledge by description. Church's appeal to translation serves to illuminate this fundamental point. Frege's explanation of the indirect sense of (0) in English as the ordinary sense of (5) rather than that of (4) almost certainly reflects his own grasp of this same fundamental point, and with it a repudiation of Dummett's Thesis.

Insofar as understanding requires acquaintance, what is it that we are required to be acquainted with? If there is no comprehension without identification, what is it that we are identifying? Not just sentences in a language, as we have seen. But if not just sentences in a language, then what else?

The Translation Argument has been with us a long time—longer even than I have. Yet many language theorists today have missed its principal point. If nothing else, there is an important lesson to be learned from the failure of the argument's critics to appreciate that point, and from the decisive collapse of at least one theory that has been proffered in defiance of the argument.

Nathan Salmon
Department of Philosophy
University of California
Santa Barbara, California 93106 USA
Email: nsalmon@humanitas.ucsb.edu

[48]This notation is normally used for definitions within a single language. For present purposes we may think of the union of all of the separate languages spoken by a particular speaker as constituting a single comprehensive language. (Any resulting lexical ambiguities may be resolved by means of disambiguating subscripts.) Alternatively, the notation is easily extendible to accommodate inter-language definitions, for example by writing: $[e_1, l_1] =_{\text{def}} [e_2, l_2]$.

INDEX

SYNTHESE LIBRARY

1. J. M. Bochénski, *A Precis of Mathematical Logic*. Translated from French and German by O. Bird. 1959 ISBN 90-277-0073-7
2. P. Guiraud, *Problèmes et méthodes de la statistique linguistique*. 1959 ISBN 90-277-0025-7
3. H. Freudenthal (ed.), *The Concept and the Role of the Model in Mathematics and Natural and Social Sciences*. 1961 ISBN 90-277-0017-6
4. E. W. Beth, *Formal Methods*. An Introduction to Symbolic Logic and to the Study of Effective Operations in Arithmetic and Logic. 1962 ISBN 90-277-0069-9
5. B. H. Kazemier and D. Vuysje (eds.), *Logic and Language*. Studies dedicated to Professor Rudolf Carnap on the Occasion of His 70th Birthday. 1962 ISBN 90-277-0019-2
6. M. W. Wartofsky (ed.), *Proceedings of the Boston Colloquium for the Philosophy of Science, 1961–1962*. [Boston Studies in the Philosophy of Science, Vol. I] 1963 ISBN 90-277-0021-4
7. A. A. Zinov'ev, *Philosophical Problems of Many-valued Logic*. A revised edition, edited and translated (from Russian) by G. Küng and D.D. Comey. 1963 ISBN 90-277-0091-5
8. G. Gurvitch, *The Spectrum of Social Time*. Translated from French and edited by M. Korenbaum and P. Bosserman. 1964 ISBN 90-277-0006-0
9. P. Lorenzen, *Formal Logic*. Translated from German by F.J. Crosson. 1965 ISBN 90-277-0080-X
10. R. S. Cohen and M. W. Wartofsky (eds.), *Proceedings of the Boston Colloquium for the Philosophy of Science, 1962–1964*. In Honor of Philipp Frank. [Boston Studies in the Philosophy of Science, Vol. II] 1965 ISBN 90-277-9004-0
11. E. W. Beth, *Mathematical Thought*. An Introduction to the Philosophy of Mathematics. 1965 ISBN 90-277-0070-2
12. E. W. Beth and J. Piaget, *Mathematical Epistemology and Psychology*. Translated from French by W. Mays. 1966 ISBN 90-277-0071-0
13. G. Küng, *Ontology and the Logistic Analysis of Language*. An Enquiry into the Contemporary Views on Universals. Revised ed., translated from German. 1967 ISBN 90-277-0028-1
14. R. S. Cohen and M. W. Wartofsky (eds.), *Proceedings of the Boston Colloquium for the Philosophy of Sciences, 1964–1966*. In Memory of Norwood Russell Hanson. [Boston Studies in the Philosophy of Science, Vol. III] 1967 ISBN 90-277-0013-3
15. C. D. Broad, *Induction, Probability, and Causation*. Selected Papers. 1968 ISBN 90-277-0012-5
16. G. Patzig, *Aristotle's Theory of the Syllogism*. A Logical-philosophical Study of *Book A* of the *Prior Analytics*. Translated from German by J. Barnes. 1968 ISBN 90-277-0030-3
17. N. Rescher, *Topics in Philosophical Logic*. 1968 ISBN 90-277-0084-2
18. R. S. Cohen and M. W. Wartofsky (eds.), *Proceedings of the Boston Colloquium for the Philosophy of Science, 1966–1968, Part I*. [Boston Studies in the Philosophy of Science, Vol. IV] 1969 ISBN 90-277-0014-1
19. R. S. Cohen and M. W. Wartofsky (eds.), *Proceedings of the Boston Colloquium for the Philosophy of Science, 1966–1968, Part II*. [Boston Studies in the Philosophy of Science, Vol. V] 1969 ISBN 90-277-0015-X
20. J. W. Davis, D. J. Hockney and W. K. Wilson (eds.), *Philosophical Logic*. 1969 ISBN 90-277-0075-3
21. D. Davidson and J. Hintikka (eds.), *Words and Objections*. Essays on the Work of W. V. Quine. 1969, rev. ed. 1975 ISBN 90-277-0074-5; Pb 90-277-0602-6
22. P. Suppes, *Studies in the Methodology and Foundations of Science. Selected Papers from 1951 to 1969*. 1969 ISBN 90-277-0020-6
23. J. Hintikka, *Models for Modalities*. Selected Essays. 1969 ISBN 90-277-0078-8; Pb 90-277-0598-4

24. N. Rescher *et al.* (eds.), *Essays in Honor of Carl G. Hempel.* A Tribute on the Occasion of His 65th Birthday. 1969 ISBN 90-277-0085-0
25. P. V. Tavanec (ed.), *Problems of the Logic of Scientific Knowledge.* Translated from Russian. 1970 ISBN 90-277-0087-7
26. M. Swain (ed.), *Induction, Acceptance, and Rational Belief.* 1970 ISBN 90-277-0086-9
27. R. S. Cohen and R. J. Seeger (eds.), *Ernst Mach: Physicist and Philosopher.* [Boston Studies in the Philosophy of Science, Vol. VI]. 1970 ISBN 90-277-0016-8
28. J. Hintikka and P. Suppes, *Information and Inference.* 1970 ISBN 90-277-0155-5
29. K. Lambert, *Philosophical Problems in Logic.* Some Recent Developments. 1970 ISBN 90-277-0079-6
30. R. A. Eberle, *Nominalistic Systems.* 1970 ISBN 90-277-0161-X
31. P. Weingartner and G. Zecha (eds.), *Induction, Physics, and Ethics.* 1970 ISBN 90-277-0158-X
32. E. W. Beth, *Aspects of Modern Logic.* Translated from Dutch. 1970 ISBN 90-277-0173-3
33. R. Hilpinen (ed.), *Deontic Logic.* Introductory and Systematic Readings. 1971 *See also* No. 152. ISBN Pb (1981 rev.) 90-277-1302-2
34. J.-L. Krivine, *Introduction to Axiomatic Set Theory.* Translated from French. 1971 ISBN 90-277-0169-5; Pb 90-277-0411-2
35. J. D. Sneed, *The Logical Structure of Mathematical Physics.* 2nd rev. ed., 1979 ISBN 90-277-1056-2; Pb 90-277-1059-7
36. C. R. Kordig, *The Justification of Scientific Change.* 1971 ISBN 90-277-0181-4; Pb 90-277-0475-9
37. M. Čapek, *Bergson and Modern Physics.* A Reinterpretation and Re-evaluation. [Boston Studies in the Philosophy of Science, Vol. VII] 1971 ISBN 90-277-0186-5
38. N. R. Hanson, *What I Do Not Believe, and Other Essays.* Ed. by S. Toulmin and H. Woolf. 1971 ISBN 90-277-0191-1
39. R. C. Buck and R. S. Cohen (eds.), *PSA 1970.* Proceedings of the Second Biennial Meeting of the Philosophy of Science Association, Boston, Fall 1970. In Memory of Rudolf Carnap. [Boston Studies in the Philosophy of Science, Vol. VIII] 1971 ISBN 90-277-0187-3; Pb 90-277-0309-4
40. D. Davidson and G. Harman (eds.), *Semantics of Natural Language.* 1972 ISBN 90-277-0304-3; Pb 90-277-0310-8
41. Y. Bar-Hillel (ed.), *Pragmatics of Natural Languages.* 1971 ISBN 90-277-0194-6; Pb 90-277-0599-2
42. S. Stenlund, *Combinators, γ Terms and Proof Theory.* 1972 ISBN 90-277-0305-1
43. M. Strauss, *Modern Physics and Its Philosophy.* Selected Paper in the Logic, History, and Philosophy of Science. 1972 ISBN 90-277-0230-6
44. M. Bunge, *Method, Model and Matter.* 1973 ISBN 90-277-0252-7
45. M. Bunge, *Philosophy of Physics.* 1973 ISBN 90-277-0253-5
46. A. A. Zinov'ev, *Foundations of the Logical Theory of Scientific Knowledge (Complex Logic).* Revised and enlarged English edition with an appendix by G. A. Smirnov, E. A. Sidorenka, A. M. Fedina and L. A. Bobrova. [Boston Studies in the Philosophy of Science, Vol. IX] 1973 ISBN 90-277-0193-8; Pb 90-277-0324-8
47. L. Tondl, *Scientific Procedures.* A Contribution concerning the Methodological Problems of Scientific Concepts and Scientific Explanation. Translated from Czech by D. Short. Edited by R.S. Cohen and M.W. Wartofsky. [Boston Studies in the Philosophy of Science, Vol. X] 1973 ISBN 90-277-0147-4; Pb 90-277-0323-X
48. N. R. Hanson, *Constellations and Conjectures.* 1973 ISBN 90-277-0192-X

49. K. J. J. Hintikka, J. M. E. Moravcsik and P. Suppes (eds.), *Approaches to Natural Language.* 1973 ISBN 90-277-0220-9; Pb 90-277-0233-0
50. M. Bunge (ed.), *Exact Philosophy.* Problems, Tools and Goals. 1973 ISBN 90-277-0251-9
51. R. J. Bogdan and I. Niiniluoto (eds.), *Logic, Language and Probability.* 1973 ISBN 90-277-0312-4
52. G. Pearce and P. Maynard (eds.), *Conceptual Change.* 1973 ISBN 90-277-0287-X; Pb 90-277-0339-6
53. I. Niiniluoto and R. Tuomela, *Theoretical Concepts and Hypothetico-inductive Inference.* 1973 ISBN 90-277-0343-4
54. R. Fraissé, *Course of Mathematical Logic* – Volume 1: *Relation and Logical Formula.* Translated from French. 1973 ISBN 90-277-0268-3; Pb 90-277-0403-1 (For *Volume 2* see under No. 69).
55. A. Grünbaum, *Philosophical Problems of Space and Time.* Edited by R.S. Cohen and M.W. Wartofsky. 2nd enlarged ed. [Boston Studies in the Philosophy of Science, Vol. XII] 1973 ISBN 90-277-0357-4; Pb 90-277-0358-2
56. P. Suppes (ed.), *Space, Time and Geometry.* 1973 ISBN 90-277-0386-8; Pb 90-277-0442-2
57. H. Kelsen, *Essays in Legal and Moral Philosophy.* Selected and introduced by O. Weinberger. Translated from German by P. Heath. 1973 ISBN 90-277-0388-4
58. R. J. Seeger and R. S. Cohen (eds.), *Philosophical Foundations of Science.* [Boston Studies in the Philosophy of Science, Vol. XI] 1974 ISBN 90-277-0390-6; Pb 90-277-0376-0
59. R. S. Cohen and M. W. Wartofsky (eds.), *Logical and Epistemological Studies in Contemporary Physics.* [Boston Studies in the Philosophy of Science, Vol. XIII] 1973 ISBN 90-277-0391-4; Pb 90-277-0377-9
60. R. S. Cohen and M. W. Wartofsky (eds.), *Methodological and Historical Essays in the Natural and Social Sciences. Proceedings of the Boston Colloquium for the Philosophy of Science, 1969–1972.* [Boston Studies in the Philosophy of Science, Vol. XIV] 1974 ISBN 90-277-0392-2; Pb 90-277-0378-7
61. R. S. Cohen, J. J. Stachel and M. W. Wartofsky (eds.), *For Dirk Struik. Scientific, Historical and Political Essays.* [Boston Studies in the Philosophy of Science, Vol. XV] 1974 ISBN 90-277-0393-0; Pb 90-277-0379-5
62. K. Ajdukiewicz, *Pragmatic Logic.* Translated from Polish by O. Wojtasiewicz. 1974 ISBN 90-277-0326-4
63. S. Stenlund (ed.), *Logical Theory and Semantic Analysis.* Essays dedicated to Stig Kanger on His 50th Birthday. 1974 ISBN 90-277-0438-4
64. K. F. Schaffner and R. S. Cohen (eds.), *PSA 1972. Proceedings of the Third Biennial Meeting of the Philosophy of Science Association.* [Boston Studies in the Philosophy of Science, Vol. XX] 1974 ISBN 90-277-0408-2; Pb 90-277-0409-0
65. H. E. Kyburg, Jr., *The Logical Foundations of Statistical Inference.* 1974 ISBN 90-277-0330-2; Pb 90-277-0430-9
66. M. Grene, *The Understanding of Nature.* Essays in the Philosophy of Biology. [Boston Studies in the Philosophy of Science, Vol. XXIII] 1974 ISBN 90-277-0462-7; Pb 90-277-0463-5
67. J. M. Broekman, *Structuralism: Moscow, Prague, Paris.* Translated from German. 1974 ISBN 90-277-0478-3
68. N. Geschwind, *Selected Papers on Language and the Brain.* [Boston Studies in the Philosophy of Science, Vol. XVI] 1974 ISBN 90-277-0262-4; Pb 90-277-0263-2
69. R. Fraissé, *Course of Mathematical Logic* – Volume 2: *Model Theory.* Translated from French. 1974 ISBN 90-277-0269-1; Pb 90-277-0510-0 (For *Volume 1* see under No. 54)

70. A. Grzegorczyk, *An Outline of Mathematical Logic*. Fundamental Results and Notions explained with all Details. Translated from Polish. 1974 ISBN 90-277-0359-0; Pb 90-277-0447-3
71. F. von Kutschera, *Philosophy of Language*. 1975 ISBN 90-277-0591-7
72. J. Manninen and R. Tuomela (eds.), *Essays on Explanation and Understanding*. Studies in the Foundations of Humanities and Social Sciences. 1976 ISBN 90-277-0592-5
73. J. Hintikka (ed.), *Rudolf Carnap, Logical Empiricist*. Materials and Perspectives. 1975 ISBN 90-277-0583-6
74. M. Čapek (ed.), *The Concepts of Space and Time*. Their Structure and Their Development. [Boston Studies in the Philosophy of Science, Vol. XXII] 1976 ISBN 90-277-0355-8; Pb 90-277-0375-2
75. J. Hintikka and U. Remes, *The Method of Analysis*. Its Geometrical Origin and Its General Significance. [Boston Studies in the Philosophy of Science, Vol. XXV] 1974 ISBN 90-277-0532-1; Pb 90-277-0543-7
76. J. E. Murdoch and E. D. Sylla (eds.), *The Cultural Context of Medieval Learning*. [Boston Studies in the Philosophy of Science, Vol. XXVI] 1975 ISBN 90-277-0560-7; Pb 90-277-0587-9
77. S. Amsterdamski, *Between Experience and Metaphysics*. Philosophical Problems of the Evolution of Science. [Boston Studies in the Philosophy of Science, Vol. XXXV] 1975 ISBN 90-277-0568-2; Pb 90-277-0580-1
78. P. Suppes (ed.), *Logic and Probability in Quantum Mechanics*. 1976 ISBN 90-277-0570-4; Pb 90-277-1200-X
79. H. von Helmholtz: *Epistemological Writings. The Paul Hertz / Moritz Schlick Centenary Edition of 1921 with Notes and Commentary by the Editors*. Newly translated from German by M. F. Lowe. Edited, with an Introduction and Bibliography, by R. S. Cohen and Y. Elkana. [Boston Studies in the Philosophy of Science, Vol. XXXVII] 1975 ISBN 90-277-0290-X; Pb 90-277-0582-8
80. J. Agassi, *Science in Flux*. [Boston Studies in the Philosophy of Science, Vol. XXVIII] 1975 ISBN 90-277-0584-4; Pb 90-277-0612-2
81. S. G. Harding (ed.), *Can Theories Be Refuted?* Essays on the Duhem-Quine Thesis. 1976 ISBN 90-277-0629-8; Pb 90-277-0630-1
82. S. Nowak, *Methodology of Sociological Research*. General Problems. 1977 ISBN 90-277-0486-4
83. J. Piaget, J.-B. Grize, A. Szeminśska and V. Bang, *Epistemology and Psychology of Functions*. Translated from French. 1977 ISBN 90-277-0804-5
84. M. Grene and E. Mendelsohn (eds.), *Topics in the Philosophy of Biology*. [Boston Studies in the Philosophy of Science, Vol. XXVII] 1976 ISBN 90-277-0595-X; Pb 90-277-0596-8
85. E. Fischbein, *The Intuitive Sources of Probabilistic Thinking in Children*. 1975 ISBN 90-277-0626-3; Pb 90-277-1190-9
86. E. W. Adams, *The Logic of Conditionals*. An Application of Probability to Deductive Logic. 1975 ISBN 90-277-0631-X
87. M. Przełęcki and R. Wójcicki (eds.), *Twenty-Five Years of Logical Methodology in Poland*. Translated from Polish. 1976 ISBN 90-277-0601-8
88. J. Topolski, *The Methodology of History*. Translated from Polish by O. Wojtasiewicz. 1976 ISBN 90-277-0550-X
89. A. Kasher (ed.), *Language in Focus: Foundations, Methods and Systems*. Essays dedicated to Yehoshua Bar-Hillel. [Boston Studies in the Philosophy of Science, Vol. XLIII] 1976 ISBN 90-277-0644-1; Pb 90-277-0645-X

90. J. Hintikka, *The Intentions of Intentionality and Other New Models for Modalities.* 1975
ISBN 90-277-0633-6; Pb 90-277-0634-4
91. W. Stegmüller, *Collected Papers on Epistemology, Philosophy of Science and History of Philosophy.* 2 Volumes. 1977 Set ISBN 90-277-0767-7
92. D. M. Gabbay, *Investigations in Modal and Tense Logics with Applications to Problems in Philosophy and Linguistics.* 1976 ISBN 90-277-0656-5
93. R. J. Bogdan, *Local Induction.* 1976 ISBN 90-277-0649-2
94. S. Nowak, *Understanding and Prediction.* Essays in the Methodology of Social and Behavioral Theories. 1976 ISBN 90-277-0558-5; Pb 90-277-1199-2
95. P. Mittelstaedt, *Philosophical Problems of Modern Physics.* [Boston Studies in the Philosophy of Science, Vol. XVIII] 1976 ISBN 90-277-0285-3; Pb 90-277-0506-2
96. G. Holton and W. A. Blanpied (eds.), *Science and Its Public: The Changing Relationship.* [Boston Studies in the Philosophy of Science, Vol. XXXIII] 1976
ISBN 90-277-0657-3; Pb 90-277-0658-1
97. M. Brand and D. Walton (eds.), *Action Theory.* 1976 ISBN 90-277-0671-9
98. P. Gochet, *Outline of a Nominalist Theory of Propositions.* An Essay in the Theory of Meaning and in the Philosophy of Logic. 1980 ISBN 90-277-1031-7
99. R. S. Cohen, P. K. Feyerabend, and M. W. Wartofsky (eds.), *Essays in Memory of Imre Lakatos.* [Boston Studies in the Philosophy of Science, Vol. XXXIX] 1976
ISBN 90-277-0654-9; Pb 90-277-0655-7
100. R. S. Cohen and J. J. Stachel (eds.), *Selected Papers of Léon Rosenfield.* [Boston Studies in the Philosophy of Science, Vol. XXI] 1979 ISBN 90-277-0651-4; Pb 90-277-0652-2
101. R. S. Cohen, C. A. Hooker, A. C. Michalos and J. W. van Evra (eds.), *PSA 1974. Proceedings of the 1974 Biennial Meeting of the Philosophy of Science Association.* [Boston Studies in the Philosophy of Science, Vol. XXXII] 1976 ISBN 90-277-0647-6; Pb 90-277-0648-4
102. Y. Fried and J. Agassi, *Paranoia.* A Study in Diagnosis. [Boston Studies in the Philosophy of Science, Vol. L] 1976 ISBN 90-277-0704-9; Pb 90-277-0705-7
103. M. Przełęcki, K. Szaniawski and R. Wójcicki (eds.), *Formal Methods in the Methodology of Empirical Sciences.* 1976 ISBN 90-277-0698-0
104. J. M. Vickers, *Belief and Probability.* 1976 ISBN 90-277-0744-8
105. K. H. Wolff, *Surrender and Catch.* Experience and Inquiry Today. [Boston Studies in the Philosophy of Science, Vol. LI] 1976 ISBN 90-277-0758-8; Pb 90-277-0765-0
106. K. Kosík, *Dialectics of the Concrete.* A Study on Problems of Man and World. [Boston Studies in the Philosophy of Science, Vol. LII] 1976 ISBN 90-277-0761-8; Pb 90-277-0764-2
107. N. Goodman, *The Structure of Appearance.* 3rd ed. with an Introduction by G. Hellman. [Boston Studies in the Philosophy of Science, Vol. LIII] 1977
ISBN 90-277-0773-1; Pb 90-277-0774-X
108. K. Ajdukiewicz, *The Scientific World-Perspective and Other Essays, 1931-1963.* Translated from Polish. Edited and with an Introduction by J. Giedymin. 1978 ISBN 90-277-0527-5
109. R. L. Causey, *Unity of Science.* 1977 ISBN 90-277-0779-0
110. R. E. Grandy, *Advanced Logic for Applications.* 1977 ISBN 90-277-0781-2
111. R. P. McArthur, *Tense Logic.* 1976 ISBN 90-277-0697-2
112. L. Lindahl, *Position and Change.* A Study in Law and Logic. Translated from Swedish by P. Needham. 1977 ISBN 90-277-0787-1
113. R. Tuomela, *Dispositions.* 1978 ISBN 90-277-0810-X
114. H. A. Simon, *Models of Discovery and Other Topics in the Methods of Science.* [Boston Studies in the Philosophy of Science, Vol. LIV] 1977 ISBN 90-277-0812-6; Pb 90-277-0858-4

115. R. D. Rosenkrantz, *Inference, Method and Decision.* Towards a Bayesian Philosophy of Science. 1977 ISBN 90-277-0817-7; Pb 90-277-0818-5
116. R. Tuomela, *Human Action and Its Explanation.* A Study on the Philosophical Foundations of Psychology. 1977 ISBN 90-277-0824-X
117. M. Lazerowitz, *The Language of Philosophy.* Freud and Wittgenstein. [Boston Studies in the Philosophy of Science, Vol. LV] 1977 ISBN 90-277-0826-6; Pb 90-277-0862-2
118. Not published 119. J. Pelc (ed.), *Semiotics in Poland, 1894–1969.* Translated from Polish. 1979 ISBN 90-277-0811-8
120. I. Pörn, *Action Theory and Social Science.* Some Formal Models. 1977 ISBN 90-277-0846-0
121. J. Margolis, *Persons and Mind.* The Prospects of Nonreductive Materialism. [Boston Studies in the Philosophy of Science, Vol. LVII] 1977 ISBN 90-277-0854-1; Pb 90-277-0863-0
122. J. Hintikka, I. Niiniluoto, and E. Saarinen (eds.), *Essays on Mathematical and Philosophical Logic.* 1979 ISBN 90-277-0879-7
123. T. A. F. Kuipers, *Studies in Inductive Probability and Rational Expectation.* 1978 ISBN 90-277-0882-7
124. E. Saarinen, R. Hilpinen, I. Niiniluoto and M. P. Hintikka (eds.), *Essays in Honour of Jaakko Hintikka on the Occasion of His 50th Birthday.* 1979 ISBN 90-277-0916-5
125. G. Radnitzky and G. Andersson (eds.), *Progress and Rationality in Science.* [Boston Studies in the Philosophy of Science, Vol. LVIII] 1978 ISBN 90-277-0921-1; Pb 90-277-0922-X
126. P. Mittelstaedt, *Quantum Logic.* 1978 ISBN 90-277-0925-4
127. K. A. Bowen, *Model Theory for Modal Logic.* Kripke Models for Modal Predicate Calculi. 1979 ISBN 90-277-0929-7
128. H. A. Bursen, *Dismantling the Memory Machine.* A Philosophical Investigation of Machine Theories of Memory. 1978 ISBN 90-277-0933-5
129. M. W. Wartofsky, *Models.* Representation and the Scientific Understanding. [Boston Studies in the Philosophy of Science, Vol. XLVIII] 1979 ISBN 90-277-0736-7; Pb 90-277-0947-5
130. D. Ihde, *Technics and Praxis.* A Philosophy of Technology. [Boston Studies in the Philosophy of Science, Vol. XXIV] 1979 ISBN 90-277-0953-X; Pb 90-277-0954-8
131. J. J. Wiatr (ed.), *Polish Essays in the Methodology of the Social Sciences.* [Boston Studies in the Philosophy of Science, Vol. XXIX] 1979 ISBN 90-277-0723-5; Pb 90-277-0956-4
132. W. C. Salmon (ed.), *Hans Reichenbach: Logical Empiricist.* 1979 ISBN 90-277-0958-0
133. P. Bieri, R.-P. Horstmann and L. Krüger (eds.), *Transcendental Arguments in Science.* Essays in Epistemology. 1979 ISBN 90-277-0963-7; Pb 90-277-0964-5
134. M. Marković and G. Petrović (eds.), *Praxis.* Yugoslav Essays in the Philosophy and Methodology of the Social Sciences. [Boston Studies in the Philosophy of Science, Vol. XXXVI] 1979 ISBN 90-277-0727-8; Pb 90-277-0968-8
135. R. Wójcicki, *Topics in the Formal Methodology of Empirical Sciences.* Translated from Polish. 1979 ISBN 90-277-1004-X
136. G. Radnitzky and G. Andersson (eds.), *The Structure and Development of Science.* [Boston Studies in the Philosophy of Science, Vol. LIX] 1979 ISBN 90-277-0994-7; Pb 90-277-0995-5
137. J. C. Webb, *Mechanism, Mentalism and Metamathematics.* An Essay on Finitism. 1980 ISBN 90-277-1046-5
138. D. F. Gustafson and B. L. Tapscott (eds.), *Body, Mind and Method.* Essays in Honor of Virgil C. Aldrich. 1979 ISBN 90-277-1013-9
139. L. Nowak, *The Structure of Idealization.* Towards a Systematic Interpretation of the Marxian Idea of Science. 1980 ISBN 90-277-1014-7

140. C. Perelman, *The New Rhetoric and the Humanities*. Essays on Rhetoric and Its Applications. Translated from French and German. With an Introduction by H. Zyskind. 1979 ISBN 90-277-1018-X; Pb 90-277-1019-8
141. W. Rabinowicz, *Universalizability*. A Study in Morals and Metaphysics. 1979 ISBN 90-277-1020-2
142. C. Perelman, *Justice, Law and Argument*. Essays on Moral and Legal Reasoning. Translated from French and German. With an Introduction by H.J. Berman. 1980 ISBN 90-277-1089-9; Pb 90-277-1090-2
143. S. Kanger and S. Öhman (eds.), *Philosophy and Grammar*. Papers on the Occasion of the Quincentennial of Uppsala University. 1981 ISBN 90-277-1091-0
144. T. Pawlowski, *Concept Formation in the Humanities and the Social Sciences*. 1980 ISBN 90-277-1096-1
145. J. Hintikka, D. Gruender and E. Agazzi (eds.), *Theory Change, Ancient Axiomatics and Galileo's Methodology*. Proceedings of the 1978 Pisa Conference on the History and Philosophy of Science, Volume I. 1981 ISBN 90-277-1126-7
146. J. Hintikka, D. Gruender and E. Agazzi (eds.), *Probabilistic Thinking, Thermodynamics, and the Interaction of the History and Philosophy of Science*. Proceedings of the 1978 Pisa Conference on the History and Philosophy of Science, Volume II. 1981 ISBN 90-277-1127-5
147. U. Mönnich (ed.), *Aspects of Philosophical Logic*. Some Logical Forays into Central Notions of Linguistics and Philosophy. 1981 ISBN 90-277-1201-8
148. D. M. Gabbay, *Semantical Investigations in Heyting's Intuitionistic Logic*. 1981 ISBN 90-277-1202-6
149. E. Agazzi (ed.), *Modern Logic – A Survey*. Historical, Philosophical, and Mathematical Aspects of Modern Logic and Its Applications. 1981 ISBN 90-277-1137-2
150. A. F. Parker-Rhodes, *The Theory of Indistinguishables*. A Search for Explanatory Principles below the Level of Physics. 1981 ISBN 90-277-1214-X
151. J. C. Pitt, *Pictures, Images, and Conceptual Change*. An Analysis of Wilfrid Sellars' Philosophy of Science. 1981 ISBN 90-277-1276-X; Pb 90-277-1277-8
152. R. Hilpinen (ed.), *New Studies in Deontic Logic*. Norms, Actions, and the Foundations of Ethics. 1981 ISBN 90-277-1278-6; Pb 90-277-1346-4
153. C. Dilworth, *Scientific Progress*. A Study Concerning the Nature of the Relation between Successive Scientific Theories. 3rd rev. ed., 1994 ISBN 0-7923-2487-0; Pb 0-7923-2488-9
154. D. Woodruff Smith and R. McIntyre, *Husserl and Intentionality*. A Study of Mind, Meaning, and Language. 1982 ISBN 90-277-1392-8; Pb 90-277-1730-3
155. R. J. Nelson, *The Logic of Mind*. 2nd. ed., 1989 ISBN 90-277-2819-4; Pb 90-277-2822-4
156. J. F. A. K. van Benthem, *The Logic of Time*. A Model-Theoretic Investigation into the Varieties of Temporal Ontology, and Temporal Discourse. 1983; 2nd ed., 1991 ISBN 0-7923-1081-0
157. R. Swinburne (ed.), *Space, Time and Causality*. 1983 ISBN 90-277-1437-1
158. E. T. Jaynes, *Papers on Probability, Statistics and Statistical Physics*. Ed. by R. D. Rozenkrantz. 1983 ISBN 90-277-1448-7; Pb (1989) 0-7923-0213-3
159. T. Chapman, *Time: A Philosophical Analysis*. 1982 ISBN 90-277-1465-7
160. E. N. Zalta, *Abstract Objects*. An Introduction to Axiomatic Metaphysics. 1983 ISBN 90-277-1474-6
161. S. Harding and M. B. Hintikka (eds.), *Discovering Reality*. Feminist Perspectives on Epistemology, Metaphysics, Methodology, and Philosophy of Science. 1983 ISBN 90-277-1496-7; Pb 90-277-1538-6
162. M. A. Stewart (ed.), *Law, Morality and Rights*. 1983 ISBN 90-277-1519-X

163. D. Mayr and G. Süssmann (eds.), *Space, Time, and Mechanics*. Basic Structures of a Physical Theory. 1983 ISBN 90-277-1525-4
164. D. Gabbay and F. Guenthner (eds.), *Handbook of Philosophical Logic*. Vol. I: Elements of Classical Logic. 1983 ISBN 90-277-1542-4
165. D. Gabbay and F. Guenthner (eds.), *Handbook of Philosophical Logic*. Vol. II: Extensions of Classical Logic. 1984 ISBN 90-277-1604-8
166. D. Gabbay and F. Guenthner (eds.), *Handbook of Philosophical Logic*. Vol. III: Alternative to Classical Logic. 1986 ISBN 90-277-1605-6
167. D. Gabbay and F. Guenthner (eds.), *Handbook of Philosophical Logic*. Vol. IV: Topics in the Philosophy of Language. 1989 ISBN 90-277-1606-4
168. A. J. I. Jones, *Communication and Meaning*. An Essay in Applied Modal Logic. 1983 ISBN 90-277-1543-2
169. M. Fitting, *Proof Methods for Modal and Intuitionistic Logics*. 1983 ISBN 90-277-1573-4
170. J. Margolis, *Culture and Cultural Entities*. Toward a New Unity of Science. 1984 ISBN 90-277-1574-2
171. R. Tuomela, *A Theory of Social Action*. 1984 ISBN 90-277-1703-6
172. J. J. E. Gracia, E. Rabossi, E. Villanueva and M. Dascal (eds.), *Philosophical Analysis in Latin America*. 1984 ISBN 90-277-1749-4
173. P. Ziff, *Epistemic Analysis*. A Coherence Theory of Knowledge. 1984 ISBN 90-277-1751-7
174. P. Ziff, *Antiaesthetics*. An Appreciation of the Cow with the Subtile Nose. 1984 ISBN 90-277-1773-7
175. W. Balzer, D. A. Pearce, and H.-J. Schmidt (eds.), *Reduction in Science*. Structure, Examples, Philosophical Problems. 1984 ISBN 90-277-1811-3
176. A. Peczenik, L. Lindahl and B. van Roermund (eds.), *Theory of Legal Science*. Proceedings of the Conference on Legal Theory and Philosophy of Science (Lund, Sweden, December 1983). 1984 ISBN 90-277-1834-2
177. I. Niiniluoto, *Is Science Progressive?* 1984 ISBN 90-277-1835-0
178. B. K. Matilal and J. L. Shaw (eds.), *Analytical Philosophy in Comparative Perspective*. Exploratory Essays in Current Theories and Classical Indian Theories of Meaning and Reference. 1985 ISBN 90-277-1870-9
179. P. Kroes, *Time: Its Structure and Role in Physical Theories*. 1985 ISBN 90-277-1894-6
180. J. H. Fetzer, *Sociobiology and Epistemology*. 1985 ISBN 90-277-2005-3; Pb 90-277-2006-1
181. L. Haaparanta and J. Hintikka (eds.), *Frege Synthesized*. Essays on the Philosophical and Foundational Work of Gottlob Frege. 1986 ISBN 90-277-2126-2
182. M. Detlefsen, *Hilbert's Program*. An Essay on Mathematical Instrumentalism. 1986 ISBN 90-277-2151-3
183. J. L. Golden and J. J. Pilotta (eds.), *Practical Reasoning in Human Affairs*. Studies in Honor of Chaim Perelman. 1986 ISBN 90-277-2255-2
184. H. Zandvoort, *Models of Scientific Development and the Case of Nuclear Magnetic Resonance*. 1986 ISBN 90-277-2351-6
185. I. Niiniluoto, *Truthlikeness*. 1987 ISBN 90-277-2354-0
186. W. Balzer, C. U. Moulines and J. D. Sneed, *An Architectonic for Science*. The Structuralist Program. 1987 ISBN 90-277-2403-2
187. D. Pearce, *Roads to Commensurability*. 1987 ISBN 90-277-2414-8
188. L. M. Vaina (ed.), *Matters of Intelligence*. Conceptual Structures in Cognitive Neuroscience. 1987 ISBN 90-277-2460-1

189. H. Siegel, *Relativism Refuted.* A Critique of Contemporary Epistemological Relativism. 1987 ISBN 90-277-2469-5
190. W. Callebaut and R. Pinxten, *Evolutionary Epistemology.* A Multiparadigm Program, with a Complete Evolutionary Epistemology Bibliograph. 1987 ISBN 90-277-2582-9
191. J. Kmita, *Problems in Historical Epistemology.* 1988 ISBN 90-277-2199-8
192. J. H. Fetzer (ed.), *Probability and Causality.* Essays in Honor of Wesley C. Salmon, with an Annotated Bibliography. 1988 ISBN 90-277-2607-8; Pb 1-5560-8052-2
193. A. Donovan, L. Laudan and R. Laudan (eds.), *Scrutinizing Science.* Empirical Studies of Scientific Change. 1988 ISBN 90-277-2608-6
194. H.R. Otto and J.A. Tuedio (eds.), *Perspectives on Mind.* 1988 ISBN 90-277-2640-X
195. D. Batens and J.P. van Bendegem (eds.), *Theory and Experiment.* Recent Insights and New Perspectives on Their Relation. 1988 ISBN 90-277-2645-0
196. J. Österberg, *Self and Others.* A Study of Ethical Egoism. 1988 ISBN 90-277-2648-5
197. D.H. Helman (ed.), *Analogical Reasoning.* Perspectives of Artificial Intelligence, Cognitive Science, and Philosophy. 1988 ISBN 90-277-2711-2
198. J. Woleński, *Logic and Philosophy in the Lvov-Warsaw School.* 1989 ISBN 90-277-2749-X
199. R. Wójcicki, *Theory of Logical Calculi.* Basic Theory of Consequence Operations. 1988 ISBN 90-277-2785-6
200. J. Hintikka and M.B. Hintikka, *The Logic of Epistemology and the Epistemology of Logic.* Selected Essays. 1989 ISBN 0-7923-0040-8; Pb 0-7923-0041-6
201. E. Agazzi (ed.), *Probability in the Sciences.* 1988 ISBN 90-277-2808-9
202. M. Meyer (ed.), *From Metaphysics to Rhetoric.* 1989 ISBN 90-277-2814-3
203. R.L. Tieszen, *Mathematical Intuition.* Phenomenology and Mathematical Knowledge. 1989 ISBN 0-7923-0131-5
204. A. Melnick, *Space, Time, and Thought in Kant.* 1989 ISBN 0-7923-0135-8
205. D.W. Smith, *The Circle of Acquaintance.* Perception, Consciousness, and Empathy. 1989 ISBN 0-7923-0252-4
206. M.H. Salmon (ed.), *The Philosophy of Logical Mechanism.* Essays in Honor of Arthur W. Burks. With his Responses, and with a Bibliography of Burk's Work. 1990 ISBN 0-7923-0325-3
207. M. Kusch, *Language as Calculus vs. Language as Universal Medium.* A Study in Husserl, Heidegger, and Gadamer. 1989 ISBN 0-7923-0333-4
208. T.C. Meyering, *Historical Roots of Cognitive Science.* The Rise of a Cognitive Theory of Perception from Antiquity to the Nineteenth Century. 1989 ISBN 0-7923-0349-0
209. P. Kosso, *Observability and Observation in Physical Science.* 1989 ISBN 0-7923-0389-X
210. J. Kmita, *Essays on the Theory of Scientific Cognition.* 1990 ISBN 0-7923-0441-1
211. W. Sieg (ed.), *Acting and Reflecting.* The Interdisciplinary Turn in Philosophy. 1990 ISBN 0-7923-0512-4
212. J. Karpiński, *Causality in Sociological Research.* 1990 ISBN 0-7923-0546-9
213. H.A. Lewis (ed.), *Peter Geach: Philosophical Encounters.* 1991 ISBN 0-7923-0823-9
214. M. Ter Hark, *Beyond the Inner and the Outer.* Wittgenstein's Philosophy of Psychology. 1990 ISBN 0-7923-0850-6
215. M. Gosselin, *Nominalism and Contemporary Nominalism.* Ontological and Epistemological Implications of the Work of W.V.O. Quine and of N. Goodman. 1990 ISBN 0-7923-0904-9
216. J.H. Fetzer, D. Shatz and G. Schlesinger (eds.), *Definitions and Definability.* Philosophical Perspectives. 1991 ISBN 0-7923-1046-2
217. E. Agazzi and A. Cordero (eds.), *Philosophy and the Origin and Evolution of the Universe.* 1991 ISBN 0-7923-1322-4

218. M. Kusch, *Foucault's Strata and Fields*. An Investigation into Archaeological and Genealogical Science Studies. 1991 ISBN 0-7923-1462-X
219. C.J. Posy, *Kant's Philosophy of Mathematics*. Modern Essays. 1992 ISBN 0-7923-1495-6
220. G. Van de Vijver, *New Perspectives on Cybernetics*. Self-Organization, Autonomy and Connectionism. 1992 ISBN 0-7923-1519-7
221. J.C. Nyíri, *Tradition and Individuality*. Essays. 1992 ISBN 0-7923-1566-9
222. R. Howell, *Kant's Transcendental Deduction*. An Analysis of Main Themes in His Critical Philosophy. 1992 ISBN 0-7923-1571-5
223. A. García de la Sienra, *The Logical Foundations of the Marxian Theory of Value*. 1992 ISBN 0-7923-1778-5
224. D.S. Shwayder, *Statement and Referent*. An Inquiry into the Foundations of Our Conceptual Order. 1992 ISBN 0-7923-1803-X
225. M. Rosen, *Problems of the Hegelian Dialectic*. Dialectic Reconstructed as a Logic of Human Reality. 1993 ISBN 0-7923-2047-6
226. P. Suppes, *Models and Methods in the Philosophy of Science: Selected Essays*. 1993 ISBN 0-7923-2211-8
227. R. M. Dancy (ed.), *Kant and Critique: New Essays in Honor of W. H. Werkmeister*. 1993 ISBN 0-7923-2244-4
228. J. Woleński (ed.), *Philosophical Logic in Poland*. 1993 ISBN 0-7923-2293-2
229. M. De Rijke (ed.), *Diamonds and Defaults*. Studies in Pure and Applied Intensional Logic. 1993 ISBN 0-7923-2342-4
230. B.K. Matilal and A. Chakrabarti (eds.), *Knowing from Words*. Western and Indian Philosophical Analysis of Understanding and Testimony. 1994 ISBN 0-7923-2345-9
231. S.A. Kleiner, *The Logic of Discovery*. A Theory of the Rationality of Scientific Research. 1993 ISBN 0-7923-2371-8
232. R. Festa, *Optimum Inductive Methods*. A Study in Inductive Probability, Bayesian Statistics, and Verisimilitude. 1993 ISBN 0-7923-2460-9
233. P. Humphreys (ed.), *Patrick Suppes: Scientific Philosopher*. Vol. 1: Probability and Probabilistic Causality. 1994 ISBN 0-7923-2552-4
234. P. Humphreys (ed.), *Patrick Suppes: Scientific Philosopher*. Vol. 2: Philosophy of Physics, Theory Structure, and Measurement Theory. 1994 ISBN 0-7923-2553-2
235. P. Humphreys (ed.), *Patrick Suppes: Scientific Philosopher*. Vol. 3: Language, Logic, and Psychology. 1994 ISBN 0-7923-2862-0
Set ISBN (Vols 233–235) 0-7923-2554-0
236. D. Prawitz and D. Westerståhl (eds.), *Logic and Philosophy of Science in Uppsala*. Papers from the 9th International Congress of Logic, Methodology, and Philosophy of Science. 1994 ISBN 0-7923-2702-0
237. L. Haaparanta (ed.), *Mind, Meaning and Mathematics*. Essays on the Philosophical Views of Husserl and Frege. 1994 ISBN 0-7923-2703-9
238. J. Hintikka (ed.), *Aspects of Metaphor*. 1994 ISBN 0-7923-2786-1
239. B. McGuinness and G. Oliveri (eds.), *The Philosophy of Michael Dummett*. With Replies from Michael Dummett. 1994 ISBN 0-7923-2804-3
240. D. Jamieson (ed.), *Language, Mind, and Art*. Essays in Appreciation and Analysis, In Honor of Paul Ziff. 1994 ISBN 0-7923-2810-8
241. G. Preyer, F. Siebelt and A. Ulfig (eds.), *Language, Mind and Epistemology*. On Donald Davidson's Philosophy. 1994 ISBN 0-7923-2811-6
242. P. Ehrlich (ed.), *Real Numbers, Generalizations of the Reals, and Theories of Continua*. 1994 ISBN 0-7923-2689-X

243. G. Debrock and M. Hulswit (eds.), *Living Doubt*. Essays concerning the epistemology of Charles Sanders Peirce. 1994 ISBN 0-7923-2898-1
244. J. Srzednicki, *To Know or Not to Know*. Beyond Realism and Anti-Realism. 1994 ISBN 0-7923-2909-0
245. R. Egidi (ed.), *Wittgenstein: Mind and Language*. 1995 ISBN 0-7923-3171-0
246. A. Hyslop, *Other Minds*. 1995 ISBN 0-7923-3245-8
247. L. Pólos and M. Masuch (eds.), *Applied Logic: How, What and Why*. Logical Approaches to Natural Language. 1995 ISBN 0-7923-3432-9
248. M. Krynicki, M. Mostowski and L.M. Szczerba (eds.), *Quantifiers: Logics, Models and Computation*. Volume One: Surveys. 1995 ISBN 0-7923-3448-5
249. M. Krynicki, M. Mostowski and L.M. Szczerba (eds.), *Quantifiers: Logics, Models and Computation*. Volume Two: Contributions. 1995 ISBN 0-7923-3449-3
Set ISBN (Vols 248 + 249) 0-7923-3450-7
250. R.A. Watson, *Representational Ideas from Plato to Patricia Churchland*. 1995 ISBN 0-7923-3453-1
251. J. Hintikka (ed.), *From Dedekind to Gödel*. Essays on the Development of the Foundations of Mathematics. 1995 ISBN 0-7923-3484-1
252. A. Wiśniewski, *The Posing of Questions*. Logical Foundations of Erotetic Inferences. 1995 ISBN 0-7923-3637-2
253. J. Peregrin, *Doing Worlds with Words*. Formal Semantics without Formal Metaphysics. 1995 ISBN 0-7923-3742-5
254. I.A. Kieseppä, *Truthlikeness for Multidimensional, Quantitative Cognitive Problems*. 1996 ISBN 0-7923-4005-1
255. P. Hugly and C. Sayward: *Intensionality and Truth*. An Essay on the Philosophy of A.N. Prior. 1996 ISBN 0-7923-4119-8
256. L. Hankinson Nelson and J. Nelson (eds.): *Feminism, Science, and the Philosophy of Science*. 1997 ISBN 0-7923-4162-7
257. P.I. Bystrov and V.N. Sadovsky (eds.): *Philosophical Logic and Logical Philosophy*. Essays in Honour of Vladimir A. Smirnov. 1996 ISBN 0-7923-4270-4
258. Å.E. Andersson and N-E. Sahlin (eds.): *The Complexity of Creativity*. 1996 ISBN 0-7923-4346-8
259. M.L. Dalla Chiara, K. Doets, D. Mundici and J. van Benthem (eds.): *Logic and Scientific Methods*. Volume One of the Tenth International Congress of Logic, Methodology and Philosophy of Science, Florence, August 1995. 1997 ISBN 0-7923-4383-2
260. M.L. Dalla Chiara, K. Doets, D. Mundici and J. van Benthem (eds.): *Structures and Norms in Science*. Volume Two of the Tenth International Congress of Logic, Methodology and Philosophy of Science, Florence, August 1995. 1997 ISBN 0-7923-4384-0
Set ISBN (Vols 259 + 260) 0-7923-4385-9
261. A. Chakrabarti: *Denying Existence*. The Logic, Epistemology and Pragmatics of Negative Existentials and Fictional Discourse. 1997 ISBN 0-7923-4388-3
262. A. Biletzki: *Talking Wolves*. Thomas Hobbes on the Language of Politics and the Politics of Language. 1997 ISBN 0-7923-4425-1
263. D. Nute (ed.): *Defeasible Deontic Logic*. 1997 ISBN 0-7923-4630-0
264. U. Meixner: *Axiomatic Formal Ontology*. 1997 ISBN 0-7923-4747-X
265. I. Brinck: *The Indexical 'I'*. The First Person in Thought and Language. 1997 ISBN 0-7923-4741-2
266. G. Hölmström-Hintikka and R. Tuomela (eds.): *Contemporary Action Theory*. Volume 1: Individual Action. 1997 ISBN 0-7923-4753-6; Set: 0-7923-4754-4

267. G. Hölmström-Hintikka and R. Tuomela (eds.): *Contemporary Action Theory.* Volume 2: Social Action. 1997 ISBN 0-7923-4752-8; Set: 0-7923-4754-4
268. B.-C. Park: *Phenomenological Aspects of Wittgenstein's Philosophy.* 1998 ISBN 0-7923-4813-3
269. J. Paśniczek: *The Logic of Intentional Objects.* A Meinongian Version of Classical Logic. 1998 Hb ISBN 0-7923-4880-X; Pb ISBN 0-7923-5578-4
270. P.W. Humphreys and J.H. Fetzer (eds.): *The New Theory of Reference.* Kripke, Marcus, and Its Origins. 1998 ISBN 0-7923-4898-2
271. K. Szaniawski, A. Chmielewski and J. Woleński (eds.): *On Science, Inference, Information and Decision Making.* Selected Essays in the Philosophy of Science. 1998 ISBN 0-7923-4922-9
272. G.H. von Wright: *In the Shadow of Descartes.* Essays in the Philosophy of Mind. 1998 ISBN 0-7923-4992-X
273. K. Kijania-Placek and J. Woleński (eds.): *The Lvov–Warsaw School and Contemporary Philosophy.* 1998 ISBN 0-7923-5105-3
274. D. Dedrick: *Naming the Rainbow.* Colour Language, Colour Science, and Culture. 1998 ISBN 0-7923-5239-4
275. L. Albertazzi (ed.): *Shapes of Forms.* From Gestalt Psychology and Phenomenology to Ontology and Mathematics. 1999 ISBN 0-7923-5246-7
276. P. Fletcher: *Truth, Proof and Infinity.* A Theory of Constructions and Constructive Reasoning. 1998 ISBN 0-7923-5262-9
277. M. Fitting and R.L. Mendelsohn (eds.): *First-Order Modal Logic.* 1998 Hb ISBN 0-7923-5334-X; Pb ISBN 0-7923-5335-8
278. J.N. Mohanty: *Logic, Truth and the Modalities from a Phenomenological Perspective.* 1999 ISBN 0-7923-5550-4
279. T. Placek: *Mathematical Intiutionism and Intersubjectivity.* A Critical Exposition of Arguments for Intuitionism. 1999 ISBN 0-7923-5630-6
280. A. Cantini, E. Casari and P. Minari (eds.): *Logic and Foundations of Mathematics.* 1999 ISBN 0-7923-5659-4 set ISBN 0-7923-5867-8
281. M.L. Dalla Chiara, R. Giuntini and F. Laudisa (eds.): *Language, Quantum, Music.* 1999 ISBN 0-7923-5727-2; set ISBN 0-7923-5867-8
282. R. Egidi (ed.): *In Search of a New Humanism.* The Philosophy of Georg Hendrik von Wright. 1999 ISBN 0-7923-5810-4
283. F. Vollmer: *Agent Causality.* 1999 ISBN 0-7923-5848-1
284. J. Peregrin (ed.): *Truth and Its Nature (if Any).* 1999 ISBN 0-7923-5865-1
285. M. De Caro (ed.): *Interpretations and Causes.* New Perspectives on Donald Davidson's Philosophy. 1999 ISBN 0-7923-5869-4
286. R. Murawski: *Recursive Functions and Metamathematics.* Problems of Completeness and Decidability, Gödel's Theorems. 1999 ISBN 0-7923-5904-6
287. T.A.F. Kuipers: *From Instrumentalism to Constructive Realism.* On Some Relations between Confirmation, Empirical Progress, and Truth Approximation. 2000 ISBN 0-7923-6086-9
288. G. Holmström-Hintikka (ed.): *Medieval Philosophy and Modern Times.* 2000 ISBN 0-7923-6102-4
289. E. Grosholz and H. Breger (eds.): *The Growth of Mathematical Knowledge.* 2000 ISBN 0-7923-6151-2

290. G. Sommaruga: *History and Philosophy of Constructive Type Theory.* 2000 ISBN 0-7923-6180-6
291. J. Gasser (ed.): *A Boole Anthology.* Recent and Classical Studies in the Logic of George Boole. 2000 ISBN 0-7923-6380-9
292. V.F. Hendricks, S.A. Pedersen and K.F. Jørgensen (eds.): *Proof Theory.* History and Philosophical Significance. 2000 ISBN 0-7923-6544-5
293. W.L. Craig: *The Tensed Theory of Time.* A Critical Examination. 2000 ISBN 0-7923-6634-4
294. W.L. Craig: *The Tenseless Theory of Time.* A Critical Examination. 2000 ISBN 0-7923-6635-2
295. L. Albertazzi (ed.): *The Dawn of Cognitive Science.* Early European Contributors. 2001 ISBN 0-7923-6799-5
296. G. Forrai: *Reference, Truth and Conceptual Schemes.* A Defense of Internal Realism. 2001 ISBN 0-7923-6885-1
297. V.F. Hendricks, S.A. Pedersen and K.F. Jørgensen (eds.): *Probability Theory.* Philosophy, Recent History and Relations to Science. 2001 ISBN 0-7923-6952-1
298. M. Esfeld: *Holism in Philosophy of Mind and Philosophy of Physics.* 2001 ISBN 0-7923-7003-1
299. E.C. Steinhart: *The Logic of Metaphor.* Analogous Parts of Possible Worlds. 2001 ISBN 0-7923-7004-X
300. To be published.
301. T.A.F. Kuipers: *Structures in Science Heuristic Patterns Based on Cognitive Structures.* An Advanced Textbook in Neo-Classical Philosophy of Science. 2001 ISBN 0-7923-7117-8
302. G. Hon and S.S. Rakover (eds.): *Explanation.* Theoretical Approaches and Applications. 2001 ISBN 1-4020-0017-0
303. G. Holmström-Hintikka, S. Lindström and R. Sliwinski (eds.): *Collected Papers of Stig Kanger with Essays on his Life and Work.* Vol. I. 2001 ISBN 1-4020-0021-9; Pb ISBN 1-4020-0022-7
304. G. Holmström-Hintikka, S. Lindström and R. Sliwinski (eds.): *Collected Papers of Stig Kanger with Essays on his Life and Work.* Vol. II. 2001 ISBN 1-4020-0111-8; Pb ISBN 1-4020-0112-6
305. C.A. Anderson and M. Zelëny (eds.): *Logic, Meaning and Computation.* Essays in Memory of Alonzo Church. 2001 ISBN 1-4020-0141-X
306. P. Schuster, U. Berger and H. Osswald (eds.): *Reuniting the Antipodes – Constructive and Nonstandard Views of the Continuum.* 2001 ISBN 1-4020-0152-5

Previous volumes are still available.

KLUWER ACADEMIC PUBLISHERS – DORDRECHT / BOSTON / LONDON

MIX
Papier aus verantwortungsvollen Quellen
Paper from responsible sources
FSC® C105338

If you have any concerns about our products,
you can contact us on
ProductSafety@springernature.com

In case Publisher is established outside the EU,
the EU authorized representative is:
Springer Nature Customer Service Center GmbH
Europaplatz 3, 69115 Heidelberg, Germany

Printed by Libri Plureos GmbH
in Hamburg, Germany